SCHAUM'S OUTLINE OF

THEORY AND PROBLEMS

of

First Year COLLEGE MATHEMATICS

College Algebra
Plane Trigonometry
Plane and Solid Analytic Geometry
Introduction to Calculus

BY

FRANK AYRES, JR., Ph.D.

Professor and Head, Department of Mathematics
Dickinson College

SCHAUM'S OUTLINE SERIES

McGRAW-HILL BOOK COMPANY

New York, St. Louis, San Francisco, Toronto, Sydney

ISBN 07-002650-5

10 11 12 13 14 15 SH SH 7 5 4 3

Preface

This book is designed primarily to assist students in acquiring a more thorough knowledge and proficiency in basic college mathematics. It includes a thorough coverage of algebra, plane trigonometry, and plane analytic geometry together with selected topics in solid analytic geometry and a brief introduction to the calculus, in that order. In addition to the use of the book by students taking a formal course in first year college mathematics, it should also be of considerable value to those who wish to review the fundamental principles and applications in anticipation of further work in mathematics.

Each chapter begins with a clear statement of the pertinent definitions, principles and theorems, together with illustrative and descriptive material. This is followed by carefully graded sets of solved and supplementary problems. The solved problems have been selected and solutions arranged so that a study of each will be rewarding. They serve to illustrate and amplify the theory, provide the repetition of basic principles so vital to effective teaching, and bring into sharp focus those fine points without which the student continually feels himself on unsafe ground. Derivations of formulas and proofs of theorems are included among the solved problems. The supplementary problems offer a complete review of the material of each chapter.

Although in many texts some degree of unification of the material has been achieved, it seemed best to make no attempt in that direction here. However, the reader will find that the material has been so divided into chapters and the problems in these chapters so arranged as to make the book a useful supplement to all current standard texts.

Considerably more material has been included here than can be covered in most first courses. This has been done to make the book more flexible, to provide a more useful book of reference, and to stimulate further interest in the topics.

The author gratefully acknowledges his indebtedness to Mr. Henry Hayden for painstaking work in the preparation of all drawings and for typographical arrangement.

FRANK AYRES, JR.

Carlisle, Pa.

June, 1958

Contents

PART I. COLLEGE ALGEBRA

PART II. PLANE TRIGONOMETRY

PART III. PLANE AND SOLID ANALYTIC GEOMETRY

PART IV. INTRODUCTION TO CALCULUS

APPENDIX

CHAPTER 1

The Number System of Algebra

ELEMENTARY MATHEMATICS is concerned mainly with certain elements called *numbers* and with certain operations defined on them.

The unending set of symbols 1, 2, 3, 4, 5, 6, 7, 8, 9, 10, 11, 12, ... used in counting are called *natural numbers*.

In adding two of these numbers, say 5 and 7, we begin with 5 (or with 7) and count to the *right* seven (or five) numbers to get 12. The sum of two natural numbers is a natural number, that is, the sum of two members of the above set is a member of the set.

In subtracting 5 from 7, we begin with 7 and count to the *left* five numbers to 2. It is clear, however, that 7 cannot be subtracted from 5 since there are only four numbers to the left of 5.

INTEGERS. In order that subtraction be always possible, it is necessary to increase our set of numbers. We prefix each natural number with a + sign (in practice, it is more convenient not to write the sign) to form the *positive integers*, we prefix each natural number with a − sign (the sign must always be written) to form the *negative integers*, and we create a new symbol 0, read *zero*. On the set of *integers*

$$\ldots, -8, -7, -6, -5, -4, -3, -2, -1, 0, +1, +2, +3, +4, +5, +6, +7, +8, \ldots$$

the operations of addition and subtraction are possible without exception.

To add two integers as +7 and −5, we begin with +7 and count to the left (indicated by the sign of −5) five numbers to +2 or we begin with −5 and count to the right (indicated by the sign of +7) seven numbers to +2. How would you add −7 and −5?

To subtract +7 from −5 we begin with −5 and count to the left (opposite to the direction indicated by +7) seven numbers to −12. To subtract −5 from +7 we begin with +7 and count to the right (opposite to the direction indicated by −5) five numbers to +12. How would you subtract +7 from +5? −7 from −5? −5 from −7?

If one is to reckon easily with integers it is necessary to avoid the process of counting. To do this we memorize an addition table and establish certain rules of procedure. We note that each of the numbers +7 and −7 is seven steps from 0 and indicate this fact by saying that the *numerical value* of each of the numbers +7 and −7 is 7. We may state:

Rule 1. To add two numbers having like signs, add their numerical values and prefix their common sign.

Rule 2. To add two numbers having unlike signs, subtract the smaller numerical value from the larger, and prefix the sign of the number having the larger numerical value.

Rule 3. To subtract a number, change its sign and add.

Since $3\cdot 2 = 2+2+2 = 3+3 = 6$, we assume

$$(+3)(+2) = +6, \quad (-3)(+2) = (+3)(-2) = -6, \quad \text{and} \quad (-3)(-2) = +6.$$

Rule 4. To multiply or divide two numbers (never divide by 0!), multiply or divide the numerical values, prefixing a + sign if the two numbers have like signs and a − sign if the two numbers have unlike signs. See Problem 1.

If m and n are integers then $m+n$, $m-n$, and $m\cdot n$ are integers but $m \div n$ may not be an integer. (Common fractions will be treated in the next section.) Moreover, there exists a unique integer x such that $m+x=n$. If $x=0$, then $m=n$; if x is positive ($x>0$), then m is less than n ($m<n$); if x is negative ($x<0$), then m is greater than n ($m>n$).

The integers may be made to correspond one-to-one with equally spaced points on a straight line as in Fig. 1 below. Then $m>n$ indicates that the point on the scale corresponding to m lies to the right of the point corresponding to n. There will be no possibility of confusion if we write the point m rather than the point which corresponds to m and we shall do so hereafter. Then $m<n$ indicates that the point m lies to the left of n. See Problems 2-4.

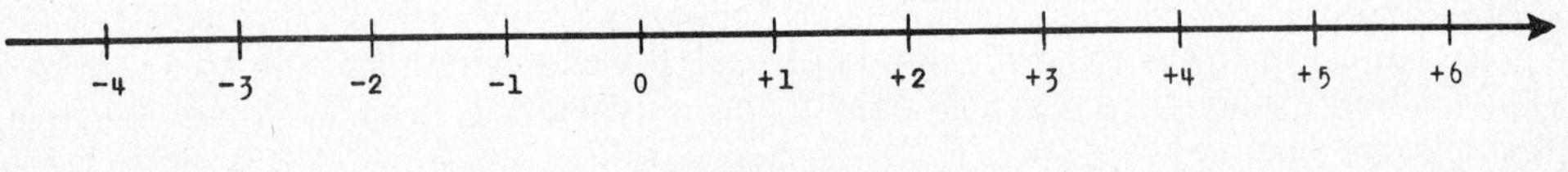

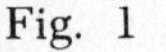

Fig. 1

Every positive integer m is divisible by ± 1 and $\pm m$. A positive integer $m>1$ is called a prime if its only factors or divisors are ± 1 and $\pm m$; otherwise, m is called *composite*. For example 2, 7, 19, are primes while $6=2\cdot 3$, $18=2\cdot 3\cdot 3$ and $30=2\cdot 3\cdot 5$ are composites. In these examples, the composite numbers have been expressed as products of prime factors, that is, factors which are prime numbers. Clearly, if $m=r\cdot s\cdot t$ is such a factorization of m, then $-m=(-1)\,r\cdot s\cdot t$ is a factorization of $-m$.
See Problems 5-6.

THE RATIONAL NUMBERS. The set of rational numbers consists of all numbers of the form m/n, where m and $n \neq 0$ are integers. Thus, the rational numbers include the integers and common fractions.

Every rational number has an infinitude of representations; for example, the integer 1 may be represented by $1/1, 2/2, 3/3, 4/4, \ldots$ and the fraction $2/3$ may be represented by $4/6, 6/9, 8/12, \ldots$ A fraction is said to be expressed in lowest terms by the representation m/n when m and n have no common prime factor. The most useful rule concerning rational numbers is therefore

Rule 5. The value of a rational number is unchanged if both the numerator and denominator are multiplied or divided by the same non-zero number.

Caution. We use Rule 5 with division to reduce a fraction to lowest terms. For example, we write $\frac{15}{21}=\frac{\not{3}\cdot 5}{\not{3}\cdot 7}=\frac{5}{7}$ and speak of canceling the 3's. Now canceling is not an operation on numbers. We cancel or strike out the 3's as a safety measure, that is, to be sure that they will not be used in computing the final result. The operation is division and Rule 5 states that we may divide the numerator by 3 provided we also divide the denominator by 3. This point is labored here because of the all too common error $\frac{12\not{a}-5}{7\not{a}}$. The fact is that $\frac{12a-5}{7a}$ cannot be further simplified for if we divide $7a$ by a we must also divide $12a$ and 5 by a. This would lead to the more cumbersome $\dfrac{12-5/a}{7}$. See Problems 7-8.

The rational numbers may be associated in a one-to-one manner with points on a straight line as in Fig. 2 below. Here the point associated with the rational number m is m units from that point (called the origin) associated with 0, the distance between the points 0 and 1 being the unit of measure.

If two rational numbers have representations r/n and s/n, where n is a positive integer, then $r/n > s/n$ if $r > s$, $r/n = s/n$ if $r = s$, and $r/n < s/n$ if $r < s$. Thus, in comparing two rational numbers it is necessary to express them with the same denominator. Of the many denominators (positive integers) there is always a least one, called the *least common denominator*. For the fractions 3/5 and 2/3, the least common denominator is 15. We conclude that $3/5 < 2/3$ since $3/5 = 9/15 < 10/15 = 2/3$.
See Problems 9-10.

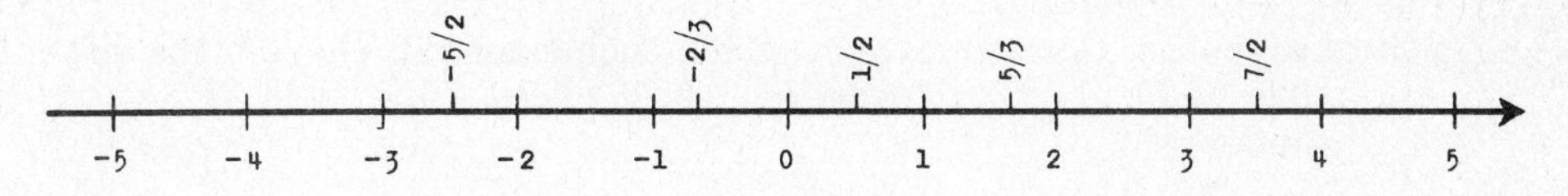

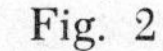
Fig. 2

Rule 6. The sum (difference) of two rational numbers expressed with the same denominator is a rational number whose denominator is the common denominator and whose numerator is the sum (difference) of the numerators.

Rule 7. The product of two or more rational numbers is a rational number whose numerator is the product of the numerators and whose denominator is the product of the denominators of the several factors.

Rule 8. The quotient of two rational numbers can be evaluated by the use of Rule 5 with the least common denominator of the two numbers as the multiplier.

See Problems 11-13.

If a and b are rational numbers, $a+b$, $a-b$ and $a \cdot b$ are rational numbers. Moreover, if a and b are $\neq 0$, there exists a rational number x, unique except for its representation, such that

$$ax = b \tag{1}$$

When a or b or both are zero, we have the following situations –

$b = 0$ and $a \neq 0$: (*1*) becomes $a \cdot x = 0$ and $x = 0$, that is, $0/a = 0$ when $a \neq 0$.

$a = 0$ and $b \neq 0$: (*1*) becomes $0 \cdot x = b$; then $b/0$, when $b \neq 0$, is without meaning since $0 \cdot x = 0$.

$a = 0$ and $b = 0$: (*1*) becomes $0 \cdot x = 0$; then $0/0$ is indeterminate since every number x satisfies the equation.

In brief: $0/a = 0$ when $a \neq 0$, but division by 0 is never permitted.

DECIMALS. In writing numbers we use a positional system, that is, the value given any particular digit depends upon its position in the sequence. For example, in 423 the the positional value of the digit 4 is 4(100) while in 234 the positional value of the digit 4 is 4(1). Since the positional value of a digit involves the number 10, this system of notation is called the *decimal system*. In this system the number 4238.75 means

$$4(1000) + 2(100) + 3(10) + 8(1) + 7(1/10) + 5(1/100)$$

It is interesting to note that from this example certain definitions to be made in a later study of exponents may be anticipated. Since $1000 = 10^3$, $100 = 10^2$, $10 = 10^1$ it would seem natural to define $1 = 10^0$, $1/10 = 10^{-1}$, $1/100 = 10^{-2}$.

By the process of division, any rational number can be expressed as a decimal; for example, $70/33 = 2.121212\ldots$ This is termed a *repeating decimal* since the digits 12, called the cycle, are repeated without end. It will be seen later that every repeating decimal represents a rational number.

In reckoning with decimals, it is necessary to "round off" a decimal representation to a prescribed number of decimal places. For example, $1/3 = 0.3333\ldots$ is written as 0.33 to two decimal places and $2/3 = 0.6666\ldots$ is written as 0.667 to three decimal places. In rounding off, use will be made of the Computer's Rule:

(*a*) Increase the last digit retained by 1 if the digits rejected exceed the sequence 50000...; for example, 2.384629... becomes 2.385 to three decimal places.

(*b*) Leave the last digit retained unchanged if the digits rejected are less than 5000...; for example, 2.384629... becomes 2.38 to two decimal places.

(c) Make the last digit retained even if the digit rejected is exactly 5; for example, to three decimal places 11.3865 becomes 11.386 and 9.3815 becomes 9.382.

See Problem 14.

PERCENTAGE. The symbol %, read per cent, means per hundred; thus 5% is equivalent to 5/100 or 0.05.

Any number, when expressed in decimal notation, can be written as a per cent by multiplying by 100 and adding the symbol %. For example, $0.0125 = 100(0.0125)\% = 1.25\% = 1\frac{1}{4}\%$, $2.3 = 230\%$, and $7/20 = 0.35 = 35\%$.

Conversely, any percentage may be expressed in decimal form by dropping the symbol % and dividing by 100. For example, $42.5\% = 42.5/100 = 0.425$, $3.25\% = 0.0325$, and $2000\% = 20$.

When reckoning percentages, express the percent as a decimal and, when possible, as a simple fraction. For example, $4\frac{1}{4}\%$ of $48 = 0.0425 \times 48 = 2.04$ and $12\frac{1}{2}\%$ of $5.28 = 1/8$ of $5.28 = 0.66$.

See Problems 15-18.

THE IRRATIONAL NUMBERS. The existence of numbers other than the rational numbers may be inferred from either of the following considerations:

(*a*) We may conceive of a non-repeating decimal constructed in endless time by setting down a succession of digits chosen at random.

(*b*) The length of the diagonal of a square of side 1 is not a rational number, that is, there exists no rational number a such that $a^2 = 2$. Numbers such as $\sqrt{2}$, $\sqrt[3]{2}$, $\sqrt[5]{-3}$, and π (but not $\sqrt{-3}$ or $\sqrt[4]{-5}$) are called *irrational numbers*. The first three of these are called *radicals*. The radical $\sqrt[n]{a}$ is said to be of order n; n is called the *index* and a is called the *radicand*.

See Problems 19-21.

THE REAL NUMBERS. The set of *real numbers* consists of the rational and irrational numbers. The real numbers may be ordered by comparing their decimal representations. For example, $\sqrt{2} = 1.4142\ldots$; then $7/5 = 1.4 < \sqrt{2}$, $3/2 = 1.5 > \sqrt{2}$, etc.

We assume that the totality of real numbers may be placed in one-to-one correspondence with the totality of points on a straight line.

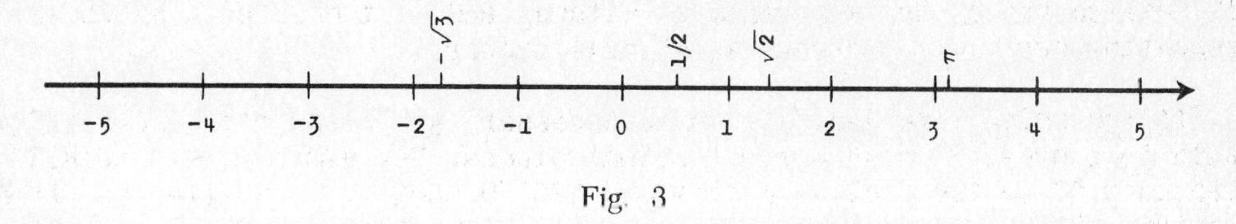

Fig. 3

The number associated with a point on the line, called the *coordinate* of the point, gives its distance and direction from that point (called the origin) associated with the number 0. If a point A has coordinate a, we shall speak of it as the point $A(a)$.

The directed distance from the point $A(a)$ to the point $B(b)$ on the real number scale is given by $AB = b - a$. The midpoint of the segment AB has coordinate $\frac{1}{2}(a+b)$.

See Problems 22-25.

THE COMPLEX NUMBERS. In the set of real numbers there is no number whose square is -2. If there is to be such a number, say $\sqrt{-2}$ then by definition $(\sqrt{-2})^2 = -2$. Note carefully that $(\sqrt{-2})^2 = \sqrt{-2}\,\sqrt{-2} = \sqrt{(-2)(-2)} = \sqrt{4} = 2$ is incorrect; also $\sqrt{-2}\,\sqrt{-3} = \sqrt{6}$ is incorrect. In order to avoid this error, the symbol i with the following properties is used:

$$\text{If } a > 0,\ \sqrt{-a} = i\sqrt{a};\quad i^2 = -1.$$

$$\text{Then}\quad (\sqrt{-2})^2 = \sqrt{-2}\,\sqrt{-2} = (i\sqrt{2})(i\sqrt{2}) = i^2\cdot 2 = -2 \quad\text{and}$$

$$\sqrt{-2}\,\sqrt{-3} = (i\sqrt{2})(i\sqrt{3}) = i^2\sqrt{6} = -\sqrt{6}.$$

Numbers of the form $a + bi$, where a and b are real numbers, are called *complex numbers*. In the complex number $a + bi$, a is called the *real part* and bi is called the *imaginary part*.

The complex number $a + bi$ is a real number when $b = 0$ and a pure imaginary number when $a = 0$. When a complex number is not a real number it is called *imaginary*.

Complex numbers will be considered in more detail in a later chapter. Only the following operations will be considered here:

To add (subtract) two complex numbers, add (subtract) the real parts and add (subtract) the pure imaginary parts.

To multiply two complex numbers, form the product treating i as an ordinary number and then replace i^2 by -1.

See Problems 26-27.

SOLVED PROBLEMS

1. Give the results when the following operations are performed on each of the numbers −9, −6, −3, 0, 3, 6, 9, 12, 15: (*a*) add −4, (*b*) subtract 6, (*c*) subtract −2, (*d*) multiply by −5, (*e*) divide by 3, (*f*) divide by −1, (*g*) divide by −3.

a) −13, −10, −7, −4, −1, 2, 5, 8, 11
b) −15, −12, −9, −6, −3, 0, 3, 6, 9
c) −7, −4, −1, 2, 5, 8, 11, 14, 17
d) 45, 30, 15, 0, −15, −30, −45, −60, −75
e) −3, −2, −1, 0, 1, 2, 3, 4, 5
f) 9, 6, 3, 0, −3, −6, −9, −12, −15
g) 3, 2, 1, 0, −1, −2, −3, −4, −5

2. Arrange the integers in each set so that they may be separated by $<$ and again so that they may be separated by $>$.

a) 3, 15, 12, 20, 0 *Ans.* $0 < 3 < 12 < 15 < 20$; $20 > 15 > 12 > 3 > 0$
b) 3, −3, 5, 0, −2 *Ans.* $-3 < -2 < 0 < 3 < 5$; $5 > 3 > 0 > -2 > -3$
c) −7, −5, −10, −8 *Ans.* $-10 < -8 < -7 < -5$; $-5 > -7 > -8 > -10$

3. Let x be an integer. By means of Fig. 1, interpret each of the following:

a) $x < 10$ *Ans.* x is to the left of 10.
b) $x > -2$ *Ans.* x is to the right of −2.
c) $x \geqq 5$ *Ans.* x is 5 or is to the right of 5.
d) $2 < x < 6$ *Ans.* x is to the right of 2 but to the left of 6.
e) $10 > x > -3$ *Ans.* x is to the left of 10 but to the right of −3.

4. List all integral values of x when:

a) $2 < x < 6$ *Ans.* 3, 4, 5
b) $2 > x > -3$ *Ans.* −2, −1, 0, 1
c) $-5 < x < 0$ *Ans.* −4, −3, −2, −1
d) $2 \leqq x < 5$ *Ans.* 2, 3, 4
e) $-4 < x \leqq -1$ *Ans.* −3, −2, −1
f) $2 \geqq x \geqq -3$ *Ans.* −3, −2, −1, 0, 1, 2

5. List the first 15 primes. *Ans.* 2, 3, 5, 7, 11, 13, 17, 19, 23, 29, 31, 37, 41, 43, 47.

6. Express each of the following integers as a product of primes: (*a*) 6930, (*b*) 23,595.

a) A systematic procedure is to test the primes 2, 3, 5... in order. When a factor is found, we then repeat the procedure, using in order all primes not already rejected, on the quotient. Thus, $6930 = 2\cdot 3465$. Since 3465 is not divisible by 2, we try 3 and obtain $6930 = 2\cdot 3\cdot 1155$. Using 3 again, we find $6930 = 2\cdot 3\cdot 3\cdot 385$. Since 385 is not divisible by 3, we try 5 and obtain $6930 = 2\cdot 3\cdot 3\cdot 5\cdot 77 = 2\cdot 3\cdot 3\cdot 5\cdot 7\cdot 11$.

b) $23{,}595 = 3\cdot 5\cdot 11\cdot 11\cdot 13$.

7. Express 5/6 as a fraction having denominator (*a*) 12, (*b*) 36, (*c*) 84, (*d*) 126.

a) $\frac{5}{6} = \frac{5\cdot 2}{6\cdot 2} = \frac{10}{12}$ *b*) $\frac{5}{6} = \frac{5\cdot 6}{6\cdot 6} = \frac{30}{36}$ *c*) $\frac{5}{6} = \frac{5\cdot 14}{6\cdot 14} = \frac{70}{84}$ *d*) $\frac{5}{6} = \frac{5\cdot 21}{6\cdot 21} = \frac{105}{126}$

8. Reduce to lowest terms: (*a*) 6/24, (*b*) 30/42, (*c*) 27/45, (*d*) 60/96.

a) $\frac{6}{24} = \frac{2\cdot 3}{2\cdot 2\cdot 2\cdot 3} = \frac{1}{2\cdot 2} = \frac{1}{4}$

c) $\frac{27}{45} = \frac{3\cdot 3\cdot 3}{3\cdot 3\cdot 5} = \frac{3}{5}$

b) $\frac{30}{42} = \frac{2\cdot 3\cdot 5}{2\cdot 3\cdot 7} = \frac{5}{7}$

d) $\frac{60}{96} = \frac{2\cdot 2\cdot 3\cdot 5}{2\cdot 2\cdot 2\cdot 2\cdot 2\cdot 3} = \frac{5}{2\cdot 2\cdot 2} = \frac{5}{8}$

9. In each of the following find the lowest common denominator (L.C.D.) of the several fractions: (*a*) 3/4, 5/6, (*b*) 1/6, 2/9, 5/24, (*c*) 1/12, 7/60, 2/25, (*d*) 7/72, 4/75, 9/80.

To find the L.C.D.: Express each of the several denominators as the product of prime factors, write each distinct factor the greatest number of times it occurs in any denominator, and form the product.

a) Here $4 = 2 \cdot 2$ and $6 = 2 \cdot 3$; L.C.D. $= (2 \cdot 2)(3) = 12$.
b) Here $6 = 2 \cdot 3$, $9 = 3 \cdot 3$, $24 = 2 \cdot 2 \cdot 2 \cdot 3$; L.C.D. $= (2 \cdot 2 \cdot 2)(3 \cdot 3) = 72$.
c) Here $12 = 2 \cdot 2 \cdot 3$, $60 = 2 \cdot 2 \cdot 3 \cdot 5$, $25 = 5 \cdot 5$; L.C.D. $= (2 \cdot 2)(3)(5 \cdot 5) = 300$.
d) Here $72 = 2 \cdot 2 \cdot 2 \cdot 3 \cdot 3$, $75 = 3 \cdot 5 \cdot 5$, $80 = 2 \cdot 2 \cdot 2 \cdot 2 \cdot 5$; L.C.D. $= 2 \cdot 2 \cdot 2 \cdot 2 \cdot 3 \cdot 3 \cdot 5 \cdot 5 = 3600$.

10. Arrange each set of rational numbers so that they may be separated by $<$: (*a*) $1, -1/2, 1/3, 3/4, -5/3$; (*b*) $-2/3, -7/8, -5/6, -11/6, -5/12$.

a) Since $1 = 12/12$, $-1/2 = -6/12$, $1/3 = 4/12$, $3/4 = 9/12$, $-5/3 = -20/12$, then $-5/3 < -1/2 < 1/3 < 3/4 < 1$.
b) Since $-2/3 = -16/24$, $-7/8 = -21/24$, $-5/6 = -20/24$, $-11/6 = -44/24$, $-5/12 = -10/24$, then $-11/6 < -7/8 < -5/6 < -2/3 < -5/12$.

11. Perform the indicated operations:

a) $\frac{1}{2} + \frac{2}{5} = \frac{5}{10} + \frac{4}{10} = \frac{5+4}{10} = \frac{9}{10}$

b) $\frac{2}{3} - \frac{1}{2} = \frac{4-3}{6} = \frac{1}{6}$

c) $\frac{11}{12} + \frac{4}{3} - \frac{1}{6} = \frac{11+16-2}{12} = \frac{25}{12}$

d) $\frac{15}{64} - \frac{17}{32} + \frac{1}{8} = \frac{15-34+8}{64} = -\frac{11}{64}$

e) $\frac{2}{3} \times \frac{4}{5} = \frac{2 \cdot 4}{3 \cdot 5} = \frac{8}{15}$

f) $\frac{5}{8} \times \frac{16}{15} = \frac{5 \cdot 16}{8 \cdot 15} = \frac{2}{3}$

g) $\frac{15}{7}(-\frac{21}{10})(\frac{4}{9}) = -\frac{15 \cdot 21 \cdot 4}{7 \cdot 10 \cdot 9} = -\frac{3 \cdot 5 \cdot 3 \cdot 7 \cdot 2 \cdot 2}{7 \cdot 2 \cdot 5 \cdot 3 \cdot 3} = -2$

h) $\frac{5}{7} \div \frac{11}{14} = 14\frac{5}{7} \div 14\frac{11}{14} = 10 \div 11 = 10/11$

i) $\frac{16}{5} \div \frac{7}{10} = 10\frac{16}{5} \div 10\frac{7}{10} = \frac{32}{7}$

j) $8 \div (-\frac{2}{3}) = 3 \cdot 8 \div 3(-\frac{2}{3}) = -\frac{24}{2} = -12$

k) $\frac{4 - 3/4}{2 + 1/2} = \frac{4 \cdot 4 - 4(3/4)}{4 \cdot 2 + 4(1/2)} = \frac{16-3}{8+2} = \frac{13}{10}$

l) $\frac{1/2 - 7/4}{1 - 3/8} = \frac{8(1/2) - 8(7/4)}{8 \cdot 1 - 8(3/8)} = \frac{4-14}{8-3} = \frac{-10}{5} = -2$

12. The product of 5 factors each equal to 2, that is, $2 \cdot 2 \cdot 2 \cdot 2 \cdot 2$, is denoted by 2^5 and read the fifth power of 2. We call 2 the *base* and 5 the *exponent*. Show that the solutions of Solved Problem 8(*a*), (*c*), (*d*) may be written as follows using exponents:

a) $\frac{6}{24} = \frac{2 \cdot 3}{2^3 \cdot 3} = \frac{1}{2^2} = \frac{1}{4}$ *c*) $\frac{27}{45} = \frac{3^3}{3^2 \cdot 5} = \frac{3}{5}$ *d*) $\frac{60}{96} = \frac{2^2 \cdot 3 \cdot 5}{2^5 \cdot 3} = \frac{5}{2^3} = \frac{5}{8}$

13. Verify: (*a*) $a^4 \cdot a^2 = (a \cdot a \cdot a \cdot a)(a \cdot a) = a^{4+2} = a^6$, (*b*) $\frac{a^5}{a^3} = \frac{a \cdot a \cdot a \cdot a \cdot a}{a \cdot a \cdot a} = a^{5-3} = a^2$, (*c*) $\frac{a^3}{a^5} = \frac{a \cdot a \cdot a}{a \cdot a \cdot a \cdot a \cdot a} = \frac{1}{a^{5-3}} = \frac{1}{a^2}$, (*d*) $(a \cdot b)^4 = (a \cdot b)(a \cdot b)(a \cdot b)(a \cdot b) = (a \cdot a \cdot a \cdot a)(b \cdot b \cdot b \cdot b) = a^4 b^4$.

The general rules are: If m and n are positive integers, then $a^m a^n = a^{m+n}$; $a^m/a^n = a^{m-n}$ if $m > n$; $a^m/a^n = 1$ if $m = n$; $a^m/a^n = 1/a^{n-m}$ if $m < n$; $(a \cdot b)^m = a^m b^m$.

14. Express 2/7 as a decimal to (*a*) 5, (*b*) 4, (*c*) 3, (*d*) 2 decimal places.

By division, $2/7 = 0.285714\ldots$ Then we have for *a*) 0.28571 , *b*) 0.2857 , *c*) 0.286 , *d*) 0.29.

15. Compute:

a) 6% of 400 $= 0.06 \times 400 = 24$
b) $4\frac{1}{2}$% of 1200 $= 0.045 \times 1200 = 54$
c) 135% of 500 $= 1.35 \times 500 = 675$
d) 2% of 6% of 8000 $= 0.02 \times 0.06 \times 8000 = 9.6$

16. What percent: (*a*) of 75 is 15? (*b*) of 112 is 14? (*c*) of 72 is 3.96? (*d*) of 0.44 is 1.034?

a) $15/75 = 1/5 = 20\%$ *c*) $3.96/72 = 0.055 = 5\frac{1}{2}\%$
b) $14/112 = 1/8 = 12\frac{1}{2}\%$ *d*) $1.034/0.44 = 2.35 = 235\%$

17. Find the number, given: (*a*) 5% of it is 32, (*b*) 8% of it is 8.4, (*c*) 210% of it is 54.6, (*d*) 0.5% of it is 2.3.

a) 1% of the number is $32/5 = 6.4$; 100% of the number is $100 \times 6.4 = 640$ or $32/0.05 = 640$.
b) $8.4/0.08 = 105$ *c*) $54.6/2.1 = 26$ *d*) $2.3/0.005 = 460$

18. Express the percentage strength of each of the following solutions (by a 10% silver nitrate solution is meant 10 grams of silver nitrate in 100 grams of solution): (*a*) 200 grams of solution containing a 0.5 gram tablet of bichloride of mercury; (*b*) 50 grams of solution containing 0.8 gram of salt.

a) $0.5/200 = 0.0025 = 0.25\% = \frac{1}{4}\%$ *b*) $0.8/50 = 0.016 = 1.6\%$

19. Simplify each of the following radicals:

a) $\sqrt{700} = \sqrt{100\cdot 7} = 10\sqrt{7}$

b) $\sqrt[3]{24} = \sqrt[3]{8\cdot 3} = 2\sqrt[3]{3}$

c) $\sqrt{72} = \sqrt{36\cdot 2} = 6\sqrt{2}$

d) $\sqrt[5]{-64} = \sqrt[5]{-32\cdot 2} = -2\sqrt[5]{2}$

e) $\dfrac{2}{\sqrt{3}} = \dfrac{2\sqrt{3}}{\sqrt{3}\,\sqrt{3}} = \dfrac{2}{3}\sqrt{3}$

f) $\sqrt{\dfrac{2}{3}} = \dfrac{\sqrt{2}}{\sqrt{3}} = \dfrac{\sqrt{2}\,\sqrt{3}}{\sqrt{3}\,\sqrt{3}} = \dfrac{\sqrt{2\cdot 3}}{3} = \dfrac{\sqrt{6}}{3}$

g) $\sqrt{\dfrac{5}{8}} = \sqrt{\dfrac{5\cdot 2}{8\cdot 2}} = \sqrt{\dfrac{10}{16}} = \dfrac{\sqrt{10}}{4}$

20. Perform the indicated operations.

a) $4\sqrt{2} + 3\sqrt{2} - 2\sqrt{2} = (4+3-2)\sqrt{2} = 5\sqrt{2}$

b) $6\sqrt{3} - \sqrt{27} = 6\sqrt{3} - 3\sqrt{3} = 3\sqrt{3}$

c) $2\sqrt[3]{5} - \sqrt[3]{135} + 4\sqrt[6]{25} = 2\sqrt[3]{5} - 3\sqrt[3]{5} + 4\sqrt[3]{5} = 3\sqrt[3]{5}$

d) $2\cdot\dfrac{1}{\sqrt{7}} + 3\sqrt{28} - \sqrt{63} = \dfrac{2}{7}\sqrt{7} + 6\sqrt{7} - 3\sqrt{7} = \dfrac{23}{7}\sqrt{7}$

e) $\sqrt[4]{64} - 5\sqrt[6]{\dfrac{1}{8}} = 2\sqrt{2} - \dfrac{5}{2}\sqrt{2} = -\dfrac{1}{2}\sqrt{2}$

f) $\sqrt{3}\cdot\sqrt{15} = \sqrt{45} = 3\sqrt{5}$

g) $\sqrt[3]{18}\cdot\sqrt[3]{4} = \sqrt[3]{72} = 2\sqrt[3]{9}$

h) $(2\sqrt{5} + 3)(3\sqrt{5} - 4) = 30 - 8\sqrt{5} + 9\sqrt{5} - 12 = 18 + \sqrt{5}$

i) $(2\sqrt{3} - 3\sqrt{2})(5\sqrt{3} + \sqrt{2}) = 30 + 2\sqrt{6} - 15\sqrt{6} - 6 = 24 - 13\sqrt{6}$

21. Simplify each of the following.

a) $\dfrac{3 + 5\sqrt{2}}{2\sqrt{2}} = \dfrac{(3+5\sqrt{2})\sqrt{2}}{(2\sqrt{2})\sqrt{2}} = \dfrac{3\sqrt{2} + 10}{4}$

b) $\dfrac{4}{\sqrt{2} + \sqrt{3}} = \dfrac{4}{\sqrt{2} + \sqrt{3}}\cdot\dfrac{\sqrt{2} - \sqrt{3}}{\sqrt{2} - \sqrt{3}} = \dfrac{4\sqrt{2} - 4\sqrt{3}}{2 - 3} = 4\sqrt{3} - 4\sqrt{2}$

c) $\dfrac{2\sqrt{3} + 3\sqrt{2}}{3\sqrt{5} - 5\sqrt{3}} = \dfrac{2\sqrt{3} + 3\sqrt{2}}{3\sqrt{5} - 5\sqrt{3}}\cdot\dfrac{3\sqrt{5} + 5\sqrt{3}}{3\sqrt{5} + 5\sqrt{3}} = \dfrac{6\sqrt{15} + 30 + 9\sqrt{10} + 15\sqrt{6}}{45 - 75} = \dfrac{30 + 15\sqrt{6} + 9\sqrt{10} + 6\sqrt{15}}{-30}$

$= -\dfrac{10 + 5\sqrt{6} + 3\sqrt{10} + 2\sqrt{15}}{10}$

d) $\dfrac{1 + 4\sqrt[3]{2}}{\sqrt[3]{2}} = \dfrac{(1+4\sqrt[3]{2})\sqrt[3]{4}}{\sqrt[3]{2}\cdot\sqrt[3]{4}} = \dfrac{\sqrt[3]{4} + 4\sqrt[3]{8}}{\sqrt[3]{8}} = \dfrac{\sqrt[3]{4} + 8}{2}$

22. The numerical value of a real number N ($|N|$) is defined as follows:

$|N| = N$ if $N > 0$; $|N| = 0$ if $N = 0$, $|N| = -N$ if $N < 0$.

Arrange each set of numbers so that they may be separated by <. (*a*) $|-12|$, $|5/2|$, $|-9|$, $|1|$, $|-2|$, $|6|$; (*b*) $|3+4|$, $|9-6|$, $|2-8|$, $|-3-6|$.

a) $|1| < |-2| < |5/2| < |6| < |-9| < |-12|$ since $1 < 2 < 5/2 < 6 < 9 < 12$.

b) $|9-6| < |2-8| < |3+4| < |-3-6|$ since $3 < 6 < 7 < 9$.

23. Find the directed distance AB, given: (*a*) $A(2)$, $B(6)$; (*b*) $A(3)$, $B(-7)$; (*c*) $A(-2)$, $B(-8)$; (*d*) $A(-10)$, $B(2)$; (*e*) $A(-9)$, $B(-2)$; (*f*) $A(1)$, $B(x)$; (*g*) $A(x_1)$, $B(3)$; (*h*) $A(x_1)$, $B(x_2)$.

a) $AB = 6-2 = 4$ *c*) $AB = -8-(-2) = -6$ *e*) $AB = -2-(-9) = 7$ *g*) $AB = 3-x_1$

b) $AB = -7-3 = -10$ *d*) $AB = 2-(-10) = 12$ *f*) $AB = x-1$ *h*) $AB = x_2-x_1$

24. (*a*) On a number scale locate the points $A(-5)$, $B(1)$, $C(7)$ and show that $AB + BC + CA = 0$. (*b*) Re-label the above points, reading from left to right, $B(-5)$, $C(1)$, $A(7)$ and show that $AB+BC+CA = 0$.

a) $AB + BC + CA = [1-(-5)] + (7-1) + (-5-7) = 6 + 6 - 12 = 0$

b) $AB + BC + CA = (-5-7) + [1-(-5)] + (7-1) = 0$

25. Find the coordinate of the midpoint of the segments AB in Problem 23.

a) $\frac{1}{2}(2+6) = 4$ *c*) $\frac{1}{2}[-2+(-8)] = -5$ *e*) $\frac{1}{2}(-9-2) = -11/2$ *g*) $\frac{1}{2}(x_1+3)$

b) $\frac{1}{2}[3+(-7)] = -2$ *d*) $\frac{1}{2}(-10+2) = -4$ *f*) $\frac{1}{2}(1+x)$ *h*) $\frac{1}{2}(x_1+x_2)$

26. Rewrite each of the following, using i:

a) $\sqrt{-5} = i\sqrt{5}$ *b*) $\sqrt{-4} = 2i$ *c*) $\sqrt{-a^2} = ai$ *d*) $\sqrt{-32} = 4i\sqrt{2}$ *e*) $3-\sqrt{-9} = 3-3i$

f) $\dfrac{8+\sqrt{-16}}{2} = \dfrac{8+4i}{2} = 4+2i$ *g*) $\dfrac{6-\sqrt{-128}}{12} = \dfrac{6-8i\sqrt{2}}{12} = \dfrac{1}{2} - \dfrac{2\sqrt{2}}{3}i$

27. Perform the indicated operations:

a) $(2-5i) + (4+3i) = (2+4) + (-5+3)i = 6-2i$

b) $(3+2i) - (-6-3i) = [3-(-6)] + [2-(-3)]i = 9+5i$

c) $(-5+2\sqrt{-4}) + (1-\sqrt{-9}) = (-5+4i) + (1-3i) = -4+i$

d) $(2+\sqrt{-27}) - (4-\sqrt{-3}) = (2+3i\sqrt{3}) - (4-i\sqrt{3}) = -2+4i\sqrt{3}$

e) $(-2-\sqrt{-8}) - (5+\sqrt{-27}) = (-2-2i\sqrt{2}) - (5+3i\sqrt{3}) = -7-(2\sqrt{2}+3\sqrt{3})i$

f) $(2+3i) + (2-3i) = 4$

g) $(2+3i) - (2-3i) = 6i$

h) $(2-5i)(4+3i) = 8+6i-20i-15i^2 = 8-14i+15 = 23-14i$

i) $(2+3i)(2-3i) = 4-6i+6i-9i^2 = 13$

SUPPLEMENTARY PROBLEMS

28. Arrange each of the following so that they may be separated by $<$.
(a) 2/3, −3/4, 5/6, −1, 4/5, −4/3, −1/4 (b) 3/2, 2, 7/5, 4/3, 3 (c) 3/2, $\sqrt{3}$, −1/2, $-\sqrt{5}$, 0

29. Determine the greater of each pair.
(a) $|4+(-2)|$ and $|-4|+|-2|$ (b) $|4+(-2)|$ and $|4|+|-2|$ (c) $|4-(-2)|$ and $|4|-|-2|$

30. Convert each of the following fractions into equivalent fractions having the indicated denominator.
(a) 3/5, 15 (b) −3/5, 20 (c) 7/3, 42 (d) 5/7, 35 (e) 12/13, 156

31. Perform the indicated operations.

(a) $(-2)(3)(-5)$
(b) $3(-2)(4) + (-5)(2)(0)$
(c) $-8-(-6)+2$
(d) $3/4 + 2/3$
(e) $3/4 - 2/3$
(f) $5/6 - 1/2 - 2/3$
(g) $3/4 - 7/12 - 1/3$
(h) $(1/2)(8/9)(6/5)$
(i) $\frac{3}{8} \times 5\frac{1}{3}$
(j) $2\frac{1}{4} \times 2\frac{2}{3} \times 1\frac{2}{5} \times 2\frac{1}{7}$
(k) $\frac{25}{32} \div \frac{35}{64}$
(l) $3\frac{1}{3} \div \frac{7}{10}$
(m) $(1\frac{1}{2} \times 2\frac{1}{4}) \div 1\frac{1}{8}$
(n) $\dfrac{3 - 2/3}{5 + 5/6}$
(o) $\dfrac{2/3 + 3/4}{5/6 - 7/8}$
(p) $\dfrac{1\frac{1}{2} - 2\frac{2}{3}}{3\frac{1}{5} - 1\frac{1}{4}}$

32. Perform the indicated operations.

(a) $5\sqrt{3} + 2\sqrt{3} - 8\sqrt{3}$
(b) $5\sqrt{2} + \sqrt{32} - 3\sqrt{8}$
(c) $\sqrt[3]{12}\cdot\sqrt[3]{36}$
(d) $(1+\sqrt{2})(3-\sqrt{2})$
(e) $(2\sqrt{3}+3\sqrt{2})(2\sqrt{3}-3\sqrt{2})$
(f) $(4\sqrt{3}-3\sqrt{5})(2\sqrt{3}+\sqrt{5})$
(g) $\dfrac{4-2\sqrt{3}}{5\sqrt{3}}$
(h) $\dfrac{2\sqrt{5}-3\sqrt{2}}{3\sqrt{5}+4\sqrt{2}}$
(i) $\dfrac{3\sqrt{2}-4\sqrt{3}}{4\sqrt{2}-3\sqrt{3}}$

33. Perform the indicated operations.

(a) $i\sqrt{12} + i\sqrt{75} - i\sqrt{108}$
(b) $i\sqrt{50} - i\sqrt{32} + i\sqrt{8}$
(c) $4\sqrt{-27} + 2\sqrt{-12} - 5\sqrt{-48}$
(d) $3\sqrt{-20} - 5\sqrt{-80} - \sqrt{-45}$
(e) $\sqrt{-9}\cdot\sqrt{-16}$
(f) $\sqrt{-12}\cdot\sqrt{-27}$
(g) $3i(2+i)$
(h) $(-3+5i)+(4-2i)$
(i) $(3+5i)-(-4-2i)$
(j) $(3+5i)(2-7i)$
(k) $(2\sqrt{3}+i\sqrt{2})(3\sqrt{3}-5i\sqrt{2})$
(l) $(3-2i)(1+5i)(-2-i)$

ANSWERS TO SUPPLEMENTARY PROBLEMS.

28. (a) −4/3, −1, −3/4, −1/4, 2/3, 4/5, 5/6 (b) 4/3, 7/5, 3/2, 2, 3 (c) $-\sqrt{5}$, −1/2, 0, 3/2, $\sqrt{3}$

29. (a) second (b) second (c) first

30. (a) 9/15 (b) −12/20 (c) 98/42 (d) 25/35 (e) 144/156

31. (a) 30 (b) −24 (c) 0 (d) 17/12 (e) 1/12 (f) −1/3 (g) −1/6 (h) 8/15 (i) 2 (j) 18 (k) 10/7 (l) 100/21 (m) 3 (n) 2/5 (o) −34 (p) −70/117

32. (a) $-\sqrt{3}$ (b) $3\sqrt{2}$ (c) $6\sqrt[3]{2}$ (d) $1+2\sqrt{2}$ (e) −6 (f) $9-2\sqrt{15}$ (g) $\dfrac{4\sqrt{3}-6}{15}$ (h) $\dfrac{54-17\sqrt{10}}{13}$ (i) $-\dfrac{12+7\sqrt{6}}{5}$

33. (a) $i\sqrt{3}$ (b) $3i\sqrt{2}$ (c) $-4i\sqrt{3}$ (d) $-17i\sqrt{5}$ (e) −12 (f) −18 (g) $-3+6i$ (h) $1+3i$ (i) $7+7i$ (j) $41-11i$ (k) $28-7i\sqrt{6}$ (l) $-13-39i$

CHAPTER 2

Elements of Algebra

IN ARITHMETIC the numbers used are always known numbers; a typical problem is to convert 5 hours and 35 minutes to minutes. This is done by multiplying 5 by 60 and adding 35; thus, $5\cdot 60+35 = 335$ minutes.

In algebra some of the numbers used may be known but others are either unknown or not specified, that is, they are represented by letters. For example, convert h hours and m minutes into minutes. This is done in precisely the same manner as in the paragraph above, by multiplying h by 60 and adding m, thus, $h\cdot 60+m = 60h+m$. We call $60h+m$ an *algebraic expression.*

See Problem 1.

Since algebraic expressions are numbers, they may be added, subtracted, etc., following the same laws as govern these operations on known numbers. For example, the sum of $5\cdot 60+35$ and $2\cdot 60+35$ is $(5+2)\cdot 60+2\cdot 35$; similarly, the sum of $h\cdot 60+m$ and $k\cdot 60+m$ is $(h+k)\cdot 60+2m$.

See Problems 2-6.

POSITIVE INTEGRAL EXPONENTS. If a is any number and n is any positive integer, the product of the n factors $a\cdot a\cdot a\ldots a$ is denoted by a^n. To distinguish between the letters, a is called the *base* and n is called the *exponent.*

If a and b are any bases and m and n are any positive integers, we have the following laws of exponents:

(1) $a^m\cdot a^n = a^{m+n}$ (2) $(a^m)^n = a^{mn}$

(3) $\dfrac{a^m}{a^n} = a^{m-n},\ a\neq 0,\ m>n;\quad \dfrac{a^m}{a^n} = \dfrac{1}{a^{n-m}},\ a\neq 0,\ m<n.$

(4) $(a\cdot b)^n = a^n b^n$ (5) $(\dfrac{a}{b})^n = \dfrac{a^n}{b^n},\ b\neq 0.$

See Problem 7.

LET n BE A POSITIVE INTEGER and a and b be two numbers such that $b^n = a$; then b is called an nth *root of a*. Every number $a\neq 0$ has exactly n distinct nth roots.

If a is imaginary, all of its nth roots are imaginary; this case will be excluded here and treated later.

If a is real and n is odd, then exactly one of the nth roots of a is real. For example, 2 is the real cube root of 8, $(2^3 = 8)$, and -3 is the real fifth root of -243, $[(-3)^5 = -243]$.

If a is real and n is even, then there are exactly two real nth roots of a when $a>0$ but no real nth roots of a when $a<0$. For example, +3 and −3 are the square roots of 9; +2 and −2 are the real sixth roots of 64.

THE PRINCIPAL nth **ROOT OF** a is the positive real nth root of a when a is positive and the real nth root of a, if any, when a is negative. The principal nth root of a is denoted by $\sqrt[n]{a}$, called a *radical*. The integer n is called the *index* of the radical and a is called the *radicand*. For example,

$$\sqrt{9} = 3, \quad \sqrt[6]{64} = 2, \quad \sqrt[5]{-243} = -3$$

See Problem 8.

ZERO, FRACTIONAL, AND NEGATIVE EXPONENTS. When r and s are positive integers and p is any rational number, the following extend the definition of a^n in such a way that the laws (*1*)-(*5*) are satisfied when n is any rational number.

	DEFINITIONS	EXAMPLES
(*6*)	$a^0 = 1,\ a \neq 0$	$2^0 = 1,\quad (1/100)^0 = 1,\quad (-8)^0 = 1$
(*7*)	$a^{r/s} = \sqrt[s]{a^r} = (\sqrt[s]{a})^r$	$3^{1/2} = \sqrt{3},\quad (64)^{5/6} = (\sqrt[6]{64})^5 = 2^5 = 32$
(*8*)	$a^{-p} = 1/a^p$	$2^{-1} = 1/2,\quad 3^{-1/2} = 1/\sqrt{3}$

Note: Without attempting to define them, we shall assume the existence of numbers as $a^{\sqrt{2}}$, a^{π}, ..., in which the exponent is irrational. We shall also assume that these numbers have been defined in such a way that the laws (*1*)-(*5*) are satisfied.

See Problems 9-10.

SOLVED PROBLEMS

1. For each of the following statements, write the equivalent algebraic expression: (*a*) the sum of x and 2, (*b*) the sum of a and $-b$, (*c*) the sum of $5a$ and $3b$, (*d*) the product of $2a$ and $3a$, (*e*) the product of $2a$ and $5b$, (*f*) the number which is 4 more than 3 times x, (*g*) the number which is 5 less than twice y, (*h*) the time required to travel 250 miles at x miles per hour, (*i*) the cost (in cents) of x eggs at 65¢ per dozen.

a) $x + 2$ *c*) $5a + 3b$ *e*) $(2a)(5b) = 10ab$ *g*) $2y - 5$ *i*) $65(x/12)$

b) $a + (-b) = a - b$ *d*) $(2a)(3a) = 6a^2$ *f*) $3x + 4$ *h*) $250/x$

2. Let x be the present age of a father. (*a*) Express the present age of his son who 2 years ago was one-third his father's age. (*b*) Express the age of his daughter who 5 years from today will be one-fourth her father's age.

a) Two years ago the father's age was $x - 2$ and the son's age was $(x-2)/3$. Today the son's age is $2 + (x-2)/3$.

b) Five years from today the father's age will be $x + 5$ and his daughter's age will be $\frac{1}{4}(x+5)$. Today the daughter's age is $\frac{1}{4}(x+5) - 5$.

3 A pair of parentheses may be inserted or removed at will in an algebraic expression if the first parenthesis of the pair is preceded by a + sign. If, however, this sign is −, the signs of all terms within the parentheses must be changed.

a) $5a + 3a - 6a = (5+3-6)a = 2a$ *b*) $\frac{1}{2}a + \frac{1}{4}b - \frac{1}{4}a + \frac{3}{4}b = \frac{1}{4}a + b$

c) $(13a^2 - b^2) + (-4a^2 + 3b^2) - (6a^2 - 5b^2) = 13a^2 - b^2 - 4a^2 + 3b^2 - 6a^2 + 5b^2 = 3a^2 + 7b^2$

d) $(2ab - 3bc) - [5 - (4ab - 2bc)] = 2ab - 3bc - [5 - 4ab + 2bc] = 2ab - 3bc - 5 + 4ab - 2bc = 6ab - 5bc - 5$

e) $(2x+5y-4)3x = (2x)(3x) + (5y)(3x) - 4(3x) = 6x^2 + 15xy - 12x$

f)
$$\begin{array}{r} 5a - 2 \\ 3a + 4 \\ \hline 15a^2 - 6a \\ (+) \quad 20a - 8 \\ \hline 15a^2 + 14a - 8 \end{array}$$

g)
$$\begin{array}{r} 2x - 3y \\ 5x + 6y \\ \hline 10x^2 - 15xy \\ (+) \quad 12xy - 18y^2 \\ \hline 10x^2 - 3xy - 18y^2 \end{array}$$

h)
$$\begin{array}{r} 3a^2 + 2a - 1 \\ 2a - 3 \\ \hline 6a^3 + 4a^2 - 2a \\ (+) \quad - 9a^2 - 6a + 3 \\ \hline 6a^3 - 5a^2 - 8a + 3 \end{array}$$

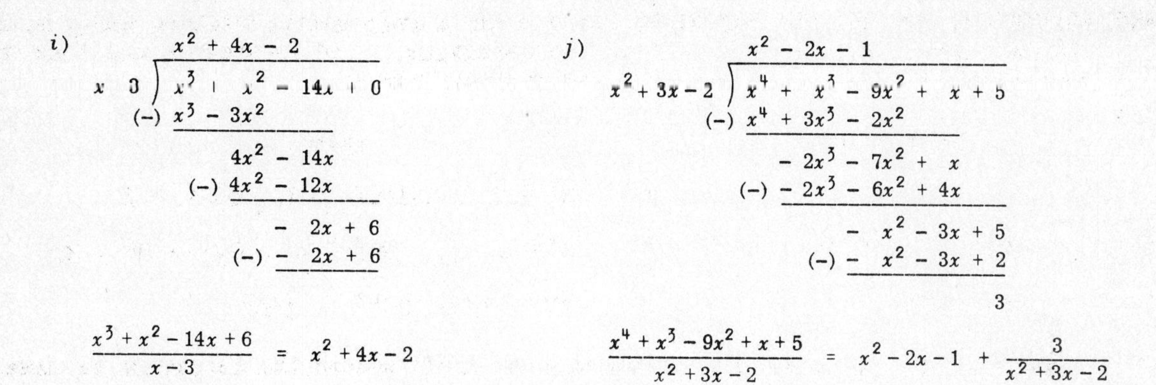

4. The problems below involve the following types of factoring:

$ab + ac - ad = a(b + c - d)$ $\qquad a^2 \pm 2ab + b^2 = (a \pm b)^2$ $\qquad a^3 + b^3 = (a+b)(a^2 - ab + b^2)$

$a^2 - b^2 = (a-b)(a+b)$ $\qquad acx^2 + (ad + bc)x + bd = (ax + b)(cx + d)$ $\qquad a^3 - b^3 = (a-b)(a^2 + ab + b^2)$

a) $5x - 10y = 5(x - 2y)$

b) $\frac{1}{2}gt^2 - \frac{1}{2}g^2t = \frac{1}{2}gt(t - g)$

c) $x^2 + 4x + 4 = (x+2)^2$

d) $x^2 + 5x + 4 = (x+1)(x+4)$

e) $x^2 - 3x - 4 = (x-4)(x+1)$

f) $4x^2 - 12x + 9 = (2x-3)^2$

g) $12x^2 + 7x - 10 = (4x+5)(3x-2)$

h) $x^3 - 8 = (x-2)(x^2 + 2x + 4)$

i) $2x^4 - 12x^3 + 10x^2 = 2x^2(x^2 - 6x + 5) = 2x^2(x-1)(x-5)$

5. Simplify:

a) $\dfrac{8}{12x + 20} = \dfrac{4 \cdot 2}{4 \cdot 3x + 4 \cdot 5} = \dfrac{2}{3x + 5}$

b) $\dfrac{9x^2}{12xy - 15xz} = \dfrac{3x \cdot 3x}{3x \cdot 4y - 3x \cdot 5z} = \dfrac{3x}{4y - 5z}$

c) $\dfrac{5x - 10}{7x - 14} = \dfrac{5(x-2)}{7(x-2)} = \dfrac{5}{7}$

d) $\dfrac{4x - 12}{15 - 5x} = \dfrac{4(x-3)}{5(3-x)} = \dfrac{4(x-3)}{-5(x-3)} = -\dfrac{4}{5}$

e) $\dfrac{x^2 - x - 6}{x^2 + 7x + 10} = \dfrac{(x+2)(x-3)}{(x+2)(x+5)} = \dfrac{x-3}{x+5}$

f) $\dfrac{6x^2 + 5x - 6}{2x^2 - 3x - 9} = \dfrac{(2x+3)(3x-2)}{(2x+3)(x-3)} = \dfrac{3x-2}{x-3}$

g) $\dfrac{3a^2 - 11a + 6}{a^2 - a - 6} \cdot \dfrac{4 - 4a - 3a^2}{36a^2 - 16} = \dfrac{(3a-2)(a-3)(2-3a)(2+a)}{(a-3)(a+2)4(3a+2)(3a-2)} = -\dfrac{3a-2}{4(3a+2)}$

6. Combine as indicated:

a) $\dfrac{2a + b}{10} + \dfrac{a - 6b}{15} = \dfrac{3(2a+b) + 2(a-6b)}{30} = \dfrac{8a - 9b}{30}$

b) $\dfrac{2}{x} - \dfrac{3}{2x} + \dfrac{5}{4} = \dfrac{2 \cdot 4 - 3 \cdot 2 + 5 \cdot x}{4x} = \dfrac{2 + 5x}{4x}$

c) $\dfrac{2}{3a - 1} - \dfrac{3}{2a + 1} = \dfrac{2(2a+1) - 3(3a-1)}{(3a-1)(2a+1)} = \dfrac{5 - 5a}{(3a-1)(2a+1)}$

d) $\dfrac{3}{x+y} - \dfrac{5}{x^2-y^2} = \dfrac{3}{x+y} - \dfrac{5}{(x+y)(x-y)} = \dfrac{3(x-y)-5}{(x+y)(x-y)} = \dfrac{3x-3y-5}{(x+y)(x-y)}$

e) $\dfrac{a-2}{6a^2-5a-6} + \dfrac{2a+1}{9a^2-4} = \dfrac{a-2}{(2a-3)(3a+2)} + \dfrac{2a+1}{(3a+2)(3a-2)}$

$= \dfrac{(a-2)(3a-2) + (2a+1)(2a-3)}{(2a-3)(3a+2)(3a-2)} = \dfrac{7a^2-12a+1}{(2a-3)(3a+2)(3a-2)}$

7. Perform the indicated operations.

a) $3^4 = 3\cdot3\cdot3\cdot3 = 81$

b) $-3^4 = -81$

c) $(-3)^4 = 81$

d) $-(-3)^4 = -81$

e) $-(-3)^3 = 27$

f) $2^6\cdot2^4 = 2^{6+4} = 2^{10} = 1024$

g) $(1/2)^3(1/2)^2 = (1/2)^5$

h) $a^{n+3}a^{m+2} = a^{m+n+5}$

i) $(a^2)^5 = a^{2\cdot5} = a^{10}$

j) $(a^{2n})^3 = a^{6n}$

k) $a^{10}/a^4 = a^{10-4} = a^6$

l) $a^4/a^{10} = 1/a^{10-4} = 1/a^6$

m) $(-2)^8/(-2)^5 = (-2)^3 = -8$

n) $a^{2n}b^{5m}/a^{3n}b^{2m} = b^{3m}/a^n$

o) $36^{x+3}/6^{x-1} = 6^{2x+6}/6^{x-1} = 6^{x+7}$

8. Evaluate.

a) $81^{1/2} = \sqrt{81} = 9$

b) $81^{3/4} = (\sqrt[4]{81})^3 = 3^3 = 27$

c) $(\frac{16}{49})^{3/2} = (\sqrt{\frac{16}{49}})^3 = (\frac{4}{7})^3 = 64/343$

d) $(-27)^{1/3} = \sqrt[3]{-27} = -3$

e) $(-32)^{4/5} = (\sqrt[5]{-32})^4 = (-2)^4 = 16$

f) $-400^{1/2} = -\sqrt{400} = -20$

9. Evaluate.

a) $4^0 = 1$

b) $4a^0 = 4\cdot1 = 4$

c) $(4a)^0 = 1$

d) $4(3+a)^0 = 4\cdot1 = 4$

e) $4^{-1} = 1/4$

f) $5^{-2} = 1/5^2 = 1/25$

g) $125^{-1/3} = 1/125^{1/3} = 1/5$

h) $(-125)^{-1/3} = -1/5$

i) $-(1/64)^{5/6} = -(\sqrt[6]{1/64})^5 = -(1/2)^5 = -1/32$

j) $-(-1/32)^{4/5} = -(\sqrt[5]{-1/32})^4 = -(-1/2)^4 = -1/16$

10. Perform each of the following operations and express the result without negative or zero exponents.

a) $(\dfrac{81a^4}{b^8})^{-1/4} = \dfrac{3^{-1}a^{-1}}{b^{-2}} = \dfrac{b^2}{3a}$

b) $(a^{1/2} + a^{-1/2})^2 = a + 2a^0 + a^{-1} = a + 2 + \dfrac{1}{a}$

c) $(a-3b^{-2})(2a^{-1}-b^2) = 2a^0 - ab^2 - 6a^{-1}b^{-2} + 3b^0 = 5 - ab^2 - 6/ab^2$

d) $\dfrac{a^{-2}+b^{-2}}{a^{-1}-b^{-1}} = \dfrac{(a^{-2}+b^{-2})(a^2b^2)}{(a^{-1}-b^{-1})(a^2b^2)} = \dfrac{b^2+a^2}{ab^2-a^2b}$

e) $(\dfrac{a^2}{b})^7(-\dfrac{b^2}{a^3})^6 = \dfrac{a^{14}\cdot b^{12}}{b^7\cdot a^{18}} = \dfrac{b^5}{a^4}$

f) $(\dfrac{a^{1/2}b^{2/3}}{c^{3/4}})^6(\dfrac{c^{1/2}}{a^{1/4}b^{1/3}})^9 = \dfrac{a^3b^4}{c^{9/2}}\cdot\dfrac{c^{9/2}}{a^{9/4}b^3} = a^{3/4}b$

SUPPLEMENTARY PROBLEMS

11. Combine. (*a*) $2x + (3x - 4y)$ (*b*) $5a + 4b - (-2a + 3b)$ (*c*) $\{(s + 2t) - (s + 3t)\} - \{(2s + 3t) - (-4s + 5t)\}$ (*d*) $8x^2y - \{3x^2y + [2xy^2 + 4x^2y - (3xy^2 - 4x^2y)]\}$

12. Perform the indicated operations.
(*a*) $4x(x - y + 2)$ (*b*) $(5x + 2)(3x - 4)$ (*c*) $(5x^2 - 4y^2)(-x^2 + 3y^2)$ (*d*) $(x^2 - 3x + 5)(2x - 7)$ (*e*) $(2x^3 + 5x^2 - 33x + 20) \div (2x - 5)$ (*f*) $(2x^3 + 5x^2 - 22x + 10) \div (2x - 3)$

13. Factor. (*a*) $8x + 12y$ (*b*) $4ax + 6ay - 24az$ (*c*) $a^2 - 4b^2$ (*d*) $50ab^4 - 98a^3b^2$ (*e*) $16a^2 - 8ab + b^2$ (*f*) $25x^2 + 30xy + 9y^2$ (*g*) $x^2 - 4x - 12$ (*h*) $a^2 + 23ab - 50b^2$ (*i*) $(x - y)^2 + 6(x - y) + 5$ (*j*) $4x^2 - 8x - 5$ (*k*) $40a^2 + ab - 6b^2$ (*l*) $x^4 + 24x^2y^2 - 25y^4$

14. Simplify. (*a*) $\dfrac{a^2 - b^2}{2ax + 2bx}$ (*b*) $\dfrac{x^2 + 4x + 3}{1 - x^2}$ (*c*) $\dfrac{1 - x - 12x^2}{1 + x - 6x^2}$

(*d*) $\dfrac{16a^2 - 25}{2a - 10} \times \dfrac{a^2 - 10a + 25}{4a + 5}$ (*e*) $\dfrac{x^2 + xy - 6y^2}{2x^3 + 6x^2y} \times \dfrac{8x^2y}{x^2 - 5xy + 6y^2}$

15. Perform the indicated operations.
(*a*) $\dfrac{5x}{18} + \dfrac{4x}{18}$ (*b*) $\dfrac{3a}{x} + \dfrac{5a}{2x}$ (*c*) $\dfrac{3a}{4b} - \dfrac{4b}{3a}$ (*d*) $\dfrac{2a - 3b}{a^2 - b^2} + \dfrac{1}{a - b}$ (*e*) $x + 5 - \dfrac{x^2}{x - 5}$ (*f*) $\dfrac{a + 2}{2a - 6} - \dfrac{a - 2}{2a + 6}$ (*g*) $\dfrac{2x + 3}{18x^2 - 27x} - \dfrac{2x - 3}{18x^2 + 27x}$

16. Simplify. (*a*) $\dfrac{a}{2 - \dfrac{3}{a}}$ (*b*) $\dfrac{4 - \dfrac{x}{3}}{\dfrac{x}{6}}$ (*c*) $\dfrac{x - \dfrac{4}{x}}{1 - \dfrac{2}{x}}$ (*d*) $\dfrac{\dfrac{1}{x} + \dfrac{1}{y}}{\dfrac{x + y}{y} + \dfrac{x + y}{x}}$

ANSWERS TO SUPPLEMENTARY PROBLEMS

11. (*a*) $5x - 4y$ (*b*) $7a + b$ (*c*) $t - 6s$ (*d*) $xy(y - 3x)$

12. (*a*) $4x^2 - 4xy + 8x$ (*b*) $15x^2 - 14x - 8$ (*c*) $-5x^4 + 19x^2y^2 - 12y^4$ (*d*) $2x^3 - 13x^2 + 31x - 35$ (*e*) $x^2 + 5x - 4$ (*f*) $x^2 + 4x - 5 - 5/(2x - 3)$

13. (*a*) $4(2x + 3y)$ (*b*) $2a(2x + 3y - 12z)$ (*c*) $(a - 2b)(a + 2b)$ (*d*) $2ab^2(5b - 7a)(5b + 7a)$ (*e*) $(4a - b)^2$ (*f*) $(5x + 3y)^2$ (*g*) $(x - 6)(x + 2)$ (*h*) $(a + 25b)(a - 2b)$ (*i*) $(x - y + 1)(x - y + 5)$ (*j*) $(2x + 1)(2x - 5)$ (*k*) $(5a + 2b)(8a - 3b)$ (*l*) $(x - y)(x + y)(x^2 + 25y^2)$

14. (*a*) $\dfrac{a - b}{2x}$ (*b*) $-\dfrac{x + 3}{x - 1}$ (*c*) $\dfrac{4x - 1}{2x - 1}$ (*d*) $\dfrac{1}{2}(4a^2 - 25a + 25)$ (*e*) $\dfrac{4y}{x - 3y}$

15. (*a*) $\dfrac{1}{2}x$ (*b*) $\dfrac{11a}{2x}$ (*c*) $\dfrac{9a^2 - 16b^2}{12ab}$ (*d*) $\dfrac{3a - 2b}{a^2 - b^2}$ (*e*) $-\dfrac{25}{x - 5}$ (*f*) $\dfrac{5a}{a^2 - 9}$ (*g*) $\dfrac{8}{12x^2 - 27}$

16. (*a*) $\dfrac{a^2}{2a - 3}$ (*b*) $\dfrac{24 - 2x}{x}$ (*c*) $x + 2$ (*d*) $\dfrac{1}{x + y}$

CHAPTER 3

Functions

A VARIABLE IS A SYMBOL selected to represent any one of a given set of numbers, here assumed to be real numbers. Should the set consist of just one number, the symbol representing it is called a *constant*.

The *range* of a variable consists of the totality of numbers of the set which it represents. For example, if x is a day in September, the range of x is the set of positive integers $1, 2, 3, \ldots, 30$; if x (ft) is the length of rope cut from a piece 50 ft long, the range of x is the set of numbers greater than 0 and less than 50.

Examples of ranges of a real variable, together with special notations and graphical representations, are given in Solved Problem 1.

FUNCTION. A correspondence (x, y) between two sets of numbers which pairs to an arbitrary number x of the first set one or more numbers y of the second is called a *function*. In this case, it is customary to speak of y as a *function of* x. The variable x is called the *independent variable* and y is called the *dependent variable*.

A function may be defined

(*a*) by a table of correspondents or table of values, as

x	1	2	3	4	5	6	7	8	9	10
y	3	4	5	6	7	8	9	10	11	12

(*b*) by an equation or formula, as $y = x + 2$.

For each value assigned to x, the above relation yields a correspondent. Note that the table above is a table of values for this function.

A FUNCTION IS CALLED *single-valued* if, to each value of x in its range, there corresponds just one value of y; otherwise, the function is called *multi-valued*. For example, $y = x^2 - 5x + 4$ defines y as a single-valued function of x while $x^2 + y^2 = 1$ or $y = \pm\sqrt{1 - x^2}$ defines y as a multi-valued (here, two-valued) function of x.

At times it will be more convenient to label a given function of x as $f(x)$, to be read "the f function of x" or simply "f of x". (Note carefully that this is not to be confused with "f times x".) If there are two functions one may be labeled $f(x)$ and the other $g(x)$. Also, if $y = f(x) = x^2 - 5x + 4$ the statement "the value of the function is -2 when $x = 3$" can be replaced by "$f(3) = -2$". See Problem 2.

Let $y = f(x)$. The range of the independent variable x is called the *domain of definition* of the function while the range of the dependent variable is called the *range of the function*. For example, $y = x^2$ defines a function whose domain of definition consists of all (real) numbers and whose range is all non-negative numbers, that is, zero and the positive numbers; $f(x) = 3/(x - 2)$ defines a function whose domain of definition consists of all numbers except 2 (why?) and whose range is all numbers except 0.

See Problems 3-8.

A VARIABLE w (dependent) is said to be a function of the (independent) variables $x, y, z, \ldots$ if, when a value of each of the variables $x, y, z, \ldots$ is known, there corresponds one or more values of w. For example, the volume (V) of a rectangular parallelipiped of dimensions x, y, z is given by $V = xyz$. Here V is a function of three independent variables.

See Problems 9-10.

SOLVED PROBLEMS

1. Represent graphically each of the following ranges:

a) $x > -2$ b) $x < 5$ c) $x \leqq -1$ d) $-3 < x < 4$ e) $-2 < x < 2$ or $|x| < 2$ f) $|x| > 3$ g) $-3 \leqq x \leqq 5$ h) $x \leqq -3,\ x \geqq 4$

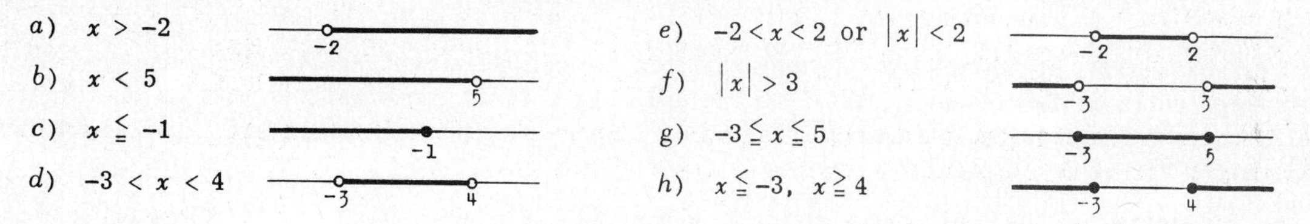

2. Given $f(x) = x^2 - 5x + 4$, find:

a) $f(0) = 0^2 - 5\cdot 0 + 4 = 4$

b) $f(2) = 2^2 - 5\cdot 2 + 4 = -2$

c) $f(-3) = (-3)^2 - 5(-3) + 4 = 28$

d) $f(a) = a^2 - 5a + 4$

e) $f(-x) = x^2 + 5x + 4$

f) $f(b+1) = (b+1)^2 - 5(b+1) + 4 = b^2 - 3b$

g) $f(3x) = (3x)^2 - 5(3x) + 4 = 9x^2 - 15x + 4$

h) $f(x+a) - f(a) = [(x+a)^2 - 5(x+a) + 4] - (a^2 - 5a + 4) = x^2 + 2ax - 5x$

i) $$\frac{f(x+a) - f(x)}{a} = \frac{[(x+a)^2 - 5(x+a) + 4] - (x^2 - 5x + 4)}{a} = \frac{2ax - 5a + a^2}{a} = 2x - 5 + a$$

3. In each of the following, state the domain of definition:

(a) $y = 5x$

(b) $y = -5x$

(c) $y = \dfrac{1}{x+5}$

(d) $y = \dfrac{x-2}{(x-3)(x+4)}$

(e) $y = \dfrac{1}{x}$

(f) $y = \sqrt{25 - x^2}$

(g) $y = \sqrt{x^2 - 9}$

(h) $y = \dfrac{1}{16 - x^2}$

(i) $y = \dfrac{1}{16 + x^2}$

Ans. a), b), all real numbers; c) $x \neq -5$; d) $x \neq 3, -4$; e) $x \neq 0$; f) $-5 \leqq x \leqq 5$ or $|x| \leqq 5$; g) $x \leqq -3,\ x \geqq 3$ or $|x| \geqq 3$; h) $x \neq \pm 4$; i) all real numbers.

4. A piece of wire 30 in. long is bent to form a rectangle. If one of its dimensions is x in., express the area as a function of x.

Since the semi-perimeter of the rectangle is $\frac{1}{2}\cdot 30 = 15$ in. and one dimension is x in., the other is $(15 - x)$ in. Thus, $A = x(15 - x)$.

5. An open box is to be formed from a rectangular sheet of tin 20×32 in. by cutting equal squares, x in. on a side, from the four corners and turning up the sides. Express the volume of the box as a function of x.

From Fig. (a) below, it is seen that the base of the box has dimensions $(20 - 2x)$ by $(32 - 2x)$ in. and the height is x in. Then

$$V = x(20 - 2x)(32 - 2x) = 4x(10 - x)(16 - x)$$

6. A closed box is to be formed from the sheet of tin of Problem 5 by cutting equal squares, x in. on a side, from two corners of the short side and two equal rectangles of width x in. from the other two corners, and folding along the dotted lines shown in Fig. (b) below. Express the volume of the box as a function of x.

One dimension of the base of the box is $(20-2x)$ in.; let y in. be the other. Then $2x+2y = 32$ and $y = 16-x$. Thus,

$$V = x(20-2x)(16-x) = 2x(10-x)(16-x)$$

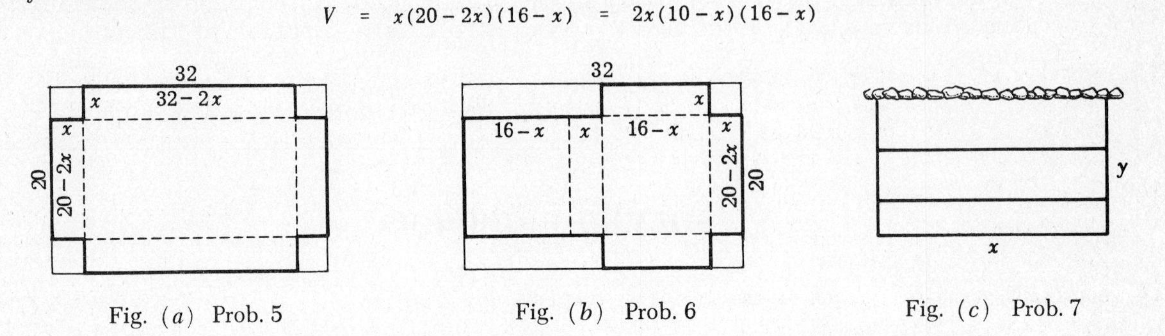

Fig. (a) Prob. 5 Fig. (b) Prob. 6 Fig. (c) Prob. 7

7. A farmer has 600 ft of woven wire fencing available to enclose a rectangular field and to divide it into three parts by two fences parallel to one end. If x ft of stone wall is used as one side of the field, express the area enclosed as a function of x when the dividing fences are parallel to the stone wall. Refer to Fig. (c) above.

The dimensions of the field are x and y ft where $3x+2y = 600$. Then $y = \frac{1}{2}(600-3x)$ and the required area is

$$A = xy = x\cdot\frac{1}{2}(600-3x) = \frac{3}{2}x(200-x)$$

8. A right circular cylinder is said to be inscribed in a sphere if the circumference of the bases of the cylinder are in the surface of the sphere. If the sphere has radius R, express the volume of the inscribed right circular cylinder as a function of the radius r of its base.

Let the altitude of the cylinder be denoted by $2h$. From the adjoining figure, $h = \sqrt{R^2-r^2}$ and the required volume is

$$V = \pi r^2\cdot 2h = 2\pi r^2\sqrt{R^2-r^2}$$

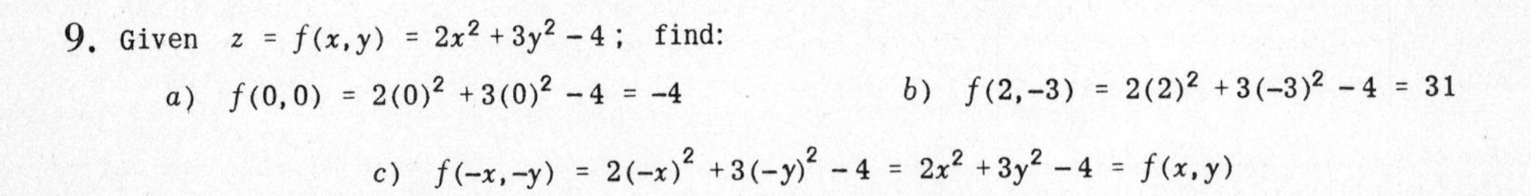

9. Given $z = f(x,y) = 2x^2+3y^2-4$; find:

a) $f(0,0) = 2(0)^2+3(0)^2-4 = -4$

b) $f(2,-3) = 2(2)^2+3(-3)^2-4 = 31$

c) $f(-x,-y) = 2(-x)^2+3(-y)^2-4 = 2x^2+3y^2-4 = f(x,y)$

10. Given $f(x,y) = \dfrac{x^2+y^2}{x^2-y^2}$, find:

a) $f(x,-y) = \dfrac{x^2+(-y)^2}{x^2-(-y)^2} = \dfrac{x^2+y^2}{x^2-y^2} = f(x,y)$

b) $f(\frac{1}{x},\frac{1}{y}) = \dfrac{(1/x)^2+(1/y)^2}{(1/x)^2-(1/y)^2} = \dfrac{1/x^2+1/y^2}{1/x^2-1/y^2} = \dfrac{y^2+x^2}{y^2-x^2} = -f(x,y)$

SUPPLEMENTARY PROBLEMS

11. Represent graphically each of the ranges.

(a) $x > -3$ (c) $x \geq 0$ (e) $|x| < 2$ (g) $-4 \leq x \leq 4$
(b) $x < 5$ (d) $-3 < x < -1$ (f) $|x| \geq 0$ (h) $x < -3,\ x \geq 5$

12. In the three angles A, B, C of a triangle, angle B exceeds twice angle A by $15°$. Express angle C in terms of angle A. *Ans.* $C = 165° - 3A$

13. A grocer has two grades of coffee, selling at 90¢ and $1.05 per pound respectively. In making a mixture of 100 pounds, he uses x pounds of the $1.05 coffee. (a) How many pounds of the 90¢ coffee does he use? (b) What is the value in cents of the mixture? (c) At what price per pound should he offer the mixture? *Ans.* (a) $100 - x$ (b) $90(100 - x) + 105x$ (c) $90 + 0.15x$

14. In a purse are nickles, dimes, and quarters. The number of dimes is twice the number of quarters and the number of nickles is 3 less than twice the number of dimes. If there are x quarters, find the sum (in cents) in the purse. *Ans.* $65x - 15$

15. A and B start from the same place. A walks 4 mi/hr and B walks 5 mi/hr. (a) How far (in mi) will each walk in x hr? (b) How far apart will they be after x hours if they leave at the same time and move in opposite directions? (c) How far apart will they be after A has walked $x > 2$ hours if they move in the same direction but B leaves 2 hr after A? (d) In (c), for how many hours would B have to walk in order to overtake A? *Ans.* (a) A, $4x$; B, $5x$ (b) $9x$ (c) $|4x - 5(x-2)|$ (d) 8

16. A motor boat, which moves at x mi/hr in still water, is on a river whose current is $y < x$ mi/hr. (a) What is the rate (mi/hr) of the boat when moving upstream? (b) What is the rate of the boat when moving downstream? (c) How far (mi) will the boat travel upstream in 8 hr? (d) How long (hr) will it take the boat moving downstream to cover 20 mi if the motor dies after the first 15 mi?

Ans. (a) $x - y$ (b) $x + y$ (c) $8(x - y)$ (d) $\dfrac{15}{x+y} + \dfrac{5}{y}$

17. Given $f(x) = \dfrac{x-3}{x+2}$; find $f(0)$, $f(1)$, $f(-3)$, $f(3)$, $f(a)$, $f(3y)$, $f(x+a)$, $\dfrac{f(x+a) - f(x)}{a}$.

Ans. $-3/2$, $-2/3$, 6, 0, $\dfrac{a-3}{a+2}$, $\dfrac{3y-3}{3y+2}$, $\dfrac{x+a-3}{x+a+2}$, $\dfrac{5}{(x+2)(x+a+2)}$

18. A ladder 25 ft long leans against a vertical wall with its foot on level ground 7 ft from the base of the wall. If the foot is pulled away from the wall at the rate 2 ft/sec, express the distance (y ft) of the top of the ladder above the ground as a function of the time t sec in moving.
Ans. $y = 2\sqrt{144 - 7t - t^2}$

19. A boat is tied to a dock by means of a cable 60 ft long. If the dock is 20 ft above the water and if the cable is being drawn in at the rate 10 ft/min, express the distance y ft of the boat from the dock after t min. *Ans.* $y = 10\sqrt{t^2 - 12t + 32}$

20. A train leaves a station at noon and travels east at the rate 30 mi/hr. At 2 P.M. of the same day a second train leaves the station and travels south at the rate 25 mi/hr. Express the distance d (mi) between the trains as a function of t(hr), the time the second train has been traveling.
Ans. $d = 5\sqrt{61t^2 + 144t + 144}$

CHAPTER 4

Functions and Loci

A FUNCTION $y = f(x)$, by definition, yields a collection of number pairs $(x, f(x))$ or (x, y) in which x is any element in the domain of definition of the function and $f(x)$ or y is the corresponding value of the function.

Example 1. Obtain ten number pairs for the function $y = 3x - 2$.

The domain of definition of the function is the set of real numbers. We may choose at random any ten real numbers as values of x. For one such choice, we obtain

x	-2	-4/3	-1/2	0	1/3	1	2	5/2	3	4
y	-8	-6	-7/2	-2	-1	1	4	11/2	7	10

See Problem 1.

THE RECTANGULAR CARTESIAN COORDINATE SYSTEM in a plane is a device by which there is established a one-to-one correspondence between the points of the plane and ordered pairs of real numbers (a, b).

Consider two real number scales intersecting at right angles in O, the origin of each (see adjoining diagram), and having the positive direction on the horizontal scale (now called the *x-axis*) directed to the right and the positive direction on the vertical scale (now called the *y-axis*) directed upward.

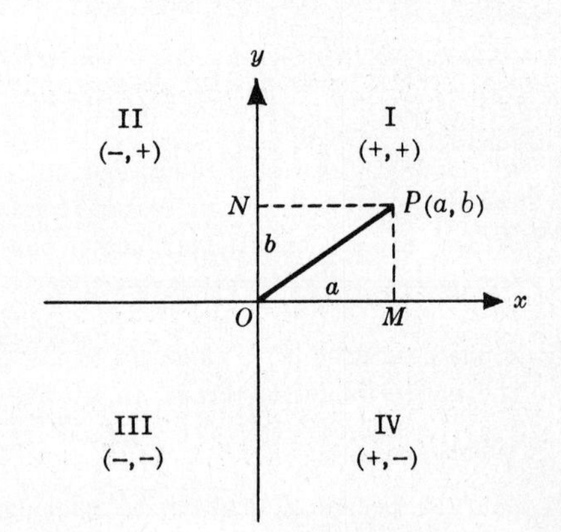

Let P be any point distinct from O in the plane of the two axes and join P to O by a straight line. Let the projection of OP on the x-axis be $OM = a$ and the projection of OP on the y-axis be $ON = b$. Then the pair of numbers (a, b) in that order are called the plane rectangular cartesian coordinates (briefly, the rectangular coordinates) of P. In particular, the coordinates of O, the *origin* of the coordinate system, are $(0,0)$.

The first coordinate, giving the directed distance of P from the y-axis, is called the *abscissa* of P while the second coordinate, giving the directed distance of P from the x-axis, is called the *ordinate* of P. Note carefully that the points $(3,4)$ and $(4,3)$ are distinct points.

The axes divide the plane into four sections, called *quadrants*. The above figure shows the customary numbering of the quadrants and the respective signs of the coordinates of a point in each quadrant.

See Problems 1-3.

THE GRAPH OF A FUNCTION $f(x)$ consists of the totality of points (x,y) whose coordinates satisfy the relation $y = f(x)$.

Example 2. Graph the function $3x - 2$.

After plotting the points whose coordinates (x,y) are given in the table of Example 1, it appears that they lie on a straight line. The adjoining sketch is not the complete locus since (1000, 2998) is one of its points and is not shown. Moreover, although we have joined the points by a straight line, we have not proved that every point on the line has as coordinates a number-pair given by the function. These matters as well as such questions as. What values of x should be chosen? How many values of x are needed? will become clearer as we proceed with the study of functions. At present,

1) build a table of values,
2) plot the corresponding points,
3) pass a smooth curve through these points, moving from left to right.

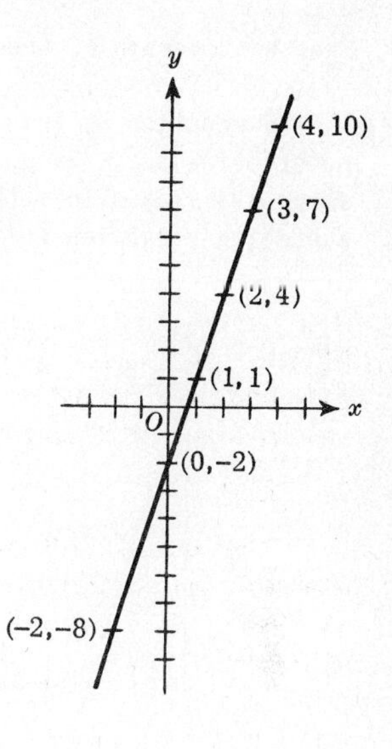

It is helpful to picture the curve in your mind before attempting to trace it on paper. If there is doubt about the curve between two plotted points, determine other points in the interval.

ANY VALUE OF x for which the corresponding value of a function $f(x)$ is zero is called a *zero* of the function. Such values of x are also called *roots* of the equation $f(x) = 0$. The real roots of an equation $f(x) = 0$ may be approximated by estimating from the graph of $f(x)$ the abscissas of its points of intersection with the x-axis.

See Problems 9-11.

Algebraic methods for finding the roots of equations will be treated in later chapters.

SOLVED PROBLEMS

1. (*a*) Show that the points $A(1,2)$, $B(0,-3)$ and $C(2,7)$ are on the graph of $y = 5x - 3$.
(*b*) Show that the points $D(0,0)$ and $E(-1,-2)$ are not on the graph of $y = 5x - 3$.

a) The point $A(1,2)$ is on the graph since $2 = 5(1) - 3$, $B(0,-3)$ is on the graph since $-3 = 5(0) - 3$, and $C(2,7)$ is on the graph since $7 = 5(2) - 3$.

b) The point $D(0,0)$ is not on the graph since $0 \neq 5(0) - 3$, and $E(-1,-2)$ is not on the graph since $-2 \neq 5(-1) - 3$.

2. Sketch the graph of the function $2x$.

This is a linear function and its graph is a straight line. For this locus only two points are necessary. Three points are used to provide a check. See adjoining figure.

x	0	1	2
$y = f(x)$	0	2	4

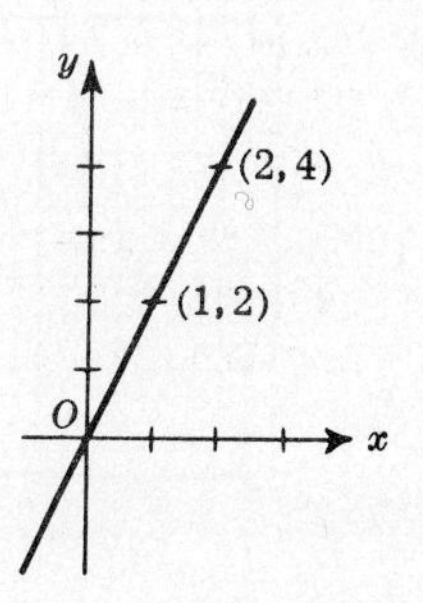

The equation of the line is $y = 2x$.

3. Sketch the graph of the function $6-3x$.

x	0	2	3
$y=f(x)$	6	0	−3

See Fig. (a) below.

The equation of the line is $y=6-3x$.

4. Sketch the graph of the function x^2.

x	3	1	0	−2	−3
$y=f(x)$	9	1	0	4	9

See Fig. (b) below.

The equation of this locus, called a *parabola*, is $y=x^2$. Note for $x\neq 0$, $x^2>0$. Thus, the curve is never below the x-axis. Moreover as $|x|$ increases, x^2 increases, that is, as we move from the origin along the x-axis in either direction, the curve moves farther and farther from the axis. Hence, in sketching parabolas sufficient points must be plotted so that its U shape can be seen.

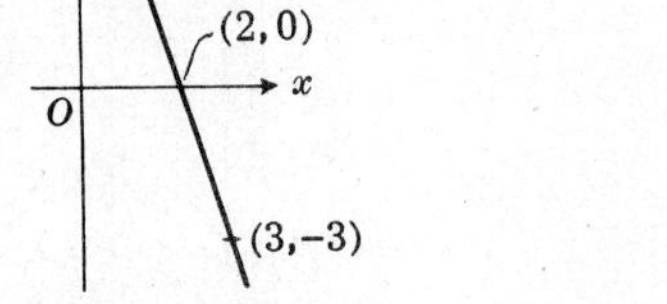

Fig. (a) Prob. 3

Fig. (b) Prob. 4

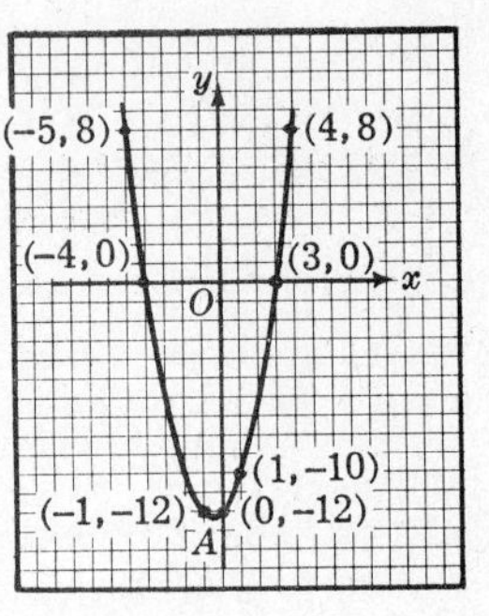

Fig. (c) Prob. 5

5. Sketch the graph of the function x^2+x-12.

x	4	3	1	0	−1	−4	−5
$y=f(x)$	8	0	−10	−12	−12	0	8

The equation of the parabola is $y=x^2+x-12$. Note that the points (0,−12) and (−1,−12) are *not* joined by a straight line segment. Check that the value of the function is $-12\frac{1}{4}$ when $x=-\frac{1}{2}$. See Fig. (c) above.

6. Sketch the graph of the function $-2x^2+4x+1$.

x	3	2	1	0	−1
$y=f(x)$	−5	1	3	1	−5

See Fig. (d) below.

7. Sketch the graph of the function $(x+1)(x-1)(x-2)$.

x	3	2	3/2	1	0	−1	−2
$y=f(x)$	8	0	−5/8	0	2	0	−12

This is a *cubic* curve of equation $y=(x+1)(x-1)(x-2)$. It crosses the x-axis where $x=-1, 1,$ and 2. See Fig. (e) below.

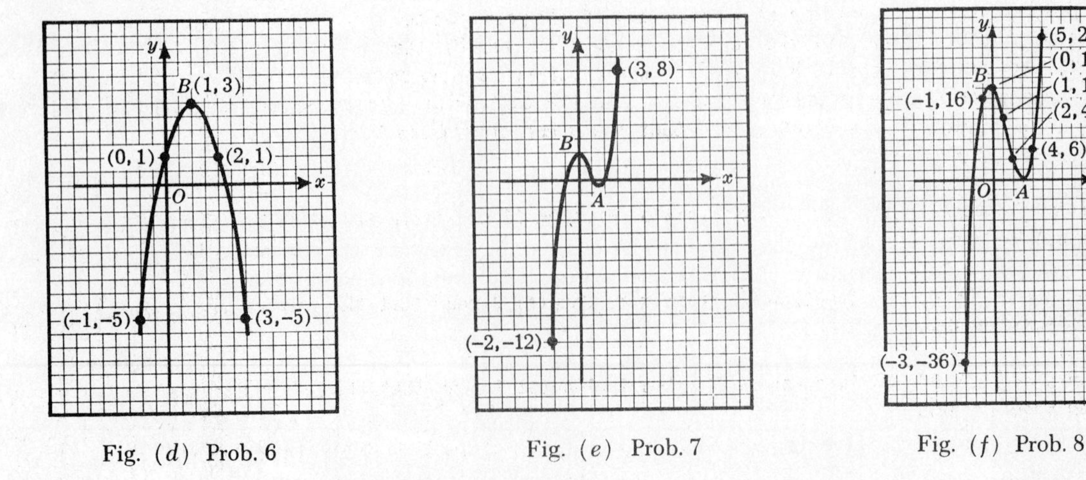

Fig. (d) Prob. 6

Fig. (e) Prob. 7

Fig. (f) Prob. 8

8. Sketch the graph of the function $(x+2)(x-3)^2$.

x	5	4	7/2	3	2	1	0	−1	−2	−3
$y=f(x)$	28	6	11/8	0	4	12	18	16	0	−36

This cubic crosses the x-axis where $x=-2$ and is tangent to the x-axis where $x=3$. Note that for $x>-2$, the value of the function is positive except for $x = 3$, where it is 0. Thus, to the right of $x=-2$, the curve is *never* below the x-axis. See Fig. (f) above.

9. Sketch the graph of the function x^2+2x-5 and by means of it determine the real roots of $x^2+2x-5=0$.

x	2	1	0	−1	−2	−3	−4
$y=f(x)$	3	−2	−5	−6	−5	−2	3

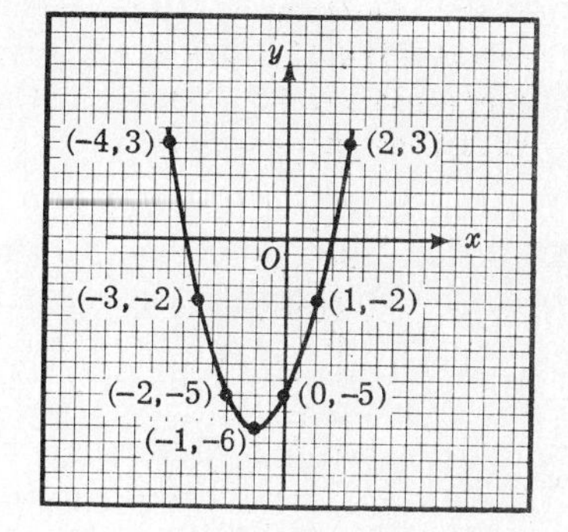

The parabola cuts the x-axis in a point whose abscissa is between 1 and 2 (the value of the function changes sign) and in a point whose abscissa is between −3 and −4.

Reading from the graph in the adjacent figure, the roots are $x=1.5$ and $x=-3.5$, approximately.

SUPPLEMENTARY PROBLEMS

10. Sketch the graph of each of the following functions.
(a) $3x-2$ (b) $2x+3$ (c) x^2-1 (d) $4-x^2$ (e) x^2-4x+4 (f) $(x+2)(x-1)(x-3)$ (g) $(x-2)(x+1)^2$

11. From the graph of each function $f(x)$ determine the real roots, if any, of $f(x)=0$.
(a) x^2-4x+3 (b) $2x^2+4x+1$ (c) x^2-2x+4 *Ans.* (a) 1, 3 (b) −0.3, −1.7 (c) none

12. If A is a point on the graph of $y=f(x)$, the function being restricted to the type considered in this chapter, and if all points of the graph sufficiently near A are higher than A (that is, lie above the horizontal drawn through A) then A is called a *relative minimum point* of the graph. (a) Verify that the origin is the relative minimum point of the graph of Problem 4. (b) Verify that the graph of Problem 5 has a relative minimum at a point whose abscissa is between $x=-1$ and $x=0$ (at $x=-\frac{1}{2}$), the graph of Problem 7 has a relative minimum at a point whose abscissa is between $x=1$ and $x=2$ (approximately $x=1.5$), and the graph of Problem 8 has (3,0) as relative minimum point.

13. If B is a point on the graph of $y=f(x)$ and if all points of the graph sufficiently near to B are lower than B (that is, lie below the horizontal drawn through B) then B is called a *relative maximum point* of the graph. (a) Verify that (1,3) is the relative maximum point of the graph of Problem 6. (b) Verify that the graph of Problem 7 has a relative maximum at a point whose abscissa is between $x=-1$ and $x=1$ (approximately $x=-0.2$) and that the graph of Problem 8 has a relative maximum between $x=-1$ and $x=0$ (at $x=-1/3$).

14. Verify that the graphs of the functions of Problem 11 have relative minimums at $x=2$, $x=-1$, and $x=1$ respectively.

15. From the graph of the function of Problem 4, Chapter 3 read that the area of the rectangle is a relative maximum when $x = 15/2$.

16. From the graph of the function of Problem 7, Chapter 3 read that the area enclosed is a relative maximum when $x=100$.

CHAPTER 5

The Linear Equation

AN EQUATION is a statement, such as (*a*) $2x - 6 = 4 - 3x$, (*b*) $y^2 + 3y = 4$, (*c*) $2x + 3y = 4xy + 1$, that two expressions are equal. The first is a *linear* equation in one unknown, the second is a *quadratic* in one unknown, the third is linear in each of the two unknowns but is of degree two in the two unknowns.

Any set of values of the unknowns for which the two members of an equation are equal is called a *solution* of the equation. Thus, $x = 2$ is a solution of (*a*) since $2(2) - 6 = 4 - 3(2)$; $y = 1$ and $y = -4$ are solutions of (*b*); and $x = 1$, $y = 1$ is a solution of (*c*). A solution of an equation in one unknown is also called a *root* of the equation.

TO SOLVE A LINEAR EQUATION in one unknown perform the same operations on both members of the equation in order to obtain the unknown alone in the left member.

Example 1. Solve: $2x - 6 = 4 - 3x$.

Add 6: $2x = 10 - 3x$
Add $3x$: $5x = 10$
Divide by 5: $x = 2$

Check: $2(2) - 6 = 4 - 3(2)$
$-2 = -2$

Example 2. Solve: $\frac{1}{3}x - \frac{1}{2} = \frac{3}{4}x + \frac{5}{6}$.

Multiply by L.C.D. = 12: $4x - 6 = 9x + 10$
Add $6 - 9x$: $-5x = 16$
Divide by -5: $x = -16/5$

Check: $\frac{1}{3}(-\frac{16}{5}) - \frac{1}{2} = \frac{3}{4}(-\frac{16}{5}) + \frac{5}{6}$
$-47/30 = -47/30$

See Problems 1-3.

An equation which contains fractions having the unknown in one or more denominators may sometimes reduce to a linear equation when cleared of fractions. When the resulting equation is solved, the solution *must* be checked since it may or may not be a root of the original equation.

See Problems 4-8.

RATIO AND PROPORTION. The ratio of two quantities is their quotient. The ratio of 1 inch to 1 foot is 1/12 or 1:12, a pure number; the ratio of 30 miles to 45 minutes is 30/45 = 2/3 miles per minute.

The expressed equality of two ratios, as $\frac{a}{b} = \frac{c}{d}$ is called a *proportion*.

See Problems 11-12.

VARIATION. A variable y is said to vary *directly* as another variable x (or y is proportional to x) if y is equal to some constant c times x, that is, if $y = cx$.

A variable y is said to vary *inversely* as another variable x if y varies directly as the reciprocal of x, that is, if $y = c/x$.

A variable z is said to vary *jointly* as x and y if z varies directly as the product xy, that is, if $z = cxy$.

See Problems 13-14.

SOLVED PROBLEMS

Solve and check the following equations. The check has been omitted in certain problems.

1. $x - 2(1-3x) = 6 + 3(4-x)$.

$$\begin{aligned} x - 2 + 6x &= 6 + 12 - 3x \\ 7x - 2 &= 18 - 3x \\ 10x &= 20 \\ x &= 2 \end{aligned}$$

2. $ay + b = cy + d$.

$$\begin{aligned} ay - cy &= d - b \\ (a-c)y &= d - b \\ y &= \frac{d-b}{a-c} \end{aligned}$$

3. $\dfrac{3x-2}{5} = 4 - \dfrac{1}{2}x$.

Multiply by 10: $6x - 4 = 40 - 5x$, $11x = 44$, $x = 4$

Check: $\dfrac{3(4)-2}{5} = 4 - \dfrac{1}{2}(4)$, $2 = 2$

4. $\dfrac{3x+1}{3x-1} = \dfrac{2x+1}{2x-3}$. Here the L.C.D. is $(3x-1)(2x-3)$.

Multiply by L.C.D.:
$$\begin{aligned} (3x+1)(2x-3) &= (2x+1)(3x-1) \\ 6x^2 - 7x - 3 &= 6x^2 + x - 1 \\ -8x &= 2 \\ x &= -\tfrac{1}{4} \end{aligned}$$

Check: $\dfrac{3(-\frac{1}{4})+1}{3(-\frac{1}{4})-1} = \dfrac{2(-\frac{1}{4})+1}{2(-\frac{1}{4})-3}$, $\dfrac{-3+4}{-3-4} = \dfrac{-2+4}{-2-12}$, $-\dfrac{1}{7} = -\dfrac{1}{7}$

5. $\dfrac{1}{x-3} - \dfrac{1}{x+1} = \dfrac{3x-2}{(x-3)(x+1)}$. Here the L.C.D. is $(x-3)(x+1)$.

$$\begin{aligned} (x+1) - (x-3) &= 3x - 2 \\ -3x &= -6 \\ x &= 2 \end{aligned}$$

Check: $-1 - \dfrac{1}{3} = \dfrac{6-2}{-3}$, $-4/3 = -4/3$

6. $\dfrac{1}{x-3} + \dfrac{1}{x-2} = \dfrac{3x-8}{(x-3)(x-2)}$. The L.C.D. is $(x-3)(x-2)$.

$$\begin{aligned} (x-2) + (x-3) &= 3x - 8 \\ 2x - 5 &= 3x - 8 \\ x &= 3 \end{aligned}$$

Check: When $x = 3$, $\dfrac{1}{x-3}$ is without meaning. The given equation has no root. The value $x = 3$ is called *extraneous*.

7. $\dfrac{x^2-2}{x-1} = x + 1 - \dfrac{1}{x-1}$.

$$\begin{aligned} x^2 - 2 &= (x+1)(x-1) - 1 \\ &= x^2 - 2 \end{aligned}$$

The given equation is satisfied by all values of x except $x = 1$. It is called an identical equation or *identity*.

8. $\dfrac{1}{x-1} + \dfrac{1}{x-3} = \dfrac{2x-5}{(x-1)(x-3)}$.

$$\begin{aligned} (x-3) + (x-1) &= 2x - 5 \\ 2x - 4 &= 2x - 5 \end{aligned}$$

There is no solution. The given equation and the resulting equation are examples of *false equations*.

9. One number is 5 more than another and the sum of the two is 71. Find the numbers.

Let x be the smaller number and $x+5$ be the larger. Then $x+(x+5) = 71$, $2x = 66$, and $x = 33$. The numbers are 33 and 38.

10. A father is now three times as old as his son. Twelve years ago he was six times as old as his son. Find the present age of each.

Let x = the age of the son and $3x$ = the age of the father. Twelve years ago, the age of the son was $x-12$ and the age of the father was $3x-12$.

Then $3x-12 = 6(x-12)$, $3x = 60$, and $x = 20$. The present age of the son is 20 and that of the father is 60.

11. When two pulleys are connected by a belt, their angular velocities (revolutions per minute) are *inversely* proportional to their diameters, that is, $\omega_1 : \omega_2 = d_2 : d_1$. Find the velocity of a pulley 15 in. in diameter when it is connected to a pulley 12 in. in diameter and rotating at 100 rev/min.

Let ω_1 be the unknown velocity; then $d_1 = 15$, $\omega_2 = 100$, and $d_2 = 12$. The given formula becomes

$$\frac{\omega_1}{100} = \frac{12}{15} \quad \text{and} \quad \omega_1 = \frac{12}{15}(100) = 80 \text{ rev/min}$$

12. Bleaching powder is obtained through the reaction of chlorine and slaked lime, 74.10 lb of lime and 70.91 lb of chlorine producing 127.00 lb of bleaching powder and 18.01 lb of water. How many lb of lime will be required to produce 1000 lb of bleaching powder ?

Let x = the number of lb of lime required. Then

$$\frac{x(\text{lb of lime})}{1000(\text{lb of powder})} = \frac{74.10(\text{lb of lime})}{127(\text{lb of powder})}, \quad 127x = 74{,}100 \quad \text{and} \quad x = 583.46 \text{ lb}$$

13. The pressure of a gas in a container at constant temperature varies inversely as the volume. If $p = 30$ when $v = 45$, find p when $v = 25$.

First Solution. Here $p = c/v$, $30 = c/45$, $c = 30 \cdot 45$; thus $p = 30 \cdot 45/v$.
When $v = 25$, $p = 30 \cdot 45/25 = 54$.

Second Solution. From $p_1 = c/v_1$ and $p_2 = c/v_2$, we obtain $\dfrac{p_1}{p_2} = \dfrac{v_2}{v_1}$.

Taking $v_1 = 25$, $p_2 = 30$, $v_2 = 45$, we obtain as before $p_1 = 54$.

14. The safe load of a horizontal beam supported at both ends varies jointly as the breadth and the square of the depth, and inversely as the length of the beam. If a 2×4 in. beam 8 ft long will support 500 lb safely, what is the safe load for a 4×8 in. beam of the same material 20 ft long ?

$$S_1 = c\,\frac{b_1 d_1^2}{l_1}, \quad S_2 = c\,\frac{b_2 d_2^2}{l_2}. \quad \text{Then} \quad \frac{S_1}{S_2} = \frac{b_1 d_1^2 l_2}{b_2 d_2^2 l_1}, \quad \frac{S_1}{500} = \frac{4 \cdot 8^2 \cdot 8}{2 \cdot 4^2 \cdot 20} \quad \text{and} \quad S_1 = 1600 \text{ lb.}$$

SUPPLEMENTARY PROBLEMS

15. Solve for x and check each of the following.

(a) $2x - 7 = 29 - 4x$ (c) $\frac{x+3}{x-3} = 3$ (e) $\frac{2x+1}{4} - \frac{1}{x-1} = \frac{x}{2}$

(b) $2(x-1) - 3(x-2) + 4(x-3) = 0$ (d) $\frac{4}{x-4} = \frac{2}{2x-5}$ (f) $a(x+3) + b(x-2) = c(x-1)$

Ans. (a) 6 (b) 8/3 (c) 6 (d) 2 (e) 5 (f) $\frac{2b-3a-c}{a+b-c}$

16. A piece of wire 11 2/3 inches long is to be divided into two parts such that one part is 2/3 that of the other. Find the length of the shorter piece. *Ans.* 4 2/3 in.

17. A train leaves a station and travels at the rate 40 mi/hr. Two hours later a second train leaves the station and travels at the rate 60 mi/hr. Where will the second train overtake the first?
Ans. 240 mi from the station

18. A tank is drained by two pipes. One pipe can empty the tank in 30 min, the other can empty it in 25 min. If the tank is 5/6 filled and both pipes are open, in what time will the tank be emptied?
Ans. 11 4/11 min

19. A man invests 1/3 of his capital at 3% and the remainder at 4%. What is his capital if his total income is \$4400 ? *Ans.* \$120,000

20. A can do a piece of work in 10 days. After he has worked 2 days, B comes to help him and together they finish it in 3 days. In how many days could B alone have done the work? *Ans.* 6 days

21. When two resistances R_1 and R_2 are placed in parallel, the resultant resistance R is given by $1/R = 1/R_1 + 1/R_2$. Find R when $R_1 = 80$ and $R_2 = 240$. *Ans.* 60

22. How soon after noon are the hands of a clock together again? *Ans.* 1 hr, 5 5/11 min

23. How much water will be produced in making the 1000 gm of bleaching powder in Problem 12?
Ans. 141.81 gm

24. The reaction of 65.4 gm of zinc and 72.9 gm of hydrochloric acid produces 136.3 gm of zinc chloride and 2 gm of hydrogen. Find the weight of hydrochloric acid necessary for a complete reaction with 300 gm of zinc and the weight of hydrogen produced. *Ans.* 334.4 gm, 9.2 gm

25. How much water must be used to prepare a 1:5000 solution of bichloride of mercury from a 0.5 gm tablet? *Ans.* 2500 grams

26. The volume of a cone varies jointly as the altitude and square of the radius. When the radius is 4 and the altitude is 6, the volume is 32π. What must the altitude be if the volume is 12π when the radius is 2? *Ans.* 9

27. Newton's law of gravitation states that the force F of attraction between two bodies varies jointly as their masses m_1 and m_2 and inversely as the square of the distance between them. Two bodies whose centers are 5000 mi apart attract each other with a force of 15 lb. What would be the force of attraction if their masses were tripled and the distance between their centers was doubled?
Ans. $33\frac{3}{4}$ lb

28. If a body weighs 20 lb on the earth's surface, what would it weigh 2000 miles above the surface? (Assume the radius of the earth to be 4000 mi) *Ans.* 8 8/9 lb

29. The horsepower of a steam engine varies jointly as the average pressure in the cylinder and the speed of rotation. When the average pressure is 400 lb/in^2 and the engine is making 750 rev/min, the horsepower is 100. What is the horsepower when the average pressure is 300 lb/in^2 and the engine is making 600 rev/min? *Ans.* 60

CHAPTER 6

Simultaneous Linear Equations

TWO LINEAR EQUATIONS IN TWO UNKNOWNS. Let the system of equations be

$$\begin{cases} a_1x + b_1y + c_1 = 0 \\ a_2x + b_2y + c_2 = 0 \end{cases}$$

Each equation has an unlimited number of solutions (x, y) corresponding to the unlimited number of points on the locus (straight line) which it represents. Our problem is to find all solutions common to the two equations or the coordinates of all points common to the two lines. There are three cases:

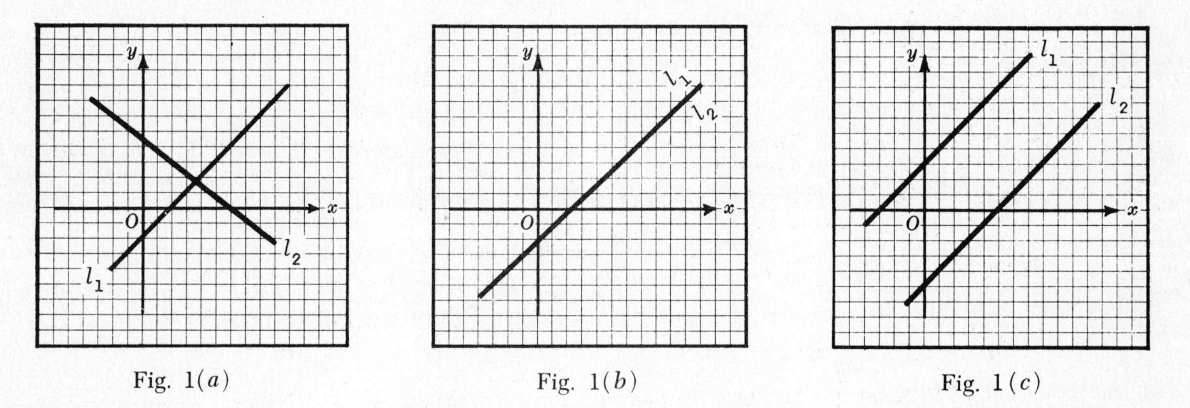

Fig. 1(a) Fig. 1(b) Fig. 1(c)

1. The system has one and only one solution, that is, the two lines have one and only one point in common. The equations are said to be *consistent* (have common solutions) *and independent*. See Fig. 1(a) above.

2. The system has an unlimited number of solutions, that is, the two equations are equivalent or the two lines are coincident. The equations are said to be *consistent and dependent*. See Fig. 1(b) above.

3. The system has no solution, that is, the two lines are parallel and distinct. The equations are said to be *inconsistent*. See Fig. 1(c) above.

GRAPHICAL SOLUTION. We plot the graphs of the two equations on the same axes and scale off the coordinates of the point of intersection. The defect of this method is that, in general, only approximate solutions are obtained.

See Problem 1.

ALGEBRAIC SOLUTION. A system of two consistent and independent equations in two unknowns may be solved algebraically by eliminating one of the unknowns.

Example 1. Solve the system $\begin{cases} 3x - 6y = 10 & (1) \\ 9x + 15y = -14 & (2) \end{cases}$

ELIMINATION BY SUBSTITUTION.

Solve (*1*) for x: $x = \frac{10}{3} + 2y$ (*3*)

Substitute in (*2*): $9(10/3 + 2y) + 15y = -14,$
$30 + 18y + 15y = -14,$ $33y = -44,$ $y = -4/3$

Substitute for y in (*3*): $x = 10/3 + 2(-4/3) = 2/3$

Check: Using (*2*), $9(2/3) + 15(-4/3) = -14$.

ELIMINATION BY ADDITION.

Multiply (*1*) by -3: $-9x + 18y = -30$
Rewrite (*2*): $9x + 15y = -14$
Add: $33y = -44$ or $y = -4/3$

Substitute for y in (*1*): $3x - 6(-4/3) = 10$ or $x = 2/3$.

Check: Using (*2*), $9(2/3) + 15(-4/3) = -14$.

See Problems 2-4.

THREE LINEAR EQUATIONS IN THREE UNKNOWNS. A system of three consistent and independent equations in three unknowns may be solved algebraically by deriving from it a system of two equations in two unknowns.

Example 2. Solve the system $\begin{cases} 2x + 3y - 4z = 1 & (1) \\ 3x - y - 2z = 4 & (2) \\ 4x - 7y - 6z = -7 & (3) \end{cases}$

We shall eliminate y.

Rewrite (*1*): $2x + 3y - 4z = 1$
$3 \times (2)$: $9x - 3y - 6z = 12$
Add: $11x - 10z = 13$ (*4*)

Rewrite (*3*): $4x - 7y - 6z = -7$
$-7 \times (2)$: $-21x + 7y + 14z = -28$
Add: $-17x + 8z = -35$ (*5*)

Next, solve (*4*) and (*5*).

$4 \times (4)$: $44x - 40z = 52$
$5 \times (5)$: $-85x + 40z = -175$
Add: $-41x = -123$
$x = 3$

From (*4*): $11(3) - 10z = 13$, $z = 2$.
From (*1*): $2(3) + 3y - 4(2) = 1$, $y = 1$.

Check: Using (*2*), $3(3) - 1 - 2(2) = 4$.

See Problems 5-6.

SOLUTIONS OF LINEAR SYSTEMS USING DETACHED COEFFICIENTS. In the example below, a variation of the method of addition and subtraction is used to solve a system of linear equations. On the left the equations themselves are used while on the right the same moves are made on the rectangular array (called a *matrix*) of the coefficients and constant terms. The numbering (*1*), (*2*), (*3*), ... refers both to the equations and to the rows of the matrices.

Example 3. Solve the system $\begin{cases} 2x - 3y = 2 \\ 4x + 7y = -9 \end{cases}$

USING EQUATIONS — USING MATRICES

$$2x - 3y = 2 \quad (1)$$
$$4x + 7y = -9 \quad (2)$$

$$\begin{pmatrix} 2 & -3 & 2 \\ 4 & 7 & -9 \end{pmatrix}$$

Multiply (1) by $\frac{1}{2}$ and write as (3). Multiply (1) by -2 and add to (2) to obtain (4).

$$x - \frac{3}{2}y = 1 \quad (3)$$
$$13y = -13 \quad (4)$$

$$\begin{pmatrix} 1 & -\frac{3}{2} & 1 \\ 0 & 13 & -13 \end{pmatrix}$$

Multiply (4) by $(3/2)/13 = 3/26$ and add to (3) to obtain (5). Multiply (4) by $1/13$ to obtain (6).

$$x = -\tfrac{1}{2} \quad (5)$$
$$y = -1 \quad (6)$$

$$\begin{pmatrix} 1 & 0 & -\frac{1}{2} \\ 0 & 1 & -1 \end{pmatrix}$$

The required solution is $x = -\frac{1}{2}$, $y = -1$.

Example 4. Solve, using matrices, the system $\begin{cases} 2x - 3y + 2z = 14 \\ 4x + 4y - 3z = 6 \\ 3x + 2y - 3z = -2 \end{cases}$

The matrix of the system $\begin{pmatrix} 2 & -3 & 2 & 14 \\ 4 & 4 & -3 & 6 \\ 3 & 2 & -3 & -2 \end{pmatrix}$ is formed by writing in order the coefficients of x, y, z, and the constant terms.

There are, in essence, two moves:

(a) Multiply the elements of a row by a non-zero number. This move is used only to obtain an element 1 in a prescribed position.

(b) Multiply the elements of a row by a non-zero number and add to the corresponding elements of another row. This move is used to obtain an element 0 in a prescribed position.

The first attack must be planned to yield a matrix of the form

$$\begin{pmatrix} 1 & * & * & * \\ 0 & * & * & * \\ 0 & * & * & * \end{pmatrix}$$

in which only the elements of the first column are prescribed.

Multiply 1st row by $\frac{1}{2}$:
Multiply 1st row by -2 and add to 2nd row:
Multiply 1st row by $-3/2$ and add to 3rd row:

$$\begin{pmatrix} 1 & -3/2 & 1 & 7 \\ 0 & 10 & -7 & -22 \\ 0 & 13/2 & -6 & -23 \end{pmatrix}$$

The second attack must be planned to yield a matrix of the form

$$\begin{pmatrix} 1 & 0 & * & * \\ 0 & 1 & * & * \\ 0 & 0 & * & * \end{pmatrix}$$

in which the elements of the first two columns are prescribed.

Multiply 2nd row by 3/20 and add to 1st row:
Multiply 2nd row by 1/10:
Multiply 2nd row by −13/20 and add to 3rd row:

$$\begin{pmatrix} 1 & 0 & -1/20 & 37/10 \\ 0 & 1 & -7/10 & -11/5 \\ 0 & 0 & -29/20 & -87/10 \end{pmatrix}$$

The final attack must be planned to yield a matrix of the form

$$\begin{pmatrix} 1 & 0 & 0 & * \\ 0 & 1 & 0 & * \\ 0 & 0 & 1 & * \end{pmatrix}$$

in which the elements of the first three columns are prescribed.

Multiply 3rd row by −1/29 and add to 1st row:
Multiply 3rd row by −14/29 and add to 2nd row:
Multiply 3rd row by −20/29 :

$$\begin{pmatrix} 1 & 0 & 0 & 4 \\ 0 & 1 & 0 & 2 \\ 0 & 0 & 1 & 6 \end{pmatrix}$$

The solution is $x = 4,\ y = 2,\ z = 6$.

SOLVED PROBLEMS

1. Solve graphically the systems: (*a*) $\begin{cases} x + 2y = 5 \\ 3x - y = 1 \end{cases}$, (*b*) $\begin{cases} x + y = 1 \\ 2x + 3y = 0 \end{cases}$, (*c*) $\begin{cases} 3x - 6y = 10 \\ 9x + 15y = -14 \end{cases}$.

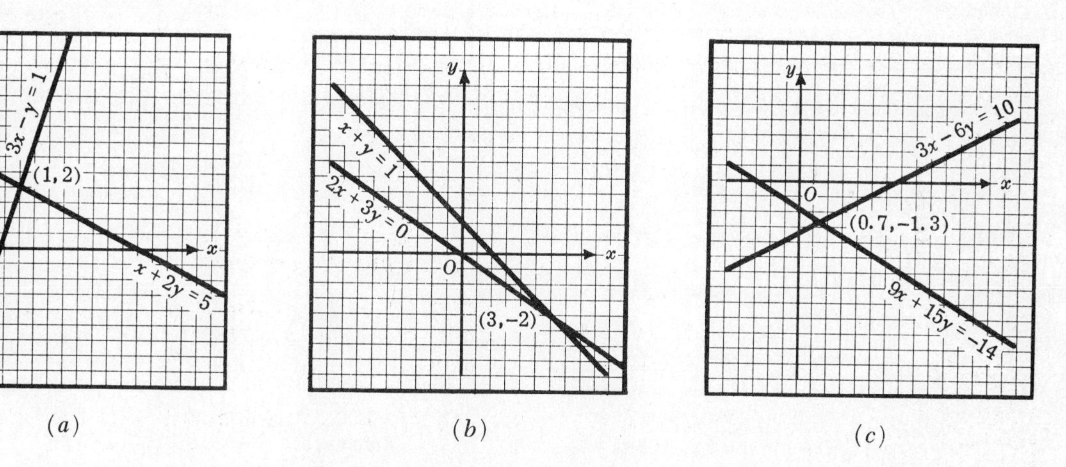

(*a*) (*b*) (*c*)

Solution: (*a*) $x = 1,\ y = 2$ (*b*) $x = 3,\ y = -2$ (*c*) $x = 0.7,\ y = -1.3$

2. Solve algebraically: (*a*) $\begin{cases} x + 2y = 5 & (1) \\ 3x - y = 1 & (2) \end{cases}$, (*b*) $\begin{cases} 3x + 2y = 2 & (1) \\ 5x + 6y = 4 & (2) \end{cases}$, (*c*) $\begin{cases} 2x + 3y = 3 & (1) \\ 5x - 9y = -4 & (2) \end{cases}$.

(*a*) Rewrite (*1*) : $x + 2y = 5$
Multiply (2) by 2: $6x - 2y = 2$
Add: $7x = 7$
$x = 1$

Substitute for x in (*1*): $1 + 2y = 5$, $y = 2$.

Check: Using (2), $3(1) - 2 = 1$.

(*b*) Multiply (*1*) by −5: $-15x - 10y = -10$
Multiply (2) by 3 : $15x + 18y = 12$
Add: $8y = 2$
$y = \frac{1}{4}$

Substitute for y in (*1*): $3x + 2(\frac{1}{4}) = 2$, $x = \frac{1}{2}$.

Check: Using (2), $5(\frac{1}{2}) + 6(\frac{1}{4}) = 4$.

(*c*) Multiply (*1*) by 3: $6x + 9y = 9$
Rewrite (2) : $5x - 9y = -4$

Add: $11x = 5$
$x = 5/11$

Substitute in (*1*): $3y = 3 - 2(5/11) = 23/11$, $y = 23/33$.

Check: Using (2), $5(5/11) - 9(23/33) = -4$.

3. If the numerator of a fraction is increased by 2, the fraction is 1/4; if the denominator is decreased by 6, the fraction is 1/6. Find the fraction.

Let x/y be the original fraction. Then

$$\frac{x+2}{y} = \frac{1}{4} \quad \text{or} \quad 4x - y = -8 \quad (1)$$

$$\frac{x}{y-6} = \frac{1}{6} \quad \text{or} \quad 6x - y = -6 \quad (2)$$

Subtract (*1*) from (2) : $2x = 2$ and $x = 1$.
Substitute $x = 1$ in (*1*): $4 - y = -8$ and $y = 12$.

The fraction is $\frac{1}{12}$.

4. A man can row downstream 6 mi in 1 hr and return in 2 hr. Find his rate in still water and the rate of the river.

Let x = rate in still water in mi/hr, y = rate of the river in mi/hr.

Then $x + y$ = rate downstream and $x - y$ = rate upstream.

Now $x + y = 6$ (*1*)
$x - y = 6/2 = 3$ (2)

Add (*1*) and (2) : $2x = 9$ and $x = 4\frac{1}{2}$
Subtract (2) from (*1*): $2y = 3$ and $y = 1\frac{1}{2}$

The rate in still water is $4\frac{1}{2}$ mi/hr and the rate of the river is $1\frac{1}{2}$ mi/hr.

5. Solve the system
$$\begin{cases} x - 5y + 3z = 9 & (1) \\ 2x - y + 4z = 6 & (2) \\ 3x - 2y + z = 2 & (3) \end{cases}$$

Eliminate z.

Rewrite (*1*) : $x - 5y + 3z = 9$
Multiply (3) by -3: $-9x + 6y - 3z = -6$

Add: $-8x + y = 3$ (4)

Rewrite (2) : $2x - y + 4z = 6$
Multiply (3) by -4: $-12x + 8y - 4z = -8$

Add: $-10x + 7y = -2$ (5)

Multiply (4) by -7: $56x - 7y = -21$
Rewrite (5) : $-10x + 7y = -2$

Add: $46x = -23$
$x = -1/2$

Substitute $x = -1/2$ in (4):
$-8(-1/2) + y = 3$ and $y = -1$.

Substitute $x = -1/2$, $y = -1$ in (*1*):
$-1/2 - 5(-1) + 3z = 9$ and $z = 3/2$.

Check: Using (2), $2(-1/2) - (-1) + 4(3/2) = -1 + 1 + 6 = 6$.

6. A parabola $y = ax^2 + bx + c$ passes through the points (1,0), (2,2), and (3,10). Determine its equation.

Since (1,0) is on the parabola : $a + b + c = 0$ (*1*)
Since (2,2) is on the parabola : $4a + 2b + c = 2$ (2)
Since (3,10) is on the parabola: $9a + 3b + c = 10$ (3)

Subtract (*1*) from (2): $3a + b = 2$ (4)
Subtract (*1*) from (3): $8a + 2b = 10$ (5)

Multiply (4) by -2 and add to (5): $2a = 6$ and $a = 3$.
Substitute $a = 3$ in (4) : $3(3) + b = 2$ and $b = -7$.
Substitute $a = 3$, $b = -7$ in (*1*) : $3 - 7 + c = 0$ and $c = 4$.

The equation of the parabola is $y = 3x^2 - 7x + 4$.

7. Solve, using matrices, the system $\begin{cases} x - 5y + 3z = 9 \\ 2x - y + 4z = 6 \\ 3x - 2y + z = 2 \end{cases}$ (See Problem 5.)

Begin with the matrix: $\begin{pmatrix} 1 & -5 & 3 & 9 \\ 2 & -1 & 4 & 6 \\ 3 & -2 & 1 & 2 \end{pmatrix}$

Rewrite 1st row (since first element is 1):
Multiply 1st row by -2 and add to 2nd row :
Multiply 1st row by -3 and add to 3rd row :
$\begin{pmatrix} 1 & -5 & 3 & 9 \\ 0 & 9 & -2 & -12 \\ 0 & 13 & -8 & -25 \end{pmatrix}$

Multiply 2nd row by 5/9 and add to 1st row:
Multiply 2nd row by 1/9 :
Multiply 2nd row by $-13/9$ and add to 3rd row:
$\begin{pmatrix} 1 & 0 & 17/9 & 7/3 \\ 0 & 1 & -2/9 & -4/3 \\ 0 & 0 & -46/9 & -23/3 \end{pmatrix}$

Multiply 3rd row by 17/46 and add to 1st row:
Multiply 3rd row by $-1/23$ and add to 2nd row:
Multiply 3rd row by $-9/46$:
$\begin{pmatrix} 1 & 0 & 0 & -1/2 \\ 0 & 1 & 0 & -1 \\ 0 & 0 & 1 & 3/2 \end{pmatrix}$

The solution is $x = -1/2$, $y = -1$, $z = 3/2$.

8. Solve, using matrices, the system $\begin{cases} 2x + 2y + 3z = 2 \\ 3x - y - 6z = 4 \\ 8x + 4y + 3z = 8 \end{cases}$

Begin with the matrix: $\begin{pmatrix} 2 & 2 & 3 & 2 \\ 3 & -1 & -6 & 4 \\ 8 & 4 & 3 & 8 \end{pmatrix}$

Multiply 1st row by 1/2 :
Multiply 1st row by $-3/2$ and add to 2nd row:
Multiply 1st row by -4 and add to 3rd row:
$\begin{pmatrix} 1 & 1 & 3/2 & 1 \\ 0 & -4 & -21/2 & 1 \\ 0 & -4 & -9 & 0 \end{pmatrix}$

Multiply 2nd row by 1/4 and add to 1st row:
Multiply 2nd row by $-1/4$:
Subtract 2nd row from 3rd row:
$\begin{pmatrix} 1 & 0 & -9/8 & 5/4 \\ 0 & 1 & 21/8 & -1/4 \\ 0 & 0 & 3/2 & -1 \end{pmatrix}$

Multiply 3rd row by 3/4 and add to 1st row:
Multiply 3rd row by $-7/4$ and add to 2nd row:
Multiply 3rd row by 2/3 :
$\begin{pmatrix} 1 & 0 & 0 & 1/2 \\ 0 & 1 & 0 & 3/2 \\ 0 & 0 & 1 & -2/3 \end{pmatrix}$

The solution is $x = 1/2$, $y = 3/2$, $z = -2/3$.

SUPPLEMENTARY PROBLEMS

9. Solve graphically the systems. (*a*) $\begin{cases} x + y = 5 \\ 2x - y = 1 \end{cases}$ (*b*) $\begin{cases} x - 3y = 1 \\ x - 2y = 0 \end{cases}$ (*c*) $\begin{cases} x + y = -1 \\ 3x - y = 3 \end{cases}$

Ans. (*a*) (2,3), (*b*) (−2,−1), (*c*) (1/2, −3/2)

10. Solve algebraically the systems.

(*a*) $\begin{cases} 3x + 2y = 2 \\ x - y = 9 \end{cases}$ (*c*) $\begin{cases} 3x - y = 1 \\ 2x + 5y = 41 \end{cases}$ (*e*) $\begin{cases} 1/x + 2/y = 2 \\ 2/x - 2/y = 1 \end{cases}$

(*b*) $\begin{cases} 3x - 5y = 5 \\ 7x + y = 75 \end{cases}$ (*d*) $\begin{cases} x + ay = b \\ 2x - by = a \end{cases}$ (*f*) $\begin{cases} 1/4x + 7/2y = 5/4 \\ 1/2x - 3/y = -5/14 \end{cases}$

Hint: In (*e*) and (*f*) solve first for $1/x$ and $1/y$.

Ans. (*a*) $x = 4,\ y = -5$ (*c*) $x = 46/17,\ y = 121/17$ (*e*) $x = 1,\ y = 2$

(*b*) $x = 10,\ y = 5$ (*d*) $x = \dfrac{b^2 + a^2}{2a + b},\ y = \dfrac{2b - a}{2a + b}$ (*f*) $x = 1,\ y = 7/2$

11. *A* and *B* are 30 miles apart. If they leave at the same time and travel in the same direction *A* overtakes *B* in 60 hours. If they walk toward each other they meet in 5 hours. What are their rates? *Ans.* *A*, $3\frac{1}{4}$ mi/hr; *B*, $2\frac{3}{4}$ mi/hr

12. Two trains, each 400 ft long, run on parallel tracks. When running in the same direction, they pass in 20 sec; when running in the opposite direction, they pass in 5 sec. Find the speed of each train. *Ans.* 100 ft/sec, 60 ft/sec

13. One alloy contains 3 times as much copper as silver, another contains 5 times as much silver as copper. How much of each alloy must be used to make 14 lb in which there is twice as much copper as silver? *Ans.* 12 lb of first, 2 lb of second

14. If a field is enlarged by making it 10 rods longer and 5 rods wider, its area is increased by 1050 square rods. If its length is decreased by 5 rods and its width is decreased by 10 rods, its area is decreased by 1050 square rods. Find the original dimensions of the field. *Ans.* 80 rods × 60 rods

15. It takes a boat $1\frac{1}{2}$ hr to go 12 miles down stream and 6 hr to return. Find the rate of the current and the rate of the boat in still water. *Ans.* 3 mi/hr, 5 mi/hr

16. Solve each of the following systems.

(*a*) $\begin{cases} x + y + z = 3 \\ 2x + y - z = -6 \\ 3x - y + z = 11 \end{cases}$ (*c*) $\begin{cases} 3x + y + 4z = 6 \\ 2x - 3y - 5z = 2 \\ 3x - 4y + 3z = 8 \end{cases}$ (*e*) $\begin{cases} 4x - 3y + 3z = 8 \\ 2x + 3y + 24z = 1 \\ 6x - y + 6z = -1 \end{cases}$

(*b*) $\begin{cases} 4x + 4y - 3z = 3 \\ 2x + 3y + 2z = -4 \\ 3x - y + 4z = 4 \end{cases}$ (*d*) $\begin{cases} 2x - 3y - 3z = 9 \\ x + 3y + 2z = 3 \\ 3x - 4y - z = 4 \end{cases}$ (*f*) $\begin{cases} 6x + 2y + 4z = 2 \\ 4x - y + 2z = -3 \\ 7x - 2y - 3z = 5 \end{cases}$

Ans. (*a*) $x = 1,\ y = -3,\ z = 5$ (*c*) $x = 3/2,\ y = -1/2,\ z = 1/2$ (*e*) $x = -3/2,\ y = -4,\ z = 2/3$
(*b*) $x = 2,\ y = -2,\ z = -1$ (*d*) $x = 3,\ y = 2,\ z = -3$ (*f*) $x = 2/3,\ y = 7/3,\ z = -5/3$

17. Find the equation of the parabola $y = ax^2 + bx + c$ which passes through the points (1,6), (4,0), (3,4). *Ans.* $y = -x^2 + 3x + 4$

18. *A* and *B* working together can do a given job in 4 4/5 days, *B* and *C* together can do it in 4 days, and *A* and *C* together can do it in 3 3/7 days. How many days would be required to do the job if the three work together? *Ans.* 2 2/3 days

CHAPTER 7

Quadratic Functions and Equations

THE GRAPH OF THE QUADRATIC FUNCTION $y = ax^2 + bx + c$, $a \neq 0$, is a parabola. If $a > 0$, the parabola opens upward (Fig. 1*a*); if $a < 0$, the parabola opens downward (Fig. 1*b*). The lowest point of the parabola of Fig. 1*a* and the highest point of the parabola of Fig. 1*b* are called *vertices*. The abscissa of the vertex is given by $x = -b/2a$.

See Problem 1.

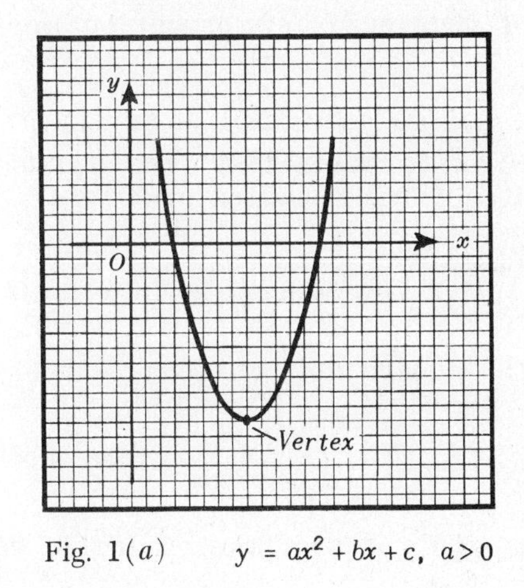

Fig. 1(*a*) $y = ax^2 + bx + c$, $a > 0$

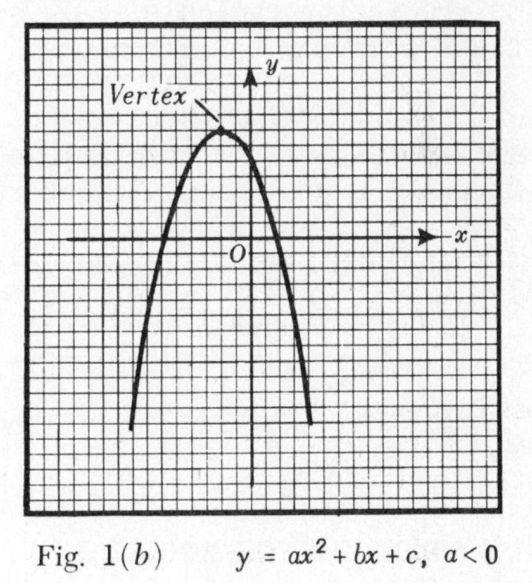

Fig. 1(*b*) $y = ax^2 + bx + c$, $a < 0$

A QUADRATIC EQUATION in one unknown x is of the form

$$ax^2 + bx + c = 0, \quad a \neq 0 \tag{1}$$

Frequently a quadratic equation may be solved by *factoring*. See Problem 2.

Every quadratic equation *(1)* can be solved by the following process, known as *completing the square*:

A) Subtract the constant term c from both members.
B) Divide both members by a, the coefficient of x^2.
C) Add to each member the square of one-half the coefficient of the term in x.
D) Set the square root of the left member (a perfect square) equal to ± the square root of the right member and solve for x.

Example 1. Solve $3x^2 - 8x - 4 = 0$ by completing the square.

A) $3x^2 - 8x = 4$, *B)* $x^2 - \frac{8}{3}x = \frac{4}{3}$,

C) $x^2 - \frac{8}{3}x + \frac{16}{9} = \frac{4}{3} + \frac{16}{9} = \frac{28}{9}$, $[\frac{1}{2}(-\frac{8}{3})]^2 = (-\frac{4}{3})^2 = \frac{16}{9}$,

D) $x - \frac{4}{3} = \pm\frac{2\sqrt{7}}{3}$. Then $x = \frac{4}{3} \pm \frac{2\sqrt{7}}{3} = \frac{4 \pm 2\sqrt{7}}{3}$.

See Problem 3.

Every quadratic equation *(1)* can be solved by means of the quadratic formula

$$x = \frac{-b \pm \sqrt{b^2 - 4ac}}{2a}$$

See Problems 4, 5.

EQUATIONS IN QUADRATIC FORM. An equation is in *quadratic form* if it is quadratic in some function of the unknown.

Example 2. Solve $x^4 + x^2 - 12 = 0$. This is a quadratic in x^2.

Factor: $x^4 + x^2 - 12 = (x^2 - 3)(x^2 + 4) = 0$

Then $x^2 - 3 = 0$ and $x = \pm\sqrt{3}$; $x^2 + 4 = 0$ and $x = \pm 2i$.

See Problems 11-12.

EQUATIONS INVOLVING RADICALS may sometimes reduce to quadratic equations after squaring to remove the radicals. All solutions of this quadratic equation *must* be tested since some may be extraneous. See Problems 13-16.

THE DISCRIMINANT of the quadratic equation *(1)* is, by definition, the quantity $b^2 - 4ac$. When a, b, c are rational numbers, the roots of the equation are

real and unequal if and only if $b^2 - 4ac > 0$,

real and equal if and only if $b^2 - 4ac = 0$,

rational if and only if $b^2 - 4ac$ is the square of a rational number,

imaginary and unequal if and only if $b^2 - 4ac < 0$.

See Problems 17-18.

SUM AND PRODUCT OF THE ROOTS. If x_1 and x_2 are the roots of the quadratic equation *(1)*, then $x_1 + x_2 = -b/a$ and $x_1 \cdot x_2 = c/a$.

A quadratic equation whose roots are x_1 and x_2 may be written in the form

$$x^2 - (x_1 + x_2)x + x_1 \cdot x_2 = 0$$

SOLVED PROBLEMS

1. Sketch the parabolas: (a) $y = x^2 - 2x - 8$, (b) $y = -2x^2 + 9x - 9$. Determine the coordinates of the vertex $V(x,y)$ of each.

Vertex: *a)* $x = -\frac{b}{2a} = -\frac{-2}{2\cdot 1} = 1$,

$y = 1^2 - 2\cdot 1 - 8 = -9$;

hence $V(1,-9)$.

b) $x = -\frac{9}{2(-2)} = \frac{9}{4}$,

$y = -2(\frac{9}{4})^2 + 9(\frac{9}{4}) - 9 = 9/8$;

hence $V(9/4, 9/8)$.

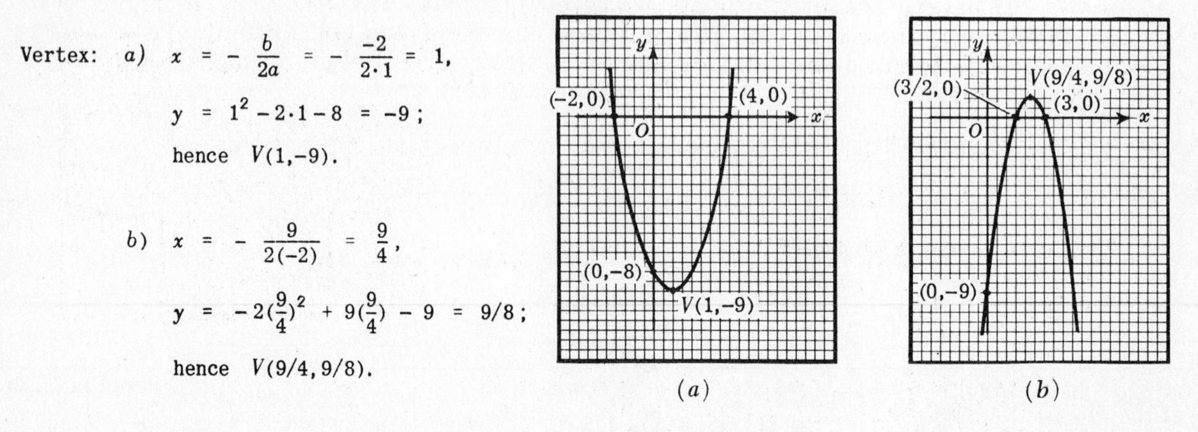

(a) (b)

2. Solve by factoring:

(*a*) $4x^2 - 5x = x(4x-5) = 0$

(*b*) $4x^2 - 9 = (2x-3)(2x+3) = 0$

(*c*) $x^2 - 4x + 3 = (x-1)(x-3) = 0$

(*d*) $x^2 - 6x + 9 = (x-3)(x-3) = 0$

(*e*) $4x^2 + 20x + 25 = (2x+5)(2x+5) = 0$

(*f*) $6x^2 + 13x + 6 = (3x+2)(2x+3) = 0$

(*g*) $3x^2 + 8ax - 3a^2 = (3x-a)(x+3a) = 0$

(*h*) $10ax^2 + (15 - 8a^2)x - 12a = (2ax+3)(5x-4a) = 0$

Ans. a) $0, 5/4$ b) $3/2, -3/2$ c) $1, 3$ d) $3, 3$ e) $-5/2, -5/2$ f) $-2/3, -3/2$ g) $a/3, -3a$ h) $-3/2a, 4a/5$

3. Solve by completing the square: (*a*) $x^2 - 2x - 1 = 0$, (*b*) $3x^2 + 8x + 7 = 0$.

a) $x^2 - 2x = 1;\ x^2 - 2x + 1 = 1 + 1 = 2;\ x - 1 = \pm\sqrt{2};\ x = 1 \pm \sqrt{2}$.

b) $3x^2 + 8x = -7;\ x^2 + \frac{8}{3}x = -\frac{7}{3};\ x^2 + \frac{8}{3}x + \frac{16}{9} = -\frac{7}{3} + \frac{16}{9} = -\frac{5}{9};$

$$x + \frac{4}{3} = \pm\sqrt{\frac{-5}{9}} = \pm\frac{i\sqrt{5}}{3};\quad x = \frac{-4 \pm i\sqrt{5}}{3}.$$

4. Solve $ax^2 + bx + c = 0$, $a \neq 0$, by completing the square.

Proceeding as in Problem 3, we have

$$x^2 + \frac{b}{a}x = -\frac{c}{a},\quad x^2 + \frac{b}{a}x + \frac{b^2}{4a^2} = \frac{b^2}{4a^2} - \frac{c}{a} = \frac{b^2 - 4ac}{4a^2},$$

$$x + \frac{b}{2a} = \pm\sqrt{\frac{b^2 - 4ac}{4a^2}} = \pm\frac{\sqrt{b^2 - 4ac}}{2a},\quad \text{and}\quad x = \frac{-b \pm \sqrt{b^2 - 4ac}}{2a}.$$

5. Solve the equations of Problem 3 using the quadratic formula.

a) $$x = \frac{-b \pm \sqrt{b^2 - 4ac}}{2a} = \frac{-(-2) \pm \sqrt{(-2)^2 - 4(1)(-1)}}{2 \cdot 1} = \frac{2 \pm \sqrt{4+4}}{2} = \frac{2 \pm 2\sqrt{2}}{2} = 1 \pm \sqrt{2}$$

b) $$x = \frac{-(8) \pm \sqrt{8^2 - 4 \cdot 3 \cdot 7}}{2 \cdot 3} = \frac{-8 \pm \sqrt{64 - 84}}{6} = \frac{-8 \pm \sqrt{-20}}{6} = \frac{-8 \pm 2i\sqrt{5}}{6} = \frac{-4 \pm i\sqrt{5}}{3}$$

6. An open box containing 24 in^3 is to be made from a square piece of tin by cutting 2 in. squares from each corner and turning up the sides. Find the dimension of the piece of tin required.

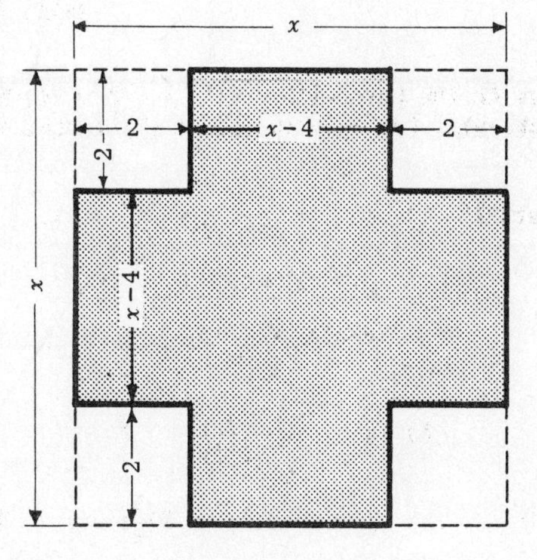

Let x = the required dimension. The resulting box will have dimensions $(x-4)$ by $(x-4)$ by 2, and its volume will be $2(x-4)(x-4)$. Then

$$2(x-4)^2 = 24,\quad x - 4 = \pm 2\sqrt{3}$$

and $x = 4 \pm 2\sqrt{3} = 7.464,\ 0.536$.

The required square of tin is 7.464 in. on a side.

7. Two pipes together can fill a reservoir in 6 hours 40 minutes. Find the time each alone will take to fill the reservoir if one of the pipes can fill it in 3 hours less time than the other.

Let x = time (hours) required by smaller pipe, $x-3$ = time required by larger pipe.

Then $\frac{1}{x}$ = part filled in 1 hr by smaller pipe, $\frac{1}{x-3}$ = part filled in 1 hr by larger pipe.

Since the two pipes together fill $\frac{1}{20/3} = 3/20$ of the reservoir in 1 hour,

$$\frac{1}{x}+\frac{1}{x-3}=\frac{3}{20},\quad 20(x-3)+20x=3x(x-3),\quad 3x^2-49x+60=(3x-4)(x-15)=0 \quad\text{and}\quad x=4/3,15.$$

The smaller pipe will fill the reservoir in 15 hr and the larger pipe in 12 hr.

8. Express each of the following in the form $a(x-h)^2 \pm b(y-k)^2 = c$.

a) $x^2+y^2-6x-9y+2=0$.

$$(x^2-6x)+(y^2-9y)=-2,\quad (x^2-6x+9)+(y^2-9y+81/4)=-2+9+81/4=109/4,$$
$$(x-3)^2+(y-9/2)^2=109/4$$

b) $3x^2+4y^2+6x-16y-21=0$.

$$3(x^2+2x)+4(y^2-4y)=21,\quad 3(x^2+2x+1)+4(y^2-4y+4)=21+3(1)+4(4)=40,$$
$$3(x+1)^2+4(y-2)^2=40$$

9. Transform each of the following into the form $a\sqrt{(x-h)^2+k}$ or $a\sqrt{k-(x-h)^2}$.

a) $\sqrt{4x^2-8x+9} = 2\sqrt{x^2-2x+9/4} = 2\sqrt{(x^2-2x+1)+5/4} = 2\sqrt{(x-1)^2+5/4}$

b) $\sqrt{8x-x^2} = \sqrt{16-(x^2-8x+16)} = \sqrt{16-(x-4)^2}$

c) $\sqrt{3-4x-2x^2} = \sqrt{2}\cdot\sqrt{3/2-2x-x^2} = \sqrt{2}\cdot\sqrt{5/2-(x^2+2x+1)} = \sqrt{2}\cdot\sqrt{5/2-(x+1)^2}$

10. If an object is thrown directly upward with initial speed v ft/sec, its distance s ft above the ground after t sec is given by

$$s = vt - \tfrac{1}{2}gt^2$$

Taking $g = 32.2$ ft/sec^2 and initial speed 120 ft/sec, find: (*a*) when the object is 60 ft above the ground, (*b*) when it is highest in its path and how high.

The equation of motion is $s = 120t - 16.1t^2$.

a) When $s = 60$: $\quad 60 = 120t-16.1t^2 \quad$ or $\quad 16.1t^2-120t+60=0$.

$$t = \frac{120 \pm \sqrt{(120)^2-4(16.1)60}}{32.2} = \frac{120\pm\sqrt{10536}}{32.2} = \frac{120\pm 102.64}{32.2} = 6.91, 0.54$$

After $t = 0.54$ sec the object is 60 ft above the ground and rising. After $t = 6.91$ sec, the object is 60 ft above the ground and falling.

b) The object is at its highest point when $t = \frac{-b}{2a} = \frac{-(-120)}{2(16.1)} = 3.73$ sec.

Its height is given by: $120t-16.1t^2 = 120(3.73) - 16.1(3.73)^2 = 223.6$ ft.

11. Solve $9x^4-10x^2+1=0$.

Factor: $(x^2-1)(9x^2-1)=0$. Then $x^2-1=0$, $9x^2-1=0$; $x=\pm 1$, $x=\pm 1/3$.

12. Solve $x^4 - 6x^3 + 12x^2 - 9x + 2 = 0$.

Complete the square on the first two terms: $(x^4 - 6x^3 + 9x^2) + 3x^2 - 9x + 2 = 0$

or $(x^2 - 3x)^2 + 3(x^2 - 3x) + 2 = 0$

Factor: $[(x^2 - 3x) + 2][(x^2 - 3x) + 1] = 0$

Then $x^2 - 3x + 2 = (x-2)(x-1) = 0$ and $x = 1, 2$

$x^2 - 3x + 1 = 0$ and $x = \dfrac{3 \pm \sqrt{9-4}}{2} = \dfrac{3 \pm \sqrt{5}}{2}$.

13. Solve $\sqrt{5x-1} - \sqrt{x} = 1$.

Transpose one of the radicals: $\sqrt{5x-1} = \sqrt{x} + 1$

Square: $5x - 1 = x + 2\sqrt{x} + 1$

Collect terms: $4x - 2 = 2\sqrt{x}$ or $2x - 1 = \sqrt{x}$

Square: $4x^2 - 4x + 1 = x$, $4x^2 - 5x + 1 = (4x-1)(x-1) = 0$, and $x = 1/4, 1$

For $x = 1/4$: $\sqrt{5(1/4) - 1} - \sqrt{1/4} = 0 \neq 1$. For $x = 1$: $\sqrt{5(1) - 1} - \sqrt{1} = 1$. The root is $x = 1$.

14. Solve $\sqrt{6x+7} - \sqrt{3x+3} = 1$.

Transpose one of the radicals: $\sqrt{6x+7} = 1 + \sqrt{3x+3}$

Square: $6x + 7 = 1 + 2\sqrt{3x+3} + 3x + 3$

Collect terms: $3x + 3 = 2\sqrt{3x+3}$

Square: $9x^2 + 18x + 9 = 4(3x+3) = 12x + 12$, $9x^2 + 6x - 3 = 3(3x-1)(x+1) = 0$, and $x = 1/3, -1$

For $x = 1/3$: $\sqrt{6(1/3) + 7} - \sqrt{3(1/3) + 3} = 3 - 2 = 1$. For $x = -1$: $1 - 0 = 1$.

The roots are $x = 1/3, -1$.

15. Solve $\dfrac{\sqrt{x+1} + \sqrt{x-1}}{\sqrt{x+1} - \sqrt{x-1}} = 3$.

Multiply the numerator and denominator of the fraction by $(\sqrt{x+1} + \sqrt{x-1})$:

$$\frac{(\sqrt{x+1} + \sqrt{x-1})(\sqrt{x+1} + \sqrt{x-1})}{(\sqrt{x+1} - \sqrt{x-1})(\sqrt{x+1} + \sqrt{x-1})} = \frac{(x+1) + 2\sqrt{x^2-1} + (x-1)}{(x+1) - (x-1)} = x + \sqrt{x^2-1} = 3$$

Then $x - 3 = -\sqrt{x^2 - 1}$, $x^2 - 6x + 9 = x^2 - 1$, and $x = 5/3$.

Check: $\dfrac{\sqrt{8/3} + \sqrt{2/3}}{\sqrt{8/3} - \sqrt{2/3}} = \dfrac{2\sqrt{2/3} + \sqrt{2/3}}{2\sqrt{2/3} - \sqrt{2/3}} = \dfrac{3\sqrt{2/3}}{\sqrt{2/3}} = 3$.

16. Solve $3x^2 - 5x + \sqrt{3x^2 - 5x + 4} = 16$.

Note that the unknown enters alike in both the expression free of radicals and under the radical.

Add 4 to both sides: $3x^2 - 5x + 4 + \sqrt{3x^2 - 5x + 4} = 20$

Let $y = \sqrt{3x^2 - 5x + 4}$. Then $y^2 + y - 20 = (y+5)(y-4) = 0$ and $y = 4, -5$.

Now $\sqrt{3x^2 - 5x + 4} = -5$ is impossible. From $\sqrt{3x^2 - 5x + 4} = 4$ we have

$3x^2 - 5x + 4 = 16$, $3x^2 - 5x - 12 = (3x+4)(x-3) = 0$ and $x = 3, -4/3$

The reader will show that both $x = 3$ and $x = -4/3$ are solutions.

17. Without solving, determine the character of the roots of:

a) $x^2 - 8x + 9 = 0$. Here $b^2 - 4ac = 28$; the roots are irrational and unequal.

b) $3x^2 - 8x + 9 = 0$. Here $b^2 - 4ac = -44$; the roots are imaginary and unequal.

c) $6x^2 - 5x - 6 = 0$. Here $b^2 - 4ac = 169$; the roots are rational and unequal.

d) $4x^2 - 4\sqrt{3}\,x + 3 = 0$. Here $b^2 - 4ac = 0$; the roots are real and equal. Note: Although the discriminant is the square of a rational number, the roots $\frac{1}{2}\sqrt{3}, \frac{1}{2}\sqrt{3}$ are not rational. Why?

18. Without sketching, state whether the graph of each of the following functions crosses the x-axis, is tangent to it, or lies wholly above or below it.

a) $3x^2 + 5x - 2$. $b^2 - 4ac = 25 + 24 > 0$; the graph crosses the x-axis.

b) $2x^2 + 5x + 4$. $b^2 - 4ac = 25 - 32 < 0$ and the graph is either wholly above or wholly below the x-axis. Since $f(0) > 0$, (the value of the function for any other value of x would do equally well), the graph lies wholly above the x-axis.

c) $4x^2 - 20x + 25$. $b^2 - 4ac = 400 - 400 = 0$; the graph is tangent to the x-axis.

d) $2x - 9 - 4x^2$. $b^2 - 4ac = 4 - 144 < 0$ and $f(0) < 0$; the graph lies wholly below the x-axis.

19. Find the sum and product of the roots of:

a) $x^2 + 5x - 8 = 0$. *Ans.* Sum $= -\frac{b}{a} = -5$, product $= \frac{c}{a} = -8$.

b) $8x^2 - x - 2 = 0$ or $x^2 - \frac{1}{8}x - \frac{1}{4} = 0$. *Ans.* Sum $= \frac{1}{8}$, product $= -\frac{1}{4}$.

c) $5 - 10x - 3x^2 = 0$ or $x^2 + \frac{10}{3}x - \frac{5}{3} = 0$. *Ans.* Sum $= -\frac{10}{3}$, product $= -\frac{5}{3}$.

20. Form the quadratic equation whose roots x_1 and x_2 are:

a) 3, 2/5. Here $x_1 + x_2 = 17/5$ and $x_1 \cdot x_2 = 6/5$.

The equation is $x^2 - \frac{17}{5}x + \frac{6}{5} = 0$ or $5x^2 - 17x + 6 = 0$.

b) $-2 + 3\sqrt{5}, -2 - 3\sqrt{5}$. Here $x_1 + x_2 = -4$ and $x_1 \cdot x_2 = 4 - 45 = -41$.

The equation is $x^2 + 4x - 41 = 0$.

c) $\frac{3 - i\sqrt{2}}{2}, \frac{3 + i\sqrt{2}}{2}$. The sum of the roots is 3 and the product is 11/4.

The equation is $x^2 - 3x + 11/4 = 0$ or $4x^2 - 12x + 11 = 0$.

21. Determine k so that the given equation will have the stated property, and write the resulting equation.

a) $x^2 + 4kx + k + 2 = 0$ has one root 0.
Since the product of the roots is to be 0, $k + 2 = 0$ and $k = -2$. The equation is $x^2 - 8x = 0$.

b) $4x^2 - 8kx - 9 = 0$ has one root the negative of the other.
Since the sum of the roots is to be 0, $2k = 0$ and $k = 0$. The equation is $4x^2 - 9 = 0$.

c) $4x^2 - 8kx + 9 = 0$ has roots whose difference is 4.
Denote the roots by r and $r + 4$. Then $r + (r + 4) = 2r + 4 = 2k$ and $r(r + 4) = 9/4$. Solving for $r = k - 2$ in the first and substituting in the second, we have $(k - 2)(k + 2) = 9/4$; then $4k^2 - 16 = 9$ and $k = \pm 5/2$. The equations are $4x^2 + 20x + 9 = 0$ and $4x^2 - 20x + 9 = 0$.

d) $2x^2 - 3kx + 5k = 0$ has one root twice the other.

Let the roots be r and $2r$. Then $r + 2r = 3r = \frac{3}{2}k$, $r = \frac{1}{2}k$, and $r(2r) = 2r^2 = \frac{5}{2}k$. Thus $k = 0, 5$.

The equations are $2x^2 = 0$ with roots 0, 0 and $2x^2 - 15x + 25 = 0$ with roots 5/2, 5.

SUPPLEMENTARY PROBLEMS

22. Locate the vertex of each of the parabolas of Problem 11, Chapter 4. Compare the results with those of Problem 14, Chapter 4.

23. Solve for x by factoring.

(*a*) $3x^2 + 4x = 0$ (*b*) $16x^2 - 25 = 0$ (*c*) $x^2 + 2x - 3 = 0$ (*d*) $2x^2 + 9x - 5 = 0$ (*e*) $10x^2 - 9x + 2 = 0$ (*f*) $2x^2 - (a + 4b)x + 2ab = 0$

Ans. (*a*) $0, -4/3$ (*b*) $\pm 5/4$ (*c*) $1, -3$ (*d*) $1/2, -5$ (*e*) $1/2, 2/5$ (*f*) $\frac{1}{2}a$, $2b$

24. Solve for x by completing the square.

(*a*) $2x^2 + x - 5 = 0$ (*b*) $2x^2 - 4x - 3 = 0$ (*c*) $3x^2 + 2x - 2 = 0$ (*d*) $5x^2 - 4x + 2 = 0$ (*e*) $15x^2 - (16m - 14)x + 4m^2 - 8m + 3 = 0$

Ans. (*a*) $\frac{1}{4}(-1 \pm \sqrt{41})$ (*b*) $\frac{1}{2}(2 \pm \sqrt{10})$ (*c*) $\frac{1}{3}(-1 \pm \sqrt{7})$ (*d*) $\frac{1}{5}(2 \pm i\sqrt{6})$ (*e*) $\frac{1}{3}(2m - 1)$, $\frac{1}{5}(2m - 3)$

25. Solve the equations of Problem 24 using the quadratic formula.

26. Solve $6x^2 + 5xy - 6y^2 + x + 8y - 2 = 0$ for (*a*) y in terms of x, (*b*) x in terms of y.

Ans. (*a*) $\frac{1}{2}(3x + 2)$, $\frac{1}{3}(1 - 2x)$ (*b*) $\frac{1}{2}(1 - 3y)$, $\frac{2}{3}(y - 1)$

27. Solve. (*a*) $x^4 - 29x^2 + 100 = 0$ (*b*) $\frac{21}{x + 2} - \frac{1}{x - 4} = 2$ (*c*) $1 - \frac{2}{2x^2 - x} = \frac{3}{(2x^2 - x)^2}$

(*d*) $\sqrt{4x + 1} - \sqrt{3x - 2} = 5$ (*e*) $\sqrt{2x + 3} - \sqrt{4 - x} = 2$ (*f*) $\sqrt{3x - 2} - \sqrt{x - 2} = 2$

Ans. (*a*) $\pm 2, \pm 5$ (*b*) 5, 7 (*c*) $-1, 3/2, \frac{1}{4}(1 \pm i\sqrt{7})$ (*d*) 342 (*e*) 3 (*f*) 2, 6

28. Form the quadratic equation whose roots are:

(*a*) the negatives of the roots of $3x^2 + 5x - 8 = 0$,
(*b*) twice the roots of $2x^2 - 5x + 2 = 0$,
(*c*) one half the roots of $2x^2 - 5x - 3 = 0$.

Ans. (*a*) $3x^2 - 5x - 8 = 0$ (*b*) $x^2 - 5x + 4 = 0$ (*c*) $8x^2 - 10x - 3 = 0$

29. The length of a rectangle is 7 in. more than its width; its area is 228 in^2. What are its dimensions?
Ans. 12×19 in.

30. A rectangular garden plot 16×24 yd is to be bordered by a strip of uniform width x yd so as to double the area. Find x. *Ans.* 4 yd

31. The interior of a cubical box is lined with insulating material $\frac{1}{2}$ in. thick. Find the original interior dimensions if the volume is thereby decreased by 271 in^3. *Ans.* 10 in.

32. What are the dimensions of the largest rectangular field which can be enclosed by 1200 ft of fencing?
Ans. 300×300 ft

CHAPTER 8

Inequalities

AN INEQUALITY is a statement that one (real) number is greater than or less than another; for example $3 > -2$, $-10 < -5$.

Two inequalities are said to have the *same sense* if their signs of inequality point in the same direction. Thus $3 > -2$ and $-5 > -10$ have the same sense; $3 > -2$ and $-10 < -5$ have opposite senses.

The sense of an equality is *not* changed:

a) if the same number is added to or subtracted from both sides.
b) if both sides are multiplied or divided by the same *positive* number.

The sense of an equality *is* changed if both sides are multiplied or divided by the same negative number.

See Problems 1-3.

AN ABSOLUTE INEQUALITY is one which is true for all real values of the letters involved; for example, $x^2 + 1 > 0$ is an absolute inequality.

A CONDITIONAL INEQUALITY is one which is true for certain values of the letters involved; for example, $x + 2 > 5$ is a conditional inequality since it is true for $x = 4$ but not for $x = 1$.

SOLUTION OF CONDITIONAL INEQUALITIES. The solution of a conditional inequality in one letter, say x, consists of all values of x for which the inequality is true. These values lie on one or more intervals of the real number scale as illustrated in the examples below.

To solve a linear inequality, proceed as in solving a linear equality keeping in mind the rules for keeping or reversing the sense.

Example 1. Solve the inequality $5x + 4 > 2x + 6$.

Subtract $2x$ from each member
Subtract 4 from each member } $3x > 2$

Divide by 3: $x > 2/3$

Graphical representation:

0 2/3 1

See Problems 5-6.

To solve a quadratic inequality, $f(x) = ax^2 + bx + c > 0$, solve the equality $f(x) = 0$, locate the roots r_1 and r_2 on a number scale, and determine the sign of $f(x)$ on each of the resulting intervals.

Example 2. Solve the inequality $3x^2 - 8x + 7 > 2x^2 - 3x + 1$.

Subtract $2x^2 - 3x + 1$ from each member: $x^2 - 5x + 6 > 0$

Solve the equality $x^2 - 5x + 6 = 0$: $x = 2$, $x = 3$

Locate the roots on a number scale: $f(x) > 0$ 2 $f(x) < 0$ 3 $f(x) > 0$

Determine the sign of $f(x) = x^2 - 5x + 6$

on the interval $x < 2$: $f(0) = 6 > 0$
on the interval $2 < x < 3$: $f(5/2) = 25/4 - 25/2 + 6 < 0$
on the interval $x > 3$: $f(4) = 16 - 20 + 6 > 0$

The given inequality is satisfied (see darkened portions of the scale) when $x < 2$ and $x > 3$.

See Problems 7-11.

SOLVED PROBLEMS

1. Given the inequality $-3 < 4$. Write the result when (a) 5 is added to both sides, (b) 5 is subtracted from both sides, (c) -6 is subtracted from both sides, (d) both sides are doubled, (e) both sides are divided by -2.

   Ans. a) $2 < 9$, b) $-8 < -1$, c) $3 < 10$, d) $-6 < 8$, e) $3/2 > -2$

2. Square each of the inequalities: (a) $-3 < 4$, (b) $-5 < 4$.

   Ans. a) $9 < 16$, b) $25 > 16$

3. If $a > 0$, $b > 0$ prove that $a^2 > b^2$ if and only if $a > b$.

   Suppose $a > b$. Since $a > 0$, $a^2 > ab$ and, since $b > 0$, $ab > b^2$. Hence, $a^2 > ab > b^2$ and $a^2 > b^2$.

   Suppose $a^2 > b^2$. Then $a^2 - b^2 = (a-b)(a+b) > 0$. Dividing by $a + b > 0$, we have $a - b > 0$ and $a > b$.

4. Prove $\dfrac{a}{b^2} + \dfrac{b}{a^2} > \dfrac{1}{a} + \dfrac{1}{b}$ if $a > 0$, $b > 0$, and $a \neq b$.

   Suppose $a > b$; then $a^2 > b^2$ and $a - b > 0$.

   Now $a^2(a-b) > b^2(a-b)$ or $a^3 - a^2b > ab^2 - b^3$ and $a^3 + b^3 > ab^2 + a^2b$. Since $a^2b^2 > 0$,

$$\frac{a^3+b^3}{a^2b^2} > \frac{ab^2+a^2b}{a^2b^2} \quad \text{and} \quad \frac{a}{b^2} + \frac{b}{a^2} > \frac{1}{a} + \frac{1}{b}$$

   Why is it necessary that $a > 0$ and $b > 0$? Hint: See Problem 3.

5. Solve $3x + 4 > 5x + 2$.

   Subtract $5x + 4$ from each member: $-2x > -2$
   Divide by -2 : $x < 1$

   Graphical representation: 0 1

6. Solve $2x - 8 < 7x + 12$.

   Subtract $7x - 8$ from each member: $-5x < 20$
   Divide by -5 : $x > -4$

   Graphical representation:

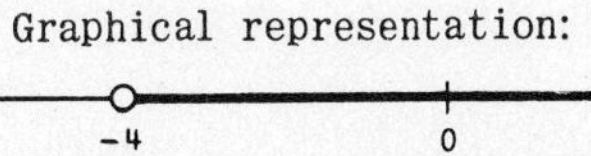

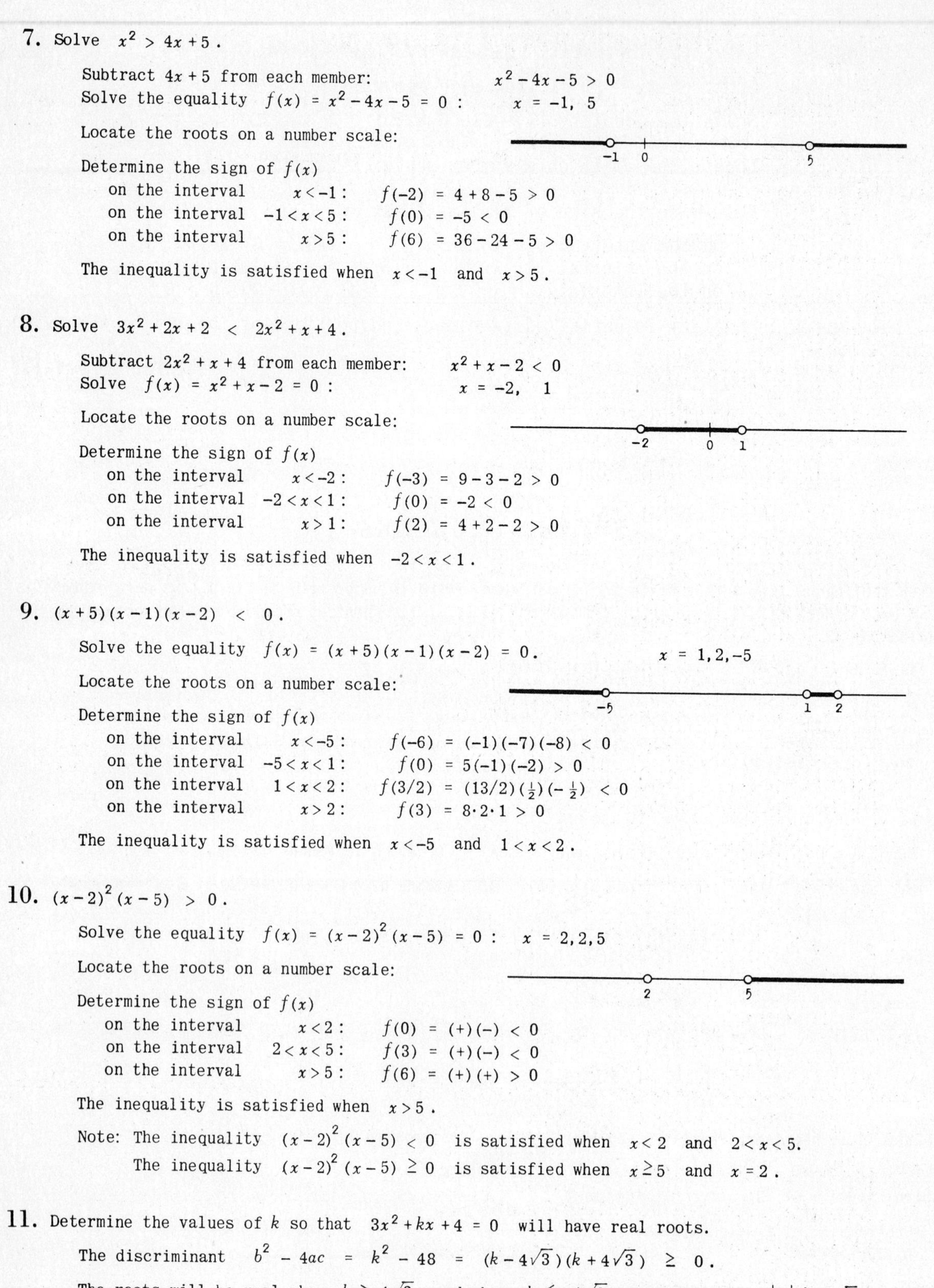

7. Solve $x^2 > 4x+5$.

Subtract $4x+5$ from each member: $x^2-4x-5 > 0$

Solve the equality $f(x) = x^2-4x-5 = 0$: $x = -1, 5$

Locate the roots on a number scale:

-1 0 5

Determine the sign of $f(x)$

on the interval $x<-1$: $f(-2) = 4+8-5 > 0$

on the interval $-1<x<5$: $f(0) = -5 < 0$

on the interval $x>5$: $f(6) = 36-24-5 > 0$

The inequality is satisfied when $x<-1$ and $x>5$.

8. Solve $3x^2+2x+2 < 2x^2+x+4$.

Subtract $2x^2+x+4$ from each member: $x^2+x-2 < 0$

Solve $f(x) = x^2+x-2 = 0$: $x = -2, 1$

Locate the roots on a number scale:

-2 0 1

Determine the sign of $f(x)$

on the interval $x<-2$: $f(-3) = 9-3-2 > 0$

on the interval $-2<x<1$: $f(0) = -2 < 0$

on the interval $x>1$: $f(2) = 4+2-2 > 0$

The inequality is satisfied when $-2<x<1$.

9. $(x+5)(x-1)(x-2) < 0$.

Solve the equality $f(x) = (x+5)(x-1)(x-2) = 0$. $x = 1, 2, -5$

Locate the roots on a number scale:

-5 1 2

Determine the sign of $f(x)$

on the interval $x<-5$: $f(-6) = (-1)(-7)(-8) < 0$

on the interval $-5<x<1$: $f(0) = 5(-1)(-2) > 0$

on the interval $1<x<2$: $f(3/2) = (13/2)(\frac{1}{2})(-\frac{1}{2}) < 0$

on the interval $x>2$: $f(3) = 8\cdot 2\cdot 1 > 0$

The inequality is satisfied when $x<-5$ and $1<x<2$.

10. $(x-2)^2(x-5) > 0$.

Solve the equality $f(x) = (x-2)^2(x-5) = 0$: $x = 2, 2, 5$

Locate the roots on a number scale:

2 5

Determine the sign of $f(x)$

on the interval $x<2$: $f(0) = (+)(-) < 0$

on the interval $2<x<5$: $f(3) = (+)(-) < 0$

on the interval $x>5$: $f(6) = (+)(+) > 0$

The inequality is satisfied when $x>5$.

Note: The inequality $(x-2)^2(x-5) < 0$ is satisfied when $x<2$ and $2<x<5$.

The inequality $(x-2)^2(x-5) \geq 0$ is satisfied when $x \geq 5$ and $x=2$.

11. Determine the values of k so that $3x^2+kx+4 = 0$ will have real roots.

The discriminant $b^2 - 4ac = k^2 - 48 = (k-4\sqrt{3})(k+4\sqrt{3}) \geq 0$.

The roots will be real when $k \geq 4\sqrt{3}$ and when $k \leq -4\sqrt{3}$, that is, when $|k| \geq 4\sqrt{3}$.

SUPPLEMENTARY PROBLEMS

12. If $2y^2 + 4xy - 3x = 0$, determine the range of values of x for which the corresponding y-roots are real.

Here $y = \dfrac{-4x \pm \sqrt{16x^2 + 24x}}{4} = \dfrac{-2x \pm \sqrt{4x^2 + 6x}}{2}$ will be real provided $4x^2 + 6x \geq 0$.

Thus y will be real for $x \leq -3/2$ and for $x \geq 0$.

13. Prove: If $a > b$ and $c > d$, then $a + c > b + d$.
Hint: $(a-b) + (c-d) = (a+c) - (b+d) > 0$.

14. Prove: If $a \neq b$ are real numbers, then $a^2 + b^2 > 2ab$.
Hint: $(a-b)^2 > 0$.

15. Prove: If $a \neq b \neq c$ are real numbers, then $a^2 + b^2 + c^2 > ab + bc + ca$.

16. Prove: If $a > 0$, $b > 0$, and $a \neq b$, then $a/b + b/a > 2$.

17. Prove: If $a^2 + b^2 = 1$ and $c^2 + d^2 = 1$, then $ac + bd \leq 1$.

18. Solve: (a) $x - 4 > -2x + 5$ (c) $x^2 - 16 > 0$ (e) $x^2 - 6x > -5$
(b) $4 + 3x < 2x + 24$ (d) $x^2 - 4x < 5$ (f) $5x^2 + 5x - 8 \leq 3x^2 + 4$

Ans. (a) $x > 3$ (b) $x < 20$ (c) $|x| > 4$ (d) $-1 < x < 5$ (e) $x < 1$, $x > 5$ (f) $-4 \leq x \leq 3/2$

19. Solve. (a) $(x+1)(x-2)(x+4) > 0$ (b) $(x-1)^3(x-3)(x+2) < 0$ (c) $(x+3)(x-2)^2(x-5)^3 < 0$
Ans. (a) $-4 < x < -1$, $x > 2$ (b) $x < -2$, $1 < x < 3$ (c) $-3 < x < 2$, $2 < x < 5$

20. In each of the following determine the range of x for which y will be real.

(a) $y = \sqrt{2x^2 - 7x + 3}$ (c) $y = \sqrt{61x^2 + 144x + 144}$ (e) $xy^2 + 3xy + 3x - 4y - 4 = 0$
(b) $y = \sqrt{6 - 5x - 4x^2}$ (d) $y^2 + 2xy + 4y + x + 14 = 0$ (f) $6x^2 + 5xy - 6y^2 + x + 8y - 2 = 0$

Ans. (a) $x \leq 1/2$, $x \geq 3$ (c) all values of x (e) $-4 \leq x \leq 4/3$
(b) $-2 \leq x \leq 3/4$ (d) $x \leq -5$, $x \geq 2$ (f) all values of x

CHAPTER 9

The Locus of an Equation

DEGENERATE LOCI. The locus of an equation $f(x,y) = 0$ is called *degenerate* if $f(x,y)$ is the product of two or more real factors $g(x,y)$, $h(x,y)$, ... The locus of $f(x,y) = 0$ then consists of the loci of $g(x,y) = 0$, $h(x,y) = 0$, ...

See Problem 1.

INTERCEPTS. The intercepts on the coordinate axes of a locus are the directed distances from the origin to the points of intersection of the locus and the coordinate axes.

To find the x-intercepts, set $y = 0$ in the equation of the locus and solve for x; to find the y-intercepts, set $x = 0$ and solve for y.

See Problem 2.

SYMMETRY. Two points P and Q are said to be symmetric with respect to a point R if R is the midpoint of the segment PQ (see Fig. 1). Each of the points is called the symmetric point of the other with respect to the point R, the *center of symmetry*.

Two points P and Q are said to be symmetric with respect to a line l if l is the perpendicular bisector of the segment PQ (see Fig. 2). Each of the points P, Q is called the symmetric point of the other with respect to l, the *axis of symmetry*.

A locus is said to be symmetric with respect to a point R or to a line l if the symmetric point with respect to R or l of every point of the locus is also a point of the locus (see Figs. 3a and 3b).

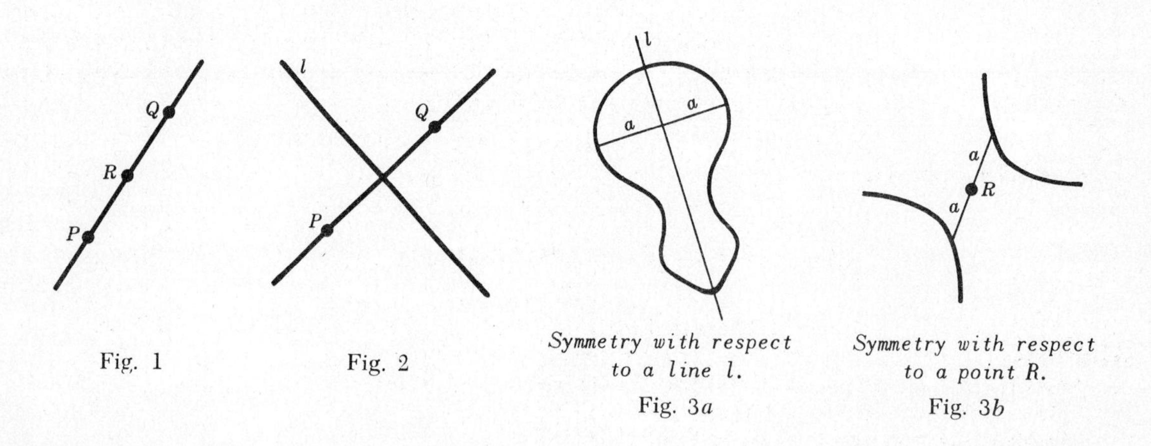

Fig. 1

Fig. 2

Symmetry with respect to a line l.
Fig. 3a

Symmetry with respect to a point R.
Fig. 3b

SYMMETRY OF A LOCUS. The locus of a given equation $f(x,y) = 0$ is symmetric with respect to the x-axis if an equivalent equation is obtained when y is replaced by $-y$, is symmetric with respect to the y-axis if an equivalent equation is obtained when x is replaced by $-x$, and is symmetric with respect to the origin if an equivalent equation is obtained when x is replaced by $-x$ and y is replaced by $-y$ simultaneously.

Example 1. Examine $x^2+2y^2+x=0$ for symmetry with respect to the coordinate axes and the origin.

When y is replaced by $-y$, we have $x^2+2y^2+x=0$; the locus is symmetric with respect to the x-axis.

When x is replaced by $-x$, we have $x^2+2y^2-x=0$; the locus is not symmetric with respect to the y-axis.

When x is replaced by $-x$ and y by $-y$, we have $x^2+2y^2-x=0$; the locus is not symmetric with respect to the origin.

See Problem 3.

EXTENT OF A LOCUS. The extent of the locus of $f(x,y)=0$ is indicated by listing either (a) the real values of x (usually intervals) which yield real values of y and the real values of y which yield real values of x or (b) the real values of x for which the values of y are imaginary or for which no value of y is defined and the real values of y for which the values of x are imaginary or for which no value of x is defined.

Example 2. Examine for extent $x^2+4y^2=4$.

Solve for x: $x^2=4-4y^2$ and $x=\pm\sqrt{4-4y^2}=\pm 2\sqrt{1-y^2}$. Now x is real provided $1-y^2\geq 0$ or $y^2\leq 1$, i.e. provided $|y|\leq 1$ or $-1\leq y\leq 1$.

Solve for y: $4y^2=4-x^2$ and $y=\pm\frac{1}{2}\sqrt{4-x^2}$. Now y is real provided $4-x^2\geq 0$ or $x^2\leq 4$, i.e. provided $|x|\leq 2$ or $-2\leq x\leq 2$.

The entire curve is confined to the ranges $-2\leq x\leq 2$ and $-1\leq y\leq 1$.

Example 3. Examine for extent $xy+3x-y-6=0$.

Solve for x: $x=\dfrac{y+6}{y+3}$

There is a value of x for every real value of y, except $y=-3$, since there the denominator is zero. The locus does not exist when $y=-3$. Note that this is not because the corresponding value of x is imaginary as in Example 2 but is due to the fact that no corresponding value of x is defined.

Solve for y: $y=\dfrac{6-3x}{x-1}$

There is a value of y for every real value of x, except $x=1$. Thus, the curve exists for $x\neq 1$ and $y\neq -3$.

See Problem 4.

ASYMPTOTES. A line l is called an asymptote of a locus if, as a point recedes indefinitely far from the origin along the locus, the distance of the point from the line decreases continually and may be made as small as we please by receding sufficiently far.

An asymptote of a locus may be horizontal, vertical, or oblique. The discussion here will be limited to horizontal and vertical asymptotes.

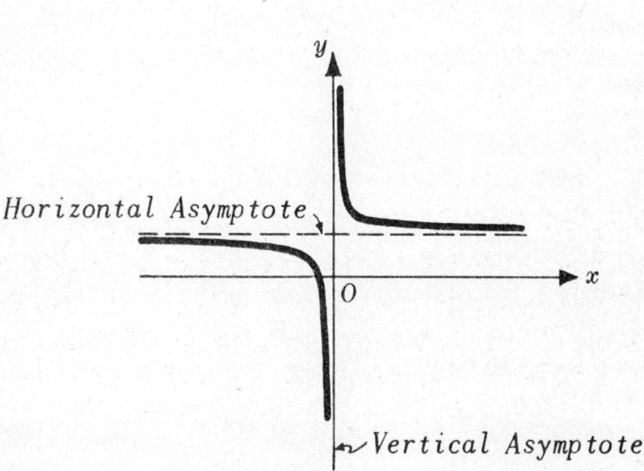

Example 4. Show that the line $y - 2 = 0$ is a horizontal asymptote of the curve $xy - 2x - 1 = 0$.

The locus exists for all $x \neq 0$ and for all $y \neq 2$. Since $y > 0$ for $x > 0$, there is a branch of the locus in the first quadrant. On this branch choose a point, say $A(1,3)$ and let a point $P(x,y)$, moving to the right from A, trace the locus.

x	y
1	3
10	2.1
100	2.01
1000	2.001
10000	2.0001

From the adjoining table which lists a few of the positions assumed by P, it is clear that as P moves to the right, (*a*) its ordinate y remains greater than 2 and (*b*) the difference $y - 2$ may be made as small as we please by taking x sufficiently large. For example, if we wish $y - 2 < 1/10^{12}$, we take $x > 10^{12}$; if we wish $y - 2 < 1/10^{999}$, we take $x > 10^{999}$, and so on. The line $y - 2 = 0$ is therefore a horizontal asymptote.

To find the horizontal asymptotes: Solve the equation of the locus for x and set each *real* linear factor of the denominator, if any, equal to zero.

To find the vertical asymptotes: Solve the equation of the locus for y and set each *real* linear factor of the denominator, if any, equal to zero.

Example 5. It is clear that the locus of Example 2 has neither horizontal nor vertical asymptotes. The locus of Example 3 has $y + 3 = 0$ as horizontal asymptote and $x - 1 = 0$ as vertical asymptote.

SOLVED PROBLEMS

1. (*a*) The locus $xy + x^2 = x(y + x) = 0$ consists of the lines $x = 0$ and $y + x = 0$.

   (*b*) The locus $y^3 + xy^2 - xy - x^2 = (x + y)(y^2 - x) = 0$ consists of the line $x + y = 0$ and the parabola $y^2 - x = 0$.

   (*c*) The locus $y^4 + y^2 - x^2 - x = (y^2 - x)(y^2 + x + 1) = 0$ consists of the parabolas

$$y^2 - x = 0 \quad \text{and} \quad y^2 + x + 1 = 0.$$

2. Examine for intercepts.

   (*a*) $4x^2 - 9y^2 = 36$.

   Set $y = 0$: $4x^2 = 36$, $x^2 = 9$; the x-intercepts are ± 3.
   Set $x = 0$: $-9y^2 = 36$, $y^2 = -4$; the y-intercepts are imaginary.

   (*b*) $x^2 - 2x = y^2 - 4y + 3$.

   Set $y = 0$: $x^2 - 2x - 3 = (x - 3)(x + 1) = 0$; the x-intercepts are -1 and 3.
   Set $x = 0$: $y^2 - 4y + 3 = (y - 1)(y - 3) = 0$; the y-intercepts are 1 and 3.

3. Examine for symmetry with respect to the coordinate axis and the origin.

   (*a*) $4x^2 - 9y^2 = 36$. Replacing y by $-y$, we have $4x^2 - 9y^2 = 36$; the locus is symmetric with respect to the x-axis.

Replacing x by $-x$, we have $4x^2 - 9y^2 = 36$; the locus is symmetric with respect to the y-axis.

Replacing x by $-x$ and y by $-y$, we have $4x^2 - 9y^2 = 36$; the locus is symmetric with respect to the origin.

Note that if a locus is symmetric with respect to the coordinate axes it is automatically symmetric with respect to the origin. It will be shown in the next problem that the converse is not true.

(*b*) $x^3 - x^2y + y^3 = 0$. Replacing y by $-y$, we have $x^3 + x^2y - y^3 = 0$; the locus is not symmetric with respect to the x-axis.

Replacing x by $-x$, we have $-x^3 - x^2y + y^3 = 0$; the locus is not symmetric with respect to the y-axis.

Replacing x by $-x$ and y by $-y$, we have $-x^3 + x^2y - y^3 = -(x^3 - x^2y + y^3) = 0$; the locus is symmetric with respect to the origin.

(*c*) $x^2 - 4y^2 - 2x + 8y = 0$. Replacing y by $-y$, we have $x^2 - 4y^2 - 2x - 8y = 0$, replacing x by $-x$, we have $x^2 - 4y^2 + 2x + 8y = 0$, replacing x by $-x$ and y by $-y$, we have $x^2 - 4y^2 + 2x - 8y = 0$; the locus is not symmetric with respect to either axis or the origin.

4. Discuss for extent.

(*a*) $3x + 4y - 12 = 0$.

Solve for y: $y = 3 - 3x/4$. Solve for x: $x = 4 - 4y/3$.

The locus exists for all values of x and for all values of y.

(*b*) $4x^2 + 5y^2 = 20$.

Solve for y: $y^2 = \frac{4}{5}(5 - x^2)$ or $y = \pm \frac{2\sqrt{5}}{5}\sqrt{5 - x^2}$.

Now y is real when $5 - x^2 \geq 0$ or $x^2 \leq 5$, that is, when $|x| \leq \sqrt{5}$ or $-\sqrt{5} \leq x \leq \sqrt{5}$.

Solve for x: $x^2 = \frac{5}{4}(4 - y^2)$ or $x = \frac{\sqrt{5}}{2}\sqrt{4 - y^2}$.

Now x is real when $4 - y^2 \geq 0$ or $y^2 \leq 4$, that is, when $|y| \leq 2$ or $-2 \leq y \leq 2$.

(*c*) $x^2 + y^2 - 4x = 0$.

Solve for y: $y = \pm\sqrt{4x - x^2}$.

The locus exists when $4x - x^2 \geq 0$ or $x(4 - x) \geq 0$, that is, when $0 \leq x \leq 4$.

Solve for x: $x^2 - 4x + 4 = 4 - y^2$, $x - 2 = \pm\sqrt{4 - y^2}$, $x = 2 \pm \sqrt{4 - y^2}$.

The locus exists when $4 - y^2 \geq 0$ or $y^2 \leq 4$, that is, when $|y| \leq 2$ or $-2 \leq y \leq 2$.

(*d*) $y^2 - 4x + 8y - 12 = 0$.

Solve for y: $y = -4 \pm 2\sqrt{x + 7}$. Solve for x: $x = \frac{1}{4}(y^2 + 8y - 12)$.

The locus exists when $x + 7 \geq 0$, that is, when $x \geq -7$ and for all values of y.

(*e*) $xy = 4$.

Solve for y: $y = 4/x$. Solve for x: $x = 4/y$.

The locus exists for all $x \neq 0$ and for all $y \neq 0$.

(*f*) $xy - y - x - 2 = 0$.

Solve for y: $y = \dfrac{x+2}{x-1}$. Solve for x: $x = \dfrac{y+2}{y-1}$.

The locus exists for all $x \neq 1$ and for all $y \neq 1$.

(*g*) $x^2y - x - 4y = 0$.

Solve for y: $y = \dfrac{x}{x^2-4}$. Solve for x: $x = \dfrac{1 \pm \sqrt{1+16y^2}}{2y}$.

The locus exists for all $x \neq \pm 2$ and for all values of y. At first glance it would seem that $y = 0$ must be excluded because of the denominator in the solution for x. This solution of the given equation as a quadratic in x assumes, however, that $y \neq 0$. When $y = 0$, the given equation becomes $-x = 0$, a linear equation, and $x = 0$.

5. Investigate for horizontal and vertical asymptotes.

(*a*) $3x + 4y - 12 = 0$. Solve for y: $y = \dfrac{12-3x}{4}$. Solve for x: $x = \dfrac{12-4y}{3}$.

Since the denominators do not involve the variables, there are neither horizontal nor vertical asymptotes.

(*b*) $xy = 8$. Solve for y: $y = 8/x$. Solve for x: $x = 8/y$.

Set each denominator equal to zero: $x = 0$ is the vertical asymptote,
$y = 0$ is the horizontal asymptote.

(*c*) $xy - y - x - 2 = 0$. (See Prob. 4(*f*).) Solve for y: $y = \dfrac{x+2}{x-1}$. Solve for x: $x = \dfrac{y+2}{y-1}$.

Set each denominator equal to zero: $x = 1$ is the vertical asymptote,
$y = 1$ is the horizontal asymptote.

(*d*) $x^2y - x - 4y = 0$. (See Prob. 4(*g*).) Solve for y: $y = \dfrac{x}{x^2-4}$. Solve for x: $x = \dfrac{1 \pm \sqrt{1+16y^2}}{2y}$.

Then $x = 2$ and $x = -2$ are vertical asymptotes and $y = 0$ is the horizontal asymptote.

(*e*) $x^2y - x^2 + 4y = 0$. Solve for y: $y = \dfrac{x^2}{x^2+4}$. Solve for x: $\pm 2\sqrt{\dfrac{y}{1-y}}$.

There are no vertical asymptotes since when $x^2 + 4 = 0$, x is imaginary. The horizontal asymptote is $y = 1$.

Discuss the following equations and sketch their loci.

6. $y^2 = -8x$.

Intercepts: When $y = 0$, $x = 0$ (x-intercept); when $x = 0$, $y = 0$ (y-intercept).

Symmetry: When y is replaced by $-y$ the equation is unchanged; the locus is symmetric with respect to the x-axis.

Extent: Solve for y: $y = \pm 2\sqrt{-2x}$. Solve for x: $x = -y^2/8$.
The locus exists when $-2x \geq 0$, i.e. when $x \leq 0$, and for all values of y.

The locus is a parabola with vertex at (0,0). It may be sketched after locating the following points: $(-1, \pm 2\sqrt{2})$, $(-2, \pm 4)$, and $(-3, \pm 2\sqrt{6})$. See Fig. (*a*) below.

7. $x^2 - 4x + 4y + 8 = 0$.

Intercepts: When $y = 0$, x is imaginary; when $x = 0$, $y = -2$ (y-intercept).

Symmetry: The locus is not symmetric with respect to the coordinate axes or the origin.

Extent: Solve for y: $y = \frac{1}{4}(4x - 8 - x^2)$. Solve for x: $x = 2 \pm 2\sqrt{-y-1}$.

The locus exists when $-y - 1 \geq 0$, i.e. when $y \leq -1$ and for all values of x.

The locus is a parabola with vertex at (2,-1). Other points on the locus are: (-2,-5), (4,-2), and (6,-5). See Fig. (b) below.

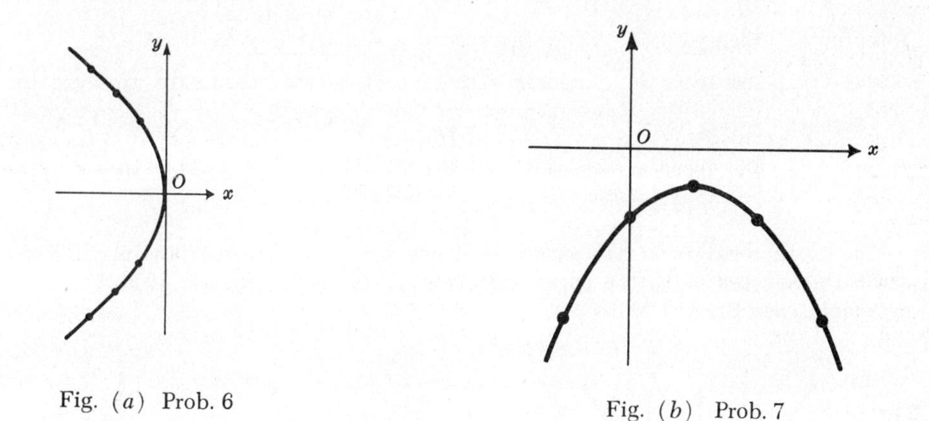

Fig. (a) Prob. 6 Fig. (b) Prob. 7

8. $x^2 + y^2 - 4x + 6y - 23 = 0$.

Intercepts: When $y = 0$, $x = \dfrac{4 \pm \sqrt{16+92}}{2} = 2 \pm 3\sqrt{3}$ (x-intercepts);

when $x = 0$, $y = \dfrac{-6 \pm \sqrt{36+92}}{2} = -3 \pm 4\sqrt{2}$ (y-intercepts).

Symmetry: There is no symmetry with respect to the coordinate axes or the origin.

Extent: Solve for y: $y = -3 \pm \sqrt{32 + 4x - x^2}$. Solve for x: $x = 2 \pm \sqrt{27 - 6y - y^2}$.
The locus exists when $32 + 4x - x^2 \geq 0$, i.e. when $x^2 - 4x - 32 = (x-8)(x+4) \leq 0$ or $-4 \leq x \leq 8$, and when $27 - 6y - y^2 \geq 0$, i.e. when $y^2 + 6y - 27 = (y-3)(y+9) \leq 0$ or $-9 \leq y \leq 3$.

Completing the squares, we have

$$(x^2 - 4x + 4) + (y^2 + 6y + 9) = 23 + 4 + 9 = 36 \quad \text{or} \quad (x-2)^2 + (y+3)^2 = 36,$$

the equation of a circle having center at $C(2,-3)$ and radius 6. See Fig. (c) below.

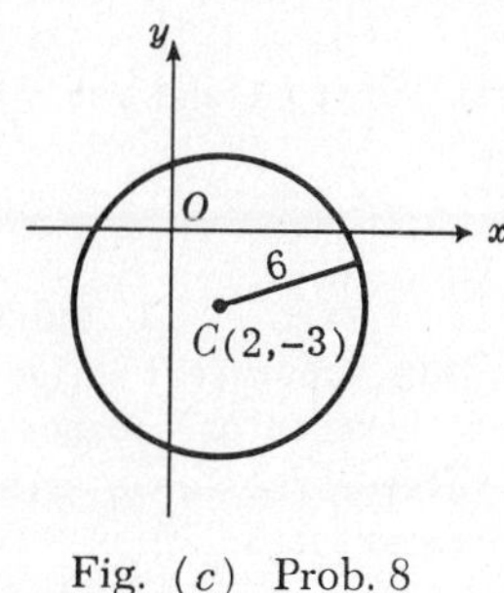

Fig. (c) Prob. 8

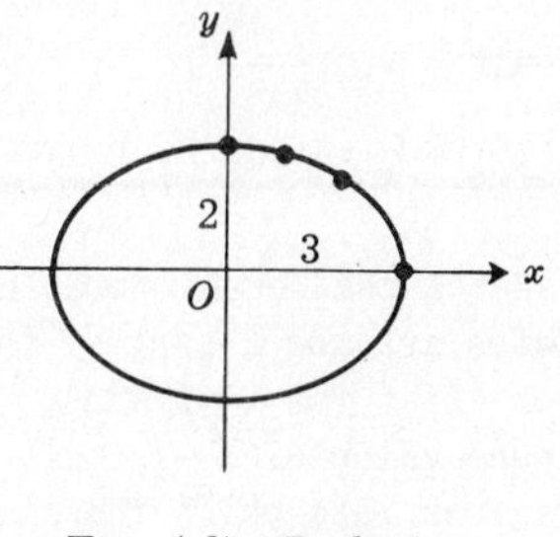

Fig. (d) Prob. 9

9. $4x^2 + 9y^2 = 36$.

Intercepts: When $y = 0$, $x = \pm 3$ (x-intercepts); when $x = 0$, $y = \pm 2$ (y-intercepts).

Symmetry: The locus is symmetric with respect to the coordinate axes and the origin.

Extent: Solve for y: $y = \pm \frac{2}{3}\sqrt{9 - x^2}$. Solve for x: $x = \pm \frac{3}{2}\sqrt{4 - y^2}$.

The locus exists when $9 - x^2 \geq 0$ or $x^2 \leq 9$, i.e. when $-3 \leq x \leq 3$,
and when $4 - y^2 \geq 0$ or $y^2 \leq 4$, i.e. when $-2 \leq y \leq 2$.

Since the locus is symmetric with respect to both the axes, only sufficient points to sketch the portion of the locus in the first quadrant are needed. Two such points are $(1, 4\sqrt{2}/3)$ and $(2, 2\sqrt{5}/3)$. The locus is called an *ellipse*. See Fig. (d) above.

10. $9x^2 - 4y^2 = 36$.

Intercepts: When $y = 0$, $x = \pm 2$ (x-intercepts); when $x = 0$, y is imaginary.

Symmetry: The locus is symmetric with respect to the coordinate axes and the origin.

Extent: Solve for y: $y = \pm \frac{3}{2}\sqrt{x^2 - 4}$. Solve for x: $x = \pm \frac{2}{3}\sqrt{y^2 + 9}$.

The locus exists when $y^2 + 9 \geq 0$, that is, for all values of y and when $x^2 - 4 \geq 0$ or $x^2 \geq 4$, that is, when $x \leq -2$ and $x \geq 2$.

The locus consists of two separate pieces and is not closed. The portion in the first quadrant has been sketched using the points $(3, 3\sqrt{5}/2)$, $(4, 3\sqrt{3})$, and $(5, 3\sqrt{21}/2)$. The locus is called a *hyperbola*. See Fig. (e) below.

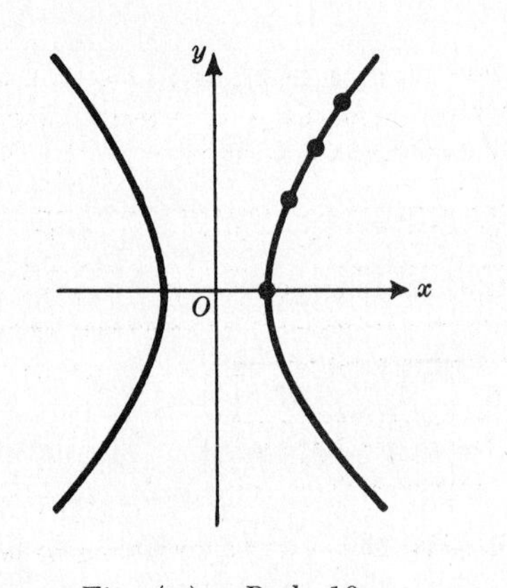

Fig. (e) Prob. 10

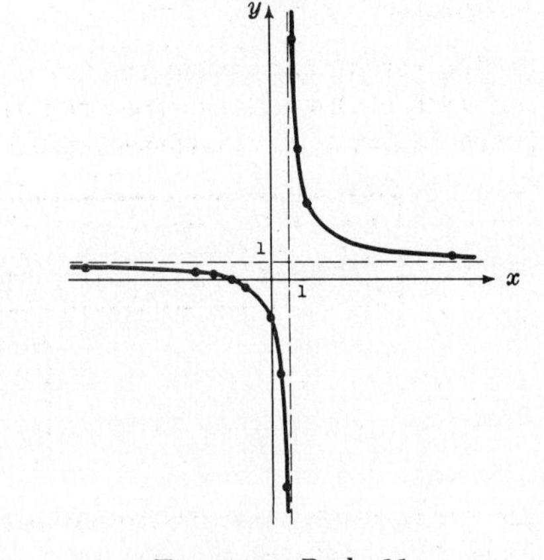

Fig. (f) Prob. 11

11. $xy - y - x - 2 = 0$.

Intercepts: The x-intercept is -2; the y-intercept is -2.

Symmetry: There is no symmetry with respect to the coordinate axes or the origin.

Extent: $y = \dfrac{x+2}{x-1}$, $x = \dfrac{y+2}{y-1}$. The locus exists for $x \neq 1$ and for $y \neq 1$.

Asymptotes: $x = 1$, $y = 1$.

To sketch the locus, first draw in the asymptotes $x = 1$ and $y = 1$ (dotted lines). While the asymptotes are *not* a part of the locus, they serve as very convenient guide lines. Since the locus does not exist for $x = 1$ and $y = 1$, it does not cross the asymptotes. Since there is one value of y for each value of $x \neq 1$, that is, since y is single valued, the locus appears in but two of the four regions into which the plane is separated by the asymptotes.

From the following table

x	-10	-4	-3	-2	-1	0	$1/2$	$3/4$	$5/4$	$3/2$	2	10
y	$8/11$	$2/5$	$1/4$	0	$-1/2$	-2	-5	-11	13	7	4	$4/3$

it is evident that the locus lies in the region to the right of the vertical asymptote and above the horizontal asymptote (see the portion of the table to the right of the double line) and in the region to the left of the vertical asymptote and below the horizontal asymptote (see the portion of the table to the left of the double line). The locus is shown in Fig. (f) above; note that it is symmetric with respect to $(1, 1)$ the point of intersection of the asymptotes.

12. $x^2y - x^2 - 4y = 0$.

Intercepts: The x-intercept is 0, the y-intercept is 0.

Symmetry: The locus is symmetric with respect to the y-axis.

Extent: $y = \dfrac{x^2}{x^2-4}$, $\quad x = \pm 2\sqrt{\dfrac{y}{y-1}}$.

The locus exists for $x \neq \pm 2$ and for all y such that $\dfrac{y}{y-1} \geq 0$. To solve this inequality, locate on the number scale the value of y for which the numerator is equal to zero, that is, $y = 0$, and the value of y for which the denominator is zero, that is, $y = 1$.

For $y < 0$, $\dfrac{y}{y-1} > 0$; for $0 < y < 1$, $\dfrac{y}{y-1} < 0$; for $y > 1$, $\dfrac{y}{y-1} > 0$.

Thus the locus exists for $y \leq 0$ and for $y > 1$.

Asymptotes: $x = \pm 2$, $y = 1$.

The asymptotes divide the plane into six regions. Since the locus does not exist when $x = \pm 2$ and when $y = 1$ (that is, does not cross an asymptote) and since y is single valued, the locus appears in but three of these regions. By means of the following table

x	-10	-5	-4	-3	-5/2	-7/4	-3/2	-1	0	1	3/2	7/4
y	25/24	25/21	4/3	9/5	25/9	-49/15	-9/7	-1/3	0	-1/3	-9/7	-49/15

x	5/2	3	4	5	10
y	25/9	9/5	4/3	25/21	25/24

the locus is sketched in Fig.(g) below. Note that only half of the table is necessary since the locus is symmetric with respect to the y-axis.

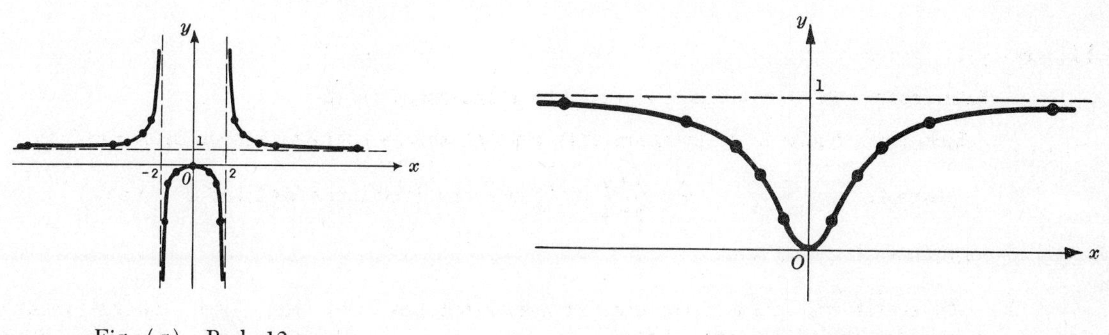

Fig. (g) Prob. 12

Fig. (h) Prob. 13

13. $x^2y - x^2 + 4y = 0$.

Intercepts: The x-intercept is 0, the y-intercept is 0.

Symmetry: The locus is symmetric with respect to the y-axis.

Extent: $y = \dfrac{x^2}{x^2+4}$, $\quad x = \pm 2\sqrt{\dfrac{y}{1-y}}$.

The locus exists for all values of x, and for y such that $\dfrac{y}{1-y} \geq 0$.

Using the diagram below (see Problem 12 above)

$\dfrac{y}{1-y} < 0 \qquad \dfrac{y}{1-y} > 0 \qquad \dfrac{y}{1-y} < 0$ the locus exists for $0 \leq y < 1$.

Asymptotes: $y = 1$.

Since the locus exists only for $0 \leq y < 1$, it lies entirely below its asymptote. The locus is sketched in Fig.(h) above using the following table:

x	−10	−5	−3	−2	−1	0	1	2	3	5	10
y	25/26	25/29	9/13	1/2	1/5	0	1/5	1/2	9/13	25/29	25/26

Only half of the table is necessary since the locus is symmetric with respect to the y-axis.

SUPPLEMENTARY PROBLEMS

14. Discuss and sketch.

(a) $x^2 - 4y^2 = 0$

(b) $x^2 + 2xy + y^2 = 4$

(c) $y = 9x^2$

(d) $y^2 = 6x - 3$

(e) $y^2 = 4 - 2x$

(f) $x^2 + y^2 = 16$

(g) $x^2 + y^2 = 0$

(h) $4x^2 + 9y^2 = 36$

(i) $9x^2 - 4y^2 = 36$

(j) $9x^2 - 4y^2 + 36 = 0$

(k) $xy = -4$

(l) $x^2y = 4$

(m) $xy^2 = -9$

(n) $y^3 + xy^2 = 2xy - 2x^2$

(o) $y = x^3$

(p) $xy - x + 4 = 0$

(q) $xy - 3x - y = 0$

(r) $x^2 + xy + y - 2 = 0$

(s) $x^2y - x - 4y = 0$

(t) $x^2y - 4xy + 3y - x - 2 = 0$

(u) $x^2y - x^2 + xy + 3x - 2 = 0$

(v) $x^3 + xy^2 - y^2 = 0$

CHAPTER 10

Simultaneous Equations Involving Quadratics

ONE LINEAR AND ONE QUADRATIC EQUATION.

Procedure: Solve the linear equation for one of the two unknowns (your choice) and substitute in the quadratic equation. Since this results in a quadratic equation in one unknown, the system can always be solved.

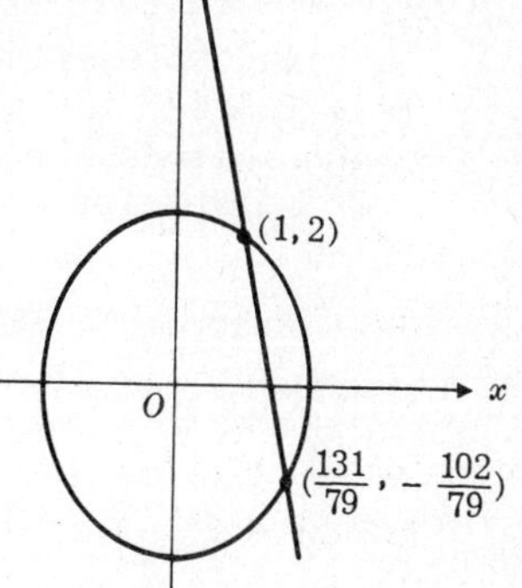

Example 1. Solve the system $\begin{cases} 4x^2 + 3y^2 = 16 \\ 5x + y = 7 \end{cases}$

Solve the linear equation for y: $y = 7 - 5x$.

Substitute in the quadratic equation:

$$4x^2 + 3(7-5x)^2 = 16,$$

$$4x^2 + 3(49 - 70x + 25x^2) = 16,$$

$$79x^2 - 210x + 131 = (x-1)(79x - 131) = 0$$

and $x = 1,\ 131/79$.

When $x = 1$, $y = 7 - 5x = 2$; when $x = 131/79$, $y = -102/79$.

The solutions are: $x = 1,\ y = 2$ and $x = 131/79,\ y = -102/79$.

The locus of the linear equation is the straight line and the locus of the quadratic equation is the ellipse of the figure above.

See Problems 1-2.

TWO QUADRATIC EQUATIONS. In general, solving a system of two quadratic equations in two unknowns involves solving an equation of the fourth degree in one of the unknowns. Since the solution of the general equation of the fourth degree in one unknown is beyond the scope of this book, only those systems which require the solution of a quadratic equation in one unknown will be treated here.

TWO QUADRATIC EQUATIONS OF THE FORM $ax^2 + by^2 = c$.

Procedure: Eliminate one of the unknowns by the method of addition of Chapter 6.

Example 2. Solve the system $\begin{cases} 4x^2 + 9y^2 = 72 & (1) \\ 3x^2 - 2y^2 = 19 & (2) \end{cases}$

Multiply (1) by 2: $8x^2 + 18y^2 = 144$

Multiply (2) by 9: $27x^2 - 18y^2 = 171$

Add: $35x^2 = 315$

Then $x^2 = 9$ and $x = \pm 3$.

When $x=3$, (1) gives $9y^2 = 72-4x^2 = 72-36 = 36$, $y^2=4$, and $y=\pm 2$.

When $x=-3$, (1) gives $9y^2 = 72-36 = 36$, $y^2=4$, and $y=\pm 2$.

The four solutions $x=3,\ y=2$; $x=3,\ y=-2$; $x=-3,\ y=2$; $x=-3,\ y=-2$ may also be written as $x=\pm 3,\ y=\pm 2$; $x=\pm 3,\ y=\mp 2$. By convention, we read the two upper signs and the two lower signs in the latter form.

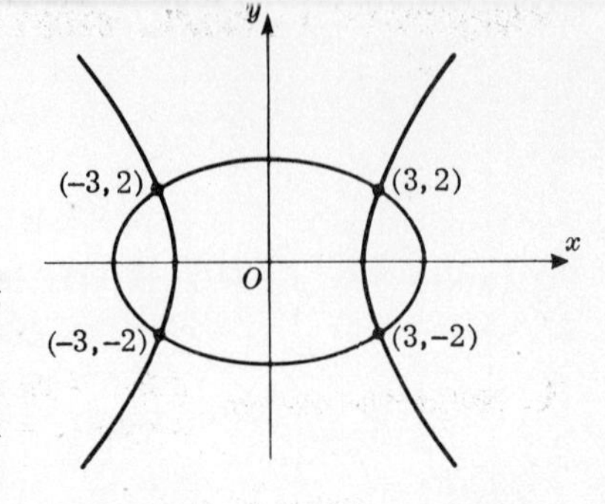

The ellipse and the hyperbola (see above figure) intersect in the points (3,2), (3,-2), (-3,2), (-3,-2).

See Problems 3-4.

TWO QUADRATIC EQUATIONS, ONE HOMOGENEOUS.

An expression, as $2x^2-3xy+y^2$, whose terms are all of the same degree in the variables, is called *homogeneous*. A homogeneous expression equated to zero is called a *homogeneous equation*. A homogeneous quadratic equation in two unknowns can always be solved for one of the unknowns in terms of the other.

Example 3. Solve the system $\begin{cases} x^2 - 3xy + 2y^2 = 0 & (1) \\ 2x^2 + 3xy - y^2 = 13 & (2) \end{cases}$

Solve (1) for x in terms of y: $(x-y)(x-2y) = 0$ and $x=y$, $x=2y$.

Solve the systems (see Example 1):

$\begin{cases} 2x^2+3xy-y^2 = 13 \\ x = y. \end{cases}$

$2y^2+3y^2-y^2 = 4y^2 = 13$,

$y^2 = 13/4$, $y = \pm\sqrt{13}/2$.

Then $x = y = \pm\sqrt{13}/2$.

$\begin{cases} 2x^2+3xy-y^2 = 13 \\ x = 2y. \end{cases}$

$8y^2+6y^2-y^2 = 13y^2 = 13$,

$y^2 = 1$, $y = \pm 1$.

Then $x = 2y = \pm 2$.

The solutions are $x = \sqrt{13}/2,\ y = \sqrt{13}/2$; $x = -\sqrt{13}/2,\ y = -\sqrt{13}/2$; $x=2,\ y=1$; $x=-2,\ y=-1$

or $x = \pm\sqrt{13}/2,\ y = \pm\sqrt{13}/2$; $x = \pm 2,\ y = \pm 1$.

See Problem 5.

TWO QUADRATIC EQUATIONS OF THE FORM $ax^2 + bxy + cy^2 = d$.

Procedure: Combine the two given equations to obtain a homogeneous equation. Solve, as in Example 3, the system consisting of this homogeneous equation and either of the given equations.

See Problems 6-7.

TWO QUADRATIC EQUATIONS, EACH SYMMETRICAL IN x AND y.

An equation, as $2x^2-3xy+2y^2+5x+5y = 1$, which is unchanged when the two unknowns are interchanged is called a *symmetrical equation*.

Procedure: Substitute $x = u+v$ and $y = u-v$ and then eliminate v^2 from the resulting equations.

See Problem 8.

FREQUENTLY A CAREFUL STUDY of a given system will reveal some special device for solving it.

See Problems 9-13.

SOLVED PROBLEMS

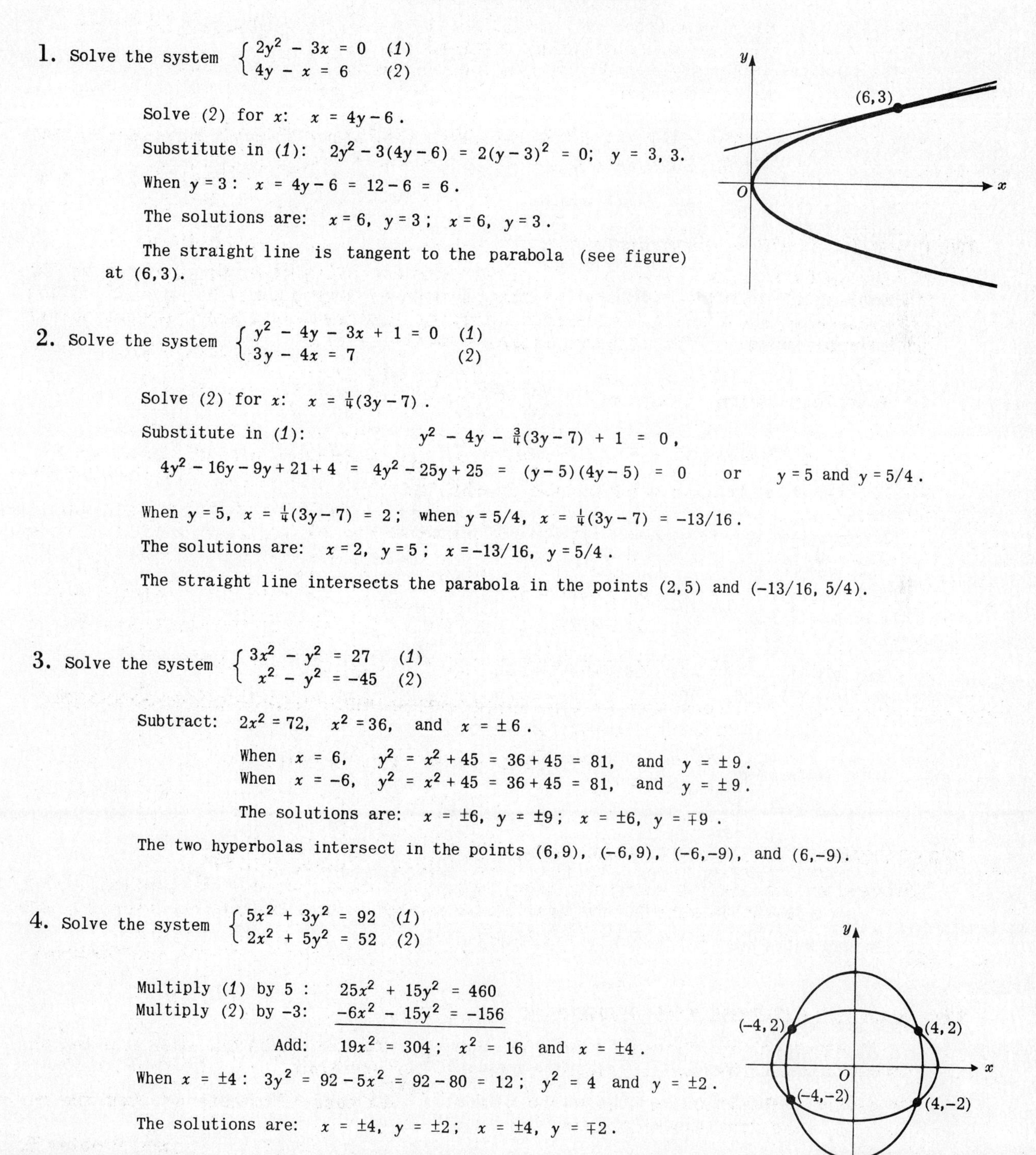

1. Solve the system $\begin{cases} 2y^2 - 3x = 0 & (1) \\ 4y - x = 6 & (2) \end{cases}$

Solve (2) for x: $x = 4y - 6$.

Substitute in (1): $2y^2 - 3(4y-6) = 2(y-3)^2 = 0$; $y = 3, 3$.

When $y = 3$: $x = 4y - 6 = 12 - 6 = 6$.

The solutions are: $x = 6,\ y = 3$; $x = 6,\ y = 3$.

The straight line is tangent to the parabola (see figure) at (6,3).

2. Solve the system $\begin{cases} y^2 - 4y - 3x + 1 = 0 & (1) \\ 3y - 4x = 7 & (2) \end{cases}$

Solve (2) for x: $x = \frac{1}{4}(3y - 7)$.

Substitute in (1): $y^2 - 4y - \frac{3}{4}(3y-7) + 1 = 0$,

$4y^2 - 16y - 9y + 21 + 4 = 4y^2 - 25y + 25 = (y-5)(4y-5) = 0$ or $y = 5$ and $y = 5/4$.

When $y = 5$, $x = \frac{1}{4}(3y-7) = 2$; when $y = 5/4$, $x = \frac{1}{4}(3y-7) = -13/16$.

The solutions are: $x = 2,\ y = 5$; $x = -13/16,\ y = 5/4$.

The straight line intersects the parabola in the points (2,5) and (−13/16, 5/4).

3. Solve the system $\begin{cases} 3x^2 - y^2 = 27 & (1) \\ x^2 - y^2 = -45 & (2) \end{cases}$

Subtract: $2x^2 = 72$, $x^2 = 36$, and $x = \pm 6$.

When $x = 6$, $y^2 = x^2 + 45 = 36 + 45 = 81$, and $y = \pm 9$.
When $x = -6$, $y^2 = x^2 + 45 = 36 + 45 = 81$, and $y = \pm 9$.

The solutions are: $x = \pm 6,\ y = \pm 9$; $x = \pm 6,\ y = \mp 9$.

The two hyperbolas intersect in the points (6,9), (−6,9), (−6,−9), and (6,−9).

4. Solve the system $\begin{cases} 5x^2 + 3y^2 = 92 & (1) \\ 2x^2 + 5y^2 = 52 & (2) \end{cases}$

Multiply (1) by 5 : $25x^2 + 15y^2 = 460$
Multiply (2) by −3: $-6x^2 - 15y^2 = -156$

Add: $19x^2 = 304$; $x^2 = 16$ and $x = \pm 4$.

When $x = \pm 4$: $3y^2 = 92 - 5x^2 = 92 - 80 = 12$; $y^2 = 4$ and $y = \pm 2$.

The solutions are: $x = \pm 4,\ y = \pm 2$; $x = \pm 4,\ y = \mp 2$.

5. Solve the system $\begin{cases} x^2 + 4xy = 0 & (1) \\ x^2 - xy + y^2 = 21 & (2) \end{cases}$

Solve (*1*) for x: $x(x+4y) = 0$ and $x = 0$, $x = -4y$.

Solve the systems

$\begin{cases} x^2 - xy + y^2 = 21 \\ x = 0. \end{cases}$ $\qquad$ $\begin{cases} x^2 - xy + y^2 = 21 \\ x = -4y. \end{cases}$

$y^2 = 21, \quad y = \pm\sqrt{21}$ $\qquad$ $y^2 = 1, \quad y = \pm 1; \quad x = -4y = \mp 4.$

The solutions are: $x = 0, \ y = \pm\sqrt{21}; \quad x = \pm 4, \ y = \mp 1.$

6. Solve the system $\begin{cases} 3x^2 + 8y^2 = 140 & (1) \\ 5x^2 + 8xy = 84 & (2) \end{cases}$

Multiply (*1*) by −3: $-9x^2 - 24y^2 = -420$
Multiply (2) by 5 : $25x^2 + 40xy = 420$
Add : $16x^2 + 40xy - 24y^2 = 0$

Then $8(2x^2 + 5xy - 3y^2) = 8(2x - y)(x + 3y) = 0$ and $x = \frac{1}{2}y, \ x = -3y$.

Solve the systems

$\begin{cases} 3x^2 + 8y^2 = 140 \\ x = \frac{1}{2}y. \end{cases}$ $\qquad$ $\begin{cases} 3x^2 + 8y^2 = 140 \\ x = -3y. \end{cases}$

$\frac{3}{4}y^2 + 8y^2 = \frac{35}{4}y^2 = 140$ $\qquad$ $27y^2 + 8y^2 = 35y^2 = 140$

$y^2 = 16, \ y = \pm 4; \quad x = \frac{1}{2}y = \pm 2.$ $\qquad$ $y^2 = 4, \ y = \pm 2; \quad x = -3y = \mp 6.$

The solutions are: $x = \pm 2, \ y = \pm 4; \quad x = \mp 6, \ y = \pm 2.$

7. Solve the system $\begin{cases} x^2 - 3xy + 2y^2 = 15 & (1) \\ 2x^2 + y^2 = 6 & (2) \end{cases}$

Multiply (*1*) by −2: $-2x^2 + 6xy - 4y^2 = -30$
Multiply (2) by 5 : $10x^2 + 5y^2 = 30$
Add : $8x^2 + 6xy + y^2 = (4x + y)(2x + y) = 0.$ Then $y = -4x$ and $y = -2x$.

Solve the systems

$\begin{cases} 2x^2 + y^2 = 6 \\ y = -4x. \end{cases}$ $\qquad$ $\begin{cases} 2x^2 + y^2 = 6 \\ y = -2x. \end{cases}$

$2x^2 + 16x^2 = 18x^2 = 6, \quad x^2 = 1/3;$ $\qquad$ $2x^2 + 4x^2 = 6x^2 = 6, \quad x^2 = 1;$

$x = \pm\sqrt{3}/3$ and $y = -4x = \mp 4\sqrt{3}/3.$ $\qquad$ $x = \pm 1$ and $y = -2x = \mp 2.$

The solutions are: $x = \pm\sqrt{3}/3, \ y = \mp 4\sqrt{3}/3; \quad x = \pm 1, \ y = \mp 2.$

8. Solve the system $\begin{cases} x^2 + y^2 + 3x + 3y = 8 \\ xy + 4x + 4y = 2 \end{cases}$

Substitute $x = u + v, \ y = u - v$ in the given system:

$$(u+v)^2 + (u-v)^2 + 3(u+v) + 3(u-v) = 2u^2 + 2v^2 + 6u = 8 \quad (1)$$
$$(u+v)(u-v) + 4(u+v) + 4(u-v) = u^2 - v^2 + 8u = 2 \quad (2)$$

Add (*1*) and 2(2): $4u^2 + 22u - 12 = 2(2u - 1)(u + 6) = 0; \quad u = \frac{1}{2}, -6.$

For $u = \frac{1}{2}$, (2) yields $v^2 = u^2 + 8u - 2 = 1/4 + 4 - 2 = 9/4$; $v = \pm 3/2$.

When $u = \frac{1}{2}$, $v = 3/2$: $x = u + v = 2$, $y = u - v = -1$.
When $u = \frac{1}{2}$, $v = -3/2$: $x = u + v = -1$, $y = u - v = 2$.

For $u = -6$, (2) yields $v^2 = u^2 + 8u - 2 = 36 - 48 - 2 = -14$; $v = \pm i\sqrt{14}$.

When $u = -6$, $v = i\sqrt{14}$: $x = u + v = -6 + i\sqrt{14}$, $y = u - v = -6 - i\sqrt{14}$.
When $u = -6$, $v = -i\sqrt{14}$: $x = u + v = -6 - i\sqrt{14}$, $y = u - v = -6 + i\sqrt{14}$.

The solutions are: $x = 2, y = -1$; $x = -1, y = 2$; $x = -6 \pm i\sqrt{14}$, $y = -6 \mp i\sqrt{14}$.

9. Solve the system $\begin{cases} x^2 + y^2 = 25 & (1) \\ xy = 12 & (2) \end{cases}$

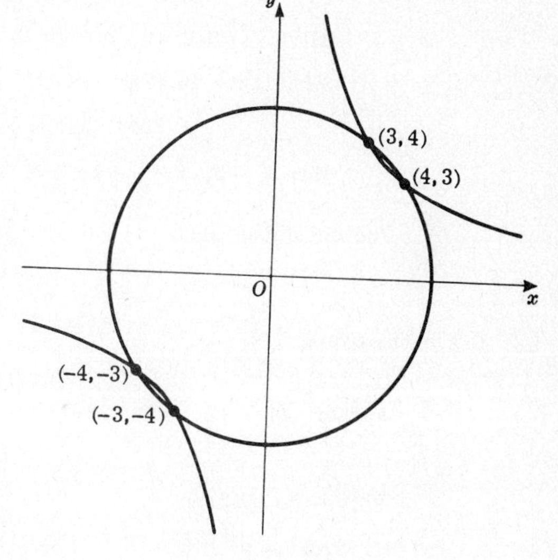

Multiply (2) by 2 and add to (1):
$$x^2 + 2xy + y^2 = 49 \quad \text{or} \quad x + y = \pm 7.$$

Multiply (2) by -2 and add to (1):
$$x^2 - 2xy + y^2 = 1 \quad \text{or} \quad x - y = \pm 1.$$

Solve the systems

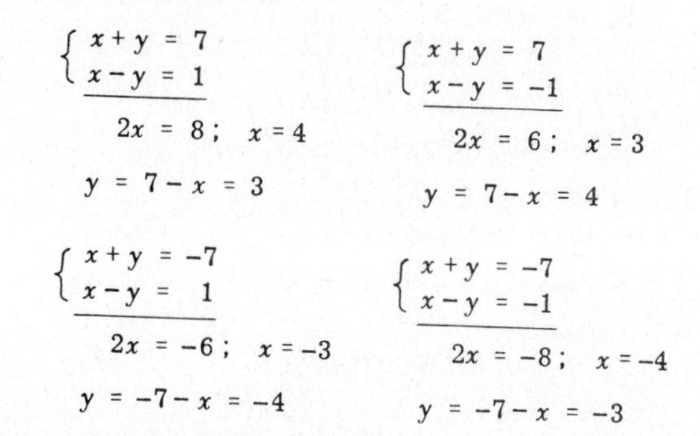

$\begin{cases} x + y = 7 \\ x - y = 1 \end{cases}$ $2x = 8$; $x = 4$; $y = 7 - x = 3$

$\begin{cases} x + y = 7 \\ x - y = -1 \end{cases}$ $2x = 6$; $x = 3$; $y = 7 - x = 4$

$\begin{cases} x + y = -7 \\ x - y = 1 \end{cases}$ $2x = -6$; $x = -3$; $y = -7 - x = -4$

$\begin{cases} x + y = -7 \\ x - y = -1 \end{cases}$ $2x = -8$; $x = -4$; $y = -7 - x = -3$

The solutions are: $x = \pm 4$, $y = \pm 3$; $x = \pm 3$, $y = \pm 4$.

Alternate Solution. Solve (2) for $y = 12/x$ and substitute in (1). The resulting quartic $x^4 - 25x^2 + 144 = 0$ can be factored readily.

10. Solve the system $\begin{cases} x^2 - xy - 12y^2 = 8 & (1) \\ x^2 + xy - 10y^2 = 20 & (2) \end{cases}$

This system may be solved by the procedure used in Problems 6 and 7. Here we give an alternate solution.

Procedure: Substitute $y = mx$ in the given equations to obtain a system in the unknowns m and x; then eliminate x to obtain a quadratic in m.

Put $y = mx$ in (1) and (2):
$$x^2 - mx^2 - 12m^2x^2 = x^2(1 - m - 12m^2) = 8$$
$$x^2 + mx^2 - 10m^2x^2 = x^2(1 + m - 10m^2) = 20$$

Now $x^2 = \dfrac{8}{1 - m - 12m^2}$ and $x^2 = \dfrac{20}{1 + m - 10m^2}$ so that $\dfrac{8}{1 - m - 12m^2} = \dfrac{20}{1 + m - 10m^2}$;

$$8 + 8m - 80m^2 = 20 - 20m - 240m^2,$$
$$160m^2 + 28m - 12 = 4(5m - 1)(8m + 3) = 0, \quad \text{and} \quad m = 1/5,\ m = -3/8.$$

When $m = 1/5$: $x^2 = \dfrac{8}{1 - m - 12m^2} = 25$; $x = \pm 5$, $y = mx = \frac{1}{5}(\pm 5) = \pm 1$.

When $m = -3/8$: $x^2 = \dfrac{8}{1-m-12m^2} = -\dfrac{128}{5}$; $x = \pm\dfrac{8i\sqrt{10}}{5}$, $y = mx = \mp\dfrac{3i\sqrt{10}}{5}$.

The solutions are: $x = \pm 5,\ y = \pm 1$; $x = \pm\dfrac{8i\sqrt{10}}{5}$, $y = \mp\dfrac{3i\sqrt{10}}{5}$.

11. Solve the system $\begin{cases} x^3 - y^3 = 19 & (1) \\ x^2 + xy + y^2 = 19 & (2) \end{cases}$

Divide (1) by (2): $x - y = 1$.

Solve the system $\begin{cases} x^2 + xy + y^2 = 19 & (2) \\ x - y = 1 & (3) \end{cases}$

Solve (3) for x: $x = y + 1$.

Substitute in (2): $(y+1)^2 + (y+1)y + y^2 = 3y^2 + 3y + 1 = 19$.

Then $3y^2 + 3y - 18 = 3(y+3)(y-2) = 0$ and $y = -3, 2$.

When $y = -3$, $x = y + 1 = -2$; when $y = 2$, $x = y + 1 = 3$.

The solutions are: $x = -2,\ y = -3$; $x = 3,\ y = 2$.

12. Solve the system $\begin{cases} (2x-y)^2 - 4(2x-y) = 5 & (1) \\ x^2 - y^2 = 3 & (2) \end{cases}$

Factor (1): $(2x-y)^2 - 4(2x-y) - 5 = (2x-y-5)(2x-y+1) = 0$. Then

$$2x - y = 5 \quad \text{and} \quad 2x - y = -1$$

Solve the systems

$\begin{cases} x^2 - y^2 = 3 \\ 2x - y = 5 \end{cases}$

$y = 2x - 5$.
$x^2 - (2x-5)^2 = 3$,
$3x^2 - 20x + 28 = (x-2)(3x-14) = 0$,
$x = 2, 14/3$.

When $x = 2$, $y = 2x - 5 = -1$.

When $x = 14/3$, $y = 2x - 5 = 13/3$.

$\begin{cases} x^2 - y^2 = 3 \\ 2x - y = -1 \end{cases}$

$y = 2x + 1$.
$x^2 - (2x+1)^2 = 3$,
$3x^2 + 4x + 4 = 0$,
$x = \dfrac{-4 \pm \sqrt{16-48}}{6} = \dfrac{-2 \pm 2i\sqrt{2}}{3}$

$y = 2x + 1 = \dfrac{-1 \pm 4i\sqrt{2}}{3}$

The solutions are: $x = 2,\ y = -1$; $x = 14/3,\ y = 13/3$; $x = \dfrac{-2 \pm 2i\sqrt{2}}{3}$, $y = \dfrac{-1 \pm 4i\sqrt{2}}{3}$.

13. Solve the system $\begin{cases} 5/x^2 + 3/y^2 = 32 & (1) \\ 4xy = 1 & (2) \end{cases}$

Write (1) as $3x^2 + 5y^2 = 32x^2y^2 = 2(4xy)^2$. Substitute (2): $3x^2 + 5y^2 = 2(1)^2 = 2$ (3)

Subtract 2(2) from (3): $3x^2 - 8xy + 5y^2 = 0$. Then $(x-y)(3x-5y) = 0$ and $x = y$, $x = 5y/3$.

Solve the systems

$\begin{cases} 4xy = 1 \\ x = y \end{cases}$

$4y^2 = 1$; $y = \pm\frac{1}{2}$,

and $x = y = \pm\frac{1}{2}$.

$\begin{cases} 4xy = 1 \\ x = 5y/3 \end{cases}$

$\dfrac{20}{3}y^2 = 1$, $y^2 = \dfrac{3}{20} = \dfrac{15}{100}$;

$y = \pm\dfrac{\sqrt{15}}{10}$ and $x = \dfrac{5}{3}y = \pm\dfrac{\sqrt{15}}{6}$.

The solutions are: $x = \pm\frac{1}{2},\ y = \pm\frac{1}{2};\quad x = \pm\dfrac{\sqrt{15}}{6},\ y = \pm\dfrac{\sqrt{15}}{10}$.

SUPPLEMENTARY PROBLEMS

Solve.

14. $\begin{cases} xy + y^2 = 5 \\ 2x + 3y = 7 \end{cases}$ *Ans.* $x = -4,\ y = 5$; $x = \frac{1}{2},\ y = 2$

15. $\begin{cases} y = x^2 - x - 1 \\ y = 2x + 3 \end{cases}$ *Ans.* $x = -1,\ y = 1$; $x = 4,\ y = 11$

16. $\begin{cases} 3x^2 - 7y^2 = 12 \\ x - 3y = -2 \end{cases}$ *Ans.* $x = -2,\ y = 0$; $x = 17/5,\ y = 9/5$

17. $\begin{cases} x^2 + 3y^2 = 43 \\ 3x^2 + y^2 = 57 \end{cases}$ *Ans.* $x = \pm 4,\ y = \pm 3$; $x = \pm 4,\ y = \mp 3$

18. $\begin{cases} 9x^2 + y^2 = 90 \\ x^2 + 9y^2 = 90 \end{cases}$ *Ans.* $x = \pm 3,\ y = \pm 3$; $x = \pm 3,\ y = \mp 3$

19. $\begin{cases} 2/x^2 - 3/y^2 = 5 \\ 1/x^2 + 2/y^2 = 6 \end{cases}$ *Ans.* $x = \pm\frac{1}{2},\ y = \pm 1$; $x = \pm\frac{1}{2},\ y = \mp 1$

20. $\begin{cases} x^2 - xy + y^2 = 28 \\ 2x^2 + 3xy - 2y^2 = 0 \end{cases}$ *Ans.* $x = \pm 4,\ y = \mp 2$; $x = \pm 2\sqrt{21}/3,\ y = \pm 4\sqrt{21}/3$

21. $\begin{cases} x^2 - xy - 12y^2 = 0 \\ x^2 + xy - 10y^2 = 20 \end{cases}$ *Ans.* $x = \pm 4\sqrt{2},\ y = \pm\sqrt{2}$; $x = \mp 3i\sqrt{5},\ y = \pm i\sqrt{5}$

22. $\begin{cases} 6x^2 + 3xy + 2y^2 = 24 \\ 3x^2 + 2xy + 2y^2 = 18 \end{cases}$ *Ans.* $x = \pm 2,\ y = \mp 3$; $x = \pm\sqrt{30}/5,\ y = \pm 2\sqrt{30}/5$

23. $\begin{cases} y^2 = 4x - 8 \\ y^2 = -6x + 32 \end{cases}$ *Ans.* $x = 4,\ y = \pm 2\sqrt{2}$

24. $\begin{cases} x^2 - y^2 = 16 \\ y^2 = 2x - 1 \end{cases}$ *Ans.* $x = 5,\ y = \pm 3$; $x = -3,\ y = \pm i\sqrt{7}$

25. $\begin{cases} 2x^2 + y^2 = 6 \\ x^2 + y^2 + 2x = 3 \end{cases}$ *Ans.* $x = -1,\ y = \pm 2$; $x = 3,\ y = \pm 2i\sqrt{3}$

26. $\begin{cases} x^2 + y^2 - 2x - 2y = 12 \\ xy = 6 \end{cases}$ *Ans.* $x = 3 \pm \sqrt{3},\ y = 3 \mp \sqrt{3}$; $x = -2 \pm i\sqrt{2},\ y = -2 \mp i\sqrt{2}$

27. $\begin{cases} x^3 - y^3 = 28 \\ x - y = 4 \end{cases}$ *Ans.* $x = 1,\ y = -3$; $x = 3,\ y = -1$

28. $\begin{cases} x + y + 3\sqrt{x+y} = 18 \\ x - y - 2\sqrt{x-y} = 15 \end{cases}$ Hint: Let $\sqrt{x+y} = u$, $\sqrt{x-y} = v$. *Ans.* $x = 17,\ y = -8$

29. Two numbers differ by 2 and their squares differ by 48. Find the numbers. *Ans.* 11, 13

30. The sum of the circumferences of two circles is 88 in. and the sum of their areas is 2200/7 in^2, when 22/7 is used for π. Find the radius of each circle. *Ans.* 6 in., 8 in.

31. A party costing $30 is planned. It is found that by adding three more to the group, the cost per person would be reduced by 50 cents. For how many was the party originally planned? *Ans.* 12

32. A truck left A at 5 AM on a trip to B 300 miles distant. At 6 AM a car left A, overtook the truck, delivered a parcel, and returned to A. If the car averaged 60 mi/hr and reached A $2\frac{1}{2}$ hours before the truck reached B, find the average speed of the truck and locate the point where the car overtook the truck. *Ans.* 40 mi/hr, 120 mi from A

CHAPTER 11

Arithmetic and Geometric Progressions

A SEQUENCE IS A SET OF NUMBERS, called *terms*, arranged in a definite order; that is, there is a rule by which the terms after the first may be formed. Sequences may be finite or infinite. Only finite sequences will be treated in this chapter.

Example 1.

(*a*) Sequence: 3, 7, 11, 15, 19, 23, 27.
Type: Finite, of 7 terms.
Rule: Add 4 to a given term to produce the next.

(*b*) Sequence: 3, 6, 12, 24, 48, 96.
Type: Finite, of 6 terms.
Rule: Multiply a given term by 2 to produce the next.

AN ARITHMETIC PROGRESSION is a sequence in which each term after the first is formed by adding a fixed amount, called the *common difference*, to the preceding term. The sequence of Example 1(*a*) is an arithmetic progression whose common difference is 4.

See Problems 1-2.

If a is the first term, d is the common difference, and n is the number of terms of an arithmetic progression, the successive terms are:

$$(1) \qquad a,\ a+d,\ a+2d,\ a+3d,\ \ldots\ldots,\ a+(n-1)d.$$

Thus, the *last term* (or nth term) l is given by

$$(2) \qquad l = a + (n-1)d$$

The *sum* S of the n terms of this progression is given by

$$(3) \qquad S = \frac{n}{2}(a+l) \qquad \text{or} \qquad S = \frac{n}{2}[2a+(n-1)d]$$

For a proof, see Problem 3.

Example 2. Find the 20th term and the sum of the first 20 terms of the arithmetic progression 4, 9, 14, 19, ...

For this progression $a = 4$, $d = 5$, and $n = 20$; then the 20th term is $l = a+(n-1)d = 4 + 19\cdot 5 = 99$ and the sum of the first 20 terms is

$$S = \frac{n}{2}(a+l) = \frac{20}{2}(4+99) = 1030$$

See Problems 4-8.

THE TERMS BETWEEN THE FIRST AND LAST TERMS of an arithmetic progression are called *arithmetic means* between these two terms. Thus, to insert k arithmetic means between two numbers is to form an arithmetic progression of $(k+2)$ terms having the two given numbers as first and last terms.

Example 3. Insert 5 arithmetic means between 4 and 22.

We have $a=4$, $l=22$, and $n=5+2=7$. Then $22=4+6d$ and $d=3$. The first mean is $4+3=7$, the second is $7+3=10$, and so on. The required means are 7, 10, 13, 16, 19 and the resulting progression is 4, 7, 10, 13, 16, 19, 22.

When just one mean is to be inserted between two numbers, it is called *the arithmetic mean* (also, the average) of the two numbers.

Example 4. Find the arithmetic mean of the two numbers a and l.

We seek the middle term of an arithmetic progression of three terms having a and l as first and third terms respectively. If d is the common difference, then $a+d = l-d$ and $d = \frac{1}{2}(l-a)$. The arithmetic mean is $a+d = a+\frac{1}{2}(l-a) = \frac{1}{2}(a+l)$.

See Problem 9.

A GEOMETRIC PROGRESSION is a sequence in which each term after the first is formed by multipling the preceding term by a fixed number, called the *common ratio*. The sequence 3, 6, 12, 24, 48, 96 of Example 1(b) is a geometric progression whose common ratio is 2.

See Problems 10-11.

If a is the first term, r is the common ratio, and n is the number of terms, the geometric progression is

(4) $$a,\ ar,\ ar^2,\ \ldots\ldots,\ ar^{n-1}$$

Thus, the last (or nth) term l is given by

(5) $$l = ar^{n-1}$$

The *sum* S of the first n terms of the geometric progression (1) is given by

(6) $$S = \frac{a-rl}{1-r} \qquad \text{or} \qquad S = \frac{a(1-r^n)}{1-r}$$

For a proof, see Problem 12.

Example 5. Find the 9th term and the sum of the first nine terms of the geometric progression 8, 4, 2, 1, ...

Here $a=8$, $r=\frac{1}{2}$, and $n=9$; the 9th term is $l = ar^{n-1} = 8(\frac{1}{2})^8 = 1/2^5 = 1/32$ and the sum of the first nine terms is

$$S = \frac{a-rl}{1-r} = \frac{8-\frac{1}{2}(1/32)}{1-\frac{1}{2}} = 16 - 1/32 = 511/32$$

See Problems 13-18.

THE TERMS BETWEEN THE FIRST AND LAST TERMS of a geometric progression are called *geometric means* between the two terms. Thus, to insert k geometric means between two numbers is to form a geometric progression of $(k+2)$ terms having the two given numbers as first and last terms.

Example 6. Insert 4 geometric means between 25 and 1/125.

We have $a=25$, $l=1/125$, and $n=4+2=6$. Using $l=ar^{n-1}$, $1/125=25r^5$; then $r^5=1/5^5$ and $r=1/5$. The first mean is $25(1/5)=5$, the second is $5(1/5)=1$, and so on. The required means are 5, 1, 1/5, 1/25 and the geometric progression is 25, 5, 1, 1/5, 1/25, 1/125.

When one mean is to be inserted between two numbers, it is called *the geometric mean* of the two numbers.

The geometric mean of two numbers a and l, having like signs, is $(\pm)\sqrt{a\cdot l}$. The sign to be used is the common sign of a and l. See Problem 19.

SOLVED PROBLEMS

ARITHMETIC PROGRESSIONS.

1. Determine which of the following sequences are arithmetic progressions (A.P.). In the case of an A.P., write three more terms.

(a) 3, 6, 9, 12, 15, 18.

Since $6-3 = 9-6 = 12-9 = 15-12 = 18-15 = 3$, the sequence is an A.P. with common difference 3. The next three terms are $18+3 = 21$, $21+3 = 24$, and $24+3 = 27$.

(b) 25, 19, 13, 7, 1, −5.

Since $19-25 = 13-19 = 7-13 = 1-7 = -5-1 = -6$, the sequence is an A.P. with $d = -6$. The next three terms are $-5+(-6) = -11$, $-11+(-6) = -17$, and $-17+(-6) = -23$.

(c) 5, 10, 14, 20, 25.

Since $10-5 \neq 14-10$, the sequence is not an A.P.

(d) $3a-2b$, $4a-b$, $5a$, $6a+b$.

Since $(4a-b)-(3a-2b) = 5a-(4a-b) = (6a+b)-5a = a+b$, the sequence is an A.P. with $d = a+b$. The next three terms are $(6a+b)+(a+b) = 7a+2b$, $8a+3b$, and $9a+4b$.

2. Find the value of k such that each sequence is an A.P.

(a) $k-1$, $k+3$, $3k-1$.

If the sequence is to form an A.P., $(k+3)-(k-1) = (3k-1)-(k+3)$. Then $k=4$ and the A.P. is 3, 7, 11.

(b) $3k^2+k+1$, $2k^2+k$, $4k^2-6k+1$.

Setting $(2k^2+k)-(3k^2+k+1) = (4k^2-6k+1)-(2k^2+k)$, we have $3k^2-7k+2 = 0$ and $k = 2, 1/3$. The progressions are 15, 10, 5 when $k=2$ and 5/3, 5/9, −5/9 when $k=1/3$.

3. Obtain the formula $S = \frac{n}{2}[2a+(n-1)d]$ for an arithmetic progression.

Write the indicated sum of the n terms in the order given by (1), then write this sum in reverse order, and sum term by term. Thus,

$$S = \{a\} + \{a+d\} + \cdots + \{a+(n-2)d\} + \{a+(n-1)d\}$$
$$S = \{a+(n-1)d\} + \{a+(n-2)d\} + \cdots + \{a+d\} + \{a\}$$
$$2S = \{2a+(n-1)d\} + \{2a+(n-1)d\} + \cdots + \{2a+(n-1)d\} + \{2a+(n-1)d\}$$
$$= n[2a+(n-1)d] \quad \text{and} \quad S = \frac{n}{2}[2a+(n-1)d].$$

4. (a) Find the 18th term and the sum of the first 18 terms of the A.P. 2, 6, 10, 14, ...

Here $a=2$, $d=4$, $n=18$. Then $l = a+(n-1)d = 2+17\cdot4 = 70$ and $S = \frac{n}{2}(a+l) = \frac{18}{2}(2+70) = 648$.

(b) Find the 49th term and the sum of the first 49 terms of the A.P. 10, 4, −2, −8, ...

Here $a=10$, $d=-6$, $n=49$. Then $l = 10+48(-6) = -278$ and $S = \frac{49}{2}(10-278) = -6566$.

(*c*) Find the 12th term and the sum of the first 15 terms of the A.P. 8, 19/3, 14/3, 3, ...

Since $a = 8$ and $d = -5/3$, the 12th term is $l = 8 + 11(-5/3) = -31/3$ and the sum of the first 15 terms is

$$S = \frac{n}{2}[2a + (n-1)d] = \frac{15}{2}[16 + 14(-5/3)] = \frac{15}{2}(-\frac{22}{3}) = -55$$

(*d*) Find the 10th term, the sum of the first 10 terms and the sum of the first 13 terms of the A.P. $2x + 3y,\ x + y,\ -y,\ \ldots$.

Here $a = 2x + 3y$ and $d = -x - 2y$. The 10th term is $l = (2x + 3y) + 9(-x - 2y) = -7x - 15y$.

Sum of first 10 terms is $S = 5[(2x + 3y) + (-7x - 15y)] = -25x - 60y$.

Sum of first 13 terms is $S = \frac{13}{2}[2(2x + 3y) + 12(-x - 2y)] = \frac{13}{2}(-8x - 18y) = -52x - 117y$.

5. The 7th term of an A.P. is 41 and the 13th term is 77. Find the 20th term.

If a is the first term and d is the common difference, then for

the 7th term $a + 6d = 41$
and for the 13th term $a + 12d = 77$

Subtracting, $6d = 36$; then $d = 6$ and $a = 41 - 6 \cdot 6 = 5$. The 20th term is $l = 5 + 19 \cdot 6 = 119$.

6. The 6th term of an A.P. is 21 and the sum of the first 17 terms is 0. Write the first three terms.

If a is the first term and d is the common difference,

$$a + 5d = 21 \qquad \text{and} \qquad 0 = \frac{17}{2}(2a + 16d) \quad \text{or} \quad a + 8d = 0.$$

Then $d = -7$ and $a = -8d = 56$. The first three terms are 56, 49, 42.

7. Obtain formulas for: (*a*) l in terms of a, n, S; (*b*) a in terms of d, n, S.

(*a*) From $S = \frac{n}{2}(a + l)$, $a + l = \frac{2S}{n}$ and $l = \frac{2S}{n} - a$.

(*b*) From $S = \frac{n}{2}[2a + (n-1)d]$, $2a + (n-1)d = \frac{2S}{n}$, $2a = \frac{2S}{n} - (n-1)d$ and $a = \frac{S}{n} - \frac{1}{2}(n-1)d$.

8. If a body is dropped, the distance (s feet) through which it falls freely in t seconds, is approximately $16t^2$. (*a*) Show that the distances through which it falls during the first, second, third, ... seconds form an A.P. (*b*) How far will the body fall in the 10th second? (*c*) How far will it fall in the first 20 seconds?

(*a*) The distance through which the body falls during the first sec is 16 ft, during the second sec is $16(2)^2 - 16 = 48$ ft, during the third sec is $16(3)^2 - 16(2)^2 = 80$ ft, during the fourth sec is $16(4)^2 - 16(3)^2 = 112$ ft, and so on. These are the first four terms of an A.P. whose common difference is 32.

(*b*) When $n = 10$, $l = 16 + 9(32) = 304$ ft.

(*c*) In the first 20 sec, the body falls $16(20)^2 = 6400$ ft.

9. (*a*) Insert six arithmetic means between 7 and 77.

For the A.P. having $a = 7$, $l = 77$ and $n = 6 + 2 = 8$, $77 = 7 + 7d$ and $d = 10$.
The required means are 17, 27, 37, 47, 57, 67 and the A.P. is 7, 17, 27, 37, 47, 57, 67, 77.

(*b*) Find the arithmetic mean of 8 and −56.

From Example 4, the arithmetic mean is $\frac{1}{2}(a + l) = \frac{1}{2}[8 + (-56)] = -24$.

GEOMETRIC PROGRESSIONS.

10. Determine which of the following sequences are geometric progressions (G.P.). In the case of a G.P., write the next three terms.

(*a*) 4, 8, 16, 32, 64.
Since $8/4 = 16/8 = 32/16 = 64/32 = 2$, the sequence is a G.P. with common ratio 2.
The next three terms are 128, 256, 512.

(*b*) 1, 1/4, 1/16, 1/48.
Since $(1/16)/(1/4) \neq (1/48)/(1/16)$, the sequence is not a G.P.

(*c*) 12, −4, 4/3, −4/9.
Since $-4/12 = (4/3)/-4 = (-4/9)/(4/3) = -1/3$, the sequence is a G.P. with common ratio −1/3.
The next three terms are 4/27, −4/81, 4/243.

11. Find the value of k so that the sequence $2k-5,\ k-4,\ 10-3k$ forms a G.P.

If the sequence is to form a G.P., $\dfrac{k-4}{2k-5} = \dfrac{10-3k}{k-4}$ or $k^2 - 8k + 16 = -6k^2 + 35k - 50$.

Then $7k^2 - 43k + 66 = (k-3)(7k-22) = 0$ and $k = 3,\ 22/7$.

The sequences are 1, −1, 1 when $k = 3$ and 9/7, −6/7, 4/7 when $k = 22/7$.

12. Obtain the formula $S = \dfrac{a(1-r^n)}{1-r}$ for a geometric progression.

Write the indicated sum of the n terms given by (4), then multiply this sum by r, and subtract term by term. Thus,

$$\begin{aligned} S &= a + ar + ar^2 + \dots\dots + ar^{n-1} \\ rS &= \phantom{a + {}} ar + ar^2 + \dots\dots + ar^{n-1} + ar^n \\ S - rS &= a \qquad\qquad\qquad\qquad\qquad\qquad - ar^n \end{aligned}$$

Then $S(1-r) = a - ar^n = a(1-r^n)$ and $S = \dfrac{a(1-r^n)}{1-r}$.

13. (*a*) Find the 7th term and the sum of the first seven terms of the G.P. 12, 16, 64/3, ...

Here $a = 12$, $r = 4/3$, $n = 7$. The 7th term is $l = ar^{n-1} = 12(4/3)^6 = 4^7/3^5 = 16384/243$ and the sum of the first seven terms is

$$S = \frac{a - rl}{1-r} = \frac{12 - \frac{4}{3}(4^7/3^5)}{1 - 4/3} = \frac{4^8}{3^5} - 36 = \frac{65536 - 8748}{243} = \frac{56788}{243}$$

(*b*) Find the 6th term and the sum of the first nine terms of the G.P. 4, −6, 9, ...

Since $a = 4$ and $r = -3/2$, the 6th term is $l = 4(-3/2)^5 = \dfrac{-3^5}{2^3} = -\dfrac{243}{8}$ and the sum of the first nine terms is

$$S = \frac{a(1-r^n)}{1-r} = \frac{4[1-(-3/2)^9]}{1-(-3/2)} = \frac{8(1+3^9/2^9)}{5} = \frac{2^9+3^9}{5\cdot 2^6} = \frac{4039}{64}$$

(*c*) Find the sum of the G.P. 8, −4, 2, ..., 1/128.

$$S = \frac{a - rl}{1-r} = \frac{8 - (-\frac{1}{2})(1/128)}{1-(-\frac{1}{2})} = \frac{2^4 + 1/2^7}{3} = \frac{2^{11}+1}{3\cdot 2^7} = \frac{683}{128}$$

14. The fourth term of a G.P. is 1 and the eighth term is 1/256. Find the tenth term.

Since the fourth term is 1, $ar^3 = 1$; since the eighth term is 1/256, $ar^7 = 1/256$. Then $ar^7/ar^3 = 1/256$, $r^4 = 1/256$ and $r = \pm\frac{1}{4}$. From $ar^3 = 1$, we have $a = \pm 64$. In each case, the tenth term is 1/4096.

15. Given $S = 3367/64$, $r = 3/4$, $l = 243/64$. Find a and n.

Since $S = \dfrac{3367}{64} = \dfrac{a - (3/4)(243/64)}{1 - 3/4} = 4a - \dfrac{729}{64}$, $4a = \dfrac{4096}{64}$ and $a = 16$.

Now $l = \dfrac{243}{64} = 16(\frac{3}{4})^{n-1}$, $(\frac{3}{4})^{n-1} = \dfrac{243}{16 \cdot 64} = (\frac{3}{4})^5$, $n - 1 = 5$ and $n = 6$.

16. Given $a = 8$, $r = 3/2$, $S = 2059/8$. Find l and n.

Since $S = \dfrac{2059}{8} = \dfrac{8 - (3/2)l}{1 - 3/2} = 3l - 16$, $3l = \dfrac{2059}{8} + 16 = \dfrac{2187}{8}$ and $l = 729/8$.

Now $l = \dfrac{729}{8} = 8(\dfrac{3}{2})^{n-1}$, $(\dfrac{3}{2})^{n-1} = \dfrac{729}{64} = (\dfrac{3}{2})^6$, $n - 1 = 6$ and $n = 7$.

17. If a boy undertakes to deposit 1¢ on Sep. 1, 2¢ on Sep. 2, 4¢ on Sep. 3, 8¢ on Sep. 4, and so on, (*a*) how much will he deposit from Sep. 1 to Sep. 15 inclusive, (*b*) how much would he deposit on Sep. 30?

Here, $a = .01$ and $r = 2$.

(*a*) When $n = 15$, $S = \dfrac{.01(1 - 2^{15})}{1 - 2} = .01(2^{15} - 1) = \$327.67$.

(*b*) When $n = 30$, $l = .01(2)^{29} = \$5,368,709.12$.

18. A rubber ball is dropped from a height of 81 feet. Each time it strikes the ground it rebounds two-thirds of the distance through which it last fell. (*a*) Through what distance did the ball fall when it struck the ground for the sixth time? (*b*) Through what distance had it traveled from the time it was dropped until it struck the ground for the sixth time?

(*a*) The successive distances through which the ball falls form a G.P. in which $a = 81$, $r = 2/3$.

When $n = 6$, $l = 81(2/3)^5 = 32/3$ ft.

(*b*) The required distance is the sum of the distances for the first six falls and the first five rebounds.

For the falls: $a = 81$, $r = 2/3$, $n = 6$, and

$$S = \frac{81[1 - (2/3)^6]}{1 - 2/3} = 81(3 - \frac{2^6}{3^5}) = \frac{3^6 - 2^6}{3} = \frac{665}{3} \text{ ft.}$$

For the rebounds: $a = 54$, $r = 2/3$, $n = 5$, and

$$S = \frac{54[1 - (2/3)^5]}{1 - 2/3} = 54(3 - \frac{2^5}{3^4}) = \frac{2}{3}(3^5 - 2^5) = \frac{422}{3} \text{ ft.}$$

Thus, the total distance is $665/3 + 422/3 = 362\ 1/3$ ft.

19. (*a*) Insert five geometric means between 8 and 1/8.

We have $a = 8$, $l = 1/8$, $n = 5 + 2 = 7$. Since $l = ar^{n-1}$, $1/8 = 8r^6$ and $r = \pm\frac{1}{2}$.

When $r = \frac{1}{2}$, the first mean is $8(\frac{1}{2}) = 4$, the second is $4(\frac{1}{2}) = 2$, and so on. The required means are $4, 2, 1, \frac{1}{2}, \frac{1}{4}$ and the G.P. is $8, 4, 2, 1, \frac{1}{2}, \frac{1}{4}, 1/8$.

When $r = -\frac{1}{2}$, the means are $-4, 2, -1, \frac{1}{2}, -\frac{1}{4}$ and the G.P. is $8, -4, 2, -1, \frac{1}{2}, -\frac{1}{4}, 1/8$.

(*b*) Insert four geometric means between 81/2 and −16/3.

Here $a = 81/2$, $l = -16/3$, $n = 6$. Then $-\frac{16}{3} = \frac{81}{2}r^5$, $r^5 = -\frac{32}{243}$ and $r = -2/3$.

The required means are −27, 18, −12, 8 and the G.P. is 81/2, −27, 18, −12, 8, −16/3.

(*c*) Find the geometric mean of 1/3 and 243. The required mean is $\sqrt{\frac{1}{3}(243)} = \sqrt{81} = 9$.

(*d*) Find the geometric mean of −2/3 and −32/27. The required mean is $-\sqrt{(-\frac{2}{3})(-\frac{32}{27})} = -\sqrt{\frac{64}{81}} = -\frac{8}{9}$.

20. A set of numbers is said to form a *harmonic progression* if their reciprocals form an arithmetic progression. To solve a problem in harmonic progressions, take the reciprocals and apply the proper formula on the arithmetic progression.
(*a*) Show that 2, 2/3, 2/5, ... to 30 terms is a harmonic progression and find its last term.
(*b*) Find the harmonic mean between 2 and 4.

(*a*) The reciprocals 1/2, 3/2, 5/2, ... form an arithmetic progression whose 30th term is $1/2 + 29\cdot 1 = 59/2$. The 30th term of the harmonic progression is 2/59.

(*b*) Let x be the required harmonic mean. Since $\frac{1}{x} = \frac{1/2 + 1/4}{2} = \frac{3}{8}$ is the arithmetic mean between 1/2 and 1/4, $x = 8/3$ is the harmonic mean between 2 and 4.

SUPPLEMENTARY PROBLEMS

ARITHMETIC PROGRESSION.

21. Find: (*a*) The 15th term and the sum of the first 15 terms of the A.P. 3, 8, 13, 18, ...
(*b*) The 12th term and the sum of the first 20 terms of the A.P. 11, 8, 5, 2, ...
(*c*) The sum of the A.P. for which $a = 6\frac{3}{4}$, $l = -3\frac{1}{4}$, $n = 17$.
(*d*) The sum of all the integers from 1 to 200 which are divisible by 3.
Ans. (*a*) 73, 570 (*b*) −22, −350 (*c*) $29\frac{3}{4}$ (*d*) 6633

22. The 4th term of an A.P. is 14 and the 9th term is 34. Find the 13th term. *Ans.* 50

23. The sum of the first 7 terms of an A.P. is 98 and the sum of the first 12 terms is 288. Find the sum of the first 20 terms. *Ans.* 800

24. Find the sum of (*a*) the first n positive integers, (*b*) the first n odd positive integers.
Ans. (*a*) $\frac{1}{2}n(n+1)$ (*b*) n^2

25. (*a*) Sum all the integers between 200 and 1000 that are divisible by 3. *Ans.* 160,200
(*b*) Sum all the even positive integers less than 200 which are not divisible by 6. *Ans.* 6534

26. In a potato race, 10 potatoes are placed 8 ft apart in a straight line. If the potatoes are to be picked up singly and returned to the basket, and if the first potato is 20 feet in front of the basket, find the total distance covered by a contestant who finishes the race. *Ans.* 1120 ft

27. In a lottery, tickets are numbered consecutively from 1 to 100. Customers draw a ticket at random and pay an amount in cents corresponding to the number on the ticket except for those tickets with numbers divisible by 5, which are free. How much is realized if 100 tickets are sold? *Ans.* $40

28. Find the arithmetic mean between (*a*) 6 and 60, (*b*) $a - 2d$ and $a + 6d$. *Ans.* (*a*) 33 (*b*) $a + 2d$

29. Insert 5 arithmetic means between 12 and 42. *Ans.* 17, 22, 27, 32, 37

30. After inserting x arithmetic means between 2 and 38, the sum of the resulting progression is 200. Find x. *Ans.* 8

GEOMETRIC PROGRESSION.

31. Find: (a) The 8th term and the sum of the first 8 terms of the G.P. 4, 12, 36, ...
(b) The 10th term and the sum of the first 12 terms of the G.P. 8, 4, 2, ...
(c) The sum of the G.P. for which $a = 64$, $l = 729$, and $n = 7$.
Ans. (a) 8748, 13120 (b) 1/64, 15 255/256 (c) 2059, 463

32. The 3rd term of a G.P. is 36 and the 5th term is 16. Find the 10th term. *Ans.* $\pm 512/243$

33. The sum of the first 3 terms of a G.P. is 21 and the sum of the first 6 terms is 20 2/9. Find the sum of the first 9 terms. *Ans.* 20 61/243

34. Given $S = 255/192$, $l = -1/64$, $r = -1/2$; find a and n. *Ans.* $a = 2$, $n = 8$

35. Find three numbers in geometric progression such that their sum is 14 and the sum of their squares is 84. *Ans.* 2, 4, 8

36. Prove: $x^n - y^n = (x-y)(x^{n-1} + x^{n-2}y + \ldots + xy^{n-2} + y^{n-1})$, n being a positive integer.

37. In a certain colony of bacteria each divides into two every hour. How many will be produced from a single baccilus if the rate of division continues for 12 hours? *Ans.* 4096

38. Find the geometric mean between (a) 2 and 32, (b) -4 and -25. *Ans.* (a) 8 (b) -10

39. Insert 5 geometric means between 6 and 384. *Ans.* 12, 24, 48, 96, 192

40. Show that for $p > q$, positive integers, their arithmetic mean A is greater than their geometric mean G. Hint: Consider $A - G$.

41. The sum of 3 numbers in A.P. is 24. If the first is decreased by 1 and the second is decreased by 2, the three numbers are in G.P. Find the A.P. *Ans.* 4, 8, 12 or 13, 8, 3

HARMONIC PROGRESSIONS.

42. Find: (a) the 16th term of the H.P. 3/4, 3/11, 1/6, ...
(b) the 20th term in the H.P. whose first term is 1 and whose 13th term is 1/19.
Ans. (a) 3/109 (b) 2/59

43. Show that the harmonic mean between a and b is $\dfrac{2ab}{a+b}$.

44. Show that the geometric mean between two positive numbers is also the geometric mean between the arithmetic and the harmonic means of the numbers.

45. Of the three means — arithmetic, geometric, harmonic — between two positive numbers $p > q$, show that the first is the greatest and the third is the least.

CHAPTER 12

Infinite Geometric Series

THE INDICATED SUM of the terms of a finite or infinite sequence is called a finite or infinite *series*. The sums of arithmetic and geometric progressions in the preceding chapter are examples of finite series.

Of course it is impossible to add up all the terms of an infinite series; that is, in the usual meaning of the word *sum*, there is no such thing as the sum of such a series. However, it is possible to associate with certain infinite series a well defined number which, for convenience, will be called the sum of the series.

Infinite series will be treated in some detail in a later chapter. For the study of the infinite geometric series here, we shall need only to examine the behavior of r^n, where $|r| < 1$, as n increases indefinitely.

Example 1. From the table of values of $(\frac{1}{2})^n$

n	1	3	5	10
$(\frac{1}{2})^n$	.5	.125	.03125	.0009765625

it appears that, as n increases indefinitely, $(\frac{1}{2})^n$ decreases indefinitely while remaining positive. Moreover, it can be made to have a value as near 0 as we please by choosing n sufficiently large. We describe this state of affairs by saying: The limit of $(\frac{1}{2})^n$, as n increases indefinitely, is 0.

By examining the behavior of r^n for other values of r, it becomes tolerably clear that

the limit of r^n, as n increases indefinitely, is 0 when $|r| < 1$.

THE SUM S of the infinite geometric series

$$a + ar + ar^2 + \dots + ar^{n-1} + \dots, \quad |r| < 1, \quad \text{is } S = \frac{a}{1-r}.$$

For a proof, see Problem 1.

Example 2. For the infinite geometric series $12 + 4 + 4/3 + 4/9 + \dots$, $a = 12$ and $r = 1/3$. The sum of the series is $S = \dfrac{a}{1-r} = \dfrac{12}{1-1/3} = 18$.

See Problems 2-5.

EVERY REPEATING DECIMAL approximates a rational number. This rational number is also called the *limiting value of the decimal*.

Example 3. Find the limiting value of the repeating decimal .727272.... We write .727272... = .72 + .0072 + .000072 + ... and note that for this infinite geometric series $a = .72$ and $r = .01$.

Then $S = \frac{a}{1-r} = \frac{.72}{1-.01} = \frac{.72}{.99} = \frac{72}{99} = \frac{8}{11}$.

See Problem 5.

SOLVED PROBLEMS

1. Prove: The sum of the infinite geometric series $a + ar + ar^2 + \ldots + ar^{n-1} + \ldots$, where $|r| < 1$, is $S = \frac{a}{1-r}$.

The sum of the first n terms of the series is $S_n = \frac{a(1-r^n)}{1-r} = \frac{a}{1-r} - \frac{a}{1-r}r^n$.

As n increases indefinitely, the first term $\frac{a}{1-r}$ remains fixed while r^n, and hence $\frac{a}{1-r}r^n$, approaches zero in value. Thus, $S = \frac{a}{1-r}$.

2. Determine the sum of each of the following infinite geometric series.

(a) $18 + 12 + 8 + \ldots$ Here $a = 18$, $r = 2/3$, and $S = \frac{a}{1-r} = \frac{18}{1-2/3} = 54$.

(b) $25 - 20 + 16 - \ldots$ Here $a = 25$, $r = -4/5$, and $S = \frac{a}{1-r} = \frac{25}{1-(-4/5)} = \frac{125}{9}$.

(c) $.6 + .06 + .006 + \ldots$ Here $a = .6$, $r = .1$, and $S = \frac{.6}{1-.1} = \frac{.6}{.9} = \frac{6}{9} = \frac{2}{3}$.

3. An equilateral triangle has a perimeter of 30 in. Another triangle is formed by joining the midpoints of the sides of the given triangle, another is formed by joining the midpoints of the sides of the second triangle, and so on. Find the sum of the perimeters of the triangles thus formed.

Since the side of each new triangle is $\frac{1}{2}$ the side of the triangle from which it is formed, the perimeters of the triangles are $30, 15, 15/2, \ldots$

Then $30 + 15 + 15/2 + \ldots = \frac{30}{1-\frac{1}{2}} = 60$ in.

4. A rubber ball is dropped from a height of 81 ft. Each time it strikes the ground it rebounds two-thirds of the distance through which it last fell. Find the total distance it travels in coming to rest.

For the falls: $a = 81$ and $r = 2/3$; $S = \frac{a}{1-r} = \frac{81}{1-2/3} = 243$ ft.

For the rebounds: $a = 54$ and $r = 2/3$; $S = \frac{54}{1-2/3} = 162$ ft.

Thus, the total distance traveled is $243 + 162 = 405$ ft.

5. For what values of x does $\frac{1}{x+1} + \frac{1}{(x+1)^2} + \frac{1}{(x+1)^3} + \ldots$ have a sum? Find the sum.

There will be a sum provided $|r| = \left|\frac{1}{x+1}\right| < 1$.

When $\left|\frac{1}{x+1}\right| = 1$, $|x+1| = 1$, $(x+1)^2 = 1$, $x^2 + 2x = 0$, and $x = -2, 0$. By examining the intervals $x < -2$, $-2 < x < 0$, and $x > 0$, we find that $\left|\frac{1}{x+1}\right| < 1$ when $x < -2$ and $x > 0$.

Thus, the series has a sum $S = \frac{1/(x+1)}{1 - 1/(x+1)} = \frac{1}{x}$ when $x < -2$ and $x > 0$.

6. Find the limiting value of each of the repeating decimals.

(a) .0123123123...

Since .0123123123... = .0123 + .0000123 + .0000000123 + ... in which $a = .0123$ and $r = .001$,

$$S = \frac{a}{1-r} = \frac{.0123}{1-.001} = \frac{.0123}{.999} = \frac{123}{9990} = \frac{41}{3330}.$$

(b) 2.373737...

The given number may be written as $2 + [.37 + .0037 + .000037 + \ldots]$. For the infinite geometric series in the brackets, $a = .37$ and $r = .01$; hence, $S = \frac{.37}{1-.01} = \frac{.37}{.99} = \frac{37}{99}$.

The limiting value is $2 + 37/99 = 235/99$.

(c) 23.1454545...

$$\text{Write } 23.1454545\ldots = 23.1 + [.045 + .00045 + .0000045 + \ldots]$$

$$= 23.1 + \frac{.045}{1-.01} = 23.1 + \frac{45}{990} = \frac{231}{10} + \frac{1}{22} = \frac{1273}{55}.$$

SUPPLEMENTARY PROBLEMS

7. Sum the following infinite geometric series.

(a) $36 + 12 + 4 + \ldots$

(b) $18 - 12 + 8 - \ldots$

(c) $5 + 3 + 1.8 + \ldots$

(d) $5.6 - 2.24 + 0.896 - \ldots$

(e) $1 + \frac{1}{2}\sqrt{2} + \frac{1}{2} + \ldots$

(f) $3 - \frac{3}{\sqrt{2}-1} + \frac{3}{3-2\sqrt{2}} - \ldots$

Ans. (a) 54 (b) 54/5 (c) $12\frac{1}{2}$ (d) 4 (e) $2+\sqrt{2}$ (f) $\frac{3}{2}(2-\sqrt{2})$

8. A swinging pendulum bob traverses the following distances: 40, 30, $22\frac{1}{2}$, ... inches. Find the distance which it travels before coming to rest. *Ans.* 160 in.

9. An unlimited sequence of squares are inscribed one within another by joining the midpoints of the sides of each preceding square. If the initial square is 8 in. on a side, find the sum of the perimeters of these squares. *Ans.* $32(2+\sqrt{2})$ in.

10. Express each repeating decimal as a rational fraction.

(a) 0.272727... (b) 1.702702... (c) 2.4242... (d) 0.076923076923...

Ans. (a) 3/11 (b) 189/111 (c) 80/33 (d) 1/13

11. Find the values of x for which each of the following geometric series may be *summed*.

(a) $3 + 3x + 3x^2 + \ldots$ Hint: $|r| = |x|$

(b) $1 + (x-1) + (x-1)^2 + \ldots$ Hint: $|r| = |x-1|$

(c) $5 + 5(x-3) + 5(x-3)^2 + \ldots$

Ans. (a) $-1 < x < 1$ (b) $0 < x < 2$ (c) $2 < x < 4$

12. Find the exact error when 1/6 is approximated as 0.1667. *Ans.* 0.000033... = 1/30,000

CHAPTER 13

Mathematical Induction

EVERYONE IS FAMILIAR with the process of reasoning, called *ordinary* or *incomplete induction,* in which a generalization is made on the basis of a number of simple observations.

Example 1. We observe that $1 = 1^2$, $1 + 3 = 4 = 2^2$, $1 + 3 + 5 = 9 = 3^2$, $1 + 3 + 5 + 7 = 16 = 4^2$, and conclude that

$$1 + 3 + 5 + \ldots. + (2n - 1) = n^2$$

or, in words, the sum of the first n odd integers is n^2.

Example 2. We observe that 2 points determine $1 = \frac{1}{2}\cdot 2(2 - 1)$ line; that 3 points, not on a line, determine $3 = \frac{1}{2}\cdot 3(3 - 1)$ lines; that 4 points, no 3 on a line, determine $6 = \frac{1}{2}\cdot 4(4 - 1)$ lines; that 5 points, no 3 on a line, determine $10 = \frac{1}{2}\cdot 5(5 - 1)$ lines; and conclude that n points, no 3 on a line, determine $\frac{1}{2}n(n - 1)$ lines.

Example 3. We observe that for $n = 1, 2, 3, 4, 5$ the values of

$$f(n) = n^4/8 - 17n^3/12 + 47n^2/8 - 103n/12 + 6$$

are 2, 3, 5, 7, 11 respectively and conclude that $f(n)$ is a prime number for every positive integral value of n.

The conclusions in Examples 1 and 2 are valid as we shall prove later. The conclusion in Example 3 is false since $f(6) = 22$ is not a prime number.

MATHEMATICAL INDUCTION or complete induction is a type of reasoning by which such conclusions as were drawn in the above examples may be proved or disproved.

The steps are:

(*1*) The verification of the proposed formula or theorem for some positive integral value of n, usually the smallest. (Of course, we would not attempt to prove an unknown theorem by mathematical induction without first verifying it for several values of n.)

(*2*) The proof that if the proposed formula or theorem is true for $n = k$, some positive integer, it is true also for $n = k + 1$.

(*3*) The conclusion that the proposed formula or theorem is true for all values of n greater than the one for which verification was made in Step (*1*).

Example 4. Prove: $1 + 3 + 5 + \ldots. + (2n - 1) = n^2$.

(*1*) The formula is true for $n = 1$ since $1 = 1^2$.

(*2*) Let us assume the formula true for $n = k$, any positive integer; that is, let us assume that

(A) $1 + 3 + 5 + ... + (2k-1) = k^2$.

We wish to show that, when (A) is true, the proposed formula is then true for $n = k+1$; that is, that

(B) $1 + 3 + 5 + ... + (2k-1) + (2k+1) = (k+1)^2$.

(Note. Statements (A) and (B) are obtained by replacing n in the proposed formula by k and $k+1$ respectively. Now it is clear that the left member of (B) can be obtained from the left member of (A) by adding $(2k+1)$. At this point the proposed formula is true or false according as we do or do not obtain the right member of (B) when $(2k+1)$ is added to the right member of (A).)

Adding $(2k+1)$ to both members of (A), we have

(C) $1+3+5+.... + (2k-1) + (2k+1) = k^2 + (2k+1) = (k+1)^2$.

Now (C) is identical with (B); thus, if the proposed formula is true for any positive integer $n = k$ it is true for the next positive integer $n = k+1$.

(3) Since the formula is true for $n = k = 1$ (Step 1), it is true for $n = k+1 = 2$; being true for $n = k = 2$, it is true for $n = k+1 = 3$; and so on. Hence, the formula is true for all positive integral values of n.

SOLVED PROBLEMS

Prove by mathematical induction.

1. $1 + 7 + 13 + ... + (6n-5) = n(3n-2)$.

(1) The proposed formula is true for $n = 1$, since $1 = 1(3-2)$.

(2) Assume the formula to be true for $n = k$, a positive integer; that is, assume

(A) $1 + 7 + 13 + ... + (6k-5) = k(3k-2)$.

Under this assumption we wish to show that

(B) $1 + 7 + 13 + ... + (6k-5) + (6k+1) = (k+1)(3k+1)$.

When $(6k+1)$ is added to both members of (A), we have on the right

$$k(3k-2) + (6k+1) = 3k^2 + 4k + 1 = (k+1)(3k+1);$$

hence, if the formula is true for $n = k$ it is true for $n = k+1$.

(3) Since the formula is true for $n = k = 1$ (Step 1), it is true for $n = k+1 = 2$; being true for $n = k = 2$ it is true for $n = k+1 = 3$; and so on, for every positive integral value of n.

2. $1 + 5 + 5^2 + ... + 5^{n-1} = \frac{1}{4}(5^n - 1)$.

(1) The proposed formula is true for $n = 1$, since $1 = \frac{1}{4}(5-1)$.

(2) Assume the formula to be true for $n = k$, a positive integer; that is, assume

(A) $1 + 5 + 5^2 + ... + 5^{k-1} = \frac{1}{4}(5^k - 1)$.

Under this assumption we wish to show that

(B) $1 + 5 + 5^2 + ... + 5^{k-1} + 5^k = \frac{1}{4}(5^{k+1} - 1)$.

When 5^k is added to both members of (A), we have on the right

$$\frac{1}{4}(5^k-1)+5^k = \frac{5}{4}(5^k)-\frac{1}{4} = \frac{1}{4}(5\cdot 5^k-1) = \frac{1}{4}(5^{k+1}-1);$$

hence, if the formula is true for $n=k$ it is true for $n=k+1$.

(3) Since the formula is true for $n=k=1$ (Step 1), it is true for $n=k+1=2$; being true for $n=k=2$ it is true for $n=k+1=3$; and so on, for every positive integral value of n.

3. $\dfrac{5}{1\cdot2\cdot3}+\dfrac{6}{2\cdot3\cdot4}+\dfrac{7}{3\cdot4\cdot5}+\ldots+\dfrac{n+4}{n(n+1)(n+2)} = \dfrac{n(3n+7)}{2(n+1)(n+2)}$.

(1) The formula is true for $n=1$, since $\dfrac{5}{1\cdot2\cdot3} = \dfrac{1(3+7)}{2\cdot2\cdot3} = \dfrac{5}{6}$.

(2) Assume the formula to be true for $n=k$, a positive integer; that is, assume

$$(A)\qquad \frac{5}{1\cdot2\cdot3}+\frac{6}{2\cdot3\cdot4}+\ldots+\frac{k+4}{k(k+1)(k+2)} = \frac{k(3k+7)}{2(k+1)(k+2)}.$$

Under this assumption we wish to show that

$$(B)\qquad \frac{5}{1\cdot2\cdot3}+\frac{6}{2\cdot3\cdot4}+\ldots+\frac{k+4}{k(k+1)(k+2)}+\frac{k+5}{(k+1)(k+2)(k+3)} = \frac{(k+1)(3k+10)}{2(k+2)(k+3)}.$$

When $\dfrac{k+5}{(k+1)(k+2)(k+3)}$ is added to both members of (A), we have on the right

$$\frac{k(3k+7)}{2(k+1)(k+2)}+\frac{k+5}{(k+1)(k+2)(k+3)} = \frac{1}{(k+1)(k+2)}\left[\frac{k(3k+7)}{2}+\frac{k+5}{k+3}\right]$$

$$= \frac{1}{(k+1)(k+2)}\,\frac{k(3k+7)(k+3)+2(k+5)}{2(k+3)} = \frac{1}{(k+1)(k+2)}\,\frac{3k^3+16k^2+23k+10}{2(k+3)}$$

$$= \frac{1}{(k+1)(k+2)}\,\frac{(k+1)^2(3k+10)}{2(k+3)} = \frac{(k+1)(3k+10)}{2(k+2)(k+3)};$$

hence, if the formula is true for $n=k$ it is true for $n=k+1$.

(3) Since the formula is true for $n=k=1$ (Step 1), it is true for $n=k+1=2$; being true for $n=k=2$, it is true for $n=k+1=3$; and so on, for all positive integral values of n.

4. $x^{2n}-y^{2n}$ is divisible by $x+y$.

(1) The theorem is true for $n=1$, since $x^2-y^2=(x-y)(x+y)$ is divisible by $x+y$.

(2) Let us assume the theorem true for $n=k$, a positive integer; that is, let us assume

$$(A)\qquad x^{2k}-y^{2k} \text{ is divisible by } x+y.$$

We wish to show that, when (A) is true,

$$(B)\qquad x^{2k+2}-y^{2k+2} \text{ is divisible by } x+y.$$

Now $x^{2k+2}-y^{2k+2} = (x^{2k+2}-x^2y^{2k})+(x^2y^{2k}-y^{2k+2}) = x^2(x^{2k}-y^{2k})+y^{2k}(x^2-y^2)$. In the first term $(x^{2k}-y^{2k})$ is divisible by $(x+y)$ by assumption, and in the second term (x^2-y^2) is divisible by $(x+y)$ by Step (1); hence, if the theorem is true for $n=k$, a positive integer, it is true for the next one $n=k+1$.

(3) Since the theorem is true for $n=k=1$, it is true for $n=k+1=2$; being true for $n=k=2$, it is true for $n=k+1=3$; and so on, for every positive integral value of n.

5. The number of straight lines determined by $n>1$ points, no 3 on the same straight line, is $\frac{1}{2}n(n-1)$.

(*1*) The theorem is true when $n = 2$, since $\frac{1}{2}\cdot 2(2-1) = 1$ and two points determine one line.

(*2*) Let us assume that k points, no 3 on the same straight line, determine $\frac{1}{2}k(k-1)$ lines.

When an additional point is added (not on any of the lines already determined) and is joined to each of the original k points, k new lines are determined. Thus, altogether we have $\frac{1}{2}k(k-1) + k = \frac{1}{2}k(k-1+2) = \frac{1}{2}k(k+1)$ lines and this agrees with the theorem when $n = k+1$.

Hence, if the theorem is true for $n = k$, a positive integer greater than 1, it is true for the next one $n = k+1$.

(*3*) Since the theorem is true for $n = k = 2$ (Step (*1*)), it is true for $n = k+1 = 3$; being true for $n = k = 3$, it is true for $n = k+1 = 4$; and so on, for every possible integral value >1 of n.

SUPPLEMENTARY PROBLEMS

Prove by mathematical induction, n being a positive integer.

6. $1 + 2 + 3 + \ldots\ldots + n \;=\; \frac{1}{2}n(n+1)$

7. $1 + 4 + 7 + \ldots\ldots + (3n-2) \;=\; \frac{1}{2}n(3n-1)$

8. $1^2 + 3^2 + 5^2 + \ldots\ldots + (2n-1)^2 \;=\; \frac{1}{3}n(4n^2-1)$

9. $1^2 + 2^2 + 3^2 + \ldots\ldots + n^2 \;=\; \frac{1}{6}n(n+1)(2n+1)$

10. $1^3 + 2^3 + 3^3 + \ldots\ldots + n^3 \;=\; \frac{1}{4}n^2(n+1)^2$

11. $1^4 + 2^4 + 3^4 + \ldots\ldots + n^4 \;=\; \frac{1}{30}n(n+1)(2n+1)(3n^2+3n-1)$

12. $1\cdot 2 + 2\cdot 3 + 3\cdot 4 + \ldots\ldots + n(n+1) \;=\; \frac{1}{3}n(n+1)(n+2)$

13. $\frac{1}{1\cdot 3} + \frac{1}{3\cdot 5} + \frac{1}{5\cdot 7} + \ldots\ldots + \frac{1}{(2n-1)(2n+1)} \;=\; \frac{n}{2n+1}$

14. $1\cdot 3 + 2\cdot 3^2 + 3\cdot 3^3 + \ldots\ldots + n\cdot 3^n \;=\; \frac{3}{4}[(2n-1)3^n + 1]$

15. $\frac{3}{1\cdot 2\cdot 2} + \frac{4}{2\cdot 3\cdot 2^2} + \frac{5}{3\cdot 4\cdot 2^3} + \ldots\ldots + \frac{n+2}{n(n+1)2^n} \;=\; 1 - \frac{1}{(n+1)2^n}$

16. A convex polygon of n sides has $\frac{1}{2}n(n-3)$ diagonals.

17. The sum of the interior angles of a regular polygon of n sides is $(n-2)180°$.

18. $f(n) \;=\; \frac{1}{24}(3n^4 - 34n^3 + 141n^2 - 206n)$ is an integer for every integral value of n.

Hint: Prove that $f(n)$ and $f(-n)$ are integers for every positive integral value of n. Also, $f(0) = 0$.

CHAPTER 14

The Binomial Theorem

BY ACTUAL MULTIPLICATION

$(a+b)^1 = a+b,\quad (a+b)^2 = a^2+2ab+b^2,\quad (a+b)^3 = a^3+3a^2b+3ab^2+b^3,$

$(a+b)^4 = a^4+4a^3b+6a^2b^2+4ab^3+b^4,$

$(a+b)^5 = a^5+5a^4b+10a^3b^2+10a^2b^3+5ab^4+b^5,$ etc.

From these cases we conclude that, when n is a positive integer,

$$(a+b)^n = a^n + na^{n-1}b + \frac{n(n-1)}{1\cdot 2}a^{n-2}b^2 + \frac{n(n-1)(n-2)}{1\cdot 2\cdot 3}a^{n-3}b^3 + \ldots + nab^{n-1} + b^n$$

and note the following properties:

(*1*) The number of terms in the expansion is $(n+1)$.

(*2*) The first term a of the binomial enters the first term of the expansion with exponent n, the second term with exponent $(n-1)$, the third term with exponent $(n-2)$, and so on.

(*3*) The second term b of the binomial enters the second term of the expansion with exponent 1, the third term with exponent 2, the fourth term with exponent 3, and so on.

(*4*) The sum of the exponents of a and b in any term is n.

(*5*) The coefficient of the first term in the expansion is 1, of the second term is $n/1$, of the third term is $\frac{n(n-1)}{1\cdot 2}$, of the fourth term is $\frac{n(n-1)(n-2)}{1\cdot 2\cdot 3}$, etc.

(*6*) The coefficients of terms equidistant from the ends of the expansion are the same. Note that the number of factors in the numerator and denominator of any coefficient except the first and last is then either the exponent of a or of b, whichever is the smaller.

The above theorem may be proved by mathematical induction.

Example 1. Expand $(3x+2y^2)^5$ and simplify term by term.

We put in first the several powers of $(3x)$, then the powers of $(2y^2)$, and finally the coefficients, recalling Property (*6*).

$$\begin{aligned}(3x+2y^2)^5 &= (3x)^5 + \frac{5}{1}(3x)^4(2y^2) + \frac{5\cdot 4}{1\cdot 2}(3x)^3(2y^2)^2 + \frac{5\cdot 4}{1\cdot 2}(3x)^2(2y^2)^3 \\ &\quad + \frac{5}{1}(3x)(2y^2)^4 + (2y^2)^5 \\ &= 3^5x^5 + 5\cdot 3^4x^4\cdot 2y^2 + 10\cdot 3^3x^3\cdot 2^2y^4 + 10\cdot 3^2x^2\cdot 2^3y^6 \\ &\quad + 5\cdot 3x\cdot 2^4y^8 + 2^5y^{10}\end{aligned}$$

$$= 243x^5 + 810x^4y^2 + 1080x^3y^4 + 720x^2y^6 + 240xy^8 + 32y^{10}.$$

See Problems 1-2.

THE rth **TERM** ($r \le n+1$) in the expansion of $(a+b)^n$ is

$$\frac{n(n-1)(n-2)\dots\dots(n-r+2)}{1\cdot 2\cdot 3\dots\dots(r-1)}\, a^{n-r+1}b^{r-1}.$$

See Problem 3.

WHEN THE LAWS ABOVE are used to expand $(a+b)^n$, where n is real but not a positive integer, an endless succession of terms is obtained. Such expansions are valid (see Problem 7(a) for a verification) when $|b| < |a|$.

Example 2. Write the first five terms in the expansion of $(a+b)^{-3}$, $|b| < |a|$.

$$(a+b)^{-3} = a^{-3} + (-3)a^{-4}b + \frac{(-3)(-4)}{1\cdot 2}a^{-5}b^2 + \frac{(-3)(-4)(-5)}{1\cdot 2\cdot 3}a^{-6}b^3$$

$$+ \frac{(-3)(-4)(-5)(-6)}{1\cdot 2\cdot 3\cdot 4}a^{-7}b^4 + \dots\dots$$

$$= \frac{1}{a^3} - \frac{3b}{a^4} + \frac{6b^2}{a^5} - \frac{10b^3}{a^6} + \frac{15b^4}{a^7} - \dots\dots$$

See Problems 5-8.

SOLVED PROBLEMS

1. Expand and simplify term by term.

(a) $$(x^2+\tfrac{1}{2}y)^6 = (x^2)^6 + \frac{6}{1}(x^2)^5(\tfrac{1}{2}y) + \frac{6\cdot 5}{1\cdot 2}(x^2)^4(\tfrac{1}{2}y)^2 + \frac{6\cdot 5\cdot 4}{1\cdot 2\cdot 3}(x^2)^3(\tfrac{1}{2}y)^3$$

$$+ \frac{6\cdot 5}{1\cdot 2}(x^2)^2(\tfrac{1}{2}y)^4 + \frac{6}{1}(x^2)(\tfrac{1}{2}y)^5 + (\tfrac{1}{2}y)^6$$

$$= x^{12} + 6(x^{10})\frac{1}{2}y + 15(x^8)\frac{1}{4}y^2 + 20(x^6)\frac{1}{8}y^3 + 15(x^4)\frac{1}{16}y^4 + 6(x^2)\frac{1}{32}y^5 + \frac{1}{64}y^6$$

$$= x^{12} + 3x^{10}y + \frac{15}{4}x^8y^2 + \frac{5}{2}x^6y^3 + \frac{15}{16}x^4y^4 + \frac{3}{16}x^2y^5 + \frac{1}{64}y^6.$$

(b) $$(x^{1/2}+2y^{1/3})^4 = (x^{1/2})^4 + \frac{4}{1}(x^{1/2})^3(2y^{1/3}) + \frac{4\cdot 3}{1\cdot 2}(x^{1/2})^2(2y^{1/3})^2$$

$$+ \frac{4}{1}(x^{1/2})(2y^{1/3})^3 + (2y^{1/3})^4$$

$$= x^2 + 4(x^{3/2})2y^{1/3} + 6(x)4y^{2/3} + 4(x^{1/2})8y + 16y^{4/3}$$

$$= x^2 + 8x^{3/2}y^{1/3} + 24xy^{2/3} + 32x^{1/2}y + 16y^{4/3}.$$

(c) $$(\tfrac{2}{3}x^{1/2} - \tfrac{1}{2x})^6 = (\tfrac{2}{3}x^{1/2})^6 + \frac{6}{1}(\tfrac{2}{3}x^{1/2})^5(-\tfrac{1}{2x}) + \frac{6\cdot 5}{1\cdot 2}(\tfrac{2}{3}x^{1/2})^4(-\tfrac{1}{2x})^2 + \frac{6\cdot 5\cdot 4}{1\cdot 2\cdot 3}(\tfrac{2}{3}x^{1/2})^3(-\tfrac{1}{2x})^3$$

$$+ \frac{6\cdot 5}{1\cdot 2}(\frac{2}{3}x^{1/2})^2(-\frac{1}{2x})^4 + \frac{6}{1}(\frac{2}{3}x^{1/2})(-\frac{1}{2x})^5 + (-\frac{1}{2x})^6$$

$$= \frac{64}{729}x^3 - \frac{32}{81}x^{3/2} + \frac{20}{27} - \frac{20x^{1/2}}{27x^2} + \frac{5}{12x^3} - \frac{x^{1/2}}{8x^5} + \frac{1}{64x^6}.$$

2. Write the first five terms in each expansion and simplify term by term.

(*a*) $$(\frac{2}{3}m^{1/2} + \frac{3}{2m^{3/2}})^{12} = (\frac{2}{3}m^{1/2})^{12} + \frac{12}{1}(\frac{2}{3}m^{1/2})^{11}(\frac{3}{2m^{3/2}}) + \frac{12\cdot 11}{1\cdot 2}(\frac{2}{3}m^{1/2})^{10}(\frac{3}{2m^{3/2}})^2$$

$$+ \frac{12\cdot 11\cdot 10}{1\cdot 2\cdot 3}(\frac{2}{3}m^{1/2})^9(\frac{3}{2m^{3/2}})^3 + \frac{12\cdot 11\cdot 10\cdot 9}{1\cdot 2\cdot 3\cdot 4}(\frac{2}{3}m^{1/2})^8(\frac{3}{2m^{3/2}})^4 + \dots$$

$$= \frac{2^{12}}{3^{12}}m^6 + 12(\frac{2^{11}}{3^{11}}m^{11/2})\frac{3}{2m^{3/2}} + 66(\frac{2^{10}}{3^{10}}m^5)\frac{3^2}{2^2m^3}$$

$$+ 220(\frac{2^9}{3^9}m^{9/2})\frac{3^3}{2^3m^{9/2}} + 495(\frac{2^8}{3^8}m^4)\frac{3^4}{2^4m^6} + \dots$$

$$= \frac{2^{12}}{3^{12}}m^6 + \frac{2^{12}}{3^9}m^4 + 11\frac{2^9}{3^7}m^2 + 55\frac{2^8}{3^6} + 55\frac{2^4}{3^2m^2} + \dots$$

(*b*) $$(\frac{x^{1/2}}{y^{2/3}z} - \frac{yz^2}{2x})^{11} = (\frac{x^{1/2}}{y^{2/3}z})^{11} + \frac{11}{1}(\frac{x^{1/2}}{y^{2/3}z})^{10}(-\frac{yz^2}{2x}) + \frac{11\cdot 10}{1\cdot 2}(\frac{x^{1/2}}{y^{2/3}z})^9(-\frac{yz^2}{2x})^2$$

$$+ \frac{11\cdot 10\cdot 9}{1\cdot 2\cdot 3}(\frac{x^{1/2}}{y^{2/3}z})^8(-\frac{yz^2}{2x})^3 + \frac{11\cdot 10\cdot 9\cdot 8}{1\cdot 2\cdot 3\cdot 4}(\frac{x^{1/2}}{y^{2/3}z})^7(-\frac{yz^2}{2x})^4 + \dots$$

$$= \frac{x^{11/2}}{y^{22/3}z^{11}} - 11(\frac{x^5}{y^{20/3}z^{10}})\frac{yz^2}{2x} + 55(\frac{x^{9/2}}{y^6z^9})\frac{y^2z^4}{2^2x^2}$$

$$- 165(\frac{x^4}{y^{16/3}z^8})\frac{y^3z^6}{2^3x^3} + 330(\frac{x^{7/2}}{y^{14/3}z^7})\frac{y^4z^8}{2^4x^4} - \dots$$

$$= \frac{x^{11/2}y^{2/3}}{y^8z^{11}} - \frac{11x^4y^{1/3}}{2y^6z^8} + \frac{55x^{5/2}}{4y^4z^5} - \frac{165xy^{2/3}}{8y^3z^2} + \frac{165x^{1/2}y^{1/3}z}{8xy} - \dots$$

3. Find the indicated term and simplify.

(*a*) The 7th term of $(a+b)^{15}$.

In the 7th term the exponent of b is $7-1=6$, the exponent of a is $15-6=9$, and the coefficient has 6 factors in the numerator and denominator. Hence, the term is

$$\frac{15\cdot 14\cdot 13\cdot 12\cdot 11\cdot 10}{1\cdot 2\cdot 3\cdot 4\cdot 5\cdot 6}a^9b^6 = 5005\,a^9b^6$$

(*b*) The 9th term of $(x - \frac{1}{x^{1/2}})^{12}$.

In the 9th term the exponent of $b = -\frac{1}{x^{1/2}}$ is $9-1=8$, the exponent of $a = x$ is $12-8=4$, and the coefficient has 4 factors in numerator and denominator. Hence the required term is

$$\frac{12\cdot 11\cdot 10\cdot 9}{1\cdot 2\cdot 3\cdot 4}(x)^4(-\frac{1}{x^{1/2}})^8 = 495(x^4)\frac{1}{x^4} = 495$$

(*c*) The 12th term of $(\frac{x^{1/2}}{4} - \frac{2y}{x^{3/2}})^{18}$. The required term is

$$\frac{18\cdot17\cdot16\cdot15\cdot14\cdot13\cdot12}{1\cdot2\cdot3\cdot4\cdot5\cdot6\cdot7}(\frac{x^{1/2}}{4})^7(-\frac{2y}{x^{3/2}})^{11} = -9\cdot17\cdot16\cdot13(\frac{x^{7/2}}{2^{14}})\frac{2^{11}y^{11}}{x^{33/2}} = -3978\frac{y^{11}}{x^{13}}$$

(*d*) The middle term in the expansion of $(x^{2/3} + \frac{1}{x^{1/2}})^{10}$.

Since there are 11 terms in all, the middle term is the 6th. This term is

$$\frac{10\cdot9\cdot8\cdot7\cdot6}{1\cdot2\cdot3\cdot4\cdot5}(x^{2/3})^5(\frac{1}{x^{1/2}})^5 = 252\frac{x^{10/3}}{x^{5/2}} = 252x^{5/6}$$

(*e*) The term involving y^{12} in the expansion of $(y^3 - \frac{x}{3})^9$.

The first term of the binomial must be raised to the 4th power to produce y^{12}; hence, the second term must be raised to the 5th power and we are to write the 6th term. This term is

$$\frac{9\cdot8\cdot7\cdot6}{1\cdot2\cdot3\cdot4}(y^3)^4(-\frac{x}{3})^5 = -3^2\cdot14y^{12}\frac{x^5}{3^5} = -\frac{14}{27}x^5y^{12}$$

(*f*) The term involving x^4 in the expansion of $(\frac{2}{x} + \frac{x^2}{4})^{14}$.

Let p and q be positive integers so that $p+q = 14$. We are required to determine p and q so that $(\frac{2}{x})^p(\frac{x^2}{4})^q$ yields a term in x^4. Then $2q-p=4$ or $2q-(14-q)=3q-14=4$ and $q=6$. The required term, the 7th in the expansion, is

$$\frac{14\cdot13\cdot12\cdot11\cdot10\cdot9}{1\cdot2\cdot3\cdot4\cdot5\cdot6}(\frac{2}{x})^8(\frac{x^2}{4})^6 = 3003\frac{2^8}{x^8}\frac{x^{12}}{2^{12}} = \frac{3003}{16}x^4$$

4. Evaluate $(1.02)^{12}$ correct to four decimal places.

$$(1.02)^{12} = (1+.02)^{12} = 1 + 12(.02) + \frac{12\cdot11}{1\cdot2}(.02)^2 + \frac{12\cdot11\cdot10}{1\cdot2\cdot3}(.02)^3 + \frac{12\cdot11\cdot10\cdot9}{1\cdot2\cdot3\cdot4}(.02)^4 + \ldots.$$

$$= 1 + .24 + .0264 + .00176 + .00008 + \ldots. = 1.26824 \text{ (approx.)}$$

Thus, $(1.02)^{12} = 1.2682$ correct to four decimal places.

5. Write the first five terms and simplify term by term.

(*a*)
$$(x^2 - \frac{2}{x^4})^{1/2} = (x^2)^{1/2} + \frac{1}{2}(x^2)^{-1/2}(-\frac{2}{x^4}) + \frac{(1/2)(-1/2)}{1\cdot2}(x^2)^{-3/2}(-\frac{2}{x^4})^2$$

$$+ \frac{(1/2)(-1/2)(-3/2)}{1\cdot2\cdot3}(x^2)^{-5/2}(-\frac{2}{x^4})^3 + \frac{(1/2)(-1/2)(-3/2)(-5/2)}{1\cdot2\cdot3\cdot4}(x^2)^{-7/2}(-\frac{2}{x^4})^4 + \ldots.$$

$$= x - \frac{1}{2}\cdot\frac{1}{x}\cdot\frac{2}{x^4} - \frac{1}{2^3}\cdot\frac{1}{x^3}\cdot\frac{2^2}{x^8} - \frac{1}{2^4}\cdot\frac{1}{x^5}\cdot\frac{2^3}{x^{12}} - \frac{5}{2^7}\cdot\frac{1}{x^7}\cdot\frac{2^4}{x^{16}} - \ldots.$$

$$= x - \frac{1}{x^5} - \frac{1}{2x^{11}} - \frac{1}{2x^{17}} - \frac{5}{8x^{23}} - \ldots., \quad (|x| > \sqrt[6]{2})$$

(b) $$(1-x^3)^{-2/3} = 1^{-2/3} + (-\frac{2}{3})1^{-5/3}(-x^3) + \frac{(-2/3)(-5/3)}{1\cdot 2}1^{-8/3}(-x^3)^2$$

$$+ \frac{(-2/3)(-5/3)(-8/3)}{1\cdot 2\cdot 3}1^{-11/3}(-x^3)^3 + \frac{(-2/3)(-5/3)(-8/3)(-11/3)}{1\cdot 2\cdot 3\cdot 4}1^{-14/3}(-x^3)^4 + \dots$$

$$= 1 + \frac{2}{3}x^3 + \frac{5}{9}x^6 + \frac{40}{81}x^9 + \frac{110}{243}x^{12} + \dots, \quad (|x| < 1)$$

6. Find and simplify the 6th term in the expansion of $(x^2 - \frac{3}{2x})^{5/3}$, $|x| > \sqrt[3]{3/2}$.

The required term is

$$\frac{(5/3)(2/3)(-1/3)(-4/3)(-7/3)}{1\cdot 2\cdot 3\cdot 4\cdot 5}(x^2)^{-10/3}(-\frac{3}{2x})^5 = \frac{7}{3^6}x^{-20/3}\frac{3^5}{2^5x^5} = \frac{7x^{1/3}}{96x^{12}}$$

7. (a) Evaluate $\sqrt{26}$ correct to four decimal places.

$$\sqrt{26} = (5^2+1)^{1/2} = (5^2)^{1/2} + \frac{1}{2}(5^2)^{-1/2}(1) + \frac{(1/2)(-1/2)}{1\cdot 2}(5^2)^{-3/2}(1)^2$$

$$+ \frac{(1/2)(-1/2)(-3/2)}{1\cdot 2\cdot 3}(5^2)^{-5/2}(1)^3 + \frac{(1/2)(-1/2)(-3/2)(-5/2)}{1\cdot 2\cdot 3\cdot 4}(5^2)^{-7/2}(1)^4 + \dots$$

$$= 5 + \frac{1}{2}\cdot\frac{1}{5} - \frac{1}{2^3}\cdot\frac{1}{5^3} + \frac{1}{2^4}\cdot\frac{1}{5^5} - \frac{1}{2^7}\cdot\frac{1}{5^6} + \dots$$

$$= 5.00000 + .10000 - .00100 + .00002 - \dots$$

$= 5.09902$ (approximately). Thus $\sqrt{26} = 5.0990$, correct to four decimal places.

Note. If we write $\sqrt{26} = (1+5^2)^{1/2} = 1 + \frac{1}{2}(5^2) - \frac{1}{8}(5^2)^2 + \frac{1}{16}(5^2)^3 - \dots$

$$= 1 + 12.5 - 78.125 + 976.5625 - \dots$$

it is clear that the expansion is not valid.

(b) Evaluate $\sqrt{23}$ correct to four decimal places.

$$\sqrt{23} = (5^2-2)^{1/2} = (5^2)^{1/2} + \frac{1}{2}(5^2)^{-1/2}(-2) + \frac{(1/2)(-1/2)}{1\cdot 2}(5^2)^{-3/2}(-2)^2$$

$$+ \frac{(1/2)(-1/2)(-3/2)}{1\cdot 2\cdot 3}(5^2)^{-5/2}(-2)^3 + \frac{(1/2)(-1/2)(-3/2)(-5/2)}{1\cdot 2\cdot 3\cdot 4}(5^2)^{-7/2}(-2)^4 + \dots$$

$$= 5 - \frac{1}{5} - \frac{1}{2}\cdot\frac{1}{5^3} - \frac{1}{2}\cdot\frac{1}{5^5} - \frac{1}{8}\cdot\frac{1}{5^6} - \dots$$

$= 4.79583$ (approximately). Thus $\sqrt{23} = 4.7958$, correct to four decimal places.

8. (a) Show that the sum of the coefficients in the expansion of $(a+b)^n$, n a positive integer, is 2^n.
(b) Show that the sum of the coefficients in the expansion of $(a-b)^n$, n a positive integer, is 0.

(a) In $(a+b)^n = a^n + na^{n-1}b + \frac{n(n-1)}{1\cdot 2}a^{n-2}b^2 + \dots + \frac{n(n-1)}{1\cdot 2}a^2b^{n-2} + nab^{n-1} + b^n$,

let $a = b = 1$; then $(1+1)^n = 2^n = 1 + n + \frac{1}{2}n(n-1) + \dots + \frac{1}{2}n(n-1) + n + 1$, as was to be proved.

(b) Similarly, let $a = b = 1$ in the expansion of $(a-b)^n$ and obtain

$$1 - n + \tfrac{1}{2}n(n-1) - \dots + (-1)^{n-2}[\tfrac{1}{2}n(n-1)] + (-1)^{n-1}n + (-1)^n = (1-1)^n = 0$$

SUPPLEMENTARY PROBLEMS

9. Expand by the binomial theorem and simplify term by term.

(a) $(a + \frac{1}{2}b)^6 = a^6 + 3a^5b + \frac{15}{4}a^4b^2 + \frac{5}{2}a^3b^3 + \frac{15}{16}a^2b^4 + \frac{3}{16}ab^5 + \frac{1}{64}b^6$

(b) $(4x + \frac{1}{4}y)^5 = 1024x^5 + 320x^4y + 40x^3y^2 + \frac{5}{2}x^2y^3 + \frac{5}{64}xy^4 + \frac{1}{1024}y^5$

(c) $(\frac{x}{4y^3} - \frac{2y}{x^2})^5 = \frac{x^5}{1024y^{15}} - \frac{5x^2}{128y^{11}} + \frac{5}{8xy^7} - \frac{5}{x^4y^3} + \frac{20y}{x^7} - \frac{32y^5}{x^{10}}$

10. Find the indicated term and simplify.

(a) 5th term of $(\frac{1}{2} + x)^{10}$

(b) 6th term of $(\frac{2}{x^{1/2}} - \frac{x^{1/4}}{4})^9$

(c) 10th term of $(\frac{27a^2}{b^3} + \frac{b^2}{6a^4})^{12}$

(d) 7th term of $(x^{1/3} - \frac{1}{2x^{2/3}})^{10}$

(e) middle term of $(\frac{1}{x} - x^2)^{12}$

(f) middle term of $(a^{1/2}b^{1/2} - \frac{a}{2b^{3/2}})^9$

(g) the term involving x^{14} in the expansion of $(2/x - x^2)^{10}$

(h) the term free of y in the expansion of $(xy^{1/6} - y^{-2/3})^{15}$

Ans. (a) $\frac{105}{32}x^4$ (b) $-\frac{63x^{1/4}}{32x}$ (c) $\frac{55b^9}{128a^{30}}$ (d) $\frac{105x^{1/3}}{32x^3}$ (e) $924x^6$ (f) $\frac{63a^{13/2}b^{1/2}}{8b^4}$, $-\frac{63a^7b^{1/2}}{16b^6}$

(g) $180x^{14}$ (h) $-455x^{12}$

11. Expand $(a+b-c)^3$. Hint: Write $(a+b-c)^3 = [(a+b)-c]^3$.
Ans. $a^3 + b^3 - c^3 + 3a^2b - 3a^2c + 3ab^2 - 3b^2c + 3ac^2 + 3bc^2 - 6abc$

12. Find the value of n if the coefficients of the 6th and 16th terms in the expansion of $(a+b)^n$ are equal. *Ans.* $n = 20$

13. Find the first five terms in the expansion and simplify term by term.

(a) $(\frac{x}{a} + \frac{a^3}{x^2})^{-3} = \frac{a^3}{x^3} - \frac{3a^7}{x^6} + \frac{6a^{11}}{x^9} - \frac{10a^{15}}{x^{12}} + \frac{15a^{19}}{x^{15}} - \dots, \quad (|x| > a^{4/3})$

(b) $(x^2 - \frac{2y^2}{x})^{3/2} = x^3 - 3y^2 + \frac{3y^4}{2x^3} + \frac{y^6}{2x^6} + \frac{3y^8}{8x^9} + \dots, \quad (|x| < \sqrt[3]{2y^2})$

(c) $(1 - \frac{3y^2}{x})^{1/3} = 1 - \frac{y^2}{x} - \frac{y^4}{x^2} - \frac{5y^6}{3x^3} - \frac{10y^8}{3x^4} - \dots, \quad (|x| > 3y^2)$

14. Find and simplify the indicated term:

(a) the 6th term in $(a^2 - 4b^2)^{1/2}$, $(|a| > 2|b|)$. *Ans.* $-28\frac{b^{10}}{a^9}$

(b) the 7th term in $(x^{1/4} - \frac{3}{x^{1/2}})^{-1}$, $(|x| > 3^{4/3})$. *Ans.* $729\frac{x^{1/4}}{x^5}$

(c) the term involving x^{20} in $(\frac{2}{x} + x^2)^{-2}$, $(|x| < \sqrt[3]{2})$. *Ans.* $\frac{7x^{20}}{256}$

CHAPTER 15

Logarithms

THE LOGARITHM OF A POSITIVE NUMBER N to a given base b (written $\log_b N$) is the exponent of the power to which b must be raised to produce N. It will be understood throughout this chapter that b is positive and different from 1.

Example 1. (a) Since $9 = 3^2$, $\log_3 9 = 2$
(b) since $64 = 4^3$, $\log_4 64 = 3$
(c) since $64 = 2^6$, $\log_2 64 = 6$
(d) since $1000 = 10^3$, $\log_{10} 1000 = 3$
(e) since $0.01 = 10^{-2}$, $\log_{10} 0.01 = -2$.

See Problems 1-3.

FUNDAMENTAL LAWS OF LOGARITHMS.

1. The logarithm of the product of two or more positive numbers is equal to the sum of the logarithms of the several numbers. For example,

$$\log_b(P \cdot Q \cdot R) = \log_b P + \log_b Q + \log_b R$$

2. The logarithm of the quotient of two positive numbers is equal to the logarithm of the dividend minus the logarithm of the divisor. For example,

$$\log_b\left(\frac{P}{Q}\right) = \log_b P - \log_b Q$$

3. The logarithm of a power of a positive number is equal to the logarithm of the number, multiplied by the exponent of the power. For example,

$$\log_b(P^n) = n \log_b P$$

4. The logarithm of a root of a positive number is equal to the logarithm of the number, divided by the index of the root. For example,

$$\log_b(\sqrt[n]{P}) = \frac{1}{n} \log_b P$$

See Problems 4-7.

IN NUMERICAL COMPUTATIONS the most useful base for a system of logarithms is 10. Such logarithms are called *common logarithms*.

The common logarithm of a positive number N (hereafter written $\log N$ instead of $\log_{10} N$) consists of two parts: an integer (positive, negative, or zero) called the *characteristic* and a positive decimal fraction called the *mantissa*.

The characteristic of the common logarithm of any number equal to or greater than 1 is one less than the number of digits to the left of the decimal point in the given number. For example, the characteristics of the common logarithms of 132, 54, and 2380.6 are 2, 1, 3 respectively.

The characteristic of the common logarithm of any positive number smaller than 1 is negative. However, because of greater convenience in computing, we shall write $-1 = 9 - 10$, $-2 = 8 - 10$, and so on. This latter form of the characteristic is obtained directly by subtracting the number of zeros immediately following the decimal point from 9 and affixing -10. For example, the characteristics of the common logarithms of 0.000436, 0.5, and 0.07086 are $6 - 10$, $9 - 10$, $8 - 10$ respectively.

See Problem 8.

The mantissa of the common logarithm of a positive number is usually a continuous decimal fraction. The discussion below refers to a four place table of mantissas.

TO FIND THE COMMON LOGARITHM OF A GIVEN POSITIVE NUMBER.

(a) Write down the characteristic.

(b_1) When the given number contains three or fewer significant digits, read the mantissa directly from the table.

Example 2. Find (a) log 2.6, (b) 32.8 .

(a) The characteristic is 0. To find the mantissa, consider the number as 2.60 and locate in the row opposite 26 the entry 4150 in the column headed 0. Thus, log 2.6 = 0.4150.

(b) The characteristic is 1. To find the mantissa, locate in the row opposite 32 the entry 5159 in the column headed 8. Thus, log 32.8 = 1.5159.

(b_2) When the given number contains four digits interpolate, using the method of proportional parts.

Example 3. Find (a) log 5462, (b) log 0.08367.

(a) The characteristic is 3. For the mantissa, we have

mantissa of log 5460 = .7372
mantissa of log 5470 = .7380

tabular difference = .0008

.2(tabular difference) = .00016
mantissa of log 5462 = .7372 + .00016 = .73736
or .7374 to four decimal places.

Hence log 5462 = 3.7374.

The essential calculation here is 7372 + .2(8) = 7373.6 or 7374.

(b) The characteristic is $8 - 10$. For the mantissa, we have

mantissa of log 8360 = .9222
mantissa of log 8370 = .9227

tabular difference = .0005

.7(tabular difference) = .00035
mantissa of log 8367 = .9222 + .00035 = .92255
or .9226 to four decimal places.

Hence log 0.08367 = 8.9226-10.

The essential calculation here is 9222 + .7(5) = 9225.5 or 9226.

The mantissas in Example 3 were found to five digits and rounded off to four. In general, a number N is said to be rounded off to p digits when it is the nearest number to N that can be written with p digits; for example, .52687 is rounded off to .5269, .38233 is rounded off to .3823.

When the rounding off process can lead to two numbers, we shall agree to use that one which ends in an even digit; for example, .52685 is rounded off to .5268, .38235 is rounded off to .3824.

See Problems 9-10.

TO FIND THE NUMBER CORRESPONDING TO A GIVEN COMMON LOGARITHM.

(*a*) When the given mantissa is found in the table, write down the row number followed by the column heading and then point off in accordance with the rule for characteristics. The resulting number is called the *antilogarithm* (antilog) of the given logarithm.

Example 4. Find: (*a*) antilog 1.6551, (*b*) antilog 8.9090-10.

(*a*) The mantissa .6551 is found in the row opposite 45 and in the column headed 2; thus, we have 452. Since the characteristic is 1, there are two digits to the left of the decimal point. Then antilog 1.6551 = 45.20.

(*b*) The mantissa .9090 is found in the row opposite 81 and under the column headed 1; thus, we have 811. Since the characteristic is 8 -10, the number is smaller than 1 with one zero immediately following the decimal point. Hence, antilog 8.9090-10 = 0.0811.

(*b*) When the given mantissa is not found in the table, interpolation must be used.

Example 5. Find: (*a*) antilog 2.7949, (*b*) antilog 9.6271-10

(*a*) Mantissa of log 6230 = .7945 | Given mantissa = .7949
Mantissa of log 6240 = .7952 | Next smaller mantissa = .7945

Tabular difference = .0007 | Difference = .0004

$$\text{Correction} = \frac{.0004}{.0007}(10) = 5.7. \text{ Add correction to smaller sequence:}$$

6230 + 5.7 = 6235.7 or 6236 to four digits.

Then antilog 2.7949 = 623.6.

The essential calculation here is $\frac{4}{7}(10) = 5.7$.

(*b*) Mantissa of log 4230 = .6263 | Given mantissa = .6271
Mantissa of log 4240 = .6274 | Next smaller mantissa = .6263

Tabular difference = .0011 | Difference = .0008

$$\text{Correction} = \frac{.0008}{.0011}(10) = 7.3. \text{ Add correction to smaller sequence:}$$

4230 + 7.3 = 4237.3 or 4237 to four digits.

Then antilog 9.6271-10 = 0.4237.

The essential calculation here is $\frac{8}{11}(10) = 7.3$.

See Problem 11.

THE COLOGARITHM of a positive number N (written, colog N) is the logarithm of its reciprocal $1/N$. Thus, colog $N = \log 1/N = \log 1 - \log N = -\log N$.

Example 6. Find (*a*) colog 38.68, (*b*) colog 0.007475.

(*a*) Since log 38.68 = 1.5875,
colog 38.68 = log 1 − log N = (10.0000−10) − 1.5875 = 8.4125−10.

(*b*) Since log 0.007475 = 7.8736−10,
colog 0.007475 = (10.0000−10) − (7.8736−10) = 2.1264.

See Problems 12-13.

AN EXPONENTIAL EQUATION is an equation involving one or more unknowns in an exponent. For example, $2^x = 7$ and $(1.03)^{-x} = 2.5$ are exponential equations. Such equations are solved by means of logarithms.

Example 7. Solve the exponential equation $2^x = 7$.

Take logarithms of both sides: $x \log 2 = \log 7$

Solve for x: $x = \dfrac{\log 7}{\log 2} = \dfrac{0.8451}{0.3010}$

Evaluate, using logarithms:

$$\begin{aligned} \log 0.8451 &= 9.9270-10 \\ -\log 0.3010 &= 9.4786-10 \\ \hline \log x &= 0.4484 \\ x &= 2.808 \end{aligned}$$

See Problem 16.

IN THE CALCULUS the most useful system of logarithms is the *natural system* in which the base is a certain irrational number $e = 2.71828$ approximately.

The natural logarithm of N, $\ln N$, and the common logarithm of N, $\log N$, are related by the formula

$$\ln N = 2.3026 \log N$$

SOLVED PROBLEMS

1. Change the following from exponential to logarithmic form:

(*a*) $7^2 = 49$, (*b*) $6^{-1} = 1/6$, (*c*) $10^\circ = 1$, (*d*) $4^\circ = 1$, (*e*) $\sqrt[3]{8} = 2$.

Ans. (*a*) $\log_7 49 = 2$, (*b*) $\log_6 (1/6) = -1$, (*c*) $\log_{10} 1 = 0$, (*d*) $\log_4 1 = 0$, (*e*) $\log_8 2 = 1/3$.

2. Change the following from logarithmic to exponential form:

(*a*) $\log_3 81 = 4$, (*b*) $\log_5 (1/625) = -4$, (*c*) $\log_{10} 10 = 1$, (*d*) $\log_9 27 = 3/2$.

Ans. (*a*) $3^4 = 81$, (*b*) $5^{-4} = 1/625$, (*c*) $10^1 = 10$, (*d*) $9^{3/2} = 27$.

3. Evaluate x, given:

(*a*) $x = \log_5 125$ (*d*) $x = \log_2 (1/16)$ (*g*) $\log_x (1/16) = -2$

(*b*) $x = \log_{10} 0.001$ (*e*) $x = \log_{\frac{1}{2}} 32$ (*h*) $\log_6 x = 2$

(*c*) $x = \log_8 2$ (*f*) $\log_x 243 = 5$ (*i*) $\log_a x = 0$.

Ans. (*a*) 3, since $5^3 = 125$ (*d*) −4, since $2^{-4} = 1/16$ (*g*) 4, since $4^{-2} = 1/16$

(*b*) −3, since $10^{-3} = 0.001$ (*e*) −5, since $(\frac{1}{2})^{-5} = 32$ (*h*) 36, since $6^2 = 36$

(*c*) 1/3, since $8^{1/3} = 2$ (*f*) 3, since $3^5 = 243$ (*i*) 1, since $a^\circ = 1$

4. Prove the four laws of logarithms.

Let $P = b^p$ and $Q = b^q$; then $\log_b P = p$ and $\log_b Q = q$.

1. Since $P \cdot Q = b^p \cdot b^q = b^{p+q}$, $\log_b (PQ) = p + q = \log_b P + \log_b Q$;
   that is, the logarithm of the product of two positive numbers is equal to the sum of the logarithms of the numbers.
2. Since $P/Q = b^p/b^q = b^{p-q}$, $\log_b (P/Q) = p - q = \log_b P - \log_b Q$;
   that is, the logarithm of the quotient of two positive numbers is the logarithm of the numerator minus the logarithm of the denominator.
3. Since $P^n = (b^p)^n = b^{np}$, $\log_b P^n = np = n \log_b P$;
   that is, the logarithm of a power of a positive number is equal to the product of the exponent and the logarithm of the number.
4. Since $\sqrt[n]{P} = P^{1/n} = b^{p/n}$, $\log_b \sqrt[n]{P} = p/n = \frac{1}{n} \log_b P$;
   that is, the logarithm of a root of a positive number is equal to the logarithm of the number divided by the index of the root.

5. Express the logarithms of the given expressions in terms of the logarithms of the individual letters involved.

(a) $\log_b \dfrac{P \cdot Q}{R} = \log_b (P \cdot Q) - \log_b R = \log_b P + \log_b Q - \log_b R$

(b) $\log_b \dfrac{P}{Q \cdot R} = \log_b P - \log_b (Q \cdot R) = \log_b P - (\log_b Q + \log_b R) = \log_b P - \log_b Q - \log_b R$

(c) $\log_b P^2 \cdot \sqrt[3]{Q} = \log_b P^2 + \log_b \sqrt[3]{Q} = 2 \log_b P + \frac{1}{3} \log_b Q$

(d) $\log_b \sqrt{\dfrac{P \cdot Q^3}{R^{1/2} \cdot S}} = \frac{1}{2} \log_b \dfrac{P \cdot Q^3}{R^{1/2} \cdot S} = \frac{1}{2}(\log_b P \cdot Q^3 - \log_b R^{1/2} \cdot S)$

$= \frac{1}{2}(\log_b P + 3 \log_b Q - \frac{1}{2} \log_b R - \log_b S)$

6. Express each of the following as a single logarithm.

(a) $\log_b x - 2 \log_b y + \log_b z = (\log_b x + \log_b z) - 2 \log_b y$

$= \log_b xz - \log_b y^2 = \log_b \dfrac{xz}{y^2}$.

(b) $\log_b 2 + \log_b \pi + \frac{1}{2} \log_b l - \frac{1}{2} \log_b g = (\log_b 2 + \log_b \pi) + \frac{1}{2}(\log_b l - \log_b g)$

$= \log_b (2\pi) + \frac{1}{2} \log_b (\frac{l}{g}) = \log_b 2\pi \sqrt{\dfrac{l}{g}}$

7. Show that $b^{3 \log_b x} = x^3$.

Let $3 \log_b x = t$. Then $\log_b x^3 = t$ and $x^3 = b^t = b^{3 \log_b x}$.

8. Determine the characteristic of the common logarithm of:

(a) 3860 (c) 7.84 (e) 5.463 (g) 0.02345 (i) 2.866 (k) 77.62
(b) 52.6 (d) 728000 (f) 0.3748 (h) 0.0001234 (j) 0.00005 (l) 0.002945

Ans. (a) 3 (c) 0 (e) 0 (g) 8 −10 (i) 0 (k) 1
(b) 1 (d) 5 (f) 9 −10 (h) 6 −10 (j) 5 −10 (l) 7 −10

9. Find

(*a*) log 3860 = 3.5866 (*c*) log 7.84 = 0.8943

(*b*) log 52.6 = 1.7210 (*d*) log 728000 = 5.8621

(*e*) log 5.463 = 0.7374 [7372 + .3(8)]

(*f*) log 0.3748 = 9.5738-10 [5729 + .8(11)]

(*g*) log 0.02345 = 8.3702-10 [3692 + .5(19)]

(*h*) log 0.0001234 = 6.0913-10 [0899 + .4(35)]

(*i*) log 2.866 = 0.4573 [4564 + .6(15)]

(*j*) log 0.00005 = 5.6990-10

(*k*) log 77.62 = 1.8900 [8899 + .2(5)]

(*l*) log 0.002945 = 7.4690-10 [4683 + .5(15)].

10. Find (*a*) $\log 2.864^3$, (*b*) $\log \sqrt{2.864}$, (*c*) $\log 0.007463^2$, (*d*) $\log \sqrt[3]{0.007463}$.

Since log 2.864 = 0.4570,

(*a*) $\log 2.864^3 = 3 \log 2.864 = 3(0.4570) = 1.3710$,

(*b*) $\log \sqrt{2.864} = \frac{1}{2} \log 2.864 = \frac{1}{2}(0.4570) = 0.2285$.

Since log 0.007463 = 7.8729-10,

(*c*) $\log 0.007463^2 = 2 \log 0.007463 = 2(7.8729-10) = 15.7458-20 = 5.7458-10$,

(*d*) $\log \sqrt[3]{0.007463} = \frac{1}{3} \log 0.007463 = \frac{1}{3}(7.8729-10) = \frac{1}{3}(27.8729-30) = 9.2910-10$.

11. Find: (*a*) antilog 1.4232 = 26.50

(*b*) antilog 7.9217-10 = 0.008350

(*c*) antilog 2.7514 = 564.1; $5640 + \frac{1}{7}(10)$

(*d*) antilog 8.6362-10 = 0.04327; $4320 + \frac{7}{10}(10)$

(*e*) antilog 0.9702 = 9.338; $9330 + \frac{3}{4}(10)$

(*f*) antilog 9.5884-10 = 0.3876; $3870 + \frac{7}{11}(10)$

(*g*) antilog 4.8353 = 68430; $6840 + \frac{2}{6}(10)$

(*h*) antilog 1.5829 = 38.27; $3820 + \frac{8}{11}(10)$

(*i*) antilog 7.9231-10 = 0.008378; $8370 + \frac{4}{5}(10)$

(*j*) antilog 0.9150 = 8.222; $8220 + \frac{1}{5}(10)$

(*k*) antilog 9.5306-10 = 0.3393; $3390 + \frac{4}{13}(10)$.

12. Find:

(*a*) colog 265 = 7.5768-10, (log 265 = 2.4232)

(*b*) colog 0.00715 = 2.1457, (log 0.00715 = 7.8543-10)

(*c*) colog 2468 = 6.6077-10, (log 2468 = 3.3923)

(*d*) colog 0.07943 = 1.1000, (log 0.07943 = 8.9000-10)

(*e*) colog 35.68 = 8.4476-10, (log 35.68 = 1.5524)

(*f*) colog 0.4466 = 0.3501, (log 0.4466 = 9.6499-10).

13. Find: (*a*) colog $(45.25)^2$, (*b*) colog $\sqrt[3]{45.25}$, (*c*) colog $(0.08742)^3$, (*d*) colog $\sqrt{0.08742}$.

(*a*) $\log (45.25)^2 = 2 \log 45.25 = 2(1.6556) = 3.3112$ and $\text{colog}\ (45.25)^2 = 6.6888-10$

(*b*) $\log \sqrt[3]{45.25} = \frac{1}{3} \log 45.25 = 0.5519$, $\text{colog} \sqrt[3]{45.25} = 9.4481-10$.

(*c*) Since $\log 0.08742 = 8.9416-10$,
$\text{colog}\ 0.08742 = 1.0584$ and $\text{colog}\ (0.08742)^3 = 3(1.0584) = 3.1752$.

(*d*) Since $\text{colog}\ 0.08742 = 1.0584$, $\text{colog} \sqrt{0.08742} = \frac{1}{2}(1.0584) = 0.5292$.

14. Evaluate, using logarithms.

(*a*) $N = 3.268 \times 0.8794$.

$$\begin{aligned} \log 3.268 &= 0.5142 \\ + \log 0.8794 &= 9.9442-10 \\ \hline \log N &= 0.4584 \\ N &= 2.873 \end{aligned}$$

(*b*) $N = \dfrac{0.8183}{0.0544}$

$$\begin{aligned} \log 0.8183 &= 9.9130-10 \\ - \log 0.0544 &= 8.7356-10 \\ \hline \log N &= 1.1774 \\ N &= 15.04 \end{aligned}$$

(*c*) $N = \dfrac{5.378 \times 92.86}{774.1 \times 0.7863}$

$$\begin{aligned} \log\ 5.378 &= 0.7306 \\ + \log\ 92.86 &= 1.9678 \\ + \text{colog}\ 774.1 &= 7.1112-10 \\ + \text{colog}\ 0.7863 &= 0.1044 \\ \hline \log N &= 9.9140-10 \\ N &= 0.8204 \end{aligned}$$

(*d*) $N = \dfrac{(0.2346)^2 \sqrt[3]{772.7}}{(12.45)^3 \sqrt{0.000382}}$

$$\begin{aligned} 2 \log 0.2346 &= 8.7406-10 \\ + \tfrac{1}{3} \log\ 772.7 &= 0.9627 \\ + 3\ \text{colog}\ 12.45 &= 6.7144-10 \\ + \tfrac{1}{2}\ \text{colog}\ 0.000382 &= 1.7090 \\ \hline \log N &= 8.1267-10 \\ N &= 0.01339 \end{aligned}$$

15. Evaluate, using five place tables of logarithms.

(*a*) $N = 36.234 \times 2.6748 \times 0.0071756$

$$\begin{aligned} \log 36.234 &= 1.55912 \\ + \log 2.6748 &= 0.42729 \\ + \log\ 0.0071756 &= 7.85586-10 \\ \hline \log N &= 9.84227-10 \\ N &= 0.69546 \end{aligned}$$

(*b*) $N = \dfrac{1.7834}{0.62315}$

$$\begin{aligned} \log 1.7834 &= 10.25125-10 \\ - \log 0.62315 &= 9.79460 \\ \hline \log N &= 0.45665 \\ N &= 2.8619 \end{aligned}$$

(*c*) $N = \sqrt{6.3794 \times 0.95327}$

$$\begin{aligned} \log 6.3794 &= 0.80478 \\ + \log 0.95327 &= 9.97922-10 \\ \hline 2 \log N &= 0.78400 \\ \log N &= 0.39200 \\ N &= 2.4661 \end{aligned}$$

(*d*) $N = \dfrac{(58.321)^3 \sqrt{0.27846}}{(7.3416)^2 \sqrt[3]{0.08423}}$

$$\begin{aligned} 3 \log\ 58.321 &= 5.29749 \\ + \tfrac{1}{2} \log 0.27846 &= 9.72238-10 \\ + 2\ \text{colog}\ 7.3416 &= 8.26842-10 \\ + \tfrac{1}{3}\ \text{colog}\ 0.08423 &= 0.35818 \\ \hline \log N &= 3.64647 \\ N &= 4430.7 \end{aligned}$$

16. Solve.

(*a*) $(1.06)^x = 3$. Taking logarithms, $x \log 1.06 = \log 3$.

$$x = \frac{\log 3}{\log 1.06} = \frac{0.4771}{0.0253}$$

$$\begin{aligned} \log 0.4771 &= 9.6786-10 \\ -\log 0.0253 &= 8.4031-10 \\ \hline \log x &= 1.2755 \\ x &= 18.86 \end{aligned}$$

(*b*) $12^{2x+5} = 55(7^{3x})$. Taking logarithms, $(2x+5)\log 12 = \log 55 + 3x \log 7$.

$$2x \log 12 - 3x \log 7 = \log 55 - 5 \log 12$$

$$x = \frac{\log 55 - 5 \log 12}{2 \log 12 - 3 \log 7} = \frac{1.7404 - 5(1.0792)}{2(1.0792) - 3(0.8451)} = \frac{3.6556}{0.3769}$$

Using a four place table, we solve $x = \dfrac{3.656}{0.3769}$.

$$\begin{aligned} \log 3.656 &= 0.5630 \\ -\log 0.3769 &= 9.5762-10 \\ \hline \log x &= 0.9868-10 \\ x &= 9.700 \end{aligned}$$

(*c*) $41.2^x = 12.6^{x-1}$. Taking logarithms, $x \log 41.2 = (x-1)\log 12.6$.

$$x \log 41.2 - x \log 12.6 = -\log 12.6 \quad \text{or} \quad x = \frac{-\log 12.6}{\log 41.2 - \log 12.6}$$

$$y = -x = \frac{\log 12.6}{\log 41.2 - \log 12.6} = \frac{1.1004}{0.5145}.$$

$$\begin{aligned} \log 1.100 &= 0.0414 \\ -\log 0.5145 &= 9.7114-10 \\ \hline \log y &= 0.3300 \\ y &= 2.138 \\ x &= -2.138 \end{aligned}$$

(*d*) $(0.8)^{2x-3} = 1.5^x$. Taking logarithms, $(2x-3)\log 0.8 = x \log 1.5$.

$$2x \log 0.8 - x \log 1.5 = 3 \log 0.8$$

$$x = \frac{3 \log 0.8}{2 \log 0.8 - \log 1.5} = \frac{-3 \log 0.8}{\log 1.5 - 2 \log 0.8} = \frac{3 \operatorname{colog} 0.8}{\log 1.5 - 2 \log 0.8} = \frac{0.2907}{0.3699}$$

$$\begin{aligned} \log 0.2907 &= 9.4634-10 \\ -\log 0.3699 &= 9.5681-10 \\ \hline \log x &= 9.8953-10 \qquad x = 0.7858 \end{aligned}$$

SUPPLEMENTARY PROBLEMS

17. Evaluate, using 4-place tables.

(*a*) $3.141 \times 0.9856 \times 58.44$ (*b*) $\dfrac{42.25}{386.2}$ (*c*) $\sqrt{76.94}$ (*d*) $\dfrac{222.6 \times 0.8988}{(5.344)^2}$

Ans. (*a*) 180.9, (*b*) 0.1094, (*c*) 8.770, (*d*) 7.005

18. Evaluate, using 5-place tables.

(*a*) $36.234 \times 2.6748 \times 0.0071756$ (*b*) $\sqrt[3]{0.48476}$ (*c*) $\dfrac{47.75 \times 8.643}{6467}$ (*d*) $\dfrac{(3.2486)^{2/3}}{\sqrt[5]{316.48}}$

Ans. (*a*) 0.69546, (*b*) 0.78554, (*c*) 0.063816, (*d*) 0.69351

19. Solve for x.

(*a*) $3^x = 30$ (*b*) $1.07^x = 3$ (*c*) $5.72^x = 8.469$ (*d*) $38.5^x = 6.5^{x-2}$

Ans. (*a*) 3.096, (*b*) 16.23, (*c*) 1.225, (*d*) −2.104

CHAPTER 16

Power, Exponential, and Logarithmic Curves

POWER FUNCTIONS in x are of the form x^n. If $n > 0$, the graph of $y = x^n$ is said to be of the *parabolic* type (the curve is a parabola for $n = 2$). If $n < 0$, the graph of $y = x^n$ is said to be of the *hyperbolic* type (the curve is a hyperbola for $n = -1$).

Example 1. Sketch the graphs of (a) $y = x^{3/2}$, (b) $y = -x^{-3/2}$.

The table below has been computed for selected values of x. We shall assume that the points corresponding to intermediate values of x lie on a smooth curve joining the points given in the table.

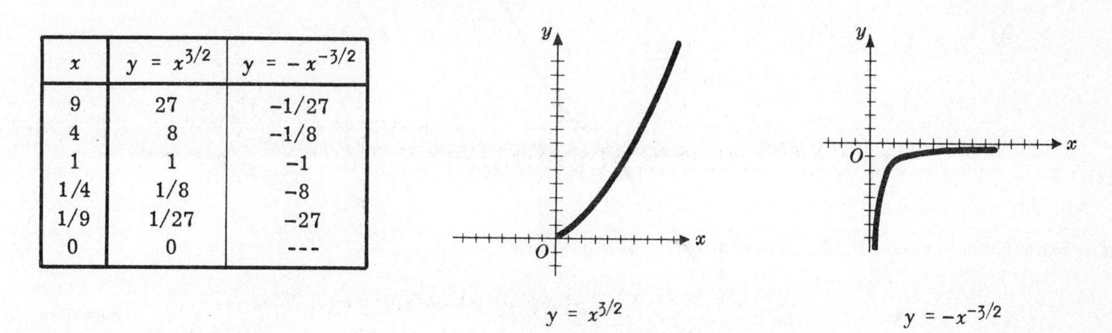

x	$y = x^{3/2}$	$y = -x^{-3/2}$
9	27	-1/27
4	8	-1/8
1	1	-1
1/4	1/8	-8
1/9	1/27	-27
0	0	---

$y = x^{3/2}$ $y = -x^{-3/2}$

See Problems 1-3.

EXPONENTIAL FUNCTIONS in x are of the form b^x where b is a constant. The discussion will be limited here to the case $b > 1$.

The curve whose equation is $y = b^x$ is called an *exponential curve*. The general properties of such curves are:

(a) the curve passes through the point (0, 1),
(b) the curve lies above the x-axis and has that axis as asymptote.

Example 2. Sketch the graphs of (a) $y = 2^x$, (b) $y = 3^x$.

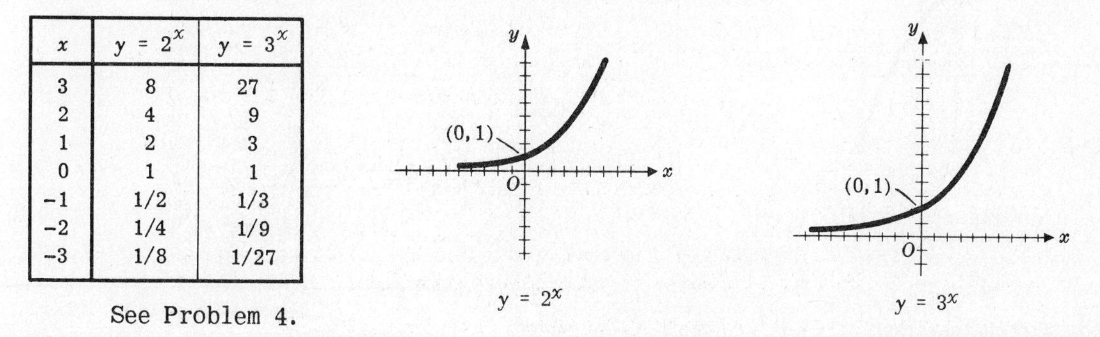

x	$y = 2^x$	$y = 3^x$
3	8	27
2	4	9
1	2	3
0	1	1
-1	1/2	1/3
-2	1/4	1/9
-3	1/8	1/27

$y = 2^x$ $y = 3^x$

See Problem 4.

The exponential equation appears frequently in the form $y = ce^{kx}$ where c and k are non-zero constants and $e = 2.71828...$ is the Naperian base.

See Problems 5-6.

THE CURVE WHOSE EQUATION IS $y = \log_b x$, $b > 1$, is called a *logarithmic curve*. The general properties are:

(*a*) the curve passes through the point (1,0),
(*b*) the curve lies to the right of the y-axis and has that axis as asymptote.

Example 3. Sketch the graph of $y = \log_2 x$.

x	8	4	2	1	1/2	1/4	1/8
y	3	2	1	0	−1	−2	−3

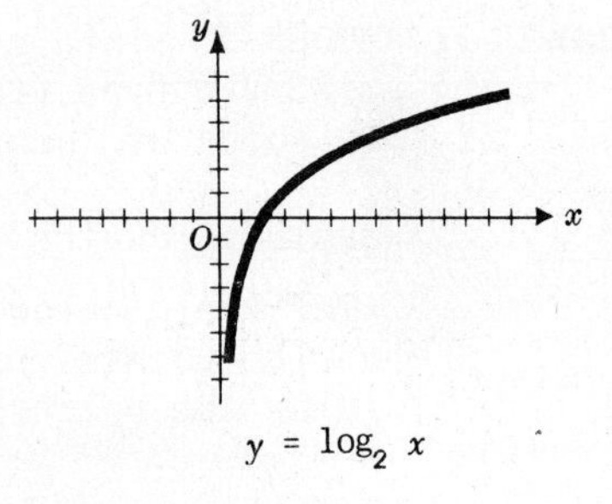

$y = \log_2 x$

Since $x = 2^y$, the above table of values may be obtained from the table for $y = 2^x$ of Example 2 by interchanging x and y.

See Problem 7.

SOLVED PROBLEMS

1. Sketch the graph of the *semi-cubic parabola* $y^2 = x^3$.

Since the given equation is equivalent to $y = \pm x^{3/2}$, the graph consists of the curve of Example 1(*a*) together with its reflection in the x-axis. See Fig. (*a*) below.

2. Sketch the graph of $y^3 = x^2$.

Refer to Fig. (*b*) below.

x	±3	±2	±1	0
y	2.1	1.6	1	0

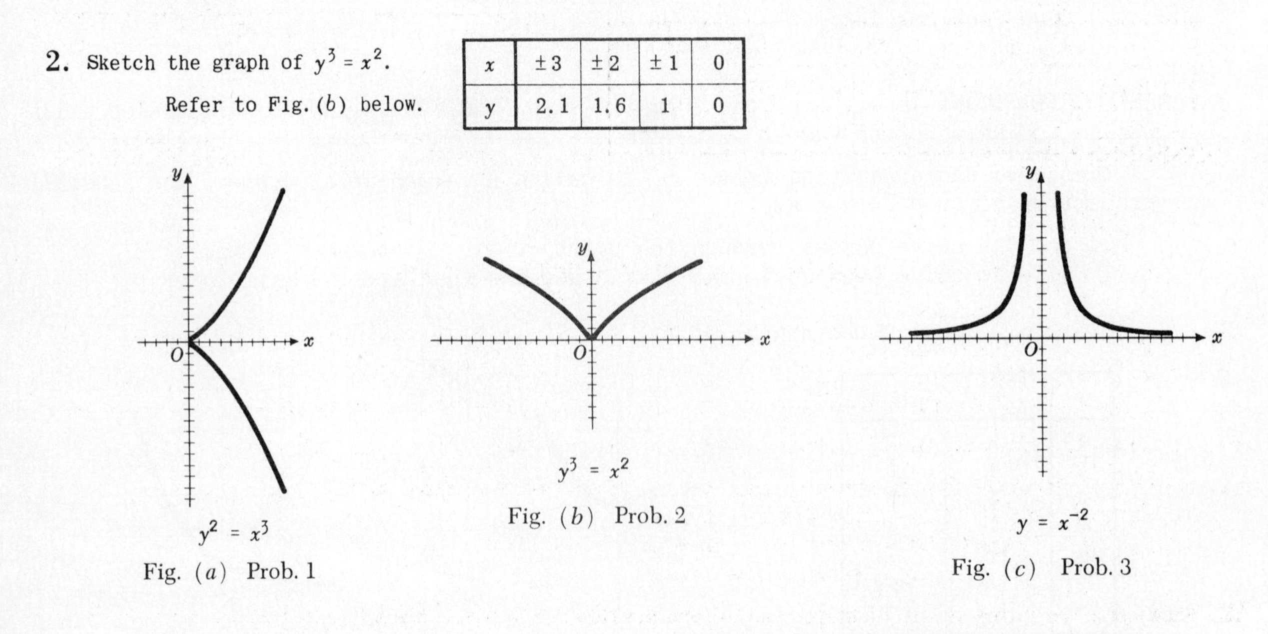

$y^2 = x^3$
Fig. (*a*) Prob. 1

$y^3 = x^2$
Fig. (*b*) Prob. 2

$y = x^{-2}$
Fig. (*c*) Prob. 3

3. Sketch the graph of $y = x^{-2}$. See Fig. (*c*) above.

x	±3	±2	±1	±1/2	±1/4	0
y	1/9	1/4	1	4	16	--

4. Sketch the graph of $y = 3^{-x}$. See Fig. (*d*) below.

Note that the graph of $y = b^{-x}$ is a reflection in the y-axis of the graph of $y = b^{x}$.

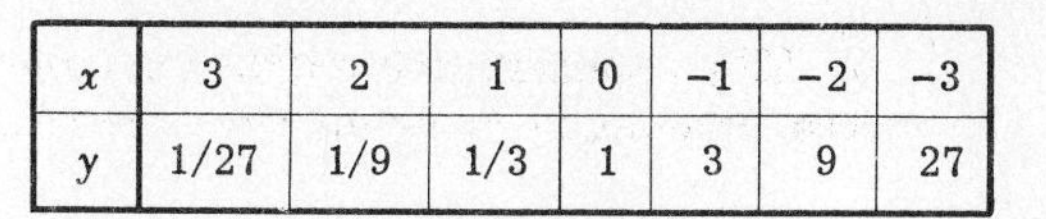

x	3	2	1	0	−1	−2	−3
y	1/27	1/9	1/3	1	3	9	27

5. Sketch the graph of $y = e^{2x}$.

See Fig. (*e*) below.

x	2	1	1/2	0	−1/2	−1	−2
$y = e^{2x}$	54.6	7.4	2.7	1	0.4	0.14	0.02

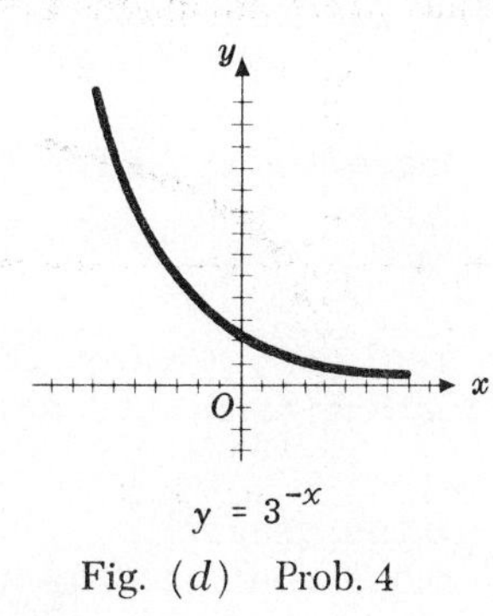

$y = 3^{-x}$

Fig. (*d*) Prob. 4

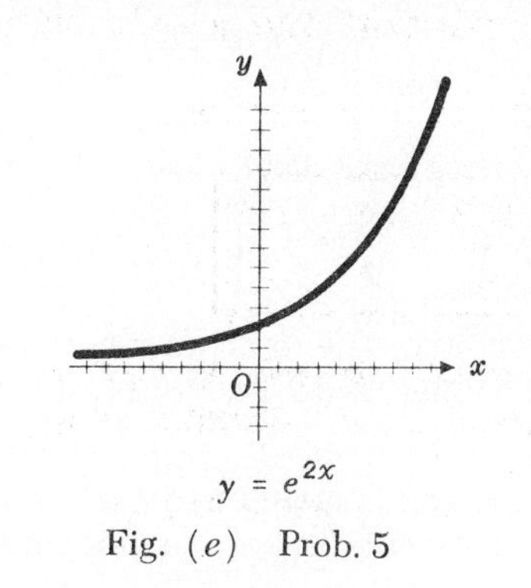

$y = e^{2x}$

Fig. (*e*) Prob. 5

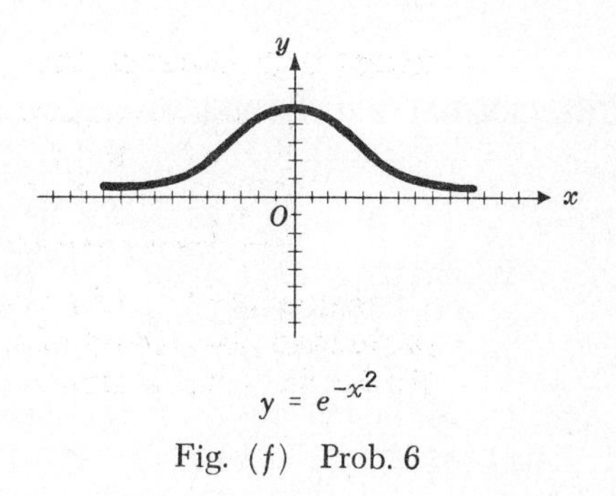

$y = e^{-x^2}$

Fig. (*f*) Prob. 6

6. Sketch the graph of $y = e^{-x^2}$. Refer to Fig. (*f*) above.

This is a simple form of the *normal probability curve* used in statistics.

x	± 2	± 3/2	± 1	± 1/2	0
y	0.02	0.1	0.4	0.8	1

7. Sketch the graphs of

(*a*) $y = \log x$

(*b*) $y = \log x^2 = 2 \log x$.

x	10	5	4	3	2	1	0.5	0.25	0.1	0.01
$y = \log x$	1	0.7	0.6	0.5	0.3	0	−0.3	−0.6	−1	−2
$y = \log x^2$	2	1.4	1.2	1	0.6	0	−0.6	−1.2	−2	−4

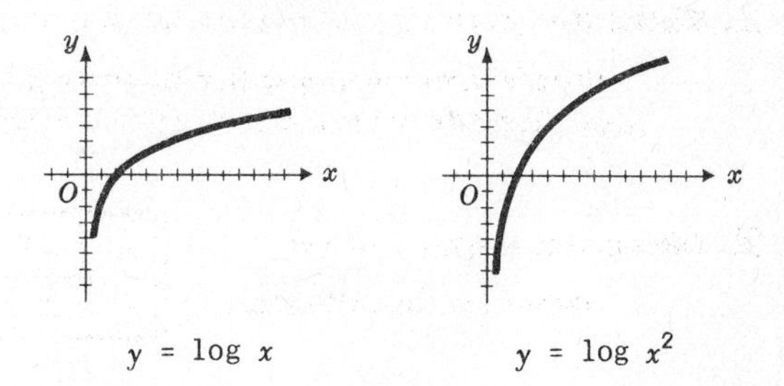

$y = \log x$ $y = \log x^2$

SUPPLEMENTARY PROBLEMS

8. Sketch the graphs of (*a*) $y^2 = x^{-3}$, (*b*) $y^3 = x^{-2}$, (*c*) $y^2 = 1/x$, (*d*) the cubical parabola $y = x^3$.

9. Sketch the graphs of (*a*) $y = (2.5)^x$ (*b*) $y = 2^{x+1}$ (*c*) $y = 2^{-1/x}$ (*d*) $y = \frac{1}{2}e^x$ (*e*) $y = e^{x/2}$ (*f*) $y = e^{-x/2}$ (*g*) $y = e^{x+2}$ (*h*) $y = xe^{-x}$

10. Sketch the graphs of (*a*) $y = \frac{1}{2} \log x$, (*b*) $y = \log (3x + 2)$, (*c*) $y = \log (x^2 + 1)$

11. Show that the curve $y^q = x^p$, where p and q are positive integers, lies entirely in:
(*a*) Quadrants I and III if p and q are both odd,
(*b*) Quadrants I and IV if p is odd and q is even,
(*c*) Quadrants I and II if p is even and q is odd.

12. Show that the curve $y^q = x^{-p}$, where p and q are positive integers, lies entirely in:
(*a*) Quadrants I and III if p and q are both odd,
(*b*) Quadrants I and II if p is even and q is odd,
(*c*) Quadrants I and IV if p is odd and q is even.

CHAPTER 17

Graphs of Polynomials

THE GENERAL POLYNOMIAL (or rational integral function) of the nth degree in x has the form

$$(1) \qquad f(x) = a_0x^n + a_1x^{n-1} + a_2x^{n-2} + \ldots. + a_{n-2}x^2 + a_{n-1}x + a_n$$

in which n is a positive integer and the a's are constants, real or complex, with $a_0 \neq 0$. The term a_0x^n is called the *leading term*, a_n the *constant term*, and a_0 the *leading coefficient*.

Although most of the theorems and statements below apply to the general polynomial, attention in this chapter will be restricted to polynomials whose coefficients (the a's) are integers.

REMAINDER THEOREM. If a polynomial $f(x)$ is divided by $x-h$ until a remainder free of x is obtained, this remainder is $f(h)$.

For a proof, see Problem 1.

Example 1. Let $f(x) = x^3 + 2x^2 - 3x - 4$ and $x-h = x-2$; then $h = 2$. By actual division

$$\frac{x^3+2x^2-3x-4}{x-2} = x^2 + 4x + 5 + \frac{6}{x-2},$$

or $x^3+2x^2-3x-4 = (x^2+4x+5)(x-2)+6$, and the remainder is 6.

By the remainder theorem, the remainder is

$$f(2) = 2^3 + 2\cdot 2^2 - 3\cdot 2 - 4 = 6$$

FACTOR THEOREM. If $x-h$ is a factor of $f(x)$ then $f(h) = 0$, and conversely.

For a proof, see Problem 2.

SYNTHETIC DIVISION. By a process known as synthetic division, the necessary work in dividing a polynomial $f(x)$ by $x-h$ may be displayed in three lines, as follows:

(*1*) Arrange the dividend $f(x)$ in descending powers of x (as usual in division) and set down in the first line the coefficients, supplying zero as coefficient whenever a term is missing.

(*2*) Place h, the synthetic divisor, in the first line to the right of the coefficients.

(*3*) Recopy the leading coefficient a_0 directly below it in the third line.

(*4*) Multiply a_0 by h; place the product a_0h in the second line under a_1 (in the first line), add to a_1, and place the sum $a_0h + a_1$ in the third line under a_1.

(5) Multiply the sum in Step 4 by h; place the product in the second line under a_2, add to a_2, and place the sum in the third line under a_2.

(6) Repeat the process of Step 5 until a product has been added to the constant term a_n.

The first n numbers in the third line are the coefficients of the quotient, a polynomial of degree $n-1$, and the last number of the third line is the remainder $f(h)$.

Example 2. Divide $5x^4 - 8x^2 - 15x - 6$ by $x-2$, using synthetic division.

Following the procedure outlined above, we have

$$\begin{array}{rrrrr|l} 5 & +\ 0 & -\ 8 & -\ 15 & -\ 6 & \underline{2} \\ & 10 & +\ 20 & +\ 24 & +\ 18 & \\ \hline 5 & +\ 10 & +\ 12 & +\ 9 & +\ 12 & \end{array}$$

The quotient is $Q(x) = 5x^3 + 10x^2 + 12x + 9$ and the remainder is $f(2) = 12$.

See Problem 4.

THE GRAPH OF A POLYNOMIAL $y = f(x)$ may be obtained by computing a table of values, locating the several points (x, y), and joining them by a smooth curve. In order to avoid unnecessary labor in constructing the table, the following systematic procedure is suggested:

(1) When $x = 0$, $y = f(0)$ is the constant term of the polynomial.

(2) Use synthetic division to find $f(1), f(2), f(3), \ldots$ stopping as soon as the numbers in the third line of the synthetic division have the same sign.

(3) Use synthetic division to find $f(-1), f(-2), f(-3), \ldots$ stopping as soon as the numbers in the third line of the synthetic division have alternating signs.

In advanced mathematics it is proved:

(a) The graph of a polynomial in x with integral coefficients is always a smooth curve without breaks or sharp corners.

(b) The number of *real* intersections of the graph of a polynomial of degree n with the x-axis is *never* greater than n.

(c) If a and b are real numbers such that $f(a)$ and $f(b)$ have opposite signs, the graph has an *odd* number of real intersections with the x-axis between $x = a$ and $x = b$.

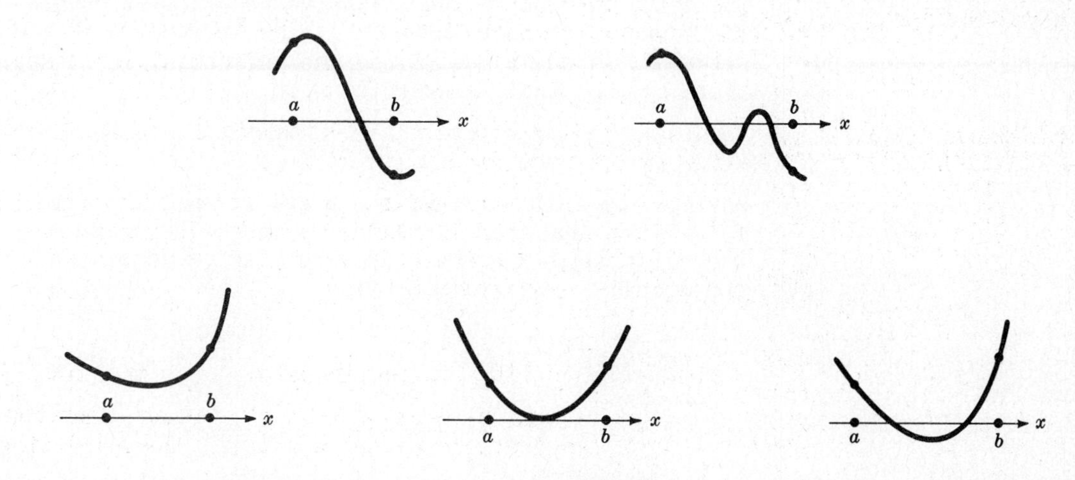

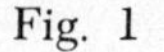
Fig. 1

(*d*) If a and b are real numbers such that $f(a)$ and $f(b)$ have the same signs, the graph either does not intersect the x-axis or intersects it an *even* number of times between $x = a$ and $x = b$.

See Figure 1 above.

Example 3. Construct the graph of $y = 2x^3 - 7x^2 - 7x + 5$.

Form the table

x	-2	-1	0	1	2	3	4	5
y	-25	3	5	-7	-21	-25	-7	45

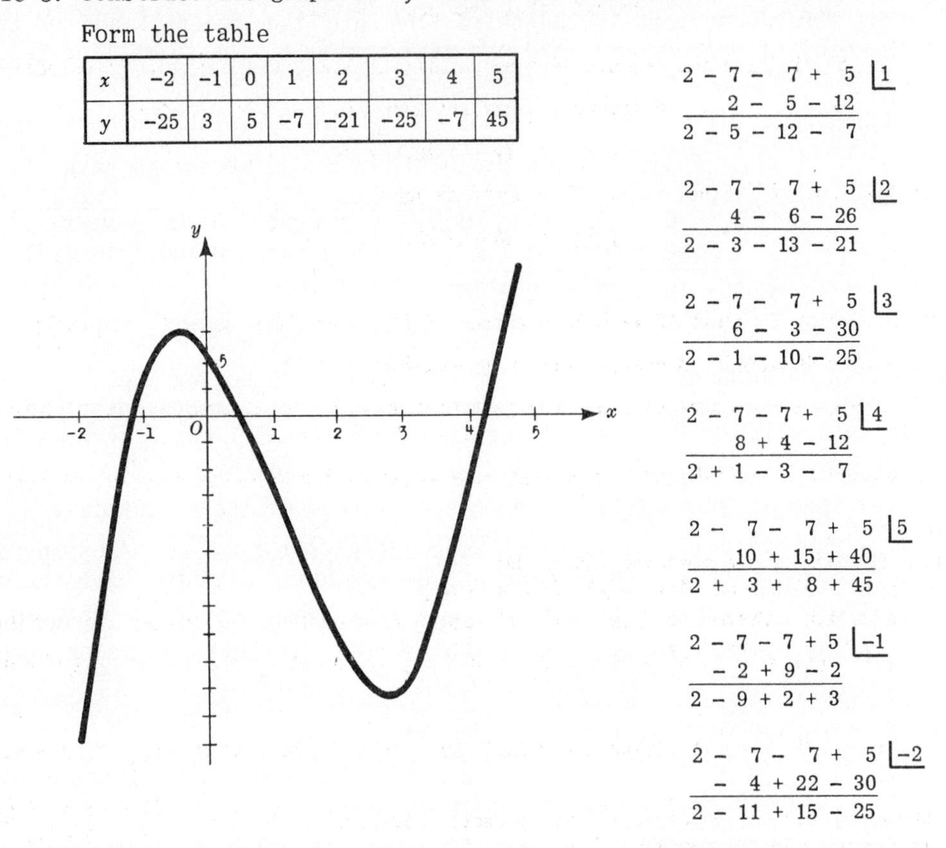

It is to be noted that:

(*a*) The numbers in the third line are all non-negative for the first time in finding $f(5)$; they alternate for the first time in finding $f(-2)$.

(*b*) The graph intersects the x-axis between $x = -2$ and $x = -1$ since $f(-2)$ and $f(-1)$ have opposite signs, between $x = 0$ and $x = 1$, and between $x = 4$ and $x = 5$. Since the polynomial is of degree three, there are no other intersections.

(*c*) Reading from the graph, the x-intercepts are *approximately* $x = -1.2$, $x = 0.5$, and $x = 4.2$.

(*d*) Moving from left to right the graph rises for a time, then falls for a time, and then rises thereafter. The problem of locating the point where a graph ceases to rise or ceases to fall will be considered in a later chapter.

See Problems 5-9.

SOLVED PROBLEMS

1. Prove the Remainder Theorem: If a polynomial $f(x)$ is divided by $x-h$ until a constant remainder is obtained, that remainder is $f(h)$.

In the division let the quotient be denoted by $Q(x)$ and the constant remainder by R. Then, since

dividend = divisor times quotient + remainder,

$$f(x) = (x-h)\,Q(x) + R$$

is true for *all* values of x. When $x=h$, we have

$$f(h) = (h-h)\,Q(h) + R = R$$

2. Prove the Factor Theorem: If $x-h$ is a factor of $f(x)$ then $f(h)=0$, and conversely.

By the Remainder Theorem, $f(x) = (x-h)\,Q(x) + f(h)$.

If $x-h$ is a factor of $f(x)$, the remainder must be zero when $f(x)$ is divided by $x-h$. Thus, $f(h) = 0$.

Conversely, if $f(h)=0$, then $f(x) = (x-h)\,Q(x)$ and $x-h$ is a factor of $f(x)$.

3. Without performing the division, show that
(*a*) $x-2$ is a factor of $f(x) = x^3 - x^2 - 14x + 24$,
(*b*) $x+a$ is not a factor of $f(x) = x^n + a^n$ for n an even positive integer and $a \neq 0$.

(*a*) $f(2) = 2^3 - 2^2 - 14\cdot 2 + 24 = 0$ (*b*) $f(-a) = (-a)^n + a^n = 2a^n \neq 0$

4. Use synthetic division to divide $4x^4 + 12x^3 - 21x^2 - 65x + 9$ by (*a*) $2x-1$, (*b*) $2x+3$.

(*a*) Write the divisor as $2(x-\frac{1}{2})$. By synthetic division with synthetic divisor $h=\frac{1}{2}$, we find

$$\begin{array}{rrrrr|l} 4 & +12 & -21 & -65 & +9 & \frac{1}{2} \\ & 2 & +7 & -7 & -36 & \\ \hline 4 & +14 & -14 & -72 & -27 & \end{array}$$

Now
$$\begin{aligned} 4x^4 + 12x^3 - 21x^2 - 65x + 9 &= (4x^3 + 14x^2 - 14x - 72)(x-\tfrac{1}{2}) - 27 \\ &= (2x^3 + 7x^2 - 7x - 36)(2x-1) - 27 \end{aligned}$$

Thus, when dividing $f(x)$ by $h=m/n$, the coefficients of the quotient have n as a common factor.

(*b*) Here $h = -3/2$. Then from

$$\begin{array}{rrrrr|l} 4 & +12 & -21 & -65 & +9 & -3/2 \\ & -6 & -9 & +45 & +30 & \\ \hline 4 & +6 & -30 & -20 & +39 & \end{array}$$

we have

$$4x^4 + 12x^3 - 21x^2 - 65x + 9 = (2x^3 + 3x^2 - 15x - 10)(2x+3) + 39$$

5. For the polynomial $y = f(x) = 4x^4 + 12x^3 - 31x^2 - 72x + 42$ form a table of values for integral values of x from $x=-5$ to $x=3$.

The required table is

x	-5	-4	-3	-2	-1	0	1	2	3
y	627	90	-21	30	75	42	-45	-66	195

$$\begin{array}{rrrrr|l} 4 & +12 & -31 & -72 & +42 & -5 \\ & -20 & +40 & -45 & +585 & \\ \hline 4 & -8 & +9 & -117 & +627 & \end{array} \qquad \begin{array}{rrrrr|l} 4 & +12 & -31 & -72 & +42 & -4 \\ & -16 & +16 & +60 & +48 & \\ \hline 4 & -4 & -15 & -12 & +90 & \end{array} \qquad \begin{array}{rrrrr|l} 4 & +12 & -31 & -72 & +42 & -3 \\ & -12 & +0 & +93 & -63 & \\ \hline 4 & +0 & -31 & +21 & -21 & \end{array}$$

$$\begin{array}{rrrrr|l} 4 & +12 & -31 & -72 & +42 & \underline{-2} \\ & -8 & -8 & +78 & -12 & \\ \hline 4 & +4 & -39 & +6 & +30 & \end{array} \qquad \begin{array}{rrrrr|l} 4 & +12 & -31 & -72 & +42 & \underline{-1} \\ & -4 & -8 & +39 & +33 & \\ \hline 4 & +8 & -39 & -33 & +75 & \end{array}$$

$f(0) = 42$, the constant term of the polynomial.

$$\begin{array}{rrrrr|l} 4 & +12 & -31 & -72 & +42 & \underline{1} \\ & 4 & +16 & -15 & -87 & \\ \hline 4 & +16 & -15 & -87 & -45 & \end{array} \qquad \begin{array}{rrrrr|l} 4 & +12 & -31 & -72 & +42 & \underline{2} \\ & 8 & +40 & +18 & -108 & \\ \hline 4 & +20 & +9 & -54 & -66 & \end{array} \qquad \begin{array}{rrrrr|l} 4 & +12 & -31 & -72 & +42 & \underline{3} \\ & 12 & +72 & +123 & +153 & \\ \hline 4 & +24 & +41 & +51 & +195 & \end{array}$$

6. Sketch the graph of $y = f(x) = x^4 - 9x^2 + 7x + 4$.

From the table

x	−4	−3	−2	−1	0	1	2	3
y	88	−17	−30	−11	4	3	−2	25

it is seen that the graph crosses the x-axis between $x=-4$ and $x=-3$, $x=-1$ and $x=0$, $x=1$ and $x=2$, and $x=2$ and $x=3$. From the graph, the points of crossing are approximately $x = -3.1$, -0.4, 1.4, and 2.1.

7. Sketch the graph of $y = f(x) = x^3 - 5x^2 + 8x - 3$.

From the table formed for integral values of x from $x=-1$ to $x=5$

x	−1	0	1	5/4	3/2	7/4	2	9/4	5/2	11/4	3	4	5
y	−17	−3	1	73/64	9/8	67/64	1	69/64	11/8	127/64	3	13	37

it is seen that the graph crosses the x-axis between $x=0$ and $x=1$. If there are two other real intersections, they are both between $x=1$ and $x=3$ since on this interval the graph rises to the right of $x=1$, then falls, and then begins to rise to the left of $x=3$.

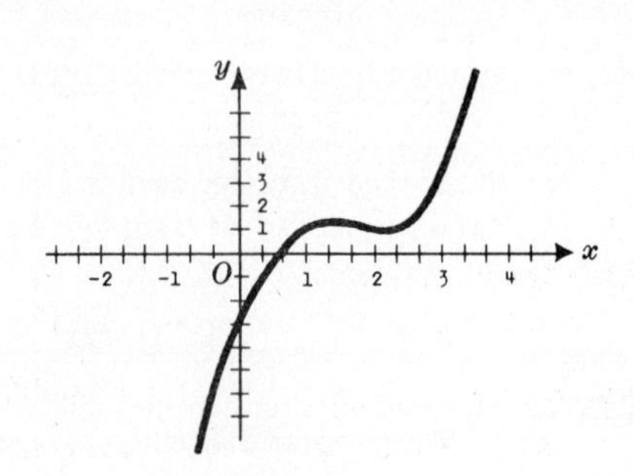

By computing additional points on the interval (some are shown in the table) we are led to suspect that there are no further intersections (see adjoining figure). Note that were we able to locate the exact points at which the graph ceases to rise or fall these additional points would not have been necessary.

8. Sketch the graph of $y = f(x) = x^3 - 4x^2 - 3x + 18$.

From the table

x	−3	−2	−1	0	1	2	3	4	5
y	−36	0	16	18	12	4	0	6	28

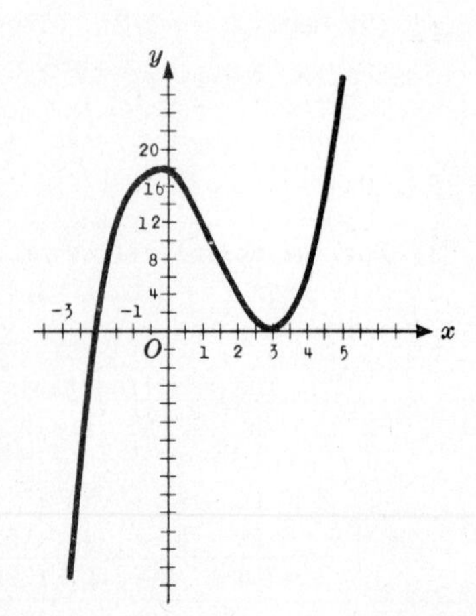

it is evident that the graph crosses the x-axis at $x=-2$ and meets it again at $x=3$. If there is a third distinct intersection, it must be between $x=2$ and $x=3$ or between $x=3$ and $x=4$. By computing additional points for x on these intervals, we are led to suspect that no such third intersection exists.

This function has been selected so that the question of intersections can be definitely settled. When $f(x)$ is divided by $x+2$, the quotient is $x^2 - 6x + 9 = (x-3)^2$ and the remainder is zero.

Thus, $f(x) = (x+2)(x-3)^2$ in factored form. It is now clear that the function is positive for $x > -2$, that is, the graph never falls below the x-axis on this interval. Thus, the graph is tangent to the x-axis at $x = 3$, the point of tangency accounting for two of its intersections with the x-axis.

9. Sketch the graph of $y = f(x) = x^3 - 6x^2 + 12x - 8$.

From the table

x	-2	-1	0	1	2	3	4	5	6
y	-64	-27	-8	-1	0	1	8	27	64

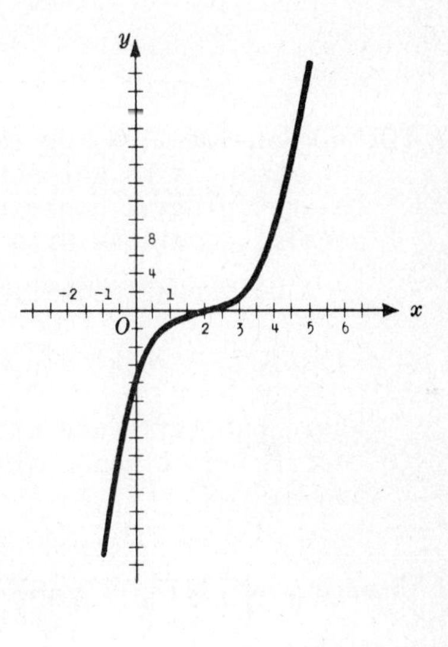

it is evident that the graph crosses the x-axis at $x = 2$ and is symmetrical with respect to that point (i.e., $f(2+h) = -f(2-h)$). Suppose the graph intersects the x-axis to the right of $x = 2$. Then, since $f(x)$ is positive for $x = 3, 4, 5$ and $x \geq 6$, the graph is either tangent at the point or crosses the axis twice between some two consecutive values of x shown in the table. But, by symmetry, the graph would then have $2 + 2 + 1 = 5$ intersections with the x-axis and this is impossible.

When $f(x)$ is divided by $x - 2$, the quotient is $x^2 - 4x + 4 = (x-2)^2$ and the remainder is zero. Thus, $f(x) = (x-2)^3$. Now it is clear that the graph lies above the x-axis when $x > 2$ and below the axis when $x < 2$. The graph crosses the x-axis at $x = 2$ and is also tangent there. In determining the intersections of the graph and the x-axis, the point $x = 2$ is to be counted three times.

SUPPLEMENTARY PROBLEMS

10. Given $f(x) = x^4 + 5x^3 - 7x^2 - 13x - 2$, use synthetic division to find:
(a) $f(2) = 0$ (b) $f(4) = 410$ (c) $f(-2) = -28$ (d) $f(-3) = -80$

11. Using synthetic division, find the quotient and remainder when:
(a) $2x^4 - 3x^3 - 6x^2 + 11x - 10$ is divided by $x - 2$; (b) $3x^4 + 11x^3 + 7x^2 - 8$ is divided by $x + 2$.
Ans. (a) $2x^3 + x^2 - 4x + 3$; -4 (b) $3x^3 + 5x^2 - 3x + 6$; -20

12. Use synthetic division to show:
(a) $x + 2$ and $3x - 2$ are factors of $3x^4 - 20x^3 + 80x - 48$.
(b) $x - 7$ and $3x + 5$ are not factors of $6x^4 - x^3 - 94x^2 + 74x + 35$.

13. Use synthetic division to form a table of values and sketch the graph of:
(a) $y = x^3 - 13x + 12$ (c) $y = 3x^3 + 5x^2 - 4x - 3$ (e) $y = -x^3 + 13x + 12$
(b) $y = 2x^3 + x^2 - 12x - 5$ (d) $y = x^4 - x^3 - 7x^2 + 13x - 6$

14. Sketch the graph of: (a) $y = x(x^2 - 4)$, (b) $y = x(4 - x^2)$, (c) $y = x(x-2)^2$, (d) $y = x(2-x)^2$.

CHAPTER 18

Polynomial Equations, Rational Roots

A POLYNOMIAL EQUATION (or rational integral equation) is obtained when any polynomial in one variable is set equal to zero. As in Chapter 17, we shall work with polynomials having integral coefficients although many of the theorems will be stated for polynomial equations with weaker restrictions on the coefficients.

A polynomial equation is said to be in *standard form* when written as

$$(1) \qquad a_0x^n + a_1x^{n-1} + a_2x^{n-2} + \ldots. + a_{n-2}x^2 + a_{n-1}x + a_n = 0$$

where the terms are arranged in descending powers of x, a zero has been inserted as coefficient of each missing term, the coefficients have no common factor except ± 1, and $a_0 \neq 0$.

See Problem 1.

A NUMBER r IS CALLED A ROOT of $f(x) = 0$ if and only if $f(r) = 0$. It follows that the abscissas of the points of intersection of the graph of $y = f(x)$ and the x-axis are roots of $f(x) = 0$.

THE FUNDAMENTAL THEOREM OF ALGEBRA. Every polynomial equation $f(x) = 0$ has at least one root, real or complex.

A polynomial equation of degree n has exactly n roots. These n roots may not all be distinct. If r is one of the roots and occurs just once, it is called a *simple root*; if r occurs exactly $m > 1$ times among the roots, it is called a *root of multiplicity* m or an *m-fold root*. If $m = 2$, r is called a *double root*; if $m = 3$, a *triple root*; and so on.

See Problems 2, 3.

IMAGINARY ROOTS. If the polynomial equation $f(x) = 0$ has real coefficients and if the imaginary $a + bi$ is a root of $f(x) = 0$, then the *conjugate* imaginary $a - bi$ is also a root.

For a proof, see Problem 11.

IRRATIONAL ROOTS. If the polynomial equation $f(x) = 0$ has rational coefficients and if the irrational number $a + \sqrt{b}$, where a and b are rational, is a root of $f(x) = 0$, then the conjugate irrational $a - \sqrt{b}$ is also a root.

See Problem 4.

LIMITS TO THE REAL ROOTS. A real number L is called an *upper limit* of the real roots of $f(x) = 0$ if no (real) root is greater than L; a real number l is called a *lower limit* if no (real) root is smaller than l.

If $L > 0$ and if, when $f(x)$ is divided by $x - L$ by synthetic division, every number in the third line is non-negative, then L is an upper limit of the real roots of $f(x) = 0$.

If $l < 0$ and if, when $f(x)$ is divided by $x - l$ by synthetic division, the numbers in the third line alternate in sign, then l is a lower limit of the real roots of $f(x) = 0$.

RATIONAL ROOTS. A polynomial equation has 0 as a root if and only if the constant term of the equation is zero.

Example 1. The roots of $x^5 - 2x^4 + 6x^3 - 5x^2 = x^2(x^3 - 2x^2 + 6x - 5) = 0$ are 0,0, and the three roots of $x^3 - 2x^2 + 6x - 5 = 0$.

If a rational fraction p/q, expressed in lowest terms, is a root of (*1*) in which $a_n \neq 0$, then p is a divisor of the constant term a_n and q is a divisor of the leading coefficient a_0 of (*1*).

For a proof, see Problem 12.

Example 2. The theorem permits us to say that 2/3 is a *possible* root of the equation $9x^4 - 5x^2 + 8x + 4 = 0$ since the numerator 2 divides the constant term 4 and the denominator 3 divides the leading coefficient 9. It does *not* assure that 2/3 is a root. However, the theorem does assure that neither 1/2 nor -4/5 is a root. In each case the denominator does not divide the leading coefficient.

If p, an integer, is a root of (*1*) then p is a divisor of its constant term.

Example 3. The *possible* rational roots of the equation

$$12x^4 - 40x^3 - 5x^2 + 45x + 18 = 0$$

are all numbers $\pm p/q$ in which the values of p are the positive divisors 1,2,3,6,9,18 of the constant term 18 and the values of q are the positive divisors 1,2,3,4,6,12 of the leading coefficient 12. Thus the rational roots, if any, of the equation are among the numbers

$$\pm1, \pm2, \pm3, \pm6, \pm9, \pm18, \pm1/2, \pm3/2, \pm9/2, \pm1/3, \pm2/3, \pm1/4, \pm3/4, \pm1/6, \pm1/12$$

THE PRINCIPAL PROBLEM OF THIS CHAPTER is to find the rational roots of a given polynomial equation. The general procedure is: Test the possible rational roots by synthetic division, accepting as roots all those for which the last number in the third line *is* zero and rejecting all those for which it is not. Certain refinements, which help to shorten the work, are pointed out in the examples and solved problems below.

Example 4. Find the rational roots of $x^5 + 2x^4 - 18x^3 - 8x^2 + 41x + 30 = 0$.

Since the leading coefficient is 1, all rational roots p/q are integers. The possible integral roots, the divisors (both positive and negative) of the constant term 30, are

$$\pm1, \ \pm2, \ \pm3, \ \pm5, \ \pm6, \ \pm10, \ \pm15, \ \pm30$$

Try 1:

$$\begin{array}{r|l} 1 + 2 - 18 - \ \ 8 + 41 + 30 & 1 \\ 1 + \ \ 3 - 15 - 23 + 18 & \\ \hline 1 + 3 - 15 - 23 + 18 + 48 & \end{array}$$

Then 1 is not a root. This number (+1) should be removed from the list of possible roots lest we forget and try it again later on.

Try 2:

$$\begin{array}{r|l} 1 + 2 - 18 - \ \ 8 + 41 + 30 & 2 \\ 2 + \ \ 8 - 20 - 56 - 30 & \\ \hline 1 + 4 - 10 - 28 - 15 + \ \ 0 & \end{array}$$

Then 2 is a root and the remaining rational roots of the given equation are the rational roots of the *depressed equation*

$$x^4 + 4x^3 - 10x^2 - 28x - 15 = 0$$

Now ±2, ±6, ±10, and ±30 cannot be roots of this equation (they are not divisors of 15) and should be removed from the list of possibilities. We return to the depressed equation.

Try 3:
$$\begin{array}{l|l} 1 + 4 - 10 - 28 - 15 & 3 \\ \quad\; 3 + 21 + 33 + 15 & \\ \hline 1 + 7 + 11 + 5 + 0 & \end{array}$$

Then 3 is a root and the new depressed equation is

$$x^3 + 7x^2 + 11x + 5 = 0$$

Since the coefficients of this equation are non-negative, it has no positive roots. We now remove +3, +5, +15 from the original list of possible roots and return to the new depressed equation.

Try −1:
$$\begin{array}{l|l} 1 + 7 + 11 + 5 & -1 \\ \quad -1 - 6 - 5 & \\ \hline 1 + 6 + 5 + 0 & \end{array}$$

Then −1 is a root and the depressed equation

$$x^2 + 6x + 5 = (x + 1)(x + 5) = 0$$

has −1 and −5 as roots.

The necessary computations may be neatly displayed as follows:

$$\begin{array}{l|l} 1 + 2 - 18 - 8 + 41 + 30 & 2 \\ \quad 2 + 8 - 20 - 56 - 30 & \\ \hline 1 + 4 - 10 - 28 - 15 & 3 \\ \quad 3 + 21 + 33 + 15 & \\ \hline 1 + 7 + 11 + 5 & -1 \\ \quad -1 - 6 - 5 & \\ \hline 1 + 6 + 5 & \end{array}$$

$$x^2 + 6x + 5 = (x + 1)(x + 5) = 0$$
$$x = -1, -5$$

The roots are 2, 3, −1, −1, −5.

Note that the roots here are numerically small numbers, that is, 3 is a root but 30 is not, −1 is a root but −15 is not. Hereafter we shall not list integers which are large numerically or fractions with large numerator or denominator among the possible roots.

See Problems 5-9.

SOLVED PROBLEMS

1. Write each of the following in standard form.

(*a*) $4x^2 + 2x^3 - 6 + 5x = 0$ *Ans.* $2x^3 + 4x^2 + 5x - 6 = 0$

(*b*) $-3x^3 + 6x - 4x^2 + 2 = 0$ *Ans.* $3x^3 + 4x^2 - 6x - 2 = 0$

(*c*) $2x^5 + x^3 + 4 = 0$ *Ans.* $2x^5 + 0\cdot x^4 + x^3 + 0\cdot x^2 + 0\cdot x + 4 = 0$

(*d*) $x^3 + \frac{1}{2}x^2 - x + 2 = 0$ *Ans.* $2x^3 + x^2 - 2x + 4 = 0$

(*e*) $4x^4 + 6x^3 - 8x^2 + 12x - 10 = 0$ *Ans.* $2x^4 + 3x^3 - 4x^2 + 6x - 5 = 0$

2. (*a*) Show that −1 and 2 are roots of $x^4 - 9x^2 + 4x + 12 = 0$.

Using synthetic division,
$$\begin{array}{l|l} 1 + 0 - 9 + 4 + 12 & -1 \\ \quad - 1 + 1 + 8 - 12 & \\ \hline 1 - 1 - 8 + 12 + 0 & \end{array} \qquad \begin{array}{l|l} 1 + 0 - 9 + 4 + 12 & 2 \\ \quad 2 + 4 - 10 - 12 & \\ \hline 1 + 2 - 5 - 6 + 0 & \end{array}$$

Since $f(-1) = 0$ and $f(2) = 0$, both −1 and 2 are roots.

(b) Show that the equation in (a) has at least two other roots by finding them.

From the synthetic division in (a) and the Factor Theorem (Chapter 17),

$$x^4 - 9x^2 + 4x + 12 = (x+1)(x^3 - x^2 - 8x + 12)$$

Since 2 is also a root of the given equation, 2 is a root of $x^3 - x^2 - 8x + 12 = 0$.

Using synthetic division
$$\begin{array}{l} 1-1-8+12 \,\lfloor 2 \\ \;\;\;\;2+2-12 \\ \hline 1+1-6+0 \end{array}$$
, we obtain $x^3 - x^2 - 8x + 12 = (x-2)(x^2 + x - 6)$.

Then $x^4 - 9x^2 + 4x + 12 = (x+1)(x-2)(x^2+x-6) = (x+1)(x-2)(x+3)(x-2)$.

Thus, the roots of the given equation are $-1, 2, -3, 2$.

Note. Since $x-2$ appears twice among the factors of $f(x)$, 2 appears twice among the roots of $f(x) = 0$ and is a double root of the equation.

3. (a) Find all of the roots of $(x+1)(x-2)^3(x+4)^2 = 0$.

The roots are $-1, 2, 2, 2, -4, -4$; thus, -1 is a simple root, 2 is a root of multiplicity three or a triple root, and -4 is a root of multiplicity two or a double root.

(b) Find all the roots of $x^2(x-2)(x-5) = 0$.

The roots are $0, 0, 2, 5$; 2 and 5 are simple roots and 0 is a double root.

4. Form the equation of lowest degree with integral coefficients having the roots $\sqrt{5}$ and $2-3i$.

To assure integral coefficients, the conjugates $-\sqrt{5}$ and $2+3i$ must also be roots. Thus the required equation is

$$(x-\sqrt{5})(x+\sqrt{5})[x-(2-3i)][x-(2+3i)] = x^4 - 4x^3 + 8x^2 + 20x - 65 = 0$$

In Problems 5-9 find the rational roots; when possible, find all the roots.

5. $2x^4 - x^3 - 11x^2 + 4x + 12 = 0$

The possible rational roots are: $\pm 1, \pm 2, \pm 3, \pm 4, \pm 1/2, \pm 3/2, \ldots$. Discarding all false trials, we find

$$\begin{array}{l} 2-1-11+\;4+12 \,\lfloor 2 \\ \;\;\;\;4+\;6-10-12 \\ \hline 2+3-\;5-\;6 \;\;\;\lfloor -1 \\ \;\;\;-2-\;1+\;6 \\ \hline 2+1-\;6 \end{array}$$

$2x^2 + x - 6 = (2x-3)(x+2) = 0$; $x = 3/2, -2$

The roots are $2, -1, 3/2, -2$.

6. $4x^4 - 3x^3 - 4x + 3 = 0$

The possible rational roots are: $\pm 1, \pm 3, \pm 1/2, \pm 3/2, \pm 1/4, \pm 3/4$. By inspection the sum of the coefficients is 0; then +1 is a root. Discarding all false trials, we find

$$\begin{array}{l} 4-3+0-4+3 \,\lfloor 1 \\ \;\;\;\;4+1+1-3 \\ \hline 4+1+1-3 \;\;\;\lfloor 3/4 \\ \;\;\;\;3+3+3 \\ \hline 4+4+4 \\ 1+1+1 \end{array}$$
(Factor out 4.)

$x^2 + x + 1 = 0$; $x = \dfrac{-1 \pm i\sqrt{3}}{2}$

The roots are $1, \dfrac{3}{4}, -\dfrac{1}{2} \pm \dfrac{i\sqrt{3}}{2}$.

7. $24x^6 - 20x^5 - 6x^4 + 9x^3 - 2x^2 = 0$

Since $24x^6 - 20x^5 - 6x^4 + 9x^3 - 2x^2 = x^2(24x^4 - 20x^3 - 6x^2 + 9x - 2)$ the roots of the given equation are 0, 0 and the roots of $24x^4 - 20x^3 - 6x^2 + 9x - 2 = 0$. Possible rational roots are: $\pm 1, \pm 2, \pm 1/2, \pm 1/3, \pm 2/3, \pm 1/4, \ldots$. Discarding all false trials, we find

$$\begin{array}{rrrrr|l}
24 & -20 & -6 & +9 & -2 & \underline{1/2} \\
 & 12 & -4 & -5 & +2 & \\
\hline
24 & -8 & -10 & +4 & & \text{(Factor out 2.)} \\
12 & -4 & -5 & +2 & & \underline{1/2} \\
 & 6 & +1 & -2 & & \\
\hline
12 & +2 & -4 & & & \text{(Factor out 2.)} \\
6 & +1 & -2 & & &
\end{array}$$

$6x^2 + x - 2 = (2x-1)(3x+2) = 0; \quad x = 1/2, -2/3$

The roots are $0, 0, 1/2, 1/2, 1/2, -2/3$.

8. $4x^5 - 32x^4 + 93x^3 - 119x^2 + 70x - 25 = 0$

Since the signs of the coefficients alternate, the rational roots (if any) are positive. Possible rational roots are: $1, 5, 1/2, 5/2, 1/4, \ldots$. Discarding all false trials, we find

$$\begin{array}{rrrrrr|l}
4 & -32 & +93 & -119 & +70 & -25 & \underline{5/2} \\
 & 10 & -55 & +95 & -60 & +25 & \\
\hline
4 & -22 & +38 & -24 & +10 & & \text{(Factor out 2.)} \\
2 & -11 & +19 & -12 & +5 & & \underline{5/2} \\
 & 5 & -15 & +10 & -5 & & \\
\hline
2 & -6 & +4 & -2 & & & \text{(Factor out 2.)} \\
1 & -3 & +2 & -1 & & &
\end{array}$$

The rational roots are $5/2, 5/2$.

The only possible rational root of $x^3 - 3x^2 + 2x - 1 = 0$ is 1; it is not a root.

The equation $f(x) = x^3 - 3x^2 + 2x - 1 = 0$ has at least one real (irrational) root since $f(0) < 0$ while for sufficiently large x $(x > 3)$, $f(x) > 0$.

9. $6x^4 + 13x^3 - 11x^2 + 5x + 1 = 0$

The possible rational roots are: $\pm 1, \pm 1/2, \pm 1/3, \pm 1/6$. After testing each possibility, we conclude that the equation has no rational roots.

10. If $ax^3 + bx^2 + cx + d = 0$, with integral coefficients, has a rational root r and if $c + ar^2 = 0$, then all of the roots are rational.

Using synthetic division to remove the known root, the depressed equation is

$$ax^2 + (b + ar)x + c + br + ar^2 = 0$$

which reduces to $ax^2 + (b + ar)x + br = 0$ when $c + ar^2 = 0$. Since its discriminant is $(b - ar)^2$, a perfect square, its roots are rational. Thus, all the roots of the given equation are rational.

11. Prove: If the polynomial $f(x)$ has real coefficients and if the imaginary $a + bi$, $b \neq 0$, is a root of $f(x) = 0$, then the conjugate imaginary $a - bi$ is also a root.

Since $a + bi$ is a root of $f(x) = 0$, $x - (a + bi)$ is a factor of $f(x)$. Similarly, if $a - bi$ is to be a root of $f(x) = 0$, $x - (a - bi)$ must be a factor of $f(x)$. We need to show then that when $a + bi$ is a root of $f(x) = 0$, it follows that

$$[x - (a + bi)][x - (a - bi)] = x^2 - 2ax + a^2 + b^2$$

is a factor of $f(x)$. By division we find

$$(A) \qquad \begin{aligned} f(x) &= [x^2 - 2ax + a^2 + b^2] \cdot Q(x) + Mx + N \\ &= [x - (a + bi)][x - (a - bi)] \cdot Q(x) + Mx + N, \end{aligned}$$

where $Q(x)$ is a polynomial of degree 2 less than that of $f(x)$ and the remainder $Mx+N$ is of degree at most one in x, that is, M and N are constants.

Since $a+bi$ is a root of $f(x)=0$, we have from (A),

$$f(a+bi) = 0\cdot Q(a+bi) + M(a+bi) + N = (aM+N) + bMi = 0$$

Then $aM+N=0$ and $bM=0$. Now $b\neq 0$; hence, $bM=0$ requires $M=0$ and then $aM+N=0$ requires $N=0$. Since $M=N=0$, (A) becomes

$$f(x) = (x^2-2ax+a^2+b^2)\cdot Q(x)$$

Then $x^2-2ax+a^2+b^2$ is a factor of $f(x)$ as was to be proved.

12. Prove: If a rational fraction p/q, expressed in lowest terms, is a root of the polynomial equation (1) whose constant term $a_n\neq 0$, then p is a divisor of the constant term a_n and q is a divisor of the leading coefficient a_0.

Let the given equation be

$$a_0x^n + a_1x^{n-1} + a_2x^{n-2} + \dots + a_{n-2}x^2 + a_{n-1}x + a_n = 0, \quad a_0a_n \neq 0$$

If p/q is a root, then

$$a_0(p/q)^n + a_1(p/q)^{n-1} + a_2(p/q)^{n-2} + \dots + a_{n-2}(p/q)^2 + a_{n-1}(p/q) + a_n = 0$$

Multiplying both members by q^n, this becomes

$$(B) \qquad a_0p^n + a_1p^{n-1}q + a_2p^{n-2}q^2 + \dots + a_{n-2}p^2q^{n-2} + a_{n-1}pq^{n-1} + a_nq^n = 0$$

When (B) is written as

$$a_0p^n + a_1p^{n-1}q + a_2p^{n-2}q^2 + \dots + a_{n-2}p^2q^{n-2} + a_{n-1}pq^{n-1} = -a_nq^n$$

it is clear that p, being a factor of every term of the left member of the equality, must divide a_nq^n. Since p/q is expressed in lowest terms, no factor of p will divide q. Hence, p must divide a_n as was to be shown.

Similarly, when (B) is written as

$$a_1p^{n-1}q + a_2p^{n-2}q^2 + \dots + a_{n-2}p^2q^{n-2} + a_{n-1}pq^{n-1} + a_nq^n = -a_0p^n$$

it follows that q must divide a_0.

SUPPLEMENTARY PROBLEMS

13. Find all the roots.

(a) $x^5 - 2x^4 - 9x^3 + 22x^2 + 4x - 24 = 0$
(b) $18x^4 - 27x^3 + x^2 + 12x - 4 = 0$
(c) $12x^4 - 40x^3 - 5x^2 + 45x + 18 = 0$
(d) $6x^4 + 5x^3 - 16x^2 - 9x - 10 = 0$
(e) $9x^4 - 19x^2 - 6x + 4 = 0$
(f) $2x^5 + 3x^4 - 3x^3 - 2x^2 = 0$

Ans. (a) $2,2,2,-1,-3$ (b) $1, 1/2, 2/3, -2/3$ (c) $-2/3, -1/2, 3/2, 3$ (d) $-2, 5/3, (-1\pm i\sqrt{7})/4$ (e) $-1, 1/3, (1\pm\sqrt{13})/3$ (f) $0,0,1,-2,-1/2$

14. Solve the inequalities. (a) $x^3-5x^2+2x+8>0$ (b) $6x^3-17x^2-5x+6<0$ (c) $x^4-3x^3+x^2+4>0$ (d) $x^5-x^4-2x^3+2x^2+x-1>0$

Ans. (a) $x>4$, $-1<x<2$ (b) $x<-2/3$, $\frac{1}{2}<x<3$ (c) all $x\neq 2$ (d) $x>1$

15. Prove: If $r\neq 1$ is a root of $f(x)=0$ then $r-1$ divides $f(1)$.

CHAPTER 19

Irrational Roots of Polynomial Equations

IF $f(x)=0$ IS A POLYNOMIAL EQUATION, the equation $f(-x)=0$ has as roots the negatives of the roots of $f(x)=0$. When $f(x)=0$ is written in standard form, the equation whose roots are the negatives of the roots of $f(x)=0$ may be obtained by changing the signs of alternate terms, beginning with the second.

Example 1. (*a*) The roots of $x^3+3x^2-4x-12 = 0$ are $2,-2,-3$; the roots of $x^3-3x^2-4x+12 = 0$ are $-2,2,3$.

(*b*) The equation $6x^4+13x^3-13x-6 = 0$ has roots $1,-1,-2/3,-3/2$; the equation $6x^4-13x^3+13x-6 = 0$ has roots $-1,1,2/3,3/2$.

See Problem 1.

VARIATION OF SIGN. If, when a polynomial is arranged in descending powers of the variable, two successive terms differ in sign, the polynomial is said to have a *variation of sign*.

Example 2. (*a*) The polynomial $x^3-3x^2-4x+12$ has two variations of sign, one from $+x^3$ to $-3x^2$ and one from $-4x$ to $+12$; the polynomial $x^3+3x^2-4x-12$ has one variation of sign.

(*b*) The polynomial $6x^4+13x^3-13x-6$ has one variation of sign; the polynomial $6x^4-13x^3+13x-6$ has three. Note that here the term with zero coefficient has not been considered.

DESCARTES' RULE OF SIGNS. The number of positive roots of a polynomial equation $f(x)=0$, with real coefficients, is equal either to the number of variations of sign in $f(x)$ or to that number diminished by an even number.

The number of negative roots of $f(x)=0$ is equal to the number of positive roots of $f(-x)=0$.

Example 3. Since $f(x) = x^3-3x^2-4x+12$ of Example 2(*a*) has two variations of sign, $f(x)=0$ has either two or no positive roots.

Since $x^3+3x^2-4x-12$ has one variation of sign, $f(-x) = 0$ has one positive root and $f(x)=0$ has one negative root.

See Problem 2.

DIMINISHING THE ROOTS OF AN EQUATION. Let

(*1*) $$f(x) = a_0x^n + a_1x^{n-1} + \dots + a_{n-1}x + a_n = 0$$

be a polynomial equation of degree n. Let

$$
(2)\quad
\begin{aligned}
f(x) &= (x-h)\cdot q_1(x) + R_1, \\
q_1(x) &= (x-h)\cdot q_2(x) + R_2, \\
&\cdots\cdots\cdots\cdots \\
&\cdots\cdots\cdots\cdots \\
q_{n-2}(x) &= (x-h)\cdot q_{n-1}(x) + R_{n-1}, \\
q_{n-1}(x) &= (x-h)\cdot q_n(x) + R_n,
\end{aligned}
$$

where each R is a constant and $q_n(x) = a_0$. Then the roots of

$$(3)\qquad g(y) = a_0 y^n + R_n y^{n-1} + R_{n-1} y^{n-2} + \ldots + R_2 y + R_1 = 0$$

are the roots of $f(x) = 0$ diminished by h.

We shall show that if r is any root of $f(x) = 0$ then $r - h$ is a root of $g(y) = 0$. Since $f(r) = 0$,

$$
\begin{aligned}
R_1 &= -(r-h)\cdot q_1(r), \\
R_2 &= q_1(r) - (r-h)\cdot q_2(r), \\
&\cdots\cdots\cdots\cdots \\
&\cdots\cdots\cdots\cdots \\
R_{n-1} &= q_{n-2}(r) - (r-h)\cdot q_{n-1}(r), \\
R_n &= q_{n-1}(r) - (r-h)a_0.
\end{aligned}
$$

When these replacements are made in (3), we have

$$
\begin{aligned}
& a_0 y^n + [q_{n-1}(r) - (r-h)a_0]y^{n-1} + [q_{n-2}(r) - (r-h)\cdot q_{n-1}(r)]y^{n-2} + \ldots \\
& \qquad \ldots + [q_1(r) - (r-h)\cdot q_2(r)]y - (r-h)\cdot q_1(r) \\
& = a_0[y - (r-h)]y^{n-1} + q_{n-1}(r)[y - (r-h)]y^{n-2} + \ldots \\
& \qquad \ldots + q_1(r)[y - (r-h)] = 0.
\end{aligned}
$$

It is clear that $r - h$ is a root of (3) as was to be proved.

Example 4. Find the equation each of whose roots is 4 less than the roots of $x^3 + 3x^2 - 4x - 12 = 0$.

```
1 +  3 -  4 - 12 |4          1 + 15 + 68 + 84 |-2
     4 + 28 + 96                 -  2 - 26 - 84
----------------             -------------------
1 +  7 + 24 + (84)           1 + 13 + 42      |-6
     4 + 44                      -  6 - 42
-----------                  -------------
1 + 11 + (68)                1 +  7           |-7
     4                           -  7
------                       --------
1 + (15)                     1
```

On the left the successive remainders have been found and circled. The resulting equation is $y^3 + 15y^2 + 68y + 84 = 0$. The given equation has roots $x = 2, -2, -3$; on the right, it is shown that $2 - 4 = -2$, $-2 - 4 = -6$, $-3 - 4 = -7$ are roots of the newly formed equation.

See Problem 3.

APPROXIMATION OF IRRATIONAL ROOTS

LET $f(x) = 0$ **BE A POLYNOMIAL EQUATION** having no rational roots. If the given equation had rational roots, we suppose that they have been found using synthetic division and that $f(x)$ is then the last third line in this process. (See Chapter 18.)

THE METHOD OF SUCCESSIVE LINEAR APPROXIMATIONS will be explained by means of examples.

Example 5. The equation $f(x) = x^3 + x - 4 = 0$ has no rational roots. By Descartes' rule of signs, it has one positive (real) root and two imaginary roots. To approximate the real root we shall first isolate it as lying between two consecutive integers. Since $f(1) = -2$ and $f(2) = 6$, the root lies between $x = 1$ and $x = 2$. Fig. 1(*a*) exhibits the portion of the graph of $f(x)$ between $(1,-2)$ and $(2,6)$.

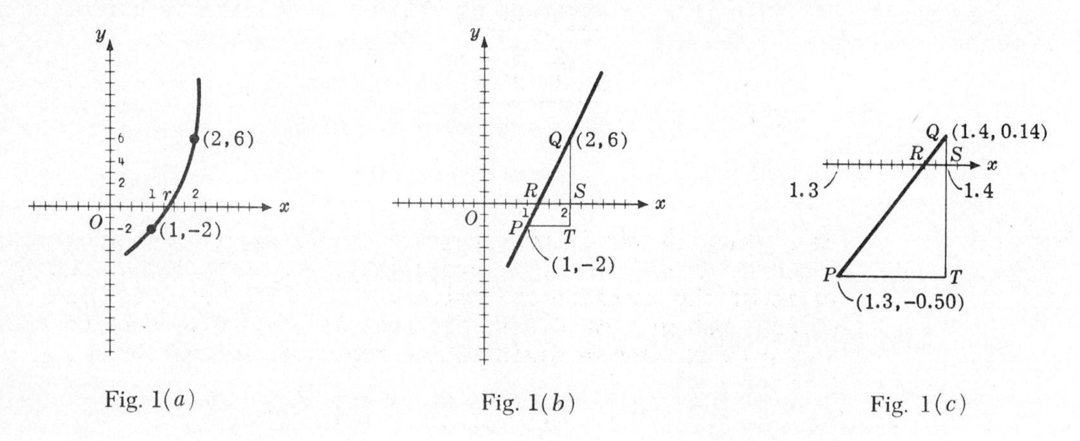

Fig. 1(*a*) Fig. 1(*b*) Fig. 1(*c*)

In Fig. 1(*b*), the curve joining the two points has been replaced by a straight line which meets the x-axis at R. We shall take OR, measured to the nearest tenth of a unit, as the first approximation of the required root and use it to isolate the root between successive tenths. From the similar triangles RSQ and PTQ,

$$\frac{RS}{PT} = \frac{SQ}{TQ} \qquad \text{or} \qquad RS = \frac{SQ}{TQ}(PT) = \frac{6}{8}(1) = \frac{3}{4} = 0.7$$

The first approximation of the root is given by $OR = OS - RS = 2 - 0.7 = 1.3$. Since $f(1.3) = -0.50$ and $f(1.4) = 0.14$, the required root lies between $x = 1.3$ and $x = 1.4$.

We now repeat the above process using the points $(1.3, -0.50)$ and $(1.4, 0.14)$ and isolate the root between successive hundredths. From Fig. 1(*c*),

$$RS = \frac{SQ}{TQ}(PT) = \frac{0.14}{0.64}(0.1) = 0.02 \quad \text{and} \quad OR = OS - RS = 1.4 - 0.02 = 1.38$$

is the next approximation. Since $f(1.38) = 0.008$ (hence, too large) and $f(1.37) = -0.059$, the root lies between $x = 1.37$ and $x = 1.38$.

Using the points $(1.37, -0.059)$ and $(1.38, 0.008)$, we isolate the root between successive thousandths. We find (no diagram needed)

$$RS = \frac{0.008}{0.067}(0.01) = 0.001 \quad \text{and} \quad OR = 1.38 - 0.001 = 1.379$$

Since $f(1.379) = 0.0012$ and $f(1.378) = -0.0054$, the root lies between $x = 1.378$ and $x = 1.379$.

For the next approximation

$$RS = \frac{0.0012}{0.0066}(0.001) = 0.0001 \quad \text{and} \quad OR = 1.379 - 0.0001 = 1.3789$$

The root correct to three decimal places is 1.379.

See Problem 4.

HORNER'S METHOD OF APPROXIMATION. This method will be explained by means of examples.

Example 6. The equation $x^3 + x^2 + x - 4 = 0$ has no rational roots. By Descartes' rule of signs, it has one positive root. Since $f(1) = -1$ and $f(2) = 10$, this root is between $x = 1$ and $x = 2$. We first diminish the roots of the given equation by 1

1	1	1	-4	1
	1	2	3	
1	2	3	-1	
	1	3		
1	3	6		
	1			
1	4			

and obtain the equation $g(y) = y^3 + 4y^2 + 6y - 1 = 0$ having a root between $y = 0$ and $y = 1$. To approximate it, we disregard the first two terms of the equation and solve $6y - 1 = 0$ for $y = 0.1$. Since $g(0.1) = -0.359$ and $g(0.2) = 0.368$, the root of $g(y) = 0$ lies between $y = 0.1$ and $y = 0.2$, and we diminish the roots of $g(y) = 0$ by 0.1.

1	4	6	-1	0.1
	0.1	0.41	0.641	
1	4.1	6.41	-0.359	
	0.1	0.42		
1	4.2	6.83		
	0.1			
1	4.3			

We obtain the equation $h(z) = z^3 + 4.3z^2 + 6.83z - 0.359 = 0$ having a root between 0 and 0.01. Disregarding the first two terms of this equation and solving $6.83z - 0.359 = 0$, we obtain $z = 0.05$ as an approximation of the root. Since $h(0.05) = -0.007$ and $h(0.06) = 0.07$, the root of $h(z) = 0$ lies between $z = 0.05$ and $z = 0.06$ and we diminish the roots by 0.05

1	4.3	6.83	-0.359	0.05
	0.05	0.2175	0.352375	
1	4.35	7.0475	-0.006625	
	0.05	0.2200		
1	4.40	7.2675		
	0.05			
1	4.45			

and obtain the equation $k(w) = w^3 + 4.45w^2 + 7.2675w - 0.006625 = 0$ having a root between $w = 0$ and $w = 0.001$. An approximation of this root, obtained by solving $7.2675w - 0.006625 = 0$ is $w = 0.0009$.

Without further computation, we are safe in stating the root of the given equation to be

$$x = 1 + 0.1 + 0.05 + 0.0009 = 1.1509$$

The complete solution may be exhibited more compactly, as follows:

1	1	1	−4	1	
	1	2	3		
1	2	3	−1		$y = \frac{1}{6} = 0.1$
	1	3			
1	3	6			
	1				
1	4	6	−1	0.1	
	0.1	0.41	0.641		
1	4.1	6.41	−0.359		$z = \frac{0.359}{6.83} = 0.05$
	0.1	0.42			
1	4.2	6.83			
	0.1				
1	4.3	6.83	−0.359	0.05	
	0.05	0.2175	0.352375		
1	4.35	7.0475	−0.006625		$w = \frac{0.006625}{7.2675} = 0.0009$
	0.05	0.2200			
1	4.40	7.2675			
	0.05				
1	4.45	7.2675	−0.006625		

See Problems 5-6.

SOLVED PROBLEMS

1. For each of the equations $f(x) = 0$, write the equation whose roots are negatives of those of $f(x) = 0$.

   (a) $x^3 - 8x^2 + x - 1 = 0$ *Ans.* $x^3 + 8x^2 + x + 1 = 0$

   (b) $x^4 + 3x^2 + 2x + 1 = 0$ *Ans.* $x^4 + 3x^2 - 2x + 1 = 0$

   (c) $2x^4 - 5x^2 + 8x - 3 = 0$ *Ans.* $2x^4 - 5x^2 - 8x - 3 = 0$

   (d) $x^5 + x + 2 = 0$ *Ans.* $x^5 + x - 2 = 0$

2. Give all the information obtainable from Descartes' rule of signs about the roots of the following equations.

   (a) $f(x) = x^3 - 8x^2 + x - 1 = 0$, (Problem 1(*a*)).

   Since there are three variations of sign in $f(x) = 0$ and no variation of sign in $f(-x) = 0$, the given equation has either three positive roots or one positive root and two imaginary roots.

   (b) $f(x) = 2x^4 - 5x^2 + 8x - 3 = 0$, (Problem 1(*c*)).

   Since there are three variations of sign in $f(x) = 0$ and one variation of sign in $f(-x) = 0$, the given equation has either three positive and one negative root or one positive, one negative, and two imaginary roots.

(c) $f(x) = x^5 + x + 2 = 0$, (Problem 1(d)).

Since there is no variation of sign in $f(x) = 0$ and one in $f(-x) = 0$, the given equation has one negative and four imaginary roots.

3. Form the equation whose roots are equal to the roots of the given equation diminished by the indicated number.

(a) $x^3 - 4x^2 + 8x - 5 = 0$; 2.

$$\begin{array}{l} 1 - 4 + 8 - 5 \;\lfloor 2 \\ \quad\; 2 - 4 + 8 \\ \hline 1 - 2 + 4 + 3 \\ \quad\; 2 + 0 \\ \hline 1 + 0 + 4 \\ \quad\; 2 \\ \hline 1 + 2 \end{array}$$

The required equation is

$y^3 + 2y^2 + 4y + 3 = 0$.

(b) $2x^3 + 9x^2 - 5x - 8 = 0$; -3.

$$\begin{array}{l} 2 + 9 - 5 - 8 \;\lfloor -3 \\ \quad - 6 - 9 + 42 \\ \hline 2 + 3 - 14 + 34 \\ \quad - 6 + 9 \\ \hline 2 - 3 - 5 \\ \quad - 6 \\ \hline 2 - 9 \end{array}$$

The required equation is

$2y^3 - 9y^2 - 5y + 34 = 0$.

(c) $x^4 - 8x^3 + 5x^2 + x + 8 = 0$; 2.

$$\begin{array}{l} 1 - 8 + 5 + 1 + 8 \;\lfloor 2 \\ \quad\; 2 - 12 - 14 - 26 \\ \hline 1 - 6 - 7 - 13 - 18 \\ \quad\; 2 - 8 - 30 \\ \hline 1 - 4 - 15 - 43 \\ \quad\; 2 - 4 \\ \hline 1 - 2 - 19 \\ \quad\; 2 \\ \hline 1 + 0 \end{array}$$

The required equation is

$y^4 - 19y^2 - 43y - 18 = 0$.

4. Use the method of successive linear approximation to approximate the irrational roots of

$$f(x) = x^3 + 3x^2 - 2x - 5 = 0.$$

By Descartes' rule of signs the equation has either one positive and two negative roots or one positive and two imaginary roots. By the location principle,

x	2	1	0	-1	-2	-3	-4
$f(x)$	11	-3	-5	-1	3	1	-13

there are roots between $x = 1$ and $x = 2$, $x = -1$ and $x = -2$, and $x = -3$ and $x = -4$.

(a) To approximate the positive root, use adjacent figure. Then

$$RS = \frac{SQ}{TQ}(PT) = \frac{11}{14}(1) = 0.7 \quad \text{and} \quad OR = 2 - 0.7 = 1.3$$

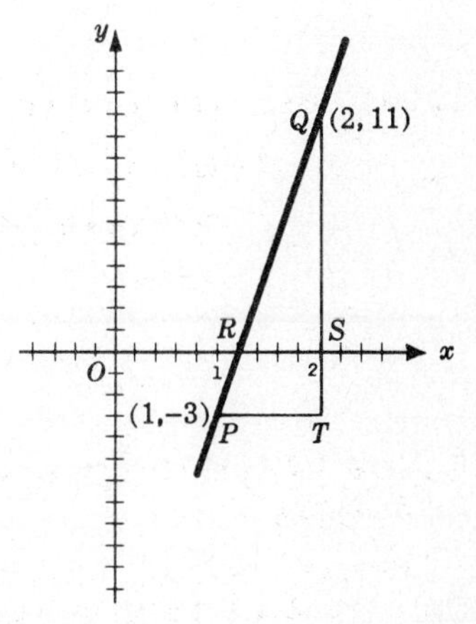

Since $f(1.3) = -0.33$ and $f(1.4) = 0.82$, the root lies between $x = 1.3$ and $x = 1.4$.

For the next approximation,

$$RS = \frac{0.82}{1.15}(0.1) = 0.07 \quad \text{and} \quad OR = 1.4 - 0.07 = 1.33$$

Since $f(1.33) = -0.0006$ and $f(1.34) = 0.113$, the root lies between $x = 1.33$ and $x = 1.34$.

For the next approximation,

$$RS = \frac{0.113}{0.1136}(0.01) = 0.009 \quad \text{and} \quad OR = 1.34 - 0.009 = 1.331$$

Now $f(1.331) > 0$ so that this approximation is too large; in fact, $f(1.3301) > 0$. Thus, the root to three decimal places is $x = 1.330$.

In approximating a negative root of $f(x) = 0$, it is more convenient to approximate the equally positive root of $f(-x) = 0$.

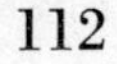

(b) To approximate the root of $f(x) = 0$ between $x = -1$ and $x = -2$, we shall approximate the positive root between $x = 1$ and $x = 2$ of $g(x) = x^3 - 3x^2 - 2x + 5 = 0$. Since $g(1) = 1$ and $g(2) = -3$, we obtain from the adjacent figure

$$SR = \frac{SP}{TP}(TQ) = \frac{1}{4}(1) = 0.2 \qquad \text{and} \qquad OR = OS + SR = 1.2$$

Since $g(1.2) = 0.01$ and $g(1.3) = -0.47$, the root is between $x = 1.2$ and $x = 1.3$.

For the next approximation,

$$SR = \frac{0.01}{0.48}(0.1) = 0.002 \qquad \text{and} \qquad OR = 1.2 + 0.002 = 1.202$$

Since $g(1.202) = -0.0018$ (hence, too large) and $g(1.201) = 0.0031$, the root is between $x = 1.201$ and $x = 1.202$.

For the next approximation,

$$SR = \frac{0.0031}{0.0049}(0.001) = 0.0006 \qquad \text{and} \qquad OR = 1.201 + 0.0006 = 1.2016$$

Since $g(1.2016) = -0.00281$ and $g(1.2015) = 0.00007$, the root of $g(x) = 0$ to three decimal places is $x = 1.202$. The corresponding root of the given equation is $x = -1.202$.

(c) The approximation of the root -3.128 between $x = -3$ and $x = -4$ is left as an exercise.

5. Use Horner's method to approximate the irrational roots of $x^3 + 2x^2 - 4 = 0$.

By Descartes' rule of signs the equation has either one positive and two negative roots or one positive and two imaginary roots. By the location principle there is one root, between $x = 1$ and $x = 2$.

Arranged in compact form, the computation is as follows:

1 + 2	+ 0	− 4	⌊1	
1	+ 3	+ 3		
1 + 3	+ 3	− 1		
1	+ 4			$y = \frac{1}{7} = 0.1$
1 + 4	+ 7			
1				
1 + 5	+ 7	− 1	⌊0.1	
0.1	+ 0.51	+ 0.751		
1 + 5.1	+ 7.51	− 0.249		$z = \frac{0.249}{8.03} = 0.03$
0.1	+ 0.52			
1 + 5.2	+ 8.03			
0.1				
1 + 5.3	+ 8.03	− 0.249	⌊0.03	
0.03	+ 0.1599	+ 0.245697		
1 + 5.33	+ 8.1899	− 0.003303		$w = \frac{0.003303}{8.3507} = 0.00039$
0.03	+ 0.1608			
1 + 5.36	+ 8.3507			
0.03				
1 + 5.39	+ 8.3507	− 0.003303		

The root, correct to four decimal places is $1 + 0.1 + 0.03 + 0.00039 = 1.1304$.

6. Use Horner's method to approximate the irrational roots of $x^3 + 3x^2 - 2x - 5 = 0$.

By the location principle there are roots between $x = 1$ and $x = 2$, $x = -1$ and $x = -2$, $x = -3$ and $x = -4$.

(a) The computation for the root between $x = 1$ and $x = 2$ is as follows:

$$
\begin{array}{llll|l}
1 & +3 & -2 & -5 & \underline{1} \\
 & 1 & +4 & +2 & \\ \hline
1 & +4 & +2 & -3 & \\
 & 1 & +5 & & \\ \hline
1 & +5 & +7 & & \\
 & 1 & & & \\ \hline
1 & +6 & +7 & -3 & \underline{0.3} \\
 & 0.3 & +1.89 & +2.667 & \\ \hline
1 & +6.3 & +8.89 & -0.333 & \\
 & 0.3 & +1.98 & & \\ \hline
1 & +6.6 & +10.87 & & \\
 & 0.3 & & & \\ \hline
1 & +6.9 & +10.87 & -0.333 & \underline{0.03} \\
 & 0.03 & +0.2079 & +0.332337 & \\ \hline
1 & +6.93 & +11.0779 & -0.000663 & \\
 & 0.03 & +0.2088 & & \\ \hline
1 & +6.96 & +11.2867 & & \\
 & 0.03 & & & \\ \hline
1 & +6.99 & +11.2867 & -0.000663 &
\end{array}
$$

$y = \dfrac{3}{7} = 0.4$ but is too large since, when used, the last number in the third line is positive.

$z = \dfrac{0.333}{10.87} = 0.03$

$w = \dfrac{0.000663}{11.2867} = 0.000058$

The root is 1.330.

When approximating a negative root of $f(x) = 0$ using Horner's method, it is more convenient to approximate the equally positive root of $f(-x) = 0$.

(b) To approximate the root between $x = -1$ and $x = -2$ of the given equation, we approximate the root between $x = 1$ and $x = 2$ of the equation $x^3 - 3x^2 - 2x + 5 = 0$. The computation is as follows:

$$
\begin{array}{llll|l}
1 & -3 & -2 & +5 & \underline{1} \\
 & 1 & -2 & -4 & \\ \hline
1 & -2 & -4 & +1 & \\
 & 1 & -1 & & \\ \hline
1 & -1 & -5 & & \\
 & 1 & & & \\ \hline
1 & +0 & -5 & +1 & \underline{0.2} \\
 & 0.2 & +0.04 & -0.992 & \\ \hline
1 & +0.2 & -4.96 & +0.008 & \\
 & 0.2 & +0.08 & & \\ \hline
1 & +0.4 & -4.88 & & \\
 & 0.2 & & & \\ \hline
1 & +0.6 & -4.88 & +0.008 & \underline{0.001} \\
 & 0.001 & +0.000601 & -0.004879399 & \\ \hline
1 & +0.601 & -4.879399 & +0.003120601 & \\
 & 0.001 & +0.000602 & & \\ \hline
1 & +0.602 & -4.878797 & & \\
 & 0.001 & & & \\ \hline
1 & +0.603 & -4.878797 & +0.003120601 &
\end{array}
$$

$y = \dfrac{1}{5} = 0.2$

$z = \dfrac{0.008}{4.88} = 0.001$

$w = \dfrac{0.003120601}{4.878797} = 0.00063$

To four decimal places the root is $x = 1.2016$; thus, the root of the given equation is $x = -1.2016$.

(c) The approximation of the root is $x = -3.1284$ between $x = -3$ and $x = -4$ is left as an exercise.

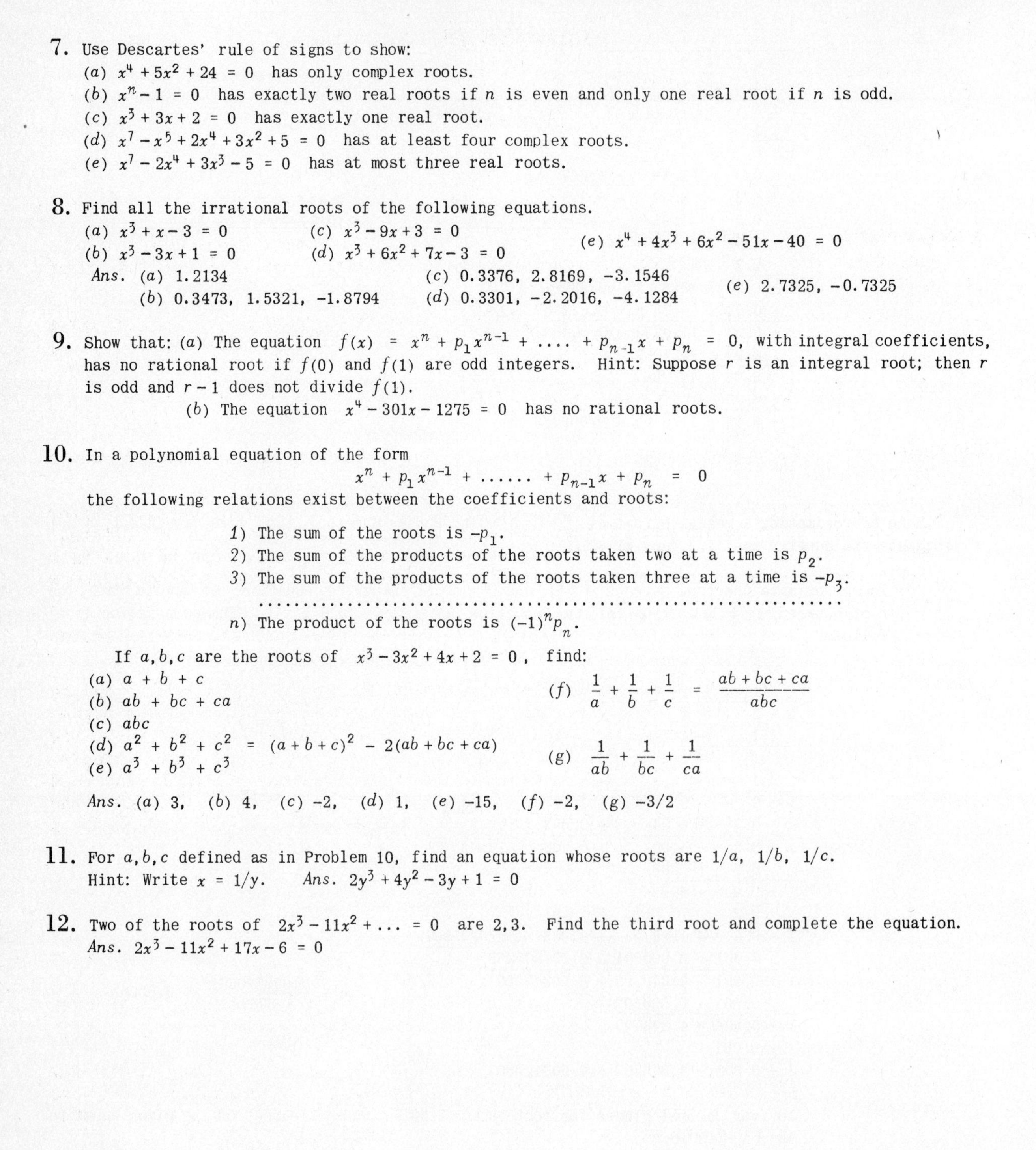

SUPPLEMENTARY PROBLEMS

7. Use Descartes' rule of signs to show:

(a) $x^4 + 5x^2 + 24 = 0$ has only complex roots.
(b) $x^n - 1 = 0$ has exactly two real roots if n is even and only one real root if n is odd.
(c) $x^3 + 3x + 2 = 0$ has exactly one real root.
(d) $x^7 - x^5 + 2x^4 + 3x^2 + 5 = 0$ has at least four complex roots.
(e) $x^7 - 2x^4 + 3x^3 - 5 = 0$ has at most three real roots.

8. Find all the irrational roots of the following equations.

(a) $x^3 + x - 3 = 0$
(b) $x^3 - 3x + 1 = 0$
(c) $x^3 - 9x + 3 = 0$
(d) $x^3 + 6x^2 + 7x - 3 = 0$
(e) $x^4 + 4x^3 + 6x^2 - 51x - 40 = 0$

Ans. (a) 1.2134
(b) 0.3473, 1.5321, −1.8794
(c) 0.3376, 2.8169, −3.1546
(d) 0.3301, −2.2016, −4.1284
(e) 2.7325, −0.7325

9. Show that: (a) The equation $f(x) = x^n + p_1 x^{n-1} + \dots + p_{n-1}x + p_n = 0$, with integral coefficients, has no rational root if $f(0)$ and $f(1)$ are odd integers. Hint: Suppose r is an integral root; then r is odd and $r-1$ does not divide $f(1)$.

(b) The equation $x^4 - 301x - 1275 = 0$ has no rational roots.

10. In a polynomial equation of the form

$$x^n + p_1 x^{n-1} + \dots\dots + p_{n-1}x + p_n = 0$$

the following relations exist between the coefficients and roots:

1) The sum of the roots is $-p_1$.
2) The sum of the products of the roots taken two at a time is p_2.
3) The sum of the products of the roots taken three at a time is $-p_3$.

....

n) The product of the roots is $(-1)^n p_n$.

If a, b, c are the roots of $x^3 - 3x^2 + 4x + 2 = 0$, find:

(a) $a + b + c$
(b) $ab + bc + ca$
(c) abc
(d) $a^2 + b^2 + c^2 = (a+b+c)^2 - 2(ab+bc+ca)$
(e) $a^3 + b^3 + c^3$
(f) $\dfrac{1}{a} + \dfrac{1}{b} + \dfrac{1}{c} = \dfrac{ab+bc+ca}{abc}$
(g) $\dfrac{1}{ab} + \dfrac{1}{bc} + \dfrac{1}{ca}$

Ans. (a) 3, (b) 4, (c) −2, (d) 1, (e) −15, (f) −2, (g) −3/2

11. For a, b, c defined as in Problem 10, find an equation whose roots are $1/a$, $1/b$, $1/c$.
Hint: Write $x = 1/y$. *Ans.* $2y^3 + 4y^2 - 3y + 1 = 0$

12. Two of the roots of $2x^3 - 11x^2 + \dots = 0$ are 2, 3. Find the third root and complete the equation.
Ans. $2x^3 - 11x^2 + 17x - 6 = 0$

CHAPTER 20

Permutations

ANY ARRANGEMENT OF A SET OF OBJECTS in a definite order is called a *permutation* of the set taken all at a time. For example: *abcd*, *acbd*, *bdca* are permutations of the set of letters a, b, c, d taken all at a time.

If a set contains n objects, any ordered arrangement of any $r \leq n$ of the objects is called a permutation of the n objects taken r at a time. For example: *ab*, *ba*, *ca*, *db* are permutations of the $n = 4$ letters a, b, c, d taken $r = 2$ at a time, while *abc*, *adb*, *bad*, *cad* are permutations of the $n = 4$ letters taken $r = 3$ at a time.

THE NUMBER OF PERMUTATIONS which may be formed in each situation can be found by means of the

FUNDAMENTAL PRINCIPLE: If one thing can be done in u different ways, if after it has been done in any one of these a second thing can be done in v different ways, if after it has been done in any one of these a third thing can be done in w different ways, ..., the several things can be done in the order stated in $u \cdot v \cdot w$ .... different ways.

Example 1. In how many ways can 6 students be assigned (a) to a row of 6 seats, (b) to a row of 8 seats?

(a) Let the seats be denoted xxxxxx. The seat on the left may be assigned to any one of the 6 students; that is, may be assigned in 6 different ways. After the assignment has been made, the next seat may be assigned to any one of the 5 remaining students. After the assignment has been made, the next seat may be assigned to any one of the 4 remaining students, and so on. Placing the number of ways in which each seat may be assigned under the x marking the seat, we have

x x x x x x
6 5 4 3 2 1 .

By the fundamental principle, the seats may be assigned in

$$6 \cdot 5 \cdot 4 \cdot 3 \cdot 2 \cdot 1 = 720 \text{ ways.}$$

The reader should assure himself that the seats might have been assigned to the students with the same result.

(b) Here each student must be assigned a seat. The first student may be assigned any one of the 8 seats, the second student any one of the 7 remaining seats, and so on. Letting x represent a student, we have

x x x x x x
8 7 6 5 4 3

and the assignment may be made in $8 \cdot 7 \cdot 6 \cdot 5 \cdot 4 \cdot 3 = 20,160$ ways.

Note. If we attempt to assign students to seats, we must first select the six seats to be used. The problem of selections will be considered under Combinations.

See Problems 1-5.

THE SYMBOL $n!$ (read, factorial n) denotes the product of the positive integers from 1 to n inclusive; for example,

$$1! = 1, \quad 3! = 1\cdot 2\cdot 3 = 6, \quad 5! = 1\cdot 2\cdot 3\cdot 4\cdot 5 = 120.$$

See Problems 6-9.

PERMUTATIONS OF OBJECTS NOT ALL DIFFERENT. If there are n objects of which k are alike while the remaining $(n-k)$ objects are different from them and from each other, it is clear that the number of different permutations of the n objects taken all together is not ${}_nP_n$.

Example 2. How many different permutations of four letters can be formed using the letters of the word bass ?

For the moment, think of the given letters as b, a, s_1, s_2 so that they are all different. Then

$$\begin{array}{ccccc} bas_1s_2 & as_1bs_2 & s_2s_1ba & s_1as_2b & bs_1s_2a \\ bas_2s_1 & as_2bs_1 & s_1s_2ba & s_2as_1b & bs_2s_1a \end{array}$$

are ten of the twenty-four permutations of the four letters taken all together. However, when the subscripts are removed, it is seen that the two permutations in each column are alike.

Thus, there are $\dfrac{1\cdot 2\cdot 3\cdot 4}{1\cdot 2} = 12$ different permutations.

In general, given n objects of which k_1 are of one sort, k_2 of another, k_3 of another, ... ; then the number of different permutations that can be made from the n objects taken all together is

$$\frac{n!}{k_1!\ k_2!\ k_3!\ \ldots.}$$

See Problem 10.

SOLVED PROBLEMS

1. Using the letters of the word MARKING and calling any arrangement a word, (*a*) how many different 7 letter words can be formed, (*b*) how many different 3 letter words can be formed ?

(*a*) We must fill each of the positions xxxxxxx with a different letter. The first position may be filled in 7 ways, the second in 6 ways, and so on.

Thus, we have $\begin{matrix} x & x & x & x & x & x & x \\ 7 & 6 & 5 & 4 & 3 & 2 & 1 \end{matrix}$ and there are $7\cdot 6\cdot 5\cdot 4\cdot 3\cdot 2\cdot 1 = 5040$ words.

(*b*) We must fill each of the positions xxx with a different letter. The first position can be filled in 7 ways, the second in 6 ways, and the third in 5 ways. Thus, there are $7\cdot 6\cdot 5 = 210$ words.

2. In forming 5 letter words using the letters of the word EQUATIONS, (*a*) how many consist only of vowels, (*b*) how many contain all of the consonants, (*c*) how many begin with E and end in S, (*d*) how many begin with a consonant, (*e*) how many contain N, (*f*) how many in which the vowels and consonants alternate, (*g*) how many in which Q is immediately followed by U?

There are 9 letters, consisting of 5 vowels and 4 consonants.

(*a*) There are five places to be filled and 5 vowels at our disposal.
Hence, we can form $5\cdot4\cdot3\cdot2\cdot1 = 120$ words.

(*b*) Each word is to contain the 4 consonants and one of the 5 vowels. There are now six things to do: first pick the vowel to be used (in 5 ways), next place the vowel (in 5 ways), and fill the remaining four positions with consonants.

We have $\begin{matrix} x & x & x & x & x & x \\ 5 & 5 & 4 & 3 & 2 & 1 \end{matrix}$; hence there are $5\cdot5\cdot4\cdot3\cdot2\cdot1 = 600$ words.

(*c*) Indicate the fact that the position of certain letters is fixed by writing $\begin{matrix} E & & & & S \\ x & x & x & x & x \end{matrix}$.
Now there are just three positions to be filled and 7 letters at our disposal. Thus, there are $7\cdot6\cdot5 = 210$ words.

(*d*) Here we have $\begin{matrix} c & & & & \\ x & x & x & x & x \\ 4 & 8 & 7 & 6 & 5 \end{matrix}$ since, after filling the first position with any one of the 4 consonants, there are 8 letters remaining. Hence, there are $4\cdot8\cdot7\cdot6\cdot5 = 6720$ words.

(*e*) There are five things to do: first, place the letter N in any one of the five positions and then fill the other four positions from among the 8 letters remaining.

We have $\begin{matrix} x & x & x & x & x \\ 5 & 8 & 7 & 6 & 5 \end{matrix}$; hence, there are $5\cdot8\cdot7\cdot6\cdot5 = 8400$ words.

(*f*) We may have $\begin{matrix} v & c & v & c & v \\ x & x & x & x & x \\ 5 & 4 & 4 & 3 & 3 \end{matrix}$ or $\begin{matrix} c & v & c & v & c \\ x & x & x & x & x \\ 4 & 5 & 3 & 4 & 2 \end{matrix}$.

Hence, there are $5\cdot4\cdot4\cdot3\cdot3 + 4\cdot5\cdot3\cdot4\cdot2 = 1200$ words.

(*g*) First we place Q so that U may follow it (Q may occupy any one of the first four positions but not the last), next we place U (in only 1 way), then we fill the three other positions from among the 7 letters remaining.

Thus, we have $\begin{matrix} x & x & x & x & x \\ 4 & 1 & 7 & 6 & 5 \end{matrix}$ and there are $4\cdot1\cdot7\cdot6\cdot5 = 840$ words.

3. If repetitions are not allowed, (*a*) how many three digit numbers can be formed with the digits 0, 1, 2, 3, 4, 5, 6, 7, 8, 9? (*b*) How many of these are odd numbers? (*c*) How many are even numbers? (*d*) How many are divisible by 5? (*e*) How many are greater than 600?

In each case we have $\begin{matrix} \neq 0 & & \\ x & x & x \end{matrix}$.

(*a*) The position on the left can be filled in 9 ways (0 cannot be used), the middle position can be filled in 9 ways (0 can be used), and the position on the right can be filled in 8 ways. Thus, there are $9\cdot9\cdot8 = 648$ numbers.

(*b*) We have $\begin{matrix} \neq 0 & & \text{odd} \\ x & x & x \end{matrix}$. Care must be exercised here in choosing the order in which to fill the positions. If the position on the left is filled first (in 9 ways), we cannot determine the number of ways in which the position on the right can be filled since, if the former is filled with an odd digit there are 4 ways of filling the latter but if the former is filled with an even digit there are 5 ways of filling the latter.

We fill first the position on the right (in 5 ways), then the position on the left (in 8 ways, since one odd digit and 0 are excluded), and the middle position (in 8 ways, since 2 digits are now excluded). Thus, there are $8\cdot8\cdot5 = 320$ numbers.

(*c*) We have $\begin{matrix} \neq 0 & & \text{even} \\ x & x & x \end{matrix}$. We note that the argument above excludes the possibility of first filling the position on the left. But if we fill the first position on the right (in 5 ways), we are unable to determine the number of ways the position on the left can be filled (9 ways if 0 was used on the right, 8 ways if 2,4,6, or 8 was used). Thus, we must separate two cases.

First, we form all numbers ending in 0; there are $9\cdot 8\cdot 1$ of them. Next, we form all numbers ending in 2,4,6, or 8; there are $8\cdot 8\cdot 4$ of them. Thus, in all, there are $9\cdot 8\cdot 1 + 8\cdot 8\cdot 4 = 328$ numbers.

As a check, we have 320 odd and 328 even numbers for a total of 648 as found in (*a*) above.

(*d*) A number is divisible by 5 if and only if it ends in 0 or 5.
There are $9\cdot 8\cdot 1$ numbers ending in 0 and $8\cdot 8\cdot 1$ numbers ending in 5.
Hence, in all, there are $9\cdot 8\cdot 1 + 8\cdot 8\cdot 1 = 136$ numbers divisible by 5.

(*e*) The position on the left can be filled in 4 ways (with 6,7,8, or 9) and the remaining positions in $9\cdot 8$ ways. Thus, there are $4\cdot 9\cdot 8 = 288$ numbers.

4. Solve Problem 3 *a*),*b*),*c*),*d*) if any digit may be used once, twice, or three times in forming the three digit number.

(*a*) The position on the left can be filled in 9 ways and each of the other positions can be filled in 10 ways. Thus, there are $9\cdot 10\cdot 10 = 900$ numbers.

(*b*) The position on the right can be filled in 5 ways, the middle position in 10 ways, and the position on the left in 9 ways. Thus, there are $9\cdot 10\cdot 5 = 450$ numbers.

(*c*) There are $9\cdot 10\cdot 5 = 450$ even numbers.

(*d*) There are $9\cdot 10\cdot 1 = 90$ numbers ending in 0 and the same number ending in 5. Thus, there are 180 numbers divisible by 5.

5. In how many ways can 10 boys be arranged (*a*) in a straight line, (*b*) in a circle?

(*a*) The boys may be arranged in a straight line in $10\cdot 9\cdot 8\cdot 7\cdot 6\cdot 5\cdot 4\cdot 3\cdot 2\cdot 1$ ways.

(*b*) We first place a boy at any point on the circle.
The other 9 boys may then be arranged in $9\cdot 8\cdot 7\cdot 6\cdot 5\cdot 4\cdot 3\cdot 2\cdot 1$ ways.
This is an example of a *circular permutation*. In general, n objects may be arranged in a circle in $(n-1)(n-2)\dots 2\cdot 1$ ways.

6. Evaluate.

(*a*) $\dfrac{8!}{3!} = \dfrac{1\cdot 2\cdot 3\cdot 4\cdot 5\cdot 6\cdot 7\cdot 8}{1\cdot 2\cdot 3} = 6720$

(*b*) $\dfrac{7!}{6!} = \dfrac{1\cdot 2\cdot 3\cdot 4\cdot 5\cdot 6\cdot 7}{1\cdot 2\cdot 3\cdot 4\cdot 5\cdot 6} = 7$

(*c*) $\dfrac{10!}{3!\,3!\,4!} = \dfrac{1\cdot 2\cdot 3\cdot 4\cdot 5\cdot 6\cdot 7\cdot 8\cdot 9\cdot 10}{1\cdot 2\cdot 3\cdot 1\cdot 2\cdot 3\cdot 1\cdot 2\cdot 3\cdot 4} = 4200$

(*d*) $\dfrac{(r+1)!}{(r-1)!} = \dfrac{1\cdot 2\dots (r-1)r(r+1)}{1\cdot 2\dots (r-1)} = r(r+1)$

7. The number of permutations ${}_nP_r$ of n different objects taken $r<n$ at a time is given by

$${}_nP_r = n(n-1)\dots(n-r+1) = \frac{n(n-1)\dots(n-r+1)\cdot(n-r)\dots 2\cdot 1}{(n-r)\dots 2\cdot 1} = \frac{n!}{(n-r)!}$$

and the number of permutations ${}_nP_n$ of n different objects taken n at a time is given by

$${}_nP_n = n(n-1)\dots 2\cdot 1 = n!$$

8. Solve for n, given: (*a*) ${}_nP_2 = 110$, (*b*) ${}_nP_4 = 30\,{}_nP_2$.

(*a*) ${}_nP_2 = n(n-1) = n^2 - n = 110$.

Then $n^2 - n - 110 = (n-11)(n+10) = 0$ and, since n is positive, $n = 11$.

(*b*) We have $n(n-1)(n-2)(n-3) = 30n(n-1)$ or $n(n-1)(n-2)(n-3) - 30n(n-1) = 0$.

Then $n(n-1)\{(n-2)(n-3) - 30\} = n(n-1)(n^2-5n-24) = n(n-1)(n-8)(n+3) = 0$.

Since $n \geq 4$, the required solution is $n = 8$.

9. Define $0! = 1$ and show that $n! = n\cdot(n-1)!$ for every positive integral value of n.

By definition, $1! = 1\cdot 0!$. For all other values of n,

$$n! = n\{(n-1)(n-2)\ldots.2\cdot 1\} = n\cdot(n-1)!$$

10. (*a*) How many permutations can be made of the letters, taken all together, of the "word" MASSESS? (*b*) In how many will the four S's be together? (*c*) How many will end in SS?

(*a*) There are seven letters of which four are S's. The number of permutations is $7!/4! = 210$.

(*b*) First, permute the non-S's in $1\cdot 2\cdot 3 = 6$ ways and then place the four S's at the ends or between any two letters in each of the six permutations. Thus, there will be $4\cdot 6 = 24$ permutations.

(*c*) After filling the last two places with S, we have to fill 5 places with 5 letters of which 2 are S's. Thus, there are $5!/2! = 60$ permutations.

SUPPLEMENTARY PROBLEMS

11. In how many different ways can 5 persons be seated on a bench? *Ans.* 120

12. In how many ways can the offices of chairman, vice-chairman, secretary, and treasurer be filled from a committee of seven? *Ans.* 840

13. How many 3 digit numbers can be formed with the digits 1,2,..,9, if no digit is repeated in any number? *Ans.* 504

14. How many 3 digit odd numbers can be formed with the digits 1,2,3,..,9, if no digit is repeated in any number? *Ans.* 280

15. How many 3 digit numbers > 300 can be formed with the digits 1,2,3,4,5,6, if no digit is repeated in any number? *Ans.* 80

16. How many 4 digit numbers > 3000 can be formed with the digits 2,3,4,5, if repetitions of digits (*a*) are not allowed, (*b*) are allowed? *Ans.* (*a*) 18, (*b*) 192

17. In how many ways can 3 girls and 3 boys be seated in a row if boys and girls alternate? *Ans.* 72

18. In how many ways can 2 letters be mailed if 5 letter boxes are available? *Ans.* 25

19. Seven letter words are formed using the letters of the word BLACKER. (*a*) How many can be formed? (*b*) How many which end in R? (*c*) How many in which E immediately follows K? (*d*) How many do not begin with B? (*e*) How many in which the vowels are separated by exactly two letters? (*f*) How many in which the vowels are separated by two or more letters?
Ans. (*a*) 5040, (*b*) 720, (*c*) 720, (*d*) 4320, (*e*) 960, (*f*) 2400

20. Eight books are to be arranged on a shelf. (*a*) In how many ways can this be done? (*b*) In how many ways if two of the books are to be placed together? (*c*) In how many ways if five of the books have red binding and three have blue binding, and the books of the same color are to be kept together? (*d*) In how many ways if four of the books belong to a numbered set and are to be kept together and in order? *Ans.* (*a*) 40,320, (*b*) 10,080, (*c*) 1440, (*d*) 120

21. How many six letter words can be formed using the letters of the word ASSIST (*a*) in which the S's alternate with other letters? (*b*) In which the three S's are together? (*c*) which begin and end with

S? (*d*) which neither begin nor end with S? *Ans.* (*a*) 12, (*b*) 24, (*c*) 24, (*d*) 24

22. (*a*) In how many ways can 8 persons be seated about a round table?
(*b*) With eight beads of different colors, how many bracelets can be formed by stringing them all together? *Ans.* (*a*) 5040, (*b*) 2520

23. How many signals can be made with 3 white, 3 green, and 2 blue flags by arranging them on a mast (*a*) all at a time? (*b*) three at a time? (*c*) five at a time? *Ans.* (*a*) 560, (*b*) 26, (*c*) 170

24. A car will hold 2 in the front seat and 1 in the rear seat. If among 6 persons only 2 can drive, in how many ways can the car be filled? *Ans.* 40

25. A chorus consists of six boys and six girls. How many arrangements of them can be made (*a*) in a row, facing front, the boys and girls alternating? (*b*) in two rows, facing front, with a boy behind each girl? (*c*) in a ring with the boys facing the center and the girls facing away from the center? (*d*) in two concentric rings, both facing the center, with a boy behind each girl?
Ans. (*a*) 1,036,800, (*b*) 518,400, (*c*) 39,916,800, (*d*) 86,400

26. (*a*) In how many ways can 10 boys take positions in a straight line if two particular boys must not stand side by side?
(*b*) In how many ways can 10 boys take positions about a round table if two particular boys must not be seated side by side?
Ans. (*a*) $8 \cdot 9!$, (*b*) $7 \cdot 8!$

27. A man has 5 large books, 7 medium-sized books, and 3 small books. In how many different ways can they be arranged on a shelf if all books of the same size are to be kept together? *Ans.* 21,772,800

28. (*a*) How many words can be made from the letters of the word MASSACHUSETTS taken all together? (*b*) Of the words in (*a*), how many begin and end with SS? (*c*) Of the words in (*a*), how many begin and end with S? (*d*) Show that there are as many words having H as middle letter as there are circular permutations, using all letters. *Ans.* (*a*) 64,864,800, (*b*) 90,720, (*c*) 4,989,600

CHAPTER 21

Combinations

THE COMBINATIONS of n objects taken r at a time consist of all possible sets of r of the objects, without regard to the order of arrangement. The number of combinations of n objects taken r at a time will be denoted by ${}_nC_r$.

For example, the combinations c the $n=4$ letters a, b, c, d taken $r=3$ at a time, are

$$abc, \quad abd, \quad acd, \quad bcd$$

Thus, ${}_4C_3 = 4$. When the letters of each combination are rearranged (in 3! ways), we obtain the ${}_4P_3$ permutations of the 4 letters taken 3 at a time. Hence, ${}_4P_3 = 3!\,({}_4C_3)$ and ${}_4C_3 = {}_4P_3/3!$.

The number of combinations of n different objects taken r at a time is equal to the number of permutations of the n objects taken r at a time divided by factorial r, or

$${}_nC_r = \frac{{}_nP_r}{r!} = \frac{n(n-1)\ldots\ldots(n-r+1)}{1\cdot 2\ldots\ldots r}$$

For a proof, see Problem 1.

Example. From a shelf containing twelve different toys, a child is permitted to select three. In how many ways can this be done?

$$\text{The required number is } {}_{12}C_3 = \frac{{}_{12}P_3}{3!} = \frac{12\cdot 11\cdot 10}{1\cdot 2\cdot 3} = 220.$$

SOLVED PROBLEMS

1. Derive the formula ${}_nC_r = \dfrac{{}_nP_r}{r!}$.

From each of the ${}_nC_r$ combinations of n objects taken r at a time, $r!$ permutations can be formed. Since two combinations differ at least in one element, the ${}_nC_r\cdot r!$ permutations thus formed are precisely the number of permutations ${}_nP_r$ of n objects taken r at a time. Thus

$${}_nC_r\cdot r! = {}_nP_r \quad \text{and} \quad {}_nC_r = \frac{{}_nP_r}{r!}$$

2. Show: (a) ${}_nC_r = \dfrac{n!}{r!\,(n-r)!}$, (b) ${}_nC_r = {}_nC_{n-r}$.

(a) $${}_nC_r = \frac{n(n-1)\ldots\ldots(n-r+1)}{1\cdot 2\ldots\ldots r} = \frac{n(n-1)\ldots\ldots(n-r+1)}{1\cdot 2\ldots\ldots r}\cdot\frac{(n-r)\ldots\ldots 2\cdot 1}{1\cdot 2\ldots\ldots(n-r)} = \frac{n!}{r!\,(n-r)!}$$

(*b*) $_nC_r = \dfrac{n!}{r!\,(n-r)!} = \dfrac{n(n-1)\ldots.(r+1)r(r-1)\ldots.2\cdot1}{1\cdot2\ldots.r\cdot(n-r)!} = \dfrac{n(n-1)\ldots.(r+1)}{(n-r)!} = \dfrac{_nP_{n-r}}{(n-r)!} = {_nC_{n-r}}$

3. Compute: (*a*) $_{10}C_2$, (*b*) $_{12}C_5$, (*c*) $_{15}C_{12}$, (*d*) $_{25}C_{21}$.

(*a*) $_{10}C_2 = \dfrac{10\cdot9}{1\cdot2} = 45$

(*b*) $_{12}C_5 = \dfrac{12\cdot11\cdot10\cdot9\cdot8}{1\cdot2\cdot3\cdot4\cdot5} = 792$

(*c*) $_{15}C_{12} = {_{15}C_3} = \dfrac{15\cdot14\cdot13}{1\cdot2\cdot3} = 455$

(*d*) $_{25}C_{21} = {_{25}C_4} = \dfrac{25\cdot24\cdot23\cdot22}{1\cdot2\cdot3\cdot4} = 12,650$

4. A lady gives a dinner party for six guests. (*a*) In how many ways may they be selected from among ten friends? (*b*) In how many ways if two of the friends will not attend the party together?

(*a*) The six guests may be selected in $_{10}C_6 = {_{10}C_4} = \dfrac{10\cdot9\cdot8\cdot7}{1\cdot2\cdot3\cdot4} = 210$ ways.

(*b*) Let A and B denote the two who will not attend together. If neither A nor B is included, the guests may be selected in $_8C_6 = {_8C_2} = \dfrac{8\cdot7}{1\cdot2} = 28$ ways. If one of A and B is included, the guests may be selected in $2\cdot{_8C_5} = 2\cdot{_8C_3} = 2\dfrac{8\cdot7\cdot6}{1\cdot2\cdot3} = 112$ ways. Thus, the six guests may be selected in $28+112 = 140$ ways.

5. A committee of five is to be selected from twelve seniors and eight juniors. In how many ways can this be done (*a*) if the committee is to consist of three seniors and two juniors, (*b*) if the committee is to contain at least three seniors and one junior?

(*a*) With each of the $_{12}C_3$ selections of three seniors, we may associate any one of the $_8C_2$ selections of two juniors. Thus, a committee can be selected in $_{12}C_3\cdot{_8C_2} = \dfrac{12\cdot11\cdot10}{1\cdot2\cdot3}\cdot\dfrac{8\cdot7}{1\cdot2} = 6160$ ways.

(*b*) The committee may consist of three seniors and two juniors or of four seniors and one junior. A committee of three seniors and two juniors can be selected in 6160 ways, and a committee of four seniors and one junior can be selected in $_{12}C_4\cdot{_8C_1} = 3960$ ways. In all, a committee may be selected in $6160+3960 = 10,120$ ways.

6. There are ten points $A, B, \ldots.$ in a plane, no three on the same straight line. (*a*) How many lines are determined by the points? (*b*) How many of the lines pass through A? (*c*) How many triangles are determined by the points? (*d*) How many of the triangles have A as a vertex? (*e*) How many of the triangles have AB as a side?

(*a*) Since any two points determine a line, there are $_{10}C_2 = 45$ lines.

(*b*) To determine a line through A, one other point must be selected. Thus there are 9 lines through A.

(*c*) Since any three of the points determine a triangle, there are $_{10}C_3 = 120$ triangles.

(*d*) Two additional points are needed to form a triangle. These points may be selected from the nine points in $_9C_2 = 36$ ways.

(*e*) One additional point is needed; there are 8 triangles having AB as a side.

7. In how many ways may 12 persons be divided into three groups (*a*) of 2, 4, and 6 persons, (*b*) of 4 persons each?

(*a*) The group of two can be selected in $_{12}C_2$ ways, then the group of four in $_{10}C_4$ ways, and the group of six in $_6C_6 = 1$ way. Thus, the division may be made in $_{12}C_2\cdot{_{10}C_4}\cdot1 = 13,860$ ways.

(*b*) One group of four can be selected in $_{12}C_4$ ways, then another in $_8C_4$ ways, and the third in 1 way.

Since the order in which the groups are formed is now immaterial, the division may be made in ${}_{12}C_4 \cdot {}_8C_4 \cdot 1 \div 3! = 5775$ ways.

8. The English alphabet consists of 21 consonants and 5 vowels.
(*a*) In how many ways can 4 consonants and two vowels be selected?
(*b*) How many words consisting of 4 consonants and 2 vowels can be formed?
(*c*) How many of the words in (*b*) begin with R?
(*d*) How many of the words in (*c*) contain E?

(*a*) The 4 consonants can be selected in ${}_{21}C_4$ ways and the 2 vowels can be selected in ${}_5C_2$ ways. Thus, the selections may be made in ${}_{21}C_4 \cdot {}_5C_2 = 59,850$ ways.

(*b*) From each of the selections in (*a*), 6! words may be formed by permuting the letters. Therefore, $59,850 \cdot 6! = 43,092,000$ words can be formed.

(*c*) Since the position of the consonant R is fixed, we must select 3 other consonants (in ${}_{20}C_3$ ways) and 2 vowels (in ${}_5C_2$ ways), and arrange each selection of 5 letters in all possible ways. Thus, there are ${}_{20}C_3 \cdot {}_5C_2 \cdot 5! = 1,368,000$ words.

(*d*) Since the position of the consonant R is fixed but the position of the vowel E is not, we must select 3 other consonants (in ${}_{20}C_3$ ways) and 1 other vowel (in 4 ways), and arrange each set of 5 letters in all possible ways. Thus, there are ${}_{20}C_3 \cdot 4 \cdot 5! = 547,200$ words.

9. From an ordinary deck of playing cards, in how many different ways can five cards be dealt (*a*) consisting of spades only, (*b*) consisting of black cards only, (*c*) containing the four aces, (*d*) consisting of three cards of one suit and two of another, (*e*) consisting of three kings and a pair, (*f*) consisting of three of one kind and two of another?

(*a*) From the 13 spades, 5 can be selected in ${}_{13}C_5 = 1287$ ways.

(*b*) From the 26 black cards, 5 can be selected in ${}_{26}C_5 = 65,780$ ways.

(*c*) One card must be selected from the 48 remaining cards. This can be done in 48 different ways.

(*d*) A suit can be selected in 4 ways and three cards from the suit can be selected in ${}_{13}C_3$ ways; a second suit can now be selected in 3 ways and two cards of this suit in ${}_{13}C_2$ ways. Thus 3 cards of one suit and 2 of another can be selected in $4 \cdot {}_{13}C_3 \cdot 3 \cdot {}_{13}C_2 = 267,696$ ways.

(*e*) Three kings can be selected from the 4 kings in ${}_4C_3$ ways, another kind can be selected in 12 ways, and two cards of this kind can be selected in ${}_4C_2$ ways. Thus, three kings and another pair can be dealt in ${}_4C_3 \cdot 12 \cdot {}_4C_2 = 288$ ways.

(*f*) A kind can be selected in 13 ways and three of this kind in ${}_4C_3$ ways; another kind can be selected in 12 ways and two of this kind can be selected in ${}_4C_2$ ways. Thus, 3 of one kind and 2 of another can be dealt in $13 \cdot {}_4C_3 \cdot 12 \cdot {}_4C_2 = 3744$ ways.

10. (*a*) Prove: The total number of combinations of n objects taken successively $1,2,3,\ldots,n$ at a time is $2^n - 1$. (*b*) In how many different ways can one invite one or more of 5 friends to the movies.

(*a*) The total number of combinations is

$$ {}_nC_1 + {}_nC_2 + {}_nC_3 + \ldots + {}_nC_n = 2^n - 1 $$

since, from Chapter 14, Problem 8(*a*),

$$ {}_nC_0 + {}_nC_1 + {}_nC_2 + \ldots + {}_nC_n = 2^n. $$

(*b*) The number is ${}_5C_1 + {}_5C_2 + {}_5C_3 + {}_5C_4 + {}_5C_5 = 2^5 - 1 = 31$.

SUPPLEMENTARY PROBLEMS

11. Evaluate: (*a*) ${}_6C_2$, (*b*) ${}_8C_6$, (*c*) ${}_nC_3$. *Ans.* (*a*) 15, (*b*) 28, (*c*) $\frac{n(n-1)(n-2)}{1\cdot 2\cdot 3}$

12. Find n if (*a*) ${}_nC_2 = 55$, (*b*) ${}_nC_3 = 84$, (*c*) ${}_{2n}C_3 = 11\cdot{}_nC_3$. *Ans.* (*a*) 11, (*b*) 9, (*c*) 6

13. Two dice can be tossed in 36 ways. In how many of these is the sum equal to (*a*) 4, (*b*) 7, (*c*) 11?
Ans. (*a*) 3, (*b*) 6, (*c*) 2

14. Four delegates are to be chosen from 8 members of a club. (*a*) How many choices are possible? (*b*) How many contain member A? (*c*) How many contain A or B but not both? *Ans.* (*a*) 70, (*b*) 35, (*c*) 40

15. A party of 8 boys and 8 girls are going on a picnic. Six of the party go in one automobile, four go in another, and the rest walk. (*a*) In how many ways can the party be distributed for the trip? (*b*) In how many ways if no girl walks? *Ans.* (*a*) 1,681,680, (*b*) 5880

16. Solve Problem 15 if the owner of each car (a boy) drives his own car. *Ans.* (*a*) 168,168, (*b*) 56

17. How many selections of five letters each can be made from the letters of the word CANADIANS?
Ans. 41

18. A bag contains 9 balls numbered 1,2,..9. In how many ways can 2 balls be drawn so that (*a*) both are odd? (*b*) their sum is odd? *Ans.* (*a*) 10, (*b*) 20

19. How many diagonals has (*a*) a hexagon, (*b*) an octagon, (*c*) an n-gon?
Ans. (*a*) 9, (*b*) 20, (*c*) $\frac{1}{2}n(n-3)$

20. (*a*) How many words consisting of 3 consonants and 2 vowels can be formed from 10 consonants and 5 vowels? (*b*) In how many of these will the consonants occupy the odd places?
Ans. (*a*) 144,000, (*b*) 14,400

21. Three balls are drawn from a bag containing 5 red, 4 white, and 3 black balls. In how many ways can this be done if (*a*) each is of a different color? (*b*) they are of the same color? (*c*) exactly 2 are red? (*d*) at least 2 are red? *Ans.* (*a*) 60, (*b*) 15, (*c*) 70, (*d*) 80

22. A squad is made up of 10 privates and 5 privates first class. (*a*) In how many ways can a detail of 4 privates and 2 privates first class be formed? (*b*) On how many of these details will private X serve? (*c*) On how many will private X but not private first class Y serve?
Ans. (*a*) 2100, (*b*) 840, (*c*) 504

23. A civic club has 60 members including 2 bankers, 4 lawers, and 5 doctors. In how many ways can a committee of 10 be formed to contain 1 banker, 2 lawyers, and 2 doctors? *Ans.* 228,826,080

24. How many committees of two or more can be selected from 10 people? *Ans.* $2^{10}-11$

25. Hands consisting of three cards are dealt from an ordinary deck. Show that a hand consisting of 3 different kinds should show 352 times as often as a hand consisting of 3 cards of the same kind.

26. Prove: (*a*) ${}_nC_r + {}_nC_{r+1} = {}_{n+1}C_{r+1}$, (*b*) ${}_{2n}C_n = 2\cdot{}_{2n-1}C_{n-1}$

CHAPTER 22

Probability

IN ESTIMATING THE PROBABILITY that a given event will or will not happen, we may, as in the case of drawing a face card from an ordinary deck, count the different ways in which this event may or may not happen. On the other hand, in the case of estimating the probability that a person who is now 25 years old will live to receive a bequest at age 30, we are forced to depend upon such knowledge of what has happened on similar occasions in the past as is available. In the first case, the result is called *mathematical* or *theoretical probability*; in the latter case, the result is called *statistical* or *empirical probability.*

MATHEMATICAL PROBABILITY. If an event must result in some one of n different but *equally likely ways* and if a certain s of these ways are considered successes and the other $f = n - s$ ways are considered failures, then the probability of success in a given trial is $p = \frac{s}{n}$ and the probability of failure is $q = \frac{f}{n}$. Since $p + q = \frac{s+f}{n} = \frac{n}{n} = 1$, $p = 1 - q$ and $q = 1 - p$.

Example 1. One card is drawn from an ordinary deck. What is the probability (*a*) that it is a red card, (*b*) that it is a spade, (*c*) that it is a king, (*d*) that it is not the ace of hearts ?

One card can be drawn from the deck in $n = 52$ different ways.

(*a*) A red card can be drawn from the deck in $s = 26$ different ways. Thus, the probability of drawing a red card is $s/n = 26/52 = 1/2$.

(*b*) A spade can be drawn from the deck in 13 different ways. The probability of drawing a spade is $13/52 = 1/4$.

(*c*) A king can be drawn in 4 ways. The required probability is $4/52 = 1/13$.

(*d*) The ace of hearts can be drawn in 1 way ; the probability of drawing the ace of hearts is $1/52$. Thus, the probability of *not* drawing the ace of hearts is $1 - 1/52 = 51/52$.

See Problems 1-4.

If p is the probability that a person will win a sum $\$S$, then $\$pS$ is called his *expectation.*

Example 2. X will win \$5 if he draws a red ball from a bag containing 3 black and 2 red balls. What is his expectation ?

The probability of drawing a red ball from the bag is $p = 2/5$; thus, his expectation is $(2/5)5 = \$2$.

See Problem 5.

Two or more events are called *mutually exclusive* if not more than one of them can occur in a single trial. Thus, the drawing of a jack and the drawing of a queen on a single draw from an ordinary deck are mutually exclusive events; however, the drawing of a jack and the drawing of a spade are not mutually exclusive.

Example 3. Find the probability of drawing a jack or a queen from an ordinary deck of cards.

Since there are four jacks and four queens, $s = 8$ and $p = 8/52 = 2/13$. Now the probability of drawing a jack is 1/13, the probability of drawing a queen is 1/13, and the required probability is $2/13 = 1/13 + 1/13$. We have verified

THEOREM *A*. The probability that some one of a set of mutually exclusive events will happen at a single trial is the sum of their separate probabilities of happening.

Two events *A* and *B* are called *independent* if the happening of one does not affect the happening of the other. Thus, in a toss of two dice, the fall of either does not affect the fall of the other, However, in drawing two cards from a deck, the probability of obtaining a red card on the second draw depends upon whether or not a red card was obtained on the draw of the first card. Two such events are called *dependent*.

Example 4. One bag contains 4 white and 4 black balls, a second bag contains 3 white and 6 black balls, and a third contains 1 white and 5 black balls. If one ball is drawn from each bag, find the probability that all are white.

A ball can be drawn from the first bag in any one of 8 ways, from the second in any one of 9 ways, and from the third in any one of 6 ways; hence, three balls can be drawn one from each bag in $8\cdot9\cdot6$ ways. A white ball can be drawn from the first bag in 4 ways, from the second in 3 ways, and from the third in 1 way; hence, three white balls can be drawn one from each bag in $4\cdot3\cdot1$ ways. Thus the required probability is

$$\frac{4\cdot3\cdot1}{8\cdot9\cdot6} = \frac{1}{36}$$

Now drawing a white ball from one bag does not affect the drawing of a white ball from another so that here we are concerned with three independent events. The probability of drawing a white ball from the first bag is 4/8, from the second is 3/9, and from the third bag is 1/6. Since the probability of drawing three white balls, one from each bag, is $\frac{4}{8}\cdot\frac{3}{9}\cdot\frac{1}{6}$, we have verified

THEOREM *B*. The probability that all of a set of independent events will happen in a single trial is the product of their separate probabilities.

THEOREM *C*, concerning dependent events. If the probability that an event will happen is p_1, and if after it has happened the probability that a second event will happen is p_2, the probability that the two events will happen in that order is p_1p_2.

Example 5. Two cards are drawn from an ordinary deck. Find the probability that both are face cards (king, queen, jack) if (*a*) the first card drawn is replaced before the second is drawn, (*b*) the first card drawn is not replaced before the second is drawn.

(*a*) Since each drawing is made from a complete deck, we have the case of two independent events. The probability of drawing a face card in a single draw is 12/52; thus, the probability of drawing two face cards, under the conditions imposed, is $\frac{12}{52}\cdot\frac{12}{52} = \frac{9}{169}$.

(*b*) Here the two events are dependent. The probability that the first drawing results in a face card is 12/52. Now, of the 51 cards remaining in the deck, there are 11 face cards; the probability that the second drawing results in a face card is 11/51. Hence, the probability of drawing two face cards is $\frac{12}{52}\cdot\frac{11}{51} = \frac{11}{221}$.

See Problems 6-10.

Example 6. Two dice are tossed six times. Find the probability (*a*) that 7 will show on the first four tosses and will not show on the other two, (*b*) that 7 will show on exactly four of the tosses.

The probability that 7 will show on a single toss is $p = 1/6$ and the probability that 7 will not show is $q = 1 - p = 5/6$.

(*a*) The probability that 7 will show on the first four tosses and will not show on the other two is

$$\frac{1}{6}\cdot\frac{1}{6}\cdot\frac{1}{6}\cdot\frac{1}{6}\cdot\frac{5}{6}\cdot\frac{5}{6} = \frac{25}{46{,}656}$$

(*b*) The four tosses on which 7 is to show may be selected in ${}_6C_4 = 15$ ways. Since these 15 ways constitute mutually exclusive events and the probability of any one of them is $(\frac{1}{6})^4(\frac{5}{6})^2$, the probability that 7 will show exactly four times in six tosses is ${}_6C_4(\frac{1}{6})^4(\frac{5}{6})^2 = \frac{125}{15{,}552}$.

We have verified

THEOREM *D*. If p is the probability that an event will happen and q is the probability that it will fail to happen at a given trial, the probability that it will happen exactly r times in n trials is

$${}_nC_r\,p^r q^{n-r}$$

See Problems 11-12.

EMPIRICAL PROBABILITY. If an event has been observed to happen s times in n trials, the ratio $p = s/n$ is defined as the *empirical probability* that the event will happen at any future trial. The confidence which can be placed in such probabilities depends in a large measure on the number of observations used. Life insurance companies, for example, base their premium rate on empirical probabilities. For this purpose they use a mortality table based on an enormous number of observations over the years.

The American Experience Table of Mortality begins with 100,000 persons all of age 10 years and indicates the number of the group who die each year thereafter. In using this table, it will be assumed that the laws stated above for mathematical probability hold also for empirical probability.

Example 7. Find the probability that a person 20 years old (*a*) will die during the year, (*b*) will die during the next 10 years, (*c*) will reach age 75.

(*a*) Of the 100,000 persons alive at age 10 years, 92,637 are alive at age

20 years. Of these 92,637 a total of 723 will die during the year. The probability that a person 20 years of age will die during the year is $\frac{723}{92,637} = 0.0078$.

(b) Of the 92,637 who reach age 20 years, 85,441 reach age 30 years; thus, 92,637 − 85,441 = 7,196 die during the 10 year period. The required probability is $\frac{7,196}{92,637} = 0.0777$.

(c) Of the 92,637 alive at age 20 years, 26,237 will reach age 75 years. The required probability is $\frac{26,237}{92,637} = 0.2832$.

See Problem 13.

SOLVED PROBLEMS

1. One ball is drawn from a bag containing 3 white, 4 red, and 5 black balls. What is the probability (a) that it is white, (b) that it is white or red, (c) that it is not red?

A ball can be drawn from the bag in $n = 3+4+5 = 12$ different ways.

(a) A white ball can be drawn in $s = 3$ different ways. Thus, the probability of drawing a white ball is $p = s/n = 3/12 = 1/4$.

(b) Here success consists in drawing either a white or a red ball; hence, $s = 3+4 = 7$ and the required probability is 7/12.

(c) The probability of drawing a red ball is $p = 4/12 = 1/3$. The probability that the ball drawn is *not* red is $1-p = 1-1/3 = 2/3$. This problem may also be solved as in (b).

2. If two dice are tossed, what is the probability (a) of throwing a total of 7, (b) of throwing a total of 8, (c) of throwing a total of 10 or more, (d) of both dice showing the same number?

Two dice may turn up in $6 \times 6 = 36$ ways.

(a) A total of 7 may result in 6 ways (1,6; 6,1; 2,5; 5,2; 3,4; 4,3). The probability of a throw of 7 is then 6/36 = 1/6.

(b) A total of 8 may result in 5 ways (2,6; 6,2; 3,5; 5,3; 4,4). The probability of a throw of 8 is 5/36.

(c) Here, success consists in throwing a total of 10, 11, or 12. Since a total of 10 may result in 3 ways, a total of 11 in 2 ways, and a total of 12 in 1 way, the probability is $(3+2+1)/36 = 1/6$.

(d) The probability that the second die will show the same number as the first is 1/6. This problem may also be solved by counting the number of successes 1,1; 2,2; etc.

3. If five coins are tossed, what is the probability (a) that all will show heads, (b) that exactly three will show heads, (c) that at least three will show heads?

Each coin can turn up in 2 ways; hence, the five coins can turn up in $2^5 = 32$ ways. (The assumption here is that HHHTT and THHHT are different results.)

(a) Five heads can turn up in only 1 way; hence, the probability of a toss of five heads is 1/32.

(b) Exactly three heads can turn up in ${}_5C_3 = 10$ ways; thus, the probability of a toss of exactly three heads is 10/32 = 5/16.

(*c*) Success here consists of a throw of exactly three, exactly four, or all heads. Exactly three heads can turn up in 10 ways, exactly four heads in 5 ways, and all heads in 1 way. Thus, the probability of throwing at least three heads is $(10+5+1)/32 = 1/2$.

4. If three cards are drawn from an ordinary deck, find the probability (*a*) that all are red cards, (*b*) that all are of the same suit, (*c*) that all are aces.

Three cards may be drawn from a deck in ${}_{52}C_3$ ways.

(*a*) Three red cards may be drawn from the 26 red cards in ${}_{26}C_3$ ways. Hence, the probability of drawing three red cards in a single draw is ${}_{26}C_3/{}_{52}C_3 = \dfrac{26\cdot25\cdot24}{1\cdot2\cdot3}\cdot\dfrac{1\cdot2\cdot3}{52\cdot51\cdot50} = \dfrac{2}{17}$.

(*b*) There are 4 ways of choosing a suit and ${}_{13}C_3$ ways of selecting three cards of that suit. Thus, the probability of drawing three cards of the same suit is $4\cdot{}_{13}C_3/{}_{52}C_3 = 22/425$.

This problem may also be solved as follows: The first card drawn determines a suit. The deck now contains 51 cards of which 12 are of that suit; hence, the probability that the next two cards drawn will be of that suit is ${}_{12}C_2/{}_{51}C_2 = 22/425$, as before.

(*c*) Three aces may be selected from the four aces in 4 ways; hence, the required probability is $4/{}_{52}C_3 = 1/5525$.

5. In a certain lottery the prize is $20 and 100 tickets have been sold. What is the expectation of a man who holds 8 tickets?

The probability that the man will win the prize is 8/100; his expectation is $\dfrac{8}{100}(\$20) = \$1.60$.

6. One bag contains 8 black balls and a second bag contains 1 white and 6 black balls. One of the bags is selected and then a ball is drawn from that bag. What is the probability that it is the white ball?

The probability that the second bag is chosen is $\frac{1}{2}$ and the probability that the white ball is drawn from this bag is 1/7. Thus, the required probability is $\frac{1}{2}(1/7) = 1/14$.

7. Two cards are drawn in succession from an ordinary deck. What is the probability (*a*) that the first will be the jack of diamonds and the second will be the queen of spades, (*b*) that the first will be a diamond and the second a spade, (*c*) that both cards are diamonds or both are spades?

(*a*) The probability that the first card is the jack of diamonds is 1/52 and the probability that the second is the queen of spades is 1/51. The required probability is $(1/52)(1/51) = 1/2652$.

(*b*) The probability that the first card is a diamond is 13/52 and the probability that the second card is a spade is 13/51. The probability of the required draw is $\dfrac{13}{52}\cdot\dfrac{13}{51} = \dfrac{13}{204}$.

(*c*) The probability that both cards are of a specified suit is $\dfrac{13}{52}\cdot\dfrac{12}{51}$. Thus, the probability that both are diamonds or both are spades is $\dfrac{13}{52}\cdot\dfrac{12}{51}+\dfrac{13}{52}\cdot\dfrac{12}{51} = \dfrac{2}{17}$.

8. *A*, *B*, and *C* work independently on a problem. If the respective probabilities that they will solve it are 1/2, 1/3, 2/5, find the probability that the problem will be solved.

The problem will be solved unless all three fail; the probability that this will happen is $\dfrac{1}{2}\cdot\dfrac{2}{3}\cdot\dfrac{3}{5} = \dfrac{1}{5}$. Thus, the probability that the problem will be solved is $1-\dfrac{1}{5} = \dfrac{4}{5}$.

9. A tosses a coin and if a head appears he wins the game; if a tail appears, B tosses the coin under the same conditions, and so on. If the stakes are \$15, find the expectation of each.

We first compute the probability that A will win. The probability that he will win on the first toss is $\frac{1}{2}$; the probability that he will win on his second toss (that is, that A first tosses a tail, B tosses a tail, and A then tosses a head) is $\frac{1}{2}\cdot\frac{1}{2}\cdot\frac{1}{2} = 1/2^3$; the probability that he will win on his third toss (that is, that A first tosses a tail, B tosses a tail, A tosses a tail, B tosses a tail, and A then tosses a head) is $\frac{1}{2}\cdot\frac{1}{2}\cdot\frac{1}{2}\cdot\frac{1}{2}\cdot\frac{1}{2} = 1/2^5$, and so on. Thus, the probability that A will win is

$$\frac{1}{2} + \frac{1}{2^3} + \frac{1}{2^5} + \ldots = \frac{1/2}{1-1/2^2} = \frac{2}{3}$$

and his expectation is $\frac{2}{3}(\$15) = \$10$. Then B's expectation is \$5.

10. On a toss of two dice, X throws a total of 5. Find the probability that he will throw another 5 before he throws 7.

X will succeed should he throw a total of 5 on the next toss, or should he not throw 5 or 7 on this toss but throw 5 on the next, or should he not throw 5 or 7 on either of these tosses but throw 5 on the next, and so on. The respective probabilities are $\frac{4}{36}$, $\frac{26}{36}\cdot\frac{4}{36}$, $\frac{26}{36}\cdot\frac{26}{36}\cdot\frac{4}{36}$, and so on. Thus, the probability that he throws 5 before 7 is

$$\frac{4}{36} + \frac{26}{36}\cdot\frac{4}{36} + \frac{26}{36}\cdot\frac{26}{36}\cdot\frac{4}{36} + \ldots = \frac{4/36}{1-26/36} = \frac{2}{5}$$

11. A bag contains 2 white and 3 black balls. A ball is drawn 5 times, each being replaced before another is drawn. Find the probability that (a) the first 4 balls drawn are white and the last is black, (b) exactly 4 of the balls drawn are white, (c) at least 4 of the balls drawn are white, (d) at least 1 ball is white.

The probability of drawing a white ball is $p = 2/5$ and the probability of drawing a black ball is 3/5.

(a) The probability that the first 4 are white and the last black is $\frac{2}{5}\cdot\frac{2}{5}\cdot\frac{2}{5}\cdot\frac{2}{5}\cdot\frac{3}{5} = \frac{48}{3125}$.

(b) Here $n = 5$, $r = 4$; the probability that exactly 4 of the balls drawn are white is

$${}_nC_r\, p^r q^{n-r} = {}_5C_4 (\tfrac{2}{5})^4 (\tfrac{3}{5}) = \frac{48}{625}$$

(c) Since success consists of drawing either 4 white and 1 black ball or 5 white balls, the probability is

$${}_5C_4 (\tfrac{2}{5})^4 (\tfrac{3}{5}) + {}_5C_5 (\tfrac{2}{5})^5 = \frac{272}{3125}$$

(d) Here failure consists of drawing 5 black balls. Since the probability of failure is $(\frac{3}{5})^5 = \frac{243}{3125}$, the probability of success is $1 - \frac{243}{3125} = \frac{2882}{3125}$. The problem may also be solved as in (c).

12. One bag contains 2 white balls and 2 black balls, and another contains 3 white balls and 5 black balls. At 5 different trials, a bag is chosen at random and one ball is drawn from that bag and replaced. Find the probability (a) that exactly 3 white balls are drawn, (b) that at least 3 white balls are drawn.

At any trial the probability that a white ball is drawn from the first bag is $\frac{1}{2}\cdot\frac{1}{2}$ and the probability that a white ball is drawn from the second bag is $\frac{1}{2}(3/8)$. Thus, the probability that a white ball is drawn at any trial is $p = \frac{1}{2}\cdot\frac{1}{2} + \frac{1}{2}(3/8) = 7/16$, and the probability that a black ball is drawn is $q = 9/16$.

(*a*) The probability of drawing exactly 3 white balls in 5 trials is $_5C_3(\frac{7}{16})^3(\frac{9}{16})^2 = \frac{138,915}{524,288}$.

(*b*) The probability of drawing at least 3 white balls in 5 trials is

$$_5C_3(\frac{7}{16})^3(\frac{9}{16})^2 + {_5C_4}(\frac{7}{16})^4(\frac{9}{16}) + {_5C_5}(\frac{7}{16})^5 = \frac{201,341}{524,288}$$

13. A husband is 35 years old and his wife is 28. Find the probability that at the end of 20 years (*a*) the husband will be alive, (*b*) the wife will be alive, (*c*) both will be alive, (*d*) both will be dead, (*e*) the wife will be alive and the husband will not, (*f*) one will be alive but not the other.

(*a*) Of the 81,822 alive at age 35, 64,563 will reach age 55. The probability that the husband will be alive at the end of 20 years is $\frac{64,563}{81,822} = 0.7890$.

(*b*) The probability that the wife will be alive at the end of 20 years is $\frac{71,627}{86,878} = 0.8245$.

(*c*) Since the survival of the husband and of the wife are independent events, the probability that both are alive after 20 years is $\frac{64,563}{81,822}\cdot\frac{71,627}{86,878} \doteq 0.6506$.

(*d*) From (*a*) 17,259 of the 81,822 alive at age 35 will not reach age 55; thus, the probability that the husband will not live for 20 years is $\frac{17,259}{81,822}$. Similarly, the probability that the wife will not live for 20 years is $\frac{15,251}{86,878}$. Hence, the probability that after 20 years both will be dead is $\frac{17,259}{81,822}\cdot\frac{15,251}{86,878} = 0.0370$.

(*e*) The probability that the husband will be dead and the wife will be alive after 20 years is

$$\frac{17,259}{81,822}\cdot\frac{71,627}{86,878} = 0.1739$$

(*f*) The probability that the wife will survive but the husband will not is found in (*e*). The probability that the husband will survive but the wife will not is $\frac{64,563}{81,822}\cdot\frac{15,251}{86,878}$. Thus, the probability that just one will survive is $\frac{17,259}{81,822}\cdot\frac{71,627}{86,878} + \frac{64,563}{81,822}\cdot\frac{15,251}{86,878} = 0.3116$.

SUPPLEMENTARY PROBLEMS

14. One ball is drawn from a bag containing 4 white and 6 black balls. Find the probability that it is (*a*) white, (*b*) black. *Ans.* (*a*) 2/5, (*b*) 3/5

15. Three balls are drawn together from a bag containing 8 white and 12 black balls. Find the probability that (*a*) all are white, (*b*) just two are white, (*c*) just one is white, (*d*) all are black.
Ans. (*a*) 14/285, (*b*) 28/95, (*c*) 44/95, (*d*) 11/57

16. Ten students are seated at random in a row. Find the probability that two particular students are not seated side by side. *Ans.* 4/5

17. If a die is cast three times, find the probability (*a*) that an even number will be thrown each time, (*b*) that an odd number will appear just once, (*c*) that the sum of the three numbers will be even.
Ans. (*a*) 1/8, (*b*) 3/8, (*c*) 1/2

18. A bag contains four \$10 bills, six \$5 bills, and ten \$1 bills. Find the expectation of a man who is permitted to draw (*a*) one bill at random, (*b*) two bills at random. *Ans.* (*a*) \$4, (*b*) \$8

19. From a box containing 10 cards numbered 1,2,3,..,10, four cards are drawn. Find the probability that their sum will be even (*a*) if the cards are drawn together, (*b*) if each card drawn is replaced before the next is drawn. *Ans.* (*a*) 11/21, (*b*) 1/2

20. A and B, having equal skill, are playing a game of three points. After A has won 2 points and B has won one point, what is the probability that A will win the game? *Ans.* 3/4

21. A,B,C, and D in that order take turns in tossing a coin. The first to toss a head wins \$15. Find the expectation of each. *Ans.* A,\$8; B,\$4; C,\$2; D,\$1.

22. One bag contains 3 white and 2 black balls, and another contains 2 white and 3 black balls. A ball is drawn from the second bag and placed in the first; then a ball is drawn from the first bag and placed in the second. When the pair of operations is repeated, what is the probability that the first bag will contain 5 white balls? *Ans.* 1/225

23. Three bags contain respectively 2 white and 1 black ball, 3 white and 3 black balls, 6 white and 2 black balls. Two bags are selected and a ball is drawn from each. Find the probability (*a*) that both balls are white, (*b*) that both balls are of the same color. *Ans.* (*a*) 29/72, (*b*) 19/36

24. If four trials are made in Problem 23, find the probability (*a*) that the first two will result in pairs of white balls and the other two in pairs of black balls, (*b*) that a pair of black balls will be obtained at least three times. *Ans.* (*a*) 841/331,776, (*b*) 125/36,864

25. Five cards numbered 1,2,3,4,5 respectively are placed in a revolving box. If the cards are drawn one at a time from the box, what is the probability that they will be drawn in their natural order? *Ans.* 1/120

26. Brown, Jones, and Smith shoot at a target in alphabetical order with probabilities 1/4, 1/3, 1/2 respectively of hitting the bull's-eye. (*a*) Find the probability that each on his first shot will be the first to hit the bull's-eye, (*b*) find the probability that the bull's-eye is not hit on the first round, (*c*) find the probability that the first to hit the bull's eye is Jones on his second shot. *Ans.* (*a*) 1/4, 1/4, 1/4, (*b*) 1/4, (*c*) 1/16

27. The probability that X will win a game of checkers is 2/5. In a five game match, what is the probability (*a*) that X will win the first, third, and fifth games, and lose the others? (*b*) that he will win exactly three games? (*c*) that he will win at least three games? *Ans.* (*a*) 72/3125, (*b*) 144/625, (*c*) 992/3125

28. Of fifty castings, two are defective. If twenty of the castings are inspected, what is the probability (*a*) that neither of the defective castings will be inspected, (*b*) that just one will be inspected, (*c*) that at least one will be inspected? *Ans.* (*a*) 0.3551, (*b*) 0.4898, (*c*) 0.6449

CHAPTER 23

Determinants of Order Two and Three

DETERMINANTS OF ORDER TWO. The symbol $\begin{vmatrix} a_1 & b_1 \\ a_2 & b_2 \end{vmatrix}$, consisting of 2^2 numbers called *elements* arranged in two rows and two columns, is called a *determinant of order two*. The elements a_1 and b_2 are said to lie along the *principal diagonal*; the elements a_2 and b_1 are said to lie along the *secondary diagonal*.

The *value* of the determinant is obtained by forming the product of the elements along the principal diagonal and subtracting from it the product of the elements along the secondary diagonal; thus,

$$\begin{vmatrix} a_1 & b_1 \\ a_2 & b_2 \end{vmatrix} = a_1b_2 - a_2b_1$$

See Problem 1.

THE SOLUTION of the consistent and independent equations

$$(1) \qquad \begin{cases} a_1x + b_1y = c_1 \\ a_2x + b_2y = c_2 \end{cases}$$

may be expressed as quotients of determinants of order two:

$$x = \frac{c_1b_2 - c_2b_1}{a_1b_2 - a_2b_1} = \frac{\begin{vmatrix} c_1 & b_1 \\ c_2 & b_2 \end{vmatrix}}{\begin{vmatrix} a_1 & b_1 \\ a_2 & b_2 \end{vmatrix}}, \qquad y = \frac{a_1c_2 - a_2c_1}{a_1b_2 - a_2b_1} = \frac{\begin{vmatrix} a_1 & c_1 \\ a_2 & c_2 \end{vmatrix}}{\begin{vmatrix} a_1 & b_1 \\ a_2 & b_2 \end{vmatrix}}$$

These equations are consistent and independent if and only if $\begin{vmatrix} a_1 & b_1 \\ a_2 & b_2 \end{vmatrix} \neq 0$.

Example 1. Solve $\begin{cases} y = 3x + 1 \\ 4x + 2y - 7 = 0 \end{cases}$ using determinants.

Arrange the equations in the form (*1*): $\begin{cases} 3x - y = -1 \\ 4x + 2y = 7 \end{cases}$. The solution requires the values of three determinants:

the denominator, D, formed by writing the coefficients of x and y in order

$$D = \begin{vmatrix} 3 & -1 \\ 4 & 2 \end{vmatrix} = 3\cdot 2 - 4(-1) = 6 + 4 = 10;$$

the numerator of x, N_x, formed from D by replacing the coefficients of x by the constant terms

$$N_x = \begin{vmatrix} -1 & -1 \\ 7 & 2 \end{vmatrix} = -1\cdot 2 - 7(-1) = -2+7 = 5;$$

the numerator of y, N_y, formed from D by replacing the coefficients of y by the constant terms

$$N_y = \begin{vmatrix} 3 & -1 \\ 4 & 7 \end{vmatrix} = 3\cdot 7 - 4(-1) = 21+4 = 25.$$

Then $x = \dfrac{N_x}{D} = \dfrac{5}{10} = \dfrac{1}{2}$ and $y = \dfrac{N_y}{D} = \dfrac{25}{10} = \dfrac{5}{2}$.

See Problem 2.

DETERMINANTS OF ORDER THREE. The symbol $\begin{vmatrix} a_1 & b_1 & c_1 \\ a_2 & b_2 & c_2 \\ a_3 & b_3 & c_3 \end{vmatrix}$, consisting of 3^2 elements arranged in three rows and three columns, is called a *determinant of order three*. Its value is

$$a_1b_2c_3 + a_2b_3c_1 + a_3b_1c_2 - a_1b_3c_2 - a_2b_1c_3 - a_3b_2c_1$$

This may be written as

$$a_1(b_2c_3 - b_3c_2) - b_1(a_2c_3 - a_3c_2) + c_1(a_2b_3 - a_3b_2)$$

or

$$(2) \qquad a_1\begin{vmatrix} b_2 & c_2 \\ b_3 & c_3 \end{vmatrix} - b_1\begin{vmatrix} a_2 & c_2 \\ a_3 & c_3 \end{vmatrix} + c_1\begin{vmatrix} a_2 & b_2 \\ a_3 & b_3 \end{vmatrix}$$

to involve three determinants of order two. Note that the elements which multiply the determinants of order two are the elements of the first row of the given determinant. In all, six such representations using the elements of each of the three rows and each of the three columns may be worked out. We shall use only the representation (2) in this chapter.

See Problem 3.

Another procedure for evaluating determinants of order three is indicated in Problem 4.

THE SOLUTION of the system of consistent and independent equations

$$\begin{cases} a_1x + b_1y + c_1z = d_1 \\ a_2x + b_2y + c_2z = d_2 \\ a_3x + b_3y + c_3z = d_3 \end{cases}$$

in determinant form is given by

$$x = \frac{N_x}{D} = \frac{\begin{vmatrix} d_1 & b_1 & c_1 \\ d_2 & b_2 & c_2 \\ d_3 & b_3 & c_3 \end{vmatrix}}{\begin{vmatrix} a_1 & b_1 & c_1 \\ a_2 & b_2 & c_2 \\ a_3 & b_3 & c_3 \end{vmatrix}}, \qquad y = \frac{N_y}{D} = \frac{\begin{vmatrix} a_1 & d_1 & c_1 \\ a_2 & d_2 & c_2 \\ a_3 & d_3 & c_3 \end{vmatrix}}{D}, \qquad z = \frac{N_z}{D} = \frac{\begin{vmatrix} a_1 & b_1 & d_1 \\ a_2 & b_2 & d_2 \\ a_3 & b_3 & d_3 \end{vmatrix}}{D}.$$

The determinant D is formed by writing the coefficients of x, y, z in order while the determinant appearing in the numerator for any unknown is obtained from D by replacing the column of coefficients of that unknown by the column of constants.

The system is consistent and independent if and only if $D \neq 0$.

Example 2. Solve, using determinants: $\begin{cases} x + 3y + 2z = -13 \\ 2x - 6y + 3z = 32 \\ 3x - 4y - z = 12 \end{cases}$

The solution requires the values of four determinants:

the denominator, $$D = \begin{vmatrix} 1 & 3 & 2 \\ 2 & 6 & 3 \\ 3 & -4 & -1 \end{vmatrix} = 1(6+12) - 3(-2-9) + 2(-8+18) = 18 + 33 + 20 = 71;$$

the numerator of x, $$N_x = \begin{vmatrix} -13 & 3 & 2 \\ 32 & -6 & 3 \\ 12 & -4 & -1 \end{vmatrix} = -13(6+12) - 3(-32-36) + 2(-128+72) = -234 + 204 - 112 = -142;$$

the numerator of y, $$N_y = \begin{vmatrix} 1 & -13 & 2 \\ 2 & 32 & 3 \\ 3 & 12 & -1 \end{vmatrix} = 1(-32-36) - (-13)(-2-9) + 2(24-96) = -68 - 143 - 144 = -355;$$

the numerator of z, $$N_z = \begin{vmatrix} 1 & 3 & -13 \\ 2 & -6 & 32 \\ 3 & -4 & 12 \end{vmatrix} = 1(-72+128) - 3(24-96) + (-13)(-8+18) = 56 + 216 - 130 = 142.$$

Then $$x = \frac{N_x}{D} = \frac{-142}{71} = -2, \quad y = \frac{N_y}{D} = \frac{-355}{71} = -5, \quad z = \frac{N_z}{D} = \frac{142}{71} = 2.$$

See Problems 5-6.

SOLVED PROBLEMS

1. Evaluate each of the following determinants.

(a) $\begin{vmatrix} 2 & 3 \\ 4 & 5 \end{vmatrix} = 2 \cdot 5 - 4 \cdot 3 = -2$ (b) $\begin{vmatrix} 5 & -2 \\ 3 & 1 \end{vmatrix} = 5 \cdot 1 - 3(-2) = 11$

2. Solve for x and y, using determinants: (a) $\begin{cases} x + 2y = -4 \\ 5x + 3y = 1 \end{cases}$, (b) $\begin{cases} ax - 2by = c \\ 2ax - 3by = 4c \end{cases}$

(a) $D = \begin{vmatrix} 1 & 2 \\ 5 & 3 \end{vmatrix} = 3 - 10 = -7$, $N_x = \begin{vmatrix} -4 & 2 \\ 1 & 3 \end{vmatrix} = -12 - 2 = -14$, $N_y = \begin{vmatrix} 1 & -4 \\ 5 & 1 \end{vmatrix} = 1 + 20 = 21$

$$x = \frac{N_x}{D} = \frac{-14}{-7} = 2, \quad y = \frac{N_y}{D} = \frac{21}{-7} = -3$$

(b) $D = \begin{vmatrix} a & -2b \\ 2a & -3b \end{vmatrix} = ab$, $N_x = \begin{vmatrix} c & -2b \\ 4c & -3b \end{vmatrix} = 5bc$, $N_y = \begin{vmatrix} a & c \\ 2a & 4c \end{vmatrix} = 2ac$

$$x = \frac{N_x}{D} = \frac{5bc}{ab} = \frac{5c}{a}, \quad y = \frac{N_y}{D} = \frac{2ac}{ab} = \frac{2c}{b}$$

3. Evaluate the following determinants.

(a) $\begin{vmatrix} 2 & -2 & -1 \\ 6 & 1 & -1 \\ 4 & 3 & 5 \end{vmatrix} = 2\begin{vmatrix} 1 & -1 \\ 3 & 5 \end{vmatrix} - (-2)\begin{vmatrix} 6 & -1 \\ 4 & 5 \end{vmatrix} + (-1)\begin{vmatrix} 6 & 1 \\ 4 & 3 \end{vmatrix}$

$= 2(5+3) + 2(30+4) - (18-4) = 2\cdot 8 + 2\cdot 34 - 14 = 70$

(b) $\begin{vmatrix} 2 & 5 & 0 \\ 0 & 3 & 4 \\ -5 & 3 & 6 \end{vmatrix} = 2\begin{vmatrix} 3 & 4 \\ 3 & 6 \end{vmatrix} - 5\begin{vmatrix} 0 & 4 \\ -5 & 6 \end{vmatrix} + 0\begin{vmatrix} 0 & 3 \\ -5 & 3 \end{vmatrix} = 2(18-12) - 5(0+20) + 0 = -88$

(c) $\begin{vmatrix} 3 & 2 & 1 \\ 1 & -2 & 4 \\ 4 & 2 & 3 \end{vmatrix} = 3(-6-8) - 2(3-16) + 1(2+8) = -6$

(d) $\begin{vmatrix} 4 & -3 & 2 \\ 5 & 9 & -7 \\ 4 & -1 & 4 \end{vmatrix} = 4(36-7) + 3(20+28) + 2(-5-36) = 178$

4. The following procedure for evaluating a determinant of order three makes use of six lines each of which joins the three elements whose product is to be formed.

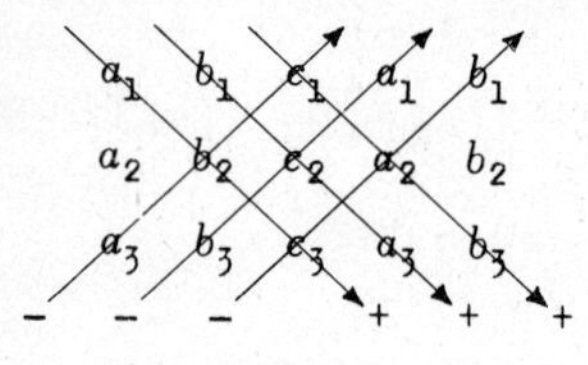

Here the first two columns are rewritten to the right of the determinant. The products formed by following the lines running down from left to right have a plus sign attached, and those formed by following the lines running up from left to right have a negative sign attached.

The algebraic sum of the products thus formed is the value of the determinant.

5. Solve, using determinants: $\begin{cases} 2x - 3y + 2z = 6 \\ x + 8y + 3z = -31 \\ 3x - 2y + z = -5 \end{cases}$

We evaluate $D = \begin{vmatrix} 2 & -3 & 2 \\ 1 & 8 & 3 \\ 3 & -2 & 1 \end{vmatrix} = 2(8+6) + 3(1-9) + 2(-2-24) = -48$

$N_x = \begin{vmatrix} 6 & -3 & 2 \\ -31 & 8 & 3 \\ -5 & -2 & 1 \end{vmatrix} = 6(8+6) + 3(-31+15) + 2(62+40) = 240$

$N_y = \begin{vmatrix} 2 & 6 & 2 \\ 1 & -31 & 3 \\ 3 & -5 & 1 \end{vmatrix} = 2(-31+15) - 6(1-9) + 2(-5+93) = 192$

$N_z = \begin{vmatrix} 2 & -3 & 6 \\ 1 & 8 & -31 \\ 3 & -2 & -5 \end{vmatrix} = 2(-40-62) + 3(-5+93) + 6(-2-24) = -96.$

Then $x = \dfrac{N_x}{D} = \dfrac{240}{-48} = -5$, $y = \dfrac{N_y}{D} = \dfrac{192}{-48} = -4$, $z = \dfrac{N_z}{D} = \dfrac{-96}{-48} = 2.$

6. Solve, using determinants: $\begin{cases} 2x + y = 2 \\ z - 4y = 0 \\ 4x + z = 6 \end{cases}$

$$D = \begin{vmatrix} 2 & 1 & 0 \\ 0 & -4 & 1 \\ 4 & 0 & 1 \end{vmatrix} = -4, \quad N_x = \begin{vmatrix} 2 & 1 & 0 \\ 0 & -4 & 1 \\ 6 & 0 & 1 \end{vmatrix} = -2, \quad N_y = \begin{vmatrix} 2 & 2 & 0 \\ 0 & 0 & 1 \\ 4 & 6 & 1 \end{vmatrix} = -4, \quad N_z = \begin{vmatrix} 2 & 1 & 2 \\ 0 & -4 & 0 \\ 4 & 0 & 6 \end{vmatrix} = -16\,.$$

Then $x = \dfrac{-2}{-4} = \dfrac{1}{2}$, $y = \dfrac{-4}{-4} = 1$, and $z = \dfrac{-16}{-4} = 4$.

SUPPLEMENTARY PROBLEMS

7. Evaluate. (*a*) $\begin{vmatrix} 1 & 2 \\ -3 & 1 \end{vmatrix}$ (*b*) $\begin{vmatrix} 2 & 0 \\ -5 & 1 \end{vmatrix}$ (*c*) $\begin{vmatrix} 3 & 2 \\ 1 & -4 \end{vmatrix}$ (*d*) $\begin{vmatrix} 2 & -10 \\ 3 & -15 \end{vmatrix}$

Ans. (*a*) 7, (*b*) 2, (*c*) −14, (*d*) 0.

8. Solve, using determinants. (*a*) $\begin{cases} 2x + y = 4 \\ 3x + 4y = 1 \end{cases}$ (*b*) $\begin{cases} 5x + 2y = 2 \\ 3x - 5y = 26 \end{cases}$ (*c*) $\begin{cases} 5x + 3y = -6 \\ 3x + 5y = -18 \end{cases}$

Ans. (*a*) $x = 3$, $y = -2$ (*b*) $x = 2$, $y = -4$ (*c*) $x = 3/2$, $y = -9/2$

9. Evaluate. (*a*) $\begin{vmatrix} 1 & 2 & 3 \\ 2 & -3 & 4 \\ -3 & 4 & 5 \end{vmatrix}$ (*b*) $\begin{vmatrix} 2 & -1 & 3 \\ 2 & 1 & 5 \\ -2 & 3 & -2 \end{vmatrix}$ (*c*) $\begin{vmatrix} -2 & 3 & 1 \\ 3 & 5 & 1 \\ 4 & -2 & 1 \end{vmatrix}$ (*d*) $\begin{vmatrix} 2 & 6 & 1 \\ 4 & -4 & 1 \\ 3 & 1 & 1 \end{vmatrix}$

Ans. (*a*) −78, (*b*) −4, (*c*) −37, (*d*) 0.

10. Solve, using determinants. (*a*) $\begin{cases} x + 2y + 2z = 4 \\ 3x - y + 4z = 25 \\ 3x + 2y - z = -4 \end{cases}$ (*b*) $\begin{cases} 2x - 3y + 5z = 4 \\ 3x + 2y + 2z = 3 \\ 4x + y - 4z = -6 \end{cases}$ (*c*) $\begin{cases} \dfrac{1}{x} + \dfrac{2}{y} + \dfrac{1}{z} = 2 \\ \dfrac{3}{x} - \dfrac{4}{y} - \dfrac{2}{z} = 1 \\ \dfrac{2}{x} + \dfrac{5}{y} - \dfrac{2}{z} = 3 \end{cases}$

Hint: In (*c*) solve first for $1/x$, $1/y$, $1/z$.

Ans. (*a*) $x = 2$, $y = -3$, $z = 4$ (*b*) $x = -1/3$, $y = 2/3$, $z = 4/3$ (*c*) $x = 1$, $y = z = 3$

11. Verify, by evaluating the determinants.

(*a*) $\begin{vmatrix} 1 & 2 & 3 \\ a & 2a & 3a \\ 8 & 9 & 10 \end{vmatrix} = 0$

(*b*) $\begin{vmatrix} 2 & 1 & -1 \\ 3 & 4 & 2 \\ -2 & -5 & 3 \end{vmatrix} = \begin{vmatrix} 0 & 1 & 0 \\ -5 & 4 & 6 \\ 8 & -5 & -2 \end{vmatrix}$

(*c*) $\begin{vmatrix} 3 & 4 & 5 \\ 7 & -2 & 3 \\ 2 & 5 & -1 \end{vmatrix} = -\begin{vmatrix} 4 & -2 & 5 \\ 3 & 7 & 2 \\ 5 & 3 & -1 \end{vmatrix}$

(*d*) $\begin{vmatrix} 4 & 2 & 5 \\ -7 & -3 & 1 \\ 9 & 4 & 8 \end{vmatrix} + \begin{vmatrix} -1 & 2 & 5 \\ 5 & -3 & 1 \\ -3 & 4 & 8 \end{vmatrix} = \begin{vmatrix} 3 & 2 & 5 \\ -2 & -3 & 1 \\ 6 & 4 & 8 \end{vmatrix}$

12. Solve, using determinants. $\begin{vmatrix} x & y & 1 \\ 1 & -1 & 1 \\ 13 & 2 & 1 \end{vmatrix} = 0$, $\begin{vmatrix} x & y & 1 \\ 3 & 2 & 1 \\ -6 & -4 & 1 \end{vmatrix} = 0$. *Ans.* $x = -3$, $y = -2$

CHAPTER 24

Determinants of Order n

A DETERMINANT of order n consists of n^2 numbers called elements arranged in n rows and columns, and enclosed by two vertical lines. For example,

$$\Delta_1 = |a_1|, \quad \Delta_2 = \begin{vmatrix} a_1 & b_1 \\ a_2 & b_2 \end{vmatrix}, \quad \Delta_3 = \begin{vmatrix} a_1 & b_1 & c_1 \\ a_2 & b_2 & c_2 \\ a_3 & b_3 & c_3 \end{vmatrix}, \quad \Delta_4 = \begin{vmatrix} a_1 & b_1 & c_1 & d_1 \\ a_2 & b_2 & c_2 & d_2 \\ a_3 & b_3 & c_3 & d_3 \\ a_4 & b_4 & c_4 & d_4 \end{vmatrix}$$

are determinants of orders one, two, three and four respectively. In this notation the letters designate columns and the subscripts designate rows. Thus, all elements with basal letter c are in the third column and all elements with subscript 2 are in the second row.

THE MINOR OF A GIVEN ELEMENT of a determinant is the determinant of the elements which remain after deleting the row and the column in which the given element stands. For example, the minor of a_1 in Δ_4 is $\begin{vmatrix} b_2 & c_2 & d_2 \\ b_3 & c_3 & d_3 \\ b_4 & c_4 & d_4 \end{vmatrix}$ and the minor of b_3 is $\begin{vmatrix} a_1 & c_1 & d_1 \\ a_2 & c_2 & d_2 \\ a_4 & c_4 & d_4 \end{vmatrix}$.

Note that the minor of a given element contains no element having either the letter or the subscript of the given element.

See Problem 1.

THE VALUE OF A DETERMINANT of order one is the single element of the determinant. A determinant of order $n > 1$ may be expressed as the sum of n products formed by multiplying each element of any chosen row (column) by its minor and prefixing a proper sign. The proper sign associated with each product is $(-1)^{i+j}$ where i is the number of the row and j is the number of the column in which the element stands. For example,

$$\Delta_3 = -a_2 \begin{vmatrix} b_1 & c_1 \\ b_3 & c_3 \end{vmatrix} + b_2 \begin{vmatrix} a_1 & c_1 \\ a_3 & c_3 \end{vmatrix} - c_2 \begin{vmatrix} a_1 & b_1 \\ a_3 & b_3 \end{vmatrix}$$

is the expansion of Δ_3 along the second row. The sign given to the first product is $-$ since a_2 stands in the second row and first column, and $(-1)^{2+1} = -1$. In all, there are six expansions of Δ_3 along its rows and columns yielding identical results when the minors are evaluated.

There are eight expansions of Δ_4 along its rows and columns, of which

$$\Delta_4 = +a_1\begin{vmatrix} b_2 & c_2 & d_2 \\ b_3 & c_3 & d_3 \\ b_4 & c_4 & d_4 \end{vmatrix} - b_1\begin{vmatrix} a_2 & c_2 & d_2 \\ a_3 & c_3 & d_3 \\ a_4 & c_4 & d_4 \end{vmatrix} + c_1\begin{vmatrix} a_2 & b_2 & d_2 \\ a_3 & b_3 & d_3 \\ a_4 & b_4 & d_4 \end{vmatrix} - d_1\begin{vmatrix} a_2 & b_2 & c_2 \\ a_3 & b_3 & c_3 \\ a_4 & b_4 & c_4 \end{vmatrix}$$

(along the first row),

$$\Delta_4 = +a_1\begin{vmatrix} b_2 & c_2 & d_2 \\ b_3 & c_3 & d_3 \\ b_4 & c_4 & d_4 \end{vmatrix} - a_2\begin{vmatrix} b_1 & c_1 & d_1 \\ b_3 & c_3 & d_3 \\ b_4 & c_4 & d_4 \end{vmatrix} + a_3\begin{vmatrix} b_1 & c_1 & d_1 \\ b_2 & c_2 & d_2 \\ b_4 & c_4 & d_4 \end{vmatrix} - a_4\begin{vmatrix} b_1 & c_1 & d_1 \\ b_2 & c_2 & d_2 \\ b_3 & c_3 & d_3 \end{vmatrix}$$

(along the first column),

$$\Delta_4 = -a_4\begin{vmatrix} b_1 & c_1 & d_1 \\ b_2 & c_2 & d \\ b_3 & c_3 & d_3 \end{vmatrix} + b_4\begin{vmatrix} a_1 & c_1 & d_1 \\ a_2 & c_2 & d_2 \\ a_3 & c_3 & d_3 \end{vmatrix} - c_4\begin{vmatrix} a_1 & b_1 & d_1 \\ a_2 & b_2 & d_2 \\ a_3 & b_3 & d_3 \end{vmatrix} + d_4\begin{vmatrix} a_1 & b_1 & c_1 \\ a_2 & b_2 & c_2 \\ a_3 & b_3 & c_3 \end{vmatrix}$$

(along the fourth row),

$$\Delta_4 = -b_1\begin{vmatrix} a_2 & c_2 & d_2 \\ a_3 & c_3 & d_3 \\ a_4 & c_4 & d_4 \end{vmatrix} + b_2\begin{vmatrix} a_1 & c_1 & d_1 \\ a_3 & c_3 & d_3 \\ a_4 & c_4 & d_4 \end{vmatrix} - b_3\begin{vmatrix} a_1 & c_1 & d_1 \\ a_2 & c_2 & d_2 \\ a_4 & c_4 & d_4 \end{vmatrix} + b_4\begin{vmatrix} a_1 & c_1 & d_1 \\ a_2 & c_2 & d_2 \\ a_3 & c_3 & d_3 \end{vmatrix}$$

(along the second column),

are examples.

See Problem 2.

THE COFACTOR OF AN ELEMENT of a determinant is the minor of that element together with the sign associated with the product of that element and its minor in the expansion of the determinant. The cofactors of the elements a_1, a_2, b_1, b_3, $c_1, \ldots$ will be denoted by A_1, A_2, B_1, B_3, $C_1, \ldots$ Thus the cofactor of c_1 in Δ_3 is $C_1 = +\begin{vmatrix} a_2 & b_2 \\ a_3 & b_3 \end{vmatrix}$ and the cofactor of b_3 is $B_3 = -\begin{vmatrix} a_1 & c_1 \\ a_2 & c_2 \end{vmatrix}$.

When cofactors are used, the expansions of Δ_4 given above take the more compact form

$$\begin{aligned} \Delta_4 &= a_1A_1 + b_1B_1 + c_1C_1 + d_1D_1 && \text{(along the first row)} \\ &= a_1A_1 + a_2A_2 + a_3A_3 + a_4A_4 && \text{(along the first column)} \\ &= a_4A_4 + b_4B_4 + c_4C_4 + d_4D_4 && \text{(along the fourth row)} \\ &= b_1B_1 + b_2B_2 + b_3B_3 + b_4B_4 && \text{(along the second column).} \end{aligned}$$

See Problems 3-4.

PROPERTIES OF DETERMINANTS. Subject always to our assumption of equivalent expansions of a determinant along any of its rows or columns, the following theorems may be proved by mathematical induction.

THEOREM I. If two rows (two columns) of a determinant are identical, the value of the determinant is zero. For example,

$$\begin{vmatrix} 2 & 3 & 2 \\ 3 & 1 & 3 \\ 1 & 4 & 1 \end{vmatrix} = 0$$

COROLLARY I. If each of the elements of a row (a column) of a determinant is multiplied by the cofactor of the corresponding element of another row (column), the sum of the products is zero.

THEOREM II. If the elements of a row (a column) of a determinant are multiplied by any number m, the determinant is multiplied by m. For example,

$$5\begin{vmatrix} 2 & 3 & 4 \\ 3 & -1 & 2 \\ 1 & 4 & -3 \end{vmatrix} = \begin{vmatrix} 10 & 3 & 4 \\ 15 & -1 & 2 \\ 5 & 4 & -3 \end{vmatrix} = \begin{vmatrix} 2 & 3 & 4 \\ 15 & -5 & 10 \\ 1 & 4 & -3 \end{vmatrix}$$

THEOREM III. If each of the elements of a row (a column) of a determinant is expressed as the sum of two or more numbers, the determinant may be written as the sum of two or more determinants. For example,

$$\begin{vmatrix} 2 & 5 & 4 \\ 4 & -2 & 3 \\ 1 & -4 & 3 \end{vmatrix} = \begin{vmatrix} -2+4 & 5 & 4 \\ 3+1 & -2 & 3 \\ 1+0 & -4 & 3 \end{vmatrix} = \begin{vmatrix} -2 & 5 & 4 \\ 3 & -2 & 3 \\ 1 & -4 & 3 \end{vmatrix} + \begin{vmatrix} 4 & 5 & 4 \\ 1 & -2 & 3 \\ 0 & -4 & 3 \end{vmatrix}$$

THEOREM IV. If to the elements of any row (any column) of a determinant there is added m times the corresponding elements of another row (another column), the value of the determinant is unchanged. For example,

$$\begin{vmatrix} -2 & 5 & 4 \\ 3 & -2 & 2 \\ 1 & -4 & 3 \end{vmatrix} = \begin{vmatrix} -2 & 5+4(-2) & 4 \\ 3 & -2+4(3) & 2 \\ 1 & -4+4(1) & 3 \end{vmatrix} = \begin{vmatrix} -2 & -3 & 4 \\ 3 & 10 & 2 \\ 1 & 0 & 3 \end{vmatrix}$$

See Problem 5.

EVALUATION OF DETERMINANTS. A determinant of any order may be evaluated by expanding it and all subsequent determinants (minors) thus obtained along a row or column. This procedure may be greatly simplified by the use of Theorem IV. In Problem 6, (*a*) and (*b*), a row (column) containing an element +1 or −1 is used to obtain an equivalent determinant having an element 0 in another row (column). In (*c*) and (*d*), the same theorem has been used to obtain an element +1 or −1; this procedure is to be followed when the given determinant is lacking in these elements.

The revised procedure consists in first obtaining an equivalent determinant in which all the elements, save one, in some row (column) are zeros and then expanding along that row (column).

Example 1. Evaluate $\begin{vmatrix} 1 & 4 & 3 & 1 \\ 2 & 8 & 2 & 5 \\ 4 & -4 & -1 & -3 \\ 2 & 5 & 3 & 3 \end{vmatrix}$

Using the first column since it contains the element 1 in the first row, we obtain an equivalent determinant all of whose elements, save the first, in the first row are zeros. We have

$$\begin{vmatrix} 1 & 4 & 3 & 1 \\ 2 & 8 & 2 & 5 \\ 4 & -4 & -1 & -3 \\ 2 & 5 & 3 & 3 \end{vmatrix} = \begin{vmatrix} 1 & 4+(-4)1 & 3+(-3)1 & 1+(-1)1 \\ 2 & 8+(-4)2 & 2+(-3)2 & 5+(-1)2 \\ 4 & -4+(-4)4 & -1+(-3)4 & -3+(-1)4 \\ 2 & 5+(-4)2 & 3+(-3)2 & 3+(-1)2 \end{vmatrix} = \begin{vmatrix} 1 & 0 & 0 & 0 \\ 2 & 0 & -4 & 3 \\ 4 & -20 & -13 & -7 \\ 2 & -3 & -3 & 1 \end{vmatrix}$$

$$= \begin{vmatrix} 0 & -4 & 3 \\ -20 & -13 & -7 \\ -3 & -3 & 1 \end{vmatrix} \quad \text{(by expanding along the first row).}$$

Expanding the resulting determinant along the first row to take full advantage of the element 0, we have

$$\begin{vmatrix} 0 & -4 & 3 \\ -20 & -13 & -7 \\ -3 & -3 & 1 \end{vmatrix} = 4(-20-21) + 3(60-39) = -101.$$

SOLVED PROBLEMS

1. Write the minors of the elements a_1, b_3, c_2 of Δ_3.

The minor of a_1 is $\begin{vmatrix} b_2 & c_2 \\ b_3 & c_3 \end{vmatrix}$, the minor of b_3 is $\begin{vmatrix} a_1 & c_1 \\ a_2 & c_2 \end{vmatrix}$, the minor of c_1 is $\begin{vmatrix} a_2 & b_2 \\ a_3 & b_3 \end{vmatrix}$.

2. Evaluate: (*a*) $\begin{vmatrix} -1 & 3 & -4 \\ 0 & 2 & 0 \\ 2 & -3 & 5 \end{vmatrix}$ by expanding along the second row,

(*b*) $\begin{vmatrix} 8 & 0 & 2 & 0 \\ 5 & 1 & -3 & 0 \\ -4 & 3 & 7 & -3 \\ 4 & 0 & 6 & 0 \end{vmatrix}$ by expanding along the fourth column.

(*a*) $$\begin{vmatrix} -1 & 3 & -4 \\ 0 & 2 & 0 \\ 2 & -3 & 5 \end{vmatrix} = -0\begin{vmatrix} 3 & -4 \\ -3 & 5 \end{vmatrix} + 2\begin{vmatrix} -1 & -4 \\ 2 & 5 \end{vmatrix} - 0\begin{vmatrix} -1 & 3 \\ 2 & -3 \end{vmatrix} = 2\begin{vmatrix} -1 & -4 \\ 2 & 5 \end{vmatrix} = 2\cdot 3 = 6$$

(*b*) $$\begin{vmatrix} 8 & 0 & 2 & 0 \\ 5 & 1 & -3 & 0 \\ -4 & 3 & 7 & -3 \\ 4 & 0 & 6 & 0 \end{vmatrix} = -(-3)\begin{vmatrix} 8 & 0 & 2 \\ 5 & 1 & -3 \\ 4 & 0 & 6 \end{vmatrix} = 3\left\{+1\begin{vmatrix} 8 & 2 \\ 4 & 6 \end{vmatrix}\right\} = 3(48-8) = 120$$

3. (*a*) Write the cofactors of the elements a_1, b_3, c_2, d_4 of Δ_4.

The cofactor of a_1 is $A_1 = +\begin{vmatrix} b_2 & c_2 & d_2 \\ b_3 & c_3 & d_3 \\ b_4 & c_4 & d_4 \end{vmatrix}$, of b_3 is $B_3 = -\begin{vmatrix} a_1 & c_1 & d_1 \\ a_2 & c_2 & d_2 \\ a_4 & c_4 & d_4 \end{vmatrix}$

of c_2 is $C_2 = -\begin{vmatrix} a_1 & b_1 & d_1 \\ a_3 & b_3 & d_3 \\ a_4 & b_4 & d_4 \end{vmatrix}$, of d_4 is $D_4 = +\begin{vmatrix} a_1 & b_1 & c_1 \\ a_2 & b_2 & c_2 \\ a_3 & b_3 & c_3 \end{vmatrix}$

(*b*) Write the expansion of Δ_4 along (1) the second row, (2) the third column, using cofactors.

$$(1)\quad \Delta_4 = a_2A_2 + b_2B_2 + c_2C_2 + d_2D_2$$

$$(2)\quad \Delta_4 = c_1C_1 + c_2C_2 + c_3C_3 + c_4C_4$$

4. Express $g_1C_1 + g_2C_2 + g_3C_3 + g_4C_4$, where the C_i are cofactors of the elements c_i of Δ_4, as a determinant.

Since the cofactors C_i contain no elements with basal letter c, we replace c_1, c_2, c_3, c_4 by g_1, g_2, g_3, g_4 respectively in Problem 3(*b*) and obtain

$$g_1C_1 + g_2C_2 + g_3C_3 + g_4C_4 = \begin{vmatrix} a_1 & b_1 & g_1 & d_1 \\ a_2 & b_2 & g_2 & d_2 \\ a_3 & b_3 & g_3 & d_3 \\ a_4 & b_4 & g_4 & d_4 \end{vmatrix}$$

5. Prove by induction: If two rows (two columns) of a determinant are identical, the value of the determinant is zero.

The theorem is true for determinants of order 2 since $\begin{vmatrix} a_1 & a_1 \\ a_2 & a_2 \end{vmatrix} = a_1a_2 - a_1a_2 = 0$.

Let us assume the theorem true for determinants of order k and consider a determinant Δ of order $(k+1)$ in which two columns are identical. When Δ is expanded along any column, other than the two with identical elements, each cofactor involved is a determinant of order k with two columns identical and, by assumption, is equal to zero. Thus, Δ is equal to zero and the theorem is proved by induction.

6. From the determinant $\begin{vmatrix} 1 & 4 & 3 & 1 \\ 2 & 8 & 2 & 5 \\ 4 & -4 & -1 & -3 \\ 2 & 5 & 3 & 3 \end{vmatrix}$ obtain an equivalent determinant

(a) by adding -4 times the elements of the first column to the corresponding elements of the second column,
(b) by adding 3 times the elements of the third row to the corresponding elements of the fourth row,
(c) by adding -1 times the elements of the third column to the corresponding elements of the second column,
(d) by adding -2 times the elements of the fourth row to the corresponding elements of the second row.

(a) $$\begin{vmatrix} 1 & 4 & 3 & 1 \\ 2 & 8 & 2 & 5 \\ 4 & -4 & -1 & -3 \\ 2 & 5 & 3 & 3 \end{vmatrix} = \begin{vmatrix} 1 & 4+(-4)1 & 3 & 1 \\ 2 & 8+(-4)2 & 2 & 5 \\ 4 & -4+(-4)4 & -1 & -3 \\ 2 & 5+(-4)2 & 3 & 3 \end{vmatrix} = \begin{vmatrix} 1 & 0 & 3 & 1 \\ 2 & 0 & 2 & 5 \\ 4 & -20 & -1 & -3 \\ 2 & -3 & 3 & 3 \end{vmatrix}$$

(b) $$\begin{vmatrix} 1 & 4 & 3 & 1 \\ 2 & 8 & 2 & 5 \\ 4 & -4 & -1 & -3 \\ 2 & 5 & 3 & 3 \end{vmatrix} = \begin{vmatrix} 1 & 4 & 3 & 1 \\ 2 & 8 & 2 & 5 \\ 4 & -4 & -1 & -3 \\ 2+(3)4 & 5+(3)(-4) & 3+(3)(-1) & 3+(3)(-3) \end{vmatrix} = \begin{vmatrix} 1 & 4 & 3 & 1 \\ 2 & 8 & 2 & 5 \\ 4 & -4 & -1 & -3 \\ 14 & -7 & 0 & -6 \end{vmatrix}$$

(c) $$\begin{vmatrix} 1 & 4 & 3 & 1 \\ 2 & 8 & 2 & 5 \\ 4 & -4 & -1 & -3 \\ 2 & 5 & 3 & 3 \end{vmatrix} = \begin{vmatrix} 1 & 4+(-1)3 & 3 & 1 \\ 2 & 8+(-1)2 & 2 & 5 \\ 4 & -4+(-1)(-1) & -1 & -3 \\ 2 & 5+(-1)3 & 3 & 3 \end{vmatrix} = \begin{vmatrix} 1 & 1 & 3 & 1 \\ 2 & 6 & 2 & 5 \\ 4 & -3 & -1 & -3 \\ 2 & 2 & 3 & 3 \end{vmatrix}$$

(d) $$\begin{vmatrix} 1 & 4 & 3 & 1 \\ 2 & 8 & 2 & 5 \\ 4 & -4 & -1 & -3 \\ 2 & 5 & 3 & 3 \end{vmatrix} = \begin{vmatrix} 1 & 4 & 3 & 1 \\ 2+(-2)2 & 8+(-2)5 & 2+(-2)3 & 5+(-2)3 \\ 4 & -4 & -1 & -3 \\ 2 & 5 & 3 & 3 \end{vmatrix} = \begin{vmatrix} 1 & 4 & 3 & 1 \\ -2 & -2 & -4 & -1 \\ 4 & -4 & -1 & -3 \\ 2 & 5 & 3 & 3 \end{vmatrix}$$

7. Evaluate:

(a) $$\begin{vmatrix} 50 & 2 & -9 \\ 250 & -10 & 45 \\ -150 & 6 & 27 \end{vmatrix} = 50\cdot2\cdot9 \begin{vmatrix} 1 & 1 & -1 \\ 5 & -5 & 5 \\ -3 & 3 & 3 \end{vmatrix} = 900\cdot5\cdot3 \begin{vmatrix} 1 & 1 & -1 \\ 1 & -1 & 1 \\ -1 & 1 & 1 \end{vmatrix} = 13{,}500 \begin{vmatrix} 2 & 0 & 0 \\ 1 & -1 & 1 \\ -1 & 1 & 1 \end{vmatrix}$$
$$= 27{,}000 \begin{vmatrix} -1 & 1 \\ 1 & 1 \end{vmatrix} = -54{,}000 .$$

(b) $$\begin{vmatrix} 3 & 2 & -3 & 1 \\ -1 & -3 & 5 & 2 \\ 2 & 1 & 6 & -3 \\ 5 & 4 & -3 & 4 \end{vmatrix} = \begin{vmatrix} 3+(-3)1 & 2+(-2)1 & -3+(3)1 & 1 \\ -1+(-3)2 & -3+(-2)2 & 5+(3)2 & 2 \\ 2+(-3)(-3) & 1+(-2)(-3) & 6+(3)(-3) & -3 \\ 5+(-3)4 & 4+(-2)4 & -3+(3)4 & 4 \end{vmatrix} = \begin{vmatrix} 0 & 0 & 0 & 1 \\ -7 & -7 & 11 & 2 \\ 11 & 7 & -3 & -3 \\ -7 & -4 & 9 & 4 \end{vmatrix}$$
$$= -\begin{vmatrix} -7 & -7 & 11 \\ 11 & 7 & -3 \\ -7 & -4 & 9 \end{vmatrix} = -\begin{vmatrix} 0 & -7 & 11 \\ 4 & 7 & -3 \\ -3 & -4 & 9 \end{vmatrix} = -[7(36-9) + 11(-16+21)] = -244 .$$

(c) $$\begin{vmatrix} 2 & -1 & -2 & 3 \\ 3 & 2 & 4 & -1 \\ 2 & 4 & 1 & -5 \\ 4 & -3 & 2 & 1 \end{vmatrix} = \begin{vmatrix} 0 & -1 & 0 & 0 \\ 7 & 2 & 0 & 5 \\ 10 & 4 & -7 & 7 \\ -2 & -3 & 8 & -8 \end{vmatrix} = \begin{vmatrix} 7 & 0 & 5 \\ 10 & -7 & 7 \\ -2 & 8 & -8 \end{vmatrix} = 5\cdot66 = 330 .$$

SUPPLEMENTARY PROBLEMS

8. Verify: The value of a determinant is unchanged if the rows are written as columns.

Hint: Show that $\begin{vmatrix} a_1 & b_1 & c_1 \\ a_2 & b_2 & c_2 \\ a_3 & b_3 & c_3 \end{vmatrix} = \begin{vmatrix} a_1 & a_2 & a_3 \\ b_1 & b_2 & b_3 \\ c_1 & c_2 & c_3 \end{vmatrix}$.

9. Prove by induction:

(a) The expansion of a determinant of order n contains $n!$ terms.

(b) If two rows (columns) of a determinant are interchanged, the sign of the determinant is changed.

Hint: Show that $\begin{vmatrix} a_1 & b_1 \\ a_2 & b_2 \end{vmatrix} = -\begin{vmatrix} b_1 & a_1 \\ b_2 & a_2 \end{vmatrix}$ and proceed as in Problem 5.

10. Show, without expanding the determinants, that:

(a) $$\begin{vmatrix} 3 & 2 & 1 & 4 \\ -1 & 5 & 2 & 6 \\ 2 & -4 & 7 & -5 \\ -2 & 1 & 3 & 5 \end{vmatrix} = -\begin{vmatrix} 3 & 2 & 1 & 4 \\ 2 & -4 & 7 & -5 \\ -1 & 5 & 2 & 6 \\ -2 & 1 & 3 & 5 \end{vmatrix} = -\begin{vmatrix} 3 & 2 & 1 & 4 \\ -2 & 1 & 3 & 5 \\ 2 & -4 & 7 & -5 \\ -1 & 5 & 2 & 6 \end{vmatrix} = -\begin{vmatrix} 3 & -2 & 2 & -1 \\ 2 & 1 & -4 & 5 \\ 1 & 3 & 7 & 2 \\ 4 & 5 & -5 & 6 \end{vmatrix}.$$

(b) $$\begin{vmatrix} 10 & 0 & -2 \\ -10 & 3 & -4 \\ -5 & -2 & 3 \end{vmatrix} = -5\begin{vmatrix} -2 & 0 & -2 \\ 2 & 3 & -4 \\ 1 & -2 & 3 \end{vmatrix} = 10\begin{vmatrix} 1 & 0 & 1 \\ 2 & 3 & -4 \\ 1 & -2 & 3 \end{vmatrix}.$$

(c) $\begin{vmatrix} -3 & -1 & 2 \\ 1 & -3 & 3 \\ 4 & -2 & 1 \end{vmatrix} = 0$. Hint: Subtract the first row from the second.

(d) $\begin{vmatrix} a-3b & a+b & a+5b & e \\ a-2b & a+2b & a+6b & f \\ a-b & a+3b & a+7b & g \\ a & a+4b & a+8b & h \end{vmatrix} = 0$. (Note. The elements of the first three columns form an A.P.)

(e) $\begin{vmatrix} 0 & a_1 & a_2 \\ -a_1 & 0 & a_3 \\ -a_2 & -a_3 & 0 \end{vmatrix} = 0$. Hint: Write the rows as columns and factor -1 from each row.

11. Verify: If the corresponding elements of two rows (columns) of a determinant are proportional, the value of the determinant is zero.

12. Evaluate each of the following determinants.

(a) $\begin{vmatrix} -3 & 6 & -1 & 1 \\ -4 & -3 & -2 & 4 \\ 5 & -4 & 1 & 3 \\ -1 & -5 & 0 & -1 \end{vmatrix}$ (b) $\begin{vmatrix} 1 & -1 & 1 & 2 \\ 3 & -2 & 4 & -3 \\ 5 & 4 & 1 & 2 \\ -3 & 0 & 3 & 1 \end{vmatrix}$ (c) $\begin{vmatrix} 3 & -1 & -1 & -4 \\ 2 & 3 & -6 & 1 \\ 4 & -1 & 3 & 1 \\ 3 & -1 & 5 & 2 \end{vmatrix}$ (d) $\begin{vmatrix} 2 & 4 & 4 & -3 & 4 \\ 1 & 3 & 1 & 0 & -1 \\ -1 & 3 & 1 & 2 & -1 \\ 4 & 8 & 11 & -10 & 9 \\ 2 & 6 & 9 & -12 & 5 \end{vmatrix}$

Ans. (a) −50, (b) −397, (c) −78, (d) −316

13. Show that $\begin{vmatrix} 1 & 1 & 1 & 1 \\ a & b & c & d \\ a^2 & b^2 & c^2 & d^2 \\ a^3 & b^3 & c^3 & d^3 \end{vmatrix} = (b-a)(c-a)(d-a)(c-b)(d-b)(d-c)$.

CHAPTER 25

Systems of Linear Equations

SYSTEMS OF n LINEAR EQUATIONS IN n UNKNOWNS. Consider, for the sake of brevity, the system of four linear equations in four unknowns

$$(1)\qquad \begin{cases} a_1x + b_1y + c_1z + d_1w = k_1 \\ a_2x + b_2y + c_2z + d_2w = k_2 \\ a_3x + b_3y + c_3z + d_3w = k_3 \\ a_4x + b_4y + c_4z + d_4w = k_4 \end{cases}$$

in which each equation is written with the unknowns x, y, z, w in that order on the left side and the constant term on the right side. Form

$$D = \begin{vmatrix} a_1 & b_1 & c_1 & d_1 \\ a_2 & b_2 & c_2 & d_2 \\ a_3 & b_3 & c_3 & d_3 \\ a_4 & b_4 & c_4 & d_4 \end{vmatrix},$$ the determinant of the coefficients of the unknowns,

and from it the determinants

$$N_x = \begin{vmatrix} k_1 & b_1 & c_1 & d_1 \\ k_2 & b_2 & c_2 & d_2 \\ k_3 & b_3 & c_3 & d_3 \\ k_4 & b_4 & c_4 & d_4 \end{vmatrix}, \quad N_y = \begin{vmatrix} a_1 & k_1 & c_1 & d_1 \\ a_2 & k_2 & c_2 & d_2 \\ a_3 & k_3 & c_3 & d_3 \\ a_4 & k_4 & c_4 & d_4 \end{vmatrix}, \quad N_z = \begin{vmatrix} a_1 & b_1 & k_1 & d_1 \\ a_2 & b_2 & k_2 & d_2 \\ a_3 & b_3 & k_3 & d_3 \\ a_4 & b_4 & k_4 & d_4 \end{vmatrix}, \quad N_w = \begin{vmatrix} a_1 & b_1 & c_1 & k_1 \\ a_2 & b_2 & c_2 & k_2 \\ a_3 & b_3 & c_3 & k_3 \\ a_4 & b_4 & c_4 & k_4 \end{vmatrix}$$

by replacing the column of coefficients of the indicated unknown by the column of constants.

(*A*) If $D \neq 0$, the system (*1*) has the unique solution

$$x = N_x/D, \quad y = N_y/D, \quad z = N_z/D, \quad w = N_w/D$$

See Problems 1-2.

(*B*) If $D = 0$ and at least one of $N_x, N_y, N_z, N_w \neq 0$, the system has no solution. For, if $D = 0$ and $N_x \neq 0$, then $x \cdot D = N_x$ leads to a contradiction. Such systems are called *inconsistent*.

See Problem 3.

(*C*) If $D = 0$ and $N_x = N_y = N_z = N_w = 0$, the system may or may not have a solution. A system having an infinite number of solutions is called *dependent*.

For systems of three or four equations, the simplest procedure is to evaluate D. If $D \neq 0$, proceed as in (*A*); if $D = 0$, proceed as in Chapter 6.

See Problem 4.

SYSTEMS OF m LINEAR EQUATIONS IN $n > m$ UNKNOWNS. Ordinarily if there are fewer equations than unknowns, the system will have an infinite number of solutions.

To solve a consistent system of m equations, solve for m of the unknowns (in certain cases for $p < m$ of the unknowns) in terms of the others. See Problem 5.

SYSTEMS OF n EQUATIONS IN $m < n$ UNKNOWNS. Ordinarily if there are more equations than unknowns the system is inconsistent. However, if $p \le m$ of the equations have a solution and if this solution satisfies each of the remaining equations, the system is consistent. See Problem 6.

A HOMOGENEOUS EQUATION is one in which all terms are of the same degree; otherwise, the equation is called *non-homogeneous*. For example, the linear equation $2x + 3y - 4z = 5$ is non-homogeneous, while $2x + 3y - 4z = 0$ is homogeneous.

Every system of homogeneous linear equations

$$\begin{array}{l} a_1x + b_1y + c_1z + \cdots\cdots = 0 \\ a_2x + b_2y + c_2z + \cdots\cdots = 0 \\ \cdots\cdots\cdots\cdots\cdots\cdots \\ \cdots\cdots\cdots\cdots\cdots\cdots \\ a_nx + b_ny + c_nz + \cdots\cdots = 0 \end{array}$$

always has the *trivial solution* $x = 0,\ y = 0,\ z = 0,\ \ldots.$

A system of n homogeneous linear equations in n unknowns has *only* the trivial solution if D, the determinant of the coefficients, is not equal to zero. If $D = 0$, the system has non-trivial solutions as well. See Problem 7.

SOLVED PROBLEMS

1. Solve the system $\begin{cases} 3x - 2y - z - 4w = 7 & (1) \\ x + 3z + 2w = -10 & (2) \\ x + 4y + 2z + w = 0 & (3) \\ 2x + 3y + 3w = 1 & (4) \end{cases}$

We find $D = \begin{vmatrix} 3 & -2 & -1 & -4 \\ 1 & 0 & 3 & 2 \\ 1 & 4 & 2 & 1 \\ 2 & 3 & 0 & 3 \end{vmatrix} = \begin{vmatrix} 3 & -2 & -10 & -10 \\ 1 & 0 & 0 & 0 \\ 1 & 4 & -1 & -1 \\ 2 & 3 & -6 & -1 \end{vmatrix} = -\begin{vmatrix} -2 & -10 & -10 \\ 4 & -1 & -1 \\ 3 & -6 & -1 \end{vmatrix} = 2\begin{vmatrix} 1 & 5 & 5 \\ 4 & -1 & -1 \\ 3 & -6 & -1 \end{vmatrix} = -210$

$$N_x = \begin{vmatrix} 7 & -2 & -1 & -4 \\ -10 & 0 & 3 & 2 \\ 0 & 4 & 2 & 1 \\ 1 & 3 & 0 & 3 \end{vmatrix} = \begin{vmatrix} 0 & -23 & -1 & -25 \\ 0 & 30 & 3 & 32 \\ 0 & 4 & 2 & 1 \\ 1 & 3 & 0 & 3 \end{vmatrix} = -\begin{vmatrix} -23 & -1 & -25 \\ 30 & 3 & 32 \\ 4 & 2 & 1 \end{vmatrix} = -\begin{vmatrix} -23 & -1 & -2 \\ 30 & 3 & 2 \\ 4 & 2 & -3 \end{vmatrix} = -105$$

$$N_y = \begin{vmatrix} 3 & 7 & -1 & -4 \\ 1 & -10 & 3 & 2 \\ 1 & 0 & 2 & 1 \\ 2 & 1 & 0 & 3 \end{vmatrix} = \begin{vmatrix} 3 & 7 & -7 & -7 \\ 1 & -10 & 1 & 1 \\ 1 & 0 & 0 & 0 \\ 2 & 1 & -4 & 1 \end{vmatrix} = \begin{vmatrix} 7 & -7 & -7 \\ -10 & 1 & 1 \\ 1 & -4 & 1 \end{vmatrix} = 7\begin{vmatrix} 1 & -1 & -1 \\ -10 & 1 & 1 \\ 1 & -4 & 1 \end{vmatrix} = -315$$

$$N_z = \begin{vmatrix} 3 & -2 & 7 & -4 \\ 1 & 0 & -10 & 2 \\ 1 & 4 & 0 & 1 \\ 2 & 3 & 1 & 3 \end{vmatrix} = \begin{vmatrix} 3 & -14 & 7 & -7 \\ 1 & -4 & -10 & 1 \\ 1 & 0 & 0 & 0 \\ 2 & -5 & 1 & 1 \end{vmatrix} = \begin{vmatrix} -14 & 7 & -7 \\ -4 & -10 & 1 \\ -5 & 1 & 1 \end{vmatrix} = -7\begin{vmatrix} 2 & -1 & 1 \\ -4 & -10 & 1 \\ -5 & 1 & 1 \end{vmatrix} = 525$$

$$N_w = \begin{vmatrix} 3 & -2 & -1 & 7 \\ 1 & 0 & 3 & -10 \\ 1 & 4 & 2 & 0 \\ 2 & 3 & 0 & 1 \end{vmatrix} = \begin{vmatrix} 3 & -14 & -7 & 7 \\ 1 & -4 & 1 & -10 \\ 1 & 0 & 0 & 0 \\ 2 & -5 & -4 & 1 \end{vmatrix} = \begin{vmatrix} -14 & -7 & 7 \\ -4 & 1 & -10 \\ -5 & -4 & 1 \end{vmatrix} = -7\begin{vmatrix} 2 & 1 & -1 \\ -4 & 1 & -10 \\ -5 & -4 & 1 \end{vmatrix} = 315 .$$

Then $x = \dfrac{N_x}{D} = \dfrac{-105}{-210} = \dfrac{1}{2}$, $y = \dfrac{N_y}{D} = \dfrac{-315}{-210} = \dfrac{3}{2}$, $z = \dfrac{N_z}{D} = \dfrac{525}{-210} = -\dfrac{5}{2}$, $w = \dfrac{N_w}{D} = \dfrac{315}{-210} = -\dfrac{3}{2}$.

Check. Using (*1*), $3(1/2) - 2(3/2) - (-5/2) - 4(-3/2) = (3-6+5+12)/2 = 7$.

Note. The above system permits some variation in procedure. For example, having found $x = 1/2$ and $y = 3/2$ using determinants, the value of w may be obtained by substituting in (*4*)

$$2(1/2) + 3(3/2) + 3w = 1, \quad 3w = -9/2, \quad w = -3/2$$

and the value of z may then be obtained by substituting in (*2*)

$$(1/2) + 3z + 2(-3/2) = -10, \quad 3z = -15/2, \quad z = -5/2 .$$

The solution may be checked by substituting in (*1*) or (*3*).

2. Solve the system $\begin{cases} 2x + y + 5z + w = 5 & (1) \\ x + y - 3z - 4w = -1 & (2) \\ 3x + 6y - 2z + w = 8 & (3) \\ 2x + 2y + 2z - 3w = 2 & (4) \end{cases}$

We have $D = \begin{vmatrix} 2 & 1 & 5 & 1 \\ 1 & 1 & -3 & -4 \\ 3 & 6 & -2 & 1 \\ 2 & 2 & 2 & -3 \end{vmatrix} = -120$, $N_x = \begin{vmatrix} 5 & 1 & 5 & 1 \\ -1 & 1 & -3 & -4 \\ 8 & 6 & -2 & 1 \\ 2 & 2 & 2 & -3 \end{vmatrix} = -240$,

$N_y = \begin{vmatrix} 2 & 5 & 5 & 1 \\ 1 & -1 & -3 & -4 \\ 3 & 8 & -2 & 1 \\ 2 & 2 & 2 & -3 \end{vmatrix} = -24$, $N_z = \begin{vmatrix} 2 & 1 & 5 & 1 \\ 1 & 1 & -1 & -4 \\ 3 & 6 & 8 & 1 \\ 2 & 2 & 2 & -3 \end{vmatrix} = 0$.

Then $x = \dfrac{N_x}{D} = \dfrac{-240}{-120} = 2$, $y = \dfrac{N_y}{D} = \dfrac{-24}{-120} = \dfrac{1}{5}$ and $z = \dfrac{N_z}{D} = \dfrac{0}{-120} = 0$.

Substituting in (*1*), $2(2) + (1/5) + 5(0) + w = 5$ and $w = 4/5$.

Check. Using (*2*), $(2) + (1/5) - 3(0) - 4(4/5) = -1$.

3. Show that the system $\begin{cases} 2x + y + 5z + w = 2 \\ x + y - z - 4w = 1 \\ 3x + 6y + 8z + w = 3 \\ 2x + 2y + 2z - 3w = 1 \end{cases}$ is inconsistent.

Since $D = \begin{vmatrix} 2 & 1 & 5 & 1 \\ 1 & 1 & -1 & -4 \\ 3 & 6 & 8 & 1 \\ 2 & 2 & 2 & -3 \end{vmatrix} = 0$ while $N_x = \begin{vmatrix} 2 & 1 & 5 & 1 \\ 1 & 1 & -1 & -4 \\ 3 & 6 & 8 & 1 \\ 1 & 2 & 2 & -3 \end{vmatrix} = -80 \neq 0$, the system is inconsistent.

4. Solve when possible:

(*a*) $\begin{cases} 2x - 3y + z = 0 & (1) \\ x + 5y - 3z = 3 & (2) \\ 5x + 12y - 8z = 9 & (3) \end{cases}$

(*b*) $\begin{cases} x + 2y + 3z = 2 & (1) \\ 2x + 4y + z = -1 & (2) \\ 3x + 6y + 5z = 2 & (3) \end{cases}$

(*c*) $\begin{cases} 6x - 2y + z = 1 & (1) \\ x - 4y + 2z = 0 & (2) \\ 4x + 6y - 3z = 0 & (3) \end{cases}$

(*d*) $\begin{cases} x + 2y - 3z + 5w = 11 & (1) \\ 4x - y + z - 2w = 0 & (2) \\ 2x + 4y - 6z + 10w = 22 & (3) \\ 5x + y - 2z + 3w = 11 & (4) \end{cases}$

(*a*) Here $D = 0$; we shall eliminate the variable x.

$$(1) - 2(2): \quad -13y + 7z = -6$$
$$(3) - 5(2): \quad -13y + 7z = -6$$

Then $y = \dfrac{7z+6}{13}$ and, from (2), $x = 3 - 5y + 3z = \dfrac{4z+9}{13}$.

The solutions may be written as $x = \dfrac{4a+9}{13}$, $y = \dfrac{7a+6}{13}$, $z = a$, where a is arbitrary.

(*b*) Here $D = 0$; we shall eliminate x.

$$(2) - 2(1): \quad -5z = -5$$
$$(3) - 3(1): \quad -4z = -4$$

Then $z = 1$ and each of the given equations reduces to $x + 2y = -1$. Note that the same situation arises when y is eliminated.

The solution may be written $x = -1 - 2y$, $z = 1$ or as $x = -1 - 2a$, $y = a$, $z = 1$, where a is arbitrary.

(*c*) Here $D = 0$; we shall eliminate z.

$$(2) - 2(1): \quad -11x = -2$$
$$(3) + 3(1): \quad 22x = 3$$

The system is inconsistent.

(*d*) Here $D = 0$; we shall eliminate x.

$$(2) - 4(1): \quad -9y + 13z - 22w = -44$$
$$(3) - 2(1): \quad 0 = 0$$
$$(4) - 5(1): \quad -9y + 13z - 22w = -44$$

Then $y = \dfrac{44 + 13z - 22w}{9}$ and, from (1), $x = \dfrac{11 + z - w}{9}$. The solutions are

$$x = \frac{11 + a - b}{9}, \quad y = \frac{44 + 13a - 22b}{9}, \quad z = a, \; w = b, \quad \text{where } a \text{ and } b \text{ are arbitrary.}$$

5. (*a*) The system of two equations in four unknowns

$$\begin{cases} x + 2y + 3z - 4w = 5 \\ 3x - y - 5z - 5w = 1 \end{cases}$$

may be solved for any two of the unknowns in terms of the others; for example, $x = 1 + z + 2w$, $y = 2 - 2z + w$.

(*b*) The system of three equations in four unknowns

$$\begin{cases} x + 2y + 3z - 4w = 5 \\ 3x - y - 5z - 5w = 1 \\ 2x + 3y + z - w = 8 \end{cases}$$

may be solved for any three of the unknowns in terms of the fourth; for example, $x = 1 + 4w$, $y = 2 - 3w$, $z = 2w$.

(*c*) The system of three equations in four unknowns

$$\begin{cases} x + 2y + 3z - 4w = 5 \\ 3x - y - 5z - 5w = 1 \\ 2x - 3y - 8z - w = -4 \end{cases}$$

may be solved for any two of the unknowns in terms of the others. Note that the third equation is the same as the second minus the first. We solve any two of these equations, say the first and second, and obtain the solution given in (*a*) above.

6. Solve when possible.

(a) $\begin{cases} 3x - 2y = 1 \\ 4x + 3y = 41 \\ 6x + 2y = 23 \end{cases}$ (b) $\begin{cases} 3x + y = 1 \\ 5x - 2y = 20 \\ 4x + 5y = -17 \end{cases}$ (c) $\begin{cases} x + y + z = 2 \\ 4x + 5y - 3z = -15 \\ 5x - 3y + 4z = 23 \\ 7x - y + 6z = 27 \end{cases}$ (d) $\begin{cases} x + y = 5 \\ y + z = 8 \\ x + z = 7 \\ 5x - 5y + z = 1 \end{cases}$

(a) The system $\begin{cases} 3x - 2y = 1 \\ 4x + 3y = 41 \end{cases}$ has solution $x = 5,\ y = 7$.

Since $6x + 2y = 6(5) + 2(7) \neq 23$, the given system is inconsistent.

(b) The system $\begin{cases} 3x + y = 1 \\ 5x - 2y = 20 \end{cases}$ has solution $x = 2,\ y = -5$.

Since $4x + 5y = 4(2) + 5(-5) = -17$, the given system is consistent with solution $x = 2$, $y = -5$.

(c) The system $\begin{cases} x + y + z = 2 \\ 4x + 5y - 3z = -15 \\ 5x - 3y + 4z = 23 \end{cases}$ has solution $x = 1,\ y = -2,\ z = 3$.

Since $7x - y + 6z = 7(1) - (-2) + 6(3) = 27$, the given system is consistent with solution $x = 1,\ y = -2,\ z = 3$.

(d) The system $\begin{cases} x + y = 5 \\ y + z = 8 \\ x + z = 7 \end{cases}$ has solution $x = 2,\ y = 3,\ z = 5$.

Since $5x - 5y + z = 5(2) - 5(3) + (5) \neq 1$, the given system is inconsistent.

Note. If the constant of the fourth equation of the system were changed from 1 to 0, the resulting system would be consistent.

7. Examine the following systems for non-trivial solutions.

(a) $\begin{cases} 2x - 3y + 3z = 0 \\ 3x - 4y + 5z = 0 \\ 5x + y + 2z = 0 \end{cases}$ (b) $\begin{cases} 4x + y - 2z = 0 \\ x - 2y + z = 0 \\ 11x - 4y - z = 0 \end{cases}$ (c) $\begin{cases} x + 2y + z = 0 \\ 3x + 6y + 3z = 0 \\ 5x + 10y + 5z = 0 \end{cases}$ (d) $\begin{cases} 2x + y = 0 \\ 3y - 2z = 0 \\ 2y + w = 0 \\ 4x - w = 0 \end{cases}$

(a) Since $D = \begin{vmatrix} 2 & -3 & 3 \\ 3 & -4 & 5 \\ 5 & 1 & 2 \end{vmatrix} \neq 0$, the system has only the trivial solution.

(b) Since $D = \begin{vmatrix} 4 & 1 & -2 \\ 1 & -2 & 1 \\ 11 & -4 & -1 \end{vmatrix} = 0$, there are non-trivial solutions.

The system $\begin{cases} 4x + y = 2z \\ x - 2y = -z \end{cases}$, for which $D = \begin{vmatrix} 4 & 1 \\ 1 & -2 \end{vmatrix} \neq 0$, has the solution $x = z/3$, $y = 2z/3$. This solution may be written as $x = a/3,\ y = 2a/3,\ z = a$ or $x = a,\ y = 2a,\ z = 3a$, where a is arbitrary, or as $x{:}y{:}z = 1{:}2{:}3$.

(c). Here $D = \begin{vmatrix} 1 & 2 & 1 \\ 3 & 6 & 3 \\ 5 & 10 & 5 \end{vmatrix} = 0$ and there are non-trivial solutions.

Since the minor of every element of D is zero, we cannot proceed as in (b). We solve the first equation for $x = -2y - z$ and write the solution as $x = -2a - b,\ y = a,\ z = b$, where a and b are arbitrary.

(d) Here $D = \begin{vmatrix} 2 & 1 & 0 & 0 \\ 0 & 3 & -2 & 0 \\ 0 & 2 & 0 & 1 \\ 4 & 0 & 0 & -1 \end{vmatrix} = 0$ and there are non-trivial solutions.

Take $x = a$, where a is arbitrary. From the first equation, $y = -2a$; from the second, $2z = 3y = -6a$ and $z = -3a$; and from the fourth equation, $w = 4a$.

Thus the solution is $x = a$, $y = -2a$, $z = -3a$, $w = 4a$, or $x:y:z:w = 1:-2:-3:4$.

SUPPLEMENTARY PROBLEMS

8. Solve, using determinants.

(a) $\begin{cases} x + y + z = 6 \\ y + z + w = 9 \\ z + w + x = 8 \\ w + x + y = 7 \end{cases}$ (b) $\begin{cases} 3x - 2y + 2z + w = 5 \\ 2x + 4y - z - 2w = 3 \\ 3x + 7y - z + 3w = 23 \\ x - 3y + 2z - 3w = -12 \end{cases}$ (c) $\begin{cases} x + y + z + w = 2 \\ 2x + 3y - 2z - w = 5 \\ 3x - 2y + z + 3w = 4 \\ 5x + 2y + 3z - 2w = -4 \end{cases}$

Ans. (a) $x = 1$, $y = 2$, $z = 3$, $w = 4$ (b) $x = 2$, $y = 1$, $z = -1$, $w = 3$ (c) $x = 1/2$, $y = 1$, $z = -3/2$, $w = 2$

9. Test for consistency, and solve when possible.

(a) $\begin{cases} 2x + 3y - 4z = 1 \\ 3x - y + 2z = -2 \\ 5x - 9y + 14z = 3 \end{cases}$ (b) $\begin{cases} x + 7y + 5z = -22 \\ x - 9y - 11z = 26 \\ x - y - 3z = 2 \end{cases}$ (c) $\begin{cases} x + y + z = 4 \\ 2x - 4y + 11z = -7 \\ 4x + 6y + z = 21 \end{cases}$

Ans. (a) Inconsistent (b) $x = 2z - 1$, $y = -z - 3$ (c) $x = \frac{1}{2}(3 - 5z)$, $y = \frac{1}{2}(5 + 3z)$

10. Solve, when possible.

(a) $\begin{cases} x - 3y + 11 = 0 \\ 3x + 2y - 33 = 0 \\ 2x - 3y + 4 = 0 \end{cases}$ (b) $\begin{cases} x - 2y - 8 = 0 \\ 3x + y - 3 = 0 \\ x - 10y + 32 = 0 \end{cases}$ (c) $\begin{cases} 2x - 3y - 7 = 0 \\ 5x + 4y + 17 = 0 \\ 4x - y + 1 = 0 \end{cases}$

Ans. (a) $x = 7$, $y = 6$ (b) No solution (c) $x = -1$, $y = -3$

11. Solve the systems.

(a) $\begin{cases} 2x - y + 3z = 8 \\ x + 3y - 2z = -3 \end{cases}$ (b) $\begin{cases} 4x + 2y + z = 13 \\ 2x + y - 2z = -6 \end{cases}$ (c) $\begin{cases} 4x + 2y - 6z + w = 10 \\ 3x - y - 9z - w = 7 \\ 7x + y - 11z - w = 13 \end{cases}$

Ans. (a) $x = 3 - z$, $y = -2 + z$ (b) $y = 4 - 2x$, $z = 5$, (c) $x = 7w/10$, $y = 2 - 23w/20$, $z = -1 + w/4$

12. Examine for non-trivial solutions.

(a) $\begin{cases} 3x + y - 9z = 0 \\ 4x - 3y + z = 0 \\ 6x - 11y + 21z = 0 \end{cases}$ (b) $\begin{cases} 2x - 3y - 5z = 0 \\ x + 2y - 13z = 0 \\ 9x - 10y - 30z = 0 \end{cases}$ (c) $\begin{cases} 2x - 3y + 2z - 9w = 0 \\ x + 4y - z + 3w = 0 \\ 3x - 2y - 2z - 6w = 0 \\ 7x + 11y + 3z - 6w = 0 \end{cases}$

Ans. (a) $x = 2z$, $y = 3z$ (b) $x = y = z = 0$ (c) $x = 2w$, $y = -w$, $z = w$

CHAPTER 26

Partial Fractions

A RATIONAL ALGEBRAIC FRACTION is the quotient of two polynomials. A rational fraction is called *proper* if the degree of the numerator is less than that of the denominator; otherwise, the fraction is called *improper*.

An improper fraction may be written as the sum of a polynomial and a proper fraction.

For example, $\frac{3}{x+2}$ is a proper fraction while $\frac{x^2+3x+5}{x+2}$ is an improper fraction.

Note that $\frac{x^2+3x+5}{x+2} = x+1+\frac{3}{x+2}$ is the sum of a polynomial and a proper fraction.

Two or more proper fractions may be summed to yield a single fraction whose denominator is the lowest common denominator of the several fractions. For example,

$$(A)\quad \frac{1}{x+2}+\frac{2}{3x-2} = \frac{5x+2}{(x+2)(3x-2)}, \qquad (B)\quad \frac{3}{x+2}+\frac{4}{(x+2)^2}-\frac{1}{x} = \frac{2x^2+6x-4}{x(x+2)^2},$$

$$(C)\quad \frac{1}{x-1}+\frac{3x-1}{x^2+2}-\frac{1}{(x^2+2)^2} = \frac{4x^4-4x^3+11x^2-9x+7}{(x-1)(x^2+2)^2}.$$

The problem of this chapter is to reverse the above process, that is, to resolve a given rational fraction into a sum of simpler proper fractions, called *partial fractions*. The cases considered here will be explained by means of examples. Proofs of the validity of the methods will be found in more advanced texts.

CASE 1. FACTORS OF THE DENOMINATOR LINEAR, NONE REPEATED.

Corresponding to each factor of the denominator form a partial fraction having an unknown constant as numerator and the factor as denominator.

Example 1. Resolve $\frac{5x+2}{(x+2)(3x-2)}$ into partial fractions.

Set $\frac{5x+2}{(x+2)(3x-2)} = \frac{A}{x+2}+\frac{B}{3x-2}$. Then

$$a)\quad \frac{5x+2}{(x+2)(3x-2)} = \frac{A(3x-2)+B(x+2)}{(x+2)(3x-2)} \quad \text{and} \quad b)\quad \frac{5x+2}{(x+2)(3x-2)} = \frac{(3A+B)x-2A+2B}{(x+2)(3x-2)}$$

are identities which hold for all values of x except possibly for $x=-2$ and $x=2/3$.

FIRST SOLUTION. Equating coefficients of like terms in the two members of the identity $5x+2 = (3A+B)x-2A+2B$, we have $3A+B=5$, $-2A+2B=2$. Then $A=1, B=2$, and

$$\frac{5x+2}{(x+2)(3x-2)} = \frac{1}{x+2}+\frac{2}{3x-2}$$

SECOND SOLUTION. Consider the identity $5x+2 = A(3x-2)+B(x+2)$. Since it is an identity between polynomials, it holds for *all* values of x. Now when

$x = -2$, coefficient of B is 0, $5(-2)+2 = A[3(-2)-2]$, and $A=1$;
$x = 2/3$, coefficient of A is 0, $5(2/3)+2 = B(2/3+2)$, and $B=2$.

Thus, as before, $\dfrac{5x+2}{(x+2)(3x-2)} = \dfrac{1}{x+2} + \dfrac{2}{3x-2}$. See Problem 1.

CASE 2. FACTORS OF THE DENOMINATOR LINEAR, SOME REPEATED.

Here (B) suggests that for each repeated factor $(ax+b)^k$, we set up a series of partial fractions

$$\frac{A}{(ax+b)^k} + \frac{B}{(ax+b)^{k-1}} + \cdots + \frac{K}{ax+b}$$

See Problem 2.

CASE 3. DENOMINATOR CONTAINS IRREDUCIBLE QUADRATIC FACTORS, NONE REPEATED.

For each irreducible quadratic factor ax^2+bx+c of the denominator set up a partial fraction of the form $\dfrac{Ax+B}{ax^2+bx+c}$.

Example 2. Resolve $\dfrac{2}{(x-1)(x^2+x-4)}$ into partial fractions.

Set $$\frac{2}{(x-1)(x^2+x-4)} = \frac{Ax+B}{x^2+x-4} + \frac{C}{x-1} = \frac{(Ax+B)(x-1)+C(x^2+x-4)}{(x-1)(x^2+x-4)}$$

$$= \frac{(A+C)x^2+(B-A+C)x-B-4C}{(x-1)(x^2+x-4)}.$$

From the identity $2 = (A+C)x^2+(B-A+C)x-B-4C$ we have

(a) $$A+C = 0, \quad B-A+C = 0, \quad -B-4C = 2$$

with solution $A=1$, $B=2$, $C=-1$. Then $\dfrac{2}{(x-1)(x^2+x-4)} = \dfrac{x+2}{x^2+x-4} - \dfrac{1}{x-1}$.

Note. The value $C=-1$ might have been obtained from the above identity $2 = (Ax+B)(x-1)+C(x^2+x-4)$ with $x=1$, and then the values of A and B found by means of (a).

See Problem 3.

CASE 4. DENOMINATOR CONTAINS IRREDUCIBLE QUADRATIC FACTORS, SOME REPEATED.

Here (C) suggests that for each repeated irreducible factor $(ax^2+bx+c)^k$ of the denominator, we set up a series of partial fractions

$$\frac{Ax+B}{(ax^2+bx+c)^k} + \frac{Cx+D}{(ax^2+bx+c)^{k-1}} + \cdots + \frac{Hx+K}{ax^2+bx+c}$$

See Problem 4.

SOLVED PROBLEMS

1. Resolve into partial fractions: (*a*) $\dfrac{54}{x^3-21x+20}$, (*b*) $\dfrac{x^4-2x^3-7x^2+5x-24}{x^2-2x-8}$.

(*a*) Using synthetic division, we find $x^3-21x+20 = (x-1)(x-4)(x+5)$. We set

$$\frac{54}{x^3-21x+20} = \frac{A}{x-1}+\frac{B}{x-4}+\frac{C}{x+5} = \frac{A(x-4)(x+5)+B(x-1)(x+5)+C(x-1)(x-4)}{x^3-21x+20}$$

and consider the identity

$$54 = A(x-4)(x+5)+B(x-1)(x+5)+C(x-1)(x-4)$$

When $x=1$, $54=A(-3)(6)$ and $A=-3$; when $x=4$, $54=B(3\cdot 9)$ and $B=2$; when $x=-5$, $54=C(-6)(-9)$ and $C=1$. Thus

$$\frac{54}{x^3-21x+20} = -\frac{3}{x-1}+\frac{2}{x-4}+\frac{1}{x+5}.$$

(*b*) The given fraction, being improper, will first be written as $x^2+1+\dfrac{7x-16}{(x-4)(x+2)}$. Using only the fractional term, we set

$$\frac{7x-16}{(x-4)(x+2)} = \frac{A}{x-4}+\frac{B}{x+2} = \frac{A(x+2)+B(x-4)}{(x-4)(x+2)}$$

and consider the identity $7x-16 = A(x+2)+B(x-4)$. Using $x=4$, we find $A=2$; using $x=-2$, we find $B=5$. Thus

$$\frac{x^4-2x^3-7x^2+5x-24}{x^2-2x-8} = x^2+1+\frac{2}{x-4}+\frac{5}{x+2}$$

2. Resolve into partial fractions: (*a*) $\dfrac{2x^2+6x-4}{x(x+2)^2}$, (*b*) $\dfrac{x^3+x^2+x-1}{(x^2-1)^2}$.

(*a*) Set $$\frac{2x^2+6x-4}{x(x+2)^2} = \frac{A}{(x+2)^2}+\frac{B}{x+2}+\frac{C}{x} = \frac{Ax+Bx(x+2)+C(x+2)^2}{x(x+2)^2}$$

$$= \frac{(B+C)x^2+(A+2B+4C)x+4C}{x(x+2)^2}.$$

Solution 1. From the identity $2x^2+6x-4 = (B+C)x^2+(A+2B+4C)x+4C$, by equating coefficients of like powers of x, we obtain $B+C=2$, $A+2B+4C=6$, $4C=-4$. Then $C=-1$, $B=2-C=3$, $A=6-2B-4C=4$, and

$$\frac{2x^2+6x-4}{x(x+2)^2} = \frac{4}{(x+2)^2}+\frac{3}{x+2}-\frac{1}{x}$$

Solution 2. Consider the identity $2x^2+6x-4 = Ax+Bx(x+2)+C(x+2)^2$. Using $x=-2$, we find $A=4$; using $x=0$, we find $C=-1$; using $x=1$, we have $2+6-4 = A+3B+9C = 4+3B-9$ and $B=3$. These are the values of A,B,C found above.

Note that only two values $x=-2$ and $x=0$ are suggested by the identity. Since three constants are to be determined, one additional value of x is needed. It may be taken at random.

(b) Set $\dfrac{x^3+x^2+x-1}{(x^2-1)^2} = \dfrac{A}{(x-1)^2} + \dfrac{B}{x-1} + \dfrac{C}{(x+1)^2} + \dfrac{D}{x+1}$

$$= \frac{A(x+1)^2 + B(x-1)(x+1)^2 + C(x-1)^2 + D(x+1)(x-1)^2}{(x^2-1)^2}$$

$$= \frac{(B+D)x^3 + (A+B+C-D)x^2 + (2A-B-2C-D)x + A - B + C + D}{(x^2-1)^2}.$$

From the identity $x^3+x^2+x-1 = (B+D)x^3 + (A+B+C-D)x^2 + (2A-B-2C-D)x + A - B + C + D$, we have

(1) $\quad B+D = 1, \quad A+B+C-D = 1, \quad 2A-B-2C-D = 1, \quad A-B+C+D = -1.$

From the identity

(2) $\quad x^3+x^2+x-1 = A(x+1)^2 + B(x-1)(x+1)^2 + C(x-1)^2 + D(x+1)(x-1)^2,$

using $x=1$ and $x=-1$, we find $A=\frac{1}{2}$ and $C=-\frac{1}{2}$. Then from (1),

$$B+D = 1, \quad B-D = 1, \quad \text{and} \quad D=0,\ B=1.$$

Thus $\dfrac{x^3+x^2+x-1}{(x^2-1)^2} = \dfrac{\frac{1}{2}}{(x-1)^2} + \dfrac{1}{x-1} - \dfrac{\frac{1}{2}}{(x+1)^2} = \dfrac{1}{2(x-1)^2} + \dfrac{1}{x-1} - \dfrac{1}{2(x+1)^2}.$

Note. The constants could have been determined by solving the four equations of (1) simultaneously or by using the values $x=1,-1,0,2$ in (2).

3. Resolve $\dfrac{x^2+2x+3}{x^4+x^3+2x^2}$ into partial fractions.

Set $\dfrac{x^2+2x+3}{x^4+x^3+2x^2} = \dfrac{A}{x^2} + \dfrac{B}{x} + \dfrac{Cx+D}{x^2+x+2} = \dfrac{A(x^2+x+2) + Bx(x^2+x+2) + (Cx+D)x^2}{x^4+x^3+2x^2}$

$$= \frac{(B+C)x^3 + (A+B+D)x^2 + (A+2B)x + 2A}{x^4+x^3+2x^2}$$

Equating coefficients of like terms in the identity

$$x^2+2x+3 = (B+C)x^3 + (A+B+D)x^2 + (A+2B)x + 2A$$

we have $B+C = 0$, $A+B+D=1$, $A+2B=2$, $2A=3$. Then $A=3/2$, $B=\frac{1}{4}$, $C=-\frac{1}{4}$, $D=-\frac{3}{4}$, and

$$\frac{x^2+2x+3}{x^4+x^3+2x^2} = \frac{3}{2x^2} + \frac{1}{4x} - \frac{x+3}{4(x^2+x+2)}$$

4. Resolve $\dfrac{x^4-x^3+8x^2-6x+7}{(x-1)(x^2+2)^2}$ into partial fractions.

Set $\dfrac{x^4-x^3+8x^2-6x+7}{(x-1)(x^2+2)^2} = \dfrac{A}{x-1} + \dfrac{Bx+C}{(x^2+2)^2} + \dfrac{Dx+E}{x^2+2}$

$$= \frac{A(x^2+2)^2 + (Bx+C)(x-1) + (Dx+E)(x^2+2)(x-1)}{(x-1)(x^2+2)^2}$$

$$= \frac{(A+D)x^4 + (-D+E)x^3 + (4A+B+2D-E)x^2 + (-B+C-2D+2E)x + 4A-C-2E}{(x-1)(x^2+2)^2}.$$

Equating coefficients of like terms in the identity

$$x^4 - x^3 + 8x^2 - 6x + 7 = (A+D)x^4 + (-D+E)x^3 + (4A+B+2D-E)x^2 + (-B+C-2D+2E)x + 4A - C - 2E,$$

we have $A+D = 1$, $-D+E = -1$, $4A+B+2D-E = 8$, $-B+C-2D+2E = -6$, $4A-C-2E = 7$.

From the identity

$$x^4 - x^3 + 8x^2 - 6x + 7 = A(x^2+2)^2 + (Bx+C)(x-1) + (Dx+E)(x^2+2)(x-1),$$

using $x = 1$, we find $A = 1$. Then $D = 1 - A = 0$, $E = -1 + D = -1$, $B = 3$, $C = -1$, and

$$\frac{x^4 - x^3 + 8x^2 - 6x + 7}{(x-1)(x^2+2)^2} = \frac{1}{x-1} + \frac{3x-1}{(x^2+2)^2} - \frac{1}{x^2+2}$$

SUPPLEMENTARY PROBLEMS

Resolve into partial fractions.

5. $\dfrac{4x+23}{(x-3)(x+2)(x+4)} = \dfrac{1}{x-3} - \dfrac{3/2}{x+2} + \dfrac{1/2}{x+4}$

6. $\dfrac{17x-45}{x^3-2x^2-15x} = \dfrac{3}{x} + \dfrac{1}{x-5} - \dfrac{4}{x+3}$

7. $\dfrac{2x^2+x+3}{x^2-9} = 2 - \dfrac{3}{x+3} + \dfrac{4}{x-3}$

8. $\dfrac{x^2+8}{x^2(x+2)} = \dfrac{4}{x^2} - \dfrac{2}{x} + \dfrac{3}{x+2}$

9. $\dfrac{x^2-1}{(x-2)^3} = \dfrac{3}{(x-2)^3} + \dfrac{4}{(x-2)^2} + \dfrac{1}{x-2}$

10. $\dfrac{3x^3-2x^2-10x+6}{(x+2)^3(2x+1)} = \dfrac{2}{(x+2)^3} - \dfrac{10}{(x+2)^2} + \dfrac{3}{2x+1}$

11. $\dfrac{x^2+5x+3}{x(x+1)^2} = \dfrac{1}{(x+1)^2} - \dfrac{2}{x+1} + \dfrac{3}{x}$

12. $\dfrac{x^2-3x}{(x-4)(x-2)^2} = \dfrac{1}{(x-2)^2} + \dfrac{1}{x-4}$

13. $\dfrac{5x^2-8}{x(x^2+2x-4)} = \dfrac{2}{x} + \dfrac{3x-4}{x^2+2x-4}$

14. $\dfrac{x^2-6x-3}{x^4+3x^2-18} = \dfrac{2x+3}{3(x^2+6)} - \dfrac{2x}{3(x^2-3)}$

15. $\dfrac{2x+7}{(x^3-1)(x^2+x+1)} = \dfrac{1}{x-1} - \dfrac{x+2}{x^2+x+1} - \dfrac{3x+4}{(x^2+x+1)^2}$

16. $\dfrac{x^3+4x^2+4}{x^3(x^2+2)^2} = \dfrac{1}{x^3} - \dfrac{x-1}{(x^2+2)^2}$

17. $\dfrac{3x^2+8}{(x^2+2)^3x^2} = \dfrac{1}{x^2} - \dfrac{1}{(x^2+2)^3} - \dfrac{2}{(x^2+2)^2} - \dfrac{1}{x^2+2}$

18. $\dfrac{x^6+5x^4+x^3+3x^2-9}{(x^2+3)^2} = x^2 - 1 + \dfrac{x}{x^2+3} - \dfrac{3x}{(x^2+3)^2}$

CHAPTER 27

Infinite Sequences

GENERAL TERM OF A SEQUENCE. Frequently the law of formation of a given sequence may be stated by giving a representative or *general term* of the sequence. This general term is a function of n, where n is the number of the term in the sequence. For this reason, it is also called the nth term of the sequence.

When the general term is given, it is a simple matter to write as many terms of the sequence as desired.

Example 1. (*a*) Write the first four terms and the tenth term of the sequence whose general term is $1/n$.

The first term ($n=1$) is $1/1=1$, the second term ($n=2$) is $1/2$, and so on. The first four terms are $1, 1/2, 1/3, 1/4$ and the tenth term is $1/10$.

(*b*) Write the first four terms and the ninth term of the sequence whose general term is $(-1)^{n-1}\dfrac{2n}{n^2+1}$.

The first term ($n=1$) is $(-1)^{1-1}\dfrac{2\cdot 1}{1^2+1} = 1$, the second term ($n=2$) is $(-1)^1\dfrac{2\cdot 2}{2^2+1} = -\dfrac{4}{5}$, and so on. The first four terms are $1, -4/5, 3/5, -8/17$ and the ninth term is $(-1)^8\dfrac{2\cdot 9}{9^2+1} = \dfrac{9}{41}$.

Note that the effect of the factor $(-1)^{n-1}$ is to produce a sequence whose terms have alternate signs, the sign of the first term being positive. The same pattern of signs is also produced by the factor $(-1)^{n+1}$. In order to produce a sequence whose terms alternate in sign, the first term being negative, the factor $(-1)^n$ is used.

When the first few terms of a sequence are given, the general term is obtained by inspection.

Example 2. Obtain the general term for each of the sequences:

(*a*) $1, 4, 9, 16, 25, \ldots$.

The terms of the sequence are the squares of the positive integers; the general term is n^2.

(*b*) $3, 7, 11, 15, 19, 23, \ldots$.

This is an arithmetic progression having $a=3$ and $d=4$. The general term is $a+(n-1)d = 4n-1$. Note, however, that the general term can be obtained about as easily by inspection.

See Problems 1-3.

LIMIT OF AN INFINITE SEQUENCE. From Example 4 of Chapter 9, the line $y = 2$ is a horizontal asymptote of $xy - 2x - 1 = 0$. To show this, let $P(x, y)$ move along the curve so that its abscissa takes on the values $10, 10^2, 10^3, \ldots, 10^n, \ldots$. Then the corresponding values of y are

(1) $$2.1,\ 2.01,\ 2.001,\ \ldots,\ 2 + 1/10^n,\ \ldots$$

and we infer that, by proceeding far enough along in this sequence, the difference between the terms of the sequence and 2 may be made as small as we please. This is equivalent to the following: Let ϵ denote a positive number, as small as we please; then there is a term of the sequence such that the difference between it and 2 is less than ϵ, and the same is true for all subsequent terms of the sequence. For example, let $\epsilon = 1/10^{25}$; then the difference between the term $2 + 1/10^{26}$ and 2, $2 + 1/10^{26} - 2 = 1/10^{26}$, is less than $\epsilon = 1/10^{25}$ and the same is true for the terms $2 + 1/10^{27}$, $2 + 1/10^{28}$, and so on.

The behavior of the terms of the sequence (1) discussed above is indicated by the statement: *The limit of the sequence (1) is 2.* In general,

If, for an infinite sequence

(2) $$s_1,\ s_2,\ s_3,\ \ldots,\ s_n,\ \ldots$$

and a positive number ϵ, however small, there exists a number s and a positive integer m such that for all $n > m$

$$|s - s_n| < \epsilon,$$

then the limit of the sequence is s.

Example 3. Show, using the above definition, that the limit of sequence (1) is 2.

Take $\epsilon = 1/10^p$, where p is a positive integer as large as we please; thus, ϵ is a positive number as small as we please. We must produce a positive integer m (in other words, a term s_m) such that for $n > m$ (that is, for all subsequent terms) $|s - s_n| < \epsilon$. Now

$$|2 - (2 + 1/10^n)| < 1/10^p \quad \text{or} \quad \frac{1}{10^n} < \frac{1}{10^p}$$

requires $n > p$. Thus $m = p$ is the required value of m.

The statement that the limit of the sequence (2) is s describes the behavior of s_n as n increases without bound over the positive integers. Since we shall repeatedly be using the phrase "as n increases without bound" or the phrase "as n becomes infinite", which we shall take to be equivalent to the former phrase, we shall introduce the notation $n \to \infty$ for it. Thus the behavior of s_n may be described briefly by

$$\lim_{n \to \infty} s_n = s,$$

(read: the limit of s_n, as n becomes infinite, is s).

We state, without proof, the following theorem:

If $\lim_{n\to\infty} s_n = s$ and $\lim_{n\to\infty} t_n = t$, then

(A) $$\lim_{n\to\infty} (s_n \pm t_n) = \lim_{n\to\infty} s_n \pm \lim_{n\to\infty} t_n = s \pm t,$$

(B) $$\lim_{n\to\infty} (s_n \cdot t_n) = \lim_{n\to\infty} s_n \cdot \lim_{n\to\infty} t_n = s \cdot t,$$

(C) $$\lim_{n\to\infty} \frac{s_n}{t_n} = \frac{\lim_{n\to\infty} s_n}{\lim_{n\to\infty} t_n} = \frac{s}{t}, \quad \text{provided } t \neq 0,$$

or, in words, if each of two variables approaches a limit, then the limits of the sum, difference, product, and quotient of the two variables are equal respectively to the sum, difference, product, and quotient of their limits provided only that, in the case of the quotient, the limit of the denominator is not zero.

This theorem makes it possible to find the limit of a sequence directly from its general term. In this connection, we shall need

(3) $\lim_{n\to\infty} a = a$, where a is any constant

(4) $\lim_{n\to\infty} (1/n^k) = 0$, $k > 0$ (See Problem 4.)

(5) $\lim_{n\to\infty} (1/b^n) = 0$, where b is a constant > 1. (See Problem 5.)

See Problem 6.

THE FOLLOWING THEOREMS are useful in establishing whether or not certain sequences have a limit.

I. Suppose M is a fixed number such that for all values of n,

$$s_n \le s_{n+1} \quad \text{and} \quad s_n \le M;$$

then $\lim_{n\to\infty} s_n$ exists and is $\le M$.

If, however, s_n eventually exceeds M, no matter how large M may be, $\lim_{n\to\infty} s_n$ does not exist.

II. Suppose M is a fixed number such that for all values of n,

$$s_n \ge s_{n+1} \quad \text{and} \quad s_n \ge M;$$

then $\lim_{n\to\infty} s_n$ exists and is $\ge M$.

If, however, s_n is eventually smaller than M, no matter how small M may be, $\lim_{n\to\infty} s_n$ does not exist.

Example 4. (*a*) For the sequence $\frac{5}{2}, 3, \frac{19}{6}, \frac{13}{4}, \ldots, (\frac{7}{2} - \frac{1}{n}), \ldots,$ $s_n < s_{n+1}$ and $s_n < 4$, for all values of n; the sequence has a limit ≤ 4. In fact, $\lim_{n\to\infty} s_n = 7/2$.

(*b*) For the sequence $3, 5, 7, 9, \ldots, 2n+1, \ldots,$ $s_n < s_{n+1}$ but s_n will eventually exceed any chosen M, however large, (if $M = 2^{1000}+1$, then $2n+1 > M$ for $n > 2^{999}$) and the sequence does not have a limit.

See Problems 7-9.

SOLVED PROBLEMS

1. Write the first five terms and the tenth term of the sequence whose general term is:

(*a*) $4n - 1$.

The first term is $4\cdot1 - 1 = 3$, the second term is $4\cdot2 - 1 = 7$, the third term is $4\cdot3 - 1 = 11$, the fourth term is $4\cdot4 - 1 = 15$, the fifth term is $4\cdot5 - 1 = 19$; the tenth term is $4\cdot10 - 1 = 39$.

(*b*) 2^{n-1}.

The first term is $2^{1-1} = 2^0 = 1$, the second term is $2^{2-1} = 2$, the third is $2^{3-1} = 2^2 = 4$, the fourth is $2^3 = 8$, the fifth is $2^4 = 16$; the tenth is $2^9 = 512$.

(*c*) $\dfrac{(-1)^{n-1}}{n+1}$.

The first term is $\dfrac{(-1)^{1-1}}{1+1} = \dfrac{(-1)^0}{2} = \dfrac{1}{2}$, the second is $\dfrac{(-1)^{2-1}}{2+1} = -\dfrac{1}{3}$, the third is $\dfrac{(-1)^2}{3+1} = \dfrac{1}{4}$, the fourth is $-\dfrac{1}{5}$, the fifth is $\dfrac{1}{6}$; the tenth is $-\dfrac{1}{11}$.

2. Write the first four terms of the sequence whose general term is:

(*a*) $\dfrac{n+1}{n!}$. The terms are: $\dfrac{1+1}{1!}, \dfrac{2+1}{2!}, \dfrac{3+1}{3!}, \dfrac{4+1}{4!}$ or $2, \dfrac{3}{2}, \dfrac{2}{3}, \dfrac{5}{24}$.

(*b*) $\dfrac{x^{2n-1}}{(2n+1)!}$. The required terms are: $\dfrac{x}{3!}, \dfrac{x^3}{5!}, \dfrac{x^5}{7!}, \dfrac{x^7}{9!}$.

(*c*) $(-1)^{n-1}\dfrac{x^{2n-2}}{(n-1)!}$. The terms are: $\dfrac{x^0}{0!}, -\dfrac{x^2}{1!}, \dfrac{x^4}{2!}, -\dfrac{x^6}{3!}$ or $1, -x^2, \dfrac{x^4}{2}, -\dfrac{x^6}{6}$.

3. Write the general term for each of the following sequences:

(*a*) $2, 4, 6, 8, 10, 12, \ldots$

The first term is $2\cdot1$, the second is $2\cdot2$, the third is $2\cdot3$, etc.; the general term is $2n$.

(*b*) $1, 3, 5, 7, 9, 11, \ldots$

Each term of the given sequence is 1 less than the corresponding term of the sequence in (*a*); the general term is $2n-1$.

(*c*) $2, 5, 8, 11, 14, \ldots$

The first term is $3\cdot1-1$, the second term is $3\cdot2-1$, the third term is $3\cdot3-1$, and so on; the general term is $3n-1$.

(*d*) $2, -5, 8, -11, 14, \ldots$

This sequence may be obtained from that in (*c*) by changing the signs of alternate terms beginning with the second; the general term is $(-1)^{n-1}(3n-1)$.

(*e*) $2.1, 2.01, 2.001, 2.0001, \ldots$

The first term is $2+1/10$, the second term is $2+1/10^2$, the third is $2+1/10^3$, and so on; the general term is $2+1/10^n$.

(*f*) $1/8, -1/27, 1/64, -1/125, \ldots$

The successive denominators are the cubes of $2, 3, 4, 5, \ldots$ or of $1+1, 2+1, 3+1, 4+1, \ldots$; the general term is $\dfrac{(-1)^{n-1}}{(n+1)^3}$.

(*g*) $\dfrac{3}{1}, \dfrac{4}{1\cdot2}, \dfrac{5}{1\cdot2\cdot3}, \ldots$

Rewriting the sequence as $\dfrac{1+2}{1!}, \dfrac{2+2}{2!}, \dfrac{3+2}{3!}, \ldots$, the general term is $\dfrac{n+2}{n!}$.

(*h*) $x/2, x^2/6, x^3/24, x^4/120, \ldots$

The denominators are $2!, 3!, \ldots, (n+1)!, \ldots$; the general term is $\dfrac{x^n}{(n+1)!}$.

(*i*) $x, -x^3/3!, x^5/5!, -x^7/7!, \ldots$

The exponents of x are $2\cdot1-1, 2\cdot2-1, 2\cdot3-1, \ldots$; the general term is $(-1)^{n-1}\dfrac{x^{2n-1}}{(2n-1)!}$.

(*j*) $3, 4, 5/2, 1, 7/24, \ldots$

Rewrite the sequence as $3/0!, 4/1!, 5/2!, 6/3!, \ldots$; the general term is $\dfrac{n+2}{(n-1)!}$.

4. Show that $\lim_{n\to\infty} \frac{1}{n^k} = 0$ when $k > 0$.

Take $\epsilon = 1/p^k$, where p is a positive integer as large as we please. We seek a positive number m such that for $n > m$, $|0 - 1/n^k| = 1/n^k < 1/p^k$. Since this inequality is satisfied when $n > p$, it is sufficient to take for m any number equal to or greater than p.

5. Show that $\lim_{n\to\infty} \frac{1}{b^n} = 0$ when $b > 1$.

Take $\epsilon = 1/b^p$, where p is a positive integer. Since $b > 1$, $b^p > 1$ and $\epsilon = 1/b^p < 1$. Thus, ϵ may be made as small as we please by taking p sufficiently large. We seek a positive number m such that for $n > m$, $|0 - 1/b^n| = 1/b^n < 1/b^p$. Since $n > p$ satisfies the inequality, it is sufficient to take for m any number equal to or greater than p.

6. Evaluate each of the following.

(a) $\lim_{n\to\infty} \left(\frac{3}{n} + \frac{5}{n^2}\right) = \lim_{n\to\infty} \frac{3}{n} + \lim_{n\to\infty} \frac{5}{n^2} = 0 + 0 = 0$

(b) $\lim_{n\to\infty} \left(\frac{n}{n+1}\right) = \lim_{n\to\infty} \left(\frac{\frac{n}{n}}{\frac{n+1}{n}}\right) = \lim_{n\to\infty} \frac{1}{1+\frac{1}{n}} = \frac{\lim_{n\to\infty} 1}{\lim_{n\to\infty} (1 + \frac{1}{n})} = \frac{1}{1+0} = 1$

(c) $\lim_{n\to\infty} \frac{n^2+2}{2n^2-3n} = \lim_{n\to\infty} \frac{1+\frac{2}{n^2}}{2-\frac{3}{n}} = \frac{1+0}{2-0} = \frac{1}{2}$

(d) $\lim_{n\to\infty} \left\{4 - \frac{2^n - 1}{2^{n+1}}\right\} = 4 - \lim_{n\to\infty} \frac{2^n-1}{2^{n+1}} = 4 - \lim_{n\to\infty} \left(\frac{1}{2} - \frac{1}{2^{n+1}}\right)$

$$= 4 - \frac{1}{2} - \lim_{n\to\infty}\left(-\frac{1}{2^{n+1}}\right) = 4 - \frac{1}{2} - 0 = 3.5$$

7. Show that every infinite arithmetic sequence fails to have a limit except when $d = 0$.

(a) If $d > 0$, then $s_n = a + (n-1)d < s_{n+1} = a + nd$; but s_n eventually exceeds any previously selected M, however large. Thus, the sequence has no limit.

(b) If $d < 0$, then $s_n > s_{n+1}$; but s_n eventually becomes smaller than any previously selected M, however small. Thus, the sequence has no limit.

(c) If $d = 0$, the sequence is $a, a, a, \ldots, a, \ldots$. with limit a.

8. Show that the infinite geometric sequence $3, 6, 12, \ldots, 3\cdot 2^{n-1}, \ldots$. does not have a limit.

Here $s_n < s_{n+1}$; but $3\cdot 2^{n-1}$ may be made to exceed any previously selected M, however large. The sequence has no limit.

9. Show that the following sequence does not have a limit:

$$1,\ 1+\frac{1}{2},\ 1+\frac{1}{2}+\frac{1}{3},\ 1+\frac{1}{2}+\frac{1}{3}+\frac{1}{4}+\ldots,\ 1+\frac{1}{2}+\frac{1}{3}+\frac{1}{4}+\ldots+\frac{1}{n}, \ldots.$$

Here $s_n < s_{n+1}$. Let M, as large as we please, be chosen. Now

$$1+\frac{1}{2}+\frac{1}{3}+\frac{1}{4}+\ldots = 1+\frac{1}{2}+\left(\frac{1}{3}+\frac{1}{4}\right)+\left(\frac{1}{5}+\frac{1}{6}+\frac{1}{7}+\frac{1}{8}\right)+\left(\frac{1}{9}+\ldots+\frac{1}{16}\right)+\left(\frac{1}{17}+\ldots+\frac{1}{32}\right)+\ldots.$$

and $\frac{1}{3}+\frac{1}{4} > \frac{1}{2}$, $\frac{1}{5}+\frac{1}{6}+\frac{1}{7}+\frac{1}{8} > \frac{1}{2}$, $\frac{1}{9}+\ldots+\frac{1}{16} > \frac{1}{2}$, and so on.

Since the sum of each group exceeds $\frac{1}{2}$ and we may add as many groups as we please, we can eventually obtain a sum of groups which exceeds M. Thus, the sequence has no limit.

SUPPLEMENTARY PROBLEMS

10. Write the first four terms of the sequence whose general term is:

(a) $\dfrac{1}{1+n}$ (b) $\dfrac{1}{1+n\sqrt{n}}$ (c) $\dfrac{1}{3^n}$ (d) $\dfrac{2n-1}{2n+3}$ (e) $\dfrac{n^2}{3n-2}$ (f) $(-1)^{n+1}\dfrac{2n+1}{2^{n+1}}$ (g) $(-1)^{n+1}\dfrac{1}{n!}$ (h) $\dfrac{n^2}{(2n)!}$

Ans. (a) 1/2, 1/3, 1/4, 1/5
(b) $\frac{1}{2}, \frac{1}{1+2\sqrt{2}}, \frac{1}{1+3\sqrt{3}}, \frac{1}{9}$
(c) 1/3, 1/9, 1/27, 1/81
(d) 1/5, 3/7, 5/9, 7/11
(e) 1, 1, 9/7, 8/5
(f) 3/4, −5/8, 7/16, −9/32
(g) 1, −1/2, 1/6, −1/24
(h) 1/2, 1/6, 1/80, 1/2520

11. Write the general term of each sequence.

(a) 1, 1/3, 1/5, 1/7, ...
(b) $\frac{4}{1\cdot 3}, \frac{5}{2\cdot 4}, \frac{6}{3\cdot 5}, \frac{7}{4\cdot 6}, \ldots$
(c) 1, 3/5, 2/5, 5/17, 3/13, ...
(d) 1/2, 3/8, 7/24, 15/64, ...
(e) 2, 1, 8/9, 1, 32/25, 16/9, ...
(f) 1/3, −1/15, 1/35, −1/63, ...
(g) 1/2, 3/8, 5/16, 9/32, ...
(h) $1/2, -x^2/4, x^4/6, -x^6/8, \ldots$

Ans. (a) $\dfrac{1}{2n-1}$ (b) $\dfrac{n+3}{n(n+2)}$ (c) $\dfrac{n+1}{n^2+1}$ (d) $\dfrac{2^n-1}{n\cdot 2^n}$ (e) $\dfrac{2^n}{n^2}$ (f) $\dfrac{(-1)^{n+1}}{(2n-1)(2n+1)}$ (g) $\dfrac{1+2^{n-1}}{2^{n+1}}$ (h) $(-1)^{n+1}\dfrac{x^{2n-2}}{2n}$

12. Evaluate. (a) $\lim_{n\to\infty}(2-\frac{1}{n})$ (b) $\lim_{n\to\infty}\dfrac{3n+1}{3n-2}$ (c) $\lim_{n\to\infty}\dfrac{2n}{(n+1)(n+2)}$ (d) $\lim_{n\to\infty}\dfrac{2n^2+5n-6}{n^2+n-1}$ (e) $\lim_{n\to\infty}\dfrac{n}{n+1}$ (f) $\lim_{n\to\infty}\dfrac{(3n)!\,n}{(3n+1)!}$ (g) $\lim_{n\to\infty}\dfrac{1}{2^n+1}$ (h) $\lim_{n\to\infty}\dfrac{2^n+1}{2^{n+1}+1}$

Ans. (a) 2, (b) 1, (c) 0, (d) 2, (e) 1, (f) 1/3, (g) 0, (h) 1/2

13. Explain why each of the following has no limit.

(a) 1, 3, 5, 7, 9, ...
(b) 1, 0, 1, 0, 1, 0, ...
(c) 1, −2, 4, −8, 16, −32, ...
(d) 1/25, 4/25, 9/25, 16/25, ...

CHAPTER 28

Infinite Series

THE INDICATED SUM of the terms of an infinite sequence is called an *infinite series*. Let

(1) $$s_1 + s_2 + s_3 + \ldots + s_n + \ldots.$$

be such a series and define the sequence of *partial sums*

$$S_1 = s_1, \quad S_2 = s_1 + s_2, \quad \ldots., \quad S_n = s_1 + s_2 + \ldots + s_n, \quad \ldots.$$

If $\lim_{n\to\infty} S_n$ exists, the series (1) is called *convergent*; if $\lim_{n\to\infty} S_n = S$, the series is said to converge to S. If $\lim_{n\to\infty} S_n$ does not exist, the series is called *divergent*.

Example 1. (a) Every infinite geometric series

$$a + ar + ar^2 + \ldots + ar^{n-1} + \ldots.$$

is convergent if $|r| < 1$ and is divergent if $|r| \geq 1$.

See Problem 1.

(b) The *harmonic series* $1 + 1/2 + 1/3 + \ldots + 1/n + \ldots.$ is divergent. See Chapter 27, Problem 9.

See Problem 2.

A NECESSARY CONDITION THAT (1) BE CONVERGENT is $\lim_{n\to\infty} s_n = 0$; that is, if (1) is convergent then $\lim_{n\to\infty} s_n = 0$. However, this condition is *not sufficient* since the harmonic series is divergent although $\lim_{n\to\infty} s_n = \lim_{n\to\infty} 1/n = 0$.

A SUFFICIENT CONDITION THAT (1) BE DIVERGENT is $\lim_{n\to\infty} s_n \neq 0$; that is, if $\lim_{n\to\infty} s_n$ exists and is different from 0, or if $\lim_{n\to\infty} s_n$ does not exist, the series is divergent. This, in turn, is not a necessary condition since the harmonic series is divergent although $\lim_{n\to\infty} s_n = 0$.

See Problem 3.

SERIES OF POSITIVE TERMS

COMPARISON TEST FOR CONVERGENCE of a series of positive terms.

I. If every term of a given series of positive terms is less than or equal to the corresponding term of a known convergent series, the given series is convergent.

II. If every term of a given series of positive terms is equal to or greater than the corresponding term of a known divergent series, the given series is divergent.

The following series will be found useful in making comparison tests:

(*a*) The geometric series $a + ar + ar^2 + \ldots + ar^n + \ldots.$ which converges when $|r| < 1$ and diverges when $|r| \geq 1$.

(*b*) The *p*-series $1 + \frac{1}{2^p} + \frac{1}{3^p} + \ldots + \frac{1}{n^p} + \ldots.$ which converges for $p > 1$ and diverges for $p \leq 1$.

(*c*) Each new series tested.

In comparing two series it is not sufficient to examine the first few terms of each series. *The general terms must be compared.*

See Problems 4-6.

THE RATIO TEST FOR CONVERGENCE. If, in a series of positive terms, the *test ratio*

$$r_n = \frac{s_{n+1}}{s_n}$$

approaches a limit R as $n \to \infty$, the series is convergent if $R < 1$ and is divergent if $R > 1$. If $R = 1$, the test fails to indicate convergency or divergency.

See Problem 7.

SERIES WITH NEGATIVE TERMS

A SERIES WITH ALL ITS TERMS NEGATIVE may be treated as the negative of a series with all of its terms positive.

ALTERNATING SERIES. A series whose terms are alternately positive and negative, as

(*2*) $$s_1 - s_2 + s_3 - \ldots + (-1)^{n-1} s_n + \ldots.$$

where each s is positive, is called an *alternating series.*

An alternating series (*2*) is convergent provided $s_n \geq s_{n+1}$, for every value of n, and $\lim_{n \to \infty} s_n = 0$.

See Problem 8.

ABSOLUTELY CONVERGENT SERIES. A series (*1*) $s_1 + s_2 + s_3 + \ldots + s_n + \ldots.$ in which some of the terms are positive and some are negative is called *absolutely convergent* if the series of absolute values of the terms

(*3*) $$|s_1| + |s_2| + |s_3| + \ldots + |s_n| + \ldots.$$

is convergent.

CONDITIONALLY CONVERGENT SERIES. A series (*1*), where some of the terms are positive and some are negative, is called *conditionally convergent* if it is convergent but the series of absolute values of its terms is divergent.

Example 2. The series $1 - 1/2 + 1/3 - 1/4 + \ldots.$ is convergent, but the series of absolute values of its terms $1 + 1/2 + 1/3 + 1/4 + \ldots.$ is divergent. Thus, the given series is conditionally convergent.

THE GENERALIZED RATIO TEST. Let (*1*) $s_1 + s_2 + s_3 + \ldots + s_n + \ldots.$ be a series some of whose terms are positive and some are negative. Let

$$\lim_{n \to \infty} \frac{|s_{n+1}|}{|s_n|} = R$$

The series (*1*) is absolutely convergent if $R < 1$ and is divergent if $R > 1$. If $R = 1$, the test fails. See Problem 9.

SOLVED PROBLEMS

1. Examine the infinite geometric series $a + ar + ar^2 + \ldots + ar^n + \ldots.$ for convergence and divergence.

The sum of the first n terms is $S_n = \dfrac{a(1-r^n)}{1-r} = \dfrac{a}{1-r}(1-r^n)$.

If $|r| < 1$, $\lim\limits_{n\to\infty} r^n = 0$; then $\lim\limits_{n\to\infty} S_n = \dfrac{a}{1-r}$ and the series is convergent.

If $|r| > 1$, $\lim\limits_{n\to\infty} r^n$ does not exist and $\lim\limits_{n\to\infty} S_n$ does not exist; the series is divergent.

If $r = 1$, the series is $a + a + a + \ldots + a + \ldots.$; then $\lim\limits_{n\to\infty} S_n = \lim\limits_{n\to\infty} na$ does not exist.

If $r = -1$, the series is $a - a + a - a + \ldots.$; then $S_n = a$ or 0 according as n is odd or even, and $\lim\limits_{n\to\infty} S_n$ does not exist.

Thus, the infinite geometric series converges to $\dfrac{a}{1-r}$ when $|r| < 1$, and diverges when $|r| \geq 1$.

2. Show that the following series are convergent.

(*a*) $1 + \dfrac{2}{3} + \dfrac{4}{9} + \dfrac{8}{27} + \ldots + \dfrac{2^{n-1}}{3^{n-1}} + \ldots.$

This is a geometric series with ratio $r = 2/3$; then $|r| < 1$ and the series is convergent.

(*b*) $2 - \dfrac{3}{2} + \dfrac{9}{8} - \dfrac{27}{32} + \ldots + (-1)^{n-1}2(\dfrac{3}{4})^{n-1} + \ldots.$

This is a geometric series with ratio $r = -3/4$; then $|r| < 1$ and the series is convergent.

(*c*) $1 + \dfrac{1}{2^p} + \dfrac{1}{2^p} + \dfrac{1}{4^p} + \dfrac{1}{4^p} + \dfrac{1}{4^p} + \dfrac{1}{4^p} + \dfrac{1}{8^p} + \ldots.,\quad p > 1.$

This series may be rewritten as $1 + \dfrac{2}{2^p} + \dfrac{4}{4^p} + \dfrac{8}{8^p} + \ldots.$, a geometric series with ratio $r = 2/2^p$. Since $|r| < 1$, when $p > 1$, the series is convergent.

3. Show that the following series are divergent.

(*a*) $2 + \dfrac{3}{2} + \dfrac{4}{3} + \ldots + \dfrac{n+1}{n} + \ldots.$

Since $\lim\limits_{n\to\infty} s_n = \lim\limits_{n\to\infty} \dfrac{n+1}{n} = \lim\limits_{n\to\infty} (1 + \dfrac{1}{n}) = 1 \neq 0$, the series is divergent.

(*b*) $\dfrac{1}{2} + \dfrac{3}{8} + \dfrac{5}{16} + \dfrac{9}{32} + \ldots + \dfrac{2^{n-1}+1}{4\cdot 2^{n-1}} + \ldots.$

Since $\lim\limits_{n\to\infty} s_n = \lim\limits_{n\to\infty} \dfrac{2^{n-1}+1}{4\cdot 2^{n-1}} = \lim\limits_{n\to\infty} \dfrac{1 + \dfrac{1}{2^{n-1}}}{4} = \dfrac{1}{4} \neq 0$, the series is divergent.

4. Show that $1 + \dfrac{1}{2^p} + \dfrac{1}{3^p} + \dfrac{1}{4^p} + \ldots + \dfrac{1}{n^p} + \ldots.$ is divergent for $p \leq 1$ and convergent for $p > 1$.

For $p = 1$, the series is the harmonic series and is divergent.

For $p < 1$, including negative values, $\frac{1}{n^p} \geq \frac{1}{n}$, for every n. Since every term of the given series is equal to or greater than the corresponding term of the harmonic series, the given series is divergent.

For $p > 1$, compare the series with the convergent series

(c) $$1 + \frac{1}{2^p} + \frac{1}{2^p} + \frac{1}{4^p} + \frac{1}{4^p} + \frac{1}{4^p} + \frac{1}{4^p} + \frac{1}{8^p} + \ldots.$$

of Solved Problem 2(c). Since each term of the given series is less than or equal to the corresponding term of series (c), the given series is convergent.

5. Use Problem 4 to determine whether the following series are convergent or divergent.

(a) $1 + \frac{1}{2\sqrt{2}} + \frac{1}{3\sqrt{3}} + \frac{1}{4\sqrt{4}} + \ldots.$ The general term is $\frac{1}{n\sqrt{n}} = \frac{1}{n^{3/2}}$.

This is a p-series with $p = 3/2 > 1$; the series is convergent.

(b) $1 + 4 + 9 + 16 + \ldots.$ The general term is $n^2 = \frac{1}{n^{-2}}$.

This is a p-series with $p = -2 < 1$; the series is divergent.

(c) $1 + \frac{\sqrt[4]{2}}{4} + \frac{\sqrt[4]{3}}{9} + \frac{\sqrt[4]{4}}{16} + \ldots.$ The general term is $\frac{\sqrt[4]{n}}{n^2} = \frac{1}{n^{7/4}}$.

The series is convergent since $p = 7/4 > 1$.

6. Use the comparison test to determine whether each of the following is convergent or divergent.

(a) $1 + \frac{1}{2!} + \frac{1}{3!} + \frac{1}{4!} + \ldots.$

The general term $\frac{1}{n!} \leq \frac{1}{n^2}$. Thus, the terms of the given series are less than or equal to the corresponding terms of the p-series with $p = 2$. The series is convergent.

(b) $\frac{1}{2} + \frac{1}{3} + \frac{1}{5} + \frac{1}{9} + \ldots.$

The general term $\frac{1}{1 + 2^{n-1}} \leq \frac{1}{2^{n-1}}$. Thus, the terms of the given series are less than or equal to the corresponding terms of the geometric series with $a = 1$ and $r = \frac{1}{2}$. The series is convergent.

(c) $\frac{2}{1} + \frac{3}{4} + \frac{4}{9} + \frac{5}{16} + \ldots.$

The general term $\frac{n+1}{n^2} = \frac{1}{n} + \frac{1}{n^2} > \frac{1}{n}$. Thus, the terms of the given series are equal to or greater than the corresponding terms of the harmonic series. The series is divergent.

(d) $\frac{1}{3} + \frac{1}{12} + \frac{1}{27} + \frac{1}{48} + \ldots.$

The general term $\frac{1}{3 \cdot n^2} \leq \frac{1}{n^2}$. Thus, the terms of the given series are less than or equal to the corresponding terms of the p-series with $p = 2$. The series is convergent.

(e) $1 + \frac{1}{2} + \frac{1}{3^2} + \frac{1}{4^3} + \frac{1}{5^4} + \ldots.$

The general term $\frac{1}{n^{n-1}} \leq \frac{1}{n^2}$ for $n \geq 3$. Thus, neglecting the first two terms, the given series is term by term less than or equal to the corresponding terms of the p-series with $p = 2$. The given series is convergent.

7. Apply the ratio test to each of the following. If it fails, use some other method to determine convergency or divergency.

(a) $\frac{1}{2} + \frac{1}{2} + \frac{3}{8} + \frac{1}{4} + \ldots$ or $\frac{1}{2} + \frac{2}{2^2} + \frac{3}{2^3} + \frac{4}{2^4} + \ldots$

For this series $s_n = \frac{n}{2^n}$, $s_{n+1} = \frac{n+1}{2^{n+1}}$, and $r_n = \frac{s_{n+1}}{s_n} = \frac{n+1}{2^{n+1}} \cdot \frac{2^n}{n} = \frac{n+1}{2n}$.

Then $R = \lim_{n\to\infty} r_n = \lim_{n\to\infty} \frac{n+1}{2n} = \lim_{n\to\infty} \frac{1+1/n}{2} = \frac{1}{2} < 1$ and the series is convergent.

(b) $3 + \frac{9}{2} + \frac{9}{2} + \frac{27}{8} + \frac{81}{40} + \ldots$ or $\frac{3}{1!} + \frac{3^2}{2!} + \frac{3^3}{3!} + \ldots$

Here $s_n = \frac{3^n}{n!}$, $s_{n+1} = \frac{3^{n+1}}{(n+1)!}$, and $r_n = \frac{3^{n+1}}{(n+1)!} \cdot \frac{n!}{3^n} = \frac{3}{n+1}$.

Then $R = \lim_{n\to\infty} \frac{3}{n+1} = 0$ and the series is convergent.

(c) $\frac{1}{1\cdot 1} + \frac{1}{2\cdot 3} + \frac{1}{3\cdot 5} + \frac{1}{4\cdot 7} + \ldots$

Here $s_n = \frac{1}{n(2n-1)}$, $s_{n+1} = \frac{1}{(n+1)(2n+1)}$, and $r_n = \frac{n(2n-1)}{(n+1)(2n+1)}$.

Then $R = \lim_{n\to\infty} \frac{n(2n-1)}{(n+1)(2n+1)} = \lim_{n\to\infty} \frac{2-1/n}{(1+1/n)(2+1/n)} = 1$ and the test fails.

Since $\frac{1}{n(2n-1)} \leq \frac{1}{n^2}$. The given series is term by term less than or equal to the convergent p-series, with $p = 2$. The given series is convergent.

(d) $\frac{2}{1^2+1} + \frac{2^3}{2^2+2} + \frac{2^5}{3^2+3} + \ldots$

Here $s_n = \frac{2^{2n-1}}{n^2+n}$, $s_{n+1} = \frac{2^{2n+1}}{(n+1)^2+(n+1)}$, and $r_n = \frac{2^{2n+1}}{(n+1)(n+2)} \cdot \frac{n(n+1)}{2^{2n-1}} = \frac{4n}{n+2}$.

Then $R = \lim_{n\to\infty} \frac{4}{1+2/n} = 4$ and the series is divergent.

(e) $\frac{1}{5} + \frac{2}{25} + \frac{6}{125} + \frac{24}{625} + \ldots$

In this series $s_n = \frac{n!}{5^n}$, $s_{n+1} = \frac{(n+1)!}{5^{n+1}}$, and $r_n = \frac{(n+1)!}{5^{n+1}} \cdot \frac{5^n}{n!} = \frac{n+1}{5}$.

Now $\lim_{n\to\infty} r_n$ does not exist. However, since $s_n \to \infty$ as $n \to \infty$, the series is divergent.

8. Test the following alternating series for convergence.

(a) $1 - 1/3 + 1/5 - 1/7 + \ldots$

$s_n > s_{n+1}$, for all values of n, and $\lim_{n\to\infty} s_n = \lim_{n\to\infty} \frac{1}{2n-1} = 0$. The series is convergent.

(b) $\frac{1}{2^3} - \frac{2}{3^3} + \frac{3}{4^3} - \frac{4}{5^3} + \ldots$

$s_n > s_{n+1}$, for all values of n, and $\lim_{n\to\infty} s_n = \lim_{n\to\infty} \frac{n}{(n+1)^3} = 0$. The series is convergent.

9. Investigate the following for absolute convergence, conditional convergence, or divergence.

(a) $1 - \frac{1}{2} + \frac{1}{4} - \frac{1}{8} + \cdots$.

Here $|s_n| = \frac{1}{2^{n-1}}$, $|s_{n+1}| = \frac{1}{2^n}$, and $R = \lim_{n\to\infty} \frac{|s_{n+1}|}{|s_n|} = \lim_{n\to\infty} \frac{2^{n-1}}{2^n} = \frac{1}{2} < 1$.

The series is absolutely convergent.

(b) $1 - \frac{4}{1!} + \frac{4^2}{2!} - \frac{4^3}{3!} + \cdots$.

Here $|s_n| = \frac{4^{n-1}}{(n-1)!}$, $|s_{n+1}| = \frac{4^n}{n!}$, and $R = \lim_{n\to\infty} \frac{4}{n} = 0$.

The series is absolutely convergent.

(c) $\frac{1}{2-\sqrt{2}} - \frac{1}{3-\sqrt{3}} + \frac{1}{4-\sqrt{4}} - \frac{1}{5-\sqrt{5}} + \cdots$. The ratio test fails here.

Since $\frac{1}{n+1-\sqrt{n+1}} > \frac{1}{n+2-\sqrt{n+2}}$ and $\lim_{n\to\infty} \frac{1}{n+1-\sqrt{n+1}} = 0$, the series is convergent.

Since $\frac{1}{n+1-\sqrt{n+1}} > \frac{1}{n+1}$ for all values of n, the series of absolute values is term by term greater than the harmonic series, and thus is divergent. The given series is conditionally convergent.

SUPPLEMENTARY PROBLEMS

10. Investigate each of the following series for convergence or divergence.

(a) $\frac{1}{3} + \frac{1}{6} + \frac{1}{11} + \cdots + \frac{1}{2^n + n} + \cdots$

(b) $\frac{1}{2} + \frac{1}{4} + \frac{1}{6} + \cdots + \frac{1}{2n} + \cdots$

(c) $1 + \frac{1}{3} + \frac{1}{5} + \cdots + \frac{1}{2n-1} + \cdots$

(d) $\frac{2}{1!} + \frac{2^2}{2!} + \frac{2^3}{3!} + \cdots + \frac{2^n}{n!} + \cdots$

(e) $2 + \frac{1}{2} + \frac{8}{27} + \frac{1}{4} + \cdots + \frac{2^n}{n^3} + \cdots$

(f) $1 + \frac{1}{\sqrt[3]{2}} + \frac{1}{\sqrt[3]{3}} + \cdots + \frac{1}{\sqrt[3]{n}} + \cdots$

(g) $\frac{1}{2} + \frac{1}{2\cdot 2^2} + \frac{1}{3\cdot 2^3} + \cdots + \frac{1}{n\cdot 2^n} + \cdots$

(h) $\frac{1}{2} + \frac{2}{3} + \frac{3}{4} + \cdots + \frac{n}{n+1} + \cdots$

(i) $\frac{1}{1\cdot 3} + \frac{1}{3\cdot 5} + \frac{1}{5\cdot 7} + \cdots + \frac{1}{(2n-1)(2n+1)} + \cdots$

(j) $1 + \frac{2^2+1}{2^3+1} + \frac{3^2+1}{3^3+1} + \cdots + \frac{n^2+1}{n^3+1} + \cdots$

Ans. (a) Convergent (b) Divergent (c) Divergent (d) Convergent (e) Divergent (f) Divergent (g) Convergent (h) Divergent (i) Convergent (j) Divergent

11. Investigate the following alternating series for convergence or divergence.

(a) 1/4 − 1/10 + 1/28 − 1/82 + ...

(b) 2 − 3/2 + 4/3 − 5/4 + ...

(c) 1 − 1/2 + 1/3 − 1/4 + ...

(d) 1/2 − 2/3 + 3/4 − 4/5 + ...

(e) 1/4 − 3/6 + 5/8 − 7/10 + ...

(f) $\frac{2}{3} - \frac{3}{4}\cdot\frac{1}{2} + \frac{4}{5}\cdot\frac{1}{3} - \frac{5}{6}\cdot\frac{1}{4} + \cdots$

(g) $\frac{2}{2\cdot 3} - \frac{2^2}{3\cdot 4} + \frac{2^3}{4\cdot 5} - \frac{2^4}{5\cdot 6} + \cdots$

(h) $2 - \frac{2^3}{3!} + \frac{2^5}{5!} - \frac{2^7}{7!} + \cdots$

Ans. (a) Abs. Conv. (b) Divergent (c) Cond. Conv. (d) Divergent (e) Divergent (f) Cond. Conv. (g) Divergent (h) Abs. Conv.

CHAPTER 29

Power Series

INFINITE SERIES OF THE FORM

(*1*) $$c_0 + c_1x + c_2x^2 + \dots + c_{n-1}x^{n-1} + \dots \quad \text{and}$$

(*2*) $$c_0 + c_1(x-a) + c_2(x-a)^2 + \dots + c_{n-1}(x-a)^{n-1} + \dots,$$

where $a, c_0, c_1, c_2, \dots$ are constants, are called *power series*. The first is called a power series in x and the second a power series in $(x-a)$.

The power series (*1*) converges for $x=0$ and (*2*) converges for $x=a$. Both series may converge for other values of x but not necessarily for every finite value of x. Our problem is to find for a given power series all values of x for which the series converges. In finding this set of values, called the *interval of convergence* of the series, the generalized ratio test of Chapter 28 will be used.

Example 1. Find the interval of convergence of the series

$$x + x^2/2 + x^3/3 + \dots$$

Since $|s_n| = |\frac{x^n}{n}|$ and $|s_{n+1}| = |\frac{x^{n+1}}{n+1}|$,

$$R = \lim_{n\to\infty} \frac{|s_{n+1}|}{|s_n|} = \lim_{n\to\infty} |\frac{x^{n+1}}{n+1} \cdot \frac{n}{x^n}| = \lim_{n\to\infty} |\frac{n}{n+1}x| = |x|.$$

Then, by the ratio test, the given series is convergent for all values of x such that $|x| < 1$, that is, for $-1 < x < 1$; the series is divergent for all values of x such that $|x| > 1$, that is, for $x < -1$ and $x > 1$; and the test fails for $x = \pm 1$.

But, when $x=1$ the series is $1 + 1/2 + 1/3 + 1/4 + \dots$ and is divergent, and when $x=-1$ the series is $-1 + 1/2 - 1/3 + 1/4 - \dots$ and is convergent.

Thus, the series converges on the interval $-1 \le x < 1$. This interval may be represented graphically as follows:

The solid line represents the interval on which the series converges, the thin line the intervals on which the series diverges. The solid circle represents the end-point for which the series converges, the open circle represents the end-point at which the series diverges.

See Problems 1-8.

SOLVED PROBLEMS

In problems 1-6, find the interval of convergence *including* the end-points.

1. $1 + \dfrac{x}{1!} + \dfrac{x^2}{2!} + \dfrac{x^3}{3!} + \ldots.$

For this series $|s_n| = \left|\dfrac{x^{n-1}}{(n-1)!}\right|$, $|s_{n+1}| = \left|\dfrac{x^n}{n!}\right|$, and

$$R = \lim_{n\to\infty} \frac{|s_{n+1}|}{|s_n|} = \lim_{n\to\infty} \left|\frac{x^n}{n!} \cdot \frac{(n-1)!}{x^{n-1}}\right| = \lim_{n\to\infty} \left|\frac{x}{n}\right| = 0$$

The series is *everywhere convergent*, that is, it is convergent for all finite values of x.

2. $1 + x + 2x^2 + 3x^3 + \ldots.$

Here $|s_n| = |(n-1)x^{n-1}|$, $|s_{n+1}| = |nx^n|$, and $R = \lim_{n\to\infty} \left|\dfrac{nx^n}{(n-1)x^{n-1}}\right| = \lim_{n\to\infty} \left|\dfrac{n}{n-1} x\right| = |x|$.

The series converges on the interval $-1 < x < 1$ and diverges on the intervals $x < -1$ and $x > 1$.

When $x = 1$, the series is $1 + 1 + 2 + 3 + \ldots.$ and is divergent.
When $x = -1$, the series is $1 - 1 + 2 - 3 + \ldots.$ and is divergent.

The interval of convergence $-1 < x < 1$ is indicated by

3. $1 + x/2 + x^2/4 + x^3/8 + \ldots.$

Here $|s_n| = \left|\dfrac{x^{n-1}}{2^{n-1}}\right|$, $|s_{n+1}| = \left|\dfrac{x^n}{2^n}\right|$, and $R = \lim_{n\to\infty} \left|\dfrac{x^n}{2^n} \cdot \dfrac{2^{n-1}}{x^{n-1}}\right| = \lim_{n\to\infty} \left|\dfrac{x}{2}\right| = \dfrac{1}{2}|x|$.

The series converges for all values of x such that $\frac{1}{2}|x| < 1$, that is, for $-2 < x < 2$; and diverges for $x < -2$ and $x > 2$.

For $x = 2$, the series is $1 + 1 + 1 + 1 + \ldots.$ and is divergent.
For $x = -2$, the series is $1 - 1 + 1 - 1 + \ldots.$ and is divergent.

The interval of convergence $-2 < x < 2$ is indicated by

4. $\dfrac{1!}{x+1} - \dfrac{2!}{(x+1)^2} + \dfrac{3!}{(x+1)^3} + \ldots.$

For this series $|s_n| = \left|\dfrac{n!}{(x+1)^n}\right|$, $|s_{n+1}| = \left|\dfrac{(n+1)!}{(x+1)^{n+1}}\right|$, and

$$R = \lim_{n\to\infty} \left|\frac{(n+1)!}{(x+1)^{n+1}} \cdot \frac{(x+1)^n}{n!}\right| = \lim_{n\to\infty} \left|\frac{n+1}{x+1}\right| \text{ does not exist.}$$

Thus, the series diverges for every value of x.

5. $\dfrac{x+3}{1\cdot 4} + \dfrac{(x+3)^2}{2\cdot 4^2} + \dfrac{(x+3)^3}{3\cdot 4^3} + \ldots.$

For this series $|s_n| = \left|\dfrac{(x+3)^n}{n\cdot 4^n}\right|$, $|s_{n+1}| = \left|\dfrac{(x+3)^{n+1}}{(n+1)\cdot 4^{n+1}}\right|$, and

$$R = \lim_{n\to\infty} \left|\frac{(x+3)^{n+1}}{(n+1)4^{n+1}} \cdot \frac{n4^n}{(x+3)^n}\right| = \lim_{n\to\infty}\left|\frac{n}{n+1}\cdot\frac{x+3}{4}\right| = \frac{1}{4}|x+3|$$

The series converges for all values of x such that $\frac{1}{4}|x+3| < 1$, that is, for $-4 < x+3 < 4$ or $-7 < x < 1$, and diverges for $x < -7$ and $x > 1$.

For $x = -7$, the series is $-1 + 1/2 - 1/3 + \ldots.$ and is convergent.
For $x = 1$, the series is $1 + 1/2 + 1/3 + \ldots.$ and is divergent.

The interval of convergence $-7 \le x < 1$ is indicated by [number line: closed dot at -7, open dot at 1]

6. $\dfrac{1}{1\cdot2\cdot3} - \dfrac{(x+2)^2}{2\cdot3\cdot4} + \dfrac{(x+2)^4}{3\cdot4\cdot5} - \ldots.$

Here $|s_n| = \left|\dfrac{(x+2)^{2n-2}}{n(n+1)(n+2)}\right|$, $|s_{n+1}| = \left|\dfrac{(x+2)^{2n}}{(n+1)(n+2)(n+3)}\right|$, and

$$R = \lim_{n\to\infty}\left|\frac{(x+2)^{2n}}{(n+1)(n+2)(n+3)}\cdot\frac{n(n+1)(n+2)}{(x+2)^{2n-2}}\right| = \lim_{n\to\infty}\left|\frac{n}{n+3}(x+2)^2\right| = (x+2)^2.$$

The series converges for all values of x such that $(x+2)^2 < 1$, that is, for $-3 < x < -1$, and diverges for $x < -3$ and $x > -1$.

For $x = -3$ and $x = -1$ the series is $\dfrac{1}{1\cdot2\cdot3} - \dfrac{1}{2\cdot3\cdot4} + \dfrac{1}{3\cdot4\cdot5} - \ldots.$ and is convergent.

The interval of convergence $-3 \le x \le -1$ is indicated by [number line: closed dots at -3 and -1]

7. Expand $(1+x)^{-1}$ as a power series in x and examine for convergence.

By division, $\dfrac{1}{x+1} = 1 - x + x^2 - x^3 + x^4 - \ldots.$ Then $R = \lim_{n\to\infty}\left|\dfrac{x^{n+1}}{x^n}\right| = |x|$.

The series converges for $-1 < x < 1$, and diverges for $x < -1$ and $x > 1$.

For $x = 1$, the series is $1-1+1-1+\ldots.$ and is divergent.
For $x = -1$, the series is $1+1+1+1+\ldots.$ and is divergent.

The interval of convergence $-1 < x < 1$ is indicated by [number line: open dots at -1 and 1]

Thus, the series $1 - x + x^2 - x^3 + x^4 - \ldots.$ *represents* the function $f(x) = (1+x)^{-1}$ for all x such that $|x| < 1$. It does not represent the function for, say, $x = -4$. Note that $f(-4) = -1/3$, while for $x = -4$ the series is $1 + 4 + 16 + 64 + \ldots.$

8. Expand $(1+x)^{1/2}$ in a power series in x and examine for convergence.

By the binomial theorem

$$(1+x)^{1/2} = 1 + \frac{1}{2}x + \frac{(1/2)(-1/2)}{1\cdot2}x^2 + \frac{(1/2)(-1/2)(-3/2)}{1\cdot2\cdot3}x^3 + \frac{(1/2)(-1/2)(-3/2)(-5/2)}{1\cdot2\cdot3\cdot4}x^4 + \ldots.$$

Except for $n = 1$, $|s_n| = \left|\dfrac{(1/2)(-1/2)(-3/2)\ldots.(-2n+5)/2}{(n-1)!}x^{n-1}\right|$,

$$|s_{n+1}| = \left|\frac{(1/2)(-1/2)(-3/2)\ldots.(-2n+3)/2}{n!}x^n\right|, \text{ and } R = \lim_{n\to\infty}\left|\frac{-2n+3}{2n}x\right| = |x|.$$

The series is convergent for $-1 < x < 1$, and divergent for $x < -1$ and $x > 1$. An investigation at the end-points is beyond the scope of this book.

SUPPLEMENTARY PROBLEMS

In Problems 9-19 find the interval of convergence including the end-points.

9. $1 + x^2 + x^4 + \dots + x^{2n-2} + \dots$ *Ans.* $-1 < x < 1$

10. $\dfrac{x}{1\cdot2} + \dfrac{x^2}{2\cdot3} + \dfrac{x^3}{3\cdot4} + \dots + \dfrac{x^n}{n(n+1)} + \dots$ *Ans.* $-1 \le x \le 1$

11. $\dfrac{x}{1\cdot3} + \dfrac{x^2}{2\cdot3^2} + \dfrac{x^3}{3\cdot3^3} + \dots + \dfrac{x^n}{n\cdot3^n} + \dots$ *Ans.* $-3 \le x < 3$

12. $\dfrac{x}{1^2+1} + \dfrac{x^2}{2^2+1} + \dfrac{x^3}{3^2+1} + \dots + \dfrac{x^n}{n^2+1} + \dots$ *Ans.* $-1 \le x \le 1$

13. $\dfrac{x^2}{4} - \dfrac{x^4}{8} + \dfrac{x^6}{16} - \dots + (-1)^{n-1}\dfrac{x^{2n}}{2^{n+1}} + \dots$ *Ans.* $-\sqrt{2} < x < \sqrt{2}$

14. $\dfrac{1\cdot x}{2\cdot1} - \dfrac{1\cdot3\cdot x^3}{2\cdot4\cdot3} + \dfrac{1\cdot3\cdot5\cdot x^5}{2\cdot4\cdot6\cdot5} - \dfrac{1\cdot3\cdot5\cdot7\cdot x^7}{2\cdot4\cdot6\cdot8\cdot7} + \dots$ *Ans.* $-1 \le x \le 1$

15. $\dfrac{1\cdot1}{3\cdot2}x + \dfrac{3\cdot2}{5\cdot3}x^2 + \dfrac{5\cdot3}{7\cdot4}x^3 + \dfrac{7\cdot4}{9\cdot5}x^4 + \dots$ *Ans.* $-1 < x < 1$

16. $(x-2) + \dfrac{1}{2}(x-2)^2 + \dfrac{1}{3}(x-2)^3 + \dfrac{1}{4}(x-2)^4 + \dots$ *Ans.* $1 \le x < 3$

17. $\dfrac{x+1}{1\cdot2} - \dfrac{(x+1)^2}{3\cdot2^2} + \dfrac{(x+1)^3}{5\cdot2^3} - \dfrac{(x+1)^4}{7\cdot2^4} + \dots$ *Ans.* $-3 < x \le 1$

18. $\dfrac{x-a}{b} + \dfrac{(x-a)^2}{b^2} + \dfrac{(x-a)^3}{b^3} + \dfrac{(x-a)^4}{b^4} + \dots$ *Ans.* $a-b < x < a+b$

19. $\dfrac{x-2}{x} + \dfrac{1}{2}\left(\dfrac{x-2}{x}\right)^2 + \dfrac{1}{3}\left(\dfrac{x-2}{x}\right)^3 + \dfrac{1}{4}\left(\dfrac{x-2}{x}\right)^4 + \dots$ *Ans.* $x \ge 1$

20. Show that the binomial series for $(1+x)^k$, k any real numbers, converges for $|x| < 1$.

Given $\lim_{n\to\infty} (1+1/n)^n = e$, where e is the base of the natural system of logarithms, find the interval of convergence, exclusive of end-points, of each of the following series.

21. $x + x^2/2^2 + x^3/3^3 + x^4/4^4 + \dots + x^n/n^n + \dots$ *Ans.* All values of x.

22. $x - 2^2x^2 + 3^3x^3 - 4^4x^4 + \dots$ *Ans.* $x = 0$

23. $x + \dfrac{2!\,x^2}{2^2} + \dfrac{3!\,x^3}{3^3} + \dfrac{4!\,x^4}{4^4} + \dots$ *Ans.* $-e < x < e$

CHAPTER 30

Angles and Arc Length

TRIGONOMETRY, as the word implies, is concerned with the measurement of the parts of a triangle. Plane trigonometry, considered in the next several chapters, is restricted to triangles lying in planes. Spherical trigonometry deals with certain triangles which lie on spheres.

The science of trigonometry is based on certain ratios, called trigonometric functions, to be defined in the next chapter. The early applications of the trigonometric functions were to surveying, navigation, and engineering. These functions also play an important role in the study of all sorts of vibratory phenomena — sound, light, electricity, etc. As a consequence, a considerable portion of the subject matter is concerned properly with a study of the properties of and relations among the trigonometric functions.

THE PLANE ANGLE XOP is formed by the two intersecting half lines OX and OP. The point O is called the *vertex* and the half lines are called the sides of the angle.

More often, a plane angle is to be thought of as generated by revolving (in a plane) a half line from the initial position OX to a terminal position OP. Then O is again the vertex, OX is called the *initial side*, and OP is called the *terminal* side of the angle.

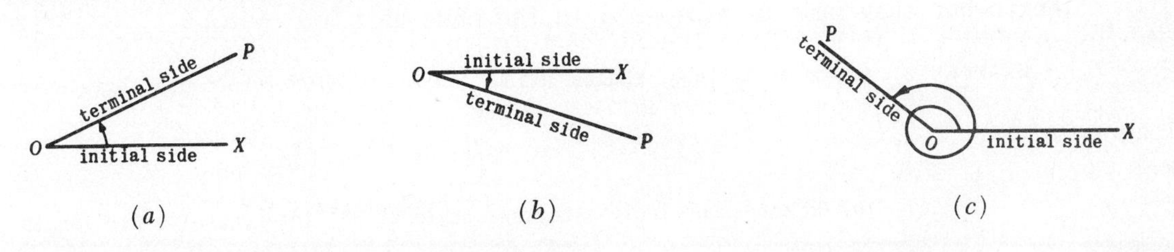

(a) (b) (c)

An angle, so generated, is called *positive* if the direction of rotation (indicated by a curved arrow) is counterclockwise and *negative* if the direction of rotation is clockwise. The angle is positive in Figures (a) and (c), and negative in Figure (b).

MEASURES OF ANGLES.

A. A *degree* (°) is defined as the measure of the central angle subtended by an arc of a circle equal to 1/360 of the circumference of the circle.

A *minute* (') is 1/60 of a degree; a *second* (") is 1/60 of a minute.

EXAMPLE 1. (a) $\frac{1}{4}(36°24') = 9°6'$ (b) $\frac{1}{2}(127°24') = \frac{1}{2}(126°84') = 63°42'$

(c) $\frac{1}{2}(81°15') = \frac{1}{2}(80°75') = 40°37.5'$ or $40°37'30''$

(d) $\frac{1}{4}(74°29'20'') = \frac{1}{4}(72°149'20'') = \frac{1}{4}(72°148'80'') = 18°37'20''$

B. A *radian* (rad) is defined as the measure of the central angle subtended by an arc of a circle equal to the radius of the circle.

The circumference of a circle = 2π(radius) and subtends an angle of 360°. Then 2π radians = 360°, from which we obtain

$$1 \text{ radian} = \frac{180°}{\pi} = 57.296° = 57°17'45''$$

and $$1 \text{ degree} = \frac{\pi}{180} \text{ radian} = 0.017453 \text{ rad, approx.,} \qquad \text{where } \pi = 3.14159.$$

EXAMPLE 2. (*a*) $\frac{7}{12}\pi \text{ rad} = \frac{7\pi}{12} \cdot \frac{180°}{\pi} = 105°$ (*b*) $50° = 50 \cdot \frac{\pi}{180} \text{ rad} = \frac{5\pi}{18} \text{ rad.}$

See Problems 1-3.

C. A *mil*, used in military science, is defined as the measure of the central angle subtended by an arc of a circle equal to 1/6400 of the circumference of the circle. The name is derived from the fact that, approximately,

$$1 \text{ mil} = \frac{1}{1000} \text{ radian.}$$

Since 6400 mils = 360°, $1 \text{ mil} = \frac{360°}{6400} = \frac{9°}{160}$ and $1° = \frac{160}{9}$ mils.

See Problems 12-13.

ARC LENGTH.

A. On a circle of radius r, a central angle of θ radians intercepts an arc of length

$$s = r\theta,$$

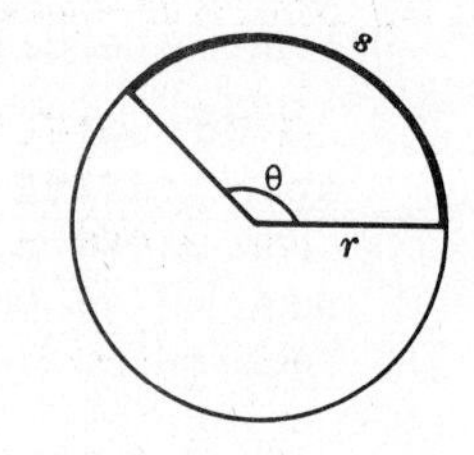

that is, arc length = radius × central angle in radians.

(Note. s and r may be measured in any convenient unit of length but they must be expressed in the same unit.)

EXAMPLE 3. (*a*) On a circle of radius 30 in., the length of arc intercepted by a central angle of 1/3 radian is

$$s = r\theta = 30(\tfrac{1}{3}) = 10 \text{ in.}$$

(*b*) On the same circle a central angle of 50° intercepts an arc length

$$s = r\theta = 30(\tfrac{5\pi}{18}) = \frac{25\pi}{3} \text{ in.}$$

(*c*) On the same circle an arc length $1\frac{1}{2}$ ft subtends a central angle

$$\theta = \frac{s}{r} = \frac{18}{30} = \frac{3}{5} \text{ rad, when } s \text{ and } r \text{ are expressed in inches,}$$

or $$\theta = \frac{s}{r} = \frac{3/2}{5/2} = \frac{3}{5} \text{ rad, when } s \text{ and } r \text{ are expressed in feet.}$$

See Problems 4-11.

B. If the central angle is relatively small, the length of the intercepted arc may be taken as a close approximation of the length of its chord.

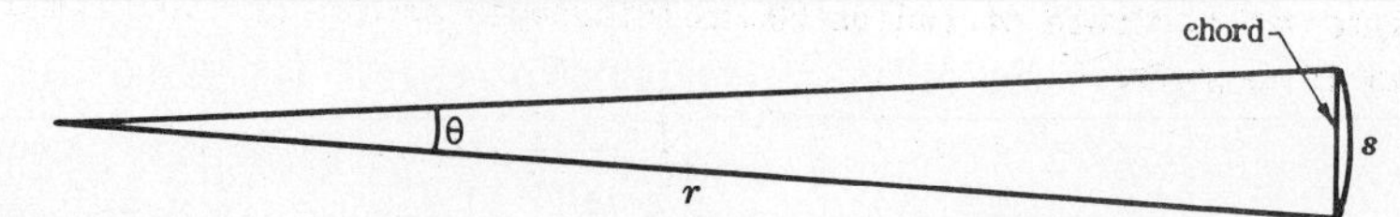

Now since θ rad = 1000 θ mils and $s = r\theta = \frac{r}{1000}(1000\,\theta)$, it follows that

$$\text{length of chord} = \frac{r}{1000}(\text{central angle in mils}), \quad \text{approx.}$$

For military purposes this is written as $W = Rm$, where m is the central angle expressed in mils, R is the radius (range) expressed in thousands of yards, and W is the chord (width) expressed in yards.

See Problems 14-16.

SOLVED PROBLEMS

1. Express each angle in radian measure; (*a*) 30°, (*b*) 135°, (*c*) 25°30', (*d*) 42°24'35".

Since $1° = \frac{\pi}{180}$ radian = 0.017453 rad,

(*a*) $30° = 30 \times \frac{\pi}{180}$ rad $= \frac{\pi}{6}$ rad or 0.5236 rad,

(*b*) $135° = 135 \times \frac{\pi}{180}$ rad $= \frac{3\pi}{4}$ rad or 2.3562 rad,

(*c*) $25°30' = 25.5° = 25.5 \times \frac{\pi}{180}$ rad = 0.4451 rad,

(*d*) $42°24'35'' = 42° + (\frac{24 \times 60 + 35}{3600})° = 42.41° = 42.41 \times \frac{\pi}{180}$ rad = 0.7402 rad.

2. Express each angle in degree measure: (*a*) $\pi/3$ rad, (*b*) $5\pi/9$ rad, (*c*) 2/5 rad, (*d*) 4/3 rad.

Since 1 rad $= \frac{180°}{\pi} = 57°17'45''$,

(*a*) $\frac{\pi}{3}$ rad $= \frac{\pi}{3} \times \frac{180°}{\pi} = 60°$,

(*b*) $\frac{5\pi}{9}$ rad $= \frac{5\pi}{9} \times \frac{180°}{\pi} = 100°$,

(*c*) $\frac{2}{5}$ rad $= \frac{2}{5} \times \frac{180°}{\pi} = \frac{72°}{\pi}$ or $\frac{2}{5}(57°17'45'') = 22°55'6''$.

(*d*) $\frac{4}{3}$ rad $= \frac{4}{3} \times \frac{180°}{\pi} = \frac{240°}{\pi}$ or $\frac{4}{3}(57°17'45'') = 76°23'40''$.

3. A wheel is turning at the rate 48 rpm (revolutions per minute or rev/min). Express this angular speed in (*a*) rev/sec, (*b*) rad/min, (*c*) rad/sec.

(*a*) 48 rev/min $= \frac{48}{60}$ rev/sec $= \frac{4}{5}$ rev/sec.

(*b*) Since 1 rev = 2π rad, 48 rev/min = $48(2\pi)$ rad/min = 301.6 rad/min.

(*c*) 48 rev/min $= \frac{4}{5}$ rev/sec $= \frac{4}{5}(2\pi)$ rad/sec = 5.03 rad/sec or

48 rev/min = 96π rad/min $= \frac{96\pi}{60}$ rad/sec = 5.03 rad/sec.

4. The minute hand of a clock is 12 in. long. How far does the tip of the hand move during 20 min?

During 20 min the hand moves through an angle $\theta = 120° = 2\pi/3$ rad and the tip of the hand moves over a distance $s = r\theta = 12(2\pi/3) = 8\pi$ in. = 25.1 in.

5. A central angle of a circle of radius 30 in. intercepts an arc of 6 in. Express the central angle θ in radians and in degrees.

$$\theta = \frac{s}{r} = \frac{6}{30} = \frac{1}{5} \text{ rad} = 11°27'33''$$

6. A railroad curve is to be laid out on a circle. What radius should be used if the track is to change direction by 25° in a distance of 120 ft ?

We are required to find the radius of a circle on which a central angle $\theta = 25° = 5\pi/36$ rad intercepts an arc of 120 ft. Then

$$r = \frac{s}{\theta} = \frac{120}{5\pi/36} = \frac{864}{\pi} \text{ ft} = 275 \text{ ft}.$$

7. Assuming the earth to be a sphere of radius 3960 miles, find the distance of a point in latitude 36°N from the equator.

Since $36° = \frac{\pi}{5}$ radian, $s = r\theta = 3960(\frac{\pi}{5}) = 2488$ miles.

8. Two cities 270 miles apart lie on the same meridian. Find their difference in latitude.

$$\theta = \frac{s}{r} = \frac{270}{3960} = \frac{3}{44} \text{ rad} \quad \text{or} \quad 3°54.4'$$

9. A wheel 4 ft in diameter is rotating at 80 rpm. Find the distance (in ft) traveled by a point on the rim in one second, that is, the linear speed of the point (in ft/sec).

$$80 \text{ rpm} = 80(\frac{2\pi}{60}) \text{ rad/sec} = \frac{8\pi}{3} \text{ rad/sec}.$$

Then in 1 sec the wheel turns through an angle $\theta = 8\pi/3$ rad and a point on the wheel will travel a distance $s = r\theta = 2(8\pi/3)$ ft $= 16.8$ ft. The linear velocity is 16.8 ft/sec.

10. Find the diameter of a pulley which is driven at 360 rpm by a belt moving at 40 ft/sec.

$$360 \text{ rev/min} = 360(\frac{2\pi}{60}) \text{ rad/sec} = 12\pi \text{ rad/sec}$$

Then in 1 sec the pulley turns through an angle $\theta = 12\pi$ rad and a point on the rim travels a distance $s = 40$ ft.

$$d = 2r = 2(\frac{s}{\theta}) = 2(\frac{40}{12\pi}) \text{ ft} = \frac{20}{3\pi} \text{ ft} = 2.12 \text{ ft}$$

11. A point on the rim of a turbine wheel of diameter 10 ft moves with a linear speed 45 ft/sec. Find the rate at which the wheel turns (angular speed) in rad/sec and in rev/sec.

In 1 sec a point on the rim travels a distance $s = 45$ ft. Then in 1 sec the wheel turns through an angle $\theta = s/r = 45/5 = 9$ radians and its angular speed is 9 rad/sec.

Since 1 rev $= 2\pi$ rad or 1 rad $= \frac{1}{2\pi}$ rev, 9 rad/sec $= 9(\frac{1}{2\pi})$ rev/sec $= 1.43$ rev/sec.

12. Express each of the following angles in mils: (*a*) 18°, (*b*) 16°20′, (*c*) 0.22 rad, (*d*) 1.6 rad.

Since $1° = \frac{160}{9}$ mils and 1 rad $= 1000$ mils,

(*a*) $18° = 18(\frac{160}{9})$ mils $= 320$ mils, (*b*) $16°20' = \frac{49}{3}(\frac{160}{9})$ mils $= 290$ mils,

(*c*) 0.22 rad $= 0.22(1000)$ mils $= 220$ mils, (*d*) 1.6 rad $= 1.6(1000)$ mils $= 1600$ mils.

13. Express each of the following angles in degrees and in radians: (*a*) 40 mils, (*b*) 100 mils.

Since 1 mil $= \frac{9°}{160} = 0.001$ rad,

(*a*) 40 mils $= 40(\frac{9°}{160}) = 2°15'$ and 40 mils $= 40(0.001)$ rad $= 0.04$ rad,

(*b*) 100 mils $= 100(\frac{9°}{160}) = 5°37.5'$ and 100 mils $= 100(0.001)$ rad $= 0.1$ rad.

14. At 5000 yd range a battery subtends an angle of 15 mils. Find the width of the battery.

$$R = \frac{5000}{1000} = 5, \quad m = 15, \quad \text{and} \quad W = Rm = 5(15) = 75 \text{ yd.}$$

15. A ship 360 ft long is found to subtend an angle of 40 mils at an observation post on shore. Find the distance from shore to ship.

$W = 360$ ft $= 120$ yd, $m = 40$, and $R = W/m = 120/40 = 3$. The required distance is 3000 yd.

16. A shell is observed to burst 200 yd to the left of the target. What angular correction should be made in aiming the gun, if the range is (*a*) 5000 yd and (*b*) 7500 yd?

(*a*) The correction is $m = W/R = 200/5 = 40$ mils, to the right.
(*b*) The correction is $m = W/R = 200/7.5 = 27$ mils, to the right.

SUPPLEMENTARY PROBLEMS

17. Express in radian measure: (*a*) 25°, (*b*) 160°, (*c*) 75°30′, (*d*) 112°40′, (*e*) 12°12′20″.
Ans. (*a*) $5\pi/36$ rad or 0.4363 rad (*b*) $8\pi/9$ rad or 2.7925 rad (*c*) $151\pi/360$ rad or 1.3177 rad (*d*) $169\pi/270$ rad or 1.9664 rad (*e*) 0.2130 rad

18. Express in degree measure: (*a*) $\pi/4$ rad, (*b*) $7\pi/10$ rad, (*c*) $5\pi/6$ rad, (*d*) 1/4 rad, (*e*) 7/5 rad.
Ans. (*a*) 45°, (*b*) 126°, (*c*) 150°, (*d*) 14°19′26″, (*e*) 80°12′51″

19. On a circle of radius 24 inches, find the length of arc subtended by a central angle (*a*) of 2/3 rad, (*b*) of $3\pi/5$ rad, (*c*) of 75°, (*d*) of 130°.
Ans. (*a*) 16 in., (*b*) 14.4π or 45.2 in., (*c*) 10π or 31.4 in., (*d*) $52\pi/3$ or 54.5 in.

20. A circle has a radius of 30 in. How many radians are there in an angle at the center subtended by an arc (*a*) of 30 in., (*b*) of 20 in., (*c*) of 50 in.? *Ans.* (*a*) 1 rad (*b*) 2/3 rad (*c*) 5/3 rad

21. Find the radius of the circle for which an arc 15 in. long subtends an angle (*a*) of 1 rad, (*b*) of 2/3 rad, (*c*) of 3 rad, (*d*) of 20°, (*e*) of 50°.
Ans. (*a*) 15 in. (*b*) 22.5 in. (*c*) 5 in. (*d*) 43.0 in. (*e*) 17.2 in.

22. The end of a 40 in. pendulum describes an arc of 5 in. Through what angle does the pendulum swing?
Ans. 1/8 rad or 7°9′43″

23. A train is traveling at the rate 12 mi/hr on a curve of radius 3000 ft. Through what angle has it turned in one minute? *Ans.* 0.352 rad or 20°10′

24. A reversed curve on a railroad track consists of two circular arcs. The central angle of one is 20° with radius 2500 ft and the central angle of the other is 25° with radius 3000 ft. Find the total length of the two arcs. *Ans.* $6250\pi/9$ ft or 2182 ft

25. A flywheel of radius 10 in. is turning at the rate 900 rpm. How fast does a point on the rim travel in ft/sec? *Ans.* 78.5 ft/sec

26. An automobile tire is 30 in. in diameter. How fast (rpm) does the wheel turn on the axle when the automobile maintains a speed of 45 mph? *Ans.* 504 rpm

27. In grinding certain tools the linear velocity of the grinding surface should not exceed 6000 ft/sec. Find the maximum number of revolutions per second (*a*) of a 12 in. (diameter) emery wheel, (*b*) of an 8 in. wheel. *Ans.* (*a*) $6000/\pi$ rev/sec or 1910 rev/sec (*b*) 2865 rev/sec

28. If an automobile wheel 32 in. in diameter rotates at 800 rpm, what is the speed of the car in mph?
Ans. 76.2 mph

29. Express each of the following angles in mils: (*a*) 45°, (*b*) 10°15′, (*c*) 0.4 rad, (*d*) 0.06 rad.
Ans. (*a*) 800 mils (*b*) 182 mils (*c*) 400 mils (*d*) 60 mils

30. Express each of the following in degree and in radian measure: (*a*) 25 mils, (*b*) 60 mils, (*c*) 110 mils.
Ans. (*a*) 1°24′ and 0.025 rad (*b*) 3°22′ and 0.06 rad (*c*) 6°11′ and 0.11 rad

31. The side of a hangar 1750 yd distant subtends an angle of 40 mils. How long is it? *Ans.* 70 yd

CHAPTER 31

Trigonometric Functions of a General Angle

ANGLES IN STANDARD POSITION. With respect to a rectangular coordinate system, an angle is said to be *in standard position* when its vertex is at the origin and its initial side coincides with the positive x-axis.

An angle is said to be a *first quadrant angle* or to be *in the first quadrant* if, when in standard position, its terminal side falls in that quadrant. Similar definitions hold for the other quadrants. For example, the angles 30°, 59°, and −330° are first quadrant angles; 119° is a second quadrant angle; −119° is a third quadrant angle; −10° and 710° are fourth quadrant angles.

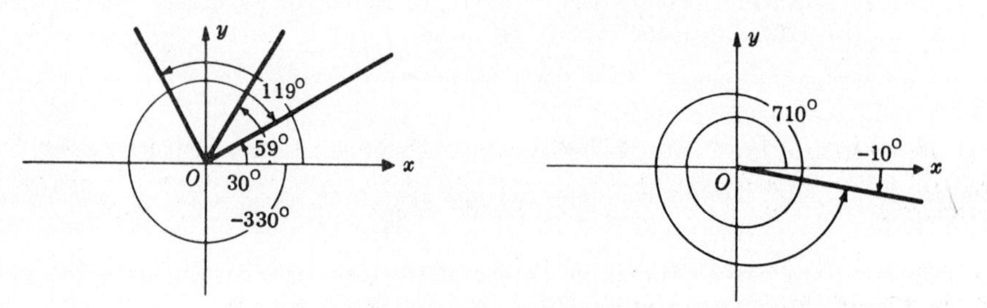

Two angles which, when placed in standard position, have coincident terminal sides are called *coterminal angles*. For example, 30° and −330°, −10° and 710° are pairs of coterminal angles. There are an unlimited number of angles coterminal with a given angle.

See Problem 1.

The angles 0°, 90°, 180°, 270°, and all angles coterminal with them are called *quadrantal angles*.

TRIGONOMETRIC FUNCTIONS OF A GENERAL ANGLE. Let θ be an angle (not quadrantal) in standard position and let $P(x,y)$ be any point, distinct from the origin, on the terminal side of the angle. The six trigonometric functions of θ are defined, in terms of the abscissa, ordinate and distance of P, as follows:

$$\text{sine } \theta = \sin\theta = \frac{\text{ordinate}}{\text{distance}} = \frac{y}{r} \qquad \text{cotangent } \theta = \cot\theta = \frac{\text{abscissa}}{\text{ordinate}} = \frac{x}{y}$$

$$\text{cosine } \theta = \cos\theta = \frac{\text{abscissa}}{\text{distance}} = \frac{x}{r} \qquad \text{secant } \theta = \sec\theta = \frac{\text{distance}}{\text{abscissa}} = \frac{r}{x}$$

$$\text{tangent } \theta = \tan\theta = \frac{\text{ordinate}}{\text{abscissa}} = \frac{y}{x} \qquad \text{cosecant } \theta = \csc\theta = \frac{\text{distance}}{\text{ordinate}} = \frac{r}{y}$$

As an immediate consequence of these definitions, we have the so-called *Reciprocal Relations*:

$\sin\theta = 1/\csc\theta$	$\tan\theta = 1/\cot\theta$	$\sec\theta = 1/\cos\theta$
$\cos\theta = 1/\sec\theta$	$\cot\theta = 1/\tan\theta$	$\csc\theta = 1/\sin\theta$

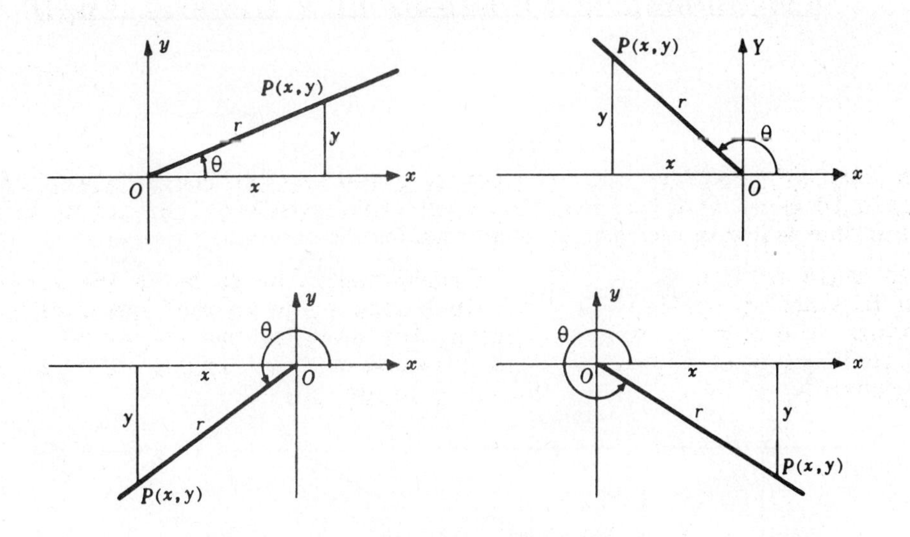

It is evident from the figures that the values of the trigonometric functions of θ change as θ changes. The values of the functions of a given angle θ are, however, independent of the choice of the point P on its terminal side.

ALGEBRAIC SIGNS OF THE FUNCTIONS. Since r is always positive, the signs of the functions in the various quadrants depend upon the signs of x and y. To determine these signs one may visualize the angle in standard position or use some device as shown in the adjacent figure in which only the functions having positive signs are listed.

II	I
$\sin\theta = +$ $\csc\theta = +$	All +
III	**IV**
$\tan\theta = +$ $\cot\theta = +$	$\cos\theta = +$ $\sec\theta = +$

When an angle is given, its trigonometric functions are uniquely determined. When, however, the value of one function of an angle is given, the angle is not uniquely determined. For example, if $\sin\theta = \frac{1}{2}$, then $\theta = 30°, 150°, 390°, 510°, \ldots\ldots$ In general, two possible positions of the terminal side are found — for example, the terminal sides of 30° and 150° in the above illustration. The exceptions to this rule occur when the angle is quadrantal.

See Problems 2-10.

TRIGONOMETRIC FUNCTIONS OF QUADRANTAL ANGLES. For a quadrantal angle, the terminal side coincides with one of the axes. A point P, distinct from the origin, on the terminal side has either $x=0, y\neq 0$ or $x\neq 0, y=0$. In either case, two of the six functions will not be defined. For example, the terminal side of the angle 0° coincides with the positive x-axis and the ordinate of P is 0. Since the ordinate occurs in the denominator of the ratio defining the cotangent and cosecant, these functions are not defined. Certain authors indicate this by writing $\cot 0° = \infty$ and others write $\cot 0° = \pm\infty$. The trigonometric functions of the quadrantal angles are:

angle θ	$\sin\theta$	$\cos\theta$	$\tan\theta$	$\cot\theta$	$\sec\theta$	$\csc\theta$
0°	0	1	0	$\pm\infty$	1	$\pm\infty$
90°	1	0	$\pm\infty$	0	$\pm\infty$	1
180°	0	−1	0	$\pm\infty$	−1	$\pm\infty$
270°	−1	0	$\pm\infty$	0	$\pm\infty$	−1

SOLVED PROBLEMS

1. (*a*) Construct the following angles in standard position and determine those which are coterminal:

$$125°,\ 210°,\ -150°,\ 385°,\ 930°,\ -370°,\ -955°,\ -870°.$$

(*b*) Give five other angles coterminal with 125°.

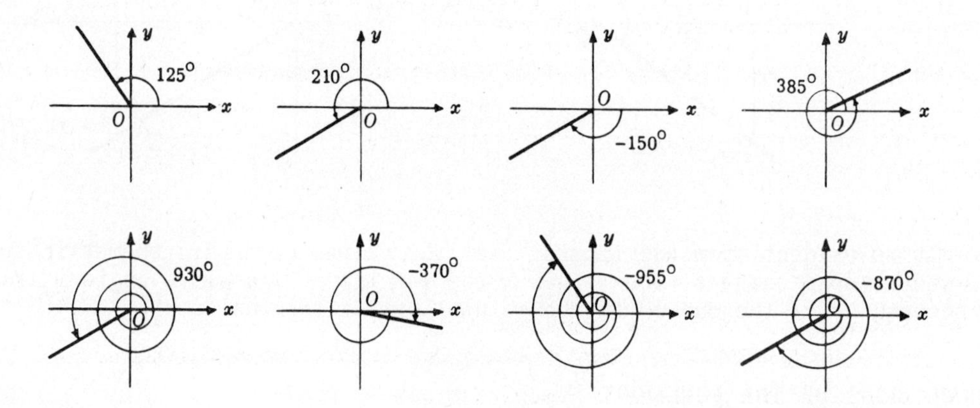

(*a*) The angles 125° and $-955° = 125° - 3\cdot 360°$ are coterminal. The angles 210°, $-150° = 210° - 360°$, $930° = 210° + 2\cdot 360°$, and $-870° = 210° - 3\cdot 360°$ are coterminal.

(*b*) $485° = 125° + 360°$, $1205° = 125° + 3\cdot 360°$, $1925° = 125° + 5\cdot 360°$, $-235° = 125° - 360°$, $-1315° = 125° - 4\cdot 360°$ are coterminal with 125°.

2. Determine the values of the trigonometric functions of angle θ (smallest positive angle in standard position) if P is a point on the terminal side of θ and the coordinates of P are:
(*a*) $P(3,4)$, (*b*) $P(-3,4)$, (*c*) $P(-1,-3)$.

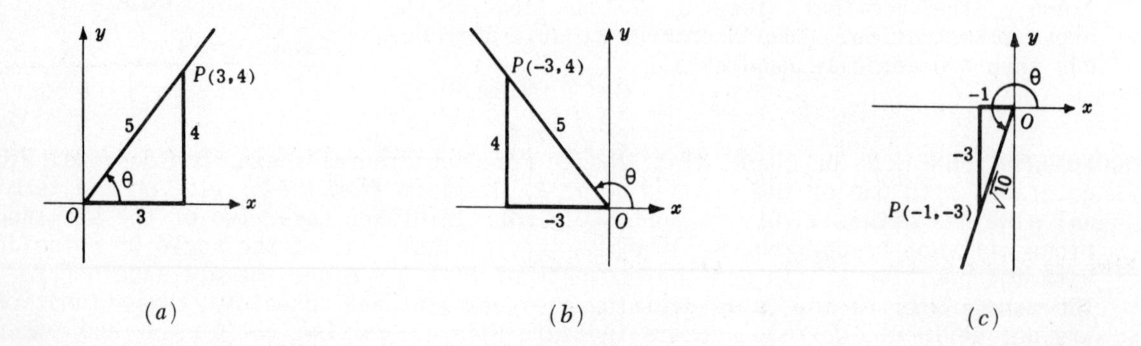

(*a*) (*b*) (*c*)

(a) $r = \sqrt{3^2 + 4^2} = 5$

$\sin\theta = y/r = 4/5$
$\cos\theta = x/r = 3/5$
$\tan\theta = y/x = 4/3$
$\cot\theta = x/y = 3/4$
$\sec\theta = r/x = 5/3$
$\csc\theta = r/y = 5/4$

(b) $r = \sqrt{(-3)^2 + 4^2} = 5$

$\sin\theta = 4/5$
$\cos\theta = -3/5$
$\tan\theta = 4/-3 = -4/3$
$\cot\theta = -3/4$
$\sec\theta = 5/-3 = -5/3$
$\csc\theta = 5/4$

(c) $r = \sqrt{(-1)^2 + (-3)^2} = \sqrt{10}$

$\sin\theta = -3/\sqrt{10} = -3\sqrt{10}/10$
$\cos\theta = -1/\sqrt{10} = -\sqrt{10}/10$
$\tan\theta = -3/-1 = 3$
$\cot\theta = -1/-3 = 1/3$
$\sec\theta = \sqrt{10}/-1 = -\sqrt{10}$
$\csc\theta = \sqrt{10}/-3 = -\sqrt{10}/3$

Note the reciprocal relationships. For example, in (b)
$\sin\theta = 1/\csc\theta = 4/5$, $\cos\theta = 1/\sec\theta = -3/5$, $\tan\theta = 1/\cot\theta = -4/3$, etc.

3. In what quadrant will θ terminate, if
(a) $\sin\theta$ and $\cos\theta$ are both negative?
(b) $\sin\theta$ and $\tan\theta$ are both positive?
(c) $\sin\theta$ is positive and secant θ is negative?
(d) $\sec\theta$ is negative and $\tan\theta$ is negative?

(a) Since $\sin\theta = y/r$ and $\cos\theta = x/r$, both x and y are negative. (Recall that r is always positive.) Thus, θ is a third quadrant angle.

(b) Since $\sin\theta$ is positive, y is positive; since $\tan\theta = y/x$ is positive, x is also positive. Thus, θ is a first quadrant angle.

(c) Since $\sin\theta$ is positive, y is positive; since $\sec\theta$ is negative, x is negative. Thus, θ is a second quadrant angle.

(d) Since $\sec\theta$ is negative, x is negative; since $\tan\theta$ is negative, y is then positive. Thus, θ is a second quadrant angle.

4. In what quadrants may θ terminate, if
(a) $\sin\theta$ is positive? (b) $\cos\theta$ is negative? (c) $\tan\theta$ is negative? (d) $\sec\theta$ is positive?

(a) Since $\sin\theta$ is positive, y is positive.
Then x may be positive or negative and θ is a first or second quadrant angle.

(b) Since $\cos\theta$ is negative, x is negative.
Then y may be positive or negative and θ is a second or third quadrant angle.

(c) Since $\tan\theta$ is negative, either y is positive and x is negative or y is negative and x is positive. Thus, θ may be a second or fourth quadrant angle.

(d) Since $\sec\theta$ is positive, x is positive. Thus, θ may be a first or fourth quadrant angle.

5. Find the values of $\cos\theta$ and $\tan\theta$, given $\sin\theta = 8/17$ and θ in quadrant I.

Let P be a point on the terminal line of θ. Since $\sin\theta = y/r = 8/17$, we take $y = 8$ and $r = 17$. Since θ is in quadrant I, x is positive; thus

$$x = \sqrt{r^2 - y^2} = \sqrt{(17)^2 - (8)^2} = 15 .$$

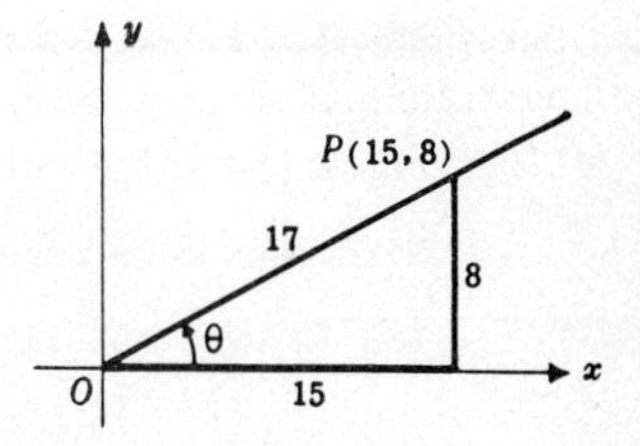

To draw the figure, locate the point $P(15,8)$, join it to the origin, and indicate the angle θ. Then

$$\cos\theta = x/r = 15/17 \quad\text{and}\quad \tan\theta = y/x = 8/15.$$

The choice of $y = 8$, $r = 17$ is one of convenience. Note that $8/17 = 16/34$ and we might have taken $y = 16$, $r = 34$. Then $x = 30$, $\cos\theta = 30/34 = 15/17$ and $\tan\theta = 16/30 = 8/15$.

6. Find the values of $\sin\theta$ and $\tan\theta$, given $\cos\theta = 5/6$.

Since $\cos\theta$ is positive, θ is in quadrant I or IV.

Since $\cos\theta = x/r = 5/6$, we take $x = 5$, $r = 6$; $y = \pm\sqrt{(6)^2 - (5)^2} = \pm\sqrt{11}$.

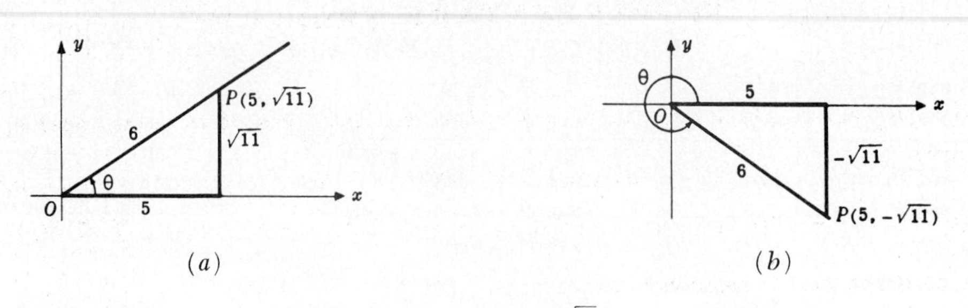

(a) For θ in quadrant I (Figure a) we have $x = 5$, $y = \sqrt{11}$, $r = 6$; then

$$\sin\theta = y/r = \sqrt{11}/6 \quad \text{and} \quad \tan\theta = y/x = \sqrt{11}/5 .$$

(b) For θ in quadrant IV (Figure b) we have $x = 5$, $y = -\sqrt{11}$, $r = 6$; then

$$\sin\theta = y/r = -\sqrt{11}/6 \quad \text{and} \quad \tan\theta = y/x = -\sqrt{11}/5 .$$

7. Find the values of $\sin\theta$ and $\cos\theta$, given $\tan\theta = -3/4$.

Since $\tan\theta = y/x$ is negative, θ is in quadrant II (take $x = -4$, $y = 3$) or in quadrant IV (take $x = 4$, $y = -3$). In either case $r = \sqrt{16+9} = 5$.

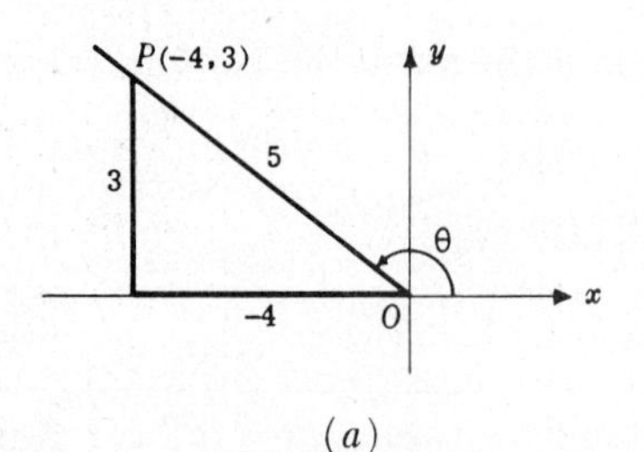

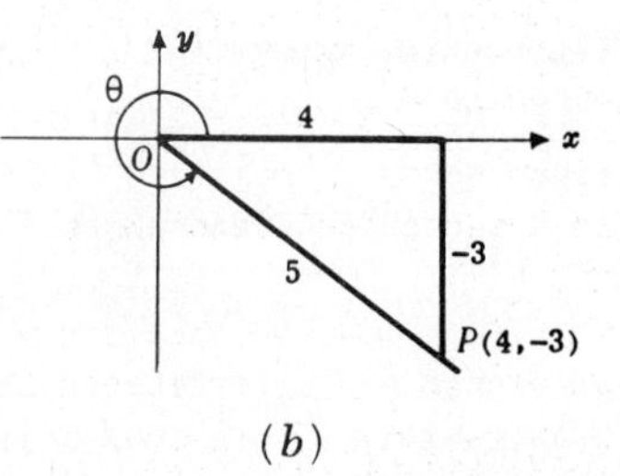

(a) For θ in quadrant II (Figure a), $\sin\theta = y/r = 3/5$ and $\cos\theta = x/r = -4/5$.
(b) For θ in quadrant IV (Figure b), $\sin\theta = y/r = -3/5$ and $\cos\theta = x/r = 4/5$.

8. Find the values of the remaining functions of θ, given $\sin\theta = \sqrt{3}/2$ and $\cos\theta = -1/2$.

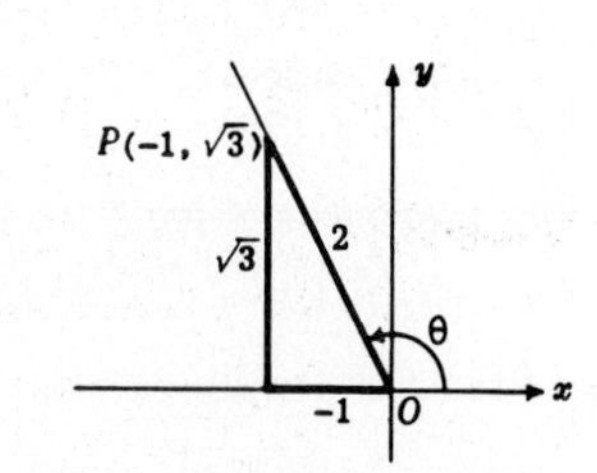

Since $\sin\theta = y/r$ is positive, y is positive. Since $\cos\theta = x/r$ is negative, x is negative. Thus, θ is in quadrant II.

Taking $x = -1$, $y = \sqrt{3}$, $r = \sqrt{(-1)^2 + (\sqrt{3})^2} = 2$, we have

$\tan\theta = y/x = \sqrt{3}/-1 = -\sqrt{3}$ $\qquad$ $\cot\theta = 1/\tan\theta = -1/\sqrt{3} = -\sqrt{3}/3$
$\sec\theta = 1/\cos\theta = -2$ $\qquad$ $\csc\theta = 1/\sin\theta = 2/\sqrt{3} = 2\sqrt{3}/3$.

9. Determine the values of $\cos\theta$ and $\tan\theta$ if $\sin\theta = m/n$, a negative fraction.

Since $\sin\theta$ is negative, θ is in quadrant III or IV.

(a) In quadrant III: Take $y = m$, $r = n$, $x = -\sqrt{n^2 - m^2}$; then

$$\cos\theta = x/r = -\sqrt{n^2-m^2}/n \quad \text{and} \quad \tan\theta = y/x = -m/\sqrt{n^2-m^2} .$$

(b) In quadrant IV: Take $y = m$, $r = n$, $x = +\sqrt{n^2 - m^2}$; then

$$\cos\theta = x/r = \sqrt{n^2-m^2}/n \quad \text{and} \quad \tan\theta = y/x = m/\sqrt{n^2-m^2} .$$

10. Evaluate: (a) $\sin 0° + 2\cos 0° + 3\sin 90° + 4\cos 90° + 5\sec 0° + 6\csc 90°$
(b) $\sin 180° + 2\cos 180° + 3\sin 270° + 4\cos 270° - 5\sec 180° - 6\csc 270°$

(a) $0 + 2(1) + 3(1) + 4(0) + 5(1) + 6(1) = 16$
(b) $0 + 2(-1) + 3(-1) + 4(0) - 5(-1) - 6(-1) = 6$

SUPPLEMENTARY PROBLEMS

11. State the quadrant in which each angle terminates and the signs of the sine, cosine, and tangent of each angle.

(*a*) $125°$ (*b*) $75°$ (*c*) $320°$ (*d*) $212°$ (*e*) $460°$ (*f*) $750°$ (*g*) $-250°$ (*h*) $-1000°$

Ans. (*a*) II; +,−,− (*b*) I; +,+,+ (*c*) IV; −,+,− (*d*) III; −,−,+ (*e*) II (*f*) I (*g*) II (*h*) I

12. In what quadrant will θ terminate if

(*a*) $\sin\theta$ and $\cos\theta$ are both positive?
(*b*) $\cos\theta$ and $\tan\theta$ are both positive?
(*c*) $\sin\theta$ and $\cos\theta$ are both negative?
(*d*) $\cos\theta$ and $\cot\theta$ are both negative?
(*e*) $\tan\theta$ is positive and $\sec\theta$ is negative?
(*f*) $\tan\theta$ is negative and $\sec\theta$ is positive?
(*g*) $\sin\theta$ is positive and $\cos\theta$ is negative?
(*h*) $\sec\theta$ is positive and $\csc\theta$ is negative?

Ans. (*a*) I (*b*) I (*c*) III (*d*) II (*e*) III (*f*) IV (*g*) II (*h*) IV

13. Denote by θ the smallest positive angle whose terminal side passes through the given point and find the trigonometric functions of θ:

(*a*) $P(-5,12)$ (*b*) $P(7,-24)$ (*c*) $P(2,3)$ (*d*) $P(-3,-5)$

Ans. (*a*) $12/13,\ -5/13,\ -12/5,\ -5/12,\ -13/5,\ 13/12$
(*b*) $-24/25,\ 7/25,\ -24/7,\ -7/24,\ 25/7,\ -25/24$
(*c*) $3/\sqrt{13},\ 2/\sqrt{13},\ 3/2,\ 2/3,\ \sqrt{13}/2,\ \sqrt{13}/3$
(*d*) $-5/\sqrt{34},\ -3/\sqrt{34},\ 5/3,\ 3/5,\ -\sqrt{34}/3,\ -\sqrt{34}/5$

14. Find the values of the trigonometric functions of θ, given:

(*a*) $\sin\theta = 7/25$
(*b*) $\cos\theta = -4/5$
(*c*) $\tan\theta = -5/12$
(*d*) $\cot\theta = 24/7$
(*e*) $\sin\theta = -2/3$
(*f*) $\cos\theta = 5/6$
(*g*) $\tan\theta = 3/5$
(*h*) $\cot\theta = \sqrt{6}/2$
(*i*) $\sec\theta = -\sqrt{5}$
(*j*) $\csc\theta = -2/\sqrt{3}$

Ans. (*a*) I: $7/25,\ 24/25,\ 7/24,\ 24/7,\ 25/24,\ 25/7$
II: $7/25,\ -24/25,\ -7/24,\ -24/7,\ -25/24,\ 25/7$

(*b*) II: $3/5,\ -4/5,\ -3/4,\ -4/3,\ -5/4,\ 5/3$; III: $-3/5,\ -4/5,\ 3/4,\ 4/3,\ -5/4,\ -5/3$

(*c*) II: $5/13,\ -12/13,\ -5/12,\ -12/5,\ -13/12,\ 13/5$
IV: $-5/13,\ 12/13,\ -5/12,\ -12/5,\ 13/12,\ -13/5$

(*d*) I: $7/25,\ 24/25,\ 7/24,\ 24/7,\ 25/24,\ 25/7$
III: $-7/25,\ -24/25,\ 7/24,\ 24/7,\ -25/24,\ -25/7$

(*e*) III: $-2/3,\ -\sqrt{5}/3,\ 2/\sqrt{5},\ \sqrt{5}/2,\ -3/\sqrt{5},\ -3/2$
IV: $-2/3,\ \sqrt{5}/3,\ -2/\sqrt{5},\ -\sqrt{5}/2,\ 3/\sqrt{5},\ -3/2$

(*f*) I: $\sqrt{11}/6,\ 5/6,\ \sqrt{11}/5,\ 5/\sqrt{11},\ 6/5,\ 6/\sqrt{11}$
IV: $-\sqrt{11}/6,\ 5/6,\ -\sqrt{11}/5,\ -5/\sqrt{11},\ 6/5,\ -6/\sqrt{11}$

(*g*) I: $3/\sqrt{34},\ 5/\sqrt{34},\ 3/5,\ 5/3,\ \sqrt{34}/5,\ \sqrt{34}/3$
III: $-3/\sqrt{34},\ -5/\sqrt{34},\ 3/5,\ 5/3,\ -\sqrt{34}/5,\ -\sqrt{34}/3$

(*h*) I: $2/\sqrt{10},\ \sqrt{3}/\sqrt{5},\ 2/\sqrt{6},\ \sqrt{6}/2,\ \sqrt{5}/\sqrt{3},\ \sqrt{10}/2$
III: $-2/\sqrt{10},\ -\sqrt{3}/\sqrt{5},\ 2/\sqrt{6},\ \sqrt{6}/2,\ -\sqrt{5}/\sqrt{3},\ -\sqrt{10}/2$

(*i*) II: $2/\sqrt{5},\ -1/\sqrt{5},\ -2,\ -1/2,\ -\sqrt{5},\ \sqrt{5}/2$; III: $-2/\sqrt{5},\ -1/\sqrt{5},\ 2,\ 1/2,\ -\sqrt{5},\ -\sqrt{5}/2$

(*j*) III: $-\sqrt{3}/2,\ -1/2,\ \sqrt{3},\ 1/\sqrt{3},\ -2,\ -2/\sqrt{3}$; IV: $-\sqrt{3}/2,\ 1/2,\ -\sqrt{3},\ -1/\sqrt{3},\ 2,\ -2/\sqrt{3}$

15. Evaluate each of the following:

(*a*) $\tan 180° - 2\cos 180° + 3\csc 270° + \sin 90° = 0$.

(*b*) $\sin 0° + 3\cot 90° + 5\sec 180° - 4\cos 270° = -5$.

CHAPTER 32

Trigonometric Functions of an Acute Angle

TRIGONOMETRIC FUNCTIONS OF AN ACUTE ANGLE. In dealing with any right triangle, it will be convenient (see Fig. 1) to denote the vertices as A, B, C = vertex of the right angle, to denote the angles of the triangle as $A, B, C = 90°$, and the sides opposite the angles as a, b, c respectively. With respect to angle A, a will be called the *opposite side* and b will be called the *adjacent side*; with respect to angle B, a will be called the *adjacent side* and b the *opposite side*. Side c will always be called the *hypotenuse*.

If now the right triangle is placed in a coordinate system (Fig. 2) so that angle A is in standard position, the point B on the terminal side of angle A has coordinates (b, a) and distance $c = \sqrt{a^2 + b^2}$. Then the trigonometric functions of angle A may be defined in terms of the sides of the right triangle, as follows:

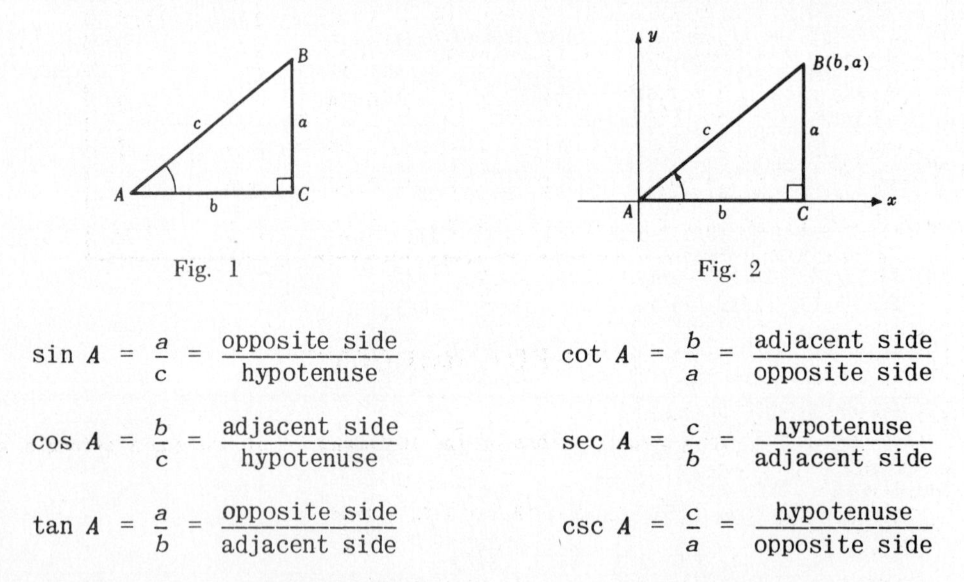

Fig. 1 Fig. 2

$$\sin A = \frac{a}{c} = \frac{\text{opposite side}}{\text{hypotenuse}} \qquad \cot A = \frac{b}{a} = \frac{\text{adjacent side}}{\text{opposite side}}$$

$$\cos A = \frac{b}{c} = \frac{\text{adjacent side}}{\text{hypotenuse}} \qquad \sec A = \frac{c}{b} = \frac{\text{hypotenuse}}{\text{adjacent side}}$$

$$\tan A = \frac{a}{b} = \frac{\text{opposite side}}{\text{adjacent side}} \qquad \csc A = \frac{c}{a} = \frac{\text{hypotenuse}}{\text{opposite side}}$$

TRIGONOMETRIC FUNCTIONS OF COMPLEMENTARY ANGLES. The acute angles A and B of the right triangle ABC are complementary, that is, $A + B = 90°$. From Fig. 1, we have

$$\sin B = b/c = \cos A \qquad \cot B = a/b = \tan A$$

$$\cos B = a/c = \sin A \qquad \sec B = c/a = \csc A$$

$$\tan B = b/a = \cot A \qquad \csc B = c/b = \sec A$$

These relations associate the functions in pairs — sine and cosine, tangent and cotangent, secant and cosecant — each function of a pair being called the *cofunction* of the other. Thus, any function of an acute angle is equal to the corresponding cofunction of the complementary angle.

TRIGONOMETRIC FUNCTIONS OF 30°, 45°, 60°. The following results are obtained in Problems 8-9.

Angle θ	$\sin\theta$	$\cos\theta$	$\tan\theta$	$\cot\theta$	$\sec\theta$	$\csc\theta$
30°	$\frac{1}{2}$	$\frac{1}{2}\sqrt{3}$	$\frac{1}{3}\sqrt{3}$	$\sqrt{3}$	$\frac{2}{3}\sqrt{3}$	2
45°	$\frac{1}{2}\sqrt{2}$	$\frac{1}{2}\sqrt{2}$	1	1	$\sqrt{2}$	$\sqrt{2}$
60°	$\frac{1}{2}\sqrt{3}$	$\frac{1}{2}$	$\sqrt{3}$	$\frac{1}{3}\sqrt{3}$	2	$\frac{2}{3}\sqrt{3}$

PROBLEMS 10-16 illustrate a number of simple applications of the trigonometric functions. For this purpose the following table will be used.

Angle θ	$\sin\theta$	$\cos\theta$	$\tan\theta$	$\cot\theta$	$\sec\theta$	$\csc\theta$
15°	0.26	0.97	0.27	3.7	1.0	3.9
20°	0.34	0.94	0.36	2.7	1.1	2.9
30°	0.50	0.87	0.58	1.7	1.2	2.0
40°	0.64	0.77	0.84	1.2	1.3	1.6
45°	0.71	0.71	1.0	1.0	1.4	1.4
50°	0.77	0.64	1.2	0.84	1.6	1.3
60°	0.87	0.50	1.7	0.58	2.0	1.2
70°	0.94	0.34	2.7	0.36	2.9	1.1
75°	0.97	0.26	3.7	0.27	3.9	1.0

SOLVED PROBLEMS

1. Find the values of the trigonometric functions of the acute angles of the right triangle ABC, given $b = 24$ and $c = 25$.

Since $a^2 = c^2 - b^2 = (25)^2 - (24)^2 = 49$, $a = 7$. Then

$$\sin A = \frac{\text{opposite side}}{\text{hypotenuse}} = \frac{7}{25} \qquad \cot A = \frac{\text{adjacent side}}{\text{opposite side}} = \frac{24}{7}$$

$$\cos A = \frac{\text{adjacent side}}{\text{hypotenuse}} = \frac{24}{25} \qquad \sec A = \frac{\text{hypotenuse}}{\text{adjacent side}} = \frac{25}{24}$$

$$\tan A = \frac{\text{opposite side}}{\text{adjacent side}} = \frac{7}{24} \qquad \csc A = \frac{\text{hypotenuse}}{\text{opposite side}} = \frac{25}{7}$$

B
c = 25
a = 7
A
C
b = 24

and

$\sin B = 24/25$ $\qquad$ $\cot B = 7/24$

$\cos B = 7/25$ $\qquad$ $\sec B = 25/7$

$\tan B = 24/7$ $\qquad$ $\csc B = 25/24$

2. Find the values of the trigonometric functions of the acute angles of the right triangle ABC, given $a = 2$, $c = 2\sqrt{5}$.

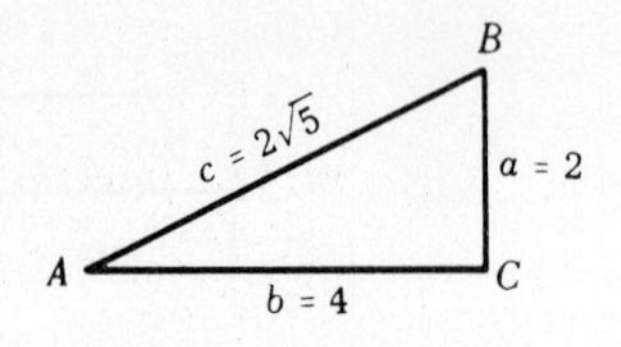

Since $b^2 = c^2 - a^2 = (2\sqrt{5})^2 - (2)^2 = 20 - 4 = 16$, $b = 4$. Then

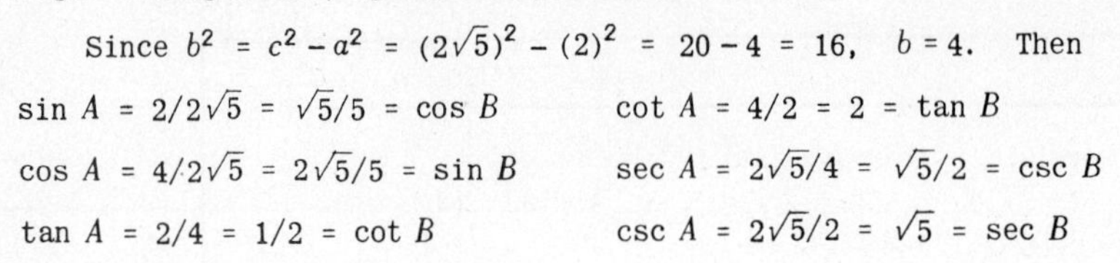

$\sin A = 2/2\sqrt{5} = \sqrt{5}/5 = \cos B$ $\qquad$ $\cot A = 4/2 = 2 = \tan B$

$\cos A = 4/2\sqrt{5} = 2\sqrt{5}/5 = \sin B$ $\qquad$ $\sec A = 2\sqrt{5}/4 = \sqrt{5}/2 = \csc B$

$\tan A = 2/4 = 1/2 = \cot B$ $\qquad$ $\csc A = 2\sqrt{5}/2 = \sqrt{5} = \sec B$

3. Find the values of the trigonometric functions of the acute angle A, given $\sin A = 3/7$.

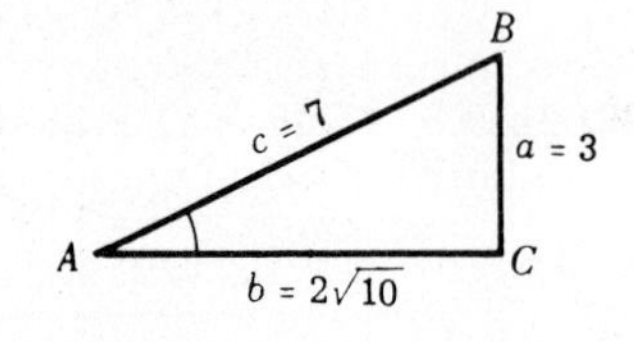

Construct the right triangle ABC having $a = 3$, $c = 7$ and $b = \sqrt{7^2 - 3^2} = 2\sqrt{10}$ units. Then

$\sin A = 3/7$ $\qquad$ $\cot A = 2\sqrt{10}/3$

$\cos A = 2\sqrt{10}/7$ $\qquad$ $\sec A = 7/2\sqrt{10} = 7\sqrt{10}/20$

$\tan A = 3/2\sqrt{10} = 3\sqrt{10}/20$ $\qquad$ $\csc A = 7/3$

4. Find the values of the trigonometric functions of the acute angle B, given $\tan B = 1.5$.

Refer to Fig. (a) below. Construct the right triangle ABC having $b = 15$ and $a = 10$ units. (Note that $1.5 = 3/2$ and a right triangle with $b = 3$, $a = 2$ will serve equally well.)

Then $c = \sqrt{a^2 + b^2} = \sqrt{10^2 + 15^2} = 5\sqrt{13}$ and

$\sin B = 15/5\sqrt{13} = 3\sqrt{13}/13$ $\qquad$ $\cot B = 2/3$

$\cos B = 10/5\sqrt{13} = 2\sqrt{13}/13$ $\qquad$ $\sec B = 5\sqrt{13}/10 = \sqrt{13}/2$

$\tan B = 15/10 = 3/2$ $\qquad$ $\csc B = 5\sqrt{13}/15 = \sqrt{13}/3$.

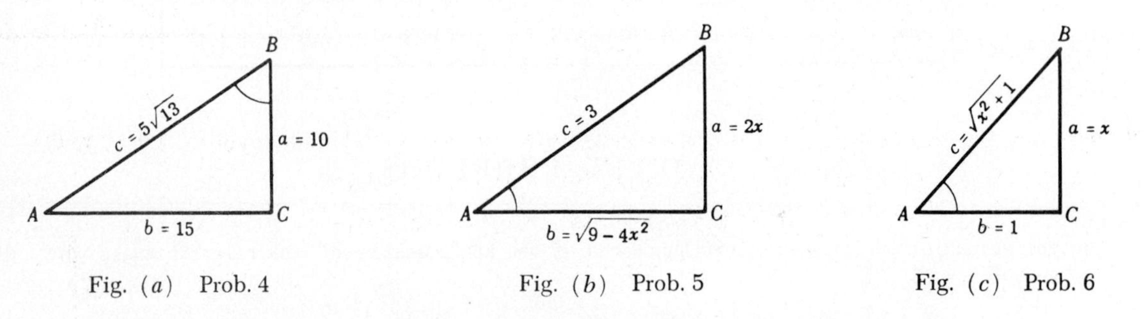

Fig. (a) Prob. 4 $\qquad$ Fig. (b) Prob. 5 $\qquad$ Fig. (c) Prob. 6

5. If A is acute and $\sin A = 2x/3$, determine the values of the remaining functions.

Construct the right triangle ABC having $a = 2x < 3$ and $c = 3$, as in Fig. (b) above.

Then $b = \sqrt{c^2 - a^2} = \sqrt{9 - 4x^2}$ and

$$\sin A = \frac{2x}{3}, \quad \cos A = \frac{\sqrt{9-4x^2}}{3}, \quad \tan A = \frac{2x}{\sqrt{9-4x^2}}, \quad \cot A = \frac{\sqrt{9-4x^2}}{2x}, \quad \sec A = \frac{3}{\sqrt{9-4x^2}}, \quad \csc A = \frac{3}{2x}.$$

6. If A is acute and $\tan A = x = x/1$, determine the values of the remaining functions.

Construct the right triangle ABC having $a = x$ and $b = 1$, as in Fig. (c) above. Then $c = \sqrt{x^2 + 1}$ and

$$\sin A = \frac{x}{\sqrt{x^2+1}}, \quad \cos A = \frac{1}{\sqrt{x^2+1}}, \quad \tan A = x, \quad \cot A = \frac{1}{x}, \quad \sec A = \sqrt{x^2+1}, \quad \csc A = \frac{\sqrt{x^2+1}}{x}.$$

7. If A is an acute angle: (*a*) Why is $\sin A < 1$? (*b*) When is $\sin A = \cos A$? (*c*) Why is $\sin A < \csc A$? (*d*) Why is $\sin A < \tan A$? (*e*) When is $\sin A < \cos A$? (*f*) When is $\tan A > 1$?

In any right triangle ABC:

(*a*) Side $a <$ side c; therefore $\sin A = a/c < 1$.
(*b*) $\sin A = \cos A$ when $a/c = b/c$; then $a = b$, $A = B$, and $A = 45°$.
(*c*) $\sin A < 1$ (above) and $\csc A = 1/\sin A > 1$.
(*d*) $\sin A = a/c$, $\tan A = a/b$, and $b < c$; therefore $a/c < a/b$ or $\sin A < \tan A$.
(*e*) $\sin A < \cos A$ when $a < b$; then $A < B$ or $A < 90° - A$, and $A < 45°$.
(*f*) $\tan A = a/b > 1$ when $a > b$; then $A > B$ and $A > 45°$.

8. Find the values of the trigonometric functions of 45°.

In any isosceles right triangle ABC, $A = B = 45°$ and $a = b$. Let $a = b = 1$; then $c = \sqrt{1+1} = \sqrt{2}$ and

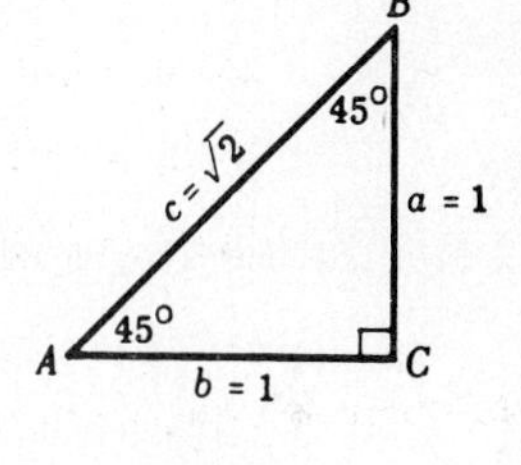

$\sin 45° = 1/\sqrt{2} = \frac{1}{2}\sqrt{2}$ $\qquad$ $\cot 45° = 1$

$\cos 45° = 1/\sqrt{2} = \frac{1}{2}\sqrt{2}$ $\qquad$ $\sec 45° = \sqrt{2}$

$\tan 45° = 1/1 = 1$ $\qquad$ $\csc 45° = \sqrt{2}$.

9. Find the values of the trigonometric functions of 30° and 60°.

In any equilateral triangle ABD, each angle is 60°. The bisector of any angle, as B, is the perpendicular bisector of the opposite side. Let the sides of the equilateral triangle be of length 2 units. Then in the right triangle ABC, $AB = 2$, $AC = 1$, and $BC = \sqrt{2^2 - 1^2} = \sqrt{3}$.

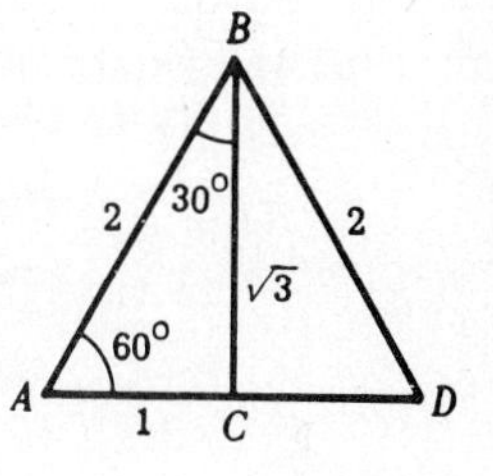

$\sin 30° = 1/2 = \cos 60°$ $\qquad$ $\cot 30° = \sqrt{3} = \tan 60°$

$\cos 30° = \sqrt{3}/2 = \sin 60°$ $\qquad$ $\sec 30° = 2/\sqrt{3} = 2\sqrt{3}/3 = \csc 60°$

$\tan 30° = 1/\sqrt{3} = \sqrt{3}/3 = \cot 60°$ $\qquad$ $\csc 30° = 2 = \sec 60°$

10. When the sun is 20° above the horizon, how long is the shadow cast by a building 150 ft high?

In Fig. (*d*) below, $A = 20°$ and $CB = 150$. Then $\cot A = AC/CB$ and

$$AC = CB \cot A = 150 \cot 20° = 150(2.7) = 405 \text{ ft.}$$

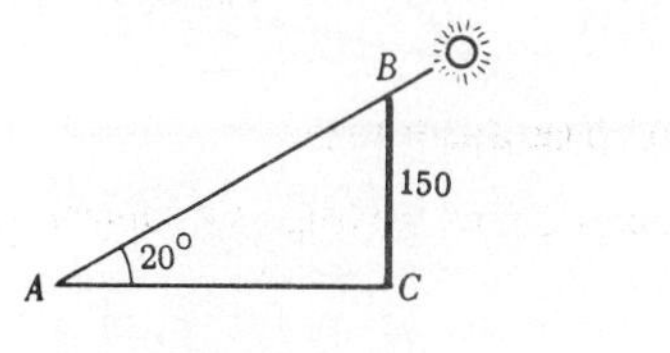

Fig. (*d*) Prob. 10

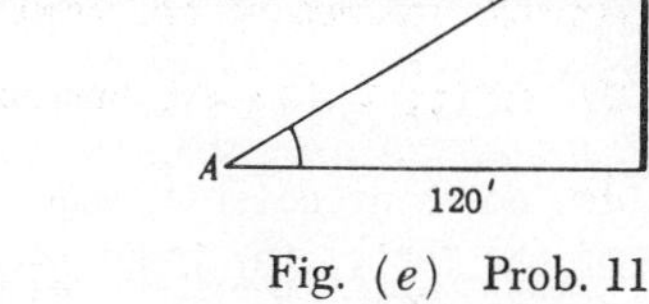

Fig. (*e*) Prob. 11

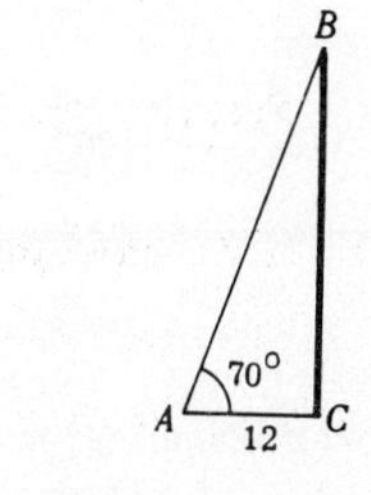

Fig. (*f*) Prob. 12

11. A tree 100 ft tall casts a shadow 120 ft long. Find the angle of elevation of the sun.

In Fig. (*e*) above, $CB = 100$ and $AC = 120$. Then $\tan A = CB/AC = 100/120 = 0.83$ and $A = 40°$.

12. A ladder leans against the side of a building with its foot 12 ft from the building. How far from the ground is the top of the ladder and how long is the ladder if it makes an angle of 70° with the ground?

From Fig. (*f*) above, $\tan A = CB/AC$; then $CB = AC \tan A = 12 \tan 70° = 12(2.7) = 32.4$. The top of the ladder is 32 ft above the ground.

$\sec A = AB/AC$; then $AB = AC \sec A = 12 \sec 70° = 12(2.9) = 34.8$. The ladder is 35 ft long.

13. From the top of a lighthouse, 120 ft above the sea, the angle of depression of a boat is 15°. How far is the boat from the lighthouse?

In Fig. (g) below, the right triangle ABC has $A = 15°$ and $CB = 120$; then

$$\cot A = AC/CB \quad \text{and} \quad AC = CB \cot A = 120 \cot 15° = 120(3.7) = 444 \text{ ft.}$$

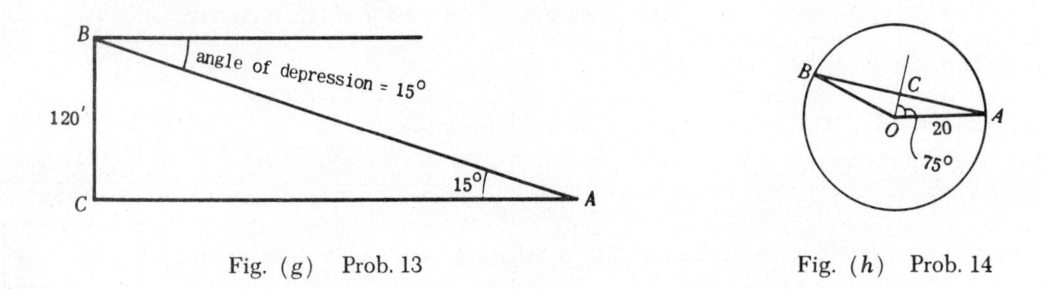

Fig. (g) Prob. 13

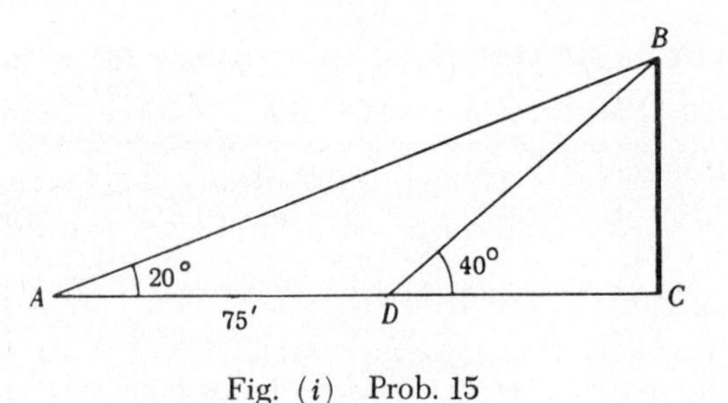

Fig. (h) Prob. 14

14. Find the length of the chord of a circle of radius 20 in. subtended by a central angle of 150°.

In Fig. (h) above, OC bisects $\angle AOB$. Then $BC = AC$ and OAC is a right triangle. In $\triangle OAC$,

$$\sin \angle COA = AC/OA \quad \text{and} \quad AC = OA \sin \angle COA = 20 \sin 75° = 20(0.97) = 19.4.$$

Then $BA = 38.8$ and the length of the chord is 39 in.

15. Find the height of a tree if the angle of elevation of its top changes from 20° to 40° as the observer advances 75 ft toward its base. See Fig. (i) below.

In the right triangle ABC, $\cot A = AC/CB$; then $AC = CB \cot A$ or $DC + 75 = CB \cot 20°$.

In the right triangle DBC, $\cot D = DC/CB$; then $DC = CB \cot 40°$.

Then $DC = CB \cot 20° - 75 = CB \cot 40°$, $CB(\cot 20° - \cot 40°) = 75$,

$$CB(2.7 - 1.2) = 75, \quad \text{and} \quad CB = 75/1.5 = 50 \text{ ft.}$$

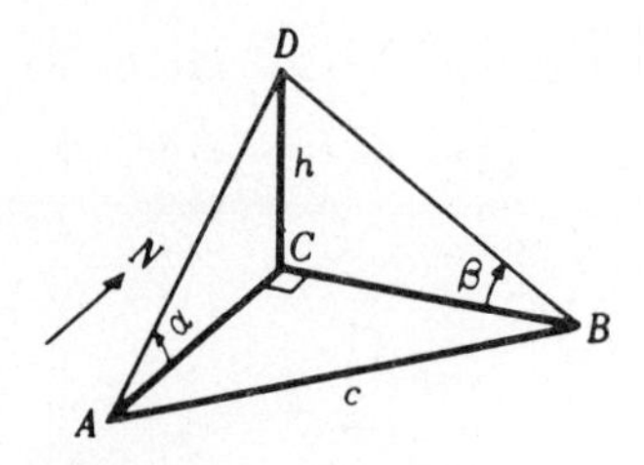

Fig. (i) Prob. 15

Fig. (j) Prob. 16

16. A tower standing on level ground is due north of point A and due west of point B, a distance c ft from A. If the angles of elevation of the top of the tower as measured from A and B are α and β respectively, find the height h of the tower.

In the right triangle ACD of Fig. (j) above, $\cot \alpha = AC/h$; and in the right triangle BCD, $\cot \beta = BC/h$. Then $AC = h \cot \alpha$ and $BC = h \cot \beta$.

Since ABC is a right triangle, $(AC)^2 + (BC)^2 = c^2 = h^2(\cot \alpha)^2 + h^2(\cot \beta)^2$ and

$$h = \frac{c}{\sqrt{(\cot \alpha)^2 + (\cot \beta)^2}}$$

17. If holes are to be spaced regularly on a circle, show that the distance d between the centers of two

successive holes is given by $d = 2r \sin \frac{180°}{n}$, where r = radius of the circle and n = number of holes. Find d when r = 20 in. and n = 4.

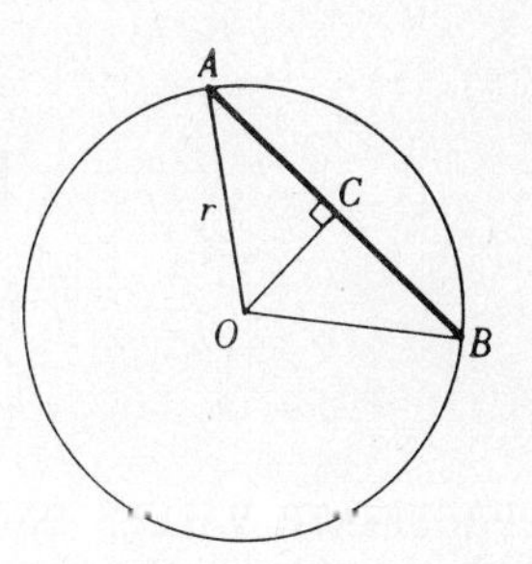

Let A and B be the centers of two consecutive holes on the circle of radius r and center O. Let the bisector of the angle O of the triangle AOB meet AB at C. In right triangle AOC,

$$\sin \angle AOC = AC/r = \tfrac{1}{2}d/r = d/2r.$$

Then $d = 2r \sin \angle AOC = 2r \sin \frac{1}{2}\angle AOB$

$$= 2r \sin \tfrac{1}{2}(360°/n) = 2r \sin \frac{180°}{n}.$$

When $r = 20$ and $n = 4$, $d = 2 \cdot 20 \sin 45° = 2 \cdot 20 \cdot \frac{\sqrt{2}}{2} = 20\sqrt{2}$ in.

SUPPLEMENTARY PROBLEMS

18. Find the values of the trigonometric functions of the acute angles of the right triangle ABC, given:
(a) $a = 3$, $b = 1$ (b) $a = 2$, $c = 5$ (c) $b = \sqrt{7}$, $c = 4$

Ans. (a) A: $3/\sqrt{10}$, $1/\sqrt{10}$, 3, $1/3$, $\sqrt{10}$, $\sqrt{10}/3$; B: $1/\sqrt{10}$, $3/\sqrt{10}$, $1/3$, 3, $\sqrt{10}/3$, $\sqrt{10}$
(b) A: $2/5$, $\sqrt{21}/5$, $2/\sqrt{21}$, $\sqrt{21}/2$, $5/\sqrt{21}$, $5/2$; B: $\sqrt{21}/5$, $2/5$, $\sqrt{21}/2$, $2/\sqrt{21}$, $5/2$, $5/\sqrt{21}$
(c) A: $3/4$, $\sqrt{7}/4$, $3/\sqrt{7}$, $\sqrt{7}/3$, $4/\sqrt{7}$, $4/3$; B: $\sqrt{7}/4$, $3/4$, $\sqrt{7}/3$, $3/\sqrt{7}$, $4/3$, $4/\sqrt{7}$

19. Which is the greater and why: (a) sin 55° or cos 55°? (b) sin 40° or cos 40°? (c) tan 15° or cot 15°? (d) sec 55° or csc 55°?
Hint: Consider a right triangle having as acute angle the given angle.
Ans. (a) sin 55° (b) cos 40° (c) cot 15° (d) sec 55°

20. Find the value of each of the following.

(a) $\sin 30° + \tan 45°$
(b) $\cot 45° + \cos 60°$
(c) $\sin 30° \cos 60° + \cos 30° \sin 60°$
(d) $\cos 30° \cos 60° - \sin 30° \sin 60°$
(e) $\dfrac{\tan 60° - \tan 30°}{1 + \tan 60° \tan 30°}$
(f) $\dfrac{\csc 30° + \csc 60° + \csc 90°}{\sec 0° + \sec 30° + \sec 60°}$

Ans. (a) 3/2 (b) 3/2 (c) 1 (d) 0 (e) $1/\sqrt{3}$ (f) 1

21. A man drives 500 ft along a road which is inclined 20° to the horizontal. How high above his starting point is he? *Ans.* 170 ft

22. A tree broken over by the wind forms a right triangle with the ground. If the broken part makes an angle of 50° with the ground and if the top of the tree is now 20 ft from its base, how tall was the tree? *Ans.* 56 ft

23. Two straight roads intersect to form an angle of 75°. Find the shortest distance from one road to a gas station on the other road 1000 ft from the junction. *Ans.* 970 ft

24. Two buildings with flat roofs are 60 ft apart. From the roof of the shorter building, 40 ft in height, the angle of elevation to the edge of the roof of the taller building is 40°. How high is the taller building? *Ans.* 90 ft

25. A ladder, with its foot in the street, makes an angle of 30° with the street when its top rests on a building on one side of the street and makes an angle of 40° with the street when its top rests on a building on the other side of the street. If the ladder is 50 ft long, how wide is the street?
Ans. 82 ft

26. Find the perimeter of an isosceles triangle whose base is 40 in. and whose base angle is 70°.
Ans. 156 in.

CHAPTER 33

Tables of Trigonometric Functions

APPROXIMATE VALUES OF THE FUNCTIONS of acute angles are given in Tables of Natural Trigonometric Functions. These tables, as published in texts, differ in several respects. Some give the values of the six functions; others are restricted to the functions sine, cosine, tangent, and cotangent; some give the values to four digits while others give values to four decimal places. We shall use the latter table here. (In case the values of secant and cosecant are not included in your table, reference to these functions is to be deleted.)

FOUR PLACE TABLE OF NATURAL TRIGONOMETRIC FUNCTIONS

WHEN THE ANGLE IS LESS THAN 45°, the angle is found in the left hand column of the table and the function is read at the top of the page. When the angle is greater than 45°, the angle is found in the right hand column and the function is read at the bottom of the page.

TO FIND THE VALUE OF A TRIGONOMETRIC FUNCTION of a given acute angle. If the angle contains a number of degrees only or a number of degrees and a multiple of 10', the value of the function is read directly from the table.

Example 1. Find sin 24°40'.

Opposite 24°40' ($<45°$) in the left hand column read the entry 0.4173 in the column labeled Sin at the top of the page.

Example 2. Find cos 72°.

Opposite 72° ($>45°$) in the right hand column read the entry 0.3090 in the column labeled Cos at the bottom of the page.

Example 3. (*a*) tan 55°20' = 1.4460. Read *up* the page since $55°20' > 45°$.
(*b*) cot 41°50' = 1.1171. Read *down* the page since $41°50' < 45°$.

If the number of minutes in the given angle is not a multiple of 10, as in 24°43', interpolate between the values of the functions of the two nearest angles (24°40' and 24°50') using the method of proportional parts.

Example 4. Find sin 24°43'.

We find

$$\begin{aligned} \sin 24°40' &= 0.4173 \\ \sin 24°50' &= \underline{0.4200} \\ \text{Difference for } 10' &= 0.0027 = \text{tabular difference} \end{aligned}$$

Correction = difference for 3' = 0.3(0.0027) = 0.00081 or 0.0008 when rounded off to four decimal places.

As the angle increases, the sine of the angle increases; thus,

$$\sin 24°43' = 0.4173 + 0.0008 = 0.4181.$$

If a five place table is available, the value 0.41813 can be read directly from the table and then rounded off to 0.4181.

Example 5. Find cos 64°26'.

We find

$$\begin{aligned} \cos 64°20' &= 0.4331 \\ \cos 64°30' &= \underline{0.4305} \\ \text{Tabular difference} &= 0.0026 \end{aligned}$$

Correction = 0.6(0.0026) = 0.00156 or 0.0016 to four decimal places.

As the angle increases, the cosine of the angle decreases. Thus

$$\cos 64°26' = 0.4331 - 0.0016 = 0.4315.$$

To save time, we should proceed as follows in Example 4.

a) Locate sin 24°40' = 0.4173. For the moment, disregard the decimal point and use only the sequence 4173.

b) Find (mentally) the tabular difference 27, that is, the difference between the sequence 4173 corresponding to 24°40' and the sequence 4200 corresponding to 24°50'.

c) Find 0.3(27) = 8.1 and round off to the nearest integer. This is the correction.

d) Add (since sine) the correction to 4173, obtaining 4181. Then

$$\sin 24°43' = 0.4181.$$

When, as in the above example, we interpolate from the smaller angle to the larger: 1) The correction is added in finding sine, tangent, and secant. 2) The correction is subtracted in finding cosine, cotangent, and cosecant.

See also Problem 1.

TO FIND THE ANGLE WHOSE FUNCTION IS GIVEN. The process is a reversal of that given above.

Example 6. Reading directly from the table, we find

$$0.2924 = \sin 17°, \qquad 2.7725 = \tan 70°10'.$$

Example 7. Find A, given $\sin A = 0.4234$.

The given value is not an entry in the table. We find, however,

$$\begin{array}{lll} & 0.4226 = \sin 25°\ 0' & 0.4226 = \sin 25°0' \\ & \underline{0.4253} = \sin 25°10' & \underline{0.4234} = \sin A \\ \text{Tabular diff.} & = 0.0027 & 0.0008 = \text{partial difference} \end{array}$$

$$\text{Correction} = \frac{0.0008}{0.0027}(10') = \frac{8}{27}(10') = 3', \text{ to the nearest minute.}$$

Adding (since sine) the correction, we have $25°0' + 3' = 25°3' = A$.

Example 8. Find A, given $\cot A = 0.6345$.

$$\begin{array}{lll} \text{We find} & 0.6330 = \cot 57°40' & 0.6330 = \cot 57°40' \\ & \underline{0.6371} = \cot 57°30 & \underline{0.6345} = \cot A \\ \text{Tabular diff.} & = 0.0041 & 0.0015 = \text{partial difference} \end{array}$$

$$\text{Correction} = \frac{0.0015}{0.0041}(10') = \frac{15}{41}(10') = 4', \text{ to the nearest minute.}$$

Subtracting (since cot) the correction, we have

$$57°40' - 4' = 57°36' = A.$$

To save time, we should proceed as follows in Example 7:

a) Locate the next smaller entry, 0.4226 = sin 25°0'. For the moment use only the sequence 4226.

b) Find the tabular difference, 27.

c) Find the partial difference, 8, between 4226 and the given sequence 4234.

d) Find $\frac{8}{27}(10') = 3'$ and add to 25°0'. See Problem 3.

ERRORS IN COMPUTED RESULTS arise from:

a) Errors in the given data. These errors are always present in data resulting from measurements.

b) The use of prepared tables. The entries in such tables are usually approximations of never ending decimals.

A measurement recorded as 35 feet means that the result is correct to the nearest foot, that is, the true length is between 34.5 and 35.5 feet. Similarly, a recorded length of 35.0 ft means that the true length is between 34.95 and 35.05 ft; a recorded length of 35.8 ft means that the true length is between 35.75 and 35.85 ft; a recorded length of 35.80 ft means that the true length is between 35.795 and 35.805 ft; and so on.

In the number 35 there are two significant digits, 3 and 5. They are also the significant digits in 3.5, 0.35, 0.035, 0.0035 but not in 35.0, 3.50, 0.350, 0.0350. In the numbers 35.0, 3.50, 0.350, 0.0350 there are three significant digits, 3, 5, and 0. This is another way of saying that 35 and 35.0 are not the same measurement.

It is impossible to determine the significant figures in a measurement recorded as 350, 3500, 35000,····. For example, 350 may mean that the true result is between 345 and 355 or between 349.5 and 350.5.

ACCURACY IN COMPUTED RESULTS. A computed result should not show more decimal places than that shown in the least accurate of the measured data. Of importance here are the following relations giving comparable degrees of accuracy in lengths and angles:

a) Distances expressed to 2 significant digits and angles expressed to the nearest degree.

b) Distances expressed to 3 significant digits and angles expressed to the nearest 10'.

c) Distances expressed to 4 significant digits and angles expressed to the nearest 1'.

d) Distances expressed to 5 significant digits and angles expressed to the nearest 0.1'.

SOLVED PROBLEMS

1. (*a*) sin 56°34' = 0.8345 ; 8339 + 0.4(16) = 8339 + 6

(*b*) cos 19°45' = 0.9412 ; 9417 − 0.5(10) = 9417 − 5

(*c*) tan 77°12' = 4.4016 ; 43897 + 0.2(597) = 43897 + 119

(*d*) cot 40°36' = 1.1667 ; 11708 − 0.6(68) = 11708 − 41

(*e*) sec 23°47' = 1.0928 ; 10918 + 0.7(14) = 10918 + 10

(*f*) csc 60°4' = 1.1539 ; 11547 − 0.4(19) = 11547 − 8

2. If the correction is 6.5, 13.5, 10.5, etc., we shall round off so that the *final* result is even.

(*a*) $\sin 28°37' = 0.4790$; $4772 + 0.7(25) = 4772 + 17.5$
(*b*) $\cot 65°53' = 0.4476$; $4487 - 0.3(35) = 4487 - 10.5$
(*c*) $\cos 35°25' = 0.8150$; $8158 - 0.5(17) = 8158 - 8.5$
(*d*) $\sec 39°35' = 1.2976$; $12960 + 0.5(31) = 12960 + 15.5$

3. (*a*) $\sin A = 0.6826$, $A = 43°3'$; $43°0' + \frac{6}{21}(10') = 43°0' + 3'$

(*b*) $\cos A = 0.5957$, $A = 53°26'$; $53°30' - \frac{9}{24}(10') = 53°30' - 4'$

(*c*) $\tan A = 0.9470$, $A = 43°26'$; $43°20' + \frac{35}{55}(10') = 43°20' + 6'$

(*d*) $\cot A = 1.7580$, $A = 29°38'$; $29°40' - \frac{24}{119}(10') = 29°40' - 2'$

(*e*) $\sec A = 2.3198$, $A = 64°28'$; $64°20' + \frac{110}{140}(10') = 64°20' + 8'$

(*f*) $\csc A = 1.5651$, $A = 39°43'$; $39°50' - \frac{40}{55}(10') = 39°50' - 7'$

4. Solve the right triangle in which $A = 35°10'$ and $c = 72.5$.

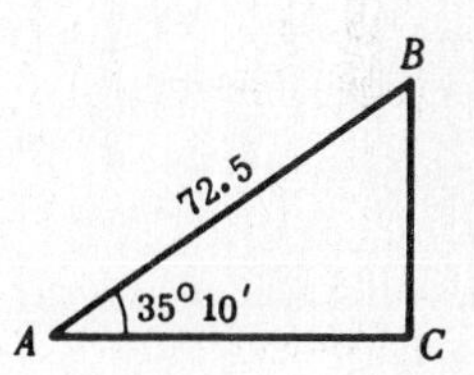

Solution: $B = 90° - 35°10' = 54°50'$.

$a/c = \sin A$, $a = c \sin A = 72.5(0.5760) = 41.8$
$b/c = \cos A$, $b = c \cos A = 72.5(0.8175) = 59.3$

Check: $a/b = \tan A$, $a = b \tan A = 59.3(0.7046) = 41.8$

5. Solve the right triangle in which $a = 24.36$, $A = 58°53'$.

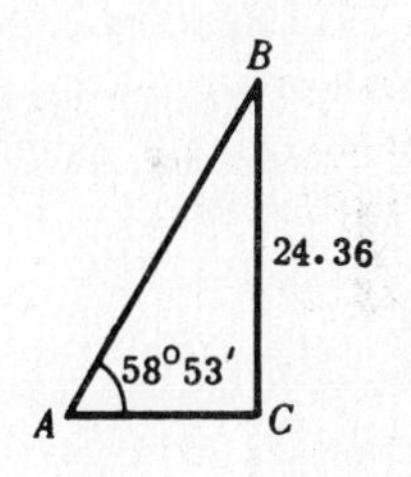

Solution: $B = 90° - 58°53' = 31°7'$.

$b/a = \cot A$, $b = a \cot A = 24.36(0.6036) = 14.70$.
$c/a = \csc A$, $c = a \csc A = 24.36(1.1681) = 28.45$, or
$a/c = \sin A$, $c = a/\sin A = 24.36/0.8562 = 28.45$.

Check: $b/c = \cos A$, $b = c \cos A = 28.45(0.5168) = 14.70$.

6. Solve the right triangle ABC in which $a = 43.9$, $b = 24.3$.

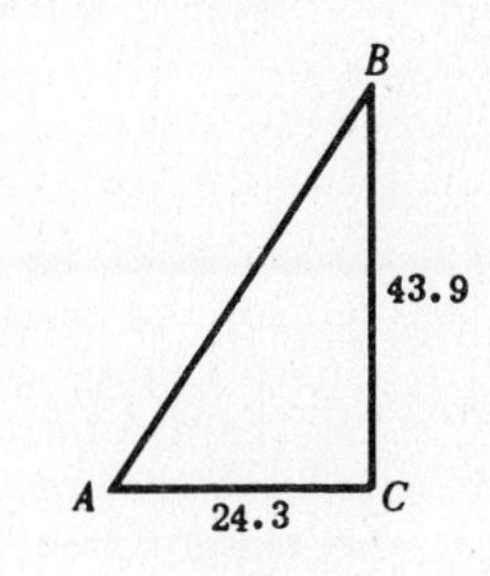

Solution: $\tan A = \frac{43.9}{24.3} = 1.8066$; $A = 61°2'$, $B = 90° - A = 28°58'$.

$c/a = \csc A$, $c = a \csc A = 43.9(1.1430) = 50.2$, or
$a/c = \sin A$, $c = a/\sin A = 43.9/0.8749 = 50.2$.

Check: $c/b = \sec A$, $c = b \sec A = 24.3(2.0649) = 50.2$, or
$b/c = \cos A$, $c = b/\cos A = 24.3/0.4843 = 50.2$

7. Solve the right triangle ABC in which $b = 15.25$, $c = 32.68$.

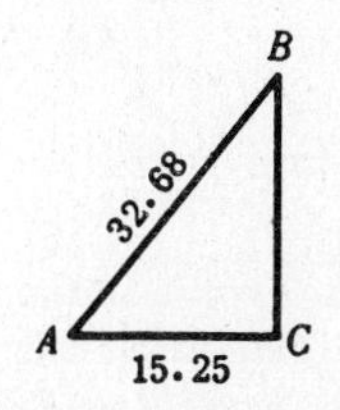

Solution: $\sin B = \frac{15.25}{32.68} = 0.4666$; $B = 27°49'$, $A = 90° - B = 62°11'$.

$a/b = \cot B$, $a = b \cot B = 15.25(1.8953) = 28.90$

Check: $a/c = \cos B$, $a = c \cos B = 32.68(0.8844) = 28.90$

8. The base of an isosceles triangle is 20.4 and the base angles are $48°40'$. Find the equal sides and the altitude of the triangle.

In the figure, BD is perpendicular to AC and bisects it.

In the right triangle ABD,

$$AB/AD = \sec A, \quad AB = 10.2(1.5141) = 15.4, \text{ or}$$
$$AD/AB = \cos A, \quad AB = 10.2/0.6604 = 15.4.$$
$$DB/AD = \tan A, \quad DB = 10.2(1.1369) = 11.6.$$

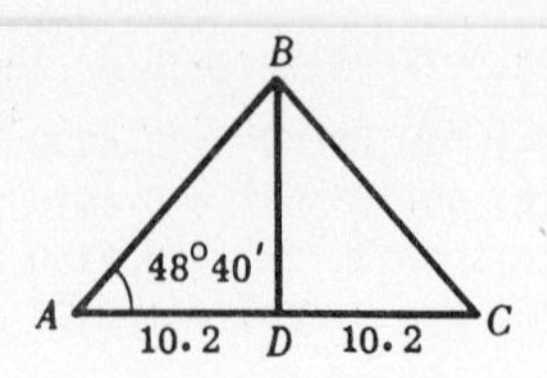

9. Considering the earth as a sphere of radius 3960 miles, find the radius r of the 40th parallel of latitude. Refer to Fig. (a) below.

In the right triangle OCB, $\angle OBC = 40°$ and $OB = 3960$.

Then $\cos \angle OBC = CB/OB$ and $r = CB = 3960 \cos 40° = 3960(0.7660) = 3030$ miles.

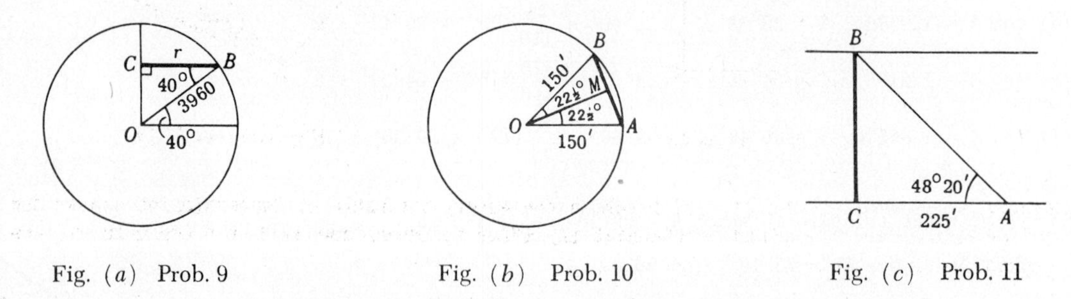

Fig. (a) Prob. 9 Fig. (b) Prob. 10 Fig. (c) Prob. 11

10. Find the perimeter of a regular octagon inscribed in a circle of radius 150 feet.

In Fig. (b) above, two consecutive vertices A and B of the octagon are joined to the center O of the circle. The triangle OAB is isosceles with equal sides 150 and $\angle AOB = 360°/8 = 45°$. As in Problem 8, we bisect $\angle AOB$ to form the right triangle MOB.

Then $MB = OB \sin \angle MOB = 150 \sin 22°30' = 150(0.3827) = 57.4$, and the perimeter of the octagon is $16MB = 16(57.4) = 918$ ft.

11. To find the width of a river, a surveyor set up his transit at C on one bank and sighted across to a point B on the opposite bank; then turning through an angle of 90°, he laid off a distance $CA = 225$ ft. Finally, setting the transit at A, he measured $\angle CAB$ as 48°20'. Find the width of the river.

See Fig. (c) above. In the right triangle ACB,

$$CB = AC \tan \angle CAB = 225 \tan 48°20' = 225(1.1237) = 253 \text{ ft.}$$

12. In the adjoining figure, the line AD crosses a swamp. In order to locate a point on this line, a surveyor turned through an angle 51°16' at A and measured 1585 feet to a point C. He then turned through an angle of 90° at C and ran a line CB. If B is on AD, how far must he measure from C to reach B ?

$$CB = AC \tan 51°16'$$
$$= 1585(1.2467) = 1976 \text{ ft.}$$

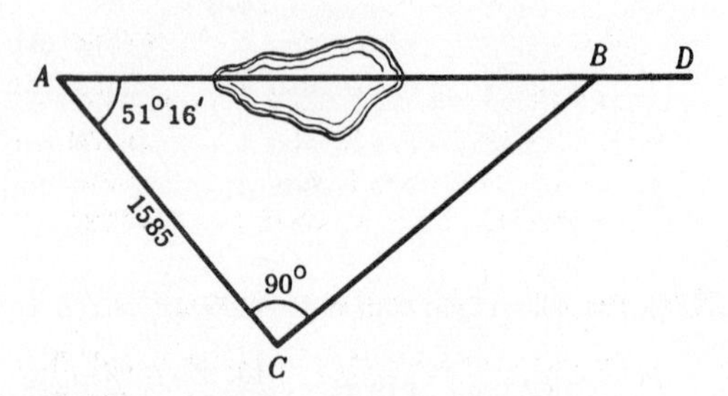

13. From a point A on level ground, the angles of elevation of the top D and bottom B of a flagpole situated on the top of a hill are measured as 47°54' and 39°45'. Find the height of the hill if the height of the flagpole is 115.5 ft. See Fig. (d) below.

Let the line of the pole meet the horizontal through A in C.

In the right triangle ACD, $AC = DC \cot 47°54' = (115.5 + BC)(0.9036)$.
In the right triangle ACB, $AC = BC \cot 39°45' = BC(1.2024)$.

Then $(115.5 + BC)(0.9036) = BC(1.2024)$

and $BC = \dfrac{115.5(0.9036)}{1.2024 - 0.9036} = 349.3$ ft.

Fig. (*d*) Prob. 13

Fig. (*e*) Prob. 14

14. From the top of a lighthouse, 175 ft above the water, the angle of depression of a boat due south is 18°50'. Calculate the speed of the boat if, after it moves due west for two minutes, the angle of depression is 14°20'.

In Fig. (*e*) above, AD is the lighthouse, C is the position of the boat when due south of the lighthouse, and B is the position two minutes later.

In the right triangle CAD, $AC = AD \cot \angle ACD = 175 \cot 18°50' = 175(2.9319) = 513$.

In the right triangle BAD, $AB = AD \cot \angle ABD = 175 \cot 14°20' = 175(3.9136) = 685$.

In the right triangle ABC, $BC = \sqrt{(AB)^2 - (AC)^2} = \sqrt{(685)^2 - (513)^2} = 454$.

The boat travels 454 ft in 2 min; its speed is 227 ft/min.

SUPPLEMENTARY PROBLEMS

15. Find the natural trigonometric functions of each of the following angles:
(*a*) 18°47′ (*b*) 32°13′ (*c*) 58°24′ (*d*) 79°45′

Ans.	sine	cosine	tangent	cotangent	secant	cosecant
(*a*) 18°47′	0.3220	0.9468	0.3401	2.9403	1.0563	3.1057
(*b*) 32°13′	0.5331	0.8460	0.6301	1.5869	1.1820	1.8757
(*c*) 58°24′	0.8517	0.5240	1.6255	0.6152	1.9084	1.1741
(*d*) 79°45′	0.9840	0.1780	5.5304	0.1808	5.6201	1.0162

16. Find (acute) angle A, given:

(*a*) $\sin A = 0.5741$ *Ans.* $A = 35°\ 2'$
(*b*) $\sin A = 0.9468$ $A = 71°13'$
(*c*) $\sin A = 0.3510$ $A = 20°33'$
(*d*) $\sin A = 0.8900$ $A = 62°52'$

(*e*) $\cos A = 0.9382$ *Ans.* $A = 20°15'$
(*f*) $\cos A = 0.6200$ $A = 51°41'$
(*g*) $\cos A = 0.7120$ $A = 44°36'$
(*h*) $\cos A = 0.4651$ $A = 62°17'$

(*i*) $\tan A = 0.2725$ $A = 15°15'$
(*j*) $\tan A = 1.1652$ $A = 49°22'$
(*k*) $\tan A = 0.5200$ $A = 27°28'$
(*l*) $\tan A = 2.7775$ $A = 70°12'$

(*m*) $\cot A = 0.2315$ $A = 76°58'$
(*n*) $\cot A = 2.9715$ $A = 18°36'$
(*o*) $\cot A = 0.7148$ $A = 54°27'$
(*p*) $\cot A = 1.7040$ $A = 30°24'$

(q) sec A = 1.1161 *Ans.* $A = 26°22'$
(r) sec A = 1.4382 $A = 45°57'$
(s) sec A = 1.2618 $A = 37°35'$
(t) sec A = 2.1584 $A = 62°24'$

(u) csc A = 3.6882 *Ans.* $A = 15°44'$
(v) csc A = 1.0547 $A = 71°28'$
(w) csc A = 1.7631 $A = 34°33'$
(x) csc A = 1.3436 $A = 48°\ 6'$

17. Solve each of the right triangles ABC, given:

(a) $A = 35°20'$, $c = 112$ *Ans.* $B = 54°40'$, $a = 64.8$, $b = 91.4$

(b) $B = 48°40'$, $c = 225$ $A = 41°20'$, $a = 149$, $b = 169$

(c) $A = 23°18'$, $c = 346.4$ $B = 66°42'$, $a = 137.0$, $b = 318.1$

(d) $B = 54°12'$, $c = 182.5$ $A = 35°48'$, $a = 106.7$, $b = 148.0$

(e) $A = 32°10'$, $a = 75.4$ $B = 57°50'$, $b = 120$, $c = 142$

(f) $A = 58°40'$, $b = 38.6$ $B = 31°20'$, $a = 63.4$, $c = 74.2$

(g) $B = 49°14'$, $b = 222.2$ $A = 40°46'$, $a = 191.6$, $c = 293.4$

(h) $A = 66°36'$, $a = 112.6$ $B = 23°24'$, $b = 48.73$, $c = 122.7$

(i) $A = 29°48'$, $b = 458.2$ $B = 60°12'$, $a = 262.4$ $c = 528.0$

(j) $a = 25.4$, $b = 38.2$ $A = 33°37'$, $B = 56°23'$, $c = 45.9$

(k) $a = 45.6$, $b = 84.8$ $A = 28°16'$, $B = 61°44'$, $c = 96.3$

(l) $a = 38.64$, $b = 48.74$ $A = 38°24'$, $B = 51°36'$, $c = 62.21$

(m) $a = 506.2$, $c = 984.8$ $A = 30°56'$, $B = 59°\ 4'$, $b = 844.7$

(n) $b = 672.9$, $c = 888.1$ $A = 40°44'$, $B = 49°16'$, $a = 579.4$

18. Find the base and altitude of an isosceles triangle whose vertical angle is 65° and whose equal sides are 415 ft. *Ans.* Base = 446 ft, altitude = 350 ft

19. The base of an isosceles triangle is 15.90 in. and the base angles are 54°28'. Find the equal sides and the altitude. *Ans.* Side = 13.68 in., altitude = 11.13 in.

20. The radius of a circle is 21.4 ft. Find (a) the length of the chord subtended by a central angle of 110°40' and (b) the distance between two parallel chords on the same side of the center subtended by central angles 118°40' and 52°20'. *Ans.* (a) 35.2 ft, (b) 8.29 ft

21. Show that the base b of an isosceles triangle whose equal sides are a and whose vertical angle is θ is given by $b = 2a \sin \frac{1}{2}\theta$.

22. Show that the perimeter P of a regular polygon of n sides inscribed in a circle of radius r is given by $P = 2nr \sin(180°/n)$.

23. A wheel, 5 ft in diameter, rolls up an incline of 18°20'. What is the height of the center of the wheel above the base of the incline when the wheel has rolled 5 ft up the incline? *Ans.* 3.95 ft

24. A wall is 15 ft high and 10 ft from a house. Find the length of the shortest ladder which will just touch the top of the wall and reach a window 20.5 ft above the ground. *Ans.* 42.5 ft

CHAPTER 34

Practical Applications

THE BEARING OF A POINT *B* FROM A POINT *A*, in a horizontal plane, is usually defined as the angle (always acute) made by the half-line drawn from *A* through *B* with the north-south line through *A*. The bearing is then read from the north or south line toward the east or west. For example,

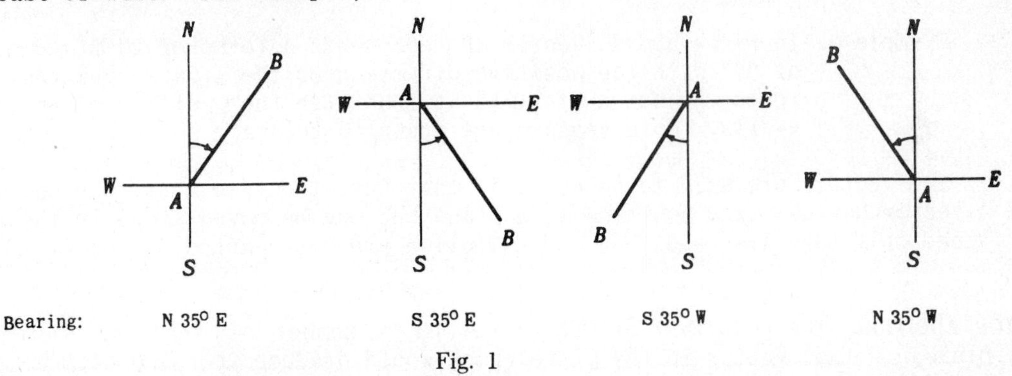

Fig. 1

In aeronautics the bearing of *B* from *A* is more often given as the angle made by the the half-line *AB* with the north line through *A*, measured clockwise from the north (i.e., from the north around through the east). For example,

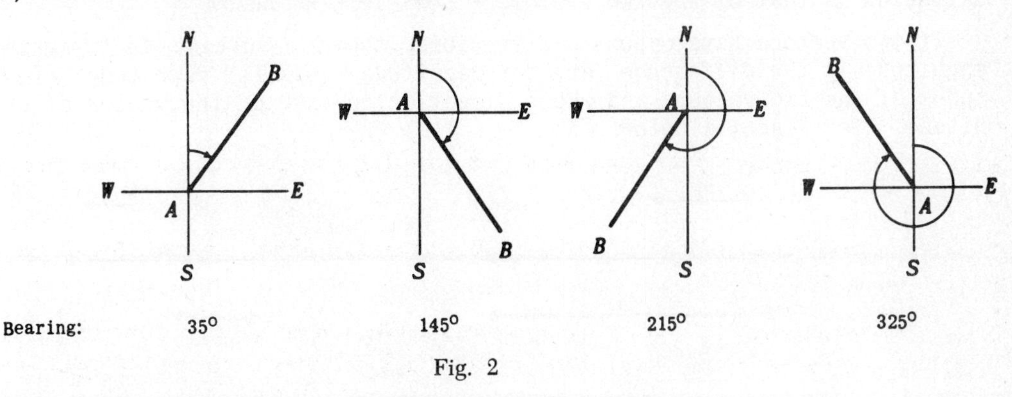

Fig. 2

VECTORS. Any physical quantity, as force or velocity, which has both magnitude and direction is called a *vector quantity.* A vector quantity may be represented by a directed line segment (arrow) called a *vector.* The *direction* of the vector is that of the given quantity and the *length* of the vector is proportional to the magnitude of the quantity.

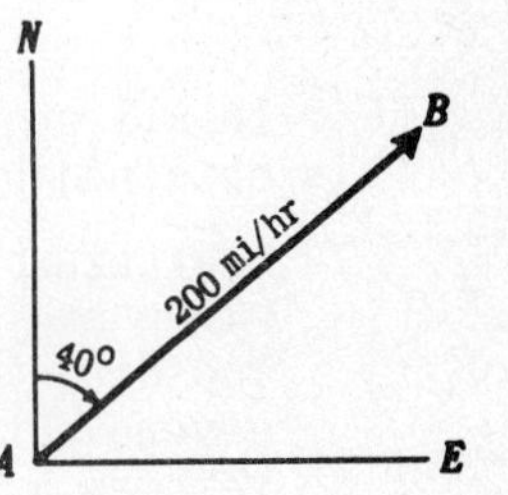

Example 1. An airplane is traveling N40°E at 200 mph. Its velocity is represented by the vector *AB* of the adjacent figure.

Example 2. A motor boat having the speed 12 mph in still water is headed directly across a river whose current is 4 mph. In Fig. 3 below, the vector *CD* represents the velocity of the current and the vector *AB* represents, to the same scale, the velocity of the boat in still water. Thus, vector *AB* is three times as long as vector *CD*.

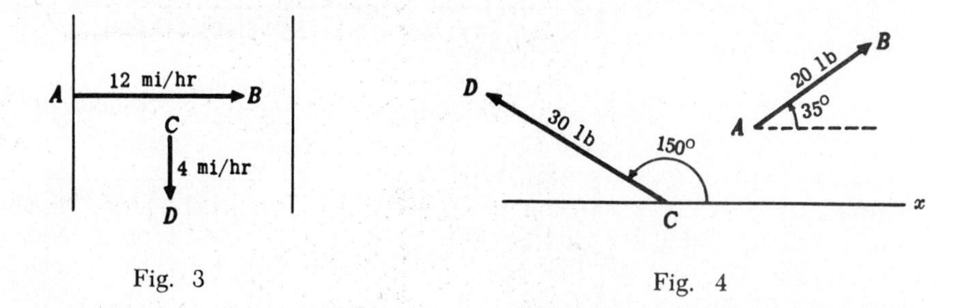

Fig. 3 Fig. 4

Example 3. In Fig. 4 above, vector *AB* represents a force of 20 lb making an angle of 35° with the positive direction on the x-axis and vector *CD* represents a force of 30 lb at 150° with the positive direction on the x-axis. Both vectors are drawn to the same scale.

Two vectors are said to be equal if they have the same magnitude and direction. A vector has no fixed position in a plane and may be moved about in the plane provided only that its magnitude and direction are not changed.

VECTOR ADDITION. The *resultant* or *vector sum* of a number of vectors, all in the same plane, is that vector in the plane which would produce the same effect as that produced by all of the original vectors acting together.

If two vectors α and β have the same direction, their resultant is a vector R whose magnitude is equal to the sum of the magnitudes of the two vectors and whose direction is that of the two vectors. See Fig. 5(*a*) below.

If two vectors have opposite directions, their resultant is a vector R whose magnitude is the difference (greater magnitude - smaller magnitude) of the magnitudes of the two vectors and whose direction is that of the vector of greater magnitude. See Fig. 5(*b*) below.

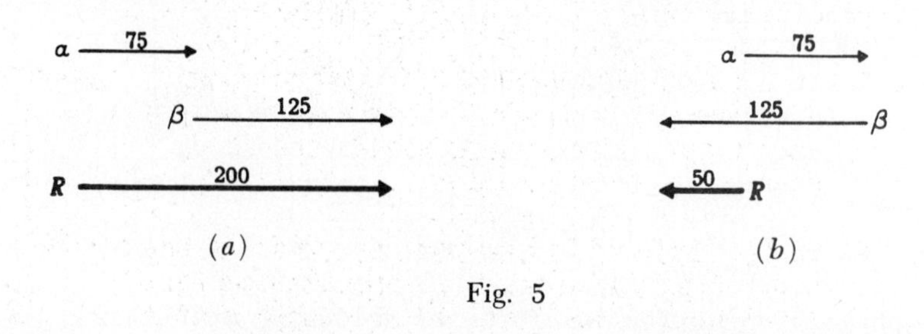

(*a*) (*b*)

Fig. 5

In all other cases, the magnitude and direction of the resultant of two vectors is obtained by either of the following two methods.

1) PARALLELOGRAM METHOD. Place the tail ends of both vectors at any point O in their plane and complete the parallelogram having these vectors as adjacent sides. The directed diagonal issuing from O is the resultant or vector sum of the two given vectors. Thus, in Fig. 6(*b*) below, the vector R is the resultant of the vectors α and β of Fig. 6(*a*).

2) TRIANGLE METHOD. Choose one of the vectors and label its tail end as O. Place

the tail end of the other vector at the arrow end of the first. The resultant is then the line segment closing the triangle and directed from O. Thus, in Fig. 6(c) and 6(d) below, R is the resultant of the vectors α and β.

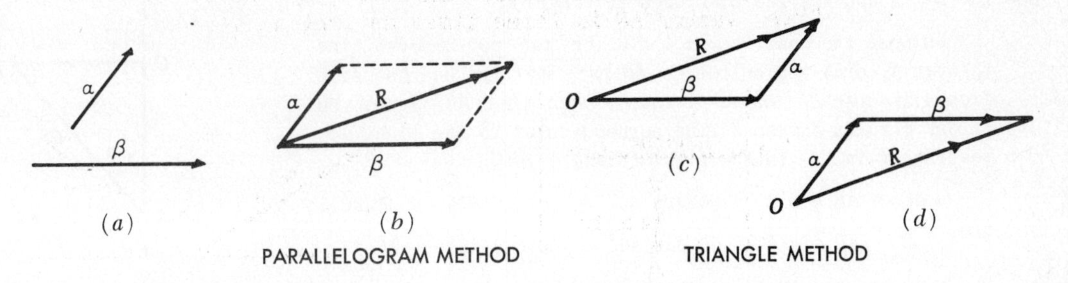

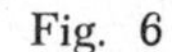
Fig. 6

Example 4. The resultant R of the two vectors of Example 2 represents the speed and direction in which the boat travels. Fig. 7(a) illustrates the parallelogram method; Fig. 7(b) and 7(c) illustrate the triangle method.

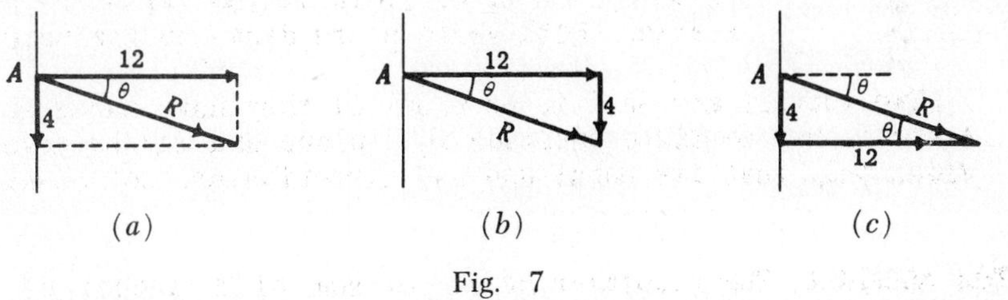

Fig. 7

The magnitude of $R = \sqrt{(12)^2 + 4^2} = 12.6$ mph.
From Fig. 7(a) or 7(b), $\tan\theta = 4/12 = 0.3333$ and $\theta = 18°30'$.
Thus, the boat moves down stream in a line making an angle $\theta = 18°30'$ with the direction in which it is headed or making an angle $90° - \theta = 71°30'$ with the bank of the river.

THE COMPONENT OF A VECTOR α along a line L is the perpendicular projection of the vector α on L. It is often very useful to resolve a vector into two components along a pair of perpendicular lines.

Example 5. In each of Fig. 7(a), (b), (c) the components of R are 1) 4 mph in the direction of the current and 2) 12 mph in the direction perpendicular to the current.

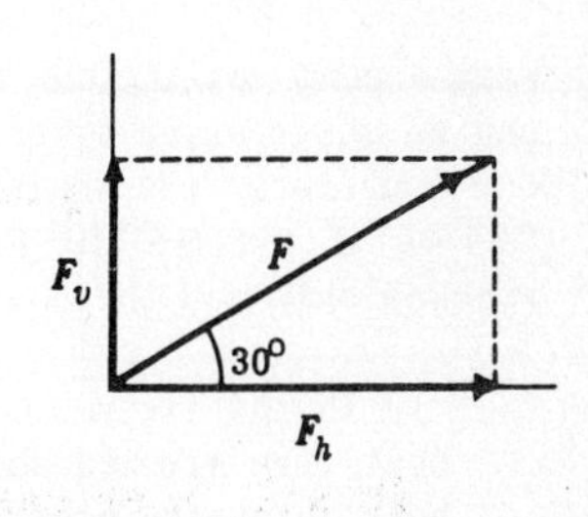

Example 6. In the adjoining diagram, the force F has horizontal component $F_h = F\cos 30°$ and vertical component $F_v = F\sin 30°$. Note that F is the vector sum or resultant of F_h and F_v.

SOLVED PROBLEMS

1. A motor boat moves in the direction N 40°E for 3 hr at 20 mph. How far north and how far east does it travel?

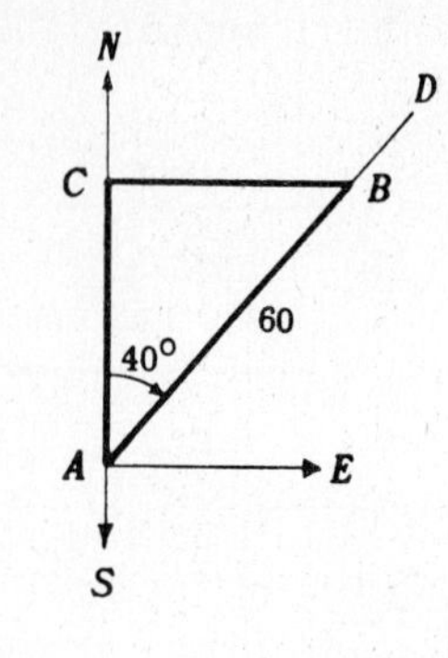

Suppose the boat leaves A. Using the north-south line through A, draw the half-line AD so that the bearing of D from A is N 40°E. On AD locate B such that $AB = 3(20) = 60$ miles. Through B pass a line perpendicular to the line NAS, meeting it in C. In the right triangle ABC,

$$AC = AB \cos A = 60 \cos 40^\circ = 60(0.7660) = 45.96$$

and

$$CB = AB \sin A = 60 \sin 40^\circ = 60(0.6428) = 38.57.$$

The boat travels 46 miles north and 39 miles east.

2. Three ships are situated as follows: A is 225 miles due north of C, and B is 375 miles due east of C. What is the bearing (*a*) of B from A, (*b*) of A from B?

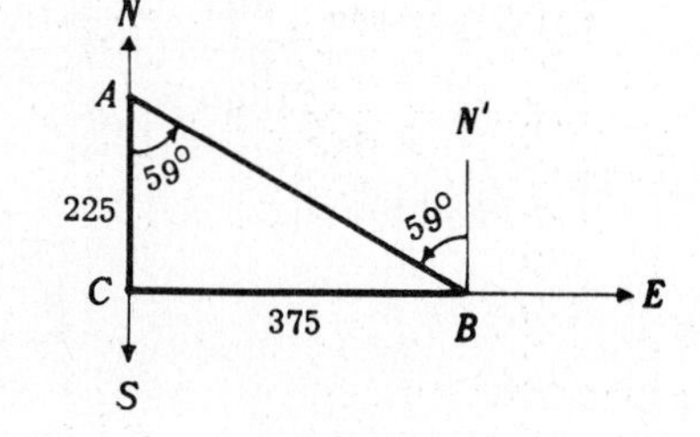

In the right triangle ABC,

$$\tan \angle CAB = 375/225 = 1.6667 \quad \text{and} \quad \angle CAB = 59^\circ 0'.$$

(*a*) The bearing of B from A (angle SAB) is S 59°0′ E.

(*b*) The bearing of A from B (angle $N'BA$) is N 59°0′ W.

3. Three ships are situated as follows: A is 225 miles west of C while B, due south of C, bears S 25°10′ E from A. (*a*) How far is B from A? (*b*) How far is B from C? (*c*) What is the bearing of A from B?

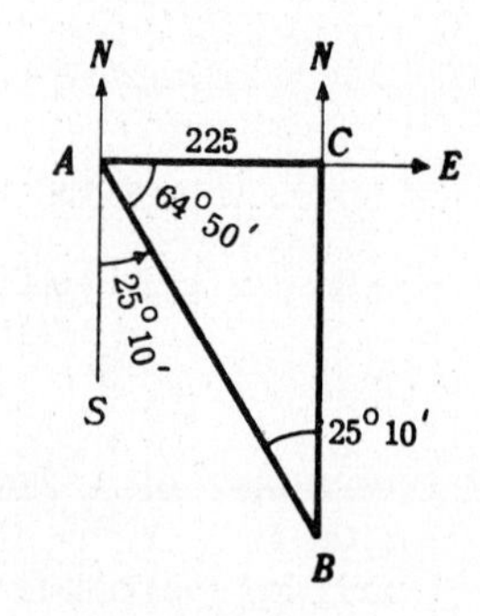

From the figure, $\angle SAB = 25^\circ 10'$ and $\angle BAC = 64^\circ 50'$. Then

$$AB = AC \sec \angle BAC = 225 \sec 64^\circ 50' = 225(2.3515) = 529.1 \quad \text{or}$$

$$AB = AC/\cos \angle BAC = 225/\cos 64^\circ 50' = 225/0.4253 = 529.0 \quad \text{and}$$

$$CB = AC \tan \angle BAC = 225 \tan 64^\circ 50' = 225(2.1283) = 478.9.$$

(*a*) B is 529 miles from A. (*b*) B is 479 miles from C.

(*c*) Since $\angle CBA = 25^\circ 10'$, the bearing of A from B is N 25°10′ W.

4. From a boat sailing due north at 16.5 mph, a wrecked ship K and an observation tower T are observed in a line due east. One hour later the wrecked ship and the tower have bearings S 34°40′ E and S 65°10′ E. Find the distance between the wrecked ship and the tower.

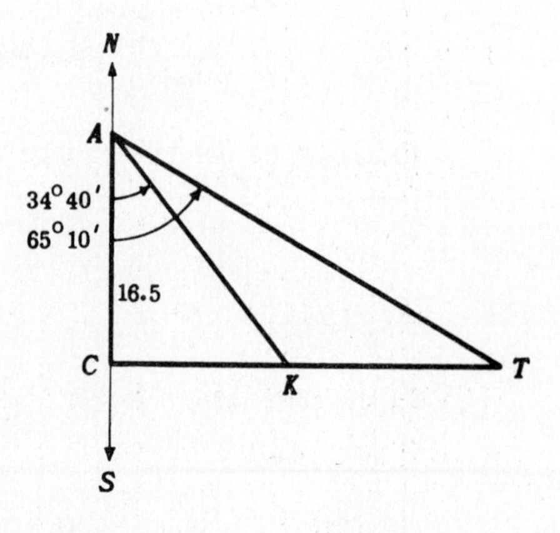

In the figure, C, K, and T represent respectively the boat, the wrecked ship, and the tower when in a line. One hour later the boat is at A, 16.5 miles due north of C. In the right triangle ACK,

$$CK = 16.5 \tan 34^\circ 40' = 16.5(0.6916).$$

In the right triangle ACT,

$$CT = 16.5 \tan 65^\circ 10' = 16.5(2.1609).$$

Then $KT = CT - CK = 16.5(2.1609 - 0.6916) = 24.2$ mi.

5. A ship is sailing due east when a light is observed bearing N $62°10'$ E. After the ship has traveled 2250 ft, the light bears N $48°25'$ E. If the course is continued, how close will the ship approach the light?

In Fig. (a) below, L is the position of the light, A is the first position of the ship, B is the second position, and C is the position when nearest L.

In the right triangle ACL, $AC = CL \cot \angle CAL = CL \cot 27°50' = 1.8940\, CL$.

In the right triangle BCL, $BC = CL \cot \angle CBL = CL \cot 41°35' = 1.1270\, CL$.

Since $AC = BC + 2250$, $1.8940\, CL = 1.1270\, CL + 2250$, and $CL = \dfrac{2250}{1.8940 - 1.1270} = 2934$ ft.

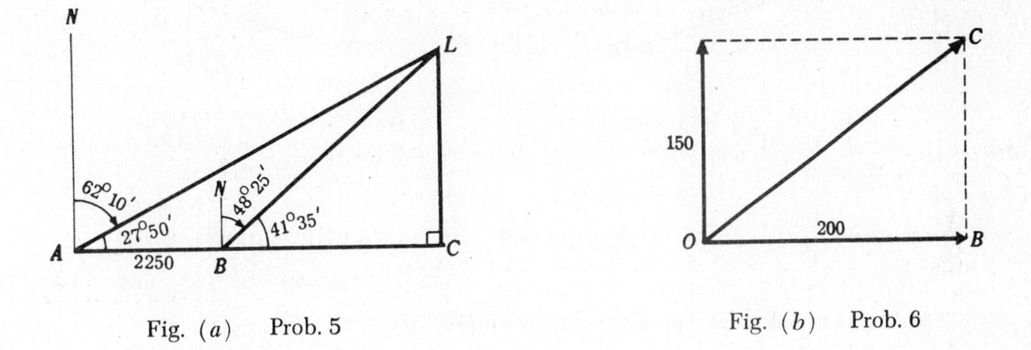

Fig. (a) Prob. 5 Fig. (b) Prob. 6

6. Refer to Fig. (b) above. A body at O is being acted upon by two forces, one of 150 lb due north and the other of 200 lb due east. Find the magnitude and direction of the resultant.

In the right triangle OBC, $OC = \sqrt{(OB)^2 + (BC)^2} = \sqrt{(200)^2 + (150)^2} = 250$ lb,

$\tan \angle BOC = 150/200 = 0.7500$ and $\angle BOC = 36°50'$.

The magnitude of the resultant force is 250 lb and its direction is N $53°10'$ E.

7. An airplane is moving horizontally at 240 mph when a bullet is shot with speed 2750 ft/sec at right angles to the path of the airplane. Find the resultant speed and direction of the bullet.

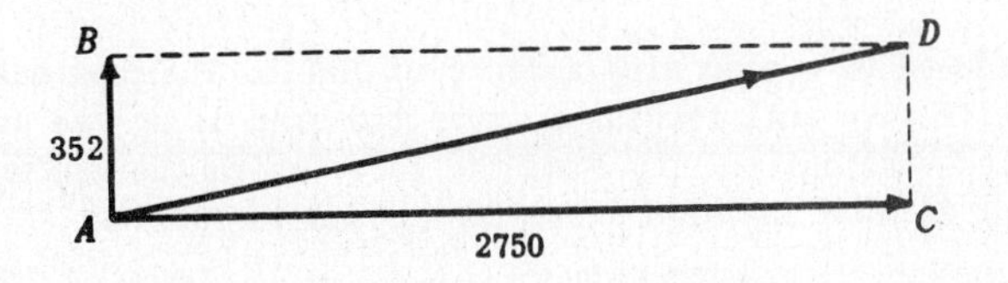

The speed of the airplane is 240 mi/hr $= \dfrac{240(5280)}{60(60)}$ ft/sec $= 352$ ft/sec.

In the figure, the vector AB represents the velocity of the airplane, the vector AC represents the initial velocity of the bullet, and the vector AD represents the resultant velocity of the bullet.

In the right triangle ACD, $AD = \sqrt{(352)^2 + (2750)^2} = 2770$ ft/sec,

$\tan \angle CAD = 352/2750 = 0.1280$ and $\angle CAD = 7°20'$.

Thus, the bullet travels at 2770 ft/sec along a path making an angle of $82°40'$ with the path of the airplane.

8. A river flows due south at 125 ft/min. A motor boat, moving at 475 ft/min in still water, is headed due east across the river. (*a*) Find the direction in which the boat moves and its speed. (*b*) In what direction must the boat be headed in order that it move due east and what is its speed in that direction?

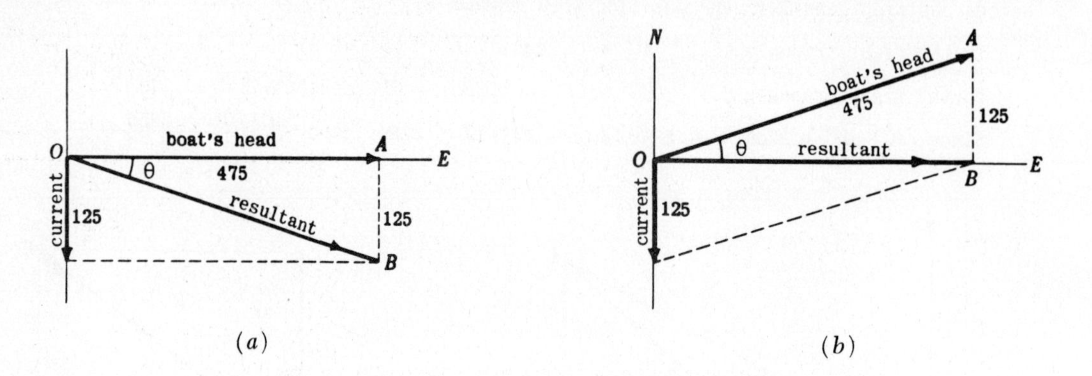

(*a*) Refer to Fig. (*a*). In right triangle OAB, $OB = \sqrt{(475)^2 + (125)^2} = 491$,

$\tan\theta = 125/475 = 0.2632$ and $\theta = 14°40'$.

Thus the boat moves at 491 ft/min in the direction S 75°20′ E.

(*b*) Refer to Fig. (*b*). In right triangle OAB, $\sin\theta = 125/475 = 0.2632$ and $\theta = 15°20'$.

Thus the boat must be headed N 74°40′ E and its speed in that direction is

$$OB = \sqrt{(475)^2 - (125)^2} = 458 \text{ ft/min}$$

9. A telegraph pole is kept vertical by a guy wire which makes an angle of 25° with the pole and which exerts a pull of F = 300 lb on the top. Find the horizontal and vertical components F_h and F_v of the pull F.

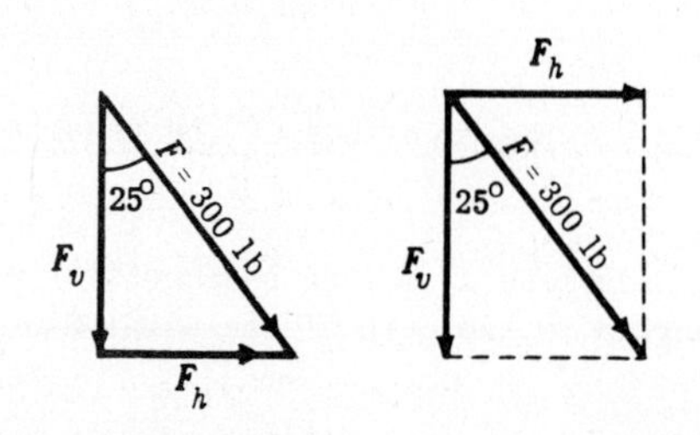

F_h = 300 sin 25° = 300(0.4226) = 127 lb

F_v = 300 cos 25° = 300(0.9063) = 272 lb

10. A man pulls a rope attached to a sled with a force of 100 lb. The rope makes an angle of 27° with the ground. (*a*) Find the effective pull tending to move the sled along the ground and the effective pull tending to lift the sled vertically. (*b*) Find the force which the man must exert in order that the effective force tending to move the sled along the ground is 100 lb.

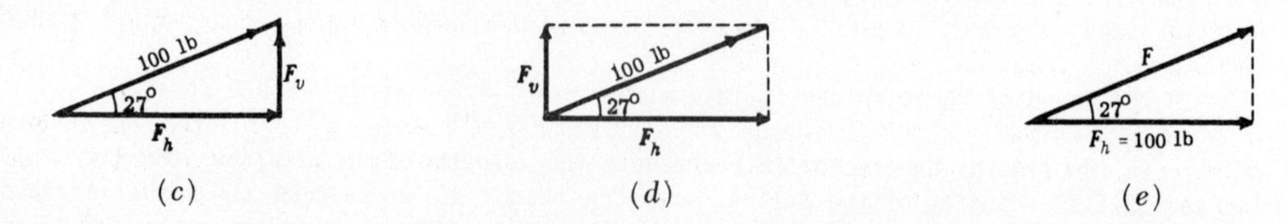

(*a*) In Fig. (*c*) and (*d*), the 100 lb pull in the rope is resolved into horizontal and vertical components, F_h and F_v respectively. Then F_h is the force tending to move the sled along the ground and F_v is the force tending to lift the sled.

F_h = 100 cos 27° = 100(0.8910) = 89 lb, F_v = 100 sin 27° = 100(0.4540) = 45 lb.

(*b*) In Fig. (*e*), the horizontal component of the required force F is F_h = 100 lb. Then

$$F = 100/\cos 27° = 100/0.8910 = 112 \text{ lb.}$$

11. A block weighing W = 500 lb rests upon a ramp inclined 29° with the horizontal. (*a*) Find the force tending to move the block down the ramp and the force of the block on the ramp. (*b*) What minimum force must be applied to keep the block from sliding down the ramp? Neglect friction.

(*a*) Refer to the adjoining figure. Resolve the weight W of the block into components F_1 and F_2, respectively parallel and perpendicular to the ramp. F_1 is the force tending to move the block down the ramp and F_2 is the force of the block on the ramp.

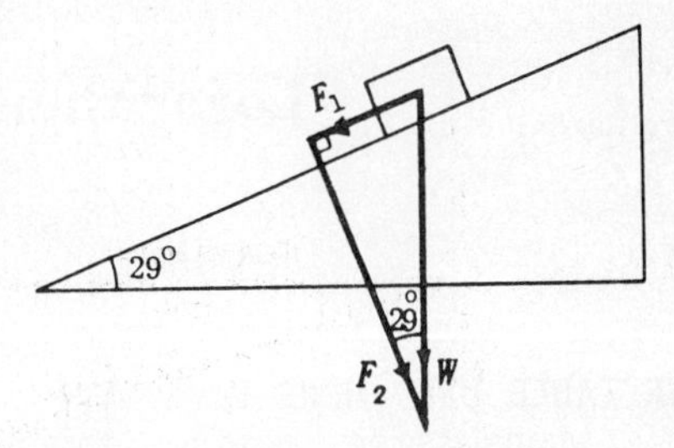

$$F_1 = W \sin 29° = 500(0.4848) = 242 \text{ lb},$$

$$F_2 = W \cos 29° = 500(0.8746) = 437 \text{ lb}.$$

(*b*) 242 lb up the ramp.

SUPPLEMENTARY PROBLEMS

12. An airplane flies 100 miles in the direction S 38°10′ E. How far south and how far east of the starting point is it? *Ans.* 78.6 mi south, 61.8 mi east

13. A plane is headed due east with airspeed 240 mph. If a wind at 40 mph from the north is blowing, find the groundspeed and track. *Ans.* Groundspeed, 243 mph; track, 99°30′ or S 80°30′ E

14. A body is acted upon by a force of 75 lb, due west, and a force of 125 lb, due north. Find the magnitude and direction of the resultant force. *Ans.* 146 lb, N 31°0′ W

15. Find the rectangular components of a force of 525.0 lb in a direction 38°25′ with the horizontal. *Ans.* 411.3 lb, 326.2 lb

16. A barge is being towed north at the rate 18 mph. A man walks across the deck from west to east at the rate 6 ft/sec. Find the magnitude and direction of the actual velocity.
Ans. 27 ft/sec, N 12°50′ E

17. A ship at A is to sail to C, 56 mi north and 258 mi east of A. After sailing N 25°10′ E for 120 mi to P, the ship is headed toward C. Find the distance of P from C and the required course to reach C.
Ans. 214 miles, S 75°40′ E

18. A guy wire 78 ft long runs from the top of a telephone pole 56 ft high to the ground and pulls on the pole with a force of 290 lb. What is the horizontal pull on the top of the pole? *Ans.* 202 lb

19. A weight of 200 lb is placed on a smooth plane inclined at an angle of 38° with the horizontal and held in place by a rope parallel to the surface and fastened to a peg in the plane. Find the pull on the string. *Ans.* 123 lb

20. A man wishes to raise a 300 lb weight to the top of a wall 20 ft high by dragging it up an incline. What is the length of the shortest inclined plane he can use if his pulling strength is 140 lb?
Ans. 43 ft

21. A 150 lb shell is dragged up a runway inclined 40° to the horizontal. Find (*a*) the force of the shell against the runway and (*b*) the force required to drag the shell.
Ans. (*a*) 115 lb, (*b*) 96 lb

CHAPTER 35

Logarithms of Trigonometric Functions

THE TABLE USED HERE is a five place table of the logarithms of the trigonometric functions (sine, cosine, tangent, and cotangent) of the angles from 0° to 90° at intervals of 1 minute.

The procedures in using such a table are essentially those for the table of natural trigonometric functions.

Example 1.

(*a*) log sin 22°34' = 9.58406 − 10

(*b*) log tan 72°18' = 0.49602

(*c*) log sin 22°34.8' = 9.58430 − 10

$$\begin{aligned} \log\sin 22°34' &= 9.58406 - 10 \\ \log\sin 22°35' &= \underline{9.58436 - 10} \\ \text{Tabular difference} &= \quad .00030 \end{aligned}$$

Correction = .8 × tabular difference = .00024

Adding the correction, since sine,
log sin 22°34.8' = 9.58406 − 10 + .00024 = 9.58430 − 10.

The essential operation here is 58406 + .8(30) = 58406 + 24 = 58430.

(*d*) log cos 66°42.4' = 9.59708 − 10

$$\begin{aligned} \log\cos 66°42' &= 9.59720 - 10 \\ \text{Tabular difference} &= 30 \end{aligned}$$

Correction = .4 × tabular difference = .4(30) = 12

Subtracting the correction, since cosine, log cos 66°42.4' = 9.59708 − 10.

Example 2.

(*a*) If log sin A = 9.66197 − 10, then A = 27°20'.

(*b*) If log cot A = 0.15262, then A = 35°8'.

(*c*) If log sin A = 9.95472 − 10, then A = 64°17.3'.

$$\begin{aligned} \log\sin 64°17' &= 9.95470 - 10 \\ \log\sin 64°18' &= \underline{9.95476 - 10} \\ \text{tabular difference} &= \quad .00006 \end{aligned} \qquad \begin{aligned} \log\sin 64°17' &= 9.95470 - 10 \\ \log\sin A &= \underline{9.95472 - 10} \\ \text{Difference} &= \quad .00002 \end{aligned}$$

$$\text{Correction} = \frac{.00002}{.00006}(1') = \frac{2}{6}(1') = .3'.$$

Adding the correction, since sine, A = 64°17.3'.

(*d*) If $\log \cos A = 9.97888 - 10$, then $A = 17°43.5'$.

$\log \cos 17°44' = 9.97886 - 10$ (next smaller logarithm)
Tabular difference = 4; difference = 2.

Correction $= \frac{2}{4}(1') = .5'$.

Subtracting the correction, since cosine, $A = 17°43.5'$.

(*e*) If $\log \tan A = 0.24372$, then $A = 60°17.6'$.

$\log \tan 60°17' = 0.24353$ (next smaller logarithm)
Tabular difference = 30; difference = 19.

Correction $= \frac{19}{30}(1') = .6'$.

Adding the correction, since tangent, $A = 60°17.6'$.

(*f*) If $\cot A = 9.41640 - 10$, then $A = 75°22.8'$.

$\log \cot 75°23' = 9.41629 - 10$
Tabular difference = 52; difference = 11.

Correction $= \frac{11}{52}(1') = .2'$.

Subtracting the correction, since cotangent, $A = 75°22.8'$.

SOLVED PROBLEMS

1. Verify each of the following.

(*a*) $\log \sin 14°28.3' = 9.39777 - 10$ $(39762 + .3 \times 39)$
(*b*) $\log \cos 66°44.8' = 9.59638 - 10$ $(59661 - .8 \times 29)$
(*c*) $\log \tan 31°26.4' = 9.78630 - 10$ $(78618 + .4 \times 29)$
(*d*) $\log \cot 45°54.6' = 9.98620 - 10$ $(98635 - .6 \times 25)$
(*e*) $\log \sin 62°29.1' = 9.94787 - 10$ $(94786 + .1 \times 7)$
(*f*) $\log \cos 23°33.7' = 9.96220 - 10$ $(96223 - .7 \times 5)$
(*g*) $\log \tan 70°20.6' = 0.44709$ $(44685 + .6 \times 40)$
(*h*) $\log \cot 11°17.3' = 0.69982$ $(70002 - .3 \times 66)$

2. Verify each of the following.

(*a*) $\log \sin A = 9.90020 - 10$, then $A = 52°37.6'$ $(\frac{6}{10} \times 1' = .6')$

(*b*) $\log \cos A = 9.93602 - 10$, then $A = 30°20.6'$ $(\frac{3}{7} \times 1' = .4')$

(*c*) $\log \tan A = 9.87150 - 10$, then $A = 36°38.7'$ $(\frac{18}{26} \times 1' = .7')$

(*d*) $\log \cot A = 0.01245$, then $A = 44°10.7'$ $(\frac{7}{25} \times 1' = .3')$

(*e*) $\log \sin A = 9.80172 - 10$, then $A = 39°18.4'$ $(\frac{6}{16} \times 1' = .4')$

(*f*) $\log \cos A = 9.55215 - 10$, then $A = 69°6.6'$ $(\frac{13}{33} \times 1' = .4')$

(*g*) log tan A = 0.44372, then A = 70°12.1' $(\frac{5}{40} \times 1' = .1')$

(*h*) log cot A = 9.31142 − 10, then A = 78°25.4' $(\frac{38}{64} \times 1' = .6')$

SUPPLEMENTARY PROBLEMS

3. Find:

(*a*) log sin 53°18' = 9.90405-10

(*b*) log cos 18°17' = 9.97750-10

(*c*) log tan 42°47' = 9.96636-10

(*d*) log cot 68°14' = 9.60130-10

(*e*) log sin 71°9.6' = 9.97608-10

(*f*) log cos 56°44.4' = 9.73913-10

(*g*) log tan 67°0.3' = 0.37226

(*h*) log cot 76°9.3' = 9.39174-10

(*i*) log sin 72°15.4' = 9.97884-10

(*j*) log cos 20° 9.2' = 9.97256-10

(*k*) log tan 84°47.1' = 1.03967

(*l*) log cot 74° 4.2' = 9.45549-10

(*m*) log sin 22°15.8' = 9.57849-10

(*n*) log cos 66°17.4' = 9.60434-10

(*o*) log tan 11°19.8' = 9.30182-10

(*p*) log cot 25°10.6' = 0.32784

4. Find acute angle A, given:

(*a*) log sin A = 9.28705-10, A = 11°10.0'

(*b*) log cos A = 9.48881-10, A = 72° 3.0'

(*c*) log tan A = 9.82325-10, A = 33°39.0'

(*d*) log cot A = 9.91765-10, A = 50°24.0'

(*e*) log sin A = 9.53928-10, A = 20°15.2'

(*f*) log cos A = 9.89900-10, A = 37°34.8'

(*g*) log tan A = 9.53042-10, A = 18°44.1'

(*h*) log cot A = 0.18960, A = 32°52.4'

(*i*) log sin A = 9.86000-10, A = 46°25.3'

(*j*) log cos A = 9.75529-10, A = 55°18.2'

(*k*) log tan A = 9.80888-10, A = 81°10.4'

(*l*) log cot A = 9.67240-10, A = 64°48.7'

(*m*) log sin A = 9.80513-10, A = 39°40.6'

(*n*) log cos A = 9.86892-10, A = 42°18.8'

(*o*) log tan A = 0.06510, A = 49°16.7'

(*p*) log cot A = 9.71700-10, A = 62°28.3'

CHAPTER 36

Logarithmic Solution of Right Triangles

ANY RIGHT TRIANGLE may be solved and partially checked by using the trigonometric functions sine, cosine, and either tangent or cotangent of one of the acute angles, together with the angle relation $A + B = 90°$. In general, a better check is obtained by using the relation $c^2 = a^2 + b^2$.

Example. Suppose the sides a and b of the right triangle ABC are given.

(1) To find angle A, use $\tan A = a/b$; then $B = 90° - A$.

(2) To find side c, use $c = a/\sin A$.

(3) To check, use $a^2 = c^2 - b^2 = (c-b)(c+b)$
or $b^2 = c^2 - a^2 = (c-a)(c+a)$.

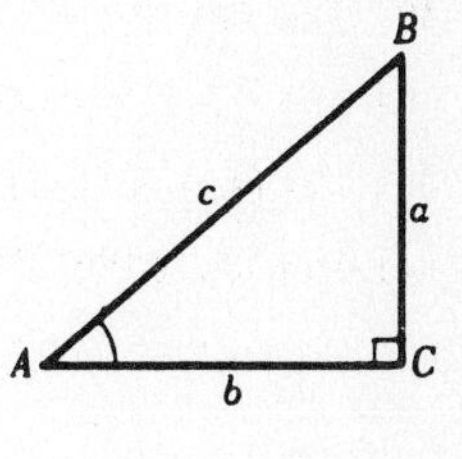

SOLVED PROBLEMS

1. Solve and check the right triangle ABC, given $a = 48.620$ and $b = 37.640$. See Fig. (a) below.

$\tan A = a/b$	$c = a/\sin A$	Check: $a^2 = (c-b)(c+b)$	
$\log a = 1.68681$	$\log a = 1.68681$	$c = 61.487$	$\log (c-b) = 1.37744$
$(-)\log b = 1.57565$	$(-)\log \sin A = 9.89803-10$	$b = 37.640$	$(+)\log (c+b) = 1.99620$
$\log \tan A = 0.11116$	$\log c = 1.78878$	$c-b = 23.847$	$2 \log a = 3.37364$
		$c+b = 99.127$	
$A = 52°15.2'$	$c = 61.487$		$\log a = 1.68682$
$B = 37°44.8'$			

Fig. (a) Prob. 1

Fig. (b) Prob. 2

2. Solve and check the right triangle ABC, given $a = 562.84$ and $A = 64°23.6'$. See Fig. (b) above.

$B = 90° - A = 25°36.4'$.

$b = a/\tan A$	$c = a/\sin A$	Check: $a^2 = (c-b)(c+b)$	
$\log a = 2.75038$	$\log a = 2.75038$	$c = 624.13$	$\log (c-b) = 2.54948$
$(-)\log \tan A = 0.31943$	$(-)\log \sin A = 9.95511-10$	$b = 269.74$	$(+)\log (c+b) = 2.95128$
$\log b = 2.43095$	$\log c = 2.79527$	$c-b = 354.39$	$2 \log a = 5.50076$
		$c+b = 893.87$	
$b = 269.74$	$c = 624.13$		$\log a = 2.75038$

3. Solve and check the right triangle ABC, given $b = 583.62$ and $c = 794.86$. See Fig. (c) below.

$\cos A = b/c$

$\log b = 2.76613$
$(-)\log c = 2.90029$
$\log \cos A = 9.86584-10$

$A = 42^\circ 45.4'$
$B = 47^\circ 14.6'$

$a = c \sin A$

$\log c = 2.90029$
$(+)\log \sin A = 9.83180-10$
$\log a = 2.73209$

$a = 539.62$

Check: $b^2 = (c-a)(c+a)$

$c = 794.86$
$a = 539.62$
$c-a = 255.24$
$c+a = 1334.48 = 1334.5$

$\log (c-a) = 2.40695$
$(+)\log (c+a) = 3.12532$
$2 \log b = 5.53227$
$\log b = 2.76614$

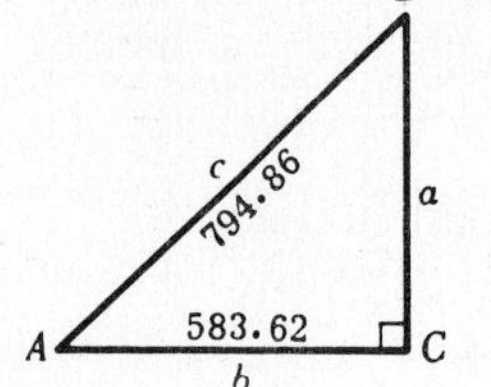

Fig. (c) Prob. 3

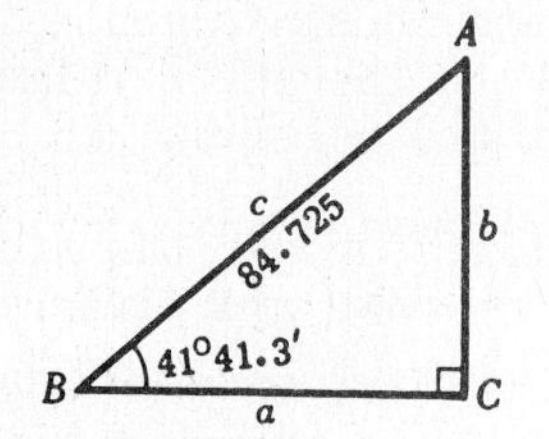

Fig. (d) Prob. 4

4. Solve and check the right triangle ABC, given $c = 84.725$ and $B = 41^\circ 41.3'$. See Fig. (d) above.
$A = 90^\circ - B = 48^\circ 18.7'$.

$b = c \sin B$

$\log c = 1.92802$
$(+)\log \sin B = 9.82287-10$
$\log b = 1.75089$

$b = 56.350$

$a = c \cos B$

$\log c = 1.92802$
$(+)\log \cos B = 9.87319-10$
$\log a = 1.80121$

$a = 63.271$

Check: $b^2 = (c-a)(c+a)$

$c = 84.725$
$a = 63.271$
$c-a = 21.454$
$c+a = 147.996 = 148.00$

$\log (c-a) = 1.33151$
$(+)\log (c+a) = 2.17026$
$2 \log b = 3.50177$
$\log b = 1.75088$

Note that this is a check of $\log b$ and not of b.

5. At a height of 23,245 ft a pilot of an airplane measures the angle of depression of a light at an airport as $28^\circ 45.2'$. How far is he from the light ?

In the adjoining figure, A is the position of the light, B is the position of the pilot, and $c = AB$ is the required distance. Then

$c = a/\sin A$

$\log a = 4.36633$
$(-)\log \sin A = 9.68218-10$

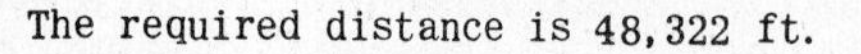

$\log c = 4.68415$

$c = 48,322$

The required distance is 48,322 ft.

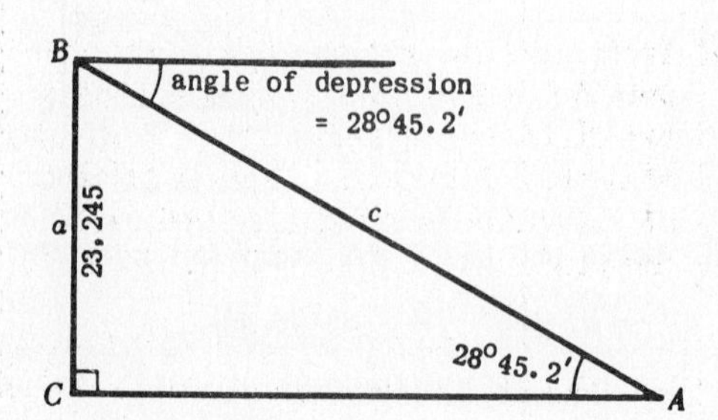

6. A shell is fired at an angle of elevation $32^\circ 14.4'$ with initial velocity 3046.8 ft/sec. Find the initial horizontal and vertical velocities.

From the figure, $v = 3046.8$, $\alpha = 32°14.4'$, and

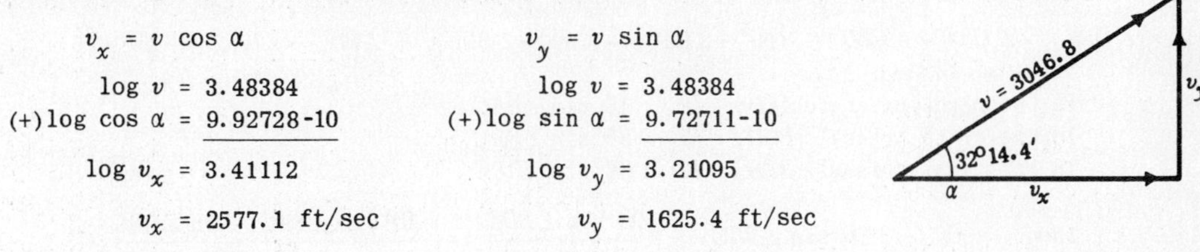

$v_x = v \cos\alpha$	$v_y = v \sin\alpha$
$\log v = 3.48384$	$\log v = 3.48384$
$(+)\log\cos\alpha = 9.92728-10$	$(+)\log\sin\alpha = 9.72711-10$
$\log v_x = 3.41112$	$\log v_y = 3.21095$
$v_x = 2577.1$ ft/sec	$v_y = 1625.4$ ft/sec

7. Two forces of 151.75 lb and 225.80 lb act at right angles. Find the magnitude of the resultant and the angle which it makes with the larger force.

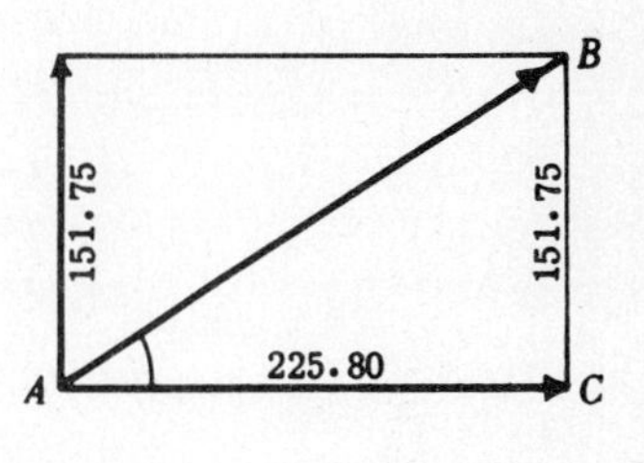

Using the right triangle ABC,

$\tan A = CB/AC$	$AB = CB/\sin A$
$\log CB = 2.18113$	$\log CB = 2.18113$
$(-)\log AC = 2.35372$	$(-)\log\sin A = 9.74648-10$
$\log\tan A = 9.82741-10$	$\log AB = 2.43465$
$A = 33°54.2'$	$AB = 272.05$

The magnitude of the resultant force is 272.05 lb and it makes an angle of 33°54.2′ with the larger force.

8. A boat travels N 28°14.6′ E for 55.375 miles and then N 61°45.4′ W for 94.625 miles. What is its distance and bearing from the starting point?

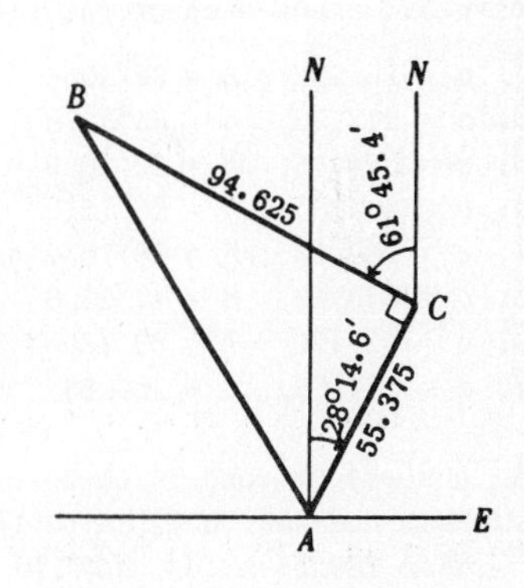

In the figure, the boat starts at A, travels to C, and then to B. In the right triangle ABC,

$\tan\angle CAB = BC/AC$	$AB = BC/\sin\angle CAB$
$\log BC = 1.97600$	$\log BC = 1.97600$
$(-)\log AC = 1.74331$	$(-)\log\sin\angle CAB = 9.93605-10$
$\log\tan\angle CAB = 0.23269$	$\log AB = 2.03995$
$\angle CAB = 59°39.8'$	$AB = 109.64$

The boat is then 109.64 miles from the starting point. Since $\angle NAB = \angle CAB - \angle CAN = 59°39.8' - 28°14.6' = 31°25.2'$, the required bearing is N 31°25.2′ W.

9. In finding the height of an inaccessible cliff CB, two points A and D, 152.75 ft apart, on a plain due west of the cliff are located. From D the angle of elevation of the top of the cliff is 44°32.4′ and from A the angle of elevation is 29°15.8′. How high is the cliff above the plain?

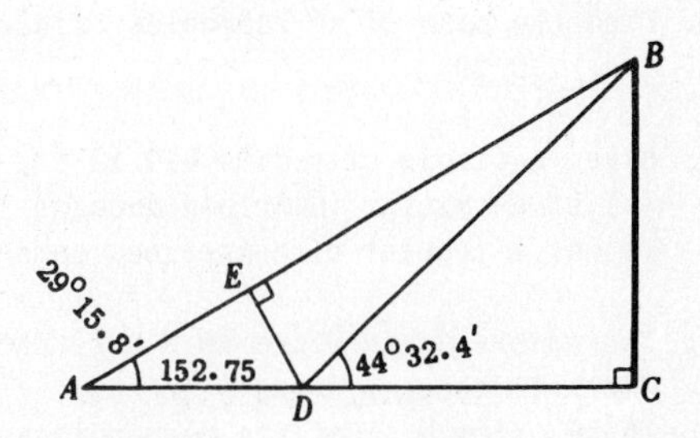

A solution of a similar problem (Problem 15, Chapter 32) made use of the relation

$$CB = \frac{AD}{\cot\angle BAC - \cot\angle BDC}.$$

In the solution here, a relation more suitable for logarithmic computation will be used.

In the figure, DE is perpendicular to AB. Then

$$\angle DBE = \angle CBA - \angle CBD = (90^\circ - \angle BAC) - (90^\circ - \angle BDC) = \angle BDC - \angle BAC = 15^\circ 16.6'.$$

In the right triangle AED, $DE = AD \sin \angle BAC$.
In the right triangle BCD, $CB = BD \sin \angle BDC$.
In the right triangle BED, $BD = DE/\sin \angle DBE$.

Then $$CB = BD \sin \angle BDC = \frac{DE \sin \angle BDC}{\sin \angle DBE} = \frac{AD \sin \angle BAC \cdot \sin \angle BDC}{\sin \angle DBE}$$

$$= \frac{152.75 \sin 29^\circ 15.8' \cdot \sin 44^\circ 32.4'}{\sin 15^\circ 16.6'}.$$

$$\begin{aligned} \log 152.75 &= 2.18398 \\ (+) \quad \log \sin 29^\circ 15.8' &= 9.68915-10 \\ (+) \quad \log \sin 44^\circ 32.4' &= 9.84597-10 \\ (+) \quad \text{colog} \sin 15^\circ 16.6' &= 0.57925 \qquad (\log \sin 15^\circ 16.6' = 9.42075-10) \\ \log CB &= 2.29835 \\ CB &= 198.77 \text{ ft} \end{aligned}$$

SUPPLEMENTARY PROBLEMS

Solve and check each of the following right triangles ABC, given:

10. $a = 25.72$, $A = 36^\circ 20'$ *Ans.* $B = 53^\circ 40'$, $b = 34.97$, $c = 43.41$
11. $a = 342.86$, $A = 55^\circ 32.8'$ *Ans.* $B = 34^\circ 27.2'$, $b = 235.23$, $c = 415.81$
12. $a = 574.16$, $B = 56^\circ 20.6'$ *Ans.* $A = 33^\circ 39.4'$, $b = 862.32$, $c = 1036.0$
13. $c = 44.26$, $A = 56^\circ 14'$ *Ans.* $B = 33^\circ 46'$, $a = 36.79$, $b = 24.60$
14. $c = 287.68$, $A = 38^\circ 10.2'$ *Ans.* $B = 51^\circ 49.8'$, $a = 177.78$, $b = 226.17$
15. $c = 67.546$, $B = 47^\circ 25.6'$ *Ans.* $A = 42^\circ 34.4'$, $a = 45.697$, $b = 49.741$
16. $a = 42.420$, $b = 58.480$ *Ans.* $A = 35^\circ 57.4'$, $B = 54^\circ 2.6'$, $c = 72.243$
17. $a = 384.66$, $b = 254.88$ *Ans.* $A = 56^\circ 28.3'$, $B = 33^\circ 31.7'$, $c = 461.44$

18. A straight road is to be constructed joining two towns A and B. If B is located 133.75 miles to the east and 256.78 miles to the north of A, find the length and direction of the road from A.
Ans. 289.53 miles, N $27^\circ 30.8'$ E

19. Two forces of 281.66 lb and 323.54 lb act at right angles. Find the magnitude of the resultant force and the angle which it makes with the larger force. *Ans.* 428.97 lb, $41^\circ 2.5'$

20. Find the base of an isosceles triangle whose vertex angle is $48^\circ 27.4'$ and whose equal legs are 168.14.
Ans. 138.00

21. Given a circle of radius 417.12 ft, find the side and area
(a) of a regular inscribed decagon. *Ans.* 257.80 ft, 511,340 ft^2
(b) of a regular circumscribed decagon. *Ans.* 271.06 ft, 565,320 ft^2

22. Two points A and D are in a horizontal line with the foot of a tower CB and on opposite sides. The distance between A and D is 535.4 ft, while the angles of elevation of the top B are $12^\circ 46'$ from A and $18^\circ 38'$ from D. Let the perpendicular through D to the line AB produced meet it at E and show that

$$CB = BD \sin \angle BDC = \frac{DE \sin \angle BDC}{\sin \angle DBE} = \frac{AD \sin \angle BAC \sin \angle BDC}{\sin \angle DBE} = 72.56 \text{ ft}.$$

CHAPTER 37

Reduction to Functions of Positive Acute Angles

COTERMINAL ANGLES. Let θ be any angle; then

$$\sin(\theta + n\,360°) = \sin\theta \qquad \cot(\theta + n\,360°) = \cot\theta$$
$$\cos(\theta + n\,360°) = \cos\theta \qquad \sec(\theta + n\,360°) = \sec\theta$$
$$\tan(\theta + n\,360°) = \tan\theta \qquad \csc(\theta + n\,360°) = \csc\theta$$

where n is any positive or negative integer or zero.

Examples. $\sin 400° = \sin(40° + 360°) = \sin 40°$
$\cos 850° = \cos(130° + 2\cdot 360°) = \cos 130°$
$\tan(-1000°) = \tan(80° - 3\cdot 360°) = \tan 80°$

FUNCTIONS OF A NEGATIVE ANGLE. Let θ be any angle; then

$$\sin(-\theta) = -\sin\theta \qquad \cot(-\theta) = -\cot\theta$$
$$\cos(-\theta) = \cos\theta \qquad \sec(-\theta) = \sec\theta$$
$$\tan(-\theta) = -\tan\theta \qquad \csc(-\theta) = -\csc\theta$$

Examples. $\sin(-50°) = -\sin 50°$, $\cos(-30°) = \cos 30°$, $\tan(-200°) = -\tan 200°$.

REDUCTION FORMULAS. Let θ be any angle; then

$$\sin(90° - \theta) = \cos\theta \qquad \sin(90° + \theta) = \cos\theta$$
$$\cos(90° - \theta) = \sin\theta \qquad \cos(90° + \theta) = -\sin\theta$$
$$\tan(90° - \theta) = \cot\theta \qquad \tan(90° + \theta) = -\cot\theta$$
$$\cot(90° - \theta) = \tan\theta \qquad \cot(90° + \theta) = -\tan\theta$$
$$\sec(90° - \theta) = \csc\theta \qquad \sec(90° + \theta) = -\csc\theta$$
$$\csc(90° - \theta) = \sec\theta \qquad \csc(90° + \theta) = \sec\theta$$

$$\sin(180° - \theta) = \sin\theta \qquad \sin(180° + \theta) = -\sin\theta$$
$$\cos(180° - \theta) = -\cos\theta \qquad \cos(180° + \theta) = -\cos\theta$$
$$\tan(180° - \theta) = -\tan\theta \qquad \tan(180° + \theta) = \tan\theta$$
$$\cot(180° - \theta) = -\cot\theta \qquad \cot(180° + \theta) = \cot\theta$$
$$\sec(180° - \theta) = -\sec\theta \qquad \sec(180° + \theta) = -\sec\theta$$
$$\csc(180° - \theta) = \csc\theta \qquad \csc(180° + \theta) = -\csc\theta$$

GENERAL REDUCTION FORMULA. Any trigonometric function of $(n\cdot 90° \pm \theta)$, where θ is any angle, is *numerically* equal

(*a*) to the same function of θ if n is an even integer,
(*b*) to the corresponding cofunction of θ if n is an odd integer.

The algebraic sign in each case is the same as the sign of the given function for that quadrant in which $n\cdot 90° \pm \theta$ lies when θ is a positive acute angle.

Examples. (*1*) $\sin(180° - \theta) = \sin(2\cdot 90° - \theta) = \sin\theta$ since 180° is an even multiple of 90° and, when θ is positive acute, the terminal side of $180° - \theta$ lies in quadrant II.

(*2*) $\cos(180° + \theta) = \cos(2\cdot 90° + \theta) = -\cos\theta$ since 180° is an even multiple of 90° and, when θ is positive acute, the terminal side of $180° + \theta$ lies in quadrant III.

(*3*) $\tan(270° - \theta) = \tan(3\cdot 90° - \theta) = \cot\theta$ since 270° is an odd multiple of 90° and, when θ is positive acute, the terminal side of $270° - \theta$ lies in quadrant III.

(*4*) $\cos(270° + \theta) = \cos(3\cdot 90° + \theta) = \sin\theta$ since 270° is an odd multiple of 90° and, when θ is positive acute, the terminal side of $270° + \theta$ lies in quadrant IV.

SOLVED PROBLEMS

1. Express each of the following in terms of a function of θ:

(*a*) $\sin(\theta - 90°)$ (*d*) $\cos(-180° + \theta)$ (*g*) $\sin(540° + \theta)$ (*j*) $\cos(-450° - \theta)$
(*b*) $\cos(\theta - 90°)$ (*e*) $\sin(-270° - \theta)$ (*h*) $\tan(720° - \theta)$ (*k*) $\csc(-900° + \theta)$
(*c*) $\sec(-\theta - 90°)$ (*f*) $\tan(\theta - 360°)$ (*i*) $\tan(720° + \theta)$ (*l*) $\sin(-540° - \theta)$

(*a*) $\sin(\theta - 90°) = \sin(-90° + \theta) = \sin(-1\cdot 90° + \theta) = -\cos\theta$, the sign being negative since, when θ is positive acute, the terminal side of $\theta - 90°$ lies in quadrant IV.

(*b*) $\cos(\theta - 90°) = \cos(-90° + \theta) = \cos(-1\cdot 90° + \theta) = \sin\theta$.

(*c*) $\sec(-\theta - 90°) = \sec(-90° - \theta) = \sec(-1\cdot 90° - \theta) = -\csc\theta$, the sign being negative since, when θ is positive acute, the terminal side of $-\theta - 90°$ lies in quadrant III.

(*d*) $\cos(-180° + \theta) = \cos(-2\cdot 90° + \theta) = -\cos\theta$. (quadrant III)

(*e*) $\sin(-270° - \theta) = \sin(-3\cdot 90° - \theta) = \cos\theta$. (quadrant I)

(*f*) $\tan(\theta - 360°) = \tan(-4\cdot 90° + \theta) = \tan\theta$. (quadrant I)

(*g*) $\sin(540° + \theta) = \sin(6\cdot 90° + \theta) = -\sin\theta$. (quadrant III)

(*h*) $\tan(720° - \theta) = \tan(8\cdot 90° - \theta) = -\tan\theta$
$= \tan(2\cdot 360° - \theta) = \tan(-\theta) = -\tan\theta$.

(*i*) $\tan(720° + \theta) = \tan(8\cdot 90° + \theta) = \tan\theta$
$= \tan(2\cdot 360° + \theta) = \tan\theta$.

(*j*) $\cos(-450° - \theta) = \cos(-5\cdot 90° - \theta) = -\sin\theta$.

(*k*) $\csc(-900° + \theta) = \csc(-10\cdot 90° + \theta) = -\csc\theta$.

(*l*) $\sin(-540° - \theta) = \sin(-6\cdot 90° - \theta) = \sin\theta$.

2. Express each of the following in terms of functions of a positive acute angle in two ways:

(*a*) sin 130° (*c*) sin 200° (*e*) tan 165° (*g*) sin 670° (*i*) csc 865° (*k*) cos(−680°)
(*b*) tan 325° (*d*) cos 310° (*f*) sec 250° (*h*) cot 930° (*j*) sin(−100°) (*l*) tan(−290°)

(*a*) $\sin 130° = \sin(2\cdot 90° - 50°) = \sin 50°$
$= \sin(1\cdot 90° + 40°) = \cos 40°$

(*c*) $\sin 200° = \sin(2\cdot 90° + 20°) = -\sin 20°$
$= \sin(3\cdot 90° - 70°) = -\cos 70°$

(*b*) $\tan 325° = \tan(4\cdot 90° - 35°) = -\tan 35°$
$= \tan(3\cdot 90° + 55°) = -\cot 55°$

(*d*) $\cos 310° = \cos(4\cdot 90° - 50°) = \cos 50°$
$= \cos(3\cdot 90° + 40°) = \sin 40°$

(*e*) $\tan 165° = \tan(2\cdot 90° - 15°) = -\tan 15°$
$= \tan(1\cdot 90° + 75°) = -\cot 75°$

(*f*) $\sec 250° = \sec(2\cdot 90° + 70°) = -\sec 70°$
$= \sec(3\cdot 90° - 20°) = -\csc 20°$

(*g*) $\sin 670° = \sin(8\cdot 90° - 50°) = -\sin 50°$
$= \sin(7\cdot 90° + 40°) = -\cos 40°$

or $\sin 670° = \sin(310° + 360°) = \sin 310° = \sin(4\cdot 90° - 50°) = -\sin 50°$

(*h*) $\cot 930° = \cot(10\cdot 90° + 30°) = \cot 30°$
$= \cot(11\cdot 90° - 60°) = \tan 60°$

or $\cot 930° = \cot(210° + 2\cdot 360°) = \cot 210° = \cot(2\cdot 90° + 30°) = \cot 30°$

(*i*) $\csc 865° = \csc(10\cdot 90° - 35°) = \csc 35°$
$= \csc(9\cdot 90° + 55°) = \sec 55°$

or $\csc 865° = \csc(145° + 2\cdot 360°) = \csc 145° = \csc(2\cdot 90° - 35°) = \csc 35°$

(*j*) $\sin(-100°) = \sin(-2\cdot 90° + 80°) = -\sin 80°$
$= \sin(-1\cdot 90° - 10°) = -\cos 10°$

or $\sin(-100°) = -\sin 100° = -\sin(2\cdot 90° - 80°) = -\sin 80°$
or $\sin(-100°) = \sin(-100° + 360°) = \sin 260° = \sin(2\cdot 90° + 80°) = -\sin 80°$

(*k*) $\cos(-680°) = \cos(-8\cdot 90° + 40°) = \cos 40°$
$= \cos(-7\cdot 90° - 50°) = \sin 50°$

or $\cos(-680°) = \cos(-680° + 2\cdot 360°) = \cos 40°$

(*l*) $\tan(-290°) = \tan(-4\cdot 90° + 70°) = \tan 70°$
$= \tan(-3\cdot 90° - 20°) = \cot 20°$

or $\tan(-290°) = \tan(-290° + 360°) = \tan 70°$

3. Find the exact values of the sine, cosine, and tangent of:
(*a*) 120°, (*b*) 210°, (*c*) 315°, (*d*) −135°, (*e*) −240°, (*f*) −330°.

Call θ, always positive acute, the *related angle* of ϕ when $\phi = 180° - \theta$, $180° + \theta$ or $360° - \theta$. Then any function of ϕ is numerically equal to the same function of θ. The algebraic sign in each case is that of the function in the quadrant in which the terminal side of ϕ lies.

(*a*) $120° = 180° - 60°$. The related angle is 60°; 120° is in quadrant II.
$\sin 120° = \sin 60° = \sqrt{3}/2$, $\cos 120° = -\cos 60° = -1/2$, $\tan 120° = -\tan 60° = -\sqrt{3}$.

(*b*) $210° = 180° + 30°$. The related angle is 30°; 210° is in quadrant III.
$\sin 210° = -\sin 30° = -1/2$, $\cos 210° = -\cos 30° = -\sqrt{3}/2$, $\tan 210° = \tan 30° = \sqrt{3}/3$.

(*c*) $315° = 360° - 45°$. The related angle is 45°; 315° is in quadrant IV.
$\sin 315° = -\sin 45° = -\sqrt{2}/2$, $\cos 315° = \cos 45° = \sqrt{2}/2$, $\tan 315° = -\tan 45° = -1$.

(*d*) Any function of −135° is the same function of $-135° + 360° = 225° = \phi$.
$225° = 180° + 45°$. The related angle is 45°; 225° is in quadrant III.
$\sin(-135°) = -\sin 45° = -\sqrt{2}/2$, $\cos(-135°) = -\cos 45° = -\sqrt{2}/2$, $\tan(-135°) = 1$.

(*e*) Any function of −240° is the same function of $-240° + 360° = 120°$.
$120° = 180° - 60°$. The related angle is 60°; 120° is in quadrant II.
$\sin(-240°) = \sin 60° = \sqrt{3}/2$, $\cos(-240°) = -\cos 60° = -1/2$, $\tan(-240°) = -\tan 60° = -\sqrt{3}$.

(*f*) Any function of −330° is the same function of $-330° + 360° = 30°$.
$\sin(-330°) = \sin 30° = 1/2$, $\cos(-330°) = \cos 30° = \sqrt{3}/2$, $\tan(-330°) = \tan 30° = \sqrt{3}/3$.

4. Using the table of natural functions, find:

(a) $\sin 125°14' = \sin(180° - 54°46') = \sin 54°46' = 0.8168$
(b) $\cos 169°40' = \cos(180° - 10°20') = -\cos 10°20' = -0.9838$
(c) $\tan 200°23' = \tan(180° + 20°23') = \tan 20°23' = 0.3716$
(d) $\cot 250°44' = \cot(180° + 70°44') = \cot 70°44' = 0.3495$
(e) $\cos 313°18' = \cos(360° - 46°42') = \cos 46°42' = 0.6858$
(f) $\sin 341°52' = \sin(360° - 18°8') = -\sin 18°8' = -0.3112$

5. If $\tan 25° = a$, find:

(a) $\dfrac{\tan 155° - \tan 115°}{1 + \tan 155° \tan 115°} = \dfrac{-\tan 25° - (-\cot 25°)}{1 + (-\tan 25°)(-\cot 25°)} = \dfrac{-a + 1/a}{1 + a(1/a)} = \dfrac{-a^2 + 1}{a + a} = \dfrac{1 - a^2}{2a}.$

(b) $\dfrac{\tan 205° - \tan 115°}{\tan 245° - \tan 335°} = \dfrac{\tan 25° - (-\cot 25°)}{\cot 25° + (-\tan 25°)} = \dfrac{a + 1/a}{1/a - a} = \dfrac{a^2 + 1}{1 - a^2}.$

6. If $A + B + C = 180°$, then

(a) $\sin(B + C) = \sin(180° - A) = \sin A$.
(b) $\sin \frac{1}{2}(B + C) = \sin \frac{1}{2}(180° - A) = \sin(90° - \frac{1}{2}A) = \cos \frac{1}{2}A$.

7. Show that $\sin \theta$ and $\tan \frac{1}{2}\theta$ have the same sign.

(a) Suppose $\theta = n \cdot 180°$. If n is even (including zero), say $2m$, then $\sin(2m \cdot 180°) = \tan(m \cdot 180°) = 0$. The case when n is odd is excluded since then $\tan \frac{1}{2}\theta$ is not defined.

(b) Suppose $\theta = n \cdot 180° + \phi$, where $0 < \phi < 180°$. If n is even, including zero, θ is in quadrant I or quadrant II and $\sin \theta$ is positive while $\frac{1}{2}\theta$ is in quadrant I or quadrant III and $\tan \frac{1}{2}\theta$ is positive. If n is odd, θ is in quadrant III or IV and $\sin \theta$ is negative while $\frac{1}{2}\theta$ is in quadrant II or IV and $\tan \frac{1}{2}\theta$ is negative.

8. Find all positive values of θ less than 360° for which $\sin \theta = -\frac{1}{2}$.

There will be two angles (see Chapter 31), one in the third quadrant and one in the fourth quadrant. The related angle of each has its sine equal to $+\frac{1}{2}$ and is 30°. Thus the required angles are $\theta = 180° + 30° = 210°$ and $\theta = 360° - 30° = 330°$.

Note. To obtain *all* values of θ for which $\sin \theta = -\frac{1}{2}$, add $n \cdot 360°$ to each of the above solutions; thus $\theta = 210° + n \cdot 360°$ and $\theta = 330° + n \cdot 360°$, where n is any integer.

9. Find all positive values of θ less than 360° for which $\cos \theta = 0.9063$.

There are two solutions, $\theta = 25°$ in quadrant I and $\theta = 360° - 25° = 335°$ in quadrant IV.

10. Find all positive values of $\frac{1}{4}\theta$ less than 360°, given $\sin \theta = 0.6428$.

The two positive angles less than 360° for which $\sin \theta = 0.6428$ are $\theta = 40°$ and $\theta = 180° - 40° = 140°$. But if $\frac{1}{4}\theta$ is to include all values less than 360°, θ must include all values less than $4 \cdot 360° = 1440°$. Hence, for θ we take the two angles above and all coterminal angles less than 1440°, that is,

$$\theta = 40°, 400°, 760°, 1120°; \ 140°, 500°, 860°, 1220° \quad \text{and}$$

$$\tfrac{1}{4}\theta = 10°, 100°, 190°, 280°; \ 35°, 125°, 215°, 305°.$$

11. Find all positive values of θ less than 360° which satisfy $\sin 2\theta = \cos \frac{1}{2}\theta$.

Since $\cos \frac{1}{2}\theta = \sin(90° - \frac{1}{2}\theta) = \sin 2\theta$, $2\theta = 90° - \frac{1}{2}\theta, 450° - \frac{1}{2}\theta, 810° - \frac{1}{2}\theta, 1170° - \frac{1}{2}\theta, \ldots$

Then $\frac{5}{2}\theta = 90°, 450°, 810°, 1170°, \ldots$ and $\theta = 36°, 180°, 324°, 468°, \ldots$

Since $\cos\frac{1}{2}\theta = \sin(90^\circ + \frac{1}{2}\theta) = \sin 2\theta$, $2\theta = 90^\circ + \frac{1}{2}\theta,\ 450^\circ + \frac{1}{2}\theta,\ 810^\circ + \frac{1}{2}\theta, \ldots.$

Then $\frac{3}{2}\theta = 90^\circ,\ 450^\circ,\ 810^\circ, \ldots.$ and $\theta = 60^\circ,\ 300^\circ,\ 540^\circ, \ldots.$

The required solutions are: $36^\circ,\ 180^\circ,\ 324^\circ;\ 60^\circ,\ 300^\circ$.

SUPPLEMENTARY PROBLEMS

12. Express each of the following in terms of functions of a positive acute angle.

(*a*) $\sin 145^\circ$ (*b*) $\cos 215^\circ$ (*c*) $\tan 440^\circ$ (*d*) $\cot 155^\circ$ (*e*) $\sec 325^\circ$ (*f*) $\csc 190^\circ$ (*g*) $\sin(-200^\circ)$ (*h*) $\cos(-760^\circ)$ (*i*) $\tan(-1385^\circ)$ (*j*) $\cot 610^\circ$ (*k*) $\sec 455^\circ$ (*l*) $\csc 825^\circ$

Ans. (*a*) $\sin 35^\circ$ or $\cos 55^\circ$
(*b*) $-\cos 35^\circ$ or $-\sin 55^\circ$
(*c*) $\tan 80^\circ$ or $\cot 10^\circ$
(*d*) $-\cot 25^\circ$ or $-\tan 65^\circ$
(*e*) $\sec 35^\circ$ or $\csc 55^\circ$
(*f*) $-\csc 10^\circ$ or $-\sec 80^\circ$
(*g*) $\sin 20^\circ$ or $\cos 70^\circ$
(*h*) $\cos 40^\circ$ or $\sin 50^\circ$
(*i*) $\tan 55^\circ$ or $\cot 35^\circ$
(*j*) $\cot 70^\circ$ or $\tan 20^\circ$
(*k*) $-\sec 85^\circ$ or $-\csc 5^\circ$
(*l*) $\csc 75^\circ$ or $\sec 15^\circ$

13. Find the exact values of the sine, cosine, and tangent of:
(*a*) 150°, (*b*) 225°, (*c*) 300°, (*d*) -120°, (*e*) -210°, (*f*) -315°.

Ans. (*a*) $1/2,\ -\sqrt{3}/2,\ -1/\sqrt{3}$
(*b*) $-\sqrt{2}/2,\ -\sqrt{2}/2,\ 1$
(*c*) $-\sqrt{3}/2,\ 1/2,\ -\sqrt{3}$
(*d*) $-\sqrt{3}/2,\ -1/2,\ \sqrt{3}$
(*e*) $1/2,\ -\sqrt{3}/2,\ -1/\sqrt{3}$
(*f*) $\sqrt{2}/2,\ \sqrt{2}/2,\ 1$

14. Using appropriate tables, find:

(*a*) $\sin 155^\circ 13' = 0.4192$
(*b*) $\cos 104^\circ 38' = -0.2526$
(*c*) $\tan 305^\circ 24' = -1.4071$
(*d*) $\sin 114^\circ 18' = 0.9114$
(*e*) $\cos 166^\circ 51' = -0.9738$
(*f*) $\log\sin 129^\circ 44.8' = 9.88586-10$
(*g*) $\log\sin 110^\circ 32.7' = 9.97146-10$
(*h*) $\log\sin 162^\circ 35.6' = 9.47589-10$
(*i*) $\log\sin 138^\circ 30.5' = 9.82119-10$
(*j*) $\log\sin 174^\circ 22.7' = 8.99104-10$

15. Find all angles, $0 \leqq \theta < 360^\circ$, for which:
(*a*) $\sin\theta = \sqrt{2}/2$ (*b*) $\cos\theta = -1$ (*c*) $\sin\theta = -0.6180$ (*d*) $\cos\theta = 0.5125$ (*e*) $\tan\theta = -1.5301$

Ans. (*a*) $45^\circ,\ 135^\circ$
(*b*) 180°
(*c*) $218^\circ 10',\ 321^\circ 50'$
(*d*) $59^\circ 10',\ 300^\circ 50'$
(*e*) $123^\circ 10',\ 303^\circ 10'$

16. When θ is a second quadrant angle for which $\tan\theta = -2/3$, show that

(*a*) $$\frac{\sin(90^\circ - \theta) - \cos(180^\circ - \theta)}{\tan(270^\circ + \theta) + \cot(360^\circ - \theta)} = -\frac{2}{\sqrt{13}}$$

(*b*) $$\frac{\tan(90^\circ + \theta) + \cos(180^\circ + \theta)}{\sin(270^\circ - \theta) - \cot(-\theta)} = \frac{2 + \sqrt{13}}{2 - \sqrt{13}}$$

CHAPTER 38

Variations and Graphs of the Trigonometric Functions

LINE REPRESENTATIONS OF THE TRIGONOMETRIC FUNCTIONS. Let θ be any given angle in standard position. (See the figures below for θ in each of the quadrants.) With the vertex O as center describe a circle of radius one unit cutting the initial side OX of θ at A, the positive y-axis at B, and the terminal side of θ at P. Draw MP perpendicular to OX; draw also the tangents to the circle at A and B meeting the terminal side of θ or its extension through O in the points Q and R respectively.

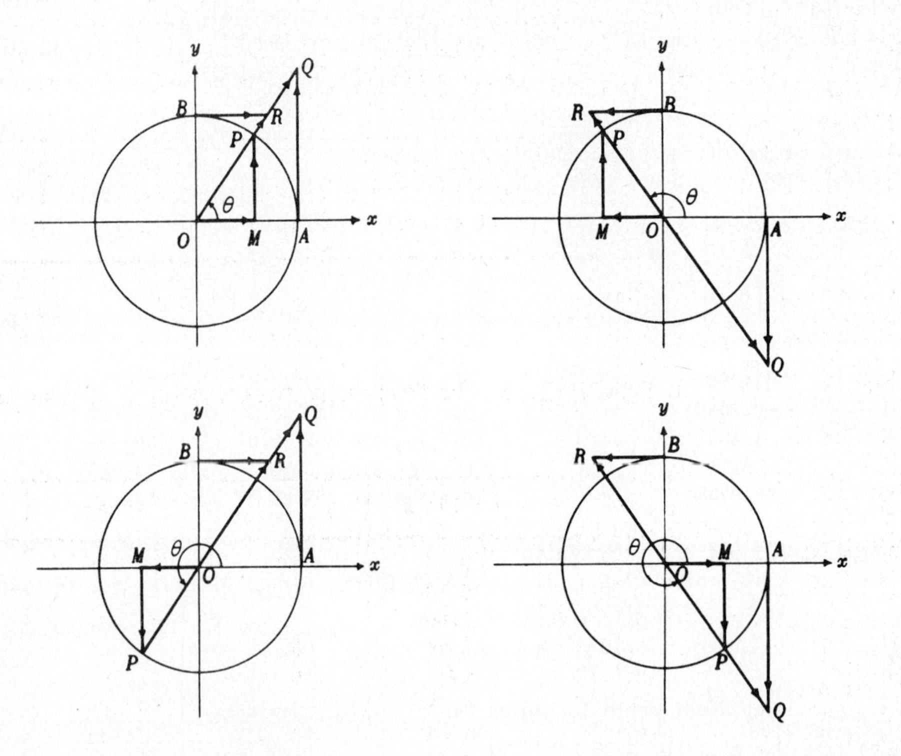

In each of the figures, the right triangles OMP, OAQ, and OBR are similar, and

$$\sin\theta = MP/OP = MP \qquad \cot\theta = OM/MP = BR/OB = BR$$
$$\cos\theta = OM/OP = OM \qquad \sec\theta = OP/OM = OQ/OA = OQ$$
$$\tan\theta = MP/OM = AQ/OA = AQ \qquad \csc\theta = OP/MP = OR/OB = OR.$$

The segments MP, OM, AQ, etc., are directed line segments, the magnitude of a function being given by the length of the corresponding segment and the sign being given by the indicated direction. The directed segments OQ and OR are to be considered positive when measured on the terminal side of the angle and negative when measured on the terminal side extended.

VARIATIONS OF THE TRIGONOMETRIC FUNCTIONS. Let P move counterclockwise about the unit circle, starting at A, so that $\theta = \angle XOP$ varies continuously from 0° to 360°. Using the figures above, it is seen that ($I.$ = increases, $D.$ = decreases):

As θ increases from	0° to 90°	90° to 180°	180° to 270°	270° to 360°
$\sin\theta$	$I.$ from 0 to 1	$D.$ from 1 to 0	$D.$ from 0 to −1	$I.$ from −1 to 0
$\cos\theta$	$D.$ from 1 to 0	$D.$ from 0 to −1	$I.$ from −1 to 0	$I.$ from 0 to 1
$\tan\theta$	$I.$ from 0 without limit (0 to $+\infty$)	$I.$ from large negative values to 0. ($-\infty$ to 0)	$I.$ from 0 without limit (0 to $+\infty$)	$I.$ from large negative values to 0. ($-\infty$ to 0)
$\cot\theta$	$D.$ from large positive values to 0. ($+\infty$ to 0)	$D.$ from 0 without limit (0 to $-\infty$)	$D.$ from large positive values to 0. ($+\infty$ to 0)	$D.$ from 0 without limit (0 to $-\infty$)
$\sec\theta$	$I.$ from 1 without limit (1 to $+\infty$)	$I.$ from large negative values to −1. ($-\infty$ to −1)	$D.$ from −1 without limit (-1 to $-\infty$)	$D.$ from large positive values to 1. ($+\infty$ to 1)
$\csc\theta$	$D.$ from large positive values to 1. ($+\infty$ to 1)	$I.$ from 1 without limit (1 to $+\infty$)	$I.$ from large negative values to −1. ($-\infty$ to −1)	$D.$ from −1 without limit (-1 to $-\infty$)

GRAPHS OF THE TRIGONOMETRIC FUNCTIONS. In the following table, values of the angle x are given in radians.

x	$y = \sin x$	$y = \cos x$	$y = \tan x$	$y = \cot x$	$y = \sec x$	$y = \csc x$
0	0	1.00	0	$\pm\infty$	1.00	$\pm\infty$
$\pi/6$	0.50	0.87	0.58	1.73	1.15	2.00
$\pi/4$	0.71	0.71	1.00	1.00	1.41	1.41
$\pi/3$	0.87	0.50	1.73	0.58	2.00	1.15
$\pi/2$	1.00	0	$\pm\infty$	0	$\pm\infty$	1.00
$2\pi/3$	0.87	−0.50	−1.73	−0.58	−2.00	1.15
$3\pi/4$	0.71	−0.71	−1.00	−1.00	−1.41	1.41
$5\pi/6$	0.50	−0.87	−0.58	−1.73	−1.15	2.00
π	0	−1.00	0	$\pm\infty$	−1.00	$\pm\infty$
$7\pi/6$	−0.50	−0.87	0.58	1.73	−1.15	−2.00
$5\pi/4$	−0.71	−0.71	1.00	1.00	−1.41	−1.41
$4\pi/3$	−0.87	−0.50	1.73	0.58	−2.00	−1.15
$3\pi/2$	−1.00	0	$\pm\infty$	0	$\pm\infty$	−1.00
$5\pi/3$	−0.87	0.50	−1.73	−0.58	2.00	−1.15
$7\pi/4$	−0.71	0.71	−1.00	−1.00	1.41	−1.41
$11\pi/6$	−0.50	0.87	−0.58	−1.73	1.15	−2.00
2π	0	1.00	0	$\pm\infty$	1.00	$\pm\infty$

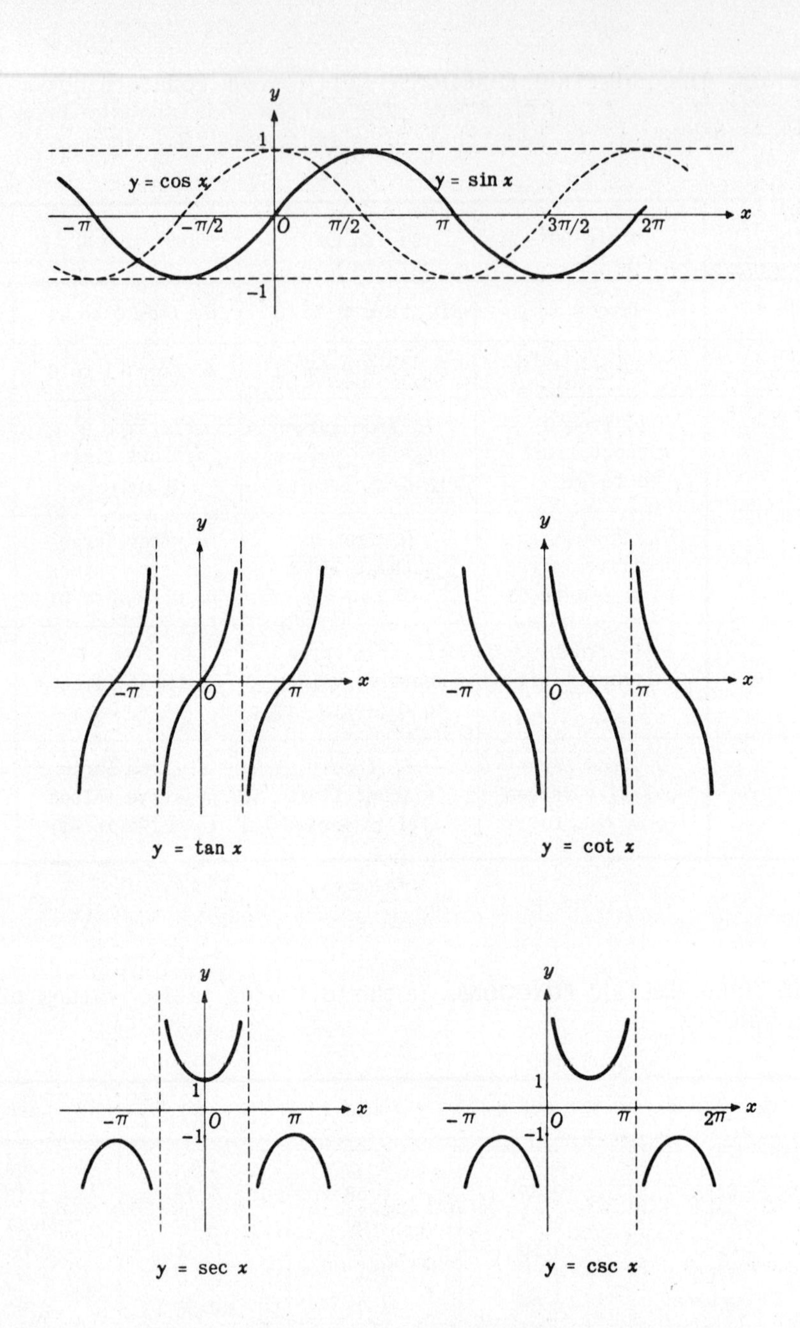

Note 1. Since $\sin(\frac{1}{2}\pi + x) = \cos x$, the graph of $y = \cos x$ may be obtained most easily by shifting the graph of $y = \sin x$ a distance $\frac{1}{2}\pi$ to the left.

Note 2. Since $\csc(\frac{1}{2}\pi + x) = \sec x$, the graph of $y = \csc x$ may be obtained by shifting the graph of $y = \sec x$ a distance $\frac{1}{2}\pi$ to the right.

PERIODIC FUNCTIONS. Any function of a variable x, $f(x)$, which repeats its values in definite cycles, is called *periodic*. The smallest range of values of x which corresponds to a complete cycle of values of the function is called the period of the function. It is evident from the graphs of the trigonometric functions that the sine, cosine, secant, and cosecant are of period 2π while the tangent and cotangent are of period π.

THE GENERAL SINE CURVE. The *amplitude* (maximum ordinate) and period (wave length) of $y = \sin x$ are respectively 1 and 2π. For a given value of x, the value of $y = a \sin x$, $a > 0$, is a times the value of $y = \sin x$. Thus, the amplitude of $y = a \sin x$ is a and the period is 2π. Since when $bx = 2\pi$, $x = 2\pi/b$, the amplitude of $y = \sin bx$, $b > 0$, is 1 and the period is $2\pi/b$.

The general sine curve (sinusoid) of equation

$$y = a \sin bx, \quad a > 0, \ b > 0,$$

has amplitude a and period $2\pi/b$. Thus the graph of $y = 3 \sin 2x$ has amplitude 3 and period $2\pi/2 = \pi$. Fig. (*a*) below exhibits the graphs of $y = \sin x$ and $y = 3 \sin 2x$ on the same axes.

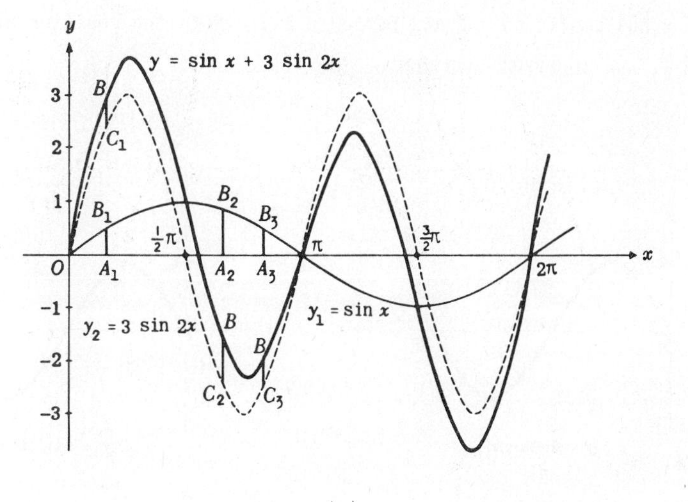

(*a*)

COMPOSITION OF SINE CURVES. More complicated forms of wave motions are obtained by combining two or more sine curves. The method of adding corresponding ordinates is illustrated in the following example.

EXAMPLE. Construct the graph of $y = \sin x + 3 \sin 2x$. See Fig. (*a*) above.

First the graphs of $y_1 = \sin x$ and $y_2 = 3 \sin 2x$ are constructed on the same axes. Then, corresponding to a given value $x = OA_1$, the ordinate A_1B of $y = \sin x + 3 \sin 2x$ is the *algebraic* sum of the ordinates A_1B_1 of $y_1 = \sin x$ and A_1C_1 of $y_2 = 3 \sin 2x$.

Also, $A_2B = A_2B_2 + A_2C_2$, $A_3B = A_3B_3 + A_3C_3$, etc.

SOLVED PROBLEMS

1. Sketch the graphs of the following for one wave length.

(a) $y = 4 \sin x$ (c) $y = 3 \sin \frac{1}{2}x$ (e) $y = 3 \cos \frac{1}{2}x = 3 \sin (\frac{1}{2}x + \frac{1}{2}\pi)$

(b) $y = \sin 3x$ (d) $y = 2 \cos x = 2 \sin (x + \frac{1}{2}\pi)$

In each case we use the same curve and then put in the y-axis and choose the units on each axis to satisfy the requirements of amplitude and period of each curve.

(a) $y = 4 \sin x$ has amplitude = 4 and period = 2π.

(b) $y = \sin 3x$ has amplitude = 1 and period = $2\pi/3$.

(c) $y = 3 \sin \frac{1}{2}x$ has amplitude = 3 and period = $2\pi/\frac{1}{2} = 4\pi$.

(d) $y = 2 \cos x$ has amplitude = 2 and period = 2π. Note the position of the y-axis.

(e) $y = 3 \cos \frac{1}{2}x$ has amplitude = 3 and period = 4π.

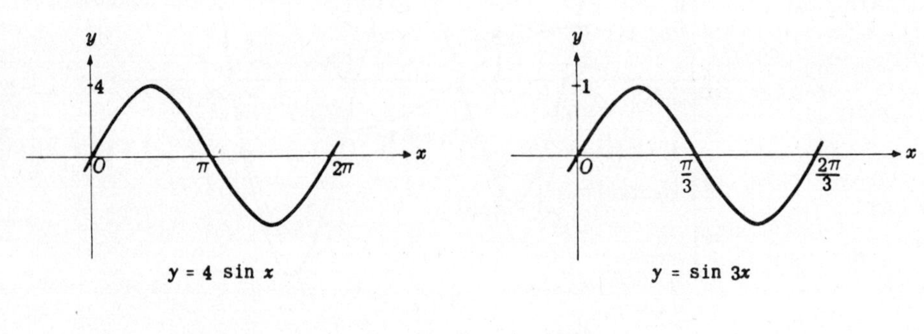

$y = 4 \sin x$ $y = \sin 3x$

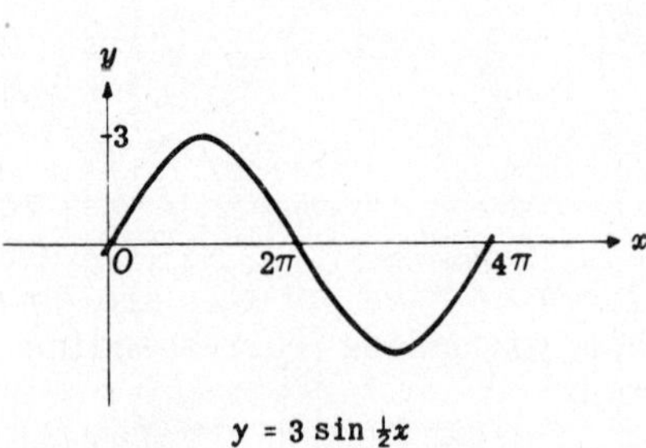

$y = 3 \sin \frac{1}{2}x$

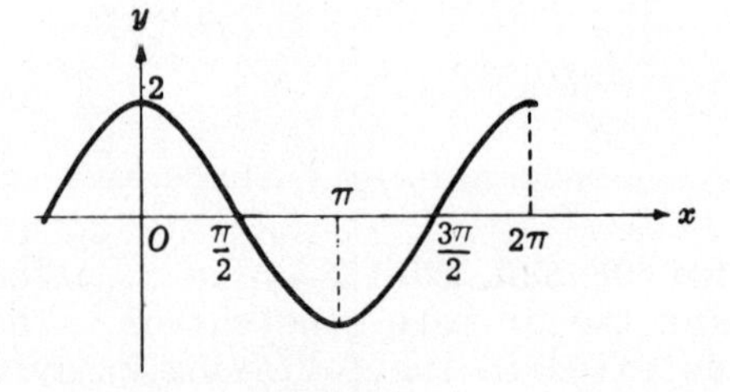

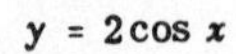

$y = 2 \cos x$

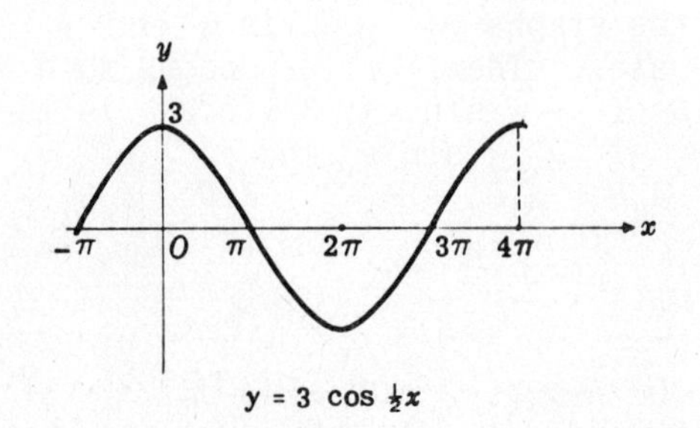

$y = 3 \cos \frac{1}{2}x$

2. Construct the graph of each of the following.

(a) $y = \sin x + \cos x$ (c) $y = \sin 2x - \cos 3x$

(b) $y = \sin 2x + \cos 3x$ (d) $y = 3 \sin 2x + 2 \cos 3x$

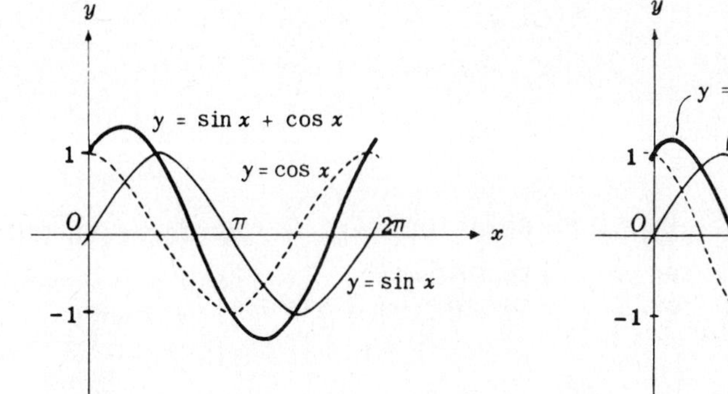

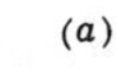

(a)

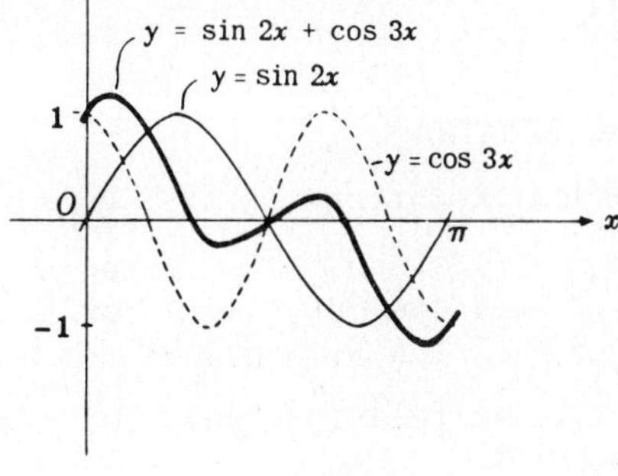

(b)

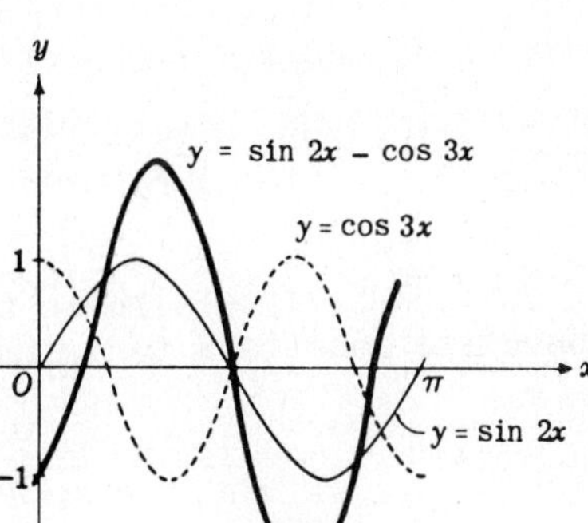

(c)

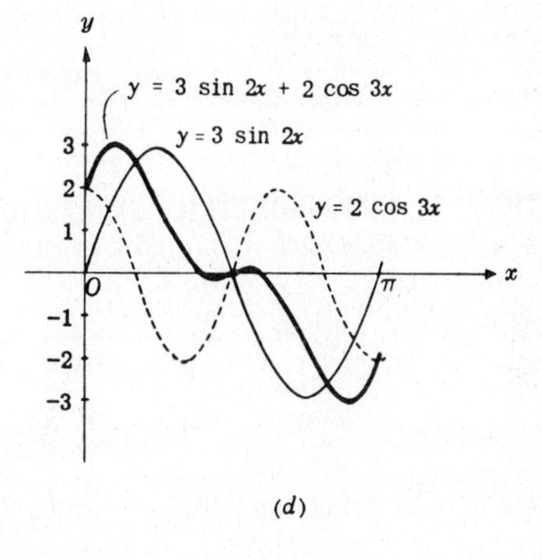

(d)

SUPPLEMENTARY PROBLEMS

3. Sketch the graph of each of the following for one wave length:
(a) $y = 3 \sin x$ (b) $y = \sin 2x$ (c) $y = 4 \sin x/2$ (d) $y = 4 \cos x$ (e) $y = 2 \cos x/3$

4. Construct the graph of each of the following for one wave length.

(a) $y = \sin x + 2 \cos x$ (d) $y = \sin 2x + \sin 3x$

(b) $y = \sin 3x + \cos 2x$ (e) $y = \sin 3x - \cos 2x$

(c) $y = \sin x + \sin 2x$ (f) $y = 2 \sin 3x + 3 \cos 2x$

CHAPTER 39

Fundamental Relations and Identities

FUNDAMENTAL RELATIONS.

Reciprocal Relations	Quotient Relations	Pythagorean Relations
$\csc\theta = 1/\sin\theta$	$\tan\theta = \sin\theta/\cos\theta$	$\sin^2\theta + \cos^2\theta = 1$
$\sec\theta = 1/\cos\theta$	$\cot\theta = \cos\theta/\sin\theta$	$1 + \tan^2\theta = \sec^2\theta$
$\cot\theta = 1/\tan\theta$		$1 + \cot^2\theta = \csc^2\theta$

The above relations hold for every value of θ for which the functions involved are defined.

Thus, $\sin^2\theta + \cos^2\theta = 1$ holds for every value of θ while $\tan\theta = \sin\theta/\cos\theta$ holds for all values of θ for which $\tan\theta$ is defined, i.e., for all $\theta \neq n\cdot 90°$ where n is odd. Note that for the excluded values of θ, $\cos\theta = 0$ and $\sin\theta \neq 0$.

For proofs of the quotient and Pythagorean relations, see Problems 1, 2. The reciprocal relations were treated in Chapter 31. (See also Problems 3-6.)

SIMPLIFICATION OF TRIGONOMETRIC EXPRESSIONS. It is frequently desirable to transform or reduce a given expression involving trigonometric functions to a simpler form.

Example 1. (*a*) Using $\csc\theta = \dfrac{1}{\sin\theta}$, $\quad \cos\theta\csc\theta = \cos\theta\dfrac{1}{\sin\theta} = \dfrac{\cos\theta}{\sin\theta} = \cot\theta$.

(*b*) Using $\tan\theta = \dfrac{\sin\theta}{\cos\theta}$, $\quad \cos\theta\tan\theta = \cos\theta\dfrac{\sin\theta}{\cos\theta} = \sin\theta$.

Example 2. Using the relation $\sin^2\theta + \cos^2\theta = 1$,

(*a*) $\sin^3\theta + \sin\theta\cos^2\theta = (\sin^2\theta + \cos^2\theta)\sin\theta = (1)\sin\theta = \sin\theta$.

(*b*) $\dfrac{\cos^2\theta}{1 - \sin\theta} = \dfrac{1 - \sin^2\theta}{1 - \sin\theta} = \dfrac{(1 - \sin\theta)(1 + \sin\theta)}{1 - \sin\theta} = 1 + \sin\theta$.

Note. The relation $\sin^2\theta + \cos^2\theta = 1$ may be written as $\sin^2\theta = 1 - \cos^2\theta$ and as $\cos^2\theta = 1 - \sin^2\theta$. Each form is equally useful.

(See Problems 7-9.)

TRIGONOMETRIC IDENTITIES. A relation involving the trigonometric functions which is valid for all values of the angle for which the functions are defined is called a trigonometric identity. The eight fundamental relations above are trigonometric identities; so also are

$$\cos\theta\csc\theta = \cot\theta \qquad \text{and} \qquad \cos\theta\tan\theta = \sin\theta$$

of Example 1 above.

A trigonometric identity is verified by transforming one member (your choice) into the other. In general, one begins with the more complicated side.

Success in verifying identities requires:
(a) complete familiarity with the fundamental relations,
(b) complete familiarity with the processes of factoring, adding fractions, etc.
(c) practice.

(See Problems 10-17.)

SOLVED PROBLEMS

1. Prove the quotient relations: $\tan\theta = \dfrac{\sin\theta}{\cos\theta}$, $\cot\theta = \dfrac{\cos\theta}{\sin\theta}$.

For any angle θ, $\sin\theta = y/r$, $\cos\theta = x/r$, $\tan\theta = y/x$, and $\cot\theta = x/y$, where $P(x,y)$ is any point on the terminal side of θ at a distance r from the origin.

Then $\tan\theta = \dfrac{y}{x} = \dfrac{y/r}{x/r} = \dfrac{\sin\theta}{\cos\theta}$ and $\cot\theta = \dfrac{x}{y} = \dfrac{x/r}{y/r} = \dfrac{\cos\theta}{\sin\theta}$. (Also, $\cot\theta = \dfrac{1}{\tan\theta} = \dfrac{\cos\theta}{\sin\theta}$.)

2. Prove the Pythagorean relations: (a) $\sin^2\theta + \cos^2\theta = 1$, (b) $1 + \tan^2\theta = \sec^2\theta$, (c) $1 + \cot^2\theta = \csc^2\theta$.

For $P(x,y)$ defined as in Problem 1, we have $A)$ $x^2 + y^2 = r^2$.

(a) Dividing $A)$ by r^2, $(x/r)^2 + (y/r)^2 = 1$ and $\sin^2\theta + \cos^2\theta = 1$.

(b) Dividing $A)$ by x^2, $1 + (y/x)^2 = (r/x)^2$ and $1 + \tan^2\theta = \sec^2\theta$.

Also, dividing $\sin^2\theta + \cos^2\theta = 1$ by $\cos^2\theta$, $\left(\dfrac{\sin\theta}{\cos\theta}\right)^2 + 1 = \left(\dfrac{1}{\cos\theta}\right)^2$ or $\tan^2\theta + 1 = \sec^2\theta$.

(c) Dividing $A)$ by y^2, $(x/y)^2 + 1 = (r/y)^2$ and $\cot^2\theta + 1 = \csc^2\theta$.

Also, dividing $\sin^2\theta + \cos^2\theta = 1$ by $\sin^2\theta$, $1 + \left(\dfrac{\cos\theta}{\sin\theta}\right)^2 = \left(\dfrac{1}{\sin\theta}\right)^2$ or $1 + \cot^2\theta = \csc^2\theta$.

3. Express each of the other functions of θ in terms of $\sin\theta$.

$$\cos^2\theta = 1 - \sin^2\theta \quad \text{and} \quad \cos\theta = \pm\sqrt{1 - \sin^2\theta},$$

$$\tan\theta = \frac{\sin\theta}{\cos\theta} = \frac{\sin\theta}{\pm\sqrt{1 - \sin^2\theta}}, \qquad \cot\theta = \frac{1}{\tan\theta} = \frac{\pm\sqrt{1 - \sin^2\theta}}{\sin\theta},$$

$$\sec\theta = \frac{1}{\cos\theta} = \frac{1}{\pm\sqrt{1 - \sin^2\theta}}, \qquad \csc\theta = \frac{1}{\sin\theta}.$$

Note that $\cos\theta = \pm\sqrt{1 - \sin^2\theta}$. Writing $\cos\theta = \sqrt{1 - \sin^2\theta}$ limits angle θ to those quadrants (first and fourth) in which the cosine is positive.

4. Express each of the other functions of θ in terms of $\tan\theta$.

$$\sec^2\theta = 1 + \tan^2\theta \quad \text{and} \quad \sec\theta = \pm\sqrt{1 + \tan^2\theta}, \qquad \cos\theta = \frac{1}{\sec\theta} = \frac{1}{\pm\sqrt{1 + \tan^2\theta}},$$

$$\frac{\sin\theta}{\cos\theta} = \tan\theta \quad \text{and} \quad \sin\theta = \tan\theta\cos\theta = \tan\theta\,\frac{1}{\pm\sqrt{1 + \tan^2\theta}} = \frac{\tan\theta}{\pm\sqrt{1 + \tan^2\theta}},$$

$$\csc\theta = \frac{1}{\sin\theta} = \frac{\pm\sqrt{1 + \tan^2\theta}}{\tan\theta}, \qquad \cot\theta = \frac{1}{\tan\theta}.$$

5. Using the fundamental relations, find the values of the functions of θ, given $\sin\theta = 3/5$.

From $\cos^2\theta = 1 - \sin^2\theta$, $\cos\theta = \pm\sqrt{1-\sin^2\theta} = \pm\sqrt{1-(3/5)^2} = \pm\sqrt{16/25} = \pm\, 4/5$.

Now $\sin\theta$ and $\cos\theta$ are both positive when θ is a first quadrant angle while $\sin\theta = +$ and $\cos\theta = -$ when θ is a second quadrant angle. Thus,

first quadrant

$\sin\theta = 3/5$ $\quad$ $\cot\theta = 4/3$

$\cos\theta = 4/5$ $\quad$ $\sec\theta = 5/4$

$\tan\theta = \dfrac{3/5}{4/5} = 3/4$ $\quad$ $\csc\theta = 5/3$

second quadrant

$\sin\theta = 3/5$ $\quad$ $\cot\theta = -4/3$

$\cos\theta = -4/5$ $\quad$ $\sec\theta = -5/4$

$\tan\theta = -3/4$ $\quad$ $\csc\theta = 5/3.$

6. Using the fundamental relations, find the values of the functions of θ, given $\tan\theta = -5/12$.

Since $\tan\theta = -$, θ is either a second or fourth quadrant angle.

second quadrant

$\tan\theta = -5/12$

$\cot\theta = 1/\tan\theta = -12/5$

$\sec\theta = -\sqrt{1+\tan^2\theta} = -13/12$

$\cos\theta = 1/\sec\theta = -12/13$

$\csc\theta = \sqrt{1+\cot^2\theta} = 13/5$

$\sin\theta = 1/\csc\theta = 5/13$

fourth quadrant

$\tan\theta = -5/12$

$\cot\theta = -12/5$

$\sec\theta = 13/12$

$\cos\theta = 12/13$

$\csc\theta = -13/5$

$\sin\theta = -5/13$

7. Perform the indicated operations.

(*a*) $(\sin\theta - \cos\theta)(\sin\theta + \cos\theta) = \sin^2\theta - \cos^2\theta$

(*b*) $(\sin A + \cos A)^2 = \sin^2 A + 2\sin A\cos A + \cos^2 A$

(*c*) $(\sin x + \cos y)(\sin y - \cos x) = \sin x\sin y - \sin x\cos x + \sin y\cos y - \cos x\cos y$

(*d*) $(\tan^2 A - \cot A)^2 = \tan^4 A - 2\tan^2 A\cot A + \cot^2 A$

(*e*) $1 + \dfrac{\cos\theta}{\sin\theta} = \dfrac{\sin\theta + \cos\theta}{\sin\theta}$

(*f*) $1 - \dfrac{\sin\theta}{\cos\theta} + \dfrac{2}{\cos^2\theta} = \dfrac{\cos^2\theta - \sin\theta\cos\theta + 2}{\cos^2\theta}$

8. Factor.

(*a*) $\sin^2\theta - \sin\theta\cos\theta = \sin\theta\,(\sin\theta - \cos\theta)$

(*b*) $\sin^2\theta + \sin^2\theta\cos^2\theta = \sin^2\theta\,(1 + \cos^2\theta)$

(*c*) $\sin^2\theta + \sin\theta\sec\theta - 6\sec^2\theta = (\sin\theta + 3\sec\theta)(\sin\theta - 2\sec\theta)$

(*d*) $\sin^3\theta\cos^2\theta - \sin^2\theta\cos^3\theta + \sin\theta\cos^2\theta = \sin\theta\cos^2\theta\,(\sin^2\theta - \sin\theta\cos\theta + 1)$

(*e*) $\sin^4\theta - \cos^4\theta = (\sin^2\theta + \cos^2\theta)(\sin^2\theta - \cos^2\theta) = (\sin^2\theta + \cos^2\theta)(\sin\theta - \cos\theta)(\sin\theta + \cos\theta)$

9. Simplify each of the following.

(*a*) $\sec\theta - \sec\theta\sin^2\theta = \sec\theta\,(1 - \sin^2\theta) = \sec\theta\cos^2\theta = \dfrac{1}{\cos\theta}\cos^2\theta = \cos\theta$

(b) $\sin\theta \sec\theta \cot\theta = \sin\theta \dfrac{1}{\cos\theta} \dfrac{\cos\theta}{\sin\theta} = \dfrac{\sin\theta\cos\theta}{\cos\theta\sin\theta} = 1$

(c) $\sin^2\theta\,(1+\cot^2\theta) = \sin^2\theta\csc^2\theta = \sin^2\theta\dfrac{1}{\sin^2\theta} = 1$

(d) $\sin^2\theta\sec^2\theta - \sec^2\theta = (\sin^2\theta - 1)\sec^2\theta = -\cos^2\theta\sec^2\theta = -\cos^2\theta\dfrac{1}{\cos^2\theta} = -1$

(e) $(\sin\theta+\cos\theta)^2 + (\sin\theta-\cos\theta)^2 = \sin^2\theta + 2\sin\theta\cos\theta + \cos^2\theta + \sin^2\theta$
$- 2\sin\theta\cos\theta + \cos^2\theta = 2(\sin^2\theta+\cos^2\theta) = 2$

(f) $\tan^2\theta\cos^2\theta + \cot^2\theta\sin^2\theta = \dfrac{\sin^2\theta}{\cos^2\theta}\cos^2\theta + \dfrac{\cos^2\theta}{\sin^2\theta}\sin^2\theta = \sin^2\theta + \cos^2\theta = 1$

(g) $\tan\theta + \dfrac{\cos\theta}{1+\sin\theta} = \dfrac{\sin\theta}{\cos\theta} + \dfrac{\cos\theta}{1+\sin\theta} = \dfrac{\sin\theta\,(1+\sin\theta) + \cos^2\theta}{\cos\theta\,(1+\sin\theta)}$

$= \dfrac{\sin\theta + \sin^2\theta + \cos^2\theta}{\cos\theta\,(1+\sin\theta)} = \dfrac{\sin\theta + 1}{\cos\theta\,(1+\sin\theta)} = \dfrac{1}{\cos\theta} = \sec\theta$

Verify the following identities.

10. $\sec^2\theta\csc^2\theta = \sec^2\theta + \csc^2\theta$

$\sec^2\theta + \csc^2\theta = \dfrac{1}{\cos^2\theta} + \dfrac{1}{\sin^2\theta} = \dfrac{\sin^2\theta+\cos^2\theta}{\sin^2\theta\cos^2\theta} = \dfrac{1}{\sin^2\theta\cos^2\theta} = \dfrac{1}{\sin^2\theta}\dfrac{1}{\cos^2\theta} = \csc^2\theta\sec^2\theta$

11. $\sec^4\theta - \sec^2\theta = \tan^4\theta + \tan^2\theta$

$\tan^4\theta + \tan^2\theta = \tan^2\theta\,(\tan^2\theta+1) = \tan^2\theta\sec^2\theta = (\sec^2\theta - 1)\sec^2\theta = \sec^4\theta - \sec^2\theta$ or

$\sec^4\theta - \sec^2\theta = \sec^2\theta\,(\sec^2\theta - 1) = \sec^2\theta\tan^2\theta = (1+\tan^2\theta)\tan^2\theta = \tan^2\theta + \tan^4\theta$

12. $2\csc x = \dfrac{\sin x}{1+\cos x} + \dfrac{1+\cos x}{\sin x}$

$\dfrac{\sin x}{1+\cos x} + \dfrac{1+\cos x}{\sin x} = \dfrac{\sin^2 x + (1+\cos x)^2}{\sin x\,(1+\cos x)} = \dfrac{\sin^2 x + 1 + 2\cos x + \cos^2 x}{\sin x\,(1+\cos x)}$

$= \dfrac{2+2\cos x}{\sin x\,(1+\cos x)} = \dfrac{2(1+\cos x)}{\sin x\,(1+\cos x)} = \dfrac{2}{\sin x} = 2\csc x$

13. $\dfrac{1-\sin x}{\cos x} = \dfrac{\cos x}{1+\sin x}$

$\dfrac{\cos x}{1+\sin x} = \dfrac{\cos^2 x}{\cos x\,(1+\sin x)} = \dfrac{1-\sin^2 x}{\cos x\,(1+\sin x)} = \dfrac{(1-\sin x)(1+\sin x)}{\cos x\,(1+\sin x)} = \dfrac{1-\sin x}{\cos x}$

14. $\dfrac{\sec A - \csc A}{\sec A + \csc A} = \dfrac{\tan A - 1}{\tan A + 1}$

$\dfrac{\sec A - \csc A}{\sec A + \csc A} = \dfrac{\dfrac{1}{\cos A} - \dfrac{1}{\sin A}}{\dfrac{1}{\cos A} + \dfrac{1}{\sin A}} = \dfrac{\dfrac{\sin A}{\cos A} - 1}{\dfrac{\sin A}{\cos A} + 1} = \dfrac{\tan A - 1}{\tan A + 1}$

15. $\dfrac{\tan x - \sin x}{\sin^3 x} = \dfrac{\sec x}{1 + \cos x}$

$$\frac{\tan x - \sin x}{\sin^3 x} = \frac{\dfrac{\sin x}{\cos x} - \sin x}{\sin^3 x} = \frac{\sin x - \sin x \cos x}{\cos x \sin^3 x} = \frac{\sin x\,(1 - \cos x)}{\cos x \sin^3 x}$$

$$= \frac{1 - \cos x}{\cos x \sin^2 x} = \frac{1 - \cos x}{\cos x\,(1 - \cos^2 x)} = \frac{1}{\cos x\,(1 + \cos x)} = \frac{\sec x}{1 + \cos x}$$

16. $\dfrac{\cos A \cot A - \sin A \tan A}{\csc A - \sec A} = 1 + \sin A \cos A$

$$\frac{\cos A \cot A - \sin A \tan A}{\csc A - \sec A} = \frac{\cos A\,\dfrac{\cos A}{\sin A} - \sin A\,\dfrac{\sin A}{\cos A}}{\dfrac{1}{\sin A} - \dfrac{1}{\cos A}} = \frac{\cos^3 A - \sin^3 A}{\cos A - \sin A}$$

$$= \frac{(\cos A - \sin A)(\cos^2 A + \cos A \sin A + \sin^2 A)}{\cos A - \sin A} = \cos^2 A + \cos A \sin A + \sin^2 A = 1 + \cos A \sin A$$

17. $\dfrac{\sin\theta - \cos\theta + 1}{\sin\theta + \cos\theta - 1} = \dfrac{\sin\theta + 1}{\cos\theta}$

$$\frac{\sin\theta + 1}{\cos\theta} = \frac{(\sin\theta + 1)(\sin\theta + \cos\theta - 1)}{\cos\theta\,(\sin\theta + \cos\theta - 1)} = \frac{\sin^2\theta + \sin\theta\cos\theta + \cos\theta - 1}{\cos\theta\,(\sin\theta + \cos\theta - 1)}$$

$$= \frac{-\cos^2\theta + \sin\theta\cos\theta + \cos\theta}{\cos\theta\,(\sin\theta + \cos\theta - 1)} = \frac{\cos\theta\,(\sin\theta - \cos\theta + 1)}{\cos\theta\,(\sin\theta + \cos\theta - 1)} = \frac{\sin\theta - \cos\theta + 1}{\sin\theta + \cos\theta - 1}$$

SUPPLEMENTARY PROBLEMS

18. Find the values of the trigonometric functions of θ, given $\sin\theta = 2/3$.

Ans. Quad I : $2/3,\ \sqrt{5}/3,\ 2/\sqrt{5},\ \sqrt{5}/2,\ 3/\sqrt{5},\ 3/2$
Quad II: $2/3,\ -\sqrt{5}/3,\ -2/\sqrt{5},\ -\sqrt{5}/2,\ -3/\sqrt{5},\ 3/2$

19. Find the values of the trigonometric functions of θ, given $\cos\theta = -5/6$.

Ans. Quad II : $\sqrt{11}/6,\ -5/6,\ -\sqrt{11}/5,\ -5/\sqrt{11},\ -6/5,\ 6/\sqrt{11}$
Quad III: $-\sqrt{11}/6,\ -5/6,\ \sqrt{11}/5,\ 5/\sqrt{11},\ -6/5,\ -6/\sqrt{11}$

20. Find the values of the trigonometric functions of θ, given $\tan\theta = 5/4$.

Ans. Quad I: $5/\sqrt{41},\ 4/\sqrt{41},\ 5/4,\ 4/5,\ \sqrt{41}/4,\ \sqrt{41}/5$
Quad III: $-5/\sqrt{41},\ -4/\sqrt{41},\ 5/4,\ 4/5,\ -\sqrt{41}/4,\ -\sqrt{41}/5$

21. Find the values of the trigonometric functions of θ, given $\cot\theta = -\sqrt{3}$.

Ans. Quad II: $1/2,\ -\sqrt{3}/2,\ -1/\sqrt{3},\ -\sqrt{3},\ -2/\sqrt{3},\ 2$
Quad IV: $-1/2,\ \sqrt{3}/2,\ -1/\sqrt{3},\ -\sqrt{3},\ 2/\sqrt{3},\ -2$

22. Find the value of $\dfrac{\sin\theta + \cos\theta - \tan\theta}{\sec\theta + \csc\theta - \cot\theta}$ when $\tan\theta = -4/3$.

Ans. Quad II: 23/5; Quad IV: 34/35

Verify the following identities.

23. $\sin\theta \sec\theta = \tan\theta$

24. $(1 - \sin^2 A)(1 + \tan^2 A) = 1$

25. $(1 - \cos\theta)(1 + \sec\theta)\cot\theta = \sin\theta$

26. $\csc^2 x\,(1 - \cos^2 x) = 1$

27. $\dfrac{\sin\theta}{\csc\theta} + \dfrac{\cos\theta}{\sec\theta} = 1$

28. $\dfrac{1 - 2\cos^2 A}{\sin A \cos A} = \tan A - \cot A$

29. $\tan^2 x \csc^2 x \cot^2 x \sin^2 x = 1$

30. $\sin A \cos A\,(\tan A + \cot A) = 1$

31. $1 - \dfrac{\cos^2\theta}{1 + \sin\theta} = \sin\theta$

32. $\dfrac{1}{\sec\theta + \tan\theta} = \sec\theta - \tan\theta$

33. $\dfrac{1}{1 - \sin A} + \dfrac{1}{1 + \sin A} = 2\sec^2 A$

34. $\dfrac{1 - \cos x}{1 + \cos x} = \dfrac{\sec x - 1}{\sec x + 1} = (\cot x - \csc x)^2$

35. $\tan\theta \sin\theta + \cos\theta = \sec\theta$

36. $\tan\theta - \csc\theta \sec\theta\,(1 - 2\cos^2\theta) = \cot\theta$

37. $\dfrac{\sin\theta}{\sin\theta + \cos\theta} = \dfrac{\sec\theta}{\sec\theta + \csc\theta}$

38. $\dfrac{\sin x + \tan x}{\cot x + \csc x} = \sin x \tan x$

39. $\dfrac{\sec x + \csc x}{\tan x + \cot x} = \sin x + \cos x$

40. $\dfrac{\sin^3\theta + \cos^3\theta}{\sin\theta + \cos\theta} = 1 - \sin\theta\cos\theta$

41. $\cot\theta + \dfrac{\sin\theta}{1 + \cos\theta} = \csc\theta$

42. $\dfrac{\sin\theta\cos\theta}{\cos^2\theta - \sin^2\theta} = \dfrac{\tan\theta}{1 - \tan^2\theta}$

43. $(\tan x + \tan y)(1 - \cot x \cot y) + (\cot x + \cot y)(1 - \tan x \tan y) = 0$

44. $(x\sin\theta - y\cos\theta)^2 + (x\cos\theta + y\sin\theta)^2 = x^2 + y^2$

45. $(2r\sin\theta\cos\theta)^2 + r^2(\cos^2\theta - \sin^2\theta)^2 = r^2$

46. $(r\sin\theta\cos\phi)^2 + (r\sin\theta\sin\phi)^2 + (r\cos\theta)^2 = r^2$

CHAPTER 40

Trigonometric Functions of Two Angles

ADDITION FORMULAS.

$$\sin(\alpha+\beta) = \sin\alpha\cos\beta + \cos\alpha\sin\beta$$

$$\cos(\alpha+\beta) = \cos\alpha\cos\beta - \sin\alpha\sin\beta$$

$$\tan(\alpha+\beta) = \frac{\tan\alpha + \tan\beta}{1 - \tan\alpha\tan\beta}$$

For a proof of these formulas, see Problems 1-2.

SUBTRACTION FORMULAS.

$$\sin(\alpha-\beta) = \sin\alpha\cos\beta - \cos\alpha\sin\beta$$

$$\cos(\alpha-\beta) = \cos\alpha\cos\beta + \sin\alpha\sin\beta$$

$$\tan(\alpha-\beta) = \frac{\tan\alpha - \tan\beta}{1 + \tan\alpha\tan\beta}$$

For a proof of these formulas, see Problem 3.

DOUBLE-ANGLE FORMULAS.

$$\sin 2\alpha = 2\sin\alpha\cos\alpha$$

$$\cos 2\alpha = \cos^2\alpha - \sin^2\alpha = 1 - 2\sin^2\alpha = 2\cos^2\alpha - 1$$

$$\tan 2\alpha = \frac{2\tan\alpha}{1 - \tan^2\alpha}$$

For a proof of these formulas, see Problem 9.

HALF-ANGLE FORMULAS.

$$\sin\tfrac{1}{2}\theta = \pm\sqrt{\frac{1-\cos\theta}{2}}$$

$$\cos\tfrac{1}{2}\theta = \pm\sqrt{\frac{1+\cos\theta}{2}}$$

$$\tan\tfrac{1}{2}\theta = \pm\sqrt{\frac{1-\cos\theta}{1+\cos\theta}} = \frac{\sin\theta}{1+\cos\theta} = \frac{1-\cos\theta}{\sin\theta}$$

For a proof of these formulas, see Problem 10

SOLVED PROBLEMS

1. Prove (1) $\sin(\alpha+\beta) = \sin\alpha\cos\beta + \cos\alpha\sin\beta$
and (2) $\cos(\alpha+\beta) = \cos\alpha\cos\beta - \sin\alpha\sin\beta$ when α and β are positive acute angles.

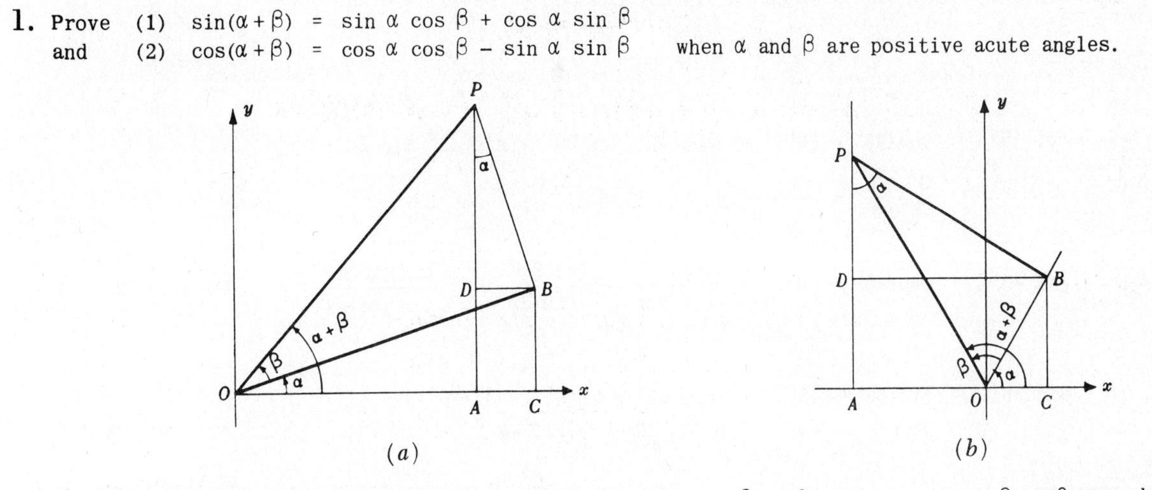

(a) (b)

Let α and β be positive acute angles such that $\alpha+\beta < 90°$ (Fig. *a*) and $\alpha+\beta > 90°$ (Fig. *b*).

To construct these figures, place angle α in standard position and then place angle β with its vertex at O and with its initial side along the terminal side of angle α. Let P be any point on the terminal side of angle $(\alpha+\beta)$: Draw PA perpendicular to OX, PB perpendicular to the terminal side of angle α, BC perpendicular to OX, and BD perpendicular to AP.

Now $\angle APB = \alpha$ since corresponding sides (OA and AP, OB and BP) are perpendicular. Then

$$\sin(\alpha+\beta) = \frac{AP}{OP} = \frac{AD+DP}{OP} = \frac{CB+DP}{OP} = \frac{CB}{OP} + \frac{DP}{OP} = \frac{CB}{OB}\cdot\frac{OB}{OP} + \frac{DP}{BP}\cdot\frac{BP}{OP}$$
$$= \sin\alpha\cos\beta + \cos\alpha\sin\beta$$

and

$$\cos(\alpha+\beta) = \frac{OA}{OP} = \frac{OC-AC}{OP} = \frac{OC-DB}{OP} = \frac{OC}{OP} - \frac{DB}{OP} = \frac{OC}{OB}\cdot\frac{OB}{OP} - \frac{DB}{BP}\cdot\frac{BP}{OP}$$
$$= \cos\alpha\cos\beta - \sin\alpha\sin\beta.$$

2. Prove: $\tan(\alpha+\beta) = \dfrac{\tan\alpha + \tan\beta}{1 - \tan\alpha\tan\beta}$.

$$\tan(\alpha+\beta) = \frac{\sin(\alpha+\beta)}{\cos(\alpha+\beta)} = \frac{\sin\alpha\cos\beta + \cos\alpha\sin\beta}{\cos\alpha\cos\beta - \sin\alpha\sin\beta} = \frac{\dfrac{\sin\alpha\cos\beta}{\cos\alpha\cos\beta} + \dfrac{\cos\alpha\sin\beta}{\cos\alpha\cos\beta}}{\dfrac{\cos\alpha\cos\beta}{\cos\alpha\cos\beta} - \dfrac{\sin\alpha\sin\beta}{\cos\alpha\cos\beta}}$$
$$= \frac{\tan\alpha + \tan\beta}{1 - \tan\alpha\tan\beta}$$

3. Prove the Subtraction formulas.

$$\sin(\alpha-\beta) = \sin[\alpha+(-\beta)] = \sin\alpha\cos(-\beta) + \cos\alpha\sin(-\beta)$$
$$= \sin\alpha(\cos\beta) + \cos\alpha(-\sin\beta) = \sin\alpha\cos\beta - \cos\alpha\sin\beta$$

$$\cos(\alpha-\beta) = \cos[\alpha+(-\beta)] = \cos\alpha\cos(-\beta) - \sin\alpha\sin(-\beta)$$
$$= \cos\alpha(\cos\beta) - \sin\alpha(-\sin\beta) = \cos\alpha\cos\beta + \sin\alpha\sin\beta$$

$$\tan(\alpha-\beta) = \tan[\alpha+(-\beta)] = \frac{\tan\alpha + \tan(-\beta)}{1 - \tan\alpha\tan(-\beta)} = \frac{\tan\alpha + (-\tan\beta)}{1 - \tan\alpha(-\tan\beta)} = \frac{\tan\alpha - \tan\beta}{1 + \tan\alpha\tan\beta}$$

4. Find the values of sine, cosine, and tangent of 15°, using (*a*) $15° = 45° - 30°$ and (*b*) $15° = 60° - 45°$.

(*a*)
$$\sin 15° = \sin(45° - 30°) = \sin 45° \cos 30° - \cos 45° \sin 30°$$
$$= \frac{1}{\sqrt{2}}\cdot\frac{\sqrt{3}}{2} - \frac{1}{\sqrt{2}}\cdot\frac{1}{2} = \frac{\sqrt{3}-1}{2\sqrt{2}} = \frac{\sqrt{2}}{4}(\sqrt{3}-1)$$

$$\cos 15° = \cos(45° - 30°) = \cos 45° \cos 30° + \sin 45° \sin 30°$$
$$= \frac{1}{\sqrt{2}}\cdot\frac{\sqrt{3}}{2} + \frac{1}{\sqrt{2}}\cdot\frac{1}{2} = \frac{\sqrt{2}}{4}(\sqrt{3}+1)$$

$$\tan 15° = \tan(45° - 30°) = \frac{\tan 45° - \tan 30°}{1 + \tan 45° \tan 30°} = \frac{1 - 1/\sqrt{3}}{1 + 1(1/\sqrt{3})} = \frac{\sqrt{3}-1}{\sqrt{3}+1} = 2 - \sqrt{3}$$

(*b*)
$$\sin 15° = \sin(60° - 45°) = \sin 60° \cos 45° - \cos 60° \sin 45°$$
$$= \frac{\sqrt{3}}{2}\cdot\frac{1}{\sqrt{2}} - \frac{1}{2}\cdot\frac{1}{\sqrt{2}} = \frac{\sqrt{2}}{4}(\sqrt{3}-1)$$

$$\cos 15° = \cos(60° - 45°) = \cos 60° \cos 45° + \sin 60° \sin 45°$$
$$= \frac{1}{2}\cdot\frac{1}{\sqrt{2}} + \frac{\sqrt{3}}{2}\cdot\frac{1}{\sqrt{2}} = \frac{\sqrt{2}}{4}(\sqrt{3}+1)$$

$$\tan 15° = \tan(60° - 45°) = \frac{\tan 60° - \tan 45°}{1 + \tan 60° \tan 45°} = \frac{\sqrt{3}-1}{\sqrt{3}+1} = 2 - \sqrt{3}$$

5. Prove: (*a*) $\sin(45° + \theta) - \sin(45° - \theta) = \sqrt{2}\sin\theta$, (*b*) $\sin(30° + \theta) + \cos(60° + \theta) = \cos\theta$.

(*a*)
$$\sin(45° + \theta) - \sin(45° - \theta) = (\sin 45° \cos\theta + \cos 45° \sin\theta) - (\sin 45° \cos\theta - \cos 45° \sin\theta)$$
$$= 2\cos 45° \sin\theta = 2\frac{1}{\sqrt{2}}\sin\theta = \sqrt{2}\sin\theta$$

(*b*)
$$\sin(30° + \theta) + \cos(60° + \theta) = (\sin 30° \cos\theta + \cos 30° \sin\theta) + (\cos 60° \cos\theta - \sin 60° \sin\theta)$$
$$= (\tfrac{1}{2}\cos\theta + \tfrac{\sqrt{3}}{2}\sin\theta) + (\tfrac{1}{2}\cos\theta - \tfrac{\sqrt{3}}{2}\sin\theta) = \cos\theta$$

6. Simplify: (*a*) $\sin(\alpha + \beta) + \sin(\alpha - \beta)$, (*b*) $\cos(\alpha + \beta) - \cos(\alpha - \beta)$, (*c*) $\dfrac{\tan(\alpha + \beta) - \tan\alpha}{1 + \tan(\alpha + \beta)\tan\alpha}$, (*d*) $(\sin\alpha\cos\beta - \cos\alpha\sin\beta)^2 + (\cos\alpha\cos\beta + \sin\alpha\sin\beta)^2$.

(*a*)
$$\sin(\alpha + \beta) + \sin(\alpha - \beta) = (\sin\alpha\cos\beta + \cos\alpha\sin\beta) + (\sin\alpha\cos\beta - \cos\alpha\sin\beta)$$
$$= 2\sin\alpha\cos\beta$$

(*b*)
$$\cos(\alpha + \beta) - \cos(\alpha - \beta) = (\cos\alpha\cos\beta - \sin\alpha\sin\beta) - (\cos\alpha\cos\beta + \sin\alpha\sin\beta)$$
$$= -2\sin\alpha\sin\beta$$

(*c*)
$$\frac{\tan(\alpha + \beta) - \tan\alpha}{1 + \tan(\alpha + \beta)\tan\alpha} = \tan[(\alpha + \beta) - \alpha] = \tan\beta$$

(*d*)
$$(\sin\alpha\cos\beta - \cos\alpha\sin\beta)^2 + (\cos\alpha\cos\beta + \sin\alpha\sin\beta)^2 = \sin^2(\alpha - \beta) + \cos^2(\alpha - \beta) = 1$$

7. Find $\sin(\alpha + \beta)$, $\cos(\alpha + \beta)$, $\sin(\alpha - \beta)$, $\cos(\alpha - \beta)$ and determine the quadrants in which $(\alpha + \beta)$ and $(\alpha - \beta)$ terminate, given (*a*) $\sin\alpha = 4/5$, $\cos\beta = 5/13$; α and β in quadrant I.
(*b*) $\sin\alpha = 2/3$, $\cos\beta = 3/4$; α in quadrant II, β in quadrant IV.

(*a*) $\cos\alpha = 3/5$ and $\sin\beta = 12/13$.

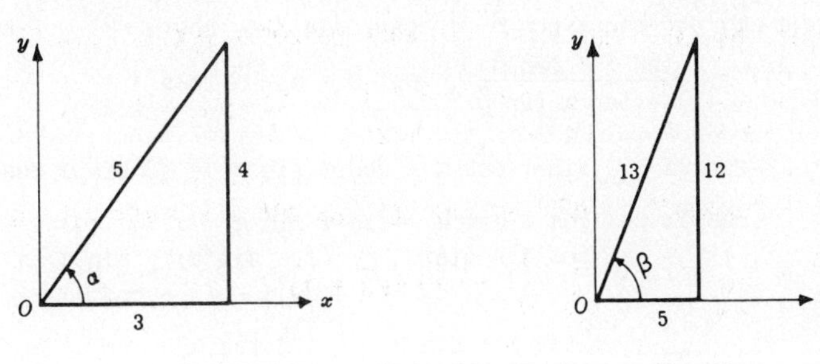

$$\sin(\alpha+\beta) = \sin\alpha\cos\beta + \cos\alpha\sin\beta = \frac{4}{5}\cdot\frac{5}{13} + \frac{3}{5}\cdot\frac{12}{13} = \frac{56}{65}$$

$$\cos(\alpha+\beta) = \cos\alpha\cos\beta - \sin\alpha\sin\beta = \frac{3}{5}\cdot\frac{5}{13} - \frac{4}{5}\cdot\frac{12}{13} = -\frac{33}{65}$$

$(\alpha+\beta)$ in quadrant II

$$\sin(\alpha-\beta) = \sin\alpha\cos\beta - \cos\alpha\sin\beta = \frac{4}{5}\cdot\frac{5}{13} - \frac{3}{5}\cdot\frac{12}{13} = -\frac{16}{65}$$

$$\cos(\alpha-\beta) = \cos\alpha\cos\beta + \sin\alpha\sin\beta = \frac{3}{5}\cdot\frac{5}{13} + \frac{4}{5}\cdot\frac{12}{13} = \frac{63}{65}$$

$(\alpha-\beta)$ in quadrant IV

(*b*) $\cos\alpha = -\sqrt{5}/3$ and $\sin\beta = -\sqrt{7}/4$.

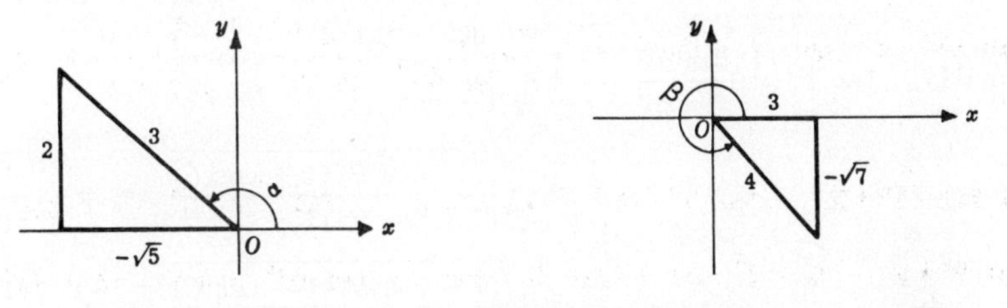

$$\sin(\alpha+\beta) = \sin\alpha\cos\beta + \cos\alpha\sin\beta = \frac{2}{3}\cdot\frac{3}{4} + \left(-\frac{\sqrt{5}}{3}\right)\left(-\frac{\sqrt{7}}{4}\right) = \frac{6+\sqrt{35}}{12}$$

$$\cos(\alpha+\beta) = \cos\alpha\cos\beta - \sin\alpha\sin\beta = \left(-\frac{\sqrt{5}}{3}\right)\frac{3}{4} - \frac{2}{3}\left(-\frac{\sqrt{7}}{4}\right) = \frac{-3\sqrt{5}+2\sqrt{7}}{12}$$

$(\alpha+\beta)$ in quadrant II

$$\sin(\alpha-\beta) = \sin\alpha\cos\beta - \cos\alpha\sin\beta = \frac{2}{3}\cdot\frac{3}{4} - \left(-\frac{\sqrt{5}}{3}\right)\left(-\frac{\sqrt{7}}{4}\right) = \frac{6-\sqrt{35}}{12}$$

$$\cos(\alpha-\beta) = \cos\alpha\cos\beta + \sin\alpha\sin\beta = \left(-\frac{\sqrt{5}}{3}\right)\frac{3}{4} + \frac{2}{3}\left(-\frac{\sqrt{7}}{4}\right) = \frac{-3\sqrt{5}-2\sqrt{7}}{12}$$

$(\alpha-\beta)$ in quadrant II

8. Prove: (*a*) $\cot(\alpha+\beta) = \dfrac{\cot\alpha\cot\beta - 1}{\cot\beta + \cot\alpha}$, (*b*) $\cot(\alpha-\beta) = \dfrac{\cot\alpha\cot\beta + 1}{\cot\beta - \cot\alpha}$.

(*a*) $$\cot(\alpha+\beta) = \frac{1}{\tan(\alpha+\beta)} = \frac{1-\tan\alpha\tan\beta}{\tan\alpha+\tan\beta} = \frac{1-\dfrac{1}{\cot\alpha\cot\beta}}{\dfrac{1}{\cot\alpha}+\dfrac{1}{\cot\beta}} = \frac{\cot\alpha\cot\beta-1}{\cot\beta+\cot\alpha}$$

(*b*) $$\cot(\alpha-\beta) = \cot[\alpha+(-\beta)] = \frac{\cot\alpha\cot(-\beta)-1}{\cot(-\beta)+\cot\alpha} = \frac{-\cot\alpha\cot\beta-1}{-\cot\beta+\cot\alpha} = \frac{\cot\alpha\cot\beta+1}{\cot\beta-\cot\alpha}$$

9. Prove the double angle formulas.

In $\sin(\alpha+\beta) = \sin\alpha\cos\beta + \cos\alpha\sin\beta$, $\cos(\alpha+\beta) = \cos\alpha\cos\beta - \sin\alpha\sin\beta$,

and $\tan(\alpha+\beta) = \dfrac{\tan\alpha + \tan\beta}{1 - \tan\alpha\tan\beta}$ put $\beta = \alpha$. Then

$$\sin 2\alpha = \sin\alpha\cos\alpha + \cos\alpha\sin\alpha = 2\sin\alpha\cos\alpha,$$

$$\begin{aligned}\cos 2\alpha &= \cos\alpha\cos\alpha - \sin\alpha\sin\alpha \\ &= \cos^2\alpha - \sin^2\alpha = (1-\sin^2\alpha) - \sin^2\alpha = 1 - 2\sin^2\alpha \\ & = \cos^2\alpha - (1-\cos^2\alpha) = 2\cos^2\alpha - 1,\end{aligned}$$

$$\tan 2\alpha = \frac{\tan\alpha + \tan\alpha}{1 - \tan\alpha\tan\alpha} = \frac{2\tan\alpha}{1-\tan^2\alpha}.$$

10. Prove the half angle formulas.

In $\cos 2\alpha = 1 - 2\sin^2\alpha$, put $\alpha = \frac{1}{2}\theta$. Then

$$\cos\theta = 1 - 2\sin^2\tfrac{1}{2}\theta, \qquad \sin^2\tfrac{1}{2}\theta = \frac{1-\cos\theta}{2} \qquad \text{and} \qquad \sin\tfrac{1}{2}\theta = \pm\sqrt{\frac{1-\cos\theta}{2}}.$$

In $\cos 2\alpha = 2\cos^2\alpha - 1$, put $\alpha = \frac{1}{2}\theta$. Then

$$\cos\theta = 2\cos^2\tfrac{1}{2}\theta - 1, \qquad \cos^2\tfrac{1}{2}\theta = \frac{1+\cos\theta}{2} \qquad \text{and} \qquad \cos\tfrac{1}{2}\theta = \pm\sqrt{\frac{1+\cos\theta}{2}}.$$

Finally, $\tan\frac{1}{2}\theta = \dfrac{\sin\frac{1}{2}\theta}{\cos\frac{1}{2}\theta} = \pm\sqrt{\dfrac{1-\cos\theta}{1+\cos\theta}}$

$$= \pm\sqrt{\frac{(1-\cos\theta)(1+\cos\theta)}{(1+\cos\theta)(1+\cos\theta)}} = \pm\sqrt{\frac{1-\cos^2\theta}{(1+\cos\theta)^2}} = \frac{\sin\theta}{1+\cos\theta}$$

$$= \pm\sqrt{\frac{(1-\cos\theta)(1-\cos\theta)}{(1+\cos\theta)(1-\cos\theta)}} = \pm\sqrt{\frac{(1-\cos\theta)^2}{1-\cos^2\theta}} = \frac{1-\cos\theta}{\sin\theta}.$$

The signs $\pm$ are not needed here since $\tan\frac{1}{2}\theta$ and $\sin\theta$ always have the same sign (Problem 7, Chapter 37) and $1-\cos\theta$ is always positive.

11. Using the half angle formulas, find the exact values of (*a*) $\sin 15°$, (*b*) $\sin 292\frac{1}{2}°$.

(*a*) $\sin 15° = \sqrt{\dfrac{1-\cos 30°}{2}} = \sqrt{\dfrac{1-\sqrt{3}/2}{2}} = \frac{1}{2}\sqrt{2-\sqrt{3}}$

(*b*) $\sin 292\frac{1}{2}° = -\sqrt{\dfrac{1-\cos 585°}{2}} = -\sqrt{\dfrac{1-\cos 225°}{2}} = -\sqrt{\dfrac{1+1/\sqrt{2}}{2}} = -\frac{1}{2}\sqrt{2+\sqrt{2}}$

12. Find the values of sine, cosine, and tangent of $\frac{1}{2}\theta$, given
(*a*) $\sin\theta = 5/13$, θ in quadrant II and
(*b*) $\cos\theta = 3/7$, θ in quadrant IV.

(*a*) $\sin\theta = 5/13$, $\cos\theta = -12/13$, and $\frac{1}{2}\theta$ in quadrant I.

$$\sin\tfrac{1}{2}\theta = \sqrt{\frac{1-\cos\theta}{2}} = \sqrt{\frac{1+12/13}{2}} = \sqrt{\frac{25}{26}} = \frac{5\sqrt{26}}{26}$$

$$\cos\tfrac{1}{2}\theta = \sqrt{\frac{1+\cos\theta}{2}} = \sqrt{\frac{1-12/13}{2}} = \sqrt{\frac{1}{26}} = \frac{\sqrt{26}}{26}$$

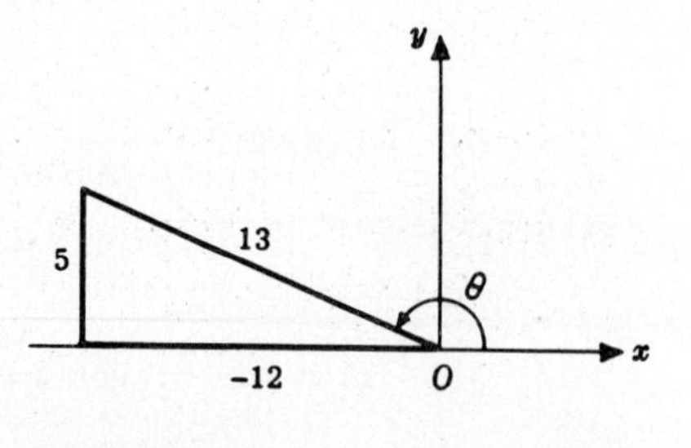

A and B are acute angles, find $A+B$ given:

) $\tan A = 1/4$, $\tan B = 3/5$. Hint: $\tan(A+B) = 1$. *Ans.* 45°

) $\tan A = 5/3$, $\tan B = 4$. *Ans.* 135°

$\tan(x+y) = 33$ and $\tan x = 3$, show that $\tan y = 0.3$.

nd the values of $\sin 2\theta$, $\cos 2\theta$, and $\tan 2\theta$, given:

) $\sin\theta = 3/5$, θ in Quadrant I. *Ans.* 24/25, 7/25, 24/7

) $\sin\theta = 3/5$, θ in Quadrant II. *Ans.* −24/25, 7/25, −24/7

) $\sin\theta = -1/2$, θ in Quadrant IV. *Ans.* $-\sqrt{3}/2,\ 1/2,\ -\sqrt{3}$

) $\tan\theta = -1/5$, θ in Quadrant II. *Ans.* 5/13, 12/13, −5/12

) $\tan\theta = u$, θ in Quadrant I. *Ans.* $\dfrac{2u}{1+u^2},\ \dfrac{1-u^2}{1+u^2},\ \dfrac{2u}{1-u^2}$

ove:

) $\tan\theta \sin 2\theta = 2\sin^2\theta$

) $\cot\theta \sin 2\theta = 1 + \cos 2\theta$

) $\dfrac{\sin^3 x - \cos^3 x}{\sin x - \cos x} = 1 + \frac{1}{2}\sin 2x$

) $\dfrac{1 - \sin 2A}{\cos 2A} = \dfrac{1 - \tan A}{1 + \tan A}$

(*e*) $\cos 2\theta = \dfrac{1 - \tan^2\theta}{1 + \tan^2\theta}$

(*f*) $\dfrac{1 + \cos 2\theta}{\sin 2\theta} = \cot\theta$

(*g*) $\cos 3\theta = 4\cos^3\theta - 3\cos\theta$

(*h*) $\cos^4 x = \dfrac{3}{8} + \dfrac{1}{2}\cos 2x + \dfrac{1}{8}\cos 4x$

nd the values of sine, cosine, and tangent of

) 30°, given $\cos 60° = 1/2$. *Ans.* $1/2,\ \sqrt{3}/2,\ 1/\sqrt{3}$

) 105°, given $\cos 210° = -\sqrt{3}/2$ *Ans.* $\frac{1}{2}\sqrt{2+\sqrt{3}},\ -\frac{1}{2}\sqrt{2-\sqrt{3}},\ -(2+\sqrt{3})$

) $\frac{1}{2}\theta$, given $\sin\theta = 3/5$, θ in Quadrant I. *Ans.* $1/\sqrt{10},\ 3/\sqrt{10},\ 1/3$

) θ, given $\cot 2\theta = 7/24$, 2θ in Quadrant I. *Ans.* 3/5, 4/5, 3/4

) θ, given $\cot 2\theta = -5/12$, 2θ in Quadrant II. *Ans.* $3/\sqrt{13},\ 2/\sqrt{13},\ 3/2$

ove:

) $\cos x = 2\cos^2\frac{1}{2}x - 1 = 1 - 2\sin^2\frac{1}{2}x$

) $\sin x = 2\sin\frac{1}{2}x\cos\frac{1}{2}x$

) $(\sin\frac{1}{2}\theta - \cos\frac{1}{2}\theta)^2 = 1 - \sin\theta$

) $\tan\frac{1}{2}\theta = \csc\theta - \cot\theta$

(*e*) $\dfrac{1 - \tan\frac{1}{2}\theta}{1 + \tan\frac{1}{2}\theta} = \dfrac{1 - \sin\theta}{\cos\theta} = \dfrac{\cos\theta}{1 + \sin\theta}$

(*f*) $\dfrac{2\tan\frac{1}{2}x}{1 + \tan^2\frac{1}{2}x} = \sin x$

the right triangle ABC, in which C is the right angle, prove:

$$\text{n } 2A = \frac{2ab}{c^2},\quad \cos 2A = \frac{b^2 - a^2}{c^2},\quad \sin\tfrac{1}{2}A = \sqrt{\frac{c-b}{2c}},\quad \cos\tfrac{1}{2}A = \sqrt{\frac{c+b}{2c}}.$$

ove: (*a*) $\dfrac{\sin 3x}{\sin x} - \dfrac{\cos 3x}{\cos x} = 2$, (*b*) $\tan 50° - \tan 40° = 2\tan 10°$.

$A+B+C = 180°$, prove:

) $\sin A + \sin B + \sin C = 4\cos\frac{1}{2}A\cos\frac{1}{2}B\cos\frac{1}{2}C$

) $\cos A + \cos B + \cos C = 1 + 4\sin\frac{1}{2}A\sin\frac{1}{2}B\sin\frac{1}{2}C$

) $\sin^2 A + \sin^2 B - \sin^2 C = 2\sin A\sin B\cos C$

) $\tan\frac{1}{2}A\tan\frac{1}{2}B + \tan\frac{1}{2}B\tan\frac{1}{2}C + \tan\frac{1}{2}C\tan\frac{1}{2}A = 1$.

$$\tan\tfrac{1}{2}\theta = \frac{1 - \cos\theta}{\sin\theta} = \frac{1 + 12/13}{5/13} = 5$$

(*b*) $\sin\theta = -2\sqrt{10}/7$, $\cos\theta = 3/7$, and $\frac{1}{2}\theta$ in quadrant II.

$$\sin\tfrac{1}{2}\theta = \sqrt{\frac{1 - \cos\theta}{2}} = \sqrt{\frac{1 - 3/7}{2}} = \frac{\sqrt{14}}{7}$$

$$\cos\tfrac{1}{2}\theta = -\sqrt{\frac{1 + \cos\theta}{2}} = -\sqrt{\frac{1 + 3/7}{2}} = -\frac{\sqrt{35}}{7}$$

$$\tan\tfrac{1}{2}\theta = \frac{1 - \cos\theta}{\sin\theta} = \frac{1 - 3/7}{-2\sqrt{10}/7} = -\frac{\sqrt{10}}{5}$$

13. Show that: (*a*) $\sin\theta = 2\sin\frac{1}{2}\theta\cos\frac{1}{2}\theta$ (*b*) $\sin A = \pm\sqrt{\dfrac{1 - \cos 2A}{2}}$ (*c*) $\tan 4x = \dfrac{\sin 8x}{1 + \cos 8x}$

(*d*) $\cos 6\theta = 1 - 2\sin^2 3\theta$ (*e*) $\sin^2\frac{1}{2}\theta = \frac{1}{2}(1 - \cos\theta)$, $\cos^2\frac{1}{2}\theta = \frac{1}{2}(1 + \cos\theta)$.

(*a*) This is obtained from $\sin 2\alpha = 2\sin\alpha\cos\alpha$ by putting $\alpha = \frac{1}{2}\theta$.

(*b*) This is obtained from $\sin\frac{1}{2}\theta = \pm\sqrt{\dfrac{1 - \cos\theta}{2}}$ by putting $\theta = 2A$.

(*c*) This is obtained from $\tan\frac{1}{2}\theta = \dfrac{\sin\theta}{1 + \cos\theta}$ by putting $\theta = 8x$.

(*d*) This is obtained from $\cos 2\alpha = 1 - 2\sin^2\alpha$ by putting $\alpha = 3\theta$.

(*e*) These formulas are obtained by squaring $\sin\frac{1}{2}\theta = \pm\sqrt{\dfrac{1 - \cos\theta}{2}}$ and $\cos\frac{1}{2}\theta = \pm\sqrt{\dfrac{1 + \cos\theta}{2}}$.

14. Express (*a*) $\sin 3\alpha$ in terms of $\sin\alpha$, (*b*) $\cos 4\alpha$ in terms of $\cos\alpha$.

(*a*)
$$\begin{aligned}\sin 3\alpha &= \sin(2\alpha + \alpha) = \sin 2\alpha\cos\alpha + \cos 2\alpha\sin\alpha\\ &= (2\sin\alpha\cos\alpha)\cos\alpha + (1 - 2\sin^2\alpha)\sin\alpha = 2\sin\alpha\cos^2\alpha + (1 - 2\sin^2\alpha)\sin\alpha\\ &= 2\sin\alpha\,(1 - \sin^2\alpha) + (1 - 2\sin^2\alpha)\sin\alpha = 3\sin\alpha - 4\sin^3\alpha.\end{aligned}$$

(*b*) $\cos 4\alpha = \cos 2(2\alpha) = 2\cos^2 2\alpha - 1 = 2(2\cos^2\alpha - 1)^2 - 1 = 8\cos^4\alpha - 8\cos^2\alpha + 1$.

15. Prove $\cos 2x = \cos^4 x - \sin^4 x$.

$$\cos^4 x - \sin^4 x = (\cos^2 x + \sin^2 x)(\cos^2 x - \sin^2 x) = \cos^2 x - \sin^2 x = \cos 2x$$

16. Prove $1 - \frac{1}{2}\sin 2x = \dfrac{\sin^3 x + \cos^3 x}{\sin x + \cos x}$.

$$\begin{aligned}\frac{\sin^3 x + \cos^3 x}{\sin x + \cos x} &= \frac{(\sin x + \cos x)(\sin^2 x - \sin x\cos x + \cos^2 x)}{\sin x + \cos x}\\ &= 1 - \sin x\cos x = 1 - \tfrac{1}{2}(2\sin x\cos x) = 1 - \tfrac{1}{2}\sin 2x\end{aligned}$$

17. Prove $\cos\theta = \sin(\theta + 30°) + \cos(\theta + 60°)$.

$$\begin{aligned}\sin(\theta + 30°) + \cos(\theta + 60°) &= (\sin\theta\cos 30° + \cos\theta\sin 30°) + (\cos\theta\cos 60° - \sin\theta\sin 60°)\\ &= \frac{\sqrt{3}}{2}\sin\theta + \frac{1}{2}\cos\theta + \frac{1}{2}\cos\theta - \frac{\sqrt{3}}{2}\sin\theta = \cos\theta\end{aligned}$$

18. Prove $\cos x = \dfrac{1-\tan^2\frac{1}{2}x}{1+\tan^2\frac{1}{2}x}$.

$$\frac{1-\tan^2\frac{1}{2}x}{1+\tan^2\frac{1}{2}x} = \frac{1-\dfrac{\sin^2\frac{1}{2}x}{\cos^2\frac{1}{2}x}}{\sec^2\frac{1}{2}x} = \frac{(1-\dfrac{\sin^2\frac{1}{2}x}{\cos^2\frac{1}{2}x})\cos^2\frac{1}{2}x}{\sec^2\frac{1}{2}x\cos^2\frac{1}{2}x} = \cos^2\tfrac{1}{2}x - \sin^2\tfrac{1}{2}x = \cos x$$

19. Prove $2\tan 2x = \dfrac{\cos x+\sin x}{\cos x-\sin x} - \dfrac{\cos x-\sin x}{\cos x+\sin x}$.

$$\frac{\cos x+\sin x}{\cos x-\sin x} - \frac{\cos x-\sin x}{\cos x+\sin x} = \frac{(\cos x+\sin x)^2-(\cos x-\sin x)^2}{(\cos x-\sin x)(\cos x+\sin x)}$$

$$= \frac{(\cos^2 x+2\sin x\cos x+\sin^2 x)-(\cos^2 x-2\sin x\cos x+\sin^2 x)}{\cos^2 x-\sin^2 x}$$

$$= \frac{4\sin x\cos x}{\cos^2 x-\sin^2 x} = \frac{2\sin 2x}{\cos 2x} = 2\tan 2x$$

20. Prove $\sin^4 A = \frac{3}{8} - \frac{1}{2}\cos 2A + \frac{1}{8}\cos 4A$.

$$\sin^4 A = (\sin^2 A)^2 = (\frac{1-\cos 2A}{2})^2 = \frac{1-2\cos 2A+\cos^2 2A}{4}$$

$$= \frac{1}{4}(1-2\cos 2A+\frac{1+\cos 4A}{2}) = \frac{3}{8}-\frac{1}{2}\cos 2A+\frac{1}{8}\cos 4A$$

21. Prove $\tan^6 x = \tan^4 x\sec^2 x - \tan^2 x\sec^2 x + \sec^2 x - 1$.

$$\tan^6 x = \tan^4 x\tan^2 x = \tan^4 x(\sec^2 x-1) = \tan^4 x\sec^2 x - \tan^2 x\tan^2 x$$

$$= \tan^4 x\sec^2 x - \tan^2 x(\sec^2 x-1) = \tan^4 x\sec^2 x - \tan^2 x\sec^2 x + \tan^2 x$$

$$= \tan^4 x\sec^2 x - \tan^2 x\sec^2 x + \sec^2 x - 1$$

22. When $A+B+C = 180°$, show that $\sin 2A + \sin 2B + \sin 2C = 4\sin A\sin B\sin C$.

Since $C = 180° - (A+B)$,

$$\sin 2A+\sin 2B+\sin 2C = \sin 2A+\sin 2B+\sin[360°-2(A+B)]$$
$$= \sin 2A+\sin 2B-\sin 2(A+B)$$
$$= \sin 2A+\sin 2B-\sin 2A\cos 2B-\cos 2A\sin 2B$$
$$= (\sin 2A)(1-\cos 2B)+(\sin 2B)(1-\cos 2A)$$
$$= 2\sin 2A\sin^2 B+2\sin 2B\sin^2 A$$
$$= 4\sin A\cos A\sin^2 B+4\sin B\cos B\sin^2 A$$
$$= 4\sin A\sin B[\sin A\cos B+\cos A\sin B]$$
$$= 4\sin A\sin B\sin(A+B)$$
$$= 4\sin A\sin B\sin[180°-(A+B)] = 4\sin A\sin B\sin C.$$

23. When $A+B+C = 180°$, show that $\tan A+\tan B+\tan C = \tan A\tan B\tan C$

Since $C = 180° - (A+B)$,

$$\tan A+\tan B+\tan C = \tan A+\tan B+\tan[180°-(A+B)] = \tan A +$$

$$= \tan A+\tan B-\frac{\tan A+\tan B}{1-\tan A\tan B} = (\tan A+\tan B)(1 -$$

$$= (\tan A+\tan B)(-\frac{\tan A\tan B}{1-\tan A\tan B}) = -\tan A\tan B\,\frac{\text{ta}}{1-}$$

$$= -\tan A\tan B\tan(A+B) = \tan A\tan B\tan[180°-(A+B)$$

SUPPLEMENTARY PROBLEMS

24. Find the values of sine, cosine, and tangent of (*a*) 75°, (*b*) 255°.

Ans. (*a*) $\frac{\sqrt{2}}{4}(\sqrt{3}+1)$, $\frac{\sqrt{2}}{4}(\sqrt{3}-1)$, $2+\sqrt{3}$ (*b*) $-\frac{\sqrt{2}}{4}(\sqrt{3}+1)$, $-\frac{\sqrt{2}}{4}(\sqrt{3}-1$

25. Find the values of $\sin(\alpha+\beta)$, $\cos(\alpha+\beta)$, and $\tan(\alpha+\beta)$, given:

(*a*) $\sin\alpha = 3/5$, $\cos\beta = 5/13$, α and β in Quadrant I. *Ans.* 63/65,

(*b*) $\sin\alpha = 8/17$, $\tan\beta = 5/12$, α and β in Quadrant I. *Ans.* 171/221

(*c*) $\cos\alpha = -12/13$, $\cot\beta = 24/7$, α in Quadrant II, β in Quadrant III. *Ans.* −36/325

(*d*) $\sin\alpha = 1/3$, $\sin\beta = 2/5$, α in Quadrant I, β in Quadrant II.

Ans. $\frac{4\sqrt{2}-\sqrt{21}}{15}$, $-\frac{2+2}{1}$

26. Find the values of $\sin(\alpha-\beta)$, $\cos(\alpha-\beta)$, and $\tan(\alpha-\beta)$, given:

(*a*) $\sin\alpha = 3/5$, $\sin\beta = 5/13$, α and β in Quadrant I. *Ans.* 16/65,

(*b*) $\sin\alpha = 8/17$, $\tan\beta = 5/12$, α and β in Quadrant I. *Ans.* 21/221,

(*c*) $\cos\alpha = -12/13$, $\cot\beta = 24/7$, α in Quadrant II, β in Quadrant I. *Ans.* 204/325

(*d*) $\sin\alpha = 1/3$, $\sin\beta = 2/5$, α in Quadrant II, β in Quadrant I.

Ans. $\frac{4\sqrt{2}+\sqrt{21}}{15}$, $-\frac{2\sqrt{42}}{1}$

27. Prove:

(*a*) $\sin(\alpha+\beta) - \sin(\alpha-\beta) = 2\cos\alpha\sin\beta$

(*b*) $\cos(\alpha+\beta) + \cos(\alpha-\beta) = 2\cos\alpha\cos\beta$

(*c*) $\tan(45°-\theta) = \dfrac{1-\tan\theta}{1+\tan\theta}$

(*d*) $\dfrac{\tan(\alpha+\beta)}{\cot(\alpha-\beta)} = \dfrac{\tan^2\alpha-\tan^2\beta}{1-\tan^2\alpha\tan^2\beta}$

(*e*) $\tan(\alpha+\beta+\gamma) = \tan[(\alpha+\beta)+\gamma] = \dfrac{\tan\alpha+\tan\beta+\tan\gamma-\tan\alpha\,\text{t}}{1-\tan\alpha\tan\beta-\tan\beta\tan\gamma-}$

(*f*) $\dfrac{\sin(x+y)}{\cos(x-y)} = \dfrac{\text{t}}{1}$

(*g*) $\tan(45°+\theta) =$

(*h*) $\sin(\alpha+\beta)\sin(\alpha$

28.

29.

30.

31.

32.

33.

34.

35.

36.

CHAPTER 41

Sum, Difference, and Product Formulas

PRODUCTS OF SINES AND COSINES.

$$\sin\alpha\cos\beta = \tfrac{1}{2}[\sin(\alpha+\beta) + \sin(\alpha-\beta)]$$

$$\cos\alpha\sin\beta = \tfrac{1}{2}[\sin(\alpha+\beta) - \sin(\alpha-\beta)]$$

$$\cos\alpha\cos\beta = \tfrac{1}{2}[\cos(\alpha+\beta) + \cos(\alpha-\beta)]$$

$$\sin\alpha\sin\beta = -\tfrac{1}{2}[\cos(\alpha+\beta) - \cos(\alpha-\beta)]$$

For proofs of these formulas, see Problem 1.

SUM AND DIFFERENCE OF SINES AND COSINES.

$$\sin A + \sin B = 2\sin\tfrac{1}{2}(A+B)\cos\tfrac{1}{2}(A-B)$$

$$\sin A - \sin B = 2\cos\tfrac{1}{2}(A+B)\sin\tfrac{1}{2}(A-B)$$

$$\cos A + \cos B = 2\cos\tfrac{1}{2}(A+B)\cos\tfrac{1}{2}(A-B)$$

$$\cos A - \cos B = -2\sin\tfrac{1}{2}(A+B)\sin\tfrac{1}{2}(A-B)$$

For proofs of these formulas, see Problem 2.

SOLVED PROBLEMS

1. Derive the product formulas.

Since $\sin(\alpha+\beta) + \sin(\alpha-\beta) = (\sin\alpha\cos\beta + \cos\alpha\cos\beta) + (\sin\alpha\cos\beta - \cos\alpha\sin\beta) = 2\sin\alpha\cos\beta$,

$$\sin\alpha\cos\beta = \tfrac{1}{2}[\sin(\alpha+\beta) + \sin(\alpha-\beta)].$$

Since $\sin(\alpha+\beta) - \sin(\alpha-\beta) = 2\cos\alpha\sin\beta$,

$$\cos\alpha\sin\beta = \tfrac{1}{2}[\sin(\alpha+\beta) - \sin(\alpha-\beta)].$$

Since $\cos(\alpha+\beta) + \cos(\alpha-\beta) = (\cos\alpha\cos\beta - \sin\alpha\sin\beta) + (\cos\alpha\cos\beta + \sin\alpha\sin\beta) = 2\cos\alpha\cos\beta$,

$$\cos\alpha\cos\beta = \tfrac{1}{2}[\cos(\alpha+\beta) + \cos(\alpha-\beta)].$$

Since $\cos(\alpha+\beta) - \cos(\alpha-\beta) = -2\sin\alpha\sin\beta$,

$$\sin\alpha\sin\beta = -\tfrac{1}{2}[\cos(\alpha+\beta) - \cos(\alpha-\beta)].$$

2. Derive the sum and difference formulas.

Let $\alpha+\beta = A$ and $\alpha-\beta = B$ so that $\alpha = \tfrac{1}{2}(A+B)$ and $\beta = \tfrac{1}{2}(A-B)$. Then (see Problem 1)

$$\sin(\alpha+\beta) + \sin(\alpha-\beta) = 2\sin\alpha\cos\beta \quad \text{becomes} \quad \sin A + \sin B = 2\sin\tfrac{1}{2}(A+B)\cos\tfrac{1}{2}(A-B),$$
$$\sin(\alpha+\beta) - \sin(\alpha-\beta) = 2\cos\alpha\sin\beta \quad \text{becomes} \quad \sin A - \sin B = 2\cos\tfrac{1}{2}(A+B)\sin\tfrac{1}{2}(A-B),$$
$$\cos(\alpha+\beta) + \cos(\alpha-\beta) = 2\cos\alpha\cos\beta \quad \text{becomes} \quad \cos A + \cos B = 2\cos\tfrac{1}{2}(A+B)\cos\tfrac{1}{2}(A-B),$$
$$\cos(\alpha+\beta) - \cos(\alpha-\beta) = -2\sin\alpha\cos\beta \quad \text{becomes} \quad \cos A - \cos B = -2\sin\tfrac{1}{2}(A+B)\sin\tfrac{1}{2}(A-B).$$

3. Express each of the following as a sum or difference:
(*a*) $\sin 40^\circ \cos 30^\circ$, (*b*) $\cos 110^\circ \sin 55^\circ$, (*c*) $\cos 50^\circ \cos 35^\circ$, (*d*) $\sin 55^\circ \sin 40^\circ$.

(*a*) $\sin 40^\circ \cos 30^\circ = \tfrac{1}{2}[\sin(40^\circ+30^\circ) + \sin(40^\circ-30^\circ)] = \tfrac{1}{2}(\sin 70^\circ + \sin 10^\circ)$

(*b*) $\cos 110^\circ \sin 55^\circ = \tfrac{1}{2}[\sin(110^\circ+55^\circ) - \sin(110^\circ-55^\circ)] = \tfrac{1}{2}(\sin 165^\circ - \sin 55^\circ)$

(*c*) $\cos 50^\circ \cos 35^\circ = \tfrac{1}{2}[\cos(50^\circ+35^\circ) + \cos(50^\circ-35^\circ)] = \tfrac{1}{2}(\cos 85^\circ + \cos 15^\circ)$

(*d*) $\sin 55^\circ \sin 40^\circ = -\tfrac{1}{2}[\cos(55^\circ+40^\circ) - \cos(55^\circ-40^\circ)] = -\tfrac{1}{2}(\cos 95^\circ - \cos 15^\circ)$

4. Express each of the following as a product:
(*a*) $\sin 50^\circ + \sin 40^\circ$, (*b*) $\sin 70^\circ - \sin 20^\circ$, (*c*) $\cos 55^\circ + \cos 25^\circ$, (*d*) $\cos 35^\circ - \cos 75^\circ$.

(*a*) $\sin 50^\circ + \sin 40^\circ = 2\sin\tfrac{1}{2}(50^\circ+40^\circ)\cos\tfrac{1}{2}(50^\circ-40^\circ) = 2\sin 45^\circ \cos 5^\circ$

(*b*) $\sin 70^\circ - \sin 20^\circ = 2\cos\tfrac{1}{2}(70^\circ+20^\circ)\sin\tfrac{1}{2}(70^\circ-20^\circ) = 2\cos 45^\circ \sin 25^\circ$

(*c*) $\cos 55^\circ + \cos 25^\circ = 2\cos\tfrac{1}{2}(55^\circ+25^\circ)\cos\tfrac{1}{2}(55^\circ-25^\circ) = 2\cos 40^\circ \cos 15^\circ$

(*d*) $\cos 35^\circ - \cos 75^\circ = -2\sin\tfrac{1}{2}(35^\circ+75^\circ)\sin\tfrac{1}{2}(35^\circ-75^\circ) = -2\sin 55^\circ \sin(-20^\circ)$
$= 2\sin 55^\circ \sin 20^\circ$

5. Prove $\dfrac{\sin 4A + \sin 2A}{\cos 4A + \cos 2A} = \tan 3A$.

$$\frac{\sin 4A + \sin 2A}{\cos 4A + \cos 2A} = \frac{2\sin\frac{1}{2}(4A+2A)\cos\frac{1}{2}(4A-2A)}{2\cos\frac{1}{2}(4A+2A)\cos\frac{1}{2}(4A-2A)} = \frac{\sin 3A}{\cos 3A} = \tan 3A$$

6. Prove $\dfrac{\sin A - \sin B}{\sin A + \sin B} = \dfrac{\tan\frac{1}{2}(A-B)}{\tan\frac{1}{2}(A+B)}$.

$$\frac{\sin A - \sin B}{\sin A + \sin B} = \frac{2\cos\frac{1}{2}(A+B)\sin\frac{1}{2}(A-B)}{2\sin\frac{1}{2}(A+B)\cos\frac{1}{2}(A-B)} = \cot\tfrac{1}{2}(A+B)\tan\tfrac{1}{2}(A-B) = \frac{\tan\frac{1}{2}(A-B)}{\tan\frac{1}{2}(A+B)}$$

7. Prove $\cos^3 x \sin^2 x = \dfrac{1}{16}(2\cos x - \cos 3x - \cos 5x)$.

$$\cos^3 x \sin^2 x = (\sin x \cos x)^2 \cos x = \frac{1}{4}\sin^2 2x \cos x = \frac{1}{4}(\sin 2x)(\sin 2x \cos x)$$
$$= \frac{1}{4}(\sin 2x)[\tfrac{1}{2}(\sin 3x + \sin x)] = \frac{1}{8}(\sin 3x \sin 2x + \sin 2x \sin x)$$
$$= \frac{1}{8}\{-\tfrac{1}{2}(\cos 5x - \cos x) + [-\tfrac{1}{2}(\cos 3x - \cos x)]\}$$
$$= \frac{1}{16}(2\cos x - \cos 3x - \cos 5x)$$

8. Prove $1 + \cos 2x + \cos 4x + \cos 6x = 4\cos x \cos 2x \cos 3x$.

$$1 + (\cos 2x + \cos 4x) + \cos 6x = 1 + 2\cos 3x \cos x + \cos 6x = (1 + \cos 6x) + 2\cos 3x \cos x$$
$$= 2\cos^2 3x + 2\cos 3x \cos x = 2\cos 3x(\cos 3x + \cos x)$$
$$= 2\cos 3x(2\cos 2x \cos x) = 4\cos x \cos 2x \cos 3x$$

9. Transform $4 \cos x + 3 \sin x$ into the form $c \cos(x - \alpha)$.

Since $c \cos(x-\alpha) = c(\cos x \cos \alpha + \sin x \sin \alpha)$, set $c \cos \alpha = 4$ and $c \sin \alpha = 3$.

Then $\cos \alpha = 4/c$ and $\sin \alpha = 3/c$. Since $\sin^2\alpha + \cos^2\alpha = 1$, $c = 5$ and -5.

Using $c = 5$, $\cos \alpha = 4/5$, $\sin \alpha = 3/5$, and $\alpha = 36°52'$. Thus,

$$4 \cos x + 3 \sin x = 5 \cos(x - 36°52').$$

Using $c = -5$, $\alpha = 216°52'$ and

$$4 \cos x + 3 \sin x = -5 \cos(x - 216°52').$$

10. Find the maximum and minimum values of $4 \cos x + 3 \sin x$ on the interval $0 \leqq x \leqq 2\pi$.

From Problem 9, $4 \cos x + 3 \sin x = 5 \cos(x - 36°52')$.

Now on the prescribed interval, $\cos \theta$ attains its maximum value 1 when $\theta = 0$ and its minimum value -1 when $\theta = \pi$. Thus, the maximum value of $4 \cos x + 3 \sin x$ is 5 which occurs when $x - 36°52' = 0$ or when $x = 36°52'$ while the minimum value is -5 which occurs when $x - 36°52' = \pi$ or when $x = 216°52'$.

SUPPLEMENTARY PROBLEMS

11. Express each of the following products as a sum or difference of sines or of cosines.

(a) $\sin 35° \cos 25° = \frac{1}{2}(\sin 60° + \sin 10°)$

(b) $\sin 25° \cos 75° = \frac{1}{2}(\sin 100° - \sin 50°)$

(c) $\cos 50° \cos 70° = \frac{1}{2}(\cos 120° + \cos 20°)$

(d) $\sin 130° \sin 55° = -\frac{1}{2}(\cos 185° - \cos 75°)$

(e) $\sin 4x \cos 2x = \frac{1}{2}(\sin 6x + \sin 2x)$

(f) $\sin x/2 \cos 3x/2 = \frac{1}{2}(\sin 2x - \sin x)$

(g) $\cos 7x \cos 4x = \frac{1}{2}(\cos 11x + \cos 3x)$

(h) $\sin 5x \sin 4x = -\frac{1}{2}(\cos 9x - \cos x)$

12. Show that

(a) $2 \sin 45° \cos 15° = \frac{1}{2}(\sqrt{3} + 1)$ and $\cos 15° = \frac{\sqrt{2}}{4}(\sqrt{3} + 1)$,

(b) $2 \sin 82\frac{1}{2}° \cos 37\frac{1}{2}° = \frac{1}{2}(\sqrt{3} + \sqrt{2})$,

(c) $2 \sin 127\frac{1}{2}° \sin 97\frac{1}{2}° = \frac{1}{2}(\sqrt{3} + \sqrt{2})$.

13. Express each of the following as a product.

(a) $\sin 50° + \sin 20° = 2 \sin 35° \cos 15°$

(b) $\sin 75° - \sin 35° = 2 \cos 55° \sin 20°$

(c) $\cos 65° + \cos 15° = 2 \cos 40° \cos 25°$

(d) $\cos 80° - \cos 70° = -2 \sin 75° \sin 5°$

(e) $\sin 4x + \sin 2x = 2 \sin 3x \cos x$

(f) $\sin 7\theta - \sin 3\theta = 2 \cos 5\theta \sin 2\theta$

(g) $\cos 6\theta + \cos 2\theta = 2 \cos 4\theta \cos 2\theta$

(h) $\cos 3x/2 - \cos 9x/2 = 2 \sin 3x \sin 3x/2$

14. Show that

(a) $\sin 40° + \sin 20° = \cos 10°$,

(b) $\sin 105° + \sin 15° = \sqrt{6}/2$,

(c) $\cos 465° + \cos 165° = -\sqrt{6}/2$,

(d) $\dfrac{\sin 75° - \sin 15°}{\cos 75° + \cos 15°} = 1/\sqrt{3}$.

15. Prove:

(a) $\dfrac{\sin A + \sin 3A}{\cos A + \cos 3A} = \tan 2A$ (b) $\dfrac{\sin 2A + \sin 4A}{\cos 2A + \cos 4A} = \tan 3A$

(c) $\dfrac{\sin A + \sin B}{\sin A - \sin B} = \dfrac{\tan \frac{1}{2}(A+B)}{\tan \frac{1}{2}(A-B)}$ (d) $\dfrac{\cos A + \cos B}{\cos A - \cos B} = -\cot \frac{1}{2}(A-B) \cot \frac{1}{2}(A+B)$

(e) $\sin\theta + \sin 2\theta + \sin 3\theta = \sin 2\theta + (\sin\theta + \sin 3\theta) = \sin 2\theta\,(1 + 2\cos\theta)$

(f) $\cos\theta + \cos 2\theta + \cos 3\theta = \cos 2\theta\,(1 + 2\cos\theta)$

(g) $\sin 2\theta + \sin 4\theta + \sin 6\theta = (\sin 2\theta + \sin 4\theta) + 2\sin 3\theta\cos 3\theta$
$= 4\cos\theta\cos 2\theta\sin 3\theta$

(h) $\dfrac{\sin 3x + \sin 5x + \sin 7x + \sin 9x}{\cos 3x + \cos 5x + \cos 7x + \cos 9x} = \dfrac{(\sin 3x + \sin 9x) + (\sin 5x + \sin 7x)}{(\cos 3x + \cos 9x) + (\cos 5x + \cos 7x)} = \tan 6x$

16. Prove:

(a) $\cos 130° + \cos 110° + \cos 10° = 0$, (b) $\cos 220° + \cos 100° + \cos 20° = 0$.

17. Prove:

(a) $\cos^2\theta\sin^3\theta = \frac{1}{16}(2\sin\theta + \sin 3\theta - \sin 5\theta)$

(b) $\cos^2\theta\sin^4\theta = \frac{1}{32}(2 - \cos 2\theta - 2\cos 4\theta + \cos 6\theta)$

(c) $\cos^5\theta = \frac{1}{16}(10\cos\theta + 5\cos 3\theta + \cos 5\theta)$

(d) $\sin^5\theta = \frac{1}{16}(10\sin\theta - 5\sin 3\theta + \sin 5\theta)$

18. Transform:

(a) $4\cos x + 3\sin x$ into the form $c\sin(x + \alpha)$. *Ans.* $5\sin(x + 53°8')$

(b) $4\cos x + 3\sin x$ into the form $c\sin(x - \alpha)$. *Ans.* $5\sin(x - 306°52')$

(c) $\sin x - \cos x$ into the form $c\sin(x - \alpha)$. *Ans.* $\sqrt{2}\sin(x - 45°)$

(d) $5\cos 3t + 12\sin 3t$ into the form $c\cos(3t - \alpha)$. *Ans.* $13\cos(3t - 67°23')$

19. Find the maximum and minimum values of each sum of Problem 18 and a value of x or t between 0 and 2π at which each occurs.

Ans. (a) Maximum = 5, when $x = 36°52'$ (i.e., when $x + 53°8' = 90°$); minimum = −5, when $x = 216°52'$.

(b) Same as (a).

(c) Maximum = $\sqrt{2}$, when $x = 135°$; minimum = $-\sqrt{2}$, when $x = 315°$.

(d) Maximum = 13, when $t = 22°28'$; minimum = −13, when $t = 82°28'$.

CHAPTER 42

Oblique Triangles. Non-logarithmic Solution

AN OBLIQUE TRIANGLE is one which does not contain a right angle. Such a triangle contains either three acute angles or two acute angles and one obtuse angle.

The convention of denoting the angles by A, B, C and the lengths of the corresponding opposite sides by a, b, c will be used here.

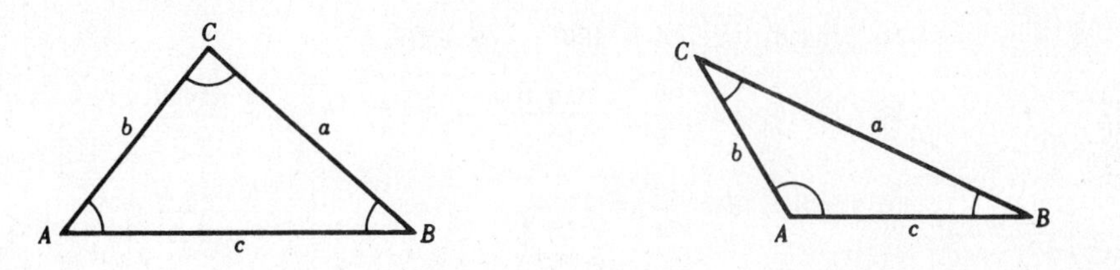

When three parts, not all angles, are known, the triangle is uniquely determined, except in one case to be noted below. The four cases of oblique triangles are:

Case I. Given one side and two angles.
Case II. Given two sides and the angle opposite one of them.
Case III. Given two sides and the included angle.
Case IV. Given the three sides.

THE LAW OF SINES. In any triangle, the sides are proportional to the sines of the opposite angles, i.e.,

$$\frac{a}{\sin A} = \frac{b}{\sin B} = \frac{c}{\sin C}.$$

The following relations follow readily:

$$\frac{a}{b} = \frac{\sin A}{\sin B}, \qquad \frac{b}{c} = \frac{\sin B}{\sin C}, \qquad \frac{c}{a} = \frac{\sin C}{\sin A}.$$

For a proof of the law of sines, see Problem 1.

MOLLWEIDE'S FORMULAS. In any triangle ABC,

$$\frac{a+b}{c} = \frac{\cos \frac{1}{2}(A-B)}{\sin \frac{1}{2}C}, \qquad \frac{a-b}{c} = \frac{\sin \frac{1}{2}(A-B)}{\cos \frac{1}{2}C}$$

together with those obtained by cyclic changes of the letters, i.e.,

$$\frac{b+c}{a} = \frac{\cos \frac{1}{2}(B-C)}{\sin \frac{1}{2}A}, \qquad \frac{b-c}{a} = \frac{\sin \frac{1}{2}(B-C)}{\cos \frac{1}{2}A}$$

$$\frac{c+a}{b} = \frac{\cos\frac{1}{2}(C-A)}{\sin\frac{1}{2}B}, \qquad \frac{c-a}{b} = \frac{\sin\frac{1}{2}(C-A)}{\cos\frac{1}{2}B}$$

and those obtained by interchanging the two letters (small and capital) in the numerators of each relation.

For derivations of these formulas, see Problem 2.

PROJECTION FORMULAS In any triangle ABC,

$$a = b\cos C + c\cos B$$
$$b = c\cos A + a\cos C$$
$$c = a\cos B + b\cos A.$$

For the derivation of these formulas, see Problem 3.

CASE I. Given one side and two angles.

Example. Suppose a, B, and C are given.

To find A, use $A = 180° - (B + C)$.

To find b, use $\frac{b}{a} = \frac{\sin B}{\sin A}$ whence $b = \frac{a\sin B}{\sin A}$.

To find c, use $\frac{c}{a} = \frac{\sin C}{\sin A}$ whence $c = \frac{a\sin C}{\sin A}$.

To check, use one of the Mollweide formulas or one of the projection formulas.

See Problems 4-6.

CASE II. Given two sides and the angle opposite one of them.

Example. Suppose b, c, and B are given.

From $\frac{\sin C}{\sin B} = \frac{c}{b}$, $\sin C = \frac{c\sin B}{b}$.

If $\sin C > 1$, no angle C is determined.

If $\sin C = 1$, $C = 90°$ and a right triangle is determined.

If $\sin C < 1$, two angles are determined: an acute angle C and an obtuse angle $C' = 180° - C$. Thus, there may be one or two triangles determined.

This case is discussed geometrically in Problem 7. The results obtained may be summarized as follows:

When the given angle is *acute*, there will be

(*a*) *one* solution if the side opposite the given angle is equal to or greater than the other given side,

(*b*) *no* solution, *one* solution (right triangle) or *two* solutions if the side opposite the given angle is less than the other given side.

When the given angle is *obtuse*, there will be

(*c*) *no* solution when the side opposite the given angle is less than or equal to the other given side,

(*d*) *one* solution if the side opposite the given angle is greater than the other given side.

Example. (*1*) When $b = 30$, $c = 20$, and $B = 40°$, there is one solution since B is acute and $b > c$.

(2) When $b = 20$, $c = 30$, and $B = 40°$, there is either no solution, one solution, or two solutions. The particular subcase is determined after computing $\sin C = \frac{c\sin B}{b}$.

(3) When $b = 30$, $c = 20$, and $B = 140°$, there is one solution.
(4) When $b = 20$, $c = 30$, and $B = 140°$, there is no solution.

This, the so-called ambiguous case, is solved by the law of sines and may be checked by either the Mollweide formulas or the projection formulas.

See Problems 8-10.

THE LAW OF COSINES. In any triangle ABC, the square of any side is equal to the sum of the squares of the other two sides diminished by twice the product of these sides and the cosine of their included angle, i.e.,

$$a^2 = b^2 + c^2 - 2bc \cos A$$
$$b^2 = c^2 + a^2 - 2ca \cos B$$
$$c^2 = a^2 + b^2 - 2ab \cos C.$$

For the derivation of these formulas, see Problem 11.

CASE III. Given two sides and the included angle.

Example. Suppose a, b, and C are given.

To find c, use $c^2 = a^2 + b^2 - 2ab \cos C$.

To find A, use $\sin A = \dfrac{a \sin C}{c}$. To find B, use $\sin B = \dfrac{b \sin C}{c}$.

To check, use $A+B+C = 180°$.

See Problems 12-14.

CASE IV. Given the three sides.

Example. With a, b, and c given, solve the law of cosines for each of the angles.

To find the angles, use $\cos A = \dfrac{b^2+c^2-a^2}{2bc}$, $\cos B = \dfrac{c^2+a^2-b^2}{2ca}$, $\cos C = \dfrac{a^2+b^2-c^2}{2ab}$.

To check, use $A+B+C = 180°$.

See Problems 15-16.

SOLVED PROBLEMS

1. Derive the law of sines.

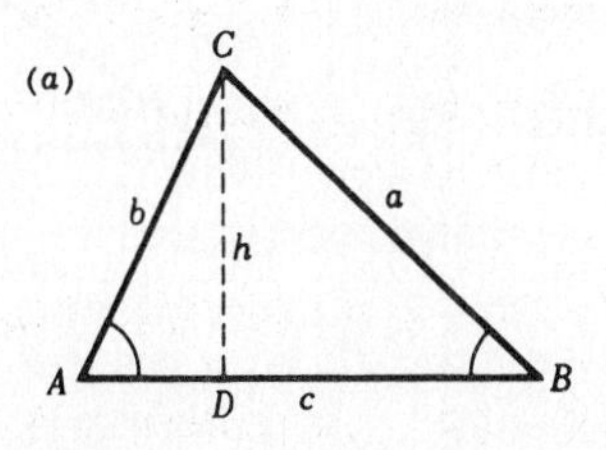

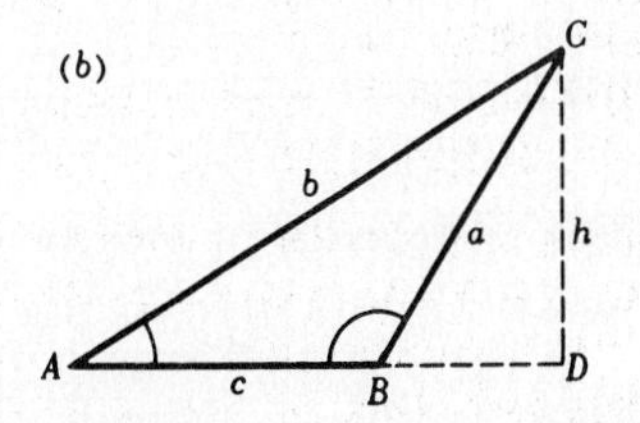

Let ABC be any oblique triangle. In Fig. (a), angles A and B are acute while in Fig. (b), angle B is obtuse. Draw CD perpendicular to AB or AB extended and denote its length by h.

In the right triangle ACD of either figure, $h = b \sin A$ while in the right triangle BCD, $h = a \sin B$ since in Fig. (b), $h = a \sin \angle DBC = a \sin(180° - B) = a \sin B$. Thus,

$$a \sin B = b \sin A \quad \text{or} \quad \frac{a}{\sin A} = \frac{b}{\sin B}.$$

In a similar manner (by drawing a perpendicular from B to AC or a perpendicular from A to BC), we obtain

$$\frac{a}{\sin A} = \frac{c}{\sin C} \quad \text{or} \quad \frac{b}{\sin B} = \frac{c}{\sin C}.$$

Thus, finally, $\dfrac{a}{\sin A} = \dfrac{b}{\sin B} = \dfrac{c}{\sin C}$.

2. Derive a pair of Mollweide's formulas.

By the law of sines, $\dfrac{a}{c} = \dfrac{\sin A}{\sin C}$ and $\dfrac{b}{c} = \dfrac{\sin B}{\sin C}$.

Then $$\frac{a+b}{c} = \frac{\sin A + \sin B}{\sin C} = \frac{2\sin\frac{1}{2}(A+B)\cos\frac{1}{2}(A-B)}{2\sin\frac{1}{2}C\cos\frac{1}{2}C} = \frac{\cos\frac{1}{2}(A-B)}{\sin\frac{1}{2}C},$$

since $\sin\frac{1}{2}(A+B) = \sin\frac{1}{2}(180° - C) = \sin(90° - \frac{1}{2}C) = \cos\frac{1}{2}C$.

Similarly, $$\frac{a-b}{c} = \frac{\sin A - \sin B}{\sin C} = \frac{2\cos\frac{1}{2}(A+B)\sin\frac{1}{2}(A-B)}{2\sin\frac{1}{2}C\cos\frac{1}{2}C} = \frac{\sin\frac{1}{2}(A-B)}{\cos\frac{1}{2}C},$$

since $\cos\frac{1}{2}(A+B) = \cos(90° - \frac{1}{2}C) = \sin\frac{1}{2}C$.

3. Derive one of the projection formulas.

Refer to the figures of Problem 1. In the right triangle ACD of either figure, $AD = b\cos A$. In the right triangle BCD of Fig. (a), $DB = a\cos B$. Thus, in Fig. (a),

$$c = AB = AD + DB = b\cos A + a\cos B = a\cos B + b\cos A.$$

In the right triangle BCD of Fig. (b), $BD = a\cos\angle DBC = a\cos(180° - B) = -a\cos B$. Thus, in Fig. ($b$),

$$c = AB = AD - BD = b\cos A - (-a\cos B) = a\cos B + b\cos A.$$

CASE I.

4. Solve the triangle ABC, given $c = 25$, $A = 35°$, and $B = 68°$.

To find C: $C = 180° - (A+B) = 180° - 103° = 77°$.

To find a: $$a = \frac{c\sin A}{\sin C} = \frac{25\sin 35°}{\sin 77°} = \frac{25(0.5736)}{0.9744} = 15.$$

To find b: $$b = \frac{c\sin B}{\sin C} = \frac{25\sin 68°}{\sin 77°} = \frac{25(0.9272)}{0.9744} = 24.$$

To check by Mollweide's formula:

$$\frac{b+a}{c} = \frac{\cos\frac{1}{2}(B-A)}{\sin\frac{1}{2}C} \quad \text{or} \quad (b+a)\sin\tfrac{1}{2}C = c\cos\tfrac{1}{2}(B-A)$$

$$(b+a)\sin\tfrac{1}{2}C = 39\sin 38°30' = 39(0.6225) = 24.3$$
$$c\cos\tfrac{1}{2}(B-A) = 25\cos 16°30' = 25(0.9588) = 24.0$$

To check by projection formula: $c = a\cos B + b\cos A = 15\cos 68° + 24\cos 35°$
$= 15(0.3746) + 24(0.8192) = 25.3.$

The required parts are $a = 15$, $b = 24$, and $C = 77°$.

5. A and B are two points on opposite banks of a river. From A a line $AC = 275$ ft is laid off and the angles $CAB = 125°40'$ and $ACB = 48°50'$ are measured. Find the length of AB.

In the triangle ABC, $B = 180° - (C+A) = 5°30'$ and

$$AB = c = \frac{b \sin C}{\sin B} = \frac{275 \sin 48°50'}{\sin 5°30'} = \frac{275(0.7528)}{0.0958} = 2160 \text{ ft.}$$

6. A tower 125 ft high is on a cliff on the bank of a river. From the top of the tower the angle of depression of a point on the opposite shore is $28°40'$ and from the base of the tower the angle of depression of the same point is $18°20'$. Find the width of the river and the height of the cliff.

In the figure BC represents the tower, DB represents the cliff, and A is the point on the opposite shore.

In triangle ABC, $C = 90° - 28°40' = 61°20'$,
$B = 90° + 18°20' = 108°20'$,
$A = 180° - (B+C) = 10°20'$.

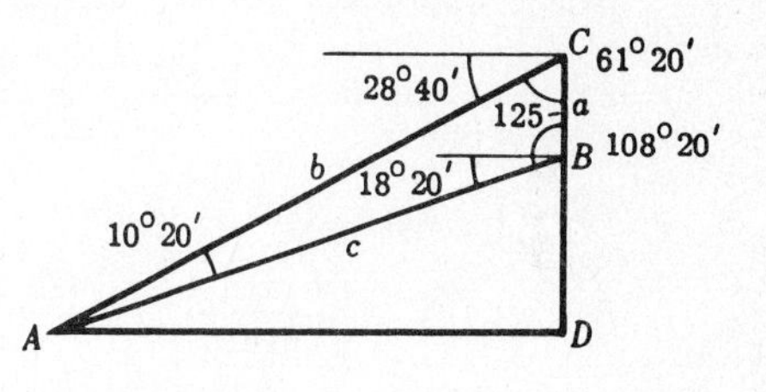

$$c = \frac{a \sin C}{\sin A} = \frac{125 \sin 61°20'}{\sin 10°20'} = \frac{125(0.8774)}{0.1794} = 611.$$

In the right triangle ABD,

$DB = c \sin 18°20' = 611(0.3145) = 192,$
$AD = c \cos 18°20' = 611(0.9492) = 580.$

The river is 580 ft wide and the cliff is 192 ft high.

7. Discuss the several special cases when two sides and the angle opposite one of them are given.

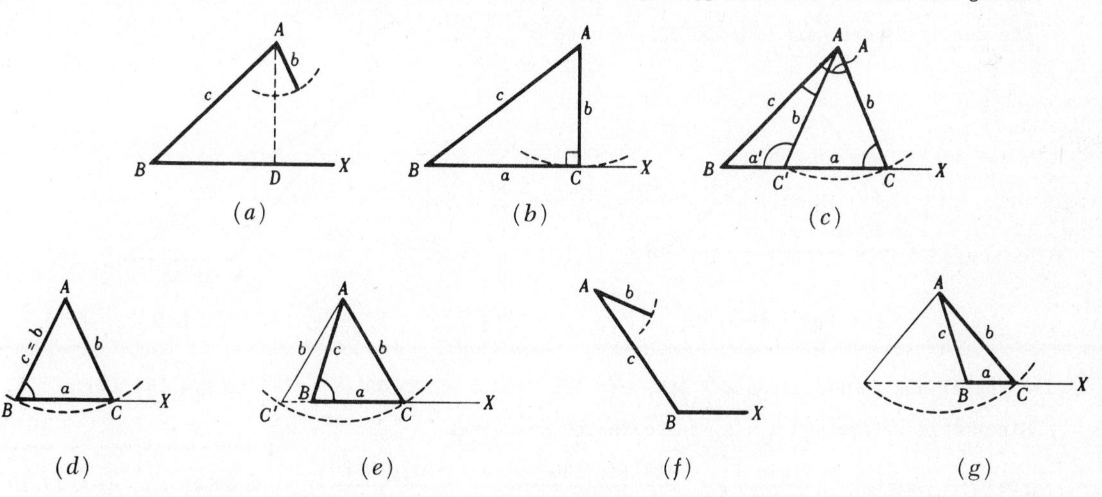

Let b, c, and B be the given parts. Construct the given angle B and lay off the side $BA = c$. With A as center and radius equal to b (the side opposite the given angle) describe an arc. Figures (a)-(e) illustrate the special cases which may occur when the given angle B is acute while Figures (f)-(g) illustrate the cases when B is obtuse.

The given angle B is acute.

Fig. (a). When $b < AD = c \sin B$, the arc does not meet BX and no triangle is determined.

Fig. (b). When $b = AD$, the arc is tangent to BX and one triangle — a right triangle with the right angle at C — is determined.

Fig. (c). When $b > AD$ and $b < c$, the arc meets BX in two points C and C' on the same side of B. Two triangles ABC, in which C is acute, and ABC' in which $C' = 180° - C$ is obtuse, are determined.

Fig. (*d*). When $b > AD$ and $b = c$, the arc meets BX in C and B. One triangle (isosceles) is determined

Fig. (*e*). When $b > c$, the arc meets BX in C and BX extended in C'. Since the triangle ABC' does not contain the given angle B, only one triangle ABC is determined.

The given angle is obtuse.

Fig. (*f*). When $b < c$ or $b = c$, no triangle is formed.

Fig. (*g*). When $b > c$, only one triangle is formed as in Fig. (*e*).

CASE II.

8. Solve the triangle ABC, given $c = 628$, $b = 480$, and $C = 55°10'$. Refer to Fig. (*a*) below.

Since C is acute and $c > b$, there is only one solution.

For B: $\sin B = \dfrac{b \sin C}{c} = \dfrac{480 \sin 55°10'}{628} = \dfrac{480(0.8208)}{628} = 0.6274$ and $B = 38°50'$.

For A: $A = 180° - (B + C) = 86°0'$.

For a: $a = \dfrac{b \sin A}{\sin B} = \dfrac{480 \sin 86°0'}{\sin 38°50'} = \dfrac{480(0.9976)}{0.6271} = 764$.

Check: $(a + b) \sin \frac{1}{2}C = c \cos \frac{1}{2}(A - B)$

$$(a + b) \sin \tfrac{1}{2}C = 1244 \sin 27°35' = 1244(0.4630) = 576.0$$
$$c \cos \tfrac{1}{2}(A - B) = 628 \cos 23°35' = 628(0.9165) = 575.6.$$

If preferred, a projection formula may be used for the check.

The required parts are $B = 38°50'$, $A = 86°0'$, and $a = 764$.

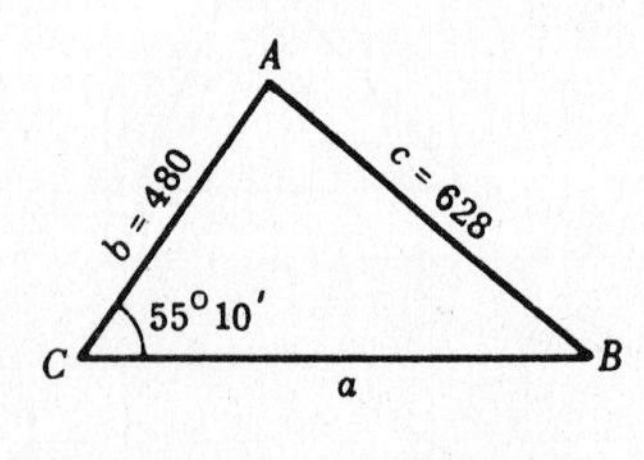

Fig. (*a*) Prob. 8

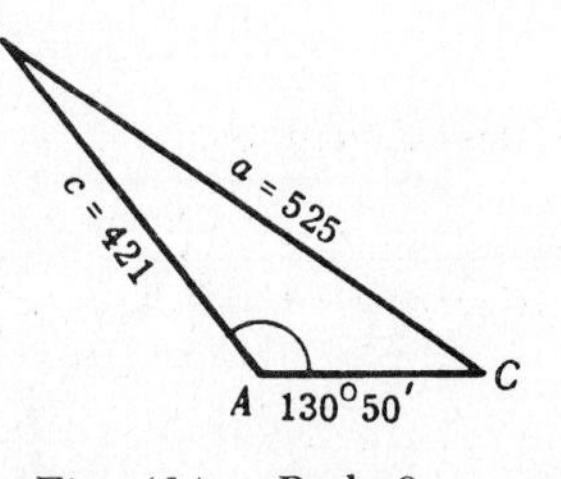

Fig. (*b*) Prob. 9

9. Solve the triangle ABC, given $a = 525$, $c = 421$, and $A = 130°50'$. Refer to Fig. (*b*) above.

Since A is obtuse and $a > c$, there is one solution.

For C: $\sin C = \dfrac{c \sin A}{a} = \dfrac{421 \sin 130°50'}{525} = \dfrac{421(0.7566)}{525} = 0.6067$ and $C = 37°20'$.

For B: $B = 180° - (C + A) = 11°50'$.

For b: $b = \dfrac{a \sin B}{\sin A} = \dfrac{525 \sin 11°50'}{\sin 130°50'} = \dfrac{525(0.2051)}{0.7566} = 142$.

Check: $(c + b) \sin \frac{1}{2}A = a \cos \frac{1}{2}(C - B)$

$$(c + b) \sin \tfrac{1}{2}A = 563 \sin 65°25' = 563(0.9094) = 512.0$$
$$a \cos \tfrac{1}{2}(C - B) = 525 \cos 12°45' = 525(0.9754) = 512.1.$$

The required parts are $C = 37°20'$, $B = 11°50'$, and $b = 142$.

10. Solve the triangle ABC, given $a = 31.5$, $b = 51.8$, and $A = 33°40'$. Refer to the adjoining figure.

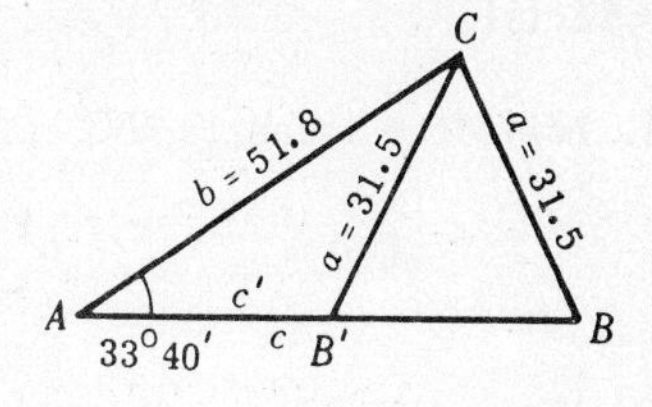

Since A is acute and $a < b$, there is the possibility of two solutions.

For B: $\sin B = \dfrac{b \sin A}{a} = \dfrac{51.8 \sin 33°40'}{31.5} = \dfrac{51.8(0.5544)}{31.5} = 0.9117.$

There are two solutions, $B = 65°40'$ and $B' = 180° - 65°40' = 114°20'$.

For C: $C = 180° - (A + B) = 80°40'$.

For c: $c = \dfrac{a \sin C}{\sin A} = \dfrac{31.5 \sin 80°40'}{\sin 33°40'} = \dfrac{31.5(0.9868)}{0.5544} = 56.1.$

Check: $(c + b) \sin \frac{1}{2}A = a \cos \frac{1}{2}(C - B)$

$(c + b) \sin \frac{1}{2}A = 107.9 \sin 16°50' = 107.9(0.2896) = 31.25$

$a \cos \frac{1}{2}(C - B) = 31.5 \cos 7°30' = 31.5(0.9914) = 31.23.$

For C': $C' = 180° - (A + B') = 32°0'$.

For c': $c' = \dfrac{a \sin C'}{\sin A} = \dfrac{31.5 \sin 32°0'}{\sin 33°40'} = \dfrac{31.5(0.5299)}{0.5544} = 30.1.$

Check: $(b + c') \sin \frac{1}{2}A = a \cos \frac{1}{2}(B' - C')$

$(b + c') \sin \frac{1}{2}A = 81.9 \sin 16°50' = 81.9(0.2896) = 23.72$

$a \cos \frac{1}{2}(B' - C') = 31.5 \cos 41°10' = 31.5(0.7528) = 23.71.$

The required parts are

for triangle ABC: $B = 65°40'$, $C = 80°40'$, and $c = 56.1$.

for triangle ABC': $B' = 114°20'$, $C' = 32°0'$, and $c' = 30.1$.

11. Derive the law of cosines.

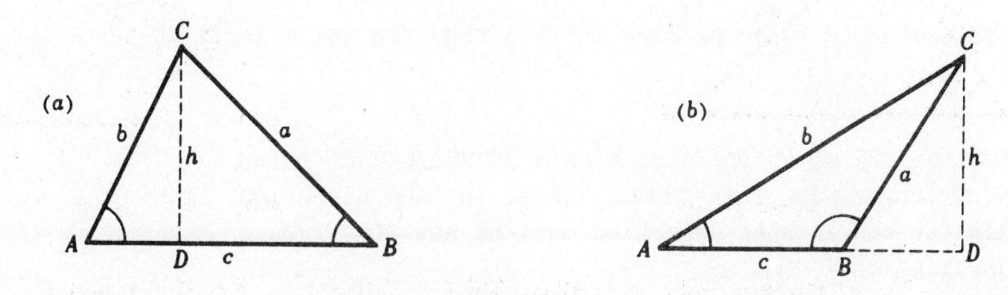

In the right triangle ACD of either figure, $b^2 = h^2 + (AD)^2$.

In the right triangle BCD of Fig. (a), $h = a \sin B$ and $DB = a \cos B$.

Then $$AD = AB - DB = c - a \cos B$$

and $$b^2 = h^2 + (AD)^2 = a^2 \sin^2 B + c^2 - 2ca \cos B + a^2 \cos^2 B$$
$$= a^2(\sin^2 B + \cos^2 B) + c^2 - 2ca \cos B = c^2 + a^2 - 2ca \cos B.$$

In the right triangle BCD of Fig. (b), $h = a \sin \angle CBD = a \sin(180° - B) = a \sin B$ and

$$BD = a \cos \angle CBD = a \cos(180° - B) = -a \cos B.$$

Then $AD = AB + BD = c - a \cos B$ and $b^2 = c^2 + a^2 - 2ca \cos B$.

The remaining equations may be obtained by cyclic changes of the letters.

CASE III.

12. Solve the triangle ABC, given $a = 132$, $b = 224$, and $C = 28°40'$.

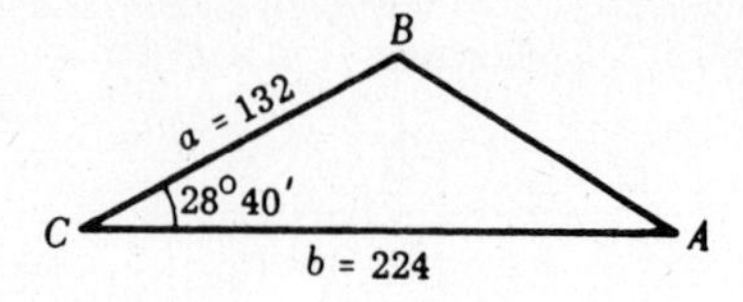

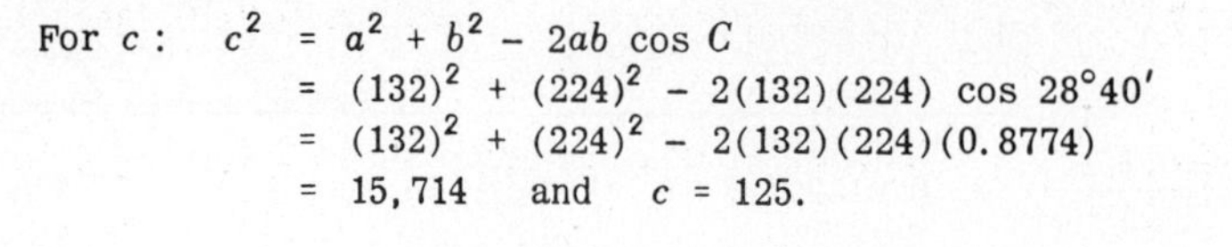

For c: $\quad c^2 = a^2 + b^2 - 2ab\cos C$
$= (132)^2 + (224)^2 - 2(132)(224)\cos 28°40'$
$= (132)^2 + (224)^2 - 2(132)(224)(0.8774)$
$= 15,714$ and $c = 125$.

For A: $\quad \sin A = \dfrac{a \sin C}{c} = \dfrac{132 \sin 28°40'}{125} = \dfrac{132(0.4797)}{125} = 0.5066$ and $A = 30°30'$.

For B: $\quad \sin B = \dfrac{b \sin C}{c} = \dfrac{224 \sin 28°40'}{125} = \dfrac{224(0.4797)}{125} = 0.8596$ and $B = 120°40'$.

(Since $b > a$, A is acute; since $A + C < 90°$, $B > 90°$.)

Check: $A + B + C = 179°50'$. The required parts are $A = 30°30'$, $B = 120°40'$, $c = 125$.

13. Two forces of 17.5 lb and 22.5 lb act on a body. If their directions make an angle of $50°10'$ with each other, find the magnitude of their resultant and the angle which it makes with the larger force.

In the parallelogram $ABCD$, $A + B = C + D = 180°$ and $B = 180° - 50°10' = 129°50'$.

In the triangle ABC,

$b^2 = c^2 + a^2 - 2ca\cos B \qquad [\cos 129°50' = -\cos(180° - 129°50') = -\cos 50°10']$
$= (22.5)^2 + (17.5)^2 - 2(22.5)(17.5)(-0.6406) = 1317$ and $b = 36.3$.

$\text{Sin } A = \dfrac{a \sin B}{b} = \dfrac{17.5 \sin 129°50'}{36.3} = \dfrac{17.5(0.7679)}{36.3} = 0.3702$ and $A = 21°40'$.

The resultant is a force of 36.3 lb; the required angle is $21°40'$.

14. From A a pilot flies 125 mi in the direction N $38°20'$ W and turns back. Through an error, he then flies 125 mi in the direction S $51°40'$ E. How far and in what direction must he now fly to reach his intended destination A?

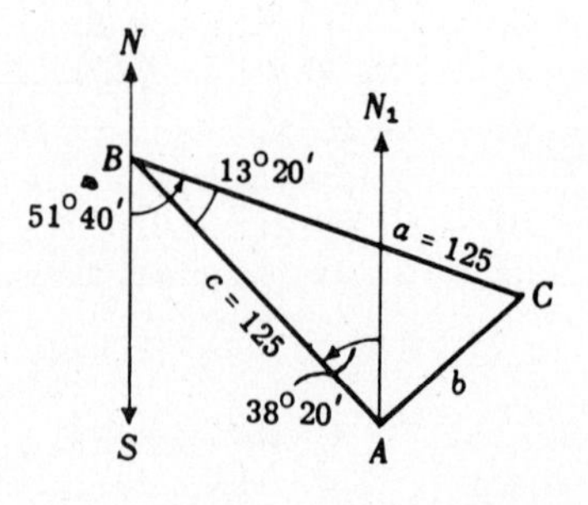

Denote the turn back point as B and his final position as C.

In the triangle ABC,

$b^2 = c^2 + a^2 - 2ca\cos B$
$= (125)^2 + (125)^2 - 2(125)(125)\cos 13°20'$
$= 2(125)^2(1 - 0.9730) = 843.7$ and $b = 29.0$.

$\text{Sin } A = \dfrac{a \sin B}{b} = \dfrac{125 \sin 13°20'}{29.0} = \dfrac{125(0.2306)}{29.0} = 0.9940$ and $A = 83°40'$.

Since $\angle CAN_1 = A - \angle N_1AB = 45°20'$, the pilot must fly a course S $45°20'$ W for 29.0 miles in going from C to A.

CASE IV.

15. Solve the triangle ABC, given $a = 30.3$, $b = 40.4$, and $c = 62.6$. Refer to Fig. (a) below.

For A: $\cos A = \dfrac{b^2 + c^2 - a^2}{2bc} = \dfrac{(40.4)^2 + (62.6)^2 - (30.3)^2}{2(40.4)(62.6)} = 0.9159$ and $A = 23°40'$.

For B: $\cos B = \dfrac{c^2 + a^2 - b^2}{2ca} = \dfrac{(62.6)^2 + (30.3)^2 - (40.4)^2}{2(62.6)(30.3)} = 0.8448$ and $B = 32°20'$.

For C: $\cos C = \dfrac{a^2 + b^2 - c^2}{2ab} = \dfrac{(30.3)^2 + (40.4)^2 - (62.6)^2}{2(30.3)(40.4)} = -0.5590$ and $C = 124°0'$.

Check: $A + B + C = 180°$.

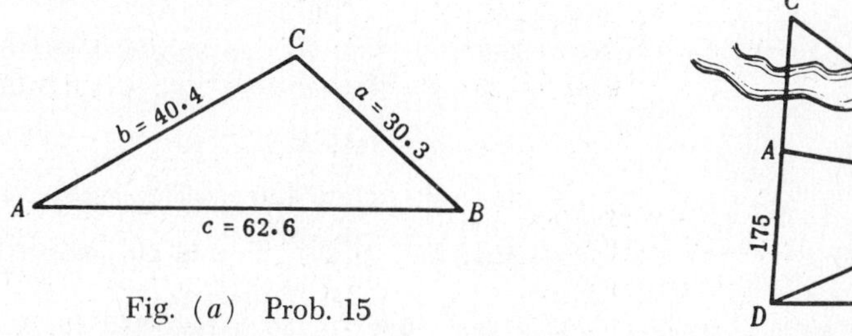

Fig. (a) Prob. 15

Fig. (b) Prob. 16

16. The distances of a point C from two points A and B, which cannot be measured directly, are required. The line CA is continued through A for a distance 175 ft to D, the line CB is continued through B for 225 ft to E, and the distances AB = 300 ft, DB = 326 ft, and DE = 488 ft are measured. Find AC and BC. Refer to Fig. (b) above.

Triangle ABC may be solved for the required parts after the angles $\angle BAC$ and $\angle ABC$ have been found. The first angle is the supplement of $\angle BAD$ and the second is the supplement of the sum of $\angle ABD$ and $\angle DBE$.

In the triangle ABD whose sides are known,

$$\cos \angle BAD = \frac{(175)^2 + (300)^2 - (326)^2}{2(175)(300)} = 0.1367 \quad \text{and} \quad \angle BAD = 82°10',$$

$$\cos \angle ABD = \frac{(300)^2 + (326)^2 - (175)^2}{2(300)(326)} = 0.8469 \quad \text{and} \quad \angle ABD = 32°10'.$$

In the triangle BDE whose sides are known,

$$\cos \angle DBE = \frac{(225)^2 + (326)^2 - (488)^2}{2(225)(326)} = -0.5538 \quad \text{and} \quad \angle DBE = 123°40'.$$

In the triangle ABC: $AB = 300$, $\angle BAC = 180° - \angle BAD = 97°50'$,
$\angle ABC = 180° - (\angle ABD + \angle DBE) = 24°10'$,
$\angle ACB = 180° - (\angle BAC + \angle ABC) = 58°0'$.

Then $$AC = \frac{AB \sin \angle ABC}{\sin \angle ACB} = \frac{300 \sin 24°10'}{\sin 58°0'} = \frac{300(0.4094)}{0.8480} = 145$$

and $$BC = \frac{AB \sin \angle BAC}{\sin \angle ACB} = \frac{300 \sin 97°50'}{\sin 58°0'} = \frac{300(0.9907)}{0.8480} = 350.$$

The required distances are AC = 145 ft and BC = 350 ft.

SUPPLEMENTARY PROBLEMS

Solve each of the following oblique triangles ABC, given:

17. $a = 125$, $A = 54°40'$, $B = 65°10'$. *Ans.* $b = 139$, $c = 133$, $C = 60°10'$

18. $b = 321$, $A = 75°20'$, $C = 38°30'$. *Ans.* $a = 339$, $c = 218$, $B = 66°10'$

19. $b = 215$, $c = 150$, $B = 42°40'$. *Ans.* $a = 300$, $A = 109°10'$, $C = 28°10'$

20. $a = 512$, $b = 426$, $A = 48°50'$. *Ans.* $c = 680$, $B = 38°50'$, $C = 92°20'$

21. $b = 50.4$, $c = 33.3$, $B = 118°30'$. *Ans.* $a = 25.1$, $A = 26°0'$, $C = 35°30'$

22. $b = 40.2$, $a = 31.5$, $B = 112°20'$. *Ans.* $c = 15.7$, $A = 46°30'$, $C = 21°10'$

23. $b = 51.5$, $a = 62.5$, $B = 40°40'$. *Ans.* $c = 78.9$, $A = 52°20'$, $C = 87°0'$
$c' = 16.0$, $A' = 127°40'$, $C' = 11°40'$

24. $a = 320$, $c = 475$, $A = 35°20'$. *Ans.* $b = 552$, $B = 85°30'$, $C = 59°10'$
$b' = 224$, $B' = 23°50'$, $C' = 120°50'$

25. $b = 120$, $c = 270$, $A = 118°40'$. *Ans.* $a = 344$, $B = 17°50'$, $C = 43°30'$

26. $a = 24.5$, $b = 18.6$, $c = 26.4$. *Ans.* $A = 63°10'$, $B = 42°40'$, $C = 74°10'$

27. $a = 6.34$, $b = 7.30$, $c = 9.98$. *Ans.* $A = 39°20'$, $B = 46°50'$, $C = 93°50'$

28. Two ships have radio equipment with a range of 200 miles. One is 155 miles N42°40′ E and the other is 165 miles N45°10′ W of a shore station. Can the two ships communicate directly?
Ans. No; they are 222 miles apart.

29. A ship sails 15.0 miles on a course S 40°10′ W and then 21.0 miles on a course N 28°20′ W. Find the distance and direction of the last position from the first.
Ans. 20.9 miles, N 70°40′ W

30. A lighthouse is 10 miles northwest of a dock. A ship leaves the dock at 9 A.M. and steams west at 12 miles per hour. At what time will it be 8 miles from the lighthouse?
Ans. 9:16 A.M. and 9:54 A.M.

31. Two forces of 115 lb and 215 lb acting on an object have a resultant of magnitude 275 lb. Find the angle between the directions in which the given forces act.
Ans. 70°50′

32. A tower 150 ft high is situated at the top of a hill. At a point 650 ft down the hill the angle between the surface of the hill and the line of sight to the top of the tower is 12°30′. Find the inclination of the hill to a horizontal plane. *Ans.* 7°50′

CHAPTER 43

Logarithmic Solution of Oblique Triangles

CASE I. Given two angles and a side.

The triangle is solved by using the angle relation, $A+B+C = 180°$, and the law of sines twice. The solution is checked by using one of the Mollweide formulas.

Example 1. Let a, A, and B be given. Then

$$C = 180° - (A+B), \qquad b = \frac{a \sin B}{\sin A}, \qquad c = \frac{a \sin C}{\sin A}.$$

Check: $(b+c) \sin \tfrac{1}{2}A = a \cos \tfrac{1}{2}(B-C)$, if $B > C$;

or $(c+b) \sin \tfrac{1}{2}A = a \cos \tfrac{1}{2}(C-B)$, if $C > B$.

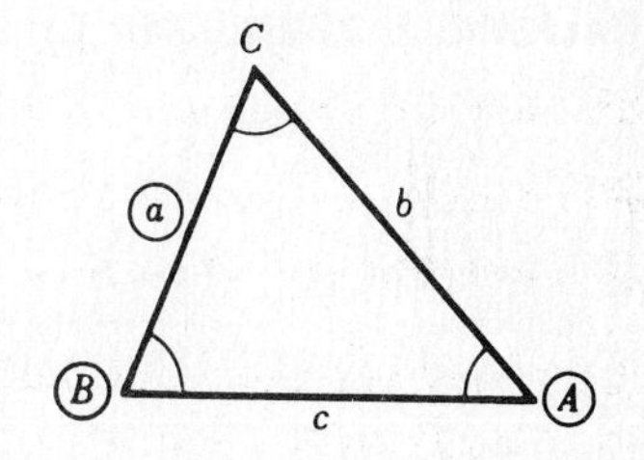

See Problems 1-3.

CASE II. Given two sides and the angle opposite one of them.

The triangle is solved by using the law of sines and the angle relation. The solution is checked by using one of the Mollweide formulas.

Example 2. Let a, b, and A be given with $a < b$. Then

$$\sin B = \frac{b \sin A}{a}, \qquad C = 180° - (A+B), \qquad c = \frac{a \sin C}{\sin A}.$$

When there are two solutions

$$B' = 180° - B, \qquad C' = 180° - (A+B'), \qquad c' = \frac{a \sin C'}{\sin A}.$$

Check: $(b+a) \sin \tfrac{1}{2}C = c \cos \tfrac{1}{2}(B-A)$,

$(b+a) \sin \tfrac{1}{2}C' = c' \cos \tfrac{1}{2}(B'-A)$.

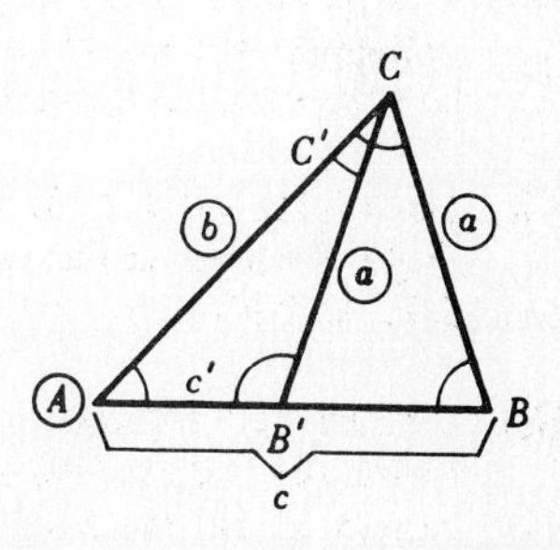

See Problems 4-5.

LAW OF TANGENTS. The law of cosines of the preceding chapter is not well adapted for logarithmic computation. In solving Case III, the law of tangents

$$\frac{a-b}{a+b} = \frac{\tan \frac{1}{2}(A-B)}{\tan \frac{1}{2}(A+B)}, \qquad \frac{b-c}{b+c} = \frac{\tan \frac{1}{2}(B-C)}{\tan \frac{1}{2}(B+C)}, \qquad \frac{c-a}{c+a} = \frac{\tan \frac{1}{2}(C-A)}{\tan \frac{1}{2}(C+A)}$$

will be used. For a proof of the law, see Problem 6.

Note. If, for example, $b > a$ it will be more convenient to write the first formula with the letters a and b (also A and B) interchanged.

CASE III. Given two sides and the included angle.

The triangle is solved by using the law of tangents to find the unknown angles and the law of sines to find the unknown side. The solution is checked by using one of the Mollweide formulas.

Example 3. Let $c > b$ and A be given. Then

$\frac{1}{2}(C+B) = 90° - \frac{1}{2}A$,

$\tan\frac{1}{2}(C-B) = \frac{c-b}{c+b}\tan\frac{1}{2}(C+B)$, $\quad a = \frac{c\sin A}{\sin C}$.

Check: $(c+b)\sin\frac{1}{2}A = a\cos\frac{1}{2}(C-B)$.

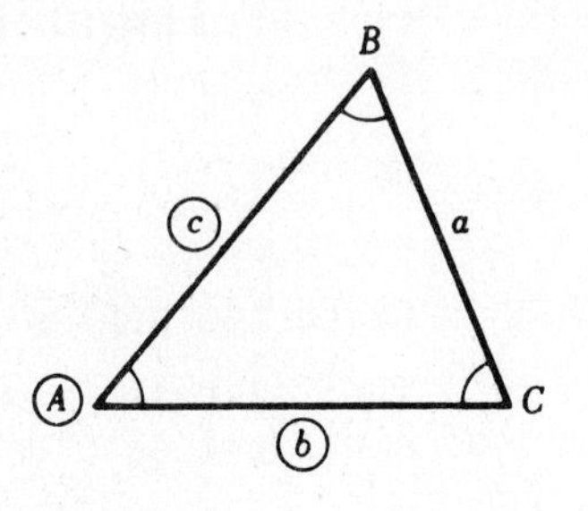

See Problems 7-8.

HALF-ANGLE FORMULAS. In any triangle *ABC*

$$\tan\tfrac{1}{2}A = \frac{r}{s-a}, \qquad \tan\tfrac{1}{2}B = \frac{r}{s-b}, \qquad \tan\tfrac{1}{2}C = \frac{r}{s-c}$$

where $s = \frac{1}{2}(a+b+c)$ is the semi-perimeter of the triangle

and $r = \sqrt{\frac{(s-a)(s-b)(s-c)}{s}}$ is the radius of the inscribed circle.

For a proof of the formulas, see Problem 9.

CASE IV. Given the three sides.

The triangle is solved by using the half-angle formulas and is checked by using the angle relation.

See Problem 10.

SOLVED PROBLEMS

CASE I.

1. Solve the triangle *ABC*, given $a = 38.124$, $A = 46°31.8'$, and $C = 79°17.4'$.

$B = 180° - (A+C) = 54°10.8'$.

$c = \frac{a\sin C}{\sin A}$ $\qquad$ $b = \frac{a\sin B}{\sin A}$

$\log a = 1.58120$	$\log a = 1.58120$
$\log\sin C = 9.99237-10$	$\log\sin B = 9.90894-10$
$\text{colog}\sin A = 0.13922$	$\text{colog}\sin A = 0.13922$
$\log c = 1.71279$	$\log b = 1.62936$
$c = 51.617$	$b = 42.595$

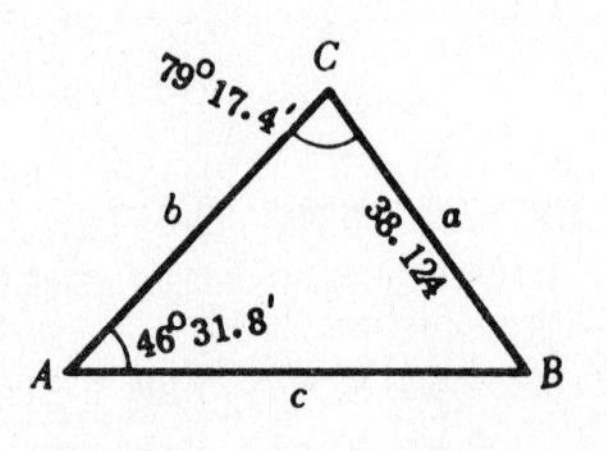

Check: $(c+b)\sin\frac{1}{2}A = a\cos\frac{1}{2}(C-B)$

$c+b = 94.212$, $\frac{1}{2}A = 23°15.9'$	$a = 38.124$, $\frac{1}{2}(C-B) = 12°33.3'$
$\log(c+b) = 1.97411$	$\log a = 1.58120$
$\log\sin\frac{1}{2}A = 9.59658-10$	$\log\cos\frac{1}{2}(C-B) = 9.98949-10$
1.57069	1.57069

2. Solve the triangle ABC, given $b = 282.66$, $A = 111°42.7'$, and $C = 24°25.8'$.

$B = 180° - (C+A) = 43°51.5'$.

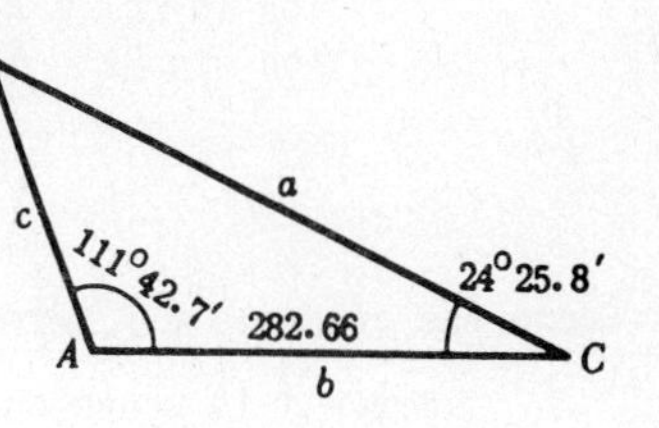

$$a = \frac{b \sin A}{\sin B} \qquad\qquad c = \frac{b \sin C}{\sin B}$$

$$\begin{aligned} \log b &= 2.45127 \\ \log \sin A &= 9.96804-10 \\ \text{colog} \sin B &= 0.15934 \\ \hline \log a &= 2.57865 \\ a &= 379.01 \end{aligned} \qquad\qquad \begin{aligned} \log b &= 2.45127 \\ \log \sin C &= 9.61656-10 \\ \text{colog} \sin B &= 0.15934 \\ \hline \log c &= 2.22717 \\ c &= 168.72 \end{aligned}$$

Check: $(a+c) \sin \frac{1}{2}B = b \cos \frac{1}{2}(A-C)$

$a+c = 547.73$, $\frac{1}{2}B = 21°55.8'$ $\qquad\qquad$ $b = 282.66$, $\frac{1}{2}(A-C) = 43°38.4'$

$$\begin{aligned} \log (a+c) &= 2.73856 \\ \log \sin \tfrac{1}{2}B &= 9.57226-10 \\ \hline &\ 2.31082 \end{aligned} \qquad\qquad \begin{aligned} \log b &= 2.45127 \\ \log \cos \tfrac{1}{2}(A-C) &= 9.85955-10 \\ \hline &\ 2.31082 \end{aligned}$$

3. In running a line PQ, S 38°42.4′ E from the point P, a surveyor encounters a swamp. At a point A, on the line and at one edge of the swamp, he changes his direction to N 61°0.0′ E for a distance of 1500.0 ft to a point B. He then sights to the other edge of the swamp, the direction being S 10°30.6′ W. If this line meets PQ in C, find the distance from B to C, the angle through which he must turn from BC to continue on the original line, and the distance AC across the swamp.

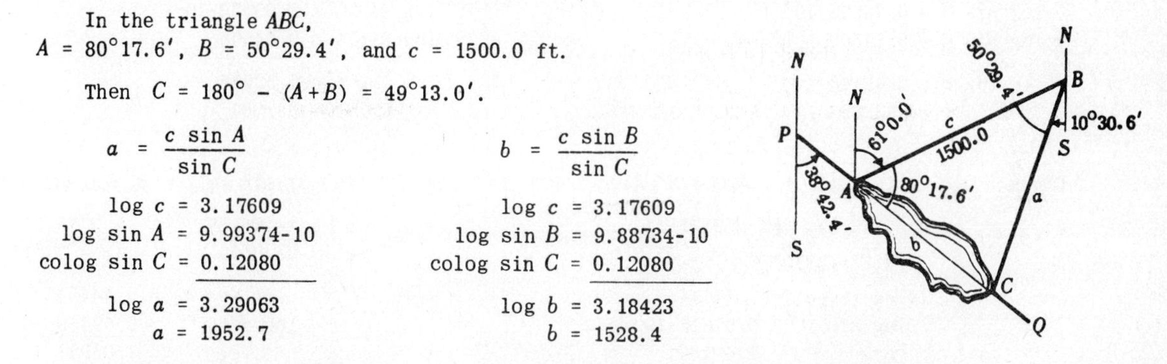

In the triangle ABC,

$A = 80°17.6'$, $B = 50°29.4'$, and $c = 1500.0$ ft.

Then $C = 180° - (A+B) = 49°13.0'$.

$$a = \frac{c \sin A}{\sin C} \qquad\qquad b = \frac{c \sin B}{\sin C}$$

$$\begin{aligned} \log c &= 3.17609 \\ \log \sin A &= 9.99374-10 \\ \text{colog} \sin C &= 0.12080 \\ \hline \log a &= 3.29063 \\ a &= 1952.7 \end{aligned} \qquad\qquad \begin{aligned} \log c &= 3.17609 \\ \log \sin B &= 9.88734-10 \\ \text{colog} \sin C &= 0.12080 \\ \hline \log b &= 3.18423 \\ b &= 1528.4 \end{aligned}$$

The distance from B to C is 1952.7 ft. The angle through which the surveyor must turn at C is $\angle BCQ = 180° - \angle ACB = 130°47.0'$. The distance across the swamp is 1528.4 ft.

CASE II.

4. Solve the triangle ABC, given $b = 67.246$, $c = 56.915$, and $B = 65°15.8'$.

Since B is acute and $b > c$, there is one solution.

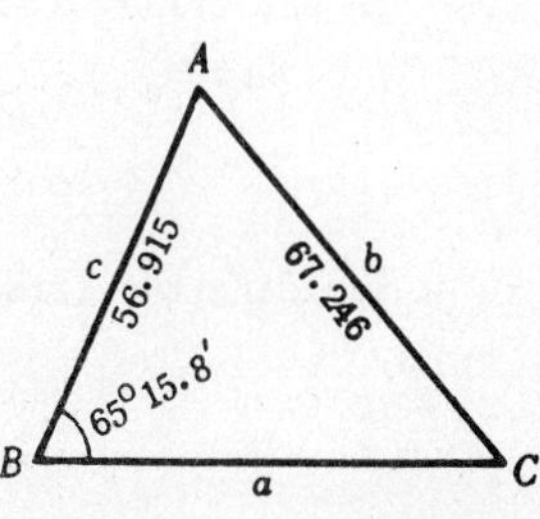

$$\sin C = \frac{c \sin B}{b} \qquad\qquad a = \frac{b \sin A}{\sin B}$$

$$\begin{aligned} \log c &= 1.75522 \\ \log \sin B &= 9.95820-10 \\ \text{colog}\ b &= 8.17233-10 \\ \hline \log \sin C &= 9.88575-10 \\ C &= 50°14.2' \\ A &= 180° - (B+C) = 64°30.0' \end{aligned} \qquad\qquad \begin{aligned} \log b &= 1.82767 \\ \log \sin A &= 9.95549-10 \\ \text{colog} \sin B &= 0.04180 \\ \hline \log a &= 1.82496 \\ a &= 66.828 \end{aligned}$$

Check: $(b+c)\sin\frac{1}{2}A = a\cos\frac{1}{2}(B-C)$

$b+c = 124.16$, $\frac{1}{2}A = 32°15.0'$ $a = 66.828$, $\frac{1}{2}(B-C) = 7°30.8'$

$\log(b+c) = 2.09398$
$\log\sin\frac{1}{2}A = 9.72723-10$
1.82121

$\log a = 1.82496$
$\log\cos\frac{1}{2}(B-C) = 9.99625-10$
1.82121

5. Solve the triangle ABC, given $a = 123.20$, $b = 155.37$, $A = 16°33.7'$.

Since A is acute and $a < b$, there may be two solutions.

$$\sin B = \frac{b\sin A}{a}$$

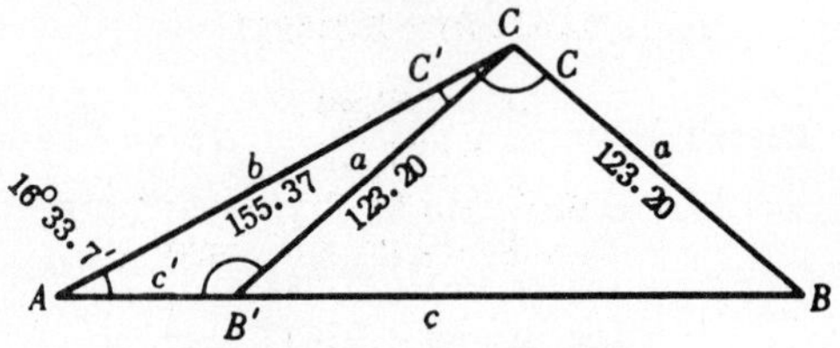

$\log b = 2.19137$
$\log\sin A = 9.45491-10$
$\text{colog } a = 7.90939-10$

$\log\sin B = 9.55567-10$
$B = 21°4.1'$
$C = 180° - (A+B) = 142°22.2'$

$B' = 180° - B = 158°55.9'$
$C' = 180° - (A+B') = 4°30.4'$

$$c = \frac{a\sin C}{\sin A}$$

$\log a = 2.09061$
$\log\sin C = 9.78573-10$
$\text{colog}\sin A = 0.54509$

$\log c = 2.42143$
$c = 263.89$

$$c' = \frac{a\sin C'}{\sin A}$$

$\log a = 2.09061$
$\log\sin C' = 8.89528-10$
$\text{colog}\sin A = 0.54509$

$\log c' = 1.53098$
$c' = 33.961$

Check: $(b+a)\sin\frac{1}{2}C = c\cos\frac{1}{2}(B-A)$

$b+a = 278.57$, $\frac{1}{2}C = 71°11.1'$
$c = 263.89$, $\frac{1}{2}(B-A) = 2°15.2'$

$\log(b+a) = 2.44494$
$\log\sin\frac{1}{2}C = 9.97615-10$
2.42109

$\log c = 2.42143$
$\log\cos\frac{1}{2}(B-A) = 9.99967-10$
2.42110

Check: $(b+a)\sin\frac{1}{2}C' = c'\cos\frac{1}{2}(B'-A)$

$b+a = 278.57$, $\frac{1}{2}C' = 2°15.2'$
$c' = 33.961$, $\frac{1}{2}(B'-A) = 71°11.1'$

$\log(b+a) = 2.44494$
$\log\sin\frac{1}{2}C' = 8.59459-10$
1.03953

$\log c' = 1.53098$
$\log\cos\frac{1}{2}(B'-A) = 9.50854-10$
1.03952

6. Derive the law of tangents.

In any triangle ABC, we have the Mollweide formulas

$$\frac{a-b}{c} = \frac{\sin\frac{1}{2}(A-B)}{\cos\frac{1}{2}C} \quad \text{and} \quad \frac{a+b}{c} = \frac{\cos\frac{1}{2}(A-B)}{\sin\frac{1}{2}C}.$$

Dividing the first by the second,

$$\frac{a-b}{c}\cdot\frac{c}{a+b} = \frac{\sin\frac{1}{2}(A-B)}{\cos\frac{1}{2}C}\cdot\frac{\sin\frac{1}{2}C}{\cos\frac{1}{2}(A-B)} = \tan\frac{1}{2}(A-B)\cdot\tan\frac{1}{2}C.$$

Since $C = 180° - (A+B)$, $\frac{1}{2}C = 90° - \frac{1}{2}(A+B)$ and $\tan\frac{1}{2}C = \cot\frac{1}{2}(A+B) = \dfrac{1}{\tan\frac{1}{2}(A+B)}$.

Thus, $$\frac{a-b}{a+b} = \tan\tfrac{1}{2}(A-B)\cdot\tan\tfrac{1}{2}C = \frac{\tan\frac{1}{2}(A-B)}{\tan\frac{1}{2}(A+B)}.$$

The two other forms may be obtained in a similar manner or by cyclic changes of letters on the above form.

CASE III.

7. Solve the triangle ABC, given $b = 472.12$, $c = 607.44$, $A = 125°14.6'$.

$C+B = 180° - A = 54°45.4'$
$\frac{1}{2}(C+B) = 27°22.7'$

$c = 607.44$
$b = 472.12$
$c-b = 135.32$
$c+b = 1079.56 = 1079.6$

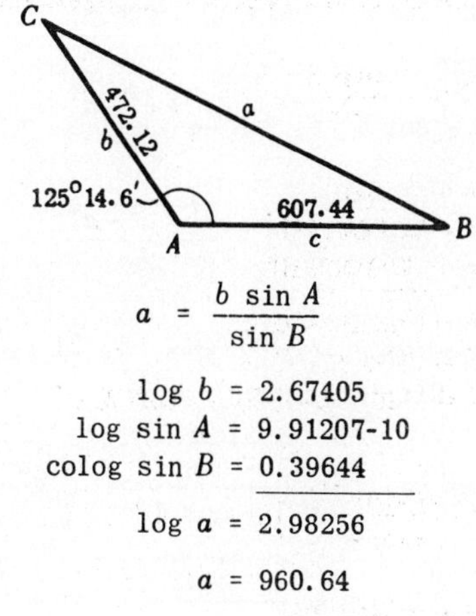

$$\tan\tfrac{1}{2}(C-B) = \frac{c-b}{c+b}\tan\tfrac{1}{2}(C+B)$$

$\log(c-b) = 2.13136$
$\text{colog}(c+b) = 6.96674-10$
$\log\tan\frac{1}{2}(C+B) = 9.71422-10$

$\log\tan\frac{1}{2}(C-B) = 8.81232-10$

$\frac{1}{2}(C-B) = 3°42.8'$
$\frac{1}{2}(C+B) = 27°22.7'$

$C = 31°\ 5.5'$
$B = 23°39.9'$

$$a = \frac{b\sin A}{\sin B}$$

$\log b = 2.67405$
$\log\sin A = 9.91207-10$
$\text{colog}\sin B = 0.39644$

$\log a = 2.98256$

$a = 960.64$

Check : To check the solution use the Mollweide formula $(c+b)\sin\frac{1}{2}A = a\cos\frac{1}{2}(C-B)$.

8. Two adjacent sides of a parallelogram are 3472.7 and 4822.3 ft respectively and the angle between them is 72°14.8′. Find the length of the longer diagonal.

In triangle ABC: $B = 180° - 72°14.8' = 107°45.2'$
$C+A = 72°14.8'$, $\frac{1}{2}(C+A) = 36°7.4'$

$c = 4822.3$
$a = 3472.7$

$c-a = 1349.6$
$c+a = 8295.0$.

$$\tan\tfrac{1}{2}(C-A) = \frac{c-a}{c+a}\tan\tfrac{1}{2}(C+A)$$

$\log(c-a) = 3.13020$
$\text{colog}(c+a) = 6.08118-10$
$\log\tan\frac{1}{2}(C+A) = 9.86322-10$

$\log\tan\frac{1}{2}(C-A) = 9.07460-10$

$\frac{1}{2}(C-A) = 6°46.3'$
$\frac{1}{2}(C+A) = 36°\ 7.4'$

$C = 42°53.7'$
$A = 29°21.1'$

$$b = \frac{c\sin B}{\sin C}$$

$\log c = 3.68326$
$\log\sin B = 9.97881-10$
$\text{colog}\sin C = 0.16707$

$\log b = 3.82914$

$b = 6747.4$ ft

Check: $(c+a)\sin\frac{1}{2}B = b\cos\frac{1}{2}(C-A)$

$$\begin{aligned} \log(c+a) &= 3.91882 \\ \log\sin\tfrac{1}{2}B &= \underline{9.90727-10} \\ & \quad 3.82609 \end{aligned} \qquad \begin{aligned} \log b &= 3.82914 \\ \log\cos\tfrac{1}{2}(C-A) &= \underline{9.99696-10} \\ & \quad 3.82610 \end{aligned}$$

9. Derive the half-angle formulas.

Let ABC be any triangle. Then $\tan\frac{1}{2}A = \sqrt{\dfrac{1-\cos A}{1+\cos A}}$ since $\frac{1}{2}A$ is always acute.

By the law of cosines, $\cos A = \dfrac{b^2+c^2-a^2}{2bc}$ so that

$$1-\cos A = 1-\frac{b^2+c^2-a^2}{2bc} = \frac{2bc-b^2-c^2+a^2}{2bc} = \frac{a^2-(b-c)^2}{2bc} = \frac{(a-b+c)(a+b-c)}{2bc}$$

and

$$1+\cos A = 1+\frac{b^2+c^2-a^2}{2bc} = \frac{2bc+b^2+c^2-a^2}{2bc} = \frac{(b+c)^2-a^2}{2bc} = \frac{(b+c+a)(b+c-a)}{2bc}.$$

Let $a+b+c = 2s$; then $a-b+c = (a+b+c)-2b = 2s-2b = 2(s-b)$, $a+b-c = 2(s-c)$, $b+c-a = 2(s-a)$, and

$$\tan\tfrac{1}{2}A = \sqrt{\frac{1-\cos A}{1+\cos A}} = \sqrt{\frac{(a-b+c)(a+b-c)}{2bc}\cdot\frac{2bc}{(b+c+a)(b+c-a)}} = \sqrt{\frac{2(s-b)\cdot 2(s-c)}{2s\cdot 2(s-a)}}$$

$$= \sqrt{\frac{(s-b)(s-c)}{s(s-a)}} = \sqrt{\frac{(s-a)(s-b)(s-c)}{s(s-a)^2}} = \frac{1}{s-a}\sqrt{\frac{(s-a)(s-b)(s-c)}{s}}.$$

Finally, setting $r = \sqrt{\dfrac{(s-a)(s-b)(s-c)}{s}}$, $\tan\frac{1}{2}A = \dfrac{r}{s-a}$. The remaining formulas may be obtained by cyclic changes of letters.

10. A triangular field has sides 2025.0, 2450.0, and 1575.0 ft respectively, as shown in the figure below. If the bearing of AB is S 35°30.4′ E, find the bearing of the other two sides.

$s = \frac{1}{2}(a+b+c)$ $\qquad r = \sqrt{\dfrac{(s-a)(s-b)(s-c)}{s}}$

$$\begin{aligned} a &= 2450.0 \\ b &= 1575.0 \\ c &= \underline{2025.0} \\ 2s &= 6050.0 \\ s &= 3025.0 \end{aligned} \qquad \begin{aligned} s-a &= 575 \\ s-b &= 1450 \\ s-c &= \underline{1000} \\ s &= 3025 \end{aligned} \qquad \begin{aligned} \log(s-a) &= 2.75967 \\ \log(s-b) &= 3.16137 \\ \log(s-c) &= 3.00000 \\ \text{colog } s &= \underline{6.51927-10} \\ 2\log r &= 5.44031 \\ \log r &= 2.72016 \end{aligned}$$

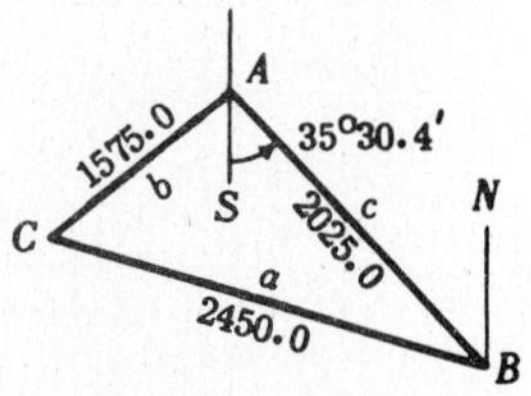

$$\tan\tfrac{1}{2}A = \frac{r}{s-a} \qquad \tan\tfrac{1}{2}B = \frac{r}{s-b} \qquad \tan\tfrac{1}{2}C = \frac{r}{s-c}$$

$$\begin{aligned} \log r &= 2.72016 \\ \log(s-a) &= \underline{2.75967} \\ \log\tan\tfrac{1}{2}A &= 9.96049-10 \\ \tfrac{1}{2}A &= 42°23.8' \\ A &= 84°47.6' \end{aligned} \qquad \begin{aligned} \log r &= 2.72016 \\ \log(s-b) &= \underline{3.16137} \\ \log\tan\tfrac{1}{2}B &= 9.55879-10 \\ \tfrac{1}{2}B &= 19°54.2' \\ B &= 39°48.4' \end{aligned} \qquad \begin{aligned} \log r &= 2.72016 \\ \log(s-c) &= \underline{3.00000} \\ \log\tan\tfrac{1}{2}C &= 9.72016-10 \\ \tfrac{1}{2}C &= 27°42.0' \\ C &= 55°24.0' \end{aligned}$$

$\angle SAC = 84°47.6' - 35°30.4' = 49°17.2'$; the bearing of AC is S 49°17.2′ W.

$\angle NBC = 35°30.4' + 39°48.4' = 75°18.8'$; the bearing of BC is N 75°18.8′ W.

SUPPLEMENTARY PROBLEMS

Solve and check each of the oblique triangles ABC, given:

11. $c = 78.753$, $A = 33°9.9'$, $C = 81°24.6'$. *Ans.* $a = 43.571$, $b = 72.432$, $B = 65°25.5'$

12. $b = 730.80$, $B = 42°12.8'$, $C = 109°32.5'$. *Ans.* $a = 514.73$, $c = 1025.0$, $A = 28°14.7'$

13. $a = 31.259$, $A = 57°59.9'$, $C = 23°36.6'$. *Ans.* $b = 36.466$, $c = 14.763$, $B = 98°23.5'$

14. $b = 13.218$, $c = 10.004$, $B = 25°57.2'$. *Ans.* $a = 21.467$, $A = 134°42.2'$, $C = 19°20.6'$

15. $b = 10.884$, $c = 35.730$, $C = 115°33.8'$. *Ans.* $a = 29.658$, $A = 48°29.2'$, $B = 15°57.0'$

16. $b = 86.425$, $c = 73.463$, $C = 49°18.9'$. *Ans.* $a = 89.534$, $B = 63°8.3'$, $A = 67°32.8'$
$a' = 23.147$, $B' = 116°51.7'$, $A' = 13°49.4'$

17. $a = 12.695$, $c = 15.873$, $A = 24°7.4'$. *Ans.* $b = 25.399$, $B = 125°8.7'$, $C = 30°43.9'$
$b' = 3.5745$, $B' = 6°36.5'$, $C' = 149°16.1'$

18. $a = 482.33$, $c = 395.71$, $B = 137°31.2'$. *Ans.* $b = 819.00$, $A = 23°26.2'$, $C = 19°2.6'$

19. $b = 561.23$, $c = 387.19$, $A = 56°43.8'$. *Ans.* $a = 475.89$, $B = 80°24.4'$, $C = 42°51.8'$

20. $a = 123.79$, $b = 264.23$, $c = 256.04$. *Ans.* $A = 27°28.2'$, $B = 79°57.0'$, $C = 72°34.8'$

21. $a = 1894.3$, $b = 2246.5$, $c = 3548.8$. *Ans.* $A = 28°11.8'$, $B = 34°4.8'$, $C = 117°43.2'$

22. A pole, which leans 10°15′ from the vertical toward the sun, casts a shadow 40.75 ft long when the angle of elevation of the sun is 40°35′. Find the length of the pole.
Ans. 41.97 ft

23. Two observers A and B, on level ground 2875 ft apart, measure the angle of elevation of an airplane as it flies over the line joining them. The angle of elevation at A is 62°45′ and at B is 50°54′. Find the distance of the airplane from A, from B, and above the earth's surface.
Ans. 2436 ft, 2790 ft, 2165 ft

24. A tunnel is to be constructed through a mountain from A to B. A point C, from which both A and B are visible, is 384.8 ft from A and 555.6 ft from B. How long is the tunnel if $\angle ACB = 35°42'$?
Ans. 330.9 ft

25. Assuming the distance of the sun from the earth to be 92,897,000 mi and the distance of the sun from Mercury to be 35,960,000 mi, find the possible distances of Mercury from the earth when the angle made by Mercury and the sun with the earth as vertex is 8°24.6′.
Ans. 58,600,000 or 125,190,000 mi

26. A point B is inaccessible and invisible from a point A. In order to find the distance AB, two points C and D on a line with A and from which B is visible are selected, and $\angle ADB = 55°18'$ and $\angle ACB = 41°36'$ are measured. If $AD = 432.3$ ft and $AC = 521.8$ ft, find AB.
Ans. 529.1 ft when A is between C and D

CHAPTER 44

Inverse Trigonometric Functions

INVERSE TRIGONOMETRIC FUNCTIONS. The equation

$$(1) \qquad x = \sin y$$

defines a unique value of x for each given angle y. But when x is given, the equation may have no solution or many solutions. For example: if $x = 2$, there is no solution, since the sine of an angle never exceeds 1; if $x = \frac{1}{2}$, there are many solutions $y = 30°, 150°, 390°, 510°, -210°, -330°, \ldots$

To express y as a function of x, we will write

$$(2) \qquad y = \text{arc} \sin x.$$

In spite of the use of the word *arc*, (2) is to be interpreted as stating that "y is an angle whose sine is x". Similarly we shall write $y = \text{arc} \cos x$ if $x = \cos y$, $y = \text{arc} \tan x$ if $x = \tan y$, etc.

The notation $y = \sin^{-1} x$, $y = \cos^{-1} x$, etc., (to be read "inverse sin of x, inverse cosine of x", etc.) are less frequently used since $\sin^{-1} x$ may be confused with $\dfrac{1}{\sin x} = (\sin x)^{-1}$.

GRAPHS OF THE INVERSE TRIGONOMETRIC FUNCTIONS. The graph of $y = \text{arc} \sin x$ is the graph of $x = \sin y$ and differs from the graph of $y = \sin x$ of Chapter 38 in that the roles of x and y are interchanged. Thus, the graph of $y = \text{arc} \sin x$ is a sine curve drawn on the y-axis instead of the x-axis.

Similarly the graphs of the remaining inverse trigonometric functions are those of the corresponding trigonometric functions except that the roles of x and y are interchanged.

PRINCIPAL VALUES. It is at times necessary to consider the inverse trigonometric functions as single valued (i.e., one value of y corresponding to each admissible value of x). To do this, we agree to select one out of the many angles corresponding to the given value of x. For example, when $x = \frac{1}{2}$, we shall agree to select the value $y = 30°$ and when $x = -\frac{1}{2}$, we shall agree to select the value $y = -30°$. This selected value is called the *principal value* of arc sin x. When only the principal value is called for, we shall write Arc sin x, Arc cos x, etc. The portions of the graphs on which the principal values of each of the inverse trigonometric functions lie are shown in the figures below by a heavier line.

When x is positive or zero and the inverse function exists, the principal value is defined as that value of y which lies between 0 and $\frac{1}{2}\pi$ inclusive. For example:

$$\text{Arc} \sin \sqrt{3}/2 = \pi/3 \quad \text{since} \quad \sin \pi/3 = \sqrt{3}/2 \quad \text{and} \quad 0 < \pi/3 < \pi/2,$$

$$\text{Arc} \cos \sqrt{3}/2 = \pi/6 \quad \text{since} \quad \cos \pi/6 = \sqrt{3}/2 \quad \text{and} \quad 0 < \pi/6 < \pi/2,$$

$$\text{Arc} \tan 1 = \pi/4 \quad \text{since} \quad \tan \pi/4 = 1 \quad \text{and} \quad 0 < \pi/4 < \pi/2.$$

When x is negative and the inverse function exists, the principal value is defined as follows:

$$-\tfrac{1}{2}\pi \leqq \text{Arc sin } x < 0 \qquad \tfrac{1}{2}\pi < \text{Arc cot } x < \pi$$

$$\tfrac{1}{2}\pi < \text{Arc cos } x \leqq \pi \qquad -\pi \leqq \text{Arc sec } x < -\tfrac{1}{2}\pi$$

$$-\tfrac{1}{2}\pi < \text{Arc tan } x < 0 \qquad -\pi < \text{Arc csc } x \leqq -\tfrac{1}{2}\pi$$

For example:

$$\text{Arc sin } (-\sqrt{3}/2) = -\pi/3 \qquad \text{Arc cot } (-1) = 3\pi/4$$

$$\text{Arc cos } (-1/2) = 2\pi/3 \qquad \text{Arc sec } (-2/\sqrt{3}) = -5\pi/6$$

$$\text{Arc tan } (-1/\sqrt{3}) = -\pi/6 \qquad \text{Arc csc } (-\sqrt{2}) = -3\pi/4$$

Note. Authors vary in defining the principal values of the inverse functions when x is negative. The definitions given above are the most convenient for the calculus.

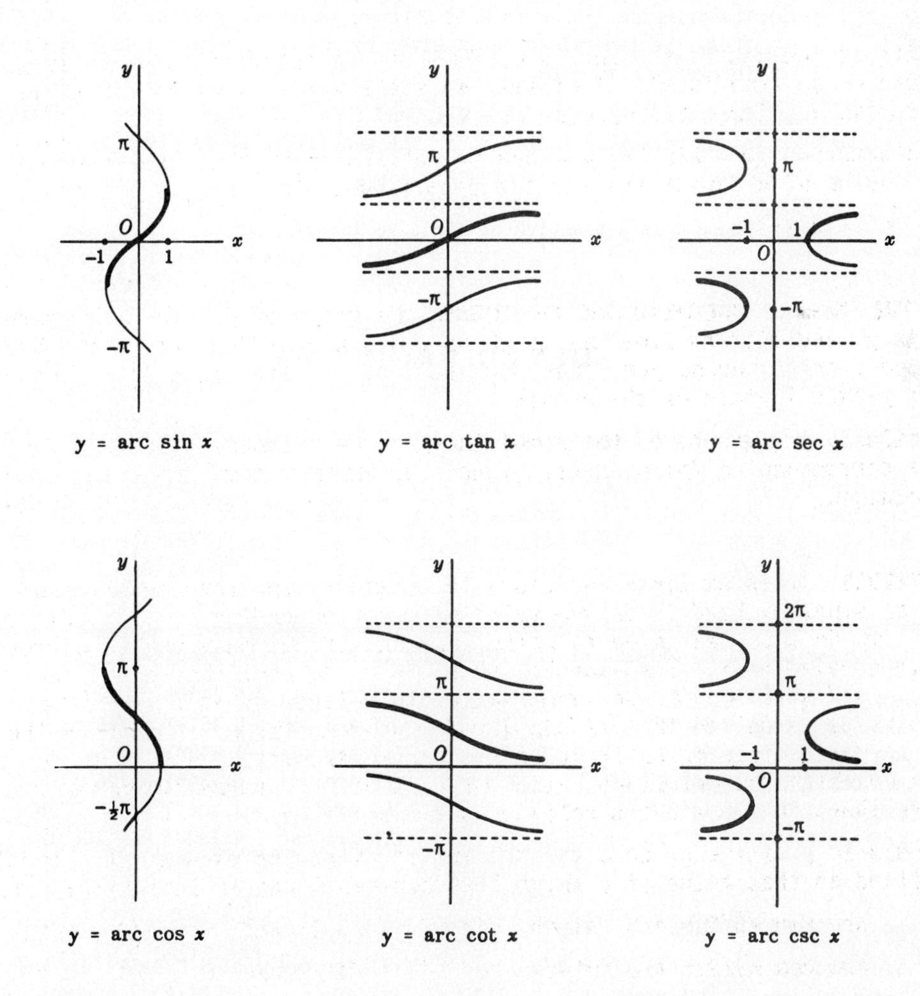

GENERAL VALUES OF THE INVERSE TRIGONOMETRIC FUNCTIONS. Let y be an inverse trigonometric function of x. Since the value of a trigonometric function of y is known, there are determined in general two positions for the terminal side of the angle y (see Chapter 31). Let y_1 and y_2 respectively be angles determined by the two positions of the terminal side. Then the totality of values of y consist of the angles y_1 and y_2, together with all angles coterminal with them, that is,

$$y_1 + 2n\pi \qquad \text{and} \qquad y_2 + 2n\pi$$

where n is any positive or negative integer, or is zero.

One of the values y_1 or y_2 may always be taken as the principal value of the inverse trigonometric function.

Example. Write expressions for the general value of (*a*) arc sin 1/2, (*b*) arc cos (–1), (*c*) arc tan (–1).

(*a*) The principal value of arc sin 1/2 is $\pi/6$, and a second value (not coterminal with the principal value) is $5\pi/6$. The general value of arc sin 1/2 is given by

$$\pi/6 + 2n\pi, \quad 5\pi/6 + 2n\pi$$

where n is any positive or negative integer, or is zero.

(*b*) The principal value is π and there is no other value not coterminal with it. Thus, the general value is given by $\pi + 2n\pi$, where n is a positive or negative integer, or is zero.

(*c*) The principal value is $-\pi/4$, and a second value (not coterminal with the principal value) is $3\pi/4$. Thus, the general value is given by

$$-\pi/4 + 2n\pi, \quad 3\pi/4 + 2n\pi$$

where n is a positive or negative integer, or is zero.

SOLVED PROBLEMS

1. Find the principal value of each of the following.

(*a*) Arc sin 0 = 0
(*b*) Arc cos (–1) = π
(*c*) Arc tan $\sqrt{3} = \pi/3$
(*d*) Arc cot $\sqrt{3} = \pi/6$
(*e*) Arc sec 2 = $\pi/3$
(*f*) Arc csc $(-\sqrt{2}) = -3\pi/4$
(*g*) Arc cos 0 = $\pi/2$
(*h*) Arc sin (–1) = $-\pi/2$
(*i*) Arc tan (–1) = $-\pi/4$
(*j*) Arc cot 0 = $\pi/2$
(*k*) Arc sec $(-\sqrt{2}) = -3\pi/4$
(*l*) Arc csc (–2) = $-5\pi/6$

2. Express the principal value of each of the following to the nearest minute.

(*a*) Arc sin 0.3333 = 19°28′
(*b*) Arc cos 0.4000 = 66°25′
(*c*) Arc tan 1.5000 = 56°19′
(*d*) Arc cot 1.1875 = 40° 6′
(*e*) Arc sec 1.0324 = 14°24′
(*f*) Arc csc 1.5082 = 41°32′
(*g*) Arc sin (–0.6439) = – 40°5′
(*h*) Arc cos (–0.4519) = 116°52′
(*i*) Arc tan (–1.4400) = – 55°13′
(*j*) Arc cot (–0.7340) = 126°17′
(*k*) Arc sec (–1.2067) = – 145°58′
(*l*) Arc csc (–4.1923) = – 166°12′

3. Verify each of the following.

(*a*) sin (Arc sin 1/2) = sin $\pi/6$ = 1/2
(*b*) cos [Arc cos (–1/2)] = cos $2\pi/3$ = –1/2
(*c*) cos [Arc sin $(-\sqrt{2}/2)$] = cos $(-\pi/4) = \sqrt{2}/2$
(*d*) Arc sin (sin $\pi/3$) = Arc sin $\sqrt{3}/2 = \pi/3$
(*e*) Arc cos [cos $(-\pi/4)$] = Arc cos $\sqrt{2}/2 = \pi/4$
(*f*) Arc sin (tan $3\pi/4$) = Arc sin (–1) = $-\pi/2$
(*g*) Arc cos [tan $(-5\pi/4)$] = Arc cos (–1) = π

4. Verify each of the following.

(*a*) $\text{Arc sin}\,\sqrt{2}/2 - \text{Arc sin}\,1/2 = \pi/4 - \pi/6 = \pi/12$

(*b*) $\text{Arc cos}\,0 + \text{Arc tan}\,(-1) = \pi/2 + (-\pi/4) = \pi/4 = \text{Arc tan}\,1$

5. Evaluate each of the following:
(*a*) $\cos(\text{Arc sin}\,3/5)$, (*b*) $\sin[\text{Arc cos}\,(-2/3)]$, (*c*) $\tan[\text{Arc sin}\,(-3/4)]$.

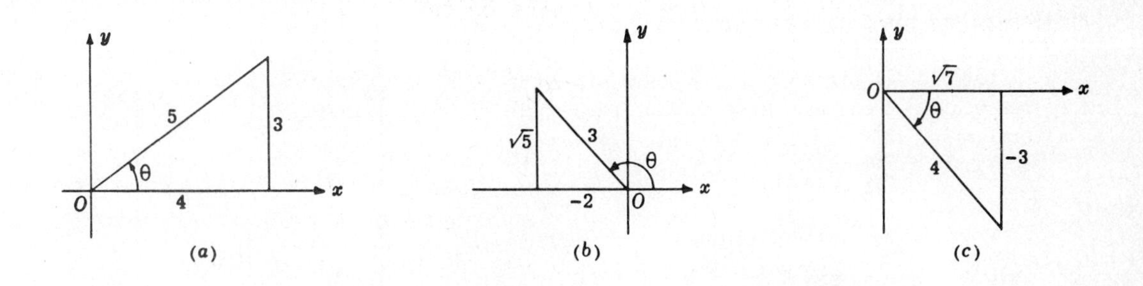

(*a*) Let $\theta = \text{Arc sin}\,3/5$; then $\sin\theta = 3/5$, θ being a first quadrant angle. From Fig. (*a*),
$$\cos(\text{Arc sin}\,3/5) = \cos\theta = 4/5.$$

(*b*) Let $\theta = \text{Arc cos}\,(-2/3)$; then $\cos\theta = -2/3$, θ being a second quadrant angle. From Fig. (*b*),
$$\sin[\text{Arc cos}\,(-2/3)] = \sin\theta = \sqrt{5}/3.$$

(*c*) Let $\theta = \text{Arc sin}\,(-3/4)$; then $\sin\theta = -3/4$, θ being a fourth quadrant angle. From Fig. (*c*),
$$\tan[\text{Arc sin}\,(-3/4)] = \tan\theta = -3/\sqrt{7} = -3\sqrt{7}/7.$$

6. Evaluate $\sin(\text{Arc sin}\,12/13 + \text{Arc sin}\,4/5)$.

Let $\theta = \text{Arc sin}\,12/13$ and
$\phi = \text{Arc sin}\,4/5$.

Then $\sin\theta = 12/13$ and $\sin\phi = 4/5$, θ and ϕ being first quadrant angles. From the adjoining figures,

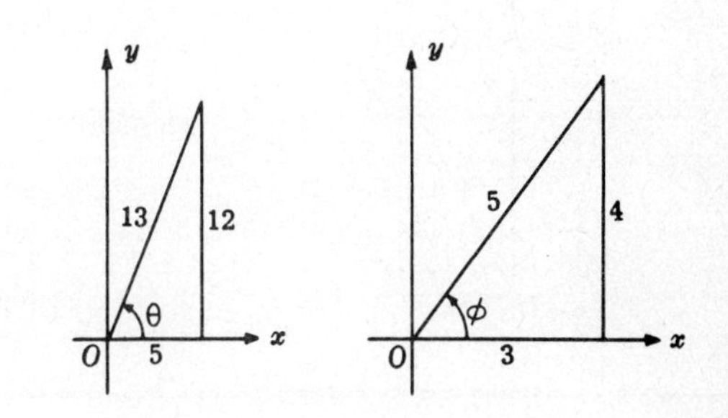

$$\sin(\text{Arc sin}\,12/13 + \text{Arc sin}\,4/5) = \sin(\theta + \phi)$$
$$= \sin\theta\cos\phi + \cos\theta\sin\phi$$
$$= \frac{12}{13}\cdot\frac{3}{5} + \frac{5}{13}\cdot\frac{4}{5} = \frac{56}{65}.$$

7. Evaluate $\cos(\text{Arc tan}\,15/8 - \text{Arc sin}\,7/25)$.

Let $\theta = \text{Arc tan}\,15/8$ and
$\phi = \text{Arc sin}\,7/25$.

Then $\tan\theta = 15/8$ and $\sin\phi = 7/25$, θ and ϕ being first quadrant angles. From the adjoining figures,

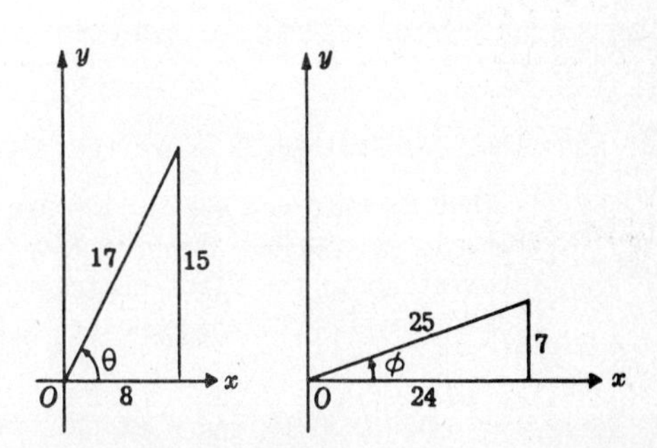

$$\cos(\text{Arc tan}\,15/8 - \text{Arc sin}\,7/25) = \cos(\theta - \phi)$$
$$= \cos\theta\cos\phi + \sin\theta\sin\phi$$
$$= \frac{8}{17}\cdot\frac{24}{25} + \frac{15}{17}\cdot\frac{7}{25} = \frac{297}{425}.$$

8. Evaluate sin (2 Arc tan 3).

Let θ = Arc tan 3; then tan θ = 3, θ being a first quadrant angle.

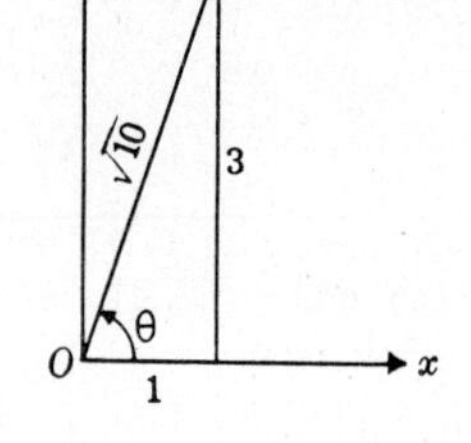

From the adjoining figure,

$$\sin(2\ \text{Arc}\tan 3) = \sin 2\theta = 2\sin\theta\cos\theta = 2(3/\sqrt{10})(1/\sqrt{10}) = 3/5.$$

9. Show that Arc sin $1/\sqrt{5}$ + Arc sin $2/\sqrt{5} = \pi/2$.

Let θ = Arc sin $1/\sqrt{5}$ and ϕ = Arc sin $2/\sqrt{5}$; then sin $\theta = 1/\sqrt{5}$ and sin $\phi = 2/\sqrt{5}$, each angle terminating in the first quadrant. We are to show that $\theta + \phi = \pi/2$ or, taking the sines of both members, that $\sin(\theta+\phi) = \sin \pi/2$.

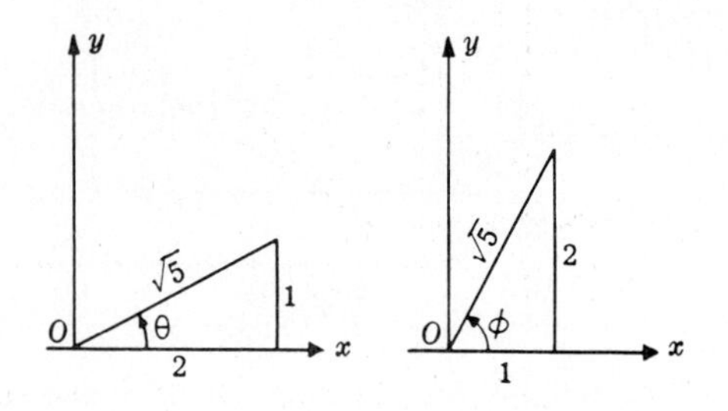

From the adjoining figures,

$$\sin(\theta+\phi) = \sin\theta\cos\phi + \cos\theta\sin\phi = \frac{1}{\sqrt{5}}\cdot\frac{1}{\sqrt{5}} + \frac{2}{\sqrt{5}}\cdot\frac{2}{\sqrt{5}} = 1 = \sin\pi/2.$$

10. Show that 2 Arc tan 1/2 = Arc tan 4/3.

Let θ = Arc tan 1/2 and ϕ = Arc tan 4/3; then tan θ = 1/2 and tan ϕ = 4/3. We are to show that $2\theta = \phi$ or, taking the tangents of both members, that tan 2θ = tan ϕ.

Now $$\tan 2\theta = \frac{2\tan\theta}{1-\tan^2\theta} = \frac{2(1/2)}{1-(1/2)^2} = 4/3 = \tan\phi.$$

11. Show that Arc sin 77/85 − Arc sin 3/5 = Arc cos 15/17.

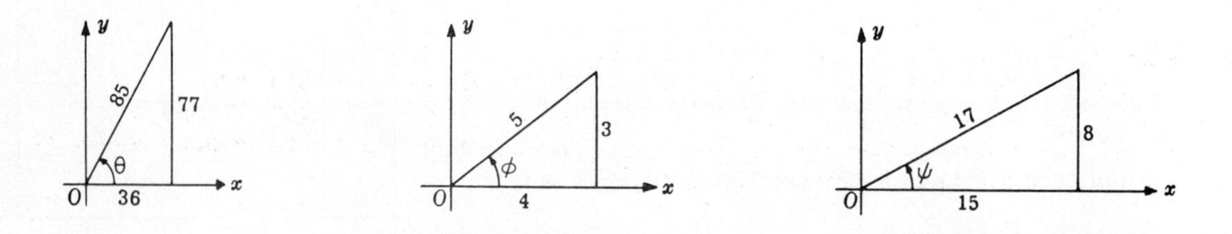

Let θ = Arc sin 77/85, ϕ = Arc sin 3/5, and ψ = Arc cos 15/17; then sin θ = 77/85, sin ϕ = 3/5, and cos ψ = 15/17, each angle terminating in the first quadrant. Taking the sine of both members of the given relation, we are to show that $\sin(\theta - \phi) = \sin\psi$. From the figures,

$$\sin(\theta-\phi) = \sin\theta\cos\phi - \cos\theta\sin\phi = \frac{77}{85}\cdot\frac{4}{5} - \frac{36}{85}\cdot\frac{3}{5} = \frac{8}{17} = \sin\psi.$$

12. Show that Arc cot 43/32 − Arc tan 1/4 = Arc cos 12/13.

Let θ = Arc cot 43/32, ϕ = Arc tan 1/4, and ψ = Arc cos 12/13; then cot θ = 43/32, tan ϕ = 1/4, and cos ψ = 12/13, each angle terminating in the first quadrant. Taking the tangent of both members of the given relation, we are to show that $\tan(\theta-\phi) = \tan\psi$.

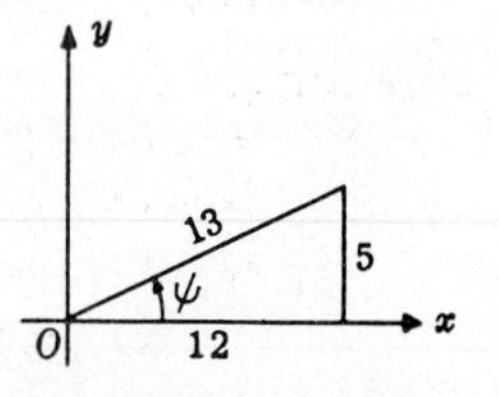

$$\tan(\theta-\phi) = \frac{\tan\theta - \tan\phi}{1+\tan\theta\tan\phi} = \frac{32/43 - 1/4}{1+(32/43)(1/4)} = \frac{5}{12} = \tan\psi.$$

13. Show that $\text{Arc tan } 1/2 + \text{Arc tan } 1/5 + \text{Arc tan } 1/8 = \pi/4$.

We shall show that $\text{Arc tan } 1/2 + \text{Arc tan } 1/5 = \pi/4 - \text{Arc tan } 1/8$.

$$\tan(\text{Arc tan } 1/2 + \text{Arc tan } 1/5) = \frac{1/2 + 1/5}{1 - (1/2)(1/5)} = \frac{7}{9}$$

and
$$\tan(\pi/4 - \text{Arc tan } 1/8) = \frac{1 - 1/8}{1 + 1/8} = \frac{7}{9}.$$

14. Show that $2 \text{ Arc tan } 1/3 + \text{Arc tan } 1/7 = \text{Arc sec } \sqrt{34}/5 + \text{Arc csc } \sqrt{17}$.

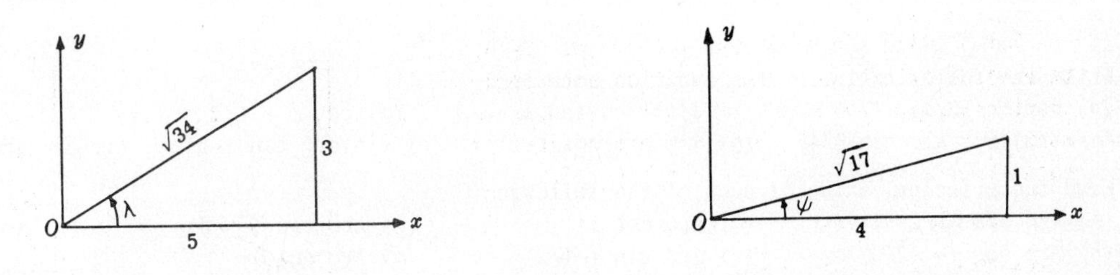

Let $\theta = \text{Arc tan } 1/3$, $\phi = \text{Arc tan } 1/7$, $\lambda = \text{Arc sec } \sqrt{34}/5$, and $\psi = \text{Arc csc } \sqrt{17}$; then $\tan\theta = 1/3$, $\tan\phi = 1/7$, $\sec\lambda = \sqrt{34}/5$, and $\csc\psi = \sqrt{17}$, each angle terminating in the first quadrant.

Taking the tangent of both members of the given relation, we are to show that
$$\tan(2\theta + \phi) = \tan(\lambda + \psi).$$

Now
$$\tan 2\theta = \frac{2\tan\theta}{1 - \tan^2\theta} = \frac{2(1/3)}{1 - (1/3)^2} = 3/4,$$

$$\tan(2\theta + \phi) = \frac{\tan 2\theta + \tan\phi}{1 - \tan 2\theta \tan\phi} = \frac{3/4 + 1/7}{1 - (3/4)(1/7)} = 1$$

and, using the figures above,
$$\tan(\lambda + \psi) = \frac{3/5 + 1/4}{1 - (3/5)(1/4)} = 1.$$

15. Find the general value of each of the following.

(*a*) $\text{arc sin } \sqrt{2}/2 = \pi/4 + 2n\pi,\ 3\pi/4 + 2n\pi$

(*b*) $\text{arc cos } 1/2 = \pi/3 + 2n\pi,\ 5\pi/3 + 2n\pi$

(*c*) $\text{arc tan } 0 = 2n\pi,\ (2n+1)\pi$

(*d*) $\text{arc sin } (-1) = -\pi/2 + 2n\pi$

(*e*) $\text{arc cos } 0 = \pi/2 + 2n\pi,\ 3\pi/2 + 2n\pi$

(*f*) $\text{arc tan } (-\sqrt{3}) = -\pi/3 + 2n\pi,\ 2\pi/3 + 2n\pi$

where n is a positive or negative integer, or is zero.

16. Show that the general value of (*a*) $\text{arc sin } x = n\pi + (-1)^n \text{ Arc sin } x$,
(*b*) $\text{arc cos } x = 2n\pi \pm \text{Arc cos } x$,
(*c*) $\text{arc tan } x = n\pi + \text{Arc tan } x$,
where n is any positive or negative integer, or is zero

(*a*) Let $\theta = \text{Arc sin } x$. Then since $\sin(\pi - \theta) = \sin\theta$, all values of $\text{arc sin } x$ are given by
$$(1)\ \theta + 2m\pi \quad \text{and} \quad (2)\ \pi - \theta + 2m\pi = (2m+1)\pi - \theta.$$

Now, when $n = 2m$, that is n is an even integer, (*1*) may be written as $n\pi + \theta = n\pi + (-1)^n\theta$; and when $n = 2m + 1$, that is n is an odd integer, (2) may be written as $n\pi - \theta = n\pi + (-1)^n\theta$. Thus, $\text{arc sin } x = n\pi + (-1)^n \text{Arc sin } x$, where n is any positive or negative integer, or is zero.

(*b*) Let $\theta = \text{Arc cos } x$. Then since $\cos(-\theta) = \cos\theta$, all values of $\text{arc cos } x$ are given by $\theta + 2n\pi$ and $-\theta + 2n\pi$ or $2n\pi \pm \theta = 2n\pi \pm \text{Arc cos } x$, where n is any positive or negative integer, or is zero.

(*c*) Let $\theta = \text{Arc tan } x$. Then since $\tan(\pi + \theta) = \tan\theta$, all values of $\text{arc tan } x$ are given by $\theta + 2m\pi$ and $(\pi + \theta) + 2m\pi = \theta + (2m+1)\pi$ or, as in (*a*), by $n\pi + \text{Arc tan } x$, where n is any positive or negative integer, or is zero.

17. Express the general value of each of the functions of Problem 15, using the form of Problem 16.

(*a*) $\text{arc sin } \sqrt{2}/2 = n\pi + (-1)^n \pi/4$
(*b*) $\text{arc cos } 1/2 = 2n\pi \pm \pi/3$
(*c*) $\text{arc tan } 0 = n\pi$
(*d*) $\text{arc sin } (-1) = n\pi + (-1)^n(-\pi/2)$
(*e*) $\text{arc cos } 0 = 2n\pi \pm \pi/2$
(*f*) $\text{arc tan } (-\sqrt{3}) = n\pi - \pi/3$

where n is any positive or negative integer, or is zero.

SUPPLEMENTARY PROBLEMS

18. Write the following in inverse function notation.
(*a*) $\sin\theta = 3/4$, (*b*) $\cos\alpha = -1$, (*c*) $\tan x = -2$, (*d*) $\cot\beta = 1/2$.
Ans. (*a*) $\theta = \text{arc sin } 3/4$, (*b*) $\alpha = \text{arc cos } (-1)$, (*c*) $x = \text{arc tan } (-2)$, (*d*) $\beta = \text{arc cot } 1/2$

19. Find the principal value of each of the following.
(*a*) $\text{Arc sin } \sqrt{3}/2$
(*b*) $\text{Arc cos } (-\sqrt{2}/2)$
(*c*) $\text{Arc tan } 1/\sqrt{3}$
(*d*) $\text{Arc cot } 1$
(*e*) $\text{Arc sin } (-1/2)$
(*f*) $\text{Arc cos } (-1/2)$
(*g*) $\text{Arc tan } (-\sqrt{3})$
(*h*) $\text{Arc cot } 0$
(*i*) $\text{Arc sec } (-\sqrt{2})$
(*j*) $\text{Arc csc } (-1)$

Ans. (*a*) $\pi/3$, (*b*) $3\pi/4$, (*c*) $\pi/6$, (*d*) $\pi/4$, (*e*) $-\pi/6$, (*f*) $2\pi/3$, (*g*) $-\pi/3$, (*h*) $\pi/2$, (*i*) $-3\pi/4$, (*j*) $-\pi/2$

20. Evaluate each of the following.
(*a*) $\sin[\text{Arc sin } (-1/2)]$
(*b*) $\cos(\text{Arc cos } \sqrt{3}/2)$
(*c*) $\tan[\text{Arc tan } (-1)]$
(*d*) $\sin[\text{Arc cos } (-\sqrt{3}/2)]$
(*e*) $\tan(\text{Arc sin } 0)$
(*f*) $\sin(\text{Arc cos } 4/5)$
(*g*) $\cos[\text{Arc sin } (-12/13)]$
(*h*) $\sin(\text{Arc tan } 2)$
(*i*) $\text{Arc cos } (\sin 220^\circ)$
(*j*) $\text{Arc sin } [\cos (-105^\circ)]$
(*k*) $\text{Arc tan } (\cot 230^\circ)$
(*l*) $\text{Arc cot } (\tan 100^\circ)$
(*m*) $\sin(2 \text{ Arc sin } 2/3)$
(*n*) $\cos(2 \text{ Arc sin } 3/5)$
(*o*) $\sin(\frac{1}{2} \text{ Arc cos } 4/5)$

Ans. (*a*) $-1/2$ (*b*) $\sqrt{3}/2$ (*c*) -1 (*d*) $1/2$ (*e*) 0 (*f*) $3/5$ (*g*) $5/13$ (*h*) $2/\sqrt{5}$ (*i*) 130° (*j*) -15° (*k*) 40° (*l*) 170° (*m*) $4\sqrt{5}/9$ (*n*) $7/25$ (*o*) $1/\sqrt{10}$

21. Show that

(*a*) $\sin(\text{Arc sin } \frac{5}{13} + \text{Arc sin } \frac{4}{5}) = \frac{63}{65}$

(*b*) $\cos(\text{Arc cos } \frac{15}{17} - \text{Arc cos } \frac{7}{25}) = \frac{297}{425}$

(*c*) $\sin(\text{Arc sin } \frac{1}{2} - \text{Arc cos } \frac{1}{3}) = \frac{1-2\sqrt{6}}{6}$

(*d*) $\tan(\text{Arc tan } \frac{3}{4} + \text{Arc cot } \frac{15}{8}) = \frac{77}{36}$

(*e*) $\cos(\text{Arc tan } \frac{-4}{3} + \text{Arc sin } \frac{12}{13}) = \frac{63}{65}$

(*f*) $\tan(\text{Arc sin } \frac{-3}{5} - \text{Arc cos } \frac{5}{13}) = \frac{63}{16}$

(*g*) $\tan(2 \text{ Arc sin } \frac{4}{5} + \text{Arc cos } \frac{12}{13}) = -\frac{253}{204}$

(*h*) $\sin(2 \text{ Arc sin } \frac{4}{5} - \text{Arc cos } \frac{12}{13}) = \frac{323}{325}$

22. Show that

(*a*) $\text{Arc tan } \frac{1}{2} + \text{Arc tan } \frac{1}{3} = \frac{\pi}{4}$

(*b*) $\text{Arc sin } \frac{4}{5} + \text{Arc tan } \frac{3}{4} = \frac{\pi}{2}$

(*c*) $\text{Arc tan } \frac{4}{3} - \text{Arc tan } \frac{1}{7} = \frac{\pi}{4}$

(*d*) $2 \text{ Arc tan } \frac{1}{3} + \text{Arc tan } \frac{1}{7} = \frac{\pi}{4}$

(*e*) $\text{Arc cos } \frac{12}{13} + \text{Arc tan } \frac{1}{4} = \text{Arc cot } \frac{43}{32}$

(*f*) $\text{Arc sin } \frac{3}{5} + \text{Arc sin } \frac{15}{17} = \text{Arc cos } \frac{-13}{85}$

(*g*) $\text{Arc tan } a + \text{Arc tan } \frac{1}{a} = \frac{\pi}{2}$ $(a > 0)$.

23. Prove: The area of the segment cut from a circle of radius r by a chord at a distance d from the center is given by $K = r^2 \text{ Arc cos } \frac{d}{r} - d\sqrt{r^2 - d^2}$.

CHAPTER 45

Trigonometric Equations

TRIGONOMETRIC EQUATIONS, i.e., equations involving trigonometric functions of unknown angles, are called:

(*a*) identical equations or *identities*, if they are satisfied by all values of the unknown angles for which the functions are defined;

(*b*) conditional equations, or equations, if they are satisfied only by particular values of the unknown angles.

For example: (*a*) $\sin x \csc x = 1$ is an identity, being satisfied by every value of x for which $\csc x$ is defined;

(*b*) $\sin x = 0$ is a conditional equation since it is not satisfied by $x = \frac{1}{4}\pi$ or $\frac{1}{2}\pi$.

Hereafter in this chapter we shall use the term "equation" instead of "conditional equation".

A SOLUTION OF A TRIGONOMETRIC EQUATION, as $\sin x = 0$, is a value of the angle x which satisfies the equation. Two solutions of $\sin x = 0$ are $x = 0$ and $x = \pi$.

If a given equation has one solution, it has in general an unlimited number of solutions. Thus, the complete solution of $\sin x = 0$ is given by

$$x = 0 + 2n\pi, \qquad x = \pi + 2n\pi$$

where n is any positive or negative integer or is zero.

In this chapter we shall list only the particular solutions for which $0 \leqq x < 2\pi$.

PROCEDURES FOR SOLVING TRIGONOMETRIC EQUATIONS. There is no general method for solving trigonometric equations. Three standard procedures are illustrated below and other procedures are introduced in the solved problems.

(*A*) The equation may be factorable.

Example 1. Solve $\sin x - 2 \sin x \cos x = 0$.

Factoring, $\sin x - 2 \sin x \cos x = \sin x\,(1 - 2 \cos x) = 0$, and setting each factor equal to zero, we have

$$\sin x = 0 \quad \text{and} \quad x = 0, \pi;$$
$$1 - 2\cos x = 0 \quad \text{or} \quad \cos x = \tfrac{1}{2} \quad \text{and} \quad x = \pi/3,\ 5\pi/3.$$

Check. For $x = 0$, $\quad \sin x - 2 \sin x \cos x = 0 - 2(0)(1) = 0$;
for $x = \pi/3$, $\quad \sin x - 2 \sin x \cos x = \frac{1}{2}\sqrt{3} - 2(\frac{1}{2}\sqrt{3})(\frac{1}{2}) = 0$;
for $x = \pi$, $\quad \sin x - 2 \sin x \cos x = 0 - 2(0)(-1) = 0$;
for $x = 5\pi/3$, $\quad \sin x - 2 \sin x \cos x = -\frac{1}{2}\sqrt{3} - 2(-\frac{1}{2}\sqrt{3})(\frac{1}{2}) = 0$.

Thus, the required solutions $(0 \leqq x < 2\pi)$ are $x = 0, \pi/3, \pi, 5\pi/3$.

(*B*) The various functions occurring in the equation may be expressed in terms of a single function.

Example 2. Solve $2\tan^2 x + \sec^2 x = 2$.

Replacing $\sec^2 x$ by $1+\tan^2 x$, we have

$$2\tan^2 x + (1+\tan^2 x) = 2, \quad 3\tan^2 x = 1, \quad \text{and} \quad \tan x = \pm 1/\sqrt{3}.$$

From $\tan x = 1/\sqrt{3}$, $x = \pi/6$ and $7\pi/6$; from $\tan x = -1/\sqrt{3}$, $x = 5\pi/6$ and $11\pi/6$. After checking each of these values in the original equation, we find that the required solutions ($0 \leqq x < 2\pi$) are $x = \pi/6, 5\pi/6, 7\pi/6, 11\pi/6$.

The necessity of the check is illustrated in

Example 3. Solve $\sec x + \tan x = 0$.

Multiplying the equation $\sec x + \tan x = \dfrac{1}{\cos x} + \dfrac{\sin x}{\cos x} = 0$ by $\cos x$, we have $1 + \sin x = 0$ or $\sin x = -1$; then $x = 3\pi/2$. However, neither $\sec x$ nor $\tan x$ is defined when $x = 3\pi/2$ and the equation has no solution.

(*C*) Both members of the equation are squared.

Example 4. Solve $\sin x + \cos x = 1$.

If the procedure of (*B*) were used, we would replace $\sin x$ by $\pm\sqrt{1-\cos^2 x}$ or $\cos x$ by $\pm\sqrt{1-\sin^2 x}$ and thereby introduce radicals. To avoid this, we write the equation in the form $\sin x = 1 - \cos x$ and square both members. We have

$$(1) \qquad \sin^2 x = 1 - 2\cos x + \cos^2 x,$$

$$1 - \cos^2 x = 1 - 2\cos x + \cos^2 x,$$

$$2\cos^2 x - 2\cos x = 2\cos x\,(\cos x - 1) = 0.$$

From $\cos x = 0$, $x = \pi/2, 3\pi/2$; from $\cos x = 1$, $x = 0$.

Check. For $x = 0$, $\sin x + \cos x = 0 + 1 = 1$;
for $x = \pi/2$, $\sin x + \cos x = 1 + 0 = 1$;
for $x = 3\pi/2$, $\sin x + \cos x = -1 + 0 \neq 1$.

Thus, the required solutions are $x = 0, \pi/2$.

The value $x = 3\pi/2$, called an *extraneous solution*, was introduced by squaring the two members. Note that (*1*) is also obtained when both members of $\sin x = \cos x - 1$ are squared and that $x = 3\pi/2$ satisfies this latter relation.

SOLVED PROBLEMS

Solve each of the trigonometric equations 1-22 for all x such that $0 \leqq x < 2\pi$. (If all solutions are required, adjoin $+2n\pi$, where n is zero or any positive or negative integer, to each result given.) In a number of solutions, the details of the check have been omitted.

1. $2\sin x - 1 = 0$.

Here $\sin x = 1/2$ and $x = \pi/6, 5\pi/6$.

2. $\sin x \cos x = 0$.

From $\sin x = 0$, $x = 0, \pi$; from $\cos x = 0$, $x = \pi/2, 3\pi/2$.
The required solutions are $x = 0, \pi/2, \pi, 3\pi/2$.

3. $(\tan x - 1)(4\sin^2 x - 3) = 0$.

From $\tan x - 1 = 0$, $\tan x = 1$ and $x = \pi/4, 5\pi/4$; from $4\sin^2 x - 3 = 0$, $\sin x = \pm\sqrt{3}/2$ and $x = \pi/3, 2\pi/3, 4\pi/3, 5\pi/3$.
The required solutions are $x = \pi/4, \pi/3, 2\pi/3, 5\pi/4, 4\pi/3, 5\pi/3$.

4. $\sin^2 x + \sin x - 2 = 0$.

Factoring, $(\sin x + 2)(\sin x - 1) = 0$.

From $\sin x + 2 = 0$, $\sin x = -2$ and there is no solution; from $\sin x - 1 = 0$, $\sin x = 1$ and $x = \pi/2$. The required solution is $x = \pi/2$.

5. $3\cos^2 x = \sin^2 x$.

First Solution. Replacing $\sin^2 x$ by $1 - \cos^2 x$, we have $3\cos^2 x = 1 - \cos^2 x$ or $4\cos^2 x = 1$. Then $\cos x = \pm 1/2$ and the required solutions are $x = \pi/3, 2\pi/3, 4\pi/3, 5\pi/3$.

Second Solution. Dividing the equation by $\cos^2 x$, we have $3 = \tan^2 x$. Then $\tan x = \pm\sqrt{3}$ and the solutions above are obtained.

6. $2\sin x - \csc x = 1$.

Multiplying the equation by $\sin x$, $2\sin^2 x - 1 = \sin x$, and rearranging, we have

$$2\sin^2 x - \sin x - 1 = (2\sin x + 1)(\sin x - 1) = 0.$$

From $2\sin x + 1 = 0$, $\sin x = -1/2$ and $x = 7\pi/6, 11\pi/6$; from $\sin x = 1$, $x = \pi/2$.

Check. For $x = \pi/2$, $2\sin x - \csc x = 2(1) - 1 = 1$;
for $x = 7\pi/6$ and $11\pi/6$, $2\sin x - \csc x = 2(-1/2) - (-2) = 1$.

The solutions are $x = \pi/2, 7\pi/6, 11\pi/6$.

7. $2\sec x = \tan x + \cot x$.

Transforming to sines and cosines, and clearing of fractions, we have

$$\frac{2}{\cos x} = \frac{\sin x}{\cos x} + \frac{\cos x}{\sin x} \quad \text{or} \quad 2\sin x = \sin^2 x + \cos^2 x = 1.$$

Then $\sin x = 1/2$ and $x = \pi/6, 5\pi/6$.

8. $\tan x + 3\cot x = 4$.

Multiplying by $\tan x$ and rearranging, $\tan^2 x - 4\tan x + 3 = (\tan x - 1)(\tan x - 3) = 0$.

From $\tan x - 1 = 0$, $\tan x = 1$ and $x = \pi/4, 5\pi/4$; from $\tan x - 3 = 0$, $\tan x = 3$ and $x = 71°34', 251°34'$.

Check. For $x = \pi/4$ and $5\pi/4$, $\tan x + 3\cot x = 1 + 3(1) = 4$;
for $x = 71°34'$ and $251°34'$, $\tan x + 3\cot x = 3 + 3(1/3) = 4$.

The solutions are $45°, 71°34', 225°, 251°34'$.

9. $\csc x + \cot x = \sqrt{3}$.

First Solution. Writing the equation in the form $\csc x = \sqrt{3} - \cot x$ and squaring, we have

$$\csc^2 x = 3 - 2\sqrt{3}\cot x + \cot^2 x.$$

Replacing $\csc^2 x$ by $1 + \cot^2 x$ and combining, this becomes $2\sqrt{3}\cot x - 2 = 0$. Then $\cot x = 1/\sqrt{3}$ and $x = \pi/3, 4\pi/3$.

Check. For $x = \pi/3$, $\csc x + \cot x = 2/\sqrt{3} + 1/\sqrt{3} = \sqrt{3}$;
for $x = 4\pi/3$, $\csc x + \cot x = -2/\sqrt{3} + 1/\sqrt{3} \neq \sqrt{3}$. The required solution is $x = \pi/3$.

Second Solution. Upon making the indicated replacement, the equation becomes

$$\frac{1}{\sin x} + \frac{\cos x}{\sin x} = \sqrt{3} \quad \text{and, clearing of fractions,} \quad 1 + \cos x = \sqrt{3}\sin x.$$

Squaring both members, we have $1 + 2\cos x + \cos^2 x = 3\sin^2 x = 3(1 - \cos^2 x)$ or

$$4\cos^2 x + 2\cos x - 2 = 2(2\cos x - 1)(\cos x + 1) = 0.$$

From $2\cos x - 1 = 0$, $\cos x = 1/2$ and $x = \pi/3, 5\pi/3$; from $\cos x + 1 = 0$, $\cos x = -1$ and $x = \pi$.

Now $x = \pi/3$ is the solution. The values $x = \pi$ and $5\pi/3$ are to be excluded since $\csc\pi$ is not defined while $\csc 5\pi/3$ and $\cot 5\pi/3$ are both negative.

10. $\cos x - \sqrt{3}\sin x = 1$.

First Solution. Putting the equation in the form $\cos x - 1 = \sqrt{3}\sin x$ and squaring, we have

$$\cos^2 x - 2\cos x + 1 = 3\sin^2 x = 3(1 - \cos^2 x);$$

then, combining and factoring,

$$4\cos^2 x - 2\cos x - 2 = 2(2\cos x + 1)(\cos x - 1) = 0.$$

From $2\cos x + 1 = 0$, $\cos x = -1/2$ and $x = 2\pi/3, 4\pi/3$; from $\cos x - 1 = 0$, $\cos x = 1$ and $x = 0$.

Check. For $x = 0$, $\cos x - \sqrt{3}\sin x = 1 - \sqrt{3}(0) = 1$;
for $x = 2\pi/3$, $\cos x - \sqrt{3}\sin x = -1/2 - \sqrt{3}(\sqrt{3}/2) \neq 1$;
for $x = 4\pi/3$, $\cos x - \sqrt{3}\sin x = -1/2 - \sqrt{3}(-\sqrt{3}/2) = 1$.

The required solutions are $x = 0, 4\pi/3$.

Second Solution. The left member of the given equation may be put in the form

$$\sin\theta\cos x + \cos\theta\sin x = \sin(\theta + x),$$

in which θ is a known angle, by dividing the given equation by $r > 0$, $\frac{1}{r}\cos x + (\frac{-\sqrt{3}}{r})\sin x = \frac{1}{r}$, and setting $\sin\theta = \frac{1}{r}$ and $\cos\theta = \frac{-\sqrt{3}}{r}$. Since $\sin^2\theta + \cos^2\theta = 1$, $(\frac{1}{r})^2 + (\frac{-\sqrt{3}}{r})^2 = 1$ and $r = 2$. Now $\sin\theta = 1/2$, $\cos\theta = -\sqrt{3}/2$ so that the given equation may be written as $\sin(\theta + x) = 1/2$ with $\theta = 5\pi/6$. Then $\theta + x = 5\pi/6 + x = \text{arc}\sin 1/2 = \pi/6, 5\pi/6, 13\pi/6, 17\pi/6, \ldots$ and $x = -2\pi/3, 0, 4\pi/3, 2\pi, \ldots$. As before, the required solutions are $x = 0, 4\pi/3$.

Note that r is the positive square root of the sum of the squares of the coefficients of $\cos x$ and $\sin x$ when the equation is written in the form $a\cos x + b\sin x = c$, that is,

$$r = \sqrt{a^2 + b^2}.$$

The equation will have no solution if $\frac{c}{\sqrt{a^2 + b^2}}$ is greater than 1 or less than -1.

11. $2\cos x = 1 - \sin x$.

First Solution. As in Problem 10, we obtain

$$4\cos^2 x = 1 - 2\sin x + \sin^2 x,$$

$$4(1 - \sin^2 x) = 1 - 2\sin x + \sin^2 x,$$

$$5\sin^2 x - 2\sin x - 3 = (5\sin x + 3)(\sin x - 1) = 0.$$

From $5\sin x + 3 = 0$, $\sin x = -3/5 = -0.6000$ and $x = 216°52', 323°8'$; from $\sin x - 1 = 0$, $\sin x = 1$ and $x = \pi/2$.

Check. For $x = \pi/2$, $2(0) = 1 - 1$;
for $x = 216°52'$, $2(-4/5) \neq 1 - (-3/5)$;
for $x = 323°8'$, $2(4/5) = 1 - (-3/5)$.

The required solutions are $x = 90°, 323°8'$.

Second Solution. Writing the equation as $2\cos x + \sin x = 1$ and dividing by $r = \sqrt{2^2 + 1^2} = \sqrt{5}$, we have

$$(1) \quad \frac{2}{\sqrt{5}} \cos x + \frac{1}{\sqrt{5}} \sin x = \frac{1}{\sqrt{5}}.$$

Let $\sin\theta = 2/\sqrt{5}$, $\cos\theta = 1/\sqrt{5}$; then (1) becomes

$$\sin\theta\cos x + \cos\theta\sin x = \sin(\theta + x) = \frac{1}{\sqrt{5}}$$

with $\theta = 63°26'$. Now $\theta + x = 63°26' + x = \arcsin(1/\sqrt{5}) = \arcsin(0.4472) = 26°34', 153°26', 386°34', \ldots$ and $x = 90°, 323°8'$ as before.

EQUATIONS INVOLVING MULTIPLE ANGLES.

12. $\sin 3x = -\frac{1}{2}\sqrt{2}$.

Since we require x such that $0 \leqq x < 2\pi$, $3x$ must be such that $0 \leqq 3x < 6\pi$.

Then $3x = 5\pi/4, 7\pi/4, 13\pi/4, 15\pi/4, 21\pi/4, 23\pi/4$ and $x = 5\pi/12, 7\pi/12, 13\pi/12, 5\pi/4, 7\pi/4, 23\pi/12$. Each of these values is a solution.

13. $\cos\frac{1}{2}x = \frac{1}{2}$.

Since we require x such that $0 \leqq x < 2\pi$, $\frac{1}{2}x$ must be such that $0 \leqq \frac{1}{2}x < \pi$.

Then $\frac{1}{2}x = \pi/3$ and $x = 2\pi/3$.

14. $\sin 2x + \cos x = 0$.

Substituting for $\sin 2x$, we have $2\sin x\cos x + \cos x = \cos x(2\sin x + 1) = 0$.

From $\cos x = 0$, $x = \pi/2, 3\pi/2$; from $\sin x = -1/2$, $x = 7\pi/6, 11\pi/6$.

The required solutions are $x = \pi/2, 7\pi/6, 3\pi/2, 11\pi/6$.

15. $2\cos^2\frac{1}{2}x = \cos^2 x$.

Substituting $1 + \cos x$ for $2\cos^2\frac{1}{2}x$, the equation becomes $\cos^2 x - \cos x - 1 = 0$; then $\cos x = \frac{1 \pm \sqrt{5}}{2} = 1.6180, -0.6180$. Since $\cos x$ cannot exceed 1, we consider $\cos x = -0.6180$ and obtain the solutions $x = 128°10', 231°50'$.

Note. To solve $\sqrt{2}\cos\frac{1}{2}x = \cos x$ and $\sqrt{2}\cos\frac{1}{2}x = -\cos x$, we square and obtain the equation of this problem. The solution of the first of these equations is $231°50'$ and the solution of the second is $128°10'$.

16. $\cos 2x + \cos x + 1 = 0$.

Substituting $2\cos^2 x - 1$ for $\cos 2x$, we have $2\cos^2 x + \cos x = \cos x(2\cos x + 1) = 0$.

From $\cos x = 0$, $x = \pi/2, 3\pi/2$; from $\cos x = -1/2$, $x = 2\pi/3, 4\pi/3$.

The required solutions are $x = \pi/2, 2\pi/3, 3\pi/2, 4\pi/3$.

17. $\tan 2x + 2\sin x = 0$.

Using $\tan 2x = \frac{\sin 2x}{\cos 2x} = \frac{2\sin x\cos x}{\cos 2x}$, we have

$$\frac{2\sin x\cos x}{\cos 2x} + 2\sin x = 2\sin x\left(\frac{\cos x}{\cos 2x} + 1\right) = 2\sin x\left(\frac{\cos x + \cos 2x}{\cos 2x}\right) = 0.$$

From $\sin x = 0$, $x = 0, \pi$; from $\cos x + \cos 2x = \cos x + 2\cos^2 x - 1 = (2\cos x - 1)(\cos x + 1) = 0$, $x = \pi/3, 5\pi/3$, and π. The required solutions are $x = 0, \pi/3, \pi, 5\pi/3$.

18. $\sin 2x = \cos 2x$.

First Solution. Let $2x = \theta$; then we are to solve $\sin\theta = \cos\theta$ for $0 \leqq \theta < 4\pi$. Then $\theta = \pi/4, 5\pi/4, 9\pi/4, 13\pi/4$ and $x = \theta/2 = \pi/8, 5\pi/8, 9\pi/8, 13\pi/8$ are the solutions.

Second Solution. Dividing by $\cos 2x$, the equation becomes $\tan 2x = 1$ for which $2x = \pi/4, 5\pi/4, 9\pi/4, 13\pi/4$ as in the first solution.

19. $\sin 2x = \cos 4x$.

Since $\cos 4x = \cos 2(2x) = 1 - 2\sin^2 2x$, the equation becomes

$$2\sin^2 2x + \sin 2x - 1 = (2\sin 2x - 1)(\sin 2x + 1) = 0.$$

From $2\sin 2x - 1 = 0$ or $\sin 2x = 1/2$, $2x = \pi/6, 5\pi/6, 13\pi/6, 17\pi/6$ and $x = \pi/12, 5\pi/12, 13\pi/12, 17\pi/12$; from $\sin 2x + 1 = 0$ or $\sin 2x = -1$, $2x = 3\pi/2, 7\pi/2$ and $x = 3\pi/4, 7\pi/4$. All of these values are solutions.

20. $\sin 3x = \cos 2x$.

To avoid the substitution for $\sin 3x$, we use one of the procedures below.

First Solution, Since $\cos 2x = \sin(\frac{1}{2}\pi - 2x)$ and also $\cos 2x = \sin(\frac{1}{2}\pi + 2x)$, we consider
(*a*) $\sin 3x = \sin(\frac{1}{2}\pi - 2x)$, obtaining $3x = \pi/2 - 2x, 5\pi/2 - 2x, 9\pi/2 - 2x, \cdots$, and
(*b*) $\sin 3x = \sin(\frac{1}{2}\pi + 2x)$, obtaining $3x = \pi/2 + 2x, 5\pi/2 + 2x, 9\pi/2 + 2x, \cdots$.

From (*a*), $5x = \pi/2, 5\pi/2, 9\pi/2, 13\pi/2, 17\pi/2$ (since $5x < 10\pi$); and from (*b*), $x = \pi/2$. The required solutions are $x = \pi/10, \pi/2, 9\pi/10, 13\pi/10, 17\pi/10$.

Second Solution. Since $\sin 3x = \cos(\frac{1}{2}\pi - 3x)$ and $\cos 2x = \cos(-2x)$, we consider
(*c*) $\cos 2x = \cos(\frac{1}{2}\pi - 3x)$, obtaining $5x = \pi/2, 5\pi/2, 9\pi/2, 13\pi/2, 17\pi/2$, and
(*d*) $\cos(-2x) = \cos(\frac{1}{2}\pi - 3x)$, obtaining $x = \pi/2$, as before.

21. $\tan 4x = \cot 6x$.

Since $\cot 6x = \tan(\frac{1}{2}\pi - 6x)$, we consider the equation $\tan 4x = \tan(\frac{1}{2}\pi - 6x)$.

Then $4x = \pi/2 - 6x, 3\pi/2 - 6x, 5\pi/2 - 6x, \cdots$, the function $\tan\theta$ being of period π.

Thus, $10x = \pi/2, 3\pi/2, 5\pi/2, 7\pi/2, 9\pi/2, \cdots, 39\pi/2$ and the required solutions are $x = \pi/20, 3\pi/20, \pi/4, 7\pi/20, \cdots, 39\pi/20$.

22. $\sin 5x - \sin 3x - \sin x = 0$.

Replacing $\sin 5x - \sin 3x$ by $2\cos 4x \sin x$ (Chapter 41), the given equation becomes

$$2\cos 4x \sin x - \sin x = \sin x\,(2\cos 4x - 1) = 0.$$

From $\sin x = 0$, $x = 0, \pi$; from $2\cos 4x - 1 = 0$ or $\cos 4x = 1/2$, $4x = \pi/3, 5\pi/3, 7\pi/3, 11\pi/3, 13\pi/3, 17\pi/3, 19\pi/3, 23\pi/3$ and $x = \pi/12, 5\pi/12, 7\pi/12, 11\pi/12, 13\pi/12, 17\pi/12, 19\pi/12, 23\pi/12$. Each of the values obtained is a solution.

23. Solve the system (*1*) $r\sin\theta = 2$, (*2*) $r\cos\theta = 3$ for $r > 0$ and $0 \leqq \theta < 2\pi$.

Squaring the two equations and adding, $r^2\sin\theta + r^2\cos\theta = r^2 = 13$ and $r = \sqrt{13} = 3.606$.

When $r > 0$, $\sin\theta$ and $\cos\theta$ are both > 0 and θ is acute.

Dividing (*1*) by (*2*), $\tan\theta = 2/3 = 0.6667$ and $\theta = 33°41'$.

24. Solve the system $\begin{matrix}(1)\ r\sin\theta = 3\\ (2)\ r = 4(1+\sin\theta)\end{matrix}$ for $r > 0$ and $0 \leqq \theta < 2\pi$.

Dividing (2) by (1), $\dfrac{1}{\sin\theta} = \dfrac{4(1+\sin\theta)}{3}$ or $4\sin^2\theta + 4\sin\theta - 3 = 0$ and

$$(2\sin\theta + 3)(2\sin\theta - 1) = 0.$$

From $2\sin\theta - 1 = 0$, $\sin\theta = 1/2$, $\theta = \pi/6$ and $5\pi/6$; using (1), $r(1/2) = 3$ and $r = 6$. Note that $2\sin\theta + 3 = 0$ is excluded since when $r > 0$, $\sin\theta > 0$ by (1).

The required solutions are $\theta = \pi/6$, $r = 6$ and $\theta = 5\pi/6$, $r = 6$.

25. Solve the system $\begin{matrix}(1)\ \sin x + \sin y = 1.2\\ (2)\ \cos x + \cos y = 1.5\end{matrix}$ for $0 \leqq x, y < 2\pi$.

Since each sum on the left is greater than 1, each of the four functions is positive and both x and y are acute.

Using the appropriate formulas of Chapter 41, we obtain

$$(1')\quad 2\sin\tfrac{1}{2}(x+y)\cos\tfrac{1}{2}(x-y) = 1.2$$
$$(2')\quad 2\cos\tfrac{1}{2}(x+y)\cos\tfrac{1}{2}(x-y) = 1.5.$$

Dividing (1′) by (2′), $\dfrac{\sin\frac{1}{2}(x+y)}{\cos\frac{1}{2}(x+y)} = \tan\tfrac{1}{2}(x+y) = \dfrac{1.2}{1.5} = 0.8000$ and $\tfrac{1}{2}(x+y) = 38°40'$ since $\tfrac{1}{2}(x+y)$ is also acute.

Substituting for $\sin\tfrac{1}{2}(x+y) = 0.6248$ in (1′), we have $\cos\tfrac{1}{2}(x-y) = \dfrac{0.6}{0.6248} = 0.9603$ and $\tfrac{1}{2}(x-y) = 16°12'$.

Then $x = \tfrac{1}{2}(x+y) + \tfrac{1}{2}(x-y) = 54°52'$ and $y = \tfrac{1}{2}(x+y) - \tfrac{1}{2}(x-y) = 22°28'$.

26. Solve $\text{Arc}\cos 2x = \text{Arc}\sin x$.

If x is positive, $\alpha = \text{Arc}\cos 2x$ and $\beta = \text{Arc}\sin x$ terminate in quadrant I; if x is negative, α terminates in quadrant II and β terminates in quadrant IV. Thus, x must be positive.

For x positive, $\sin\beta = x$ and $\cos\beta = \sqrt{1-x^2}$. Taking the cosine of both members of the given equation, we have

$$\cos(\text{Arc}\cos 2x) = \cos(\text{Arc}\sin x) = \cos\beta \quad\text{or}\quad 2x = \sqrt{1-x^2}.$$

Squaring, $4x^2 = 1 - x^2$, $5x^2 = 1$, and $x = \sqrt{5}/5 = 0.4472$.

Check. $\text{Arc}\cos 2x = \text{Arc}\cos 0.8944 = 26°30' = \text{Arc}\sin 0.4472$, approximating the angle to the nearest $10'$.

27. Solve $\text{Arc}\cos(2x^2 - 1) = 2\,\text{Arc}\cos\tfrac{1}{2}$.

Let $\alpha = \text{Arc}\cos(2x^2-1)$ and $\beta = \text{Arc}\cos\tfrac{1}{2}$; then $\cos\alpha = 2x^2 - 1$ and $\cos\beta = \tfrac{1}{2}$.

Taking the cosine of both members of the given equation,

$$\cos\alpha = 2x^2 - 1 = \cos 2\beta = 2\cos^2\beta - 1 = 2(\tfrac{1}{2})^2 - 1 = -\tfrac{1}{2}.$$

Then $2x^2 = \tfrac{1}{2}$ and $x = \pm\tfrac{1}{2}$.

Check. For $x = \pm\tfrac{1}{2}$, $\text{Arc}\cos(-\tfrac{1}{2}) = 2\,\text{Arc}\cos\tfrac{1}{2}$ or $120° = 2(60°)$.

28. Solve $\text{Arc}\cos 2x - \text{Arc}\cos x = \pi/3$.

If x is positive, $0 < \text{Arc}\cos 2x < \text{Arc}\cos x$; if x is negative, $\text{Arc}\cos 2x > \text{Arc}\cos x > 0$.

Thus, x must be negative.

Let α = Arc cos $2x$ and β = Arc cos x; then $\cos\alpha = 2x$, $\sin\alpha = \sqrt{1-4x^2}$, $\cos\beta = x$ and $\sin\beta = \sqrt{1-x^2}$ since both α and β terminate in quadrant II.

Taking the cosine of both members of the given equation,

$$\cos(\alpha-\beta) = \cos\alpha\cos\beta + \sin\alpha\sin\beta = 2x^2 + \sqrt{1-4x^2}\sqrt{1-x^2} = \cos\pi/3 = \tfrac{1}{2}$$

or

$$\sqrt{1-4x^2}\sqrt{1-x^2} = \tfrac{1}{2} - 2x^2.$$

Squaring, $1 - 5x^2 + 4x^4 = \dfrac{1}{4} - 2x^2 + 4x^4$, $3x^2 = \dfrac{3}{4}$, and $x = -\tfrac{1}{2}$.

Check. Arc cos (-1) − Arc cos $(-\tfrac{1}{2}) = \pi - 2\pi/3 = \pi/3$.

29. Solve Arc sin $2x = \tfrac{1}{4}\pi$ − Arc sin x.

Let α = Arc sin $2x$ and β = arc sin x; then $\sin\alpha = 2x$ and $\sin\beta = x$. If x is negative, α and β terminate in quadrant IV; thus, x must be positive and β acute.

Taking the sine of both members of the given equation,

$$\sin\alpha = \sin(\tfrac{1}{4}\pi - \beta) = \sin\tfrac{1}{4}\pi\cos\beta - \cos\tfrac{1}{4}\pi\sin\beta$$

or

$$2x = \tfrac{1}{2}\sqrt{2}\sqrt{1-x^2} - \tfrac{1}{2}\sqrt{2}\,x \quad\text{and}\quad (2\sqrt{2}+1)\,x = \sqrt{1-x^2}.$$

Squaring, $(8 + 4\sqrt{2} + 1)x^2 = 1 - x^2$, $x^2 = 1/(10 + 4\sqrt{2})$, and $x = 0.2527$.

Check. Arc sin 0.5054 = 30°22′; Arc sin 0.2527 = 14°38′ and $\tfrac{1}{4}\pi$ − 14°38′ = 30°22′.

30. Solve Arc tan x + Arc tan $(1-x)$ = Arc tan 4/3.

Let α = Arc tan x and β = Arc tan $(1-x)$; then $\tan\alpha = x$ and $\tan\beta = 1-x$.

Taking the tangent of both members of the given equation,

$$\tan(\alpha+\beta) = \frac{\tan\alpha + \tan\beta}{1 - \tan\alpha\tan\beta} = \frac{x + (1-x)}{1 - x(1-x)} = \frac{1}{1 - x + x^2} = \tan(\text{Arc tan } 4/3) = 4/3.$$

Then $3 = 4 - 4x + 4x^2$, $4x^2 - 4x + 1 = (2x-1)^2 = 0$, and $x = \tfrac{1}{2}$.

Check. Arc tan $\tfrac{1}{2}$ + Arc tan $(1 - \tfrac{1}{2})$ = 2 Arc tan 0.5000 = 53°8′ and
Arc tan 4/3 = Arc tan 1.3333 = 53°8′.

SUPPLEMENTARY PROBLEMS

Solve each of the following equations for all x such that $0 \leq x < 2\pi$.

31. $\sin x = \sqrt{3}/2$. *Ans.* $\pi/3$, $2\pi/3$

32. $\cos^2 x = 1/2$. *Ans.* $\pi/4$, $3\pi/4$, $5\pi/4$, $7\pi/4$

33. $\sin x \cos x = 0$. *Ans.* 0, $\pi/2$, π, $3\pi/2$

34. $(\tan x - 1)(2\sin x + 1) = 0$. *Ans.* $\pi/4$, $7\pi/6$, $5\pi/4$, $11\pi/6$

35. $2\sin^2 x - \sin x - 1 = 0$. *Ans.* $\pi/2$, $7\pi/6$, $11\pi/6$

36. $\sin 2x + \sin x = 0$. *Ans.* 0, $2\pi/3$, π, $4\pi/3$

37. $\cos x + \cos 2x = 0$. *Ans.* $\pi/3$, π, $5\pi/3$

38. $2 \tan x \sin x - \tan x = 0$. *Ans.* $0, \pi/6, 5\pi/6, \pi$

39. $2 \cos x + \sec x = 3$. *Ans.* $0, \pi/3, 5\pi/3$

40. $2 \sin x + \csc x = 3$. *Ans.* $\pi/6, \pi/2, 5\pi/6$

41. $\sin x + 1 = \cos x$. *Ans.* $0, 3\pi/2$

42. $\sec x - 1 = \tan x$. *Ans.* 0

43. $2 \cos x + 3 \sin x = 2$. *Ans.* $0°, 112°37'$

44. $3 \sin x + 5 \cos x + 5 = 0$. *Ans.* $180°, 241°56'$

45. $1 + \sin x = 2 \cos x$. *Ans.* $36°52', 270°$

46. $3 \sin x + 4 \cos x = 2$. *Ans.* $103°18', 330°27'$

47. $\sin 2x = -\sqrt{3}/2$. *Ans.* $2\pi/3, 5\pi/6, 5\pi/3, 11\pi/6$

48. $\tan 3x = 1$. *Ans.* $\pi/12, 5\pi/12, 3\pi/4, 13\pi/12, 17\pi/12, 7\pi/4$

49. $\cos x/2 = \sqrt{3}/2$. *Ans.* $\pi/3$

50. $\cot x/3 = -1/\sqrt{3}$. *Ans.* No solution in given interval

51. $\sin x \cos x = 1/2$. *Ans.* $\pi/4, 5\pi/4$

52. $\sin x/2 + \cos x = 1$. *Ans.* $0, \pi/3, 5\pi/3$

53. $\sin 3x + \sin x = 0$ *Ans.* $0, \pi/2, \pi, 3\pi/2$

54. $\cos 2x + \cos 3x = 0$. *Ans.* $\pi/5, 3\pi/5, \pi, 7\pi/5, 9\pi/5$

55. $\sin 2x + \sin 4x = 2 \sin 3x$. *Ans.* $0, \pi/3, 2\pi/3, \pi, 4\pi/3, 5\pi/3$

56. $\cos 5x + \cos x = 2 \cos 2x$. *Ans.* $0, \pi/4, 2\pi/3, 3\pi/4, 5\pi/4, 4\pi/3, 7\pi/4$

57. $\sin x + \sin 3x = \cos x + \cos 3x$. *Ans.* $\pi/8, \pi/2, 5\pi/8, 9\pi/8, 3\pi/2, 13\pi/8$

Solve each of the following systems for $r \geqq 0$ and $0 \leqq \theta < 2\pi$.

58. $r = a \sin \theta$
$r = a \cos 2\theta$
Ans. $\theta = \pi/6, r = a/2$
$\theta = 5\pi/6, r = a/2$; $\theta = 3\pi/2, r = -a$

59. $r = a \cos \theta$
$r = a \sin 2\theta$
Ans. $\theta = \pi/2, r = 0$; $\theta = 3\pi/2, r = 0$
$\theta = \pi/6, r = \sqrt{3}a/2$
$\theta = 5\pi/6, r = -\sqrt{3}a/2$

60. $r = 4(1 + \cos \theta)$
$r = 3 \sec \theta$
Ans. $\theta = \pi/3, r = 6$
$\theta = 5\pi/3, r = 6$

Solve each of the following equations.

61. $\text{Arc} \tan 2x + \text{Arc} \tan x = \pi/4$. *Ans.* $x = 0.281$

62. $\text{Arc} \sin x + \text{Arc} \tan x = \pi/2$. *Ans.* $x = 0.786$

63. $\text{Arc} \cos x + \text{Arc} \tan x = \pi/2$. *Ans.* $x = 0$

CHAPTER 46

Complex Numbers

PURE IMAGINARY NUMBERS. The square root of a negative number (i.e., $\sqrt{-1}$, $\sqrt{-5}$, $\sqrt{-9}$) is called a *pure imaginary number*. Since by definition $\sqrt{-5} = \sqrt{5}\cdot\sqrt{-1}$ and $\sqrt{-9} = \sqrt{9}\cdot\sqrt{-1} = 3\sqrt{-1}$, it is convenient to introduce the symbol $i = \sqrt{-1}$ and to adopt $\sqrt{-5} = i\sqrt{5}$ and $\sqrt{-9} = 3i$ as the standard form for these numbers.

The symbol i has the property $i^2 = -1$; and for higher integral powers we have

$$i^3 = i^2 \cdot i = (-1)i = -i, \quad i^4 = (i^2)^2 = (-1)^2 = 1, \quad i^5 = i^4 \cdot i = i, \text{ etc.}$$

The use of the standard form simplifies the operations on pure imaginaries and eliminates the possibility of certain common errors. Thus, $\sqrt{-9}\cdot\sqrt{4} = \sqrt{-36} = 6i$ since $\sqrt{-9}\cdot\sqrt{4} = 3i(2) = 6i$ but $\sqrt{-9}\cdot\sqrt{-4} \neq \sqrt{36}$ since $\sqrt{-9}\cdot\sqrt{-4} = (3i)(2i) = 6i^2 = -6$.

COMPLEX NUMBERS. A number $a + bi$, where a and b are real numbers, is called a *complex number*. The first term a is called the *real part* of the complex number and the second term bi is called the *pure imaginary part*.

Complex numbers may be thought of as including all real numbers and all pure imaginary numbers. For example, $5 = 5 + 0i$ and $3i = 0 + 3i$.

Two complex numbers $a + bi$ and $c + di$ are said to be *equal* if and only if $a = c$ and $b = d$.

The *conjugate* of a complex number $a + bi$ is the complex number $a - bi$. Thus, $2 + 3i$ and $2 - 3i$, $-3 + 4i$ and $-3 - 4i$ are pairs of conjugate complex numbers.

ALGEBRAIC OPERATIONS.

(1) *Addition.* To add two complex numbers, add the real parts and add the pure imaginary parts.

EXAMPLE 1. $(2 + 3i) + (4 - 5i) = (2 + 4) + (3 - 5)i = 6 - 2i$.

(2) *Subtraction.* To subtract two complex numbers, subtract the real parts and subtract the pure imaginary parts.

EXAMPLE 2. $(2 + 3i) - (4 - 5i) = (2 - 4) + [3 - (-5)]i = -2 + 8i$.

(3) *Multiplication.* To multiply two complex numbers, carry out the multiplication as if the numbers were ordinary binomials and replace i^2 by -1.

EXAMPLE 3. $(2 + 3i)(4 - 5i) = 8 + 2i - 15i^2 = 8 + 2i - 15(-1) = 23 + 2i$.

(4) *Division.* To divide two complex numbers, multiply both numerator and denominator of the fraction by the conjugate of the denominator.

EXAMPLE 4. $\dfrac{2 + 3i}{4 - 5i} = \dfrac{(2 + 3i)(4 + 5i)}{(4 - 5i)(4 + 5i)} = \dfrac{(8 - 15) + (10 + 12)i}{16 + 25} = -\dfrac{7}{41} + \dfrac{22}{41}i$.

(Note the form of the result; it is neither $\dfrac{-7+22i}{41}$ nor $\dfrac{1}{41}(-7+22i)$.)

See Problems 1-9.

GRAPHIC REPRESENTATION OF COMPLEX NUMBERS. The complex number $x+yi$ may be represented graphically by the point P (see Fig. 1) whose rectangular coordinates are (x, y).

The point O, having coordinates $(0,0)$ represents the complex number $0+0i = 0$. All points on the x-axis have coordinates of the form $(x,0)$ and correspond to real numbers $x+0i = x$. For this reason, the x-axis is called the *axis of reals*. All points on the y-axis have coordinates of the form $(0,y)$ and correspond to pure imaginary numbers $0+yi = yi$. The y-axis is called the *axis of imaginaries*. The plane on which the complex numbers are represented is called the *complex plane*.

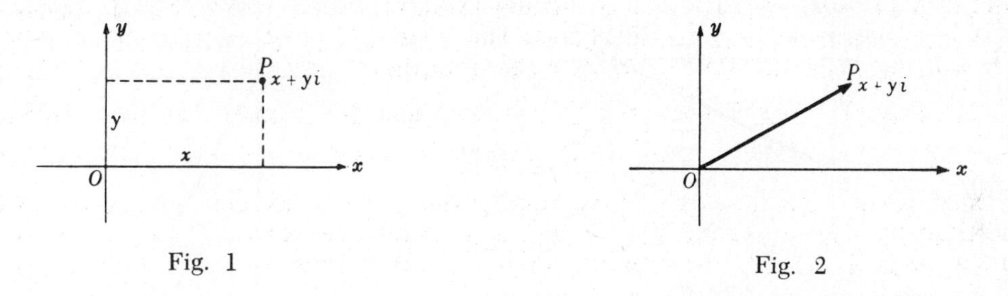

Fig. 1 Fig. 2

In addition to representing a complex number by a point P in the complex plane, the number may be represented (see Fig. 2) by the directed line segment or vector OP.

GRAPHIC REPRESENTATION OF ADDITION AND SUBTRACTION. Let $z_1 = x_1 + iy_1$ and $z_2 = x_2 + iy_2$ be two complex numbers. The vector representation of these numbers (Fig. 3) suggests the familiar parallelogram law for determining graphically the sum

$$z_1 + z_2 = (x_1 + iy_1) + (x_2 + iy_2)$$

Since $z_1 - z_2 = (x_1+iy_1) - (x_2+iy_2) = (x_1+iy_1) + (-x_2-iy_2)$, the difference $z_1 - z_2$ of the two complex numbers may be obtained graphically by applying the parallelogram law to $x_1 + iy_1$ and $-x_2 - iy_2$. (See Fig. 4)

In Fig. 5 both the sum $OR = z_1 + z_2$ and the difference $OS = z_1 - z_2$ are shown. Note that the segments OS and P_2P_1 (the other diagonal of OP_2RP_1) are equal.

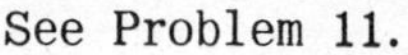
See Problem 11.

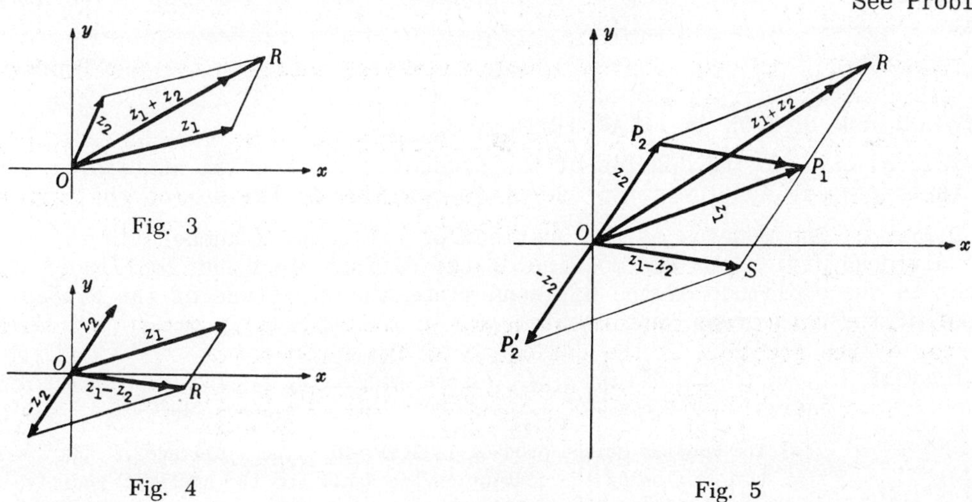

Fig. 3 Fig. 4 Fig. 5

POLAR OR TRIGONOMETRIC FORM OF COMPLEX NUMBERS. Let the complex number $x+yi$ be represented (Fig.6) by the vector OP. This vector (and hence the complex number) may be described in terms of the length r of the vector and *any* positive angle θ which the vector makes with the positive x-axis (axis of positive reals). The number $r = \sqrt{x^2+y^2}$ is called the *modulus* or *absolute value* of the complex number. The angle θ, called the *amplitude* of the complex number, is usually chosen as the smallest, positive angle for which $\tan\theta = y/x$ but at times it will be found more convenient to choose some other angle coterminal with it.

From Fig.6, $x = r\cos\theta$ and $y = r\sin\theta$; then $z = x+yi = r\cos\theta + ir\sin\theta = r(\cos\theta + i\sin\theta)$. We call $z = r(\cos\theta + i\sin\theta)$ the *polar* or *trigonometric form* and $z = x+yi$ the *rectangular form* of the complex number z.

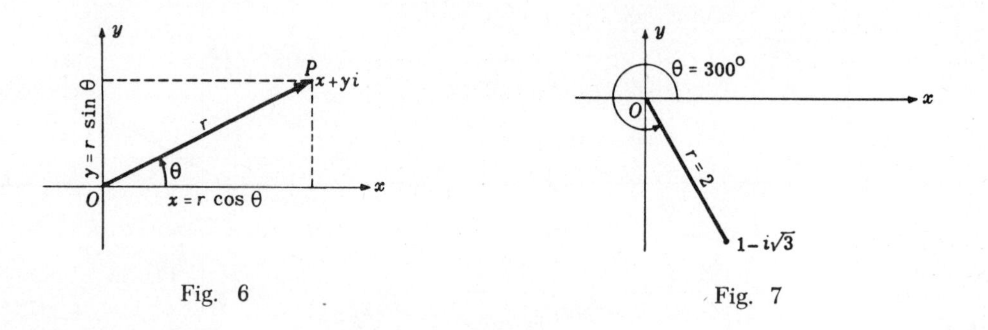

Fig. 6 Fig. 7

EXAMPLE 5. Express $z = 1 - i\sqrt{3}$ in polar form. (See Fig.7 above.)

The modulus is $r = \sqrt{(1)^2 + (-\sqrt{3})^2} = 2$. Since $\tan\theta = y/x = -\sqrt{3}/1 = -\sqrt{3}$, the amplitude θ is either 120° or 300°. Now we know that P lies in quadrant IV; hence, $\theta = 300°$ and the required polar form is

$$z = r(\cos\theta + i\sin\theta) = 2(\cos 300° + i\sin 300°).$$

Note that z may also be represented in polar form by

$$z = 2[\cos(300° + n360°) + i\sin(300° + n360°)]$$

where n is any integer.

EXAMPLE 6. Express the complex number $z = 8(\cos 210° + i\sin 210°)$ in rectangular form.

Since $\cos 210° = -\sqrt{3}/2$ and $\sin 210° = -1/2$,

$$z = 8(\cos 210° + i\sin 210°) = 8[-\sqrt{3}/2 + i(-1/2)] = -4\sqrt{3} - 4i$$

is the required rectangular form.

See Problems 12-13.

MULTIPLICATION AND DIVISION IN POLAR FORM.

Multiplication. The modulus of the product of two complex numbers is the product of their moduli, and the amplitude of the product is the sum of their amplitudes.

Division. The modulus of the quotient of two complex numbers is the modulus of the dividend divided by the modulus of the divisor, and the amplitude of the quotient is the amplitude of the dividend minus the amplitude of the divisor. For a proof of these theorems, see Problem 14.

EXAMPLE 7. Find (*a*) the product z_1z_2, (*b*) the quotient z_1/z_2, and (*c*) the quotient z_2/z_1 where $z_1 = 2(\cos 300° + i\sin 300°)$ and $z_2 = 8(\cos 210° + i\sin 210°)$.

(*a*) The modulus of the product is $2(8) = 16$. The amplitude is $300° + 210° = 510°$ but, following the convention, we shall use the smallest positive coterminal angle $510° - 360° = 150°$. Thus $z_1z_2 = 16(\cos 150° + i\sin 150°)$.

(*b*) The modulus of the quotient z_1/z_2 is $2/8 = \frac{1}{4}$ and the amplitude is $300° - 210° = 90°$. Thus $z_1/z_2 = \frac{1}{4}(\cos 90° + i \sin 90°)$.

(*c*) The modulus of the quotient z_2/z_1 is $8/2 = 4$.

The amplitude is $210° - 300° = -90°$ but we shall use the smallest positive coterminal angle $-90° + 360° = 270°$. Thus

$$z_2/z_1 = 4(\cos 270° + i \sin 270°).$$

Note. From Examples 5 and 6 the numbers are

$$z_1 = 1 - i\sqrt{3} \quad \text{and} \quad z_2 = -4\sqrt{3} - 4i$$

in rectangular form. Then

$$z_1 z_2 = (1 - i\sqrt{3})(-4\sqrt{3} - 4i) = -8\sqrt{3} + 8i = 16(\cos 150° + i \sin 150°)$$

as in (*a*), and

$$z_2/z_1 = \frac{-4\sqrt{3} - 4i}{1 - i\sqrt{3}} = \frac{(-4\sqrt{3} - 4i)(1 + i\sqrt{3})}{(1 - i\sqrt{3})(1 + i\sqrt{3})} = \frac{-16i}{4} = -4i$$

$$= 4(\cos 270° + i \sin 270°) \quad \text{as in } (c).$$

See Problems 15-16.

DE MOIVRE'S THEOREM. If n is any rational number,

$$\{r(\cos\theta + i \sin\theta)\}^n = r^n(\cos n\theta + i \sin n\theta).$$

A proof of this theorem is beyond the scope of this book; a verification for $n = 2$ and $n = 3$ is given in Problem 17.

EXAMPLE 8. $(\sqrt{3} - i)^{10} = \{2(\cos 330° + i \sin 330°)\}^{10}$

$$= 2^{10}(\cos 10 \cdot 330° + i \sin 10 \cdot 330°)$$

$$= 1024(\cos 60° + i \sin 60°) = 1024(1/2 + i\sqrt{3}/2)$$

$$= 512 + 512i\sqrt{3}.$$

See Problem 18.

ROOTS OF COMPLEX NUMBERS. We state, without proof, the theorem:

A complex number $a + bi = r(\cos\theta + i \sin\theta)$ has exactly n distinct nth roots.

The procedure for determining these roots is given in Example 9.

EXAMPLE 9. Find all fifth roots of $4 - 4i$.

The usual polar form of $4 - 4i$ is $4\sqrt{2}(\cos 315° + i \sin 315°)$ but we shall need the more general form

$$4\sqrt{2}[\cos(315° + k360°) + i \sin(315° + k360°)],$$

where k is any integer, including zero.

Using De Moivre's theorem, a fifth root of $4 - 4i$ is given by

$$\{4\sqrt{2}[\cos(315° + k360°) + i \sin(315° + k360°)]\}^{1/5}$$

$$= (4\sqrt{2})^{1/5}(\cos\frac{315° + k360°}{5} + i \sin\frac{315° + k360°}{5})$$

$$= \sqrt{2}[\cos(63° + k72°) + i \sin(63° + k72°)].$$

Assigning in turn the values $k = 0, 1, 2, \cdots$, we find

$k = 0: \sqrt{2}(\cos 63^\circ + i \sin 63^\circ) = R_1$
$k = 1: \sqrt{2}(\cos 135^\circ + i \sin 135^\circ) = R_2$
$k = 2: \sqrt{2}(\cos 207^\circ + i \sin 207^\circ) = R_3$
$k = 3: \sqrt{2}(\cos 279^\circ + i \sin 279^\circ) = R_4$
$k = 4: \sqrt{2}(\cos 351^\circ + i \sin 351^\circ) = R_5$
$k = 5: \sqrt{2}(\cos 423^\circ + i \sin 423^\circ)$
$= \sqrt{2}(\cos 63^\circ + i \sin 63^\circ) = R_1$, etc.

Thus, the five fifth roots are obtained by assigning the values 0,1,2,3,4 (i.e., $0,1,2,3,\cdots, n-1$) to k.

(See also Problem 19.)

The modulus of each of the roots is $\sqrt{2}$; hence these roots lie on a circle of radius $\sqrt{2}$ with center at the origin. The difference in amplitude of two consecutive roots is 72°; hence the roots are equally spaced on this circle, as shown in the adjoining figure.

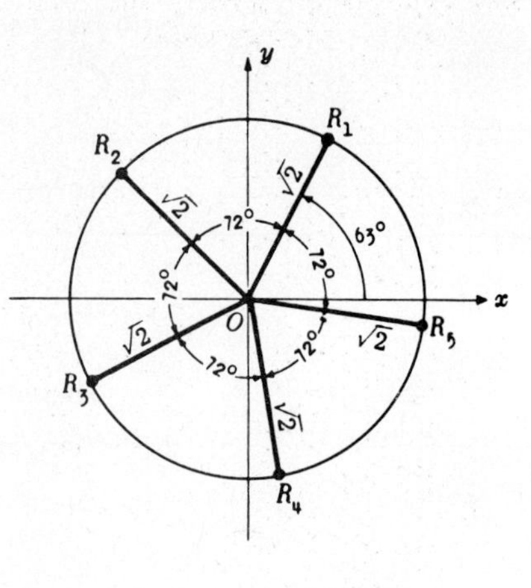

SOLVED PROBLEMS

In Problems 1-6, perform the indicated operations, simplify, and write the result in the form $a + bi$.

1. $(3-4i) + (-5+7i) = (3-5) + (-4+7)i = -2+3i$

2. $(4+2i) - (-1+3i) = [4-(-1)] + (2-3)i = 5-i$

3. $(2+i)(3-2i) = (6+2) + (-4+3)i = 8-i$

4. $(3+4i)(3-4i) = 9+16 = 25$

5. $\dfrac{1+3i}{2+i} = \dfrac{(1+3i)(2-i)}{(2+i)(2-i)} = \dfrac{(2+3) + (-1+6)i}{4+1} = 1+i$

6. $\dfrac{3-2i}{2-3i} = \dfrac{(3-2i)(2+3i)}{(2-3i)(2+3i)} = \dfrac{(6+6) + (9-4)i}{4+9} = \dfrac{12}{13} + \dfrac{5}{13}i$

7. Find x and y such that $2x - yi = 4+3i$.

Here $2x = 4$ and $-y = 3$; then $x = 2$ and $y = -3$.

8. Show that the conjugate complex numbers $2+i$ and $2-i$ are roots of the equation $x^2 - 4x + 5 = 0$.

For $x = 2+i$: $(2+i)^2 - 4(2+i) + 5 = 4 + 4i + i^2 - 8 - 4i + 5 = 0$.
For $x = 2-i$: $(2-i)^2 - 4(2-i) + 5 = 4 - 4i + i^2 - 8 + 4i + 5 = 0$.

Since each number satisfies the equation, it is a root of the equation.

9. Show that the conjugate of the sum of two complex numbers is equal to the sum of their conjugates.

Let the complex numbers be $a+bi$ and $c+di$. Their sum is $(a+c) + (b+d)i$ and the conjugate of the sum is $(a+c) - (b+d)i$.

The conjugates of the two given numbers are $a-bi$ and $c-di$, and their sum is

$$(a+c) + (-b-d)i = (a+c) - (b+d)i.$$

10. Represent graphically (as a vector) the following complex numbers:
(a) $3+2i$, (b) $2-i$, (c) $-2+i$, (d) $-1-3i$.

We locate, in turn, the points whose coordinates are (3,2), (2,−1), (−2,1), (−1,−3) and join each to the origin O.

11. Perform graphically the following operations:

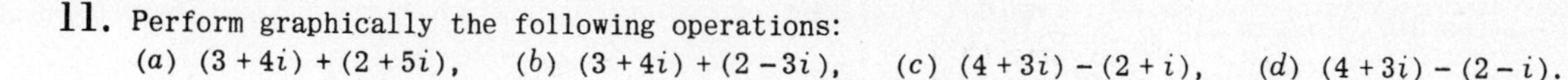

(*a*) $(3+4i)+(2+5i)$, (*b*) $(3+4i)+(2-3i)$, (*c*) $(4+3i)-(2+i)$, (*d*) $(4+3i)-(2-i)$.

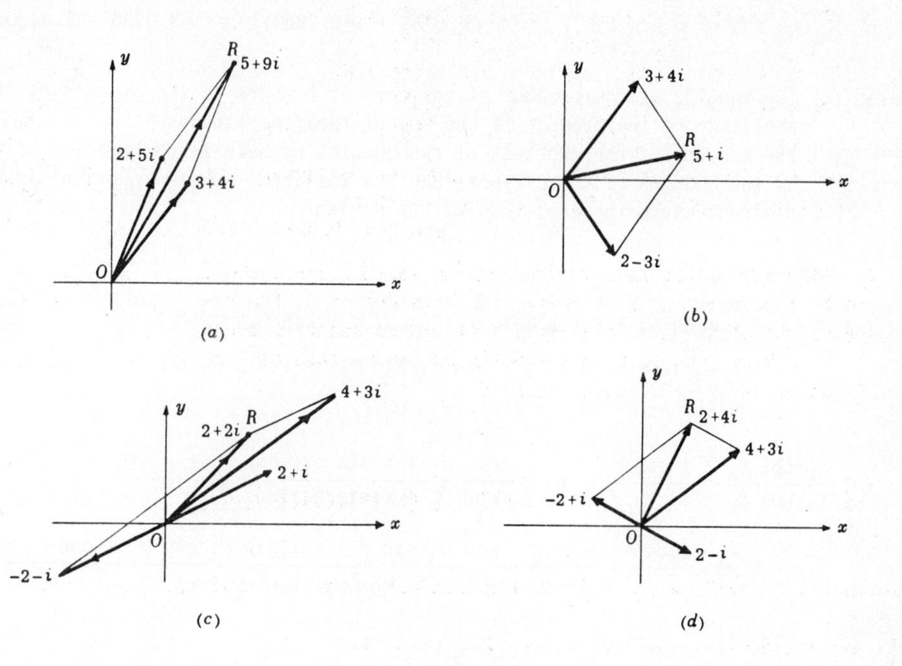

For (*a*) and (*b*), draw as in Fig. (*a*) and (*b*) the two vectors and apply the parallelogram law.

For (*c*) draw the vectors representing $4+3i$ and $-2-i$ and apply the parallelogram law as in Fig. (*c*).

For (*d*) draw the vectors representing $4+3i$ and $-2+i$ and apply the parallelogram law as in Fig. (*d*).

12. Express each of the following complex numbers z in polar form:
(*a*) $-1+i\sqrt{3}$, (*b*) $6\sqrt{3}+6i$, (*c*) $2-2i$, (*d*) $-3=-3+0i$, (*e*) $4i=0+4i$, (*f*) $-3-4i$.

(*a*) P lies in the second quadrant; $r=\sqrt{(-1)^2+(\sqrt{3})^2}=2$; $\tan\theta=\sqrt{3}/-1=-\sqrt{3}$ and $\theta=120°$.
Thus, $z=2(\cos 120°+i\sin 120°)$.

(*b*) P lies in the first quadrant; $r=\sqrt{(6\sqrt{3})^2+6^2}=12$; $\tan\theta=6/6\sqrt{3}=1/\sqrt{3}$ and $\theta=30°$.
Thus, $z=12(\cos 30°+i\sin 30°)$.

(*c*) P lies in the fourth quadrant; $r=\sqrt{2^2+(-2)^2}=2\sqrt{2}$; $\tan\theta=-2/2=-1$ and $\theta=315°$.
Thus, $z=2\sqrt{2}(\cos 315°+i\sin 315°)$.

(*d*) P lies on the negative x-axis and $\theta=180°$; $r=\sqrt{(-3)^2+0^2}=3$.
Thus, $z=3(\cos 180°+i\sin 180°)$.

(*e*) P lies on the positive y-axis and $\theta=90°$; $r=\sqrt{0^2+4^2}=4$.
Thus, $z=4(\cos 90°+i\sin 90°)$.

(*f*) P lies in the third quadrant; $r=\sqrt{(-3)^2+(-4)^2}=5$; $\tan\theta=-4/-3=1.3333$, $\theta=233°8'$.
Thus, $z=5(\cos 233°8'+i\sin 233°8')$.

13. Express each of the following complex numbers z in rectangular form:
(*a*) $4(\cos 240°+i\sin 240°)$
(*b*) $2(\cos 315°+i\sin 315°)$
(*c*) $3(\cos 90°+i\sin 90°)$
(*d*) $5(\cos 128°+i\sin 128°)$.

(*a*) $4(\cos 240°+i\sin 240°)=4[-1/2+i(-\sqrt{3}/2)]=-2-2i\sqrt{3}$

(*b*) $2(\cos 315^\circ + i \sin 315^\circ) = 2[1/\sqrt{2} + i(-1/\sqrt{2})] = \sqrt{2} - i\sqrt{2}$

(*c*) $3(\cos 90^\circ + i \sin 90^\circ) = 3[0 + i(1)] = 3i$

(*d*) $5(\cos 128^\circ + i \sin 128^\circ) = 5[-0.6157 + i(0.7880)] = -3.0785 + 3.9400i$

14. Prove: (*a*) The modulus of the product of two complex numbers is the product of their moduli, and the amplitude of the product is the sum of their amplitudes.
(*b*) The modulus of the quotient of two complex numbers is the modulus of the dividend divided by the modulus of the divisor, and the amplitude of the quotient is the amplitude of the dividend minus the amplitude of the divisor.

Let $z_1 = r_1(\cos\theta_1 + i\sin\theta_1)$ and $z_2 = r_2(\cos\theta_2 + i\sin\theta_2)$.

(*a*)
$$\begin{aligned} z_1 z_2 &= r_1(\cos\theta_1 + i\sin\theta_1)\cdot r_2(\cos\theta_2 + i\sin\theta_2) \\ &= r_1 r_2[(\cos\theta_1\cos\theta_2 - \sin\theta_1\sin\theta_2) + i(\sin\theta_1\cos\theta_2 + \cos\theta_1\sin\theta_2)] \\ &= r_1 r_2[\cos(\theta_1+\theta_2) + i\sin(\theta_1+\theta_2)]. \end{aligned}$$

(*b*)
$$\begin{aligned} \frac{r_1(\cos\theta_1 + i\sin\theta_1)}{r_2(\cos\theta_2 + i\sin\theta_2)} &= \frac{r_1(\cos\theta_1 + i\sin\theta_1)(\cos\theta_2 - i\sin\theta_2)}{r_2(\cos\theta_2 + i\sin\theta_2)(\cos\theta_2 - i\sin\theta_2)} \\ &= \frac{r_1}{r^2}\cdot\frac{(\cos\theta_1\cos\theta_2 + \sin\theta_1\sin\theta_2) + i(\sin\theta_1\cos\theta_2 - \cos\theta_1\sin\theta_2)}{\cos^2\theta_2 + \sin^2\theta_2} \\ &= \frac{r_1}{r_2}[\cos(\theta_1-\theta_2) + i\sin(\theta_1-\theta_2)]. \end{aligned}$$

15. Perform the indicated operations, giving the result in both polar and rectangular form.
(*a*) $5(\cos 170^\circ + i \sin 170^\circ)\cdot(\cos 55^\circ + i \sin 55^\circ)$
(*b*) $2(\cos 50^\circ + i \sin 50^\circ)\cdot 3(\cos 40^\circ + i \sin 40^\circ)$
(*c*) $6(\cos 110^\circ + i \sin 110^\circ)\cdot \frac{1}{2}(\cos 212^\circ + i \sin 212^\circ)$
(*d*) $10(\cos 305^\circ + i \sin 305^\circ) \div 2(\cos 65^\circ + i \sin 65^\circ)$
(*e*) $4(\cos 220^\circ + i \sin 220^\circ) \div 2(\cos 40^\circ + i \sin 40^\circ)$
(*f*) $6(\cos 230^\circ + i \sin 230^\circ) \div 3(\cos 75^\circ + i \sin 75^\circ)$

(*a*) The modulus of the product is $5(1) = 5$ and the amplitude is $170^\circ + 55^\circ = 225^\circ$.
In polar form the product is $5(\cos 225^\circ + i \sin 225^\circ)$ and in rectangular form the product is $5(-\sqrt{2}/2 - i\sqrt{2}/2) = -5\sqrt{2}/2 - 5i\sqrt{2}/2$.

(*b*) The modulus of the product is $2(3) = 6$ and the amplitude is $50^\circ + 40^\circ = 90^\circ$.
In polar form the product is $6(\cos 90^\circ + i \sin 90^\circ)$ and in rectangular form it is $6(0 + i) = 6i$.

(*c*) The modulus of the product is $6(\frac{1}{2}) = 3$ and the amplitude is $110^\circ + 212^\circ = 322^\circ$.
In polar form the product is $3(\cos 322^\circ + i \sin 322^\circ)$ and in rectangular form it is $3(0.7880 - 0.6157i) = 2.3640 - 1.8471i$.

(*d*) The modulus of the quotient is $10/2 = 5$ and the amplitude is $305^\circ - 65^\circ = 240^\circ$.
In polar form the product is $5(\cos 240^\circ + i \sin 240^\circ)$ and in rectangular form it is $5(-1/2 - i\sqrt{3}/2) = -5/2 - 5i\sqrt{3}/2$.

(*e*) The modulus of the quotient is $4/2 = 2$ and the amplitude is $220^\circ - 40^\circ = 180^\circ$.
In polar form the quotient is $2(\cos 180^\circ + i \sin 180^\circ)$ and in rectangular form it is $2(-1 + 0i) = -2$.

(*f*) The modulus of the quotient is $6/3 = 2$ and the amplitude is $230^\circ - 75^\circ = 155^\circ$.
In polar form the quotient is $2(\cos 155^\circ + i \sin 155^\circ)$ and in rectangular form it is $2(-0.9063 + 0.4226i) = -1.8126 + 0.8452i$.

16. Express each of the numbers in polar form, perform the indicated operation, and give the result in rectangular form.

(a) $(-1 + i\sqrt{3})(\sqrt{3} + i)$ (d) $-2 \div (-\sqrt{3} + i)$ (g) $(3 + 2i)(2 + i)$
(b) $(3 - 3i\sqrt{3})(-2 - 2i\sqrt{3})$ (e) $6i \div (-3 - 3i)$ (h) $(2 + 3i) \div (2 - 3i)$
(c) $(4 - 4i\sqrt{3}) \div (-2\sqrt{3} + 2i)$ (f) $(1 + i\sqrt{3})(1 + i\sqrt{3})$

(a) $(-1 + i\sqrt{3})(\sqrt{3} + i) = 2(\cos 120^\circ + i \sin 120^\circ) \cdot 2(\cos 30^\circ + i \sin 30^\circ)$
$= 4(\cos 150^\circ + i \sin 150^\circ) = 4(-\sqrt{3}/2 + \frac{1}{2}i) = -2\sqrt{3} + 2i$

(b) $(3 - 3i\sqrt{3})(-2 - 2i\sqrt{3}) = 6(\cos 300^\circ + i \sin 300^\circ) \cdot 4(\cos 240^\circ + i \sin 240^\circ)$
$= 24(\cos 540^\circ + i \sin 540^\circ) = 24(-1 + 0i) = -24$

(c) $(4 - 4i\sqrt{3}) \div (-2\sqrt{3} + 2i) = 8(\cos 300^\circ + i \sin 300^\circ) \div 4(\cos 150^\circ + i \sin 150^\circ)$
$= 2(\cos 150^\circ + i \sin 150^\circ) = 2(-\sqrt{3}/2 + \frac{1}{2}i) = -\sqrt{3} + i$

(d) $-2 \div (-\sqrt{3} + i) = 2(\cos 180^\circ + i \sin 180^\circ) \div 2(\cos 150^\circ + i \sin 150^\circ)$
$= \cos 30^\circ + i \sin 30^\circ = \frac{1}{2}\sqrt{3} + \frac{1}{2}i$

(e) $6i \div (-3 - 3i) = 6(\cos 90^\circ + i \sin 90^\circ) \div 3\sqrt{2}(\cos 225^\circ + i \sin 225^\circ)$
$= \sqrt{2}(\cos 225^\circ + i \sin 225^\circ) = -1 - i$

(f) $(1 + i\sqrt{3})(1 + i\sqrt{3}) = 2(\cos 60^\circ + i \sin 60^\circ) \cdot 2(\cos 60^\circ + i \sin 60^\circ)$
$= 4(\cos 120^\circ + i \sin 120^\circ) = 4(-\frac{1}{2} + \frac{1}{2}i\sqrt{3}) = -2 + 2i\sqrt{3}$

(g) $(3 + 2i)(2 + i) = \sqrt{13}(\cos 33^\circ 41' + i \sin 33^\circ 41') \cdot \sqrt{5}(\cos 26^\circ 34' + i \sin 26^\circ 34')$
$= \sqrt{65}(\cos 60^\circ 15' + i \sin 60^\circ 15')$
$= \sqrt{65}(0.4962 + 0.8682i) = 4.001 + 7.000i = 4 + 7i$

(h) $$\frac{2+3i}{2-3i} = \frac{\sqrt{13}(\cos 56^\circ 19' + i \sin 56^\circ 19')}{\sqrt{13}(\cos 303^\circ 41' + i \sin 303^\circ 41')} = \frac{\cos 416^\circ 19' + i \sin 416^\circ 19'}{\cos 303^\circ 41' + i \sin 303^\circ 41'}$$
$= \cos 112^\circ 38' + i \sin 112^\circ 38' = -0.3849 + 0.9230i$

17. Verify De Moivre's theorem for $n = 2$ and $n = 3$.

Let $z = r(\cos\theta + i \sin\theta)$.

For $n = 2$: $z^2 = [r(\cos\theta + i \sin\theta)][r(\cos\theta + i \sin\theta)]$
$= r^2[(\cos^2\theta - \sin^2\theta) + i(2 \sin\theta \cos\theta)] = r^2(\cos 2\theta + i \sin 2\theta)$

For $n = 3$: $z^3 = z^2 \cdot z = [r^2(\cos 2\theta + i \sin 2\theta)][r(\cos\theta + i \sin\theta)]$
$= r^3[(\cos 2\theta \cos\theta - \sin 2\theta \sin\theta) + i(\sin 2\theta \cos\theta + \cos 2\theta \sin\theta)]$
$= r^3(\cos 3\theta + i \sin 3\theta)$.

The theorem may be established for n a positive integer by mathematical induction.

18. Evaluate each of the following using De Moivre's theorem and express each result in rectangular form:

(a) $(1 + i\sqrt{3})^4$, (b) $(\sqrt{3} - i)^5$, (c) $(-1 + i)^{10}$, (d) $(2 + 3i)^4$.

(a) $(1 + i\sqrt{3})^4 = [2(\cos 60^\circ + i \sin 60^\circ)]^4 = 2^4(\cos 4\cdot 60^\circ + i \sin 4\cdot 60^\circ)$
$= 2^4(\cos 240^\circ + i \sin 240^\circ) = -8 - 8i\sqrt{3}$

(b) $(\sqrt{3} - i)^5 = [2(\cos 330^\circ + i \sin 330^\circ)]^5 = 32(\cos 1650^\circ + i \sin 1650^\circ)$
$= 32(\cos 210^\circ + i \sin 210^\circ) = -16\sqrt{3} - 16i$

(c) $(-1 + i)^{10} = [\sqrt{2}(\cos 135^\circ + i \sin 135^\circ)]^{10} = 32(\cos 270^\circ + i \sin 270^\circ) = -32i$

(d) $(2 + 3i)^4 = [\sqrt{13}(\cos 56^\circ 19' + i \sin 56^\circ 19')]^4 = 13^2(\cos 225^\circ 16' + i \sin 225^\circ 16')$
$= 169(-0.7038 - 7104i) = -118.9 - 120.1i$

19. Find the indicated roots in rectangular form, except when this would necessitate the use of tables.

(a) Square roots of $2 - 2i\sqrt{3}$
(b) Fourth roots of $-8 - 8i\sqrt{3}$
(c) Cube roots of $-4\sqrt{2} + 4i\sqrt{2}$
(d) Cube roots of 1
(e) Fourth roots of i
(f) Sixth roots of -1
(g) Fourth roots of $-16i$
(h) Fifth roots of $1+3i$

(a) $$2 - 2i\sqrt{3} = 4[\cos(300° + k360°) + i\sin(300° + k360°)]$$

and $$(2 - 2i\sqrt{3})^{1/2} = 2[\cos(150° + k180°) + i\sin(150° + k180°)].$$

Putting $k = 0$ and 1, the required roots are

$$R_1 = 2(\cos 150° + i\sin 150°) = 2(-\tfrac{1}{2}\sqrt{3} + \tfrac{1}{2}i) = -\sqrt{3} + i$$
$$R_2 = 2(\cos 330° + i\sin 330°) = 2(\tfrac{1}{2}\sqrt{3} - \tfrac{1}{2}i) = \sqrt{3} - i.$$

(b) $$-8 - 8i\sqrt{3} = 16[\cos(240° + k360°) + i\sin(240° + k360°)]$$

and $$(-8 - 8i\sqrt{3})^{1/4} = 2[\cos(60° + k90°) + i\sin(60° + k90°)].$$

Putting $k = 0,1,2,3,$ the required roots are

$$R_1 = 2(\cos 60° + i\sin 60°) = 2(\tfrac{1}{2} + i\tfrac{1}{2}\sqrt{3}) = 1 + i\sqrt{3}$$
$$R_2 = 2(\cos 150° + i\sin 150°) = 2(-\tfrac{1}{2}\sqrt{3} + \tfrac{1}{2}i) = -\sqrt{3} + i$$
$$R_3 = 2(\cos 240° + i\sin 240°) = 2(-\tfrac{1}{2} - i\tfrac{1}{2}\sqrt{3}) = -1 - i\sqrt{3}$$
$$R_4 = 2(\cos 330° + i\sin 330°) = 2(\tfrac{1}{2}\sqrt{3} - \tfrac{1}{2}i) = \sqrt{3} - i.$$

(c) $$-4\sqrt{2} + 4i\sqrt{2} = 8[\cos(135° + k360°) + i\sin(135° + k360°)]$$

and $$(-4\sqrt{2} + 4i\sqrt{2})^{1/3} = 2[\cos(45° + k120°) + i\sin(45° + k120°)].$$

Putting $k = 0,1,2,$ the required roots are

$$R_1 = 2(\cos 45° + i\sin 45°) = 2(1/\sqrt{2} + i/\sqrt{2}) = \sqrt{2} + i\sqrt{2}$$
$$R_2 = 2(\cos 165° + i\sin 165°)$$
$$R_3 = 2(\cos 285° + i\sin 285°).$$

(d) $1 = \cos(0° + k360°) + i\sin(0° + k360°)$ and $1^{1/3} = \cos(k120°) + i\sin(k120°)$.

Putting $k = 0,1,2,$ the required roots are

$$R_1 = \cos 0° + i\sin 0° = 1$$
$$R_2 = \cos 120° + i\sin 120° = -\tfrac{1}{2} + i\tfrac{1}{2}\sqrt{3}$$
$$R_3 = \cos 240° + i\sin 240° = -\tfrac{1}{2} - i\tfrac{1}{2}\sqrt{3}.$$

Note that $R_2^2 = \cos 2(120°) + i\sin 2(120°) = R_3$,

$$R_3^2 = \cos 2(240°) + i\sin 2(240°) = R_2, \text{ and}$$

$$R_2R_3 = (\cos 120° + i\sin 120°)(\cos 240° + i\sin 240°)$$
$$= \cos 0° + i\sin 0° = R_1.$$

(e) $i = \cos(90° + k360°) + i\sin(90° + k360°)$ and $i^{1/4} = \cos(22\tfrac{1}{2}° + k90°) + i\sin(22\tfrac{1}{2}° + k90°)$.

Thus, the required roots are

$$R_1 = \cos 22\tfrac{1}{2}° + i\sin 22\tfrac{1}{2}° \qquad R_3 = \cos 202\tfrac{1}{2}° + i\sin 202\tfrac{1}{2}°$$
$$R_2 = \cos 112\tfrac{1}{2}° + i\sin 112\tfrac{1}{2}° \qquad R_4 = \cos 292\tfrac{1}{2}° + i\sin 292\tfrac{1}{2}°$$

(f) $-1 = \cos(180° + k360°) + i\sin(180° + k360°)$ and $(-1)^{1/6} = \cos(30° + k60°) + i\sin(30° + k60°)$.

Thus, the required roots are

$$R_1 = \cos 30° + i\sin 30° = \tfrac{1}{2}\sqrt{3} + \tfrac{1}{2}i$$
$$R_2 = \cos 90° + i\sin 90° = i$$
$$R_3 = \cos 150° + i\sin 150° = -\tfrac{1}{2}\sqrt{3} + \tfrac{1}{2}i$$

$$R_4 = \cos 210° + i \sin 210° = -\tfrac{1}{2}\sqrt{3} - \tfrac{1}{2}i$$
$$R_5 = \cos 270° + i \sin 270° = -i$$
$$R_6 = \cos 330° + i \sin 330° = \tfrac{1}{2}\sqrt{3} - \tfrac{1}{2}i.$$

Note that $R_2^2 = R_5^2 = \cos 180° + i \sin 180°$ and thus R_2 and R_5 are the square roots of -1; that $R_1^3 = R_3^3 = R_5^3 = \cos 90° + i \sin 90° = i$ and thus R_1, R_3, R_5 are the cube roots of i; and that $R_2^3 = R_4^3 = R_6^3 = \cos 270° + i \sin 270° = -i$ and thus R_2, R_4, R_6 are the cube roots of $-i$.

(g) $-16i = 16[\cos(270° + k360°) + i \sin(270° + k360°)]$ and

$(-16i)^{1/4} = 2[\cos(67\tfrac{1}{2}° + k90°) + i \sin(67\tfrac{1}{2}° + k90°)]$. Thus, the required roots are

$$R_1 = 2(\cos 67\tfrac{1}{2}° + i \sin 67\tfrac{1}{2}°) \qquad R_3 = 2(\cos 247\tfrac{1}{2}° + i \sin 247\tfrac{1}{2}°)$$
$$R_2 = 2(\cos 157\tfrac{1}{2}° + i \sin 157\tfrac{1}{2}°) \qquad R_4 = 2(\cos 337\tfrac{1}{2}° + i \sin 337\tfrac{1}{2}°).$$

(h) $1 + 3i = \sqrt{10}\,[\cos(71°34' + k360°) + i \sin(71°34' + k360°)]$ and

$(1 + 3i)^{1/5} = \sqrt[10]{10}\,[\cos 14°19' + k72°) + i \sin(14°19' + k72°)]$. The required roots are

$$R_1 = \sqrt[10]{10}(\cos 14°19' + i \sin 14°19')$$
$$R_2 = \sqrt[10]{10}(\cos 86°19' + i \sin 86°19')$$
$$R_3 = \sqrt[10]{10}(\cos 158°19' + i \sin 158°19')$$
$$R_4 = \sqrt[10]{10}(\cos 230°19' + i \sin 230°19')$$
$$R_5 = \sqrt[10]{10}(\cos 302°19' + i \sin 302°19').$$

SUPPLEMENTARY PROBLEMS

20. Perform the indicated operations, writing the results in the form $a + bi$.

(a) $(6 - 2i) + (2 + 3i) = 8 + i$
(b) $(6 - 2i) - (2 + 3i) = 4 - 5i$
(c) $(3 + 2i) + (-4 - 3i) = -1 - i$
(d) $(3 - 2i) - (4 - 3i) = -1 + i$
(e) $3(2 - i) = 6 - 3i$
(f) $2i(3 + 4i) = -8 + 6i$
(g) $(2 + 3i)(1 + 2i) = -4 + 7i$
(h) $(2 - 3i)(5 + 2i) = 16 - 11i$
(i) $(3 - 2i)(-4 + i) = -10 + 11i$
(j) $(2 + 3i)(3 + 2i) = 13i$
(k) $(2 + \sqrt{-5})(3 - 2\sqrt{-4}) = (6 + 4\sqrt{5}) + (3\sqrt{5} - 8)i$
(l) $(1 + 2\sqrt{-3})(2 - \sqrt{-3}) = 8 + 3\sqrt{3}\,i$
(m) $(2 - i)^2 = 3 - 4i$
(n) $(4 + 2i)^2 = 12 + 16i$
(o) $(1 + i)^2(2 + 3i) = -6 + 4i$
(p) $\dfrac{2 + 3i}{1 + i} = \dfrac{5}{2} + \dfrac{1}{2}i$
(q) $\dfrac{3 - 2i}{3 - 4i} = \dfrac{17}{25} + \dfrac{6}{25}i$
(r) $\dfrac{3 - 2i}{2 + 3i} = -i$

21. Show that $3 + 2i$ and $3 - 2i$ are roots of $x^2 - 6x + 13 = 0$.

22. Perform graphically the following operations.

(a) $(2 + 3i) + (1 + 4i)$
(b) $(4 - 2i) + (2 + 3i)$
(c) $(2 + 3i) - (1 + 4i)$
(d) $(4 - 2i) - (2 + 3i)$

23. Express each of the following complex numbers in polar form.

(a) $3 + 3i = 3\sqrt{2}(\cos 45° + i \sin 45°)$
(b) $1 + \sqrt{3}\,i = 2(\cos 60° + i \sin 60°)$
(c) $-2\sqrt{3} - 2i = 4(\cos 210° + i \sin 210°)$
(d) $\sqrt{2} - i\sqrt{2} = 2(\cos 315° + i \sin 315°)$
(e) $-8 = 8(\cos 180° + i \sin 180°)$
(f) $-2i = 2(\cos 270° + i \sin 270°)$
(g) $-12 + 5i = 13(\cos 157°23' + i \sin 157°23')$
(h) $-4 - 3i = 5(\cos 216°52' + i \sin 216°52')$

24. Perform the indicated operation and express the results in the form $a+bi$.

(a) $3(\cos 25^\circ + i \sin 25^\circ)\ 8(\cos 200^\circ + i \sin 200^\circ) = -12\sqrt{2} - 12\sqrt{2}\,i$

(b) $4(\cos 50^\circ + i \sin 50^\circ)\ 2(\cos 100^\circ + i \sin 100^\circ) = -4\sqrt{3} + 4i$

(c) $\dfrac{4(\cos 190^\circ + i \sin 190^\circ)}{2(\cos 70^\circ + i \sin 70^\circ)} = -1 + i\sqrt{3}$

(d) $\dfrac{12(\cos 200^\circ + i \sin 200^\circ)}{3(\cos 350^\circ + i \sin 350^\circ)} = -2\sqrt{3} - 2i$

25. Use the polar form in finding each of the following products and quotients, and express each result in the form $a+bi$.

(a) $(1+i)(\sqrt{2} - i\sqrt{2}) = 2\sqrt{2}$

(b) $(-1 - i\sqrt{3})(-4\sqrt{3} + 4i) = 8\sqrt{3} + 8i$

(c) $\dfrac{1-i}{1+i} = -i$

(d) $\dfrac{4 + 4\sqrt{3}\,i}{\sqrt{3} + i} = 2\sqrt{3} + 2i$

(e) $\dfrac{-1 + i\sqrt{3}}{\sqrt{2} + i\sqrt{2}} = 0.2588 + 0.9659i$

(f) $\dfrac{3+i}{2+i} = 1.4 - 0.2i$

26. Use De Moivre's Theorem to evaluate each of the following and express each result in the form $a+bi$.

(a) $[2(\cos 6^\circ + i \sin 6^\circ)]^5 = 16\sqrt{3} + 16i$

(b) $[\sqrt{2}(\cos 75^\circ + i \sin 75^\circ)]^4 = 2 - 2\sqrt{3}\,i$

(c) $(1+i)^8 = 16$

(d) $(1-i)^6 = 8i$

(e) $(1/2 - i\sqrt{3}/2)^{20} = -1/2 - i\sqrt{3}/2$

(f) $(\sqrt{3}/2 + i/2)^9 = -i$

(g) $(3+4i)^4 = -526.9 - 336.1i$

(h) $\dfrac{(1-i\sqrt{3})^3}{(-2+2i)^4} = \dfrac{1}{8}$

(i) $\dfrac{(1+i)(\sqrt{3}+i)^3}{(1-i\sqrt{3})^3} = 1-i$

27. Find all the indicated roots, expressing the results in the form $a+bi$ unless tables would be needed to do so.

(a) The square roots of i. *Ans.* $\sqrt{2}/2 + i\sqrt{2}/2$, $-\sqrt{2}/2 - i\sqrt{2}/2$

(b) The square roots of $1 + i\sqrt{3}$. *Ans.* $\sqrt{6}/2 + i\sqrt{2}/2$, $-\sqrt{6}/2 - i\sqrt{2}/2$

(c) The cube roots of -8. *Ans.* $1 + i\sqrt{3}$, -2, $1 - i\sqrt{3}$

(d) The cube roots of $27i$. *Ans.* $3\sqrt{3}/2 + 3i/2$, $-3\sqrt{3}/2 + 3i/2$, $-3i$

(e) The cube roots of $-4\sqrt{3} + 4i$.
Ans. $2(\cos 50^\circ + i \sin 50^\circ)$, $2(\cos 170^\circ + i \sin 170^\circ)$, $2(\cos 290^\circ + i \sin 290^\circ)$

(f) The fifth roots of $1+i$. *Ans.* $\sqrt[10]{2}(\cos 9^\circ + i \sin 9^\circ)$, $\sqrt[10]{2}(\cos 81^\circ + i \sin 81^\circ)$, etc.

(g) The sixth roots of $-\sqrt{3}+i$. *Ans.* $\sqrt[6]{2}(\cos 25^\circ + i \sin 25^\circ)$, $\sqrt[6]{2}(\cos 85^\circ + i \sin 85^\circ)$, etc.

28. Find the tenth roots of 1 and show that the product of any two of them is again one of the tenth roots of 1.

29. Show that the reciprocal of any one of the tenth roots of 1 is again a tenth root of 1.

30. Denote either of the complex cube roots of 1 (Problem 19*d*) by ω_1 and the other by ω_2. Show that $\omega_1^2\omega_2 = \omega_1$ and $\omega_1\omega_2^2 = \omega_2$.

31. Show that $(\cos\theta + i \sin\theta)^{-n} = \cos n\theta - i \sin n\theta$.

32. Use the fact that the segments OS and P_2P_1 in Fig. 5 are equal to devise a second procedure for constructing the difference $OS = z_1 - z_2$ of two complex numbers z_1 and z_2.

CHAPTER 47

Coordinates and Loci

THE PROJECTION of a point P on a line l is the foot M of the perpendicular to l through P. The projection of the line segment PQ on a line l is the line segment MN where M is the projection of P and N is the projection of Q on l. See Fig. 1 below.

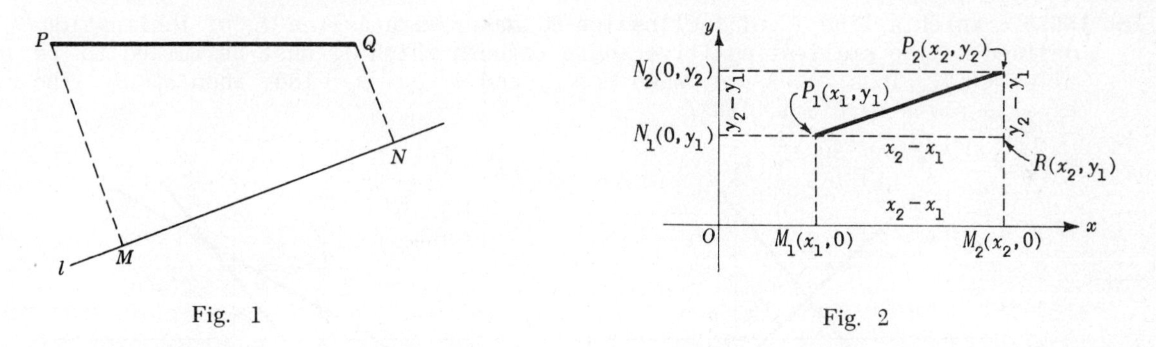

Fig. 1 Fig. 2

THE PROJECTION OF THE LINE SEGMENT joining $P_1(x_1, y_1)$ and $P_2(x_2, y_2)$ on the x-axis or on any line parallel to the x-axis is $(x_2 - x_1)$ and the projection of P_1P_2 on the y-axis or any line parallel to the y-axis is $(y_2 - y_1)$.

Thus, in Fig. 2, $M_1M_2 = P_1R = x_2 - x_1$ and $N_1N_2 = RP_2 = y_2 - y_1$. These are directed distances, that is, $M_1M_2 = -M_2M_1$.

See Problems 1-2.

THE LENGTH d OF THE LINE SEGMENT P_1P_2, assumed not parallel to a coordinate axis, is given by

$$d = \sqrt{(x_2 - x_1)^2 + (y_2 - y_1)^2}$$

In Fig. 2, $d = \sqrt{(P_1R)^2 + (RP_2)^2}$. This is not a directed distance.

See Problems 3-5.

A POINT P is said to divide a line segment P_1P_2 in the ratio $r_1 : r_2$ if P is on the line P_1P_2 and if $P_1P/PP_2 = r_1/r_2$. The point $P(x, y)$ dividing the segment joining $P_1(x_1, y_1)$ and $P_2(x_2, y_2)$ in the ratio $r_1 : r_2$ is given by

$$x = \frac{r_1x_2 + r_2x_1}{r_1 + r_2}, \qquad y = \frac{r_1y_2 + r_2y_1}{r_1 + r_2}$$

The midpoint $P(x, y)$ of the segment P_1P_2 is given by

$$x = \tfrac{1}{2}(x_1 + x_2), \qquad y = \tfrac{1}{2}(y_1 + y_2)$$

See Problems 6-8.

THE INCLINATION θ of a straight line is the smallest positive angle (measured counter-clockwise) from the positive end of the x-axis to the line. The range of θ is given by $0° \leq \theta \leq 180°$. This is equivalent to assuming upward direction along the line as positive and defining the inclination of the line as the angle between the positive directions on the line and x-axis.

THE SLOPE OF A STRAIGHT LINE is the tangent of its inclination, that is, $m = \tan\theta$. If $P_1(x_1, y_1)$ and $P_2(x_2, y_2)$ are any two distinct points on a line, its slope (see Fig. 2) is given by

$$m = \tan\theta = \tan\angle RP_1P_2 = \frac{RP_2}{P_1R} = \frac{y_2 - y_1}{x_2 - x_1}$$

The vertical line has no slope. See Problem 9.

THE ANGLE α which a line l_1 of inclination θ_1 makes with a line l_2 of inclination θ_2 is defined as the smallest positive angle through which l_2 must be turned to be parallel to l_1. Thus, $\alpha = \theta_1 - \theta_2$ when $\theta_1 > \theta_2$ and $\alpha = \theta_1 - \theta_2 + 180°$ when $\theta_1 < \theta_2$. The range of α is given by $0° \leq \alpha < 180°$.

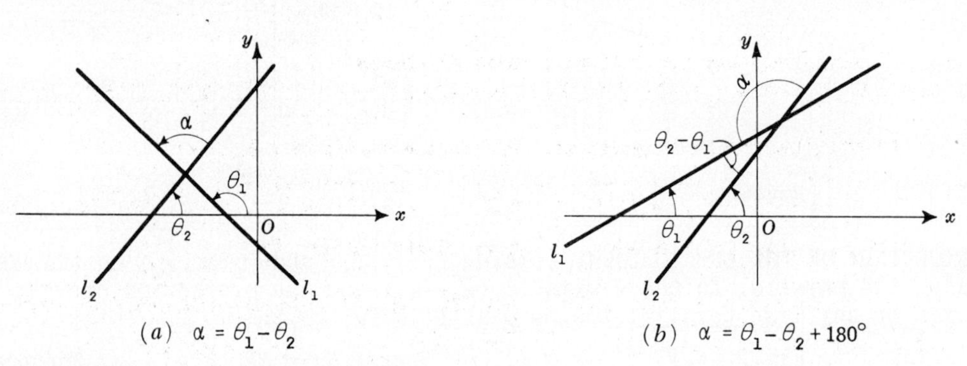

(a) $\alpha = \theta_1 - \theta_2$ (b) $\alpha = \theta_1 - \theta_2 + 180°$

The tangent of the angle α is given by

$$(1) \qquad \tan\alpha = \frac{\tan\theta_1 - \tan\theta_2}{1 + \tan\theta_1 \tan\theta_2} = \frac{m_1 - m_2}{1 + m_1 m_2}$$

(Note. The angle α is sometimes defined as $\alpha = \theta_1 - \theta_2$ with the consequent range $-180° \leq \alpha \leq 180°$.)

See Problems 10-14.

THE AREA OF A TRIANGLE whose vertices are $P_1(x_1, y_1)$, $P_2(x_2, y_2)$ and $P_3(x_3, y_3)$ is given, except possibly for sign, by

$$\tfrac{1}{2}(x_1y_2 + x_2y_3 + x_3y_1 - x_1y_3 - x_2y_1 - x_3y_2) = \frac{1}{2}\begin{vmatrix} x_1 & y_1 & 1 \\ x_2 & y_2 & 1 \\ x_3 & y_3 & 1 \end{vmatrix}$$

See Problem 15.

THE BASIC PROBLEMS of Analytic Geometry are:

1. To find the locus of a given equation. (See Chapter 9.)
2. To find the equation of a locus defined by a given geometric condition.

See Problems 17-18.

SOLVED PROBLEMS

1. For each pair of points, find the projection of P_1P_2 on the coordinate axes.

(a) $P_1(2,5)$, $P_2(6,3)$

The projection on the x-axis is $x_2 - x_1 = 6 - 2 = 4$.
The projection on the y-axis is $y_2 - y_1 = 3 - 5 = -2$.

(b) $P_1(3,4)$, $P_2(2,-1)$

The projections on the x- and y-axes are respectively $2 - 3 = -1$ and $-1 - 4 = -5$.

2. For each pair of points, find the directed length of P_1P_2.

(a) $P_1(-3,4)$, $P_2(5,4)$ Since P_1P_2 is parallel to the x-axis, $P_1P_2 = x_2 - x_1 = 5 - (-3) = 8$.

(b) $P_1(6,3)$, $P_2(-4,3)$ Here $P_1P_2 = -4 - 6 = -10$.

(c) $P_1(5,2)$, $P_2(5,8)$ Since P_1P_2 is parallel to the y-axis, $P_1P_2 = y_2 - y_1 = 8 - 2 = 6$.

(d) $P_1(-3,-2)$, $P_2(-3,-8)$ Here $P_1P_2 = -8 - (-2) = -6$.

3. Find the distance between the following pairs of points.
(a) (2,1), (8,7) (b) (−3,−2), (−5,−8) (c) (0,−2), (5,−6) (d) (0,0), (3,5) (e) (0,0), (x,y)

(a) Identifying the first point with P_1 and the second with P_2,

$$d = \sqrt{(x_2 - x_1)^2 + (y_2 - y_1)^2} = \sqrt{(8-2)^2 + (7-1)^2} = \sqrt{36+36} = 6\sqrt{2}.$$

(It is immaterial here which point is identified with P_1.)

(b) $d = \sqrt{[-5-(-3)]^2 + [(-8-(-2)]^2} = \sqrt{4+36} = 2\sqrt{10}$

(c) $d = \sqrt{(5-0)^2 + [-6-(-2)]^2} = \sqrt{25+16} = \sqrt{41}$

(d) $d = \sqrt{(3-0)^2 + (5-0)^2} = \sqrt{9+25} = \sqrt{34}$

(e) $d = \sqrt{x^2 + y^2}$

4. Show that the points $A(3,5)$, $B(1,-1)$ and $C(-4,-16)$ lie on a straight line by showing that $AB + BC = AC$.

$$AB = \sqrt{(1-3)^2 + (-1-5)^2} = \sqrt{40} = 2\sqrt{10}, \quad BC = \sqrt{(-4-1)^2 + [-16-(-1)]^2} = \sqrt{250} = 5\sqrt{10},$$

and $AC = \sqrt{(-4-3)^2 + (-16-5)^2} = \sqrt{490} = 7\sqrt{10}$. Thus, $AB + BC = 2\sqrt{10} + 5\sqrt{10} = 7\sqrt{10} = AC$.

5. Show that the triangle whose vertices are $A(4,3)$, $B(6,-2)$, $C(-11,-3)$ is a right triangle.

$AB = \sqrt{(6-4)^2 + (-2-3)^2} = \sqrt{29}$, $BC = \sqrt{(-11-6)^2 + [-3-(-2)]^2} = \sqrt{290}$, $AC = \sqrt{(-11-4)^2 + (-3-3)^2} = \sqrt{261}$

Since $(AB)^2 + (AC)^2 = 29 + 261 = 290 = (BC)^2$, the triangle is a right triangle with hypotenuse BC.

6. Derive the point of division formula: $x = \dfrac{r_1x_2 + r_2x_1}{r_1 + r_2}$, $y = \dfrac{r_1y_2 + r_2y_1}{r_1 + r_2}$.

The point $P(x,y)$ divides the segment P_1P_2, joining $P_1(x_1,y_1)$ and $P_2(x_2,y_2)$, in the ratio $P_1P/PP_2 = r_1/r_2$.

Now $\dfrac{P_1P}{PP_2} = \dfrac{P_1S}{PT} = \dfrac{x - x_1}{x_2 - x} = \dfrac{r_1}{r_2}$; then

$$x(r_1 + r_2) = r_1x_2 + r_2x_1 \quad \text{and} \quad x = \frac{r_1x_2 + r_2x_1}{r_1 + r_2}.$$

Similarly, $\dfrac{P_1P}{PP_2} = \dfrac{SP}{TP_2} = \dfrac{r_1}{r_2}$ and $y = \dfrac{r_1y_2 + r_2y_1}{r_1 + r_2}$.

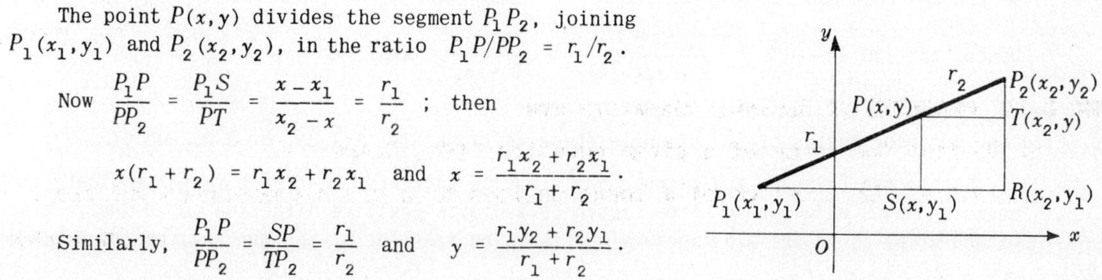

7. Find the coordinates of the point $P(x,y)$ which divides the segment P_1P_2 in the given ratio.

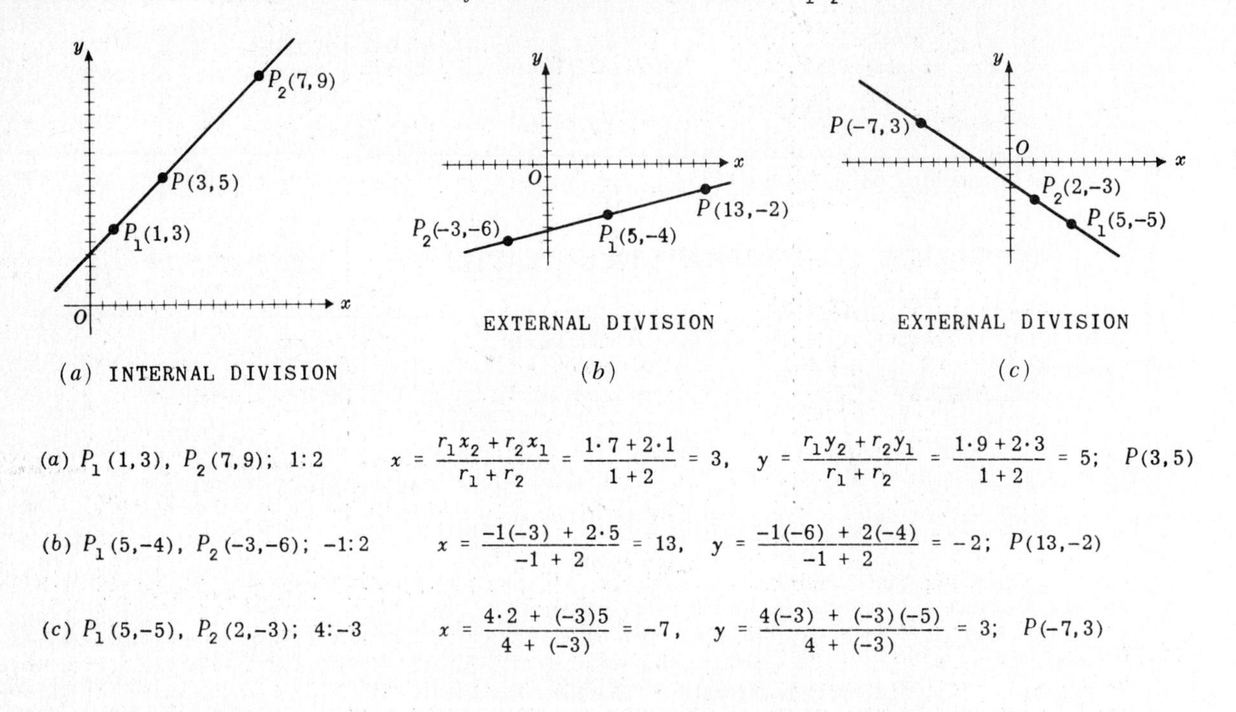

(a) INTERNAL DIVISION (b) (c)

(a) $P_1(1,3)$, $P_2(7,9)$; 1:2 $\quad x = \dfrac{r_1x_2+r_2x_1}{r_1+r_2} = \dfrac{1\cdot 7+2\cdot 1}{1+2} = 3,\quad y = \dfrac{r_1y_2+r_2y_1}{r_1+r_2} = \dfrac{1\cdot 9+2\cdot 3}{1+2} = 5;\quad P(3,5)$

(b) $P_1(5,-4)$, $P_2(-3,-6)$; −1:2 $\quad x = \dfrac{-1(-3)+2\cdot 5}{-1+2} = 13,\quad y = \dfrac{-1(-6)+2(-4)}{-1+2} = -2;\quad P(13,-2)$

(c) $P_1(5,-5)$, $P_2(2,-3)$; 4:−3 $\quad x = \dfrac{4\cdot 2+(-3)5}{4+(-3)} = -7,\quad y = \dfrac{4(-3)+(-3)(-5)}{4+(-3)} = 3;\quad P(-7,3)$

8. Find the coordinates of the midpoint M of the segment joining the pair of points.

(a) (1,2), (5,8) $\quad x = \frac{1}{2}(x_1+x_2) = \frac{1}{2}(1+5) = 3,\quad y = \frac{1}{2}(y_1+y_2) = \frac{1}{2}(2+8) = 5;\quad M(3,5)$

(b) (2,−3), (−5,7) $\quad x = \frac{1}{2}(2-5) = -3/2,\quad y = \frac{1}{2}(-3+7) = 2;\quad M(-3/2,2)$

(c) (a,0), (0,b) $\quad x = \frac{1}{2}(a+0) = \frac{1}{2}a,\quad y = \frac{1}{2}(0+b) = \frac{1}{2}b;\quad M(\frac{1}{2}a,\frac{1}{2}b)$

9. Find the slope m and inclination θ of the line passing through the points (a) (6,1) and (1,−4), (b) (−3,2) and (4,−1).

(a) Here $m = \tan\theta = \dfrac{y_2-y_1}{x_2-x_1} = \dfrac{-4-1}{1-6} = 1$ and $\theta = 45^\circ$.

(b) Here $m = \tan\theta = \dfrac{-1-2}{4-(-3)} = -\dfrac{3}{7} = -0.4286$ and $\theta = 156^\circ 48'$.

10. Find the angle α which a line whose slope is 3/5 makes with a line whose slope is 4.

Take $m_1 = \dfrac{3}{5}$ and $m_2 = 4$; then $\tan\alpha = \dfrac{m_1-m_2}{1+m_1m_2} = \dfrac{3/5-4}{1+(3/5)4} = -1$ and $\alpha = 135^\circ$.

11. Find the slope of the line which makes an angle of 120° with a line of slope −1/3.

Denote the required slope by m_1; then $\alpha = 120^\circ$ and $m_2 = -1/3$.

Hence $\tan 120^\circ = -\sqrt{3} = \dfrac{m_1-(-1/3)}{1+m_1(-1/3)} = \dfrac{3m_1+1}{3-m_1}$ and $m_1 = -\dfrac{6+5\sqrt{3}}{3}$.

12. Find the interior angles of the triangle whose vertices are $A(4,3)$, $B(-2,2)$ and $C(2,-8)$.

Locate the triangle and compute the slope of each side. At each vertex indicate the required angle by means of a curved arrow tipped to indicate positive direction about that vertex. The head of the arrow is on the side whose slope is to be taken as m_1 and the tail of the arrow is on the side whose slope is to be taken as m_2, as shown in Fig. (*d*) below. Then

$$\tan A = \frac{m_1 - m_2}{1 + m_1 \cdot m_2} = \frac{11/2 - 1/6}{1 + (11/2)(1/6)} = \frac{64}{23} = 2.7826 \quad \text{and} \quad A = 70°14',$$

$$\tan B = \frac{1/6 - (-5/2)}{1 + (1/6)(-5/2)} = \frac{32}{7} = 4.5714 \quad \text{and} \quad B = 77°40',$$

$$\tan C = \frac{-5/2 - 11/2}{1 + (-5/2)(11/2)} = \frac{32}{51} = 0.6275 \quad \text{and} \quad C = 32°6'.$$

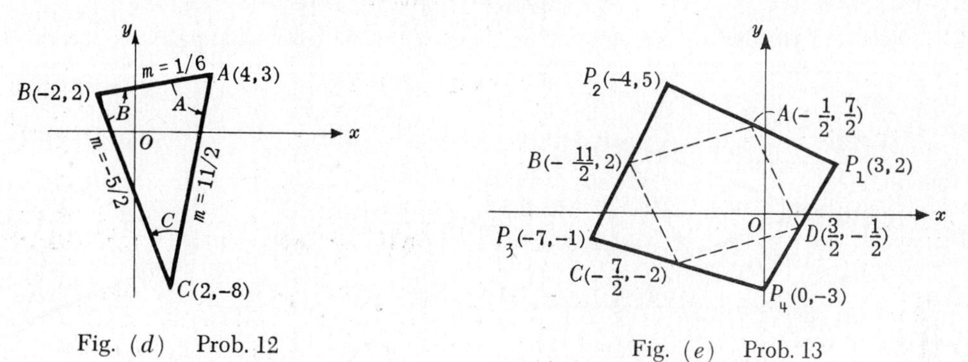

Fig. (*d*) Prob. 12 Fig. (*e*) Prob. 13

13. The vertices of a quadrilateral are $P_1(3,2)$, $P_2(-4,5)$, $P_3(-7,-1)$, $P_4(0,-3)$. Show, by finding numerical values, that the perimeter of the quadrilateral whose vertices are the midpoints of the sides is equal to the sum of the diagonals of the given quadrilateral. Refer to Fig. (*e*) above.

The midpoints of the sides $P_1P_2, P_2P_3, P_3P_4, P_4P_1$ are $A(-\frac{1}{2}, 7/2)$, $B(-11/2, 2)$, $C(-7/2, -2)$, $D(3/2, -\frac{1}{2})$ respectively. The perimeter of the quadrilateral $ABCD$ is

$$AB + BC + CD + DA = \sqrt{(-1/2 + 11/2)^2 + (7/2 - 2)^2} + \sqrt{(-11/2 + 7/2)^2 + (2+2)^2}$$
$$+ \sqrt{(-7/2 - 3/2)^2 + (-2 + 1/2)^2} + \sqrt{(3/2 + 1/2)^2 + (-1/2 - 7/2)^2}$$
$$= \sqrt{109}/2 + 2\sqrt{5} + \sqrt{109}/2 + 2\sqrt{5} = \sqrt{109} + 4\sqrt{5}.$$

(Note that the quadrilateral is a parallelogram since the opposite sides are equal.)

The sum of the diagonals of the given quadrilateral is

$$P_1P_3 + P_2P_4 = \sqrt{(3+7)^2 + (2+1)^2} + \sqrt{(-4-0)^2 + (5+3)^2} = \sqrt{109} + \sqrt{80} = \sqrt{109} + 4\sqrt{5}.$$

14. Prove: The sum of the squares of the medians of a triangle is equal to 3/4 the sum of the square of its sides.

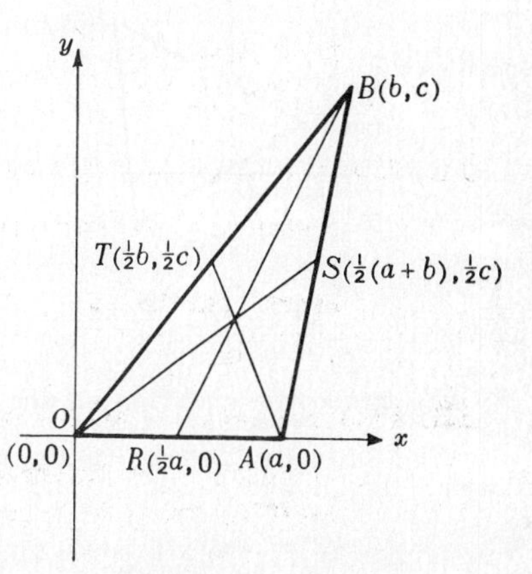

Place the triangle so that its vertices are $O(0,0)$, $A(a,0)$, and $B(b,c)$. The midpoints of the sides OA, AB, BO are respectively $R(\frac{1}{2}a, 0)$, $S[\frac{1}{2}(a+b), \frac{1}{2}c]$, $T(\frac{1}{2}b, \frac{1}{2}c)$. Then

$$(OS)^2 + (AT)^2 + (BR)^2 = \{[\tfrac{1}{2}(a+b)]^2 + (\tfrac{1}{2}c)^2\} + [(\tfrac{1}{2}b - a)^2 + (\tfrac{1}{2}c)^2]$$
$$+ [(b - \tfrac{1}{2}a)^2 + c^2] = \frac{3}{2}(a^2 - ab + b^2 + c^2)$$

$$(OA)^2 + (AB)^2 + (BO)^2 = [a^2] + [(b-a)^2 + c^2] + [b^2 + c^2]$$
$$= 2(a^2 - ab + b^2 + c^2)$$

Thus, $(OS)^2 + (AT)^2 + (BR)^2 = \frac{3}{4}[(OA)^2 + (AB)^2 + (BO)^2]$.

15. Show that the area of the triangle whose vertices are $P_1(x_1, y_1)$, $P_2(x_2, y_2)$, $P_3(x_3, y_3)$ is given, apart from sign, by $A = \frac{1}{2}(x_1y_2 + x_2y_3 + x_3y_1 - x_1y_3 - x_2y_1 - x_3y_2)$.

In Fig. (f) below,

$$\begin{aligned}\text{Area triangle } P_1P_2P_3 &= \text{area trapezoid } M_3P_3P_2M_2 - \text{area trapezoid } M_1P_1P_2M_2 \\ &\quad - \text{area trapezoid } M_3P_3P_1M_1 \\ &= \tfrac{1}{2}(y_3 + y_2)(x_2 - x_3) - \tfrac{1}{2}(y_1 + y_2)(x_2 - x_1) - \tfrac{1}{2}(y_3 + y_1)(x_1 - x_3) \\ &= \tfrac{1}{2}(x_1y_2 + x_2y_3 + x_3y_1 - x_1y_3 - x_2y_1 - x_3y_2)\end{aligned}$$

Note. In the derivation above, P_1, P_2 and P_3 have been assigned in counterclockwise order about the triangle; when so assigned, the formula yields a positive result.

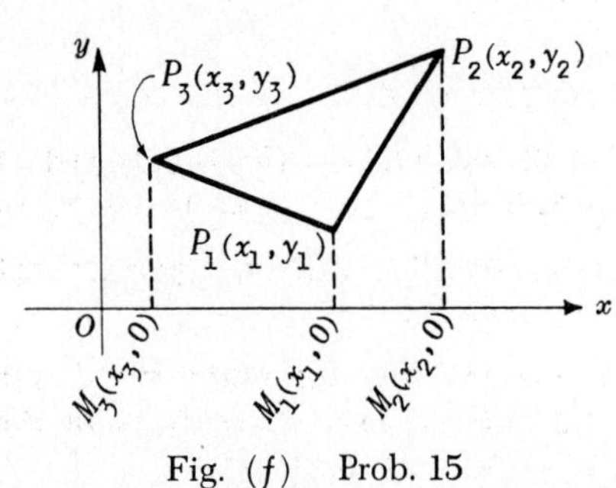

Fig. (f) Prob. 15

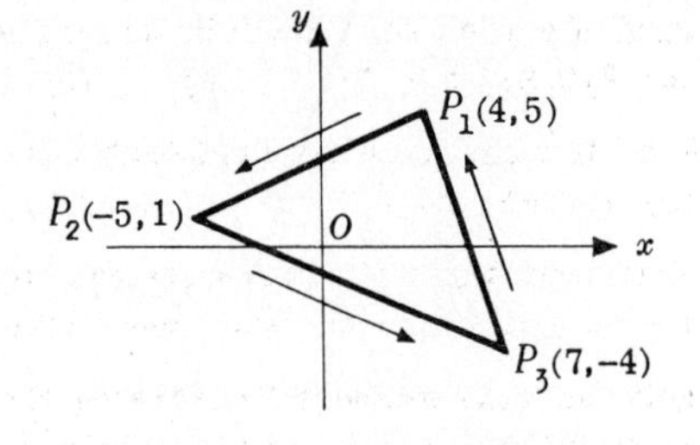

Fig. (g) Prob. 16

16. Find the area of the triangle whose vertices are $(-5,1)$, $(4,5)$ and $(7,-4)$.

To insure a positive result, we locate the points and number them so that as we describe the path $P_1P_2P_3P_1$ the area lies on the left (see Fig. (g) above). Then

$$A = \frac{1}{2}\begin{vmatrix} 4 & 5 & 1 \\ -5 & 1 & 1 \\ 7 & -4 & 1 \end{vmatrix} = 46\tfrac{1}{2}$$

17. Find the equation of the locus of a point which is (a) equidistant from the points $A(-3,2)$ and $B(5,6)$, (b) at a distance 6 from the point $C(3,4)$.

Let $P(x,y)$ be an arbitrary point on the required locus.

(a) In Fig. (h) below, $AP = BP$; hence, $(AP)^2 = (BP)^2$ or $(x+3)^2 + (y-2)^2 = (x-5)^2 + (y-6)^2$. This reduces to $2x + y = 6$, the equation of the required locus.

(b) In Fig. (i), $CP = 6$; hence, $(CP)^2 = 36$ or $(x-3)^2 + (y-4)^2 = 36$. The locus is a circle.

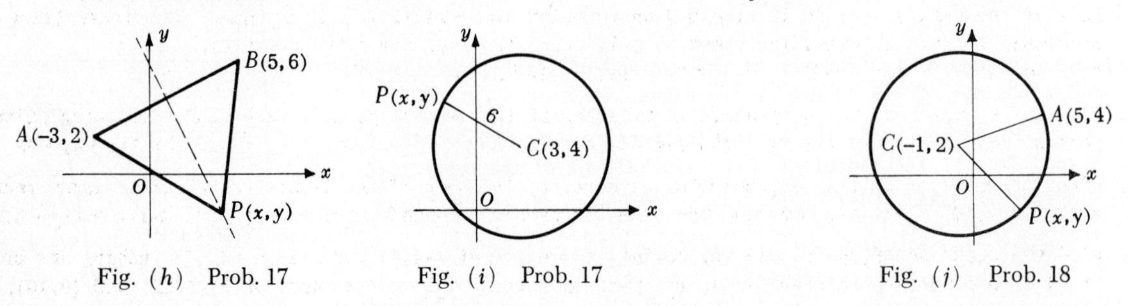

Fig. (h) Prob. 17 Fig. (i) Prob. 17 Fig. (j) Prob. 18

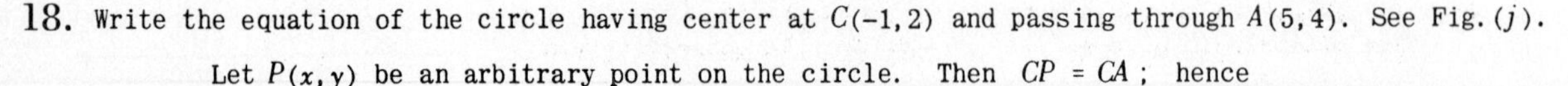

18. Write the equation of the circle having center at $C(-1,2)$ and passing through $A(5,4)$. See Fig. (j).

Let $P(x,y)$ be an arbitrary point on the circle. Then $CP = CA$; hence

$$(CP)^2 = (CA)^2 \quad \text{or} \quad (x+1)^2 + (y-2)^2 = (5+1)^2 + (4-2)^2 = 40$$

is the required equation.

SUPPLEMENTARY PROBLEMS

19. The sides of a square are 12 units in length. Locate its vertices when it is placed in a coordinate system so that: (*a*) one vertex is at the origin, two of its sides are along the coordinate axes, and the fourth vertex is in Quadrant IV; (*b*) its center is at the origin and its sides are parallel to the coordinate axes; (*c*) its center is at the origin and its vertices are on the coordinate axes.
Ans. (*a*) (0,0), (12,0), (0,−12), (12,−12); (*b*) (6,6), (−6,6), (−6,−6), (6,−6);
(*c*) $(6\sqrt{2}, 0)$, $(0, 6\sqrt{2})$, $(-6\sqrt{2}, 0)$, $(0, -6\sqrt{2})$

20. The radius of a circle is 8 units. (*a*) Locate its points of intersection with the coordinate axes if its center is at the origin, (*b*) Locate the center and the points of tangency if the circle is tangent to the coordinate axes with center in Quadrant II.
Ans. (*a*) (4,0), (0,4), (−4,0), (0,−4); (*b*) $C(-8,8)$, $T_1(0,8)$, $T_2(-8,0)$

21. Find the area of the right triangle whose vertices are:
(*a*) $A(2,3)$, $B(8,7)$, $C(8,3)$; (*b*) $A(5,4)$, $B(-3,6)$, $C(-3,4)$. *Ans.* (*a*) 12, (*b*) 8

22. Find the distance between each pair of points: (*a*) (3,6) and (7,3), (*b*) (−4,2) and (1,7), (*c*) (10,4) and (−2,−2), (*d*) (0,0), and (−3,−2). *Ans.* (*a*) 5, (*b*) $5\sqrt{2}$, (*c*) $6\sqrt{5}$, (*d*) $\sqrt{13}$

23. Show that the triangle whose vertices are (4,−1), (−2,3), and (−6,−3) is a right triangle. Show that the midpoint of its hypotenuse is equidistant from the vertices.

24. For the line segment P_1P_2 joining $P_1(-3,2)$ and $P_2(4,-4)$, locate: (*a*) the midpoint, (*b*) the points of trisection, (*c*) the point dividing it in the ratio 3:2, (*d*) the point dividing it in the ratio −3:5.
Ans. (*a*) (1/2,−1); (*b*) (−2/3,0), (5/3,−2); (*c*) (6/5,−8/5); (*d*) (−27/2,11)

25. Find the area of the triangle whose vertices are:
(*a*) (4,−1), (6,5), (−2,9); (*b*) (4,−3), (9,4), (−3,6). *Ans.* (*a*) 28; (*b*) 47

26. Prove: The sum of the squares of the distances of any point in the plane to two opposite vertices of a rectangle is equal to the sum of the squares of the distances to the other two vertices.
Hint: Take the vertices at (0,0), $(a,0)$, $(0,b)$, (a,b).

27. Prove: The segment joining the midpoints of two opposite sides of a quadrilateral and the segment joining the midpoints of its diagonals bisect each other.
Hint: Take the vertices at (0,0), $(a,0)$, (b,c), (d,e).

28. Find the slope and inclination of the line through each pair of points of Problem 22.
Ans. (*a*) −3/4, 143°8′; (*b*) 1, 45°; (*c*) 1/2, 26°34′; (*d*) 2/3, 33°41′

29. Find the interior angles of the triangle whose vertices are (3,1), (−3,−2), (−4,4).
Ans. 49°46′, 72°54′, 57°20′

30. Find the area of the triangle of Problem 29. *Ans.* 39/2

31. Write the equation of the locus of a point $P(x,y)$ which moves so that (*a*) it is always 3 units to the left of the y-axis, (*b*) it is always 4 units below the x-axis, (*c*) its ordinate is always twice its abscissa, (*d*) its abscissa increased by 2 is equal to three times its ordinate.
Ans. (*a*) $x = -3$, (*b*) $y = -4$, (*c*) $y = 2x$, (*d*) $x - 3y + 2 = 0$

32. Find the equation of the perpendicular bisector of the segment joining each of the following pairs of points: (*a*) (2,−3) and (−3,2), (*b*) (−4,−5) and (3,−3). *Ans.* (*a*) $x - y = 0$, (*b*) $14x + 4y + 23 = 0$

33. Find the equation of the perpendicular bisector of the base of the isosceles triangle whose vertices are (1,1), (10,2), (6,−3). Show that the bisector passes through the third vertex. *Ans.* $9x + y - 51 = 0$

34. Find the equation of the circle (*a*) having its center at (−2,4) and radius = 6, (*b*) having its center at (4,3) and passing through (8,0), (*c*) passing through the points (4,0), (8,0), (0,2) and (0,16), (*d*) circumscribing the right triangle whose vertices are (2,−4), (6,−4) and (6,6).
Ans. (*a*) $x^2 + y^2 + 4x - 8y - 16 = 0$ (*c*) $x^2 + y^2 - 12x - 18y + 32 = 0$
(*b*) $x^2 + y^2 - 8x - 6y = 0$ (*d*) $x^2 + y^2 - 8x - 2y - 12 = 0$

35. Find the equation of the locus of a point $P(x,y)$ which moves so that the sum of the squares of its distances from $A(2,4)$ and $B(-3,5)$ is 30. *Ans.* $x^2 + y^2 + x - 9y + 12 = 0$

36. Find the equation of the locus of a point $P(x,y)$ which moves so that the difference of the squares of its distances from $A(2,4)$ and $B(-3,5)$ is 30. *Ans.* $5x - y = -22$, $5x - y = 8$

CHAPTER 48

The Straight Line

THE EQUATION OF THE STRAIGHT LINE parallel to the y-axis at a distance a from that axis is $x = a$.

The equation of the straight line having slope m and passing through the point (x_1, y_1) is

$$y - y_1 = m(x - x_1). \qquad \textit{(Point-slope form)}$$

See Problem 1.

The equation of the line having slope m and y-intercept b is

$$y = mx + b. \qquad \textit{(Slope-intercept form)}$$

See Problem 2.

The equation of the line passing through the points (x_1, y_1) and (x_2, y_2), where $x_1 \neq x_2$, is

$$y - y_1 = \frac{y_2 - y_1}{x_2 - x_1}(x - x_1). \qquad \textit{(Two point form)}$$

See Problem 3.

The equation of the line whose x-intercept is a and whose y-intercept is b, where $ab \neq 0$, is

$$\frac{x}{a} + \frac{y}{b} = 1. \qquad \textit{(Intercept form)}$$

See Problem 4.

THE GENERAL EQUATION of the straight line is $Ax + By + C = 0$, where A, B, C are arbitrary constants except that not both A and B are zero.

If $C = 0$, the line passes through the origin.

If $B = 0$, the line is vertical; if $A = 0$, the line is horizontal.

Otherwise, the line has slope $m = -A/B$ and y-intercept $b = -C/B$.

If two non-vertical lines are parallel, their slopes are equal. Thus the lines $Ax + By + C = 0$ and $Ax + By + D = 0$ are parallel.

If two oblique lines are perpendicular, the slope of one is the negative reciprocal of the slope of the other. If m_1 and m_2 are the slopes of two perpendicular lines, then $m_1 = -1/m_2$ or $m_1 m_2 = -1$. Thus $Ax + By + C = 0$ and $Bx - Ay + D = 0$, where $AB \neq 0$, are perpendicular lines.

See Problems 5-8.

THE NORMAL FORM of the equation of a straight line, not passing through the origin, is

$$x \cos\omega + y \sin\omega - p = 0$$

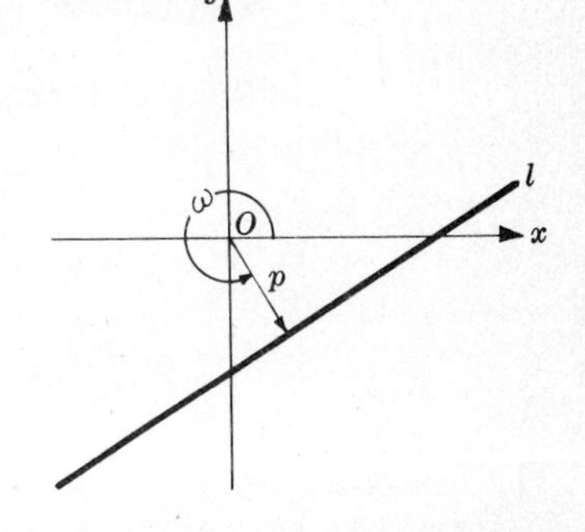

where $p > 0$ is the length of the normal (that is, the perpendicular dropped from the origin to the line) measured always away from the origin, and ω is the positive angle $< 360°$ measured from the positive end of the x-axis to the normal.

An important by-product of the normal form of the equation is that it establishes a sense of direction with respect to the line. The origin is *always* on the negative side of the line. In all figures, we shall show the normal as a directed line segment with the arrow (pointed away from the origin) indicating positive direction with respect to the line.

For a line passing through the origin, $p = 0$ and no direction with respect to the line is established. For this case, we shall agree to assume a normal directed upward from the origin; thus the range of ω is here restricted to $0° \le \omega < 180°$.

See Problems 9-10.

TO REDUCE THE GENERAL EQUATION $Ax + By + C = 0$ to normal form divide each term by $\pm\sqrt{A^2 + B^2}$, taking the sign of the radical

(*1*) opposite to that of C, if $C \neq 0$,

(*2*) to agree with that of B, if $C = 0$ and $B \neq 0$,

(*3*) to agree with that of A, if $C = B = 0$.

See Problem 11.

THE DIRECTED DISTANCE d from a straight line l to a point $P_1(x_1, y_1)$ is obtained by substituting the coordinates of the point for x and y in the left member of the normal form of the equation of the line.

Example 1. Find the directed distance from the line $3x - 4y + 20 = 0$ to the point (*a*) (0,0), (*b*) (2,8), (*c*) (-1,-3).

The left member of the normal form of the equation of the given line is

$$(A) \qquad -\frac{3}{5}x + \frac{4}{5}y - 4.$$

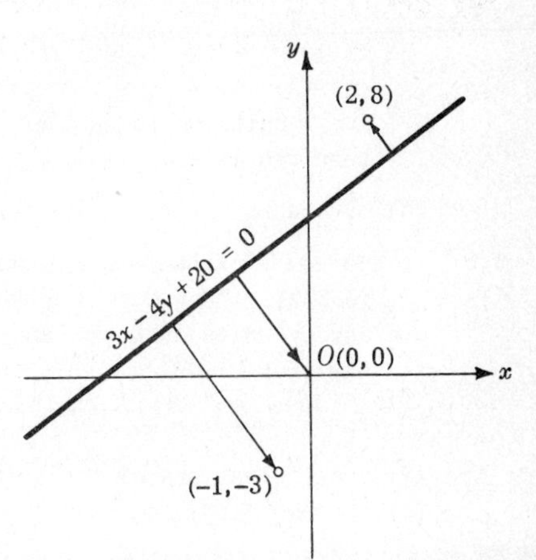

(*a*) Substitute $x = 0$, $y = 0$ in (*A*):

$$d = -\frac{3}{5}(0) + \frac{4}{5}(0) - 4 = -4$$

Note that this agrees with the fact that the distance from the line to the origin is given by $-p$.

(*b*) Substitute $x = 2$, $y = 8$ in (*A*):

$$d = -\frac{3}{5}(2) + \frac{4}{5}(8) - 4 = \frac{6}{5}$$

Thus the origin and the point (2,8) lie on opposite sides of the line.

(c) Substitute $x=-1$, $y=-3$ in (A): $d = -\frac{3}{5}(-1) + \frac{4}{5}(-3) - 4 = -\frac{29}{5}$

Thus the origin and the point (-1,-3) lie on the same side of the line.

See Problems 12-17.

SOLVED PROBLEMS

1. Construct and find the equation of the straight line which passes through the point (-1,-2) with slope (a) 3/4, (b) -4/5 .

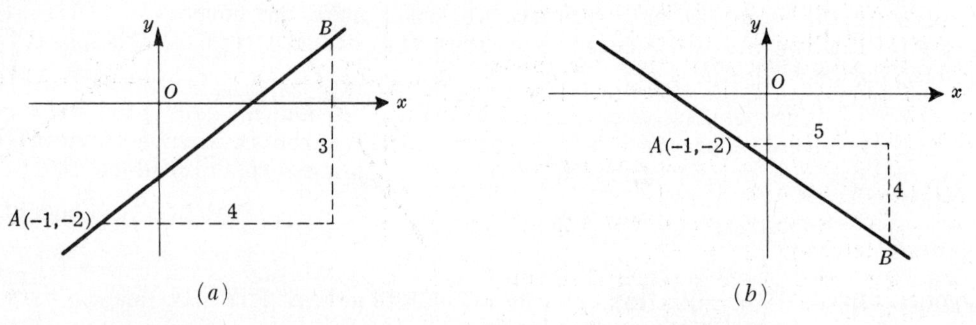

(a) (b)

(a) A point tracing the line *rises* (slope is positive) 3 units as it moves a horizontal distance of 4 units to the right. Thus, after locating the point A(-1,-2), move 4 units to the right and 3 up to the point B(3,1). The required line is AB. (See Fig. (a) above.)

Using $y - y_1 = m(x - x_1)$, the equation is $y + 2 = \frac{3}{4}(x+1)$ or $3x - 4y - 5 = 0$.

(b) A point tracing the line falls (slope is negative) 4 units as it moves a horizontal distance of 5 units to the right. Thus, after locating the point A(-1,-2), move 5 units to the right and 4 units down to the point B(4,-6). The required line is AB as shown in Fig.(b) above. Its equation is

$$y + 2 = -\frac{4}{5}(x+1) \quad \text{or} \quad 4x + 5y + 14 = 0.$$

2. Determine the slope m and y-intercept b of the following lines. Sketch each.

(a) $y = \frac{3}{2}x - 2$ *Ans.* $m = 3/2$; $b = -2$

To sketch the locus, locate the point (0, - 2). Then move 2 units to the right and 3 units up to another point on the required line.

(b) $y = -3x + 5/2$ *Ans.* $m = -3$; $b = 5/2$

To sketch the locus, locate the point (0,5/2). Then move 1 unit to the right and 3 units down to another point on the line.

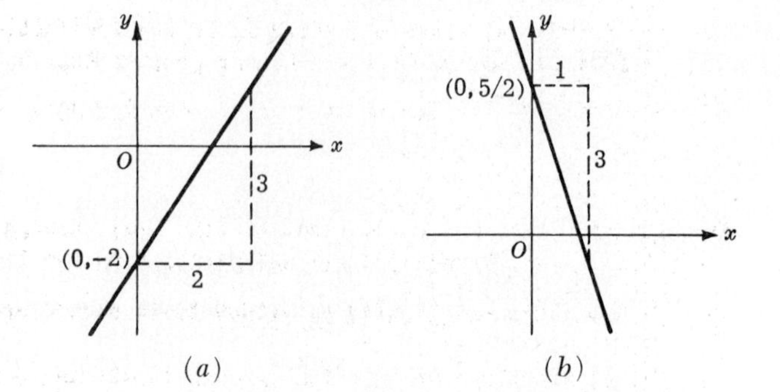

(a) (b)

3. Write the equations of the straight lines: (a) through (2,3) and (-1,4); (b) through (-7,-2) and (-2,-5); (c) through (3,3) and (3,6).

We use $y - y_1 = \frac{y_2 - y_1}{x_2 - x_1}(x - x_1)$ and label each pair of points P_1 and P_2 in the order given.

(*a*) The equation is $y-3 = \frac{4-3}{-1-2}(x-2) = -\frac{1}{3}(x-2)$ or $x+3y-11 = 0$.

(*b*) The equation is $y+2 = \frac{-5+2}{-2+7}(x+7) = -\frac{3}{5}(x+7)$ or $3x+5y+31 = 0$.

(*c*) Here $x_1 = x_2 = 3$. The required equation is $x-3 = 0$.

4. Determine the x-intercept a and the y-intercept b of the following lines. Sketch each.

(*a*) $3x-2y-4 = 0$

When $y=0$, $3x-4=0$ and $x=4/3$; the x-intercept is $a=4/3$.

When $x=0$, $-2y-4=0$ and $y=-2$; the y-intercept is $b=-2$.

To obtain the locus, join the points $(4/3,0)$ and $(0,-2)$ by a straight line.

(*b*) $3x+4y+12 = 0$

When $y=0$, $3x+12=0$ and $x=-4$; the x-intercept is $a=-4$.

When $x=0$, $4y+12=0$ and $y=-3$; the y-intercept is $b=-3$.

The locus is the straight line joining the points $(-4,0)$ and $(0,-3)$.

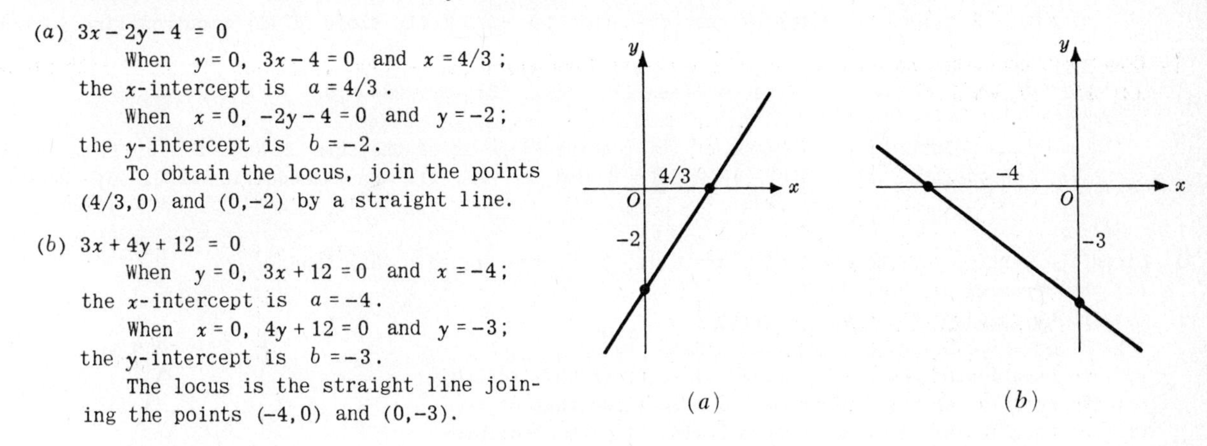

(*a*) (*b*)

5. Prove: If two oblique lines l_1 and l_2 of slope m_1 and m_2 respectively are mutually perpendicular, then $m_1 = -1/m_2$.

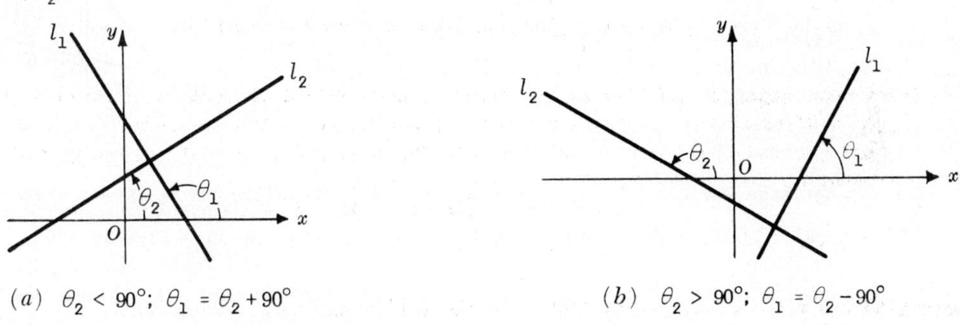

(*a*) $\theta_2 < 90°$; $\theta_1 = \theta_2 + 90°$ (*b*) $\theta_2 > 90°$; $\theta_1 = \theta_2 - 90°$

Let $m_2 = \tan\theta_2$, where θ_2 is the inclination of l_2. The inclination of l_1 is $\theta_1 = \theta_2 \pm 90°$ according as θ_2 is less than or greater than 90°. Then

$$m_1 = \tan\theta_1 = \tan(\theta_2 \pm 90°) = -\cot\theta_2 = -\frac{1}{\tan\theta_2} = -\frac{1}{m_2}$$

6. Find the equation of the straight line (*a*) through (3,1) and parallel to the line through (3,−2) and (−6,5), (*b*) through (−2,−4) and parallel to the line $8x-2y+3 = 0$.

Two lines are parallel provided their slopes are equal.

(*a*) The slope of the line through (3,−2) and (−6,5) is $m = \frac{y_2-y_1}{x_2-x_1} = \frac{5+2}{-6-3} = -\frac{7}{9}$.

The equation of the line through (3,1) with slope $-\frac{7}{9}$ is $y-1 = -\frac{7}{9}(x-3)$ or $7x+9y-30=0$.

(*b*) First Solution. From $y = 4x+3/2$, the slope of the given line is $m=4$. The equation of the line through (−2,−4) with slope 4 is $y+4 = 4(x+2)$ or $4x-y+4 = 0$.

Second Solution. The equation of the required line is of the form $8x-2y+D = 0$. If (−2,−4) is on the line, then $8(-2)-2(-4)+D = 0$ and $D=8$. The required equation is $8x-2y+8=0$ or $4x-y+4 = 0$.

7. Find the equation of the straight line (*a*) through (−1,−2) and perpendicular to the line through (−2,3) and (−5,−6), (*b*) through (2,−4) and perpendicular to the line $5x+3y-8=0$.

Two lines are perpendicular provided the slope of one is the negative reciprocal of the slope of the other.

(*a*) The slope of the line through (−2,3) and (−5,−6) is $m=3$; the slope of the required line is $-1/m=-1/3$. The required equation is

$$y+2 = -\frac{1}{3}(x+1) \quad \text{or} \quad x+3y+7 = 0$$

(*b*) First Solution. The slope of the given line is $-5/3$; the slope of the required line is 3/5. The required equation is

$$y+4 = \frac{3}{5}(x-2) \quad \text{or} \quad 3x-5y-26 = 0$$

Second Solution. The equation of the required line is of the form $3x-5y+D=0$. If (2,−4) is on the line, then $3(2)-5(-4)+D=0$ and $D=-26$. The required equation is $3x-5y-26=0$.

8. Given the vertices $A(7,9)$, $B(-5,-7)$, and $C(12,-3)$ of the triangle ABC, find:
(*a*) the equation of the side AB,
(*b*) the equation of the median through A,
(*c*) the equation of the altitude through B,
(*d*) the equation of the perpendicular bisector of the side AB,
(*e*) the equation of the line through C with slope that of AB,
(*f*) the equation of the line through C with slope the reciprocal of that of AB.

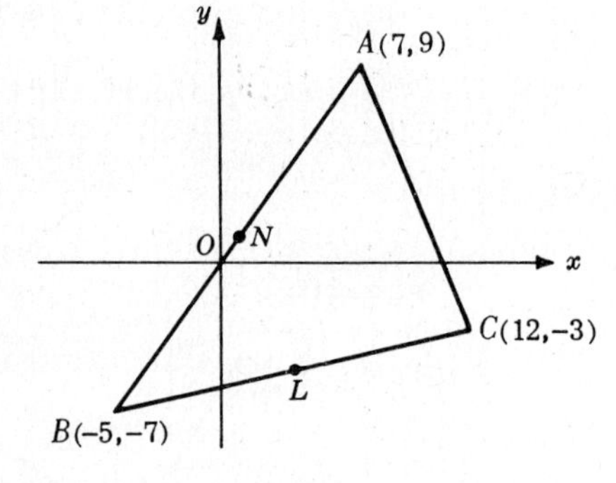

(*a*) $y+7 = \frac{9+7}{7+5}(x+5) = \frac{4}{3}(x+5)$ or $4x-3y-1=0$

(*b*) The median through a vertex bisects the opposite side. The midpoint of BC is $L(7/2,-5)$. The equation of the median through A is

$$y-9 = \frac{-5-9}{7/2-7}(x-7) = 4(x-7) \quad \text{or} \quad 4x-y-19 = 0$$

(*c*) The altitude through B is perpendicular to CA. The slope of CA is −12/5 and its negative reciprocal is 5/12. The equation is $y+7 = \frac{5}{12}(x+5)$ or $5x-12y-59=0$.

(*d*) The perpendicular bisector of AB passes through the midpoint $N(1,1)$ and has slope $-3/4$. The equation is $y-1 = -\frac{3}{4}(x-1)$ or $3x+4y-7=0$.

(*e*) $y+3 = \frac{4}{3}(x-12)$ or $4x-3y-57=0$. (*f*) $y+3 = \frac{3}{4}(x-12)$ or $3x-4y-48=0$.

9. Derive the normal form $x\cos\omega + y\sin\omega - p = 0$ of the equation of a straight line, not passing through the origin.

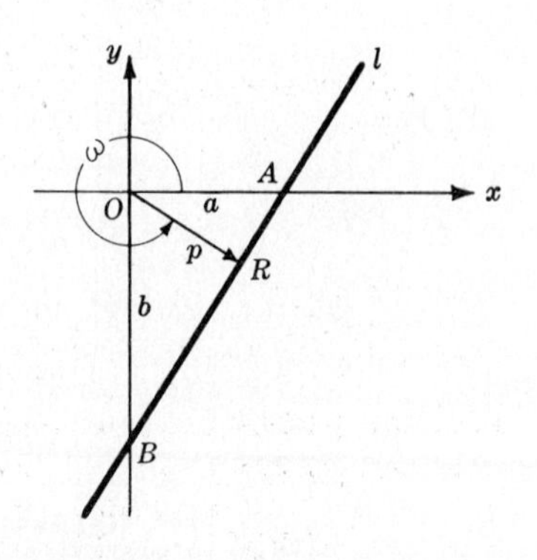

Let the given line have intercepts $x=a$ and $y=b$ so that its equation is (*A*) $\frac{x}{a}+\frac{y}{b}=1$.

Denote by p the length of the normal OR and by ω the positive angle which the normal makes with the x-axis.

From the right triangle ARO, $p = a\cos(360°-\omega) = a\cos\omega$ or $a=p/\cos\omega$. From the right triangle BRO, $p = b\cos(\omega-270°) = b\sin\omega$ or $b=p/\sin\omega$. Substituting in (*A*), we have

$$\frac{x\cos\omega}{p}+\frac{y\sin\omega}{p} = 1 \quad \text{or} \quad x\cos\omega + y\sin\omega - p = 0$$

In the above derivation the line l was chosen so that the normal extends into the fourth quadrant. It is left for the reader to consider other positions of l.

10. Construct all lines satisfying the given condition and write their equations:
(a) $\omega = 135°$; $p = 4$.
(b) inclination, $\theta = 150°$; $p = 5$
(c) x-intercept, $a = 5$; distance from the origin $= 3$.
(d) whose nearest point to the origin is $(-6, 8)$.

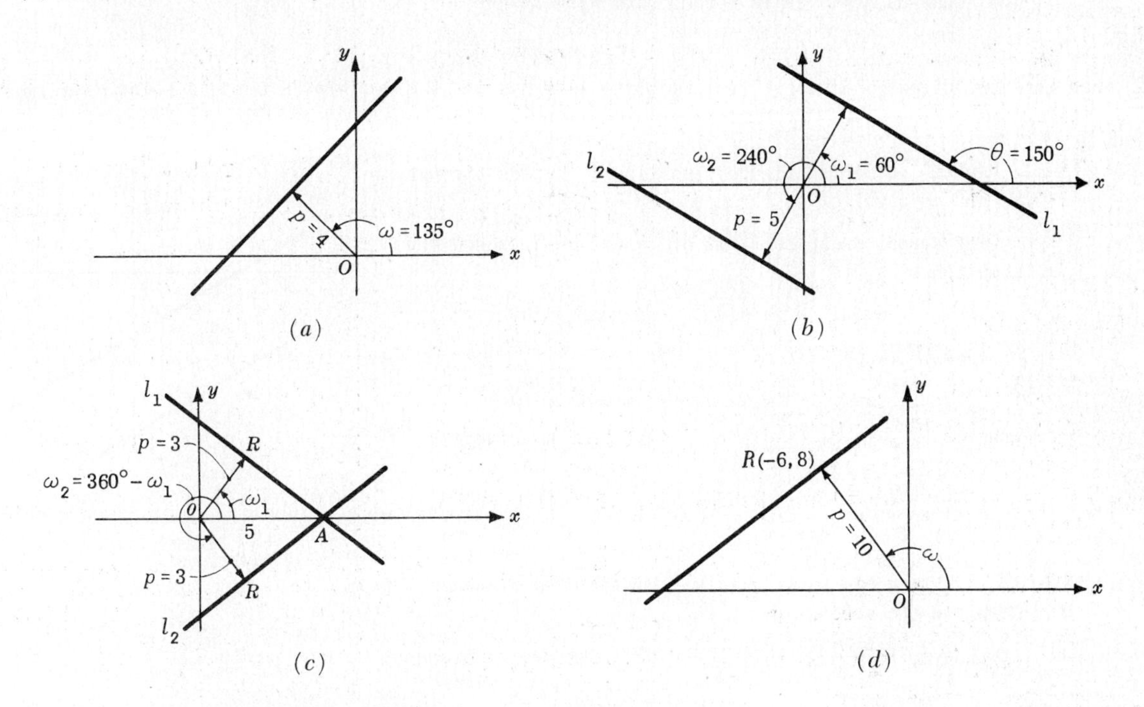

The normal form of the equation of a straight line is $x\cos\omega + y\sin\omega - p = 0$.

(a) The equation is $x\cos 135° + y\sin 135° - 4 = -\frac{1}{2}\sqrt{2}\,x + \frac{1}{2}\sqrt{2}\,y - 4 = 0$ or $x - y + 4\sqrt{2} = 0$.

(b) From Fig. (b), the line l_1 is characterized by $\omega_1 = 60°$ and $p = 5$; the equation is

$$x\cos 60° + y\sin 60° - 5 = \tfrac{1}{2}x + \tfrac{1}{2}\sqrt{3}\,y - 5 = 0 \quad\text{or}\quad x + \sqrt{3}\,y - 10 = 0$$

The line l_2 is characterized by $\omega_2 = 240°$ and $p = 5$; the equation is

$$x\cos 240° + y\sin 240° - 5 = -\tfrac{1}{2}x - \tfrac{1}{2}\sqrt{3}\,y - 5 = 0 \quad\text{or}\quad x + \sqrt{3}\,y + 10 = 0$$

(c) In Fig. (c), $OA = 5$, $OR = 3$, and $AR = 4$. Then for l_1, $\sin\omega_1 = 4/5$, $\cos\omega_1 = 3/5$ and the equation is

$$\frac{3}{5}x + \frac{4}{5}y - 3 = 0 \quad\text{or}\quad 3x + 4y - 15 = 0$$

For the line l_2, $\omega_2 = 360° - \omega_1$. Then $\sin\omega_2 = -4/5$, $\cos\omega_2 = 3/5$ and the equation is

$$\frac{3}{5}x - \frac{4}{5}y - 3 = 0 \quad\text{or}\quad 3x - 4y - 15 = 0$$

(d) Here the normal joins the origin to $(-6, 8)$. Then $p = 10$, $\sin\omega = 4/5$, $\cos\omega = -3/5$ and the equation is

$$-\frac{3}{5}x + \frac{4}{5}y - 10 = 0 \quad\text{or}\quad 3x - 4y + 50 = 0$$

11. Write each of the following equations in normal form and determine p and ω:
(a) $3x + 4y + 10 = 0$, (b) $x - y = 0$, (c) $x = 5$, (d) $y = 0$.

(a) Here $C > 0$ and we divide by $-\sqrt{(3)^2 + (4)^2} = -5$ to get $-3x/5 - 4y/5 - 2 = 0$.

Since $\sin\omega < 0$ and $\cos\omega < 0$, ω is the third quadrant angle whose tangent is $4/3 = 1.3333$. Then $\omega = 223°8'$ and $p = 2$.

(*b*) Here $C = 0$ and $B = -1$; we divide by $-\sqrt{(1)^2 + (-1)^2} = -\sqrt{2}$ to get $-x/\sqrt{2} + y/\sqrt{2} = 0$. Then ω is the second quadrant angle whose tangent is -1, that is, $\omega = 135°$ and $p = 0$.

(*c*) The equation $x - 5 = 0$ is in normal form with $\cos\omega = 1$ and $\sin\omega = 0$. Thus $\omega = 0°$ and $p = 5$.

(*d*) The equation $y = 0$ is in normal form with $\omega = 90°$ and $p = 0$.

12. Show that the directed distance from the line l: $x\cos\omega + y\sin\omega - p = 0$ to a point $P_1(x_1, y_1)$ is given by $d = x_1\cos\omega + y_1\sin\omega - p$.

In the adjoining figure, the line l_1 passes through P_1 and is parallel to l.

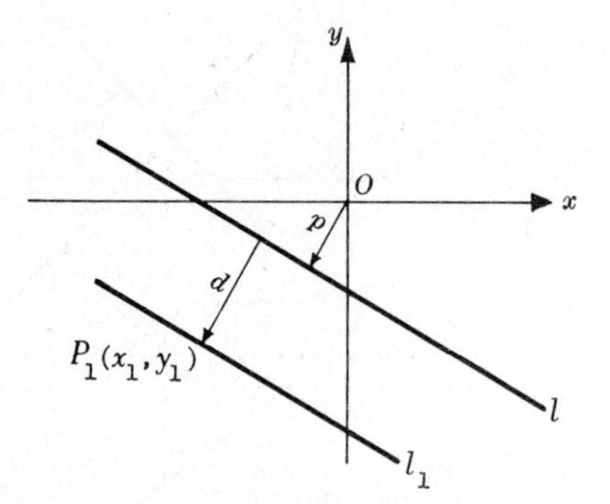

The normal drawn to l_1 is of length $p + d$; hence the equation of l_1 is

$$x\cos\omega + y\sin\omega - (p + d) = 0$$

Since $P_1(x_1, y_1)$ is on l_1,

$$x_1\cos\omega + y_1\sin\omega - (p + d) = 0$$

and
$$d = x_1\cos\omega + y_1\sin\omega - p$$

13. Show that the line l: $8x - y + 26 = 0$ is tangent to the circle, with center at $C(5,1)$, passing through $P(-2,-3)$.

If l is tangent to the circle, the directed distance d from l to C is equal, except possibly for sign, to the radius $CP = \sqrt{65}$.

Now $d = \dfrac{8(5) - 1(1) + 26}{-\sqrt{65}} = -\sqrt{65}$; hence l is tangent to the circle.

14. Find the value of k such that the line l: $4x - 3y + k = 0$ shall be tangent to the circle with center at $C(6,1)$ and radius 2.

The directed distance from l to C must be numerically equal to 2; then $\left|\dfrac{4(6) - 3(1) + k}{\pm 5}\right| = 2$, $\dfrac{21 + k}{5} = \pm 2$ and $k = -11, -31$. The corresponding equations are $4x - 3y - 11 = 0$ and $4x - 3y - 31 = 0$.

15. Find the equations of the lines which are parallel to l: $2x - 5y - 2\sqrt{29} = 0$ and 4 units distant from it.

Let (X,Y) be any point on the required line. Then $\dfrac{2X - 5Y - 2\sqrt{29}}{\sqrt{29}} = \pm 4$. Simplifying and replacing (X,Y) by the customary (x,y), we have $2x - 5y - 6\sqrt{29} = 0$ and $2x - 5y + 2\sqrt{29} = 0$ as the required equations.

16. Find the equation of the line which is equidistant from the parallel lines l_1: $2x - 3y + 14 = 0$ and l_2: $2x - 3y - 6 = 0$.

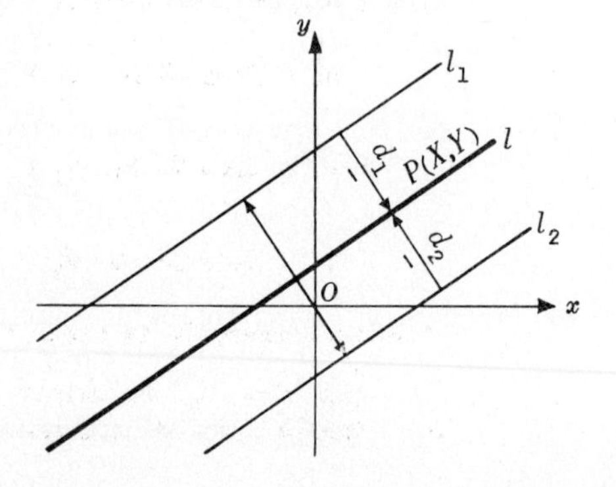

Let $P(X,Y)$ be any point on the required line. The directed distances of P from l_1 and l_2 respectively are

$$d_1 = \frac{2X - 3Y + 14}{-\sqrt{13}} \quad \text{and} \quad d_2 = \frac{2X - 3Y - 6}{\sqrt{13}}$$

From the adjoining figure it is clear that d_1 and d_2 are negative. Setting $d_1 = d_2$, simplifying, and replacing (X,Y) by (x,y), we have $2x - 3y + 4 = 0$ as the equation of the line.

17. Find the distance (undirected) between the parallel lines l_1: $5x + 12y + 4 = 0$ and l_2: $10x + 24y - 5 = 0$.

We seek the numerical value of the distance from either line to any point on the other line. Using l_2 and the point (4,−2) on l_1, we find $d = \dfrac{10(4) + 24(-2) - 5}{26} = -\dfrac{1}{2}$. The required distance is $\frac{1}{2}$.

18. The equations of the sides of a triangle are AB: $x + y - 5 = 0$, BC: $x + 7y - 7 = 0$, and CA: $7x + y + 14 = 0$. Find (a) the bisector of the interior angle at B and (b) the bisector of the exterior angle at C.

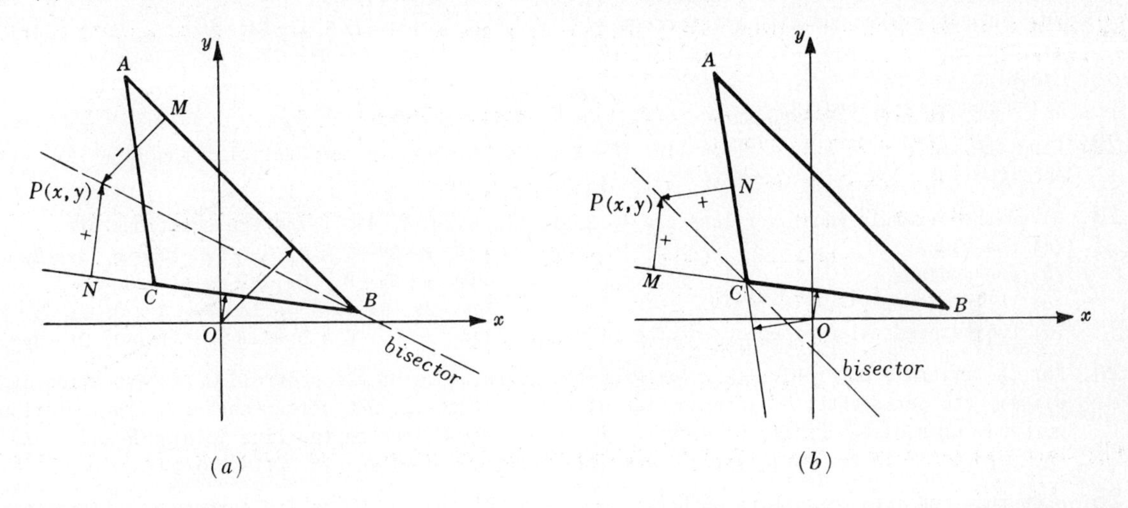

We make use of the fact that any point on the bisector of an angle is equidistant from the sides of the angle.

(a) Refer to Fig. (a). On the bisector of the interior angle at B, take any point $P(x,y)$ and drop perpendiculars MP and NP to the sides AB and BC respectively. Now the normal from the origin to AB and the segment MP (from line to point) have opposite directions while the normal to BC and the segment NP have the same direction.

Then $MP = -NP$, $\dfrac{x + y - 5}{\sqrt{2}} = -\dfrac{x + 7y - 7}{5\sqrt{2}}$ and the equation of the bisector is $3x + 6y - 16 = 0$.

(b) Refer to Fig. (b). As in (a) take $P(x,y)$ any point on the bisector of the exterior angle at C and drop the perpendiculars MP and NP to the sides BC and CA respectively.

Then $MP = NP$, $\dfrac{x + 7y - 7}{5\sqrt{2}} = \dfrac{7x + y + 14}{-5\sqrt{2}}$ and the equation of the bisector is $8x + 8y + 7 = 0$.

19. Find the center of the inscribed circle of the triangle of Problem 18.

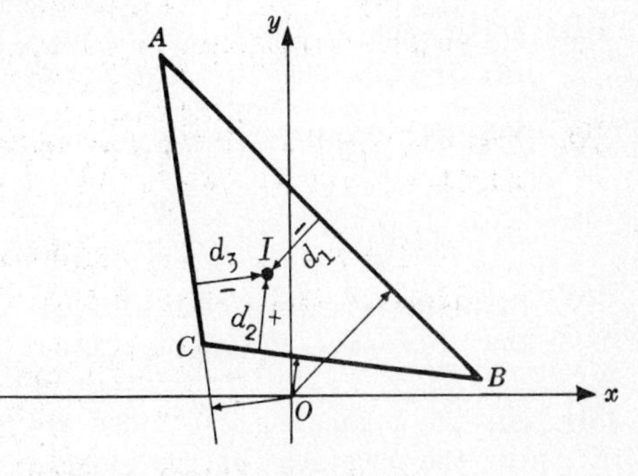

Let $I(h,k)$ be the coordinates of the center; then the directed distance

of I from AB is $d_1 = \dfrac{h + k - 5}{\sqrt{2}}$,

of I from BC is $d_2 = \dfrac{h + 7k - 7}{5\sqrt{2}}$,

of I from CA is $d_3 = \dfrac{7h + k + 14}{-5\sqrt{2}}$.

From the figure it is clear that $d_1 = -d_2 = d_3$.

From $d_1 = -d_2$, we obtain $3h + 6k - 16 = 0$, and from $d_1 = d_3$, we obtain $12h + 6k - 11 = 0$. Then $h = -5/9$, $k = 53/18$ and the incenter of the triangle has coordinates (−5/9, 53/18).

SUPPLEMENTARY PROBLEMS

20. Determine the slope and the intercepts on the coordinate axes of each of the lines (*a*) $4x - 5y + 20 = 0$, (*b*) $2x + 3y - 12 = 0$. *Ans.* (*a*) $m = 4/5$, $a = -5$, $b = 4$; (*b*) $m = -2/3$, $a = 6$, $b = 4$

21. Write the equation of each of the straight lines:

(*a*) with slope 3 and y-intercept 2. *Ans.* $3x - y + 2 = 0$
(*b*) through (5,4) and parallel to $2x + 3y - 12 = 0$. *Ans.* $2x + 3y - 22 = 0$
(*c*) through (−3,3) and with y-intercept 6. *Ans.* $x - y + 6 = 0$
(*d*) through (−3,3) and with x-intercept 4. *Ans.* $3x + 7y - 12 = 0$
(*e*) through (5,4) and (−3,3). *Ans.* $x - 8y + 27 = 0$
(*f*) with $a = 3$ and $b = -5$. *Ans.* $5x - 3y - 15 = 0$
(*g*) through (2,3) and (2,−5). *Ans.* $x - 2 = 0$

22. Find the value of k such that the line $(k-1)x + (k+1)y - 7 = 0$ is parallel to the line $3x + 5y + 7 = 0$.
Ans. $k = 4$

23. Given the triangle whose vertices are $A(-3,2)$, $B(5,6)$, $C(1,-4)$. Find the equations of:

(*a*) the sides. *Ans.* $x - 2y + 7 = 0$, $5x - 2y - 13 = 0$, $3x + 2y + 5 = 0$
(*b*) the medians. *Ans.* $x + 6y - 9 = 0$, $7x - 6y + 1 = 0$, $x = 1$
(*c*) the altitudes. *Ans.* $2x + 5y - 4 = 0$, $2x - 3y + 8 = 0$, $2x + y + 2 = 0$
(*d*) the perpendicular bisectors of the sides. *Ans.* $2x + y - 6 = 0$, $2x + 5y - 11 = 0$, $2x - 3y - 1 = 0$

24. For the triangle of Problem 23: (*a*) Find the coordinates of the centroid G (intersection of the medians), the orthocenter H (intersection of the altitudes), and circum-center C (intersection of the perpendicular bisectors of the sides). (*b*) Show that G lies on the line joining H and C and divides the segment HC in the ratio 2:1. *Ans.* (*a*) $G(1,4/3)$, $H(-7/4,3/2)$, $C(19/8,5/4)$

25. Show that the line passing through the points (5,−4) and (−2,7) is the perpendicular bisector of the line segment whose end points are (−4,−2) and (7,5).

26. Use the determinant form of the formula for the area of a triangle (Chapter 47) to show:

(*a*) The equation of the straight line through $P_1(x_1,y_1)$ and $P_2(x_2,y_2)$ is given by $\begin{vmatrix} x & y & 1 \\ x_1 & y_1 & 1 \\ x_2 & y_2 & 1 \end{vmatrix} = 0$.

(*b*) Three points $P_1(x_1,y_1)$, $P_2(x_2,y_2)$, and $P_3(x_3,y_3)$ are on a straight line provided $\begin{vmatrix} x_1 & y_1 & 1 \\ x_2 & y_2 & 1 \\ x_3 & y_3 & 1 \end{vmatrix} = 0$.

27. Draw the following straight lines and write their equations.

(*a*) $\omega = 60°$, $p = 6$
(*b*) $\omega = 180°$, $p = 5$
(*c*) $\omega = 330°$, $p = 2$
(*d*) Angle of inclination, 120°; $p = 3$
(*e*) x-intercept, 5; $p = 4$
(*f*) The line whose nearest point to the origin is (3,−2)

Ans. (*a*) $x + \sqrt{3}y - 12 = 0$ (*b*) $x + 5 = 0$ (*c*) $\sqrt{3}x - y - 4 = 0$ (*d*) $\sqrt{3}x + y \pm 6 = 0$ (*e*) $4x \pm 3y - 20 = 0$ (*f*) $3x - 2y - 13 = 0$

28. Write each of the following equations in normal form and determine p and ω.
(*a*) $5x - 12y - 39 = 0$, (*b*) $3x + 4y + 15 = 0$, (*c*) $6x - 8y + 25 = 0$, (*d*) $x + 3y = 0$

Ans. (*a*) $\frac{5}{13}x - \frac{12}{13}y - 3 = 0$; $p = 3$, $\omega = 292°40'$
(*b*) $-\frac{3}{5}x - \frac{4}{5}y - 3 = 0$; $p = 3$, $\omega = 233°10'$
(*c*) $-\frac{3}{5}x + \frac{4}{5}y - \frac{5}{2} = 0$; $p = \frac{5}{2}$, $\omega = 126°50'$
(*d*) $\frac{x}{\sqrt{10}} + \frac{3y}{\sqrt{10}} = 0$; $p = 0$, $\omega = 71°30'$

29. Find the lengths of the altitudes of the triangle of Problem 23.
Ans. $16\sqrt{5}/5$, $32\sqrt{29}/29$, $32\sqrt{13}/13$

30. For the triangle whose sides are $x + y - 6 = 0$, $7x + y - 6 = 0$, and $x - 7y - 6 = 0$, find:
(*a*) the equations of the bisectors of the interior angles, (*b*) the coordinates of the incenter.
Ans. (*a*) $2x + y - 6 = 0$, $x + 3y - 6 = 0$, $4x - 3y - 6 = 0$; (*b*) (12/5, 6/5)

CHAPTER 49

Families of Straight Lines

THE EQUATION $y = 3x + b$ represents the set of all lines having slope 3. The quantity b is a variable over the set of lines, each value of b being associated with one and only one line of the set. To distinguish it from the variables x, y which vary over the points of each line, b is called a *parameter*. Such sets of lines are also called *one-parameter systems* or *families* of lines.

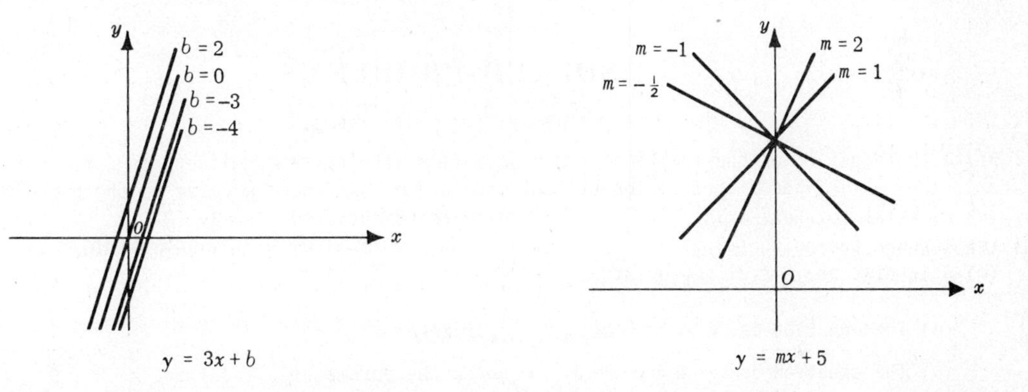

Similarly, the equation $y = mx + 5$, in which m is the parameter, represents the one-parameter family of lines having y-intercept = 5 or passing through the point (0,5). It is important to note that one line satisfying the geometric condition is not included since for no value of m does the equation $y = mx + 5$ yield the line $x = 0$.

See Problems 1-4.

THE EQUATION (*1*) $$A_1x + B_1y + C_1 + k(A_2x + B_2y + C_2) = 0$$

represents the family of all lines, except for l_2, passing through the point of intersection of the lines

$$l_1:\ A_1x + B_1y + C_1 = 0 \quad \text{and} \quad l_2:\ A_2x + B_2y + C_2 = 0$$

See Problem 5.

Equation (*1*) provides a means of finding the equation of a line which passes through the point of intersection of two given lines and satisfies one other condition, without having first to compute the coordinates of the point of intersection.

Example 1. Write the equation of the line which passes through the point of intersection of the lines

$$l_1:\ 3x + 5y + 29 = 0 \quad \text{and} \quad l_2:\ 7x - 11y - 13 = 0$$

and through the point (2,1).

Since the given point is not on l_2, the required line is among those given by

$$l_1 + kl_2: \quad 3x + 5y + 29 + k(7x - 11y - 13) = 0$$

For the line which passes through (2, 1),

$$3(2) + 5(1) + 29 + k[7(2) - 11(1) - 13] = 0 \quad \text{and} \quad k = 4;$$

hence, its equation is

$$l_1 + 4l_2: \quad 3x + 5y + 29 + 4(7x - 11y - 13) = 31x - 39y - 23 = 0$$

See Problem 6.

THE EQUATION $x \cos \omega + y \sin \omega - p = 0$, with parameters p and ω, represents the two parameter family of *all* lines of the plane. The equation $Ax + By + C = 0$ contains three parameters but only two are essential; that is, since at least one of $A, B \neq 0$, say $B \neq 0$, we may divide the equation by B to get $\frac{A}{B}x + y + \frac{C}{B} = 0$ or, letting $a = A/B$ and $c = C/B$, $ax + y + c = 0$ with only two parameters.

SOLVED PROBLEMS

1. Write the equation of the family of lines satisfying the given condition. Name the parameter and list any lines satisfying the condition but not obtained by assigning a value to the parameter.

(*a*) Parallel to the x-axis.
(*b*) Through the point $(-3, 2)$.
(*c*) At a distance 5 from the origin.
(*d*) Perpendicular to $5x + 3y - 8 = 0$.
(*e*) The sum of whose intercepts is 10.

(*a*) The equation is $y = k$, k being the parameter.

(*b*) The equation is $y - 2 = m(x + 3)$, m being the parameter. The line $x + 3 = 0$ is not included.

(*c*) The equation is $x \cos \omega + y \sin \omega - 5 = 0$, ω being the parameter.

(*d*) The slope of the given line is $-5/3$; hence the slope of the perpendicular is $3/5$.

The equation of the family is $y = \frac{3}{5}x + c$ or $3x - 5y + 5c = 0$, c being the parameter. The latter is equivalent to $3x - 5y + k = 0$, with k as parameter.

(*e*) Taking the x-intercept as $a \neq 0$, the y-intercept is $10 - a$ and the equation of the family is $\frac{x}{a} + \frac{y}{10 - a} = 1$. The parameter is a.

2. Describe each family of lines: (*a*) $x = k$, (*b*) $x \cos \omega + y \sin \omega - 10 = 0$, (*c*) $x/\cos \theta + y/\sin \theta = 1$, (*d*) $kx + \sqrt{1-k^2}\, y - 10 = 0$.

(*a*) This is the family of all vertical lines.

(*b*) This is the family of all tangents to the circle with center at the origin and radius = 10.

(*c*) This is the family of all lines the sum of the squares of whose intercepts is 1.

(*d*) Since $\sqrt{1-k^2}$ is to be assumed real, the range of $k = \cos \omega$ is $-1 \leq k \leq 1$, while the range of $\sqrt{1-k^2} = \sin \omega$ is $0 \leq \sqrt{1-k^2} \leq 1$. The equation is that of the family of tangents to the upper half circle of (*b*).

3. In each of the following write the equation of the family of lines satisfying the first condition and then obtain the equation of the line of the family satisfying the second condition: (*a*) parallel to $3x - 5y + 12 = 0$, through $P(1, -2)$; (*b*) perpendicular to $3x - 5y + 12 = 0$, through $P(1, -2)$; (*c*) through $P(-3, 2)$, at a distance 3 from the origin.

(a) The equation of the family is $3x-5y+k = 0$. Substituting $x=1$, $y=-2$ in this equation and solving for k, we find $k=-13$. The required line has equation $3x-5y-13 = 0$.

(b) The equation of the family is $5x+3y+k = 0$. (see Problem 1(d)). Proceeding as in (a) above, we find $k=1$; the required line has equation $5x+3y+1 = 0$.

(c) The equation of the family is $y-2 = k(x+3)$ or, in normal form, $\dfrac{kx-y+(3k+2)}{\pm\sqrt{k^2+1}} = 0$. Setting the undirected distance of a line of the family from the origin equal to 3, we have $\left|\dfrac{3k+2}{\pm\sqrt{k^2+1}}\right| = 3$. Then $\dfrac{(3k+2)^2}{k^2+1} = 9$ and $k = 5/12$. Thus, the required line has equation

$$y-2 = \frac{5}{12}(x+3) \qquad \text{or} \qquad 5x-12y+39 = 0$$

Now there is a second line, having equation $x+3 = 0$, satisfying the conditions but this line (see Problem 1(b)) was not included in the equation of the family.

4. Write the equation of the line which passes through the point of intersection of the lines l_1: $3x+5y+26 = 0$ and l_2: $7x-11y-13 = 0$ and satisfies the additional condition: (a) Passes through the origin. (b) Is perpendicular to the line $7x+3y-8 = 0$.

Each of the required lines is a member of the family

$$(A) \qquad l_1 + kl_2: \quad 3x+5y+26 + k(7x-11y-13) = 0$$

(a) Substituting $x=0$, $y=0$ in (A), we find $k=2$; the required line has equation

$$l_1 + 2l_2: \quad 3x+5y+26 + 2(7x-11y-13) = 0 \qquad \text{or} \qquad x-y = 0$$

(b) The slope of the given line is $-7/3$ and the slope of a line of (A) is $-\dfrac{3+7k}{5-11k}$.

Setting one slope equal to the negative reciprocal of the other, we find $-\dfrac{7}{3} = \dfrac{5-11k}{3+7k}$ and $k = -9/4$. The required line has equation

$$l_1 - \frac{9}{4}l_2: \quad 3x+5y+26 - \frac{9}{4}(7x-11y-13) = 0 \qquad \text{or} \qquad 3x-7y-13 = 0$$

5. Write the equation of the line which passes through the point of intersection of the lines l_1: $x+4y-18 = 0$ and l_2: $x+2y-2 = 0$, and satisfies the additional condition: (a) is parallel to the line $3x+8y+1 = 0$, (b) whose distance from the origin is 2.

The equation of the family of lines through the point of intersection of the given lines is

$$l_1 + kl_2: \quad x+4y-18 + k(x+2y-2) = 0 \qquad \text{or}$$

$$(A) \qquad (1+k)x + (4+2k)y - (18+2k) = 0.$$

(a) Since the required line is to have slope $-3/8$, we set $-\dfrac{1+k}{4+2k} = -\dfrac{3}{8}$.

Then $k=2$ and the line has equation

$$l_1 + 2l_2: \quad 3x+8y-22 = 0.$$

(b) The normal form of (A) is $\dfrac{(1+k)x + (4+2k)y - (18+2k)}{\pm\sqrt{17+18k+5k^2}} = 0$.

Setting $\left|\dfrac{-(18+2k)}{\pm\sqrt{17+18k+5k^2}}\right| = 2$ and squaring both members, we find $k = \pm 4$. The required lines have equations

$$l_1 + 4l_2: \quad 5x+12y-26 = 0 \qquad \text{and} \qquad l_1 - 4l_2: \quad 3x+4y+10 = 0.$$

SUPPLEMENTARY PROBLEMS

6. Write the equation of the family of lines satisfying the given condition. List any lines satisfying the condition but not obtained by assigning a value to the parameter.
(a) Perpendicular to the x-axis.
(b) Through the point (3,−1).
(c) At the distance 6 from the origin.
(d) Parallel to $2x+5y-8=0$.
(e) The product of whose intercepts on the coordinate axes is 10.
Ans. (a) $x=k$ (b) $y=kx-3k-1$, $x=3$ (c) $x\cos\omega+y\sin\omega-6=0$ (d) $2x+5y+k=0$ (e) $10x+k^2y-10k=0$

7. Write the equation of the family of lines satisfying the first condition and then obtain the equation of that line of the family satisfying the second condition.
(a) Parallel to $2x-3y+8=0$; passing through $P(2,-2)$.
(b) Perpendicular to $2x-3y+8=0$; passing through $P(2,-2)$.
(c) Sum of the intercepts on the coordinate axes is 2; forms with the coordinate axes a triangle of area 24.
(d) At a distance 3 from the origin; passes through $P(1,3)$.
(e) Slope is 3/2; product of the intercepts on the coordinate axes is −54.
(f) The y-intercept is 6; makes an angle of 135° with the line $7x-y-23=0$.
(g) Slope is 5/12; 5 units from (2,−3).
Ans. (a) $2x-3y-10=0$
(b) $3x+2y-2=0$
(c) $3x-4y-24=0$, $4x-3y+24=0$
(d) $3x+4y-15=0$, $y=3$
(e) $3x-2y\pm18=0$
(f) $3x-4y+24=0$
(g) $5x-12y+19=0$, $5x-12y-111=0$

8. Use the parametric method to find the equation of the line which passes through the point $P(3,-4)$ and has the additional property:
(a) The sum of its intercepts on the coordinate axes is −5.
(b) The product of its intercepts on the coordinate axes is the negative reciprocal of its slope.
(c) It forms with the coordinate axes a triangle of area 1.
Ans. (a) $2x-y-10=0$, $2x+3y+6=0$, (b) $x+y+1=0$, $5x+3y-3=0$ (c) $2x+y-2=0$, $8x+9y+12=0$

9. Write the equation of the line which passes through the point of intersection of the lines l_1: $x-2y-4=0$ and l_2: $4x-y-4=0$ and satisfies the additional condition:
(a) passes through the origin,
(b) passes through the point $P(4,-6)$,
(c) is parallel to the line $16x-11y+3=0$,
(d) is perpendicular to the line $9x+22y-8=0$,
(e) is 12/7 units from the origin,
(f) makes an angle of 45° with the line $9x-5y-12=0$,
(g) its y-intercept is −20/11 its slope,
(h) makes with the coordinate axes a triangle of area 12/5.
Ans. (a) $3x+y=0$
(b) $5x+4y+4=0$
(c) $16x-11y-28=0$
(d) $22x-9y-28=0$
(e) $7y+12=0$, $21x-28y-60=0$
(f) $49x+14y-4=0$
(g) $11x-8y-20=0$
(h) $6x-5y-12=0$, $15x-2y-12=0$,
$12(12\mp\sqrt{89})x+(41\pm3\sqrt{89})y-12(1\mp\sqrt{89})=0$

CHAPTER 50

The Circle

A CIRCLE IS THE LOCUS of a point which moves in a plane so that it is always at a constant distance from a fixed point in the plane. The fixed point is called the *center* and the constant distance is the length of the *radius* of the circle.

THE STANDARD FORM of the equation of the circle whose center is at the point $C(h,k)$ and whose radius is the constant r is

$$(1) \qquad (x-h)^2 + (y-k)^2 = r^2$$

See Problem 1.

THE GENERAL FORM of the equation of a circle is

$$(2) \qquad x^2 + y^2 + 2Dx + 2Ey + F = 0$$

By completing the squares this may be put in the form

$$(x+D)^2 + (x+E)^2 = D^2 + E^2 - F$$

Thus, (2) represents a circle with center at $C(-D,-E)$ and radius $\sqrt{D^2+E^2-F}$ if $(D^2+E^2-F) > 0$, a point if $(D^2+E^2-F) = 0$, and an imaginary locus if $(D^2+E^2-F) < 0$.

See Problems 2-3.

IN BOTH THE STANDARD AND GENERAL FORM the equation of a circle contains three independent arbitrary constants. It follows that a circle is uniquely determined by three independent conditions.

See Problems 4-6.

THE EQUATION OF A TANGENT to a circle may be found by making use of the fact that a tangent and the radius drawn to the point of tangency are perpendicular lines.

See Problem 7.

THE LENGTH OF A TANGENT to a circle from an external point P_1 is defined as the undirected distance from the point P_1 to the point of tangency. The two tangents from an external point are of equal length.

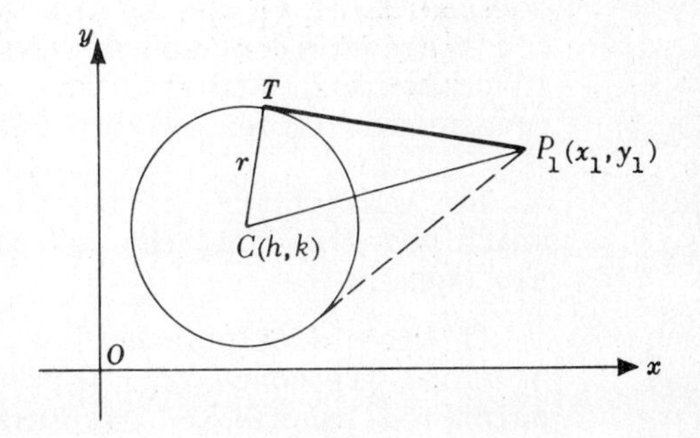

The square of the length of a tangent from the external point $P_1(x_1,y_1)$ to a circle is obtained by substituting the coordinates of the point in the left member of the equation of the circle when written in the form

$$(x-h)^2 + (y-k)^2 - r^2 = 0 \qquad \text{or} \qquad x^2 + y^2 + 2Dx + 2Ey + F = 0$$

See Problems 8-9.

THE EQUATION (3) $x^2 + y^2 + 2D_1x + 2E_1y + F_1 + k(x^2 + y^2 + 2D_2x + 2E_2y + F_2) = 0$,
where K_1: $x^2 + y^2 + 2D_1x + 2E_1y + F_1 = 0$ and K_2: $x^2 + y^2 + 2D_2x + 2E_2y + F_2 = 0$
are distinct circles and $k \neq -1$ is a parameter, represents a one-parameter family of circles.

If K_1 and K_2 are concentric, the circles of (3) are concentric with them.

If K_1 and K_2 are not concentric, the circles of (3) have a common line of centers with them and the centers of the circles (3) divide the segment joining the centers of K_1 and K_2 in the ratio k:1.

If K_1 and K_2 intersect in two distinct points P_1 and P_2, (3) consists of all circles except K_2 which pass through these points. If K_1 and K_2 are tangent to each other at the point P_1, (3) consists of all circles except K_2 which are tangent to each other at P_1. If K_1 and K_2 have no point in common, any two circles of the family (3) have no point in common with each other.

See Problems 11-13.

THE RADICAL AXIS OF TWO CIRCLES. For $k = -1$, equation (3) takes the form

(4) $$2(D_1 - D_2)x + 2(E_1 - E_2)y + F_1 - F_2 = 0$$

If K_1 and K_2 are non-concentric, equation (4) represents a straight line, called the *radical axis*, of the two circles.

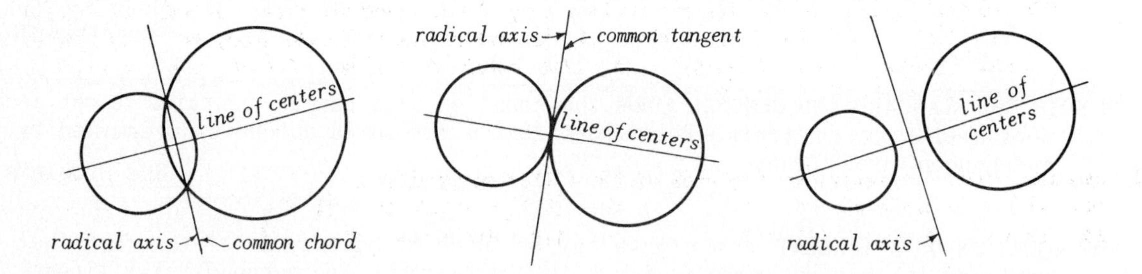

If K_1 and K_2 intersect in two distinct points, their radical axis is their common chord; if K_1 and K_2 are tangent to each other, their radical axis is their common tangent; if K_1 and K_2 have no point in common, their radical axis has no point in common with either circle. In all cases, however, the radical axis of two non-concentric circles is perpendicular to their line of centers.

See Problems 14-15.

The radical axis of two distinct non-concentric circles is the locus of a point which moves so that the lengths of the tangents drawn from it to the two circles are equal.

If three distinct circles, which do not have a common line of centers and no two of which are concentric, are taken in pairs, the three radical axes intersect in a point, called the *radical center* of the three circles.

See Problem 16.

SOLVED PROBLEMS

1. Write the equation of the circle satisfying the given conditions.

(a) $C(0,0)$, $r = 5$
(b) $C(4,-2)$, $r = 8$
(c) $C(-4,-2)$ and passing through $P(1,3)$
(d) $C(-5,6)$ and tangent to the x-axis
(e) $C(3,4)$ and tangent to $2x - y + 5 = 0$
(f) Center on $y = x$, tangent to both axes, $r = 4$

(a) Using $(x-h)^2 + (y-k)^2 = r^2$, the equation is $(x-0)^2 + (y-0)^2 = 25$ or $x^2 + y^2 = 25$.

(b) Using $(x-h)^2 + (y-k)^2 = r^2$, the equation is $(x-4)^2 + (y+2)^2 = 64$.

(c) Since the center is at $C(-4,-2)$, the equation has the form $(x+4)^2 + (y+2)^2 = r^2$. The condition that $P(1,3)$ lie on this circle is $(1+4)^2 + (3+2)^2 = r^2 = 50$. Hence the required equation is $(x+4)^2 + (y+2)^2 = 50$.

(d) The tangent to a circle is perpendicular to the radius drawn to the point of tangency; hence $r = 6$. The equation of the circle is $(x+5)^2 + (y-6)^2 = 36$.

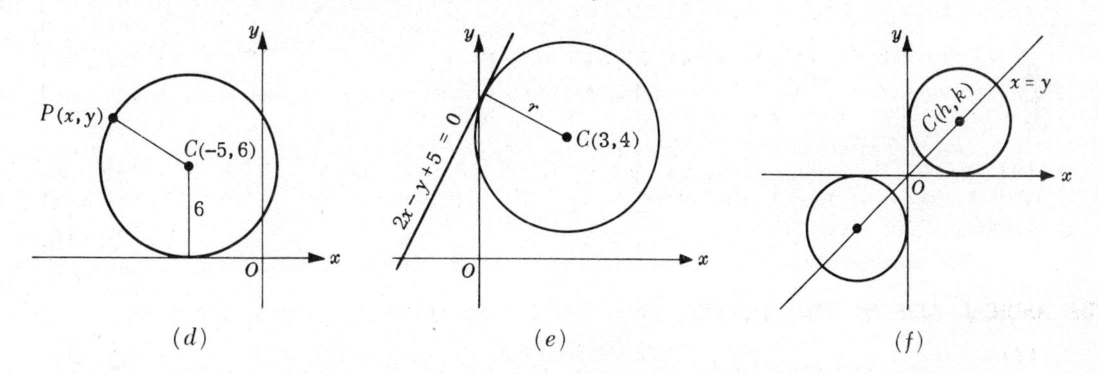

(d) (e) (f)

(e) The radius is the undirected distance of the point $C(3,4)$ from the line $2x - y + 5 = 0$; thus $r = \left|\dfrac{2\cdot 3 - 4 + 5}{-\sqrt{5}}\right| = \dfrac{7}{\sqrt{5}}$. The equation of the circle is $(x-3)^2 + (y-4)^2 = 49/5$.

(f) Since the center (h,k) lies on the line $x = y$, $h = k$; since the circle is tangent to both axes, $|h| = |k| = r$. Thus there are two circles satisfying the conditions, one with center $(4,4)$ and equation $(x-4)^2 + (y-4)^2 = 16$, the other with center $(-4,-4)$ and equation $(x+4)^2 + (y+4)^2 = 16$.

2. Describe the locus represented by each of the following equations.

(a) $x^2 + y^2 - 10x + 8y + 5 = 0$
(b) $x^2 + y^2 - 6x - 8y + 25 = 0$
(c) $x^2 + y^2 + 4x - 6y + 24 = 0$
(d) $4x^2 + 4y^2 + 80x + 12y + 265 = 0$

(a) From the standard form $(x-5)^2 + (y+4)^2 = 36$, the locus is a circle with center at $C(5,-4)$ and radius 6.

(b) From the standard form $(x-3)^2 + (y-4)^2 = 0$, the locus is a point circle or the point $(3,4)$.

(c) Here we have $(x+2)^2 + (y-3)^2 = -11$; the locus is imaginary.

(d) Dividing by 4, we have $x^2 + y^2 + 20x + 3y + 265/4 = 0$. From $(x+10)^2 + (y+3/2)^2 = 36$, the locus is a circle with center at $C(-10,-3/2)$ and radius 6.

3. Show that the circles $x^2 + y^2 - 16x - 20y + 115 = 0$ and $x^2 + y^2 + 8x - 10y + 5 = 0$ are tangent and find the point of tangency.

The first circle has center $C_1(8,10)$ and radius 7; the second has center $C_2(-4,5)$ and radius 6. The two circles are tangent externally since the distance between their centers $C_1C_2 = \sqrt{144 + 25} = 13$ is equal to the *sum* of the radii.

The point of tangency $P(x,y)$ divides the segment C_2C_1 in the ratio 6:7. Then

$$x = \frac{6\cdot 8 + 7(-4)}{6+7} = \frac{20}{13}, \qquad y = \frac{6\cdot 10 + 7\cdot 5}{6+7} = \frac{95}{13}$$

and the point of tangency has coordinates (20/13, 95/13).

4. Find the equation of the circle through the points (5,1), (4,6), and (2,-2).

Take the equation in general form $x^2 + y^2 + 2Dx + 2Ey + F = 0$. Substituting successively the coordinates of the given points, we have

$$\begin{cases} 25 + 1 + 10D + 2E + F = 0 \\ 16 + 36 + 8D + 12E + F = 0 \\ 4 + 4 + 4D - 4E + F = 0 \end{cases} \quad \text{or} \quad \begin{cases} 10D + 2E + F = -26 \\ 8D + 12E + F = -52 \\ 4D - 4E + F = -8 \end{cases}$$

with solution $D = -1/3$, $E = -8/3$, $F = -52/3$. Thus the required equation is

$$x^2 + y^2 - \frac{2}{3}x - \frac{16}{3}y - \frac{52}{3} = 0 \quad \text{or} \quad 3x^2 + 3y^2 - 2x - 16y - 52 = 0$$

5. Write the equations of the circle having radius $\sqrt{13}$ and tangent to the line $2x - 3y + 1 = 0$ at (1,1).

Let the equation of the circle be (A) $(x-h)^2 + (y-k)^2 = 13$. Since the coordinates (1,1) satisfy this equation, we have

(1) $$(1-h)^2 + (1-k)^2 = 13$$

The undirected distance from the tangent to the center of the circle is equal to the radius, that is,

(2) $$\left|\frac{2h - 3k + 1}{-\sqrt{13}}\right| = \sqrt{13} \quad \text{and} \quad \frac{2h - 3k + 1}{\sqrt{13}} = \pm\sqrt{13}$$

Finally, the radius through (1,1) is perpendicular to the tangent there, that is,

(3) $$\text{slope of radius through } (1,1) = -\frac{1}{\text{slope of given line}} \quad \text{or} \quad \frac{k-1}{h-1} = -\frac{3}{2}$$

Since there are only two unknowns, we may solve simultaneously any two of the three equations. Using (2) and (3), noting that there are two equations in (2), we find $h = 3$, $k = -2$ and $h = -1$, $k = 4$. The equations of the circles are $(x-3)^2 + (y+2)^2 = 13$ and $(x+1)^2 + (y-4)^2 = 13$.

6. Write the equations of the circles satisfying the following sets of conditions:
(a) through (2,3) and (-1,6), with center on $2x + 5y + 1 = 0$.
(b) tangent to $5x - y - 17 = 0$ at (4,3) and also tangent to $x - 5y - 5 = 0$.
(c) tangent to $x - 2y + 2 = 0$ and to $2x - y - 17 = 0$, and passing through (6,-1).
(d) with center in first quadrant and tangent to lines $y = 0$, $5x - 12y = 0$, $12x + 5y - 39 = 0$.

Take the equation of the circle in standard form $(x-h)^2 + (y-k)^2 = r^2$.

(a) We obtain the following system of equations:

(1) $2h + 5k + 1 = 0$ (center (h,k) on $2x + 5y + 1 = 0$)

(2) $(2-h)^2 + (3-k)^2 = r^2$ (point (2,3) on the circle)

(3) $(-1-h)^2 + (6-k)^2 = r^2$ (point (-1,6) on the circle)

The elimination of r between (2) and (3) yields $h - k + 4 = 0$ and when this is solved simultaneously with (1), we obtain $h = -3$, $k = 1$. By (2), $r^2 = (2+3)^2 + (3-1)^2 = 29$; the equation of the circle is $(x+3)^2 + (y-1)^2 = 29$.

(b) We obtain the following system of equations:

(1) $(4-h)^2 + (3-k)^2 = r^2$ (point (4,3) on the circle)

(2) $\left(\frac{5h-k-17}{\sqrt{26}}\right)^2 = \left(\frac{h-5k-5}{\sqrt{26}}\right)^2$ (the square of the directed distance from each tangent to the center is r^2)

(3) $\frac{k-3}{h-4} = -\frac{1}{5}$ (the radius drawn to (4,3) is perpendicular to the tangent there)

The elimination of h between (2) and (3) yields $\left(\frac{-26k+78}{\sqrt{26}}\right)^2 = \left(\frac{-10k+14}{\sqrt{26}}\right)^2$ or

$$9k^2 - 59k + 92 = (k-4)(9k-23) = 0; \quad \text{then } k = 4,\ 23/9$$

When $k = 4$, $h = -5k + 19 = -1$; then $r^2 = (4+1)^2 + (3-4)^2 = 26$ and the equation of the circle is $(x+1)^2 + (y-4)^2 = 26$. When $k = 23/9$, $h = 56/9$ and $r^2 = 416/81$; the equation of the circle is $(x - 56/9)^2 + (y - 23/9)^2 = 416/81$.

(c) Observing the directions indicated in Fig. (c) below, we obtain for each circle the following system of equations:

(1) $-\frac{h-2k+2}{-\sqrt{5}} = r$ ($x - 2y + 2 = 0$ is tangent to the circle)

(2) $-\frac{2h-k-17}{\sqrt{5}} = r$ ($2x - y - 17 = 0$ is tangent to the circle)

(3) $(6-h)^2 + (-1-k)^2 = r^2$ (point (6,−1) is on the circle)

The elimination of r between (1) and (2) yields

(4) $$h - k - 5 = 0$$

and the elimination of r between (1) and (3) yields

(5) $$(6-h)^2 + (-1-k)^2 = (h - 2k + 2)^2/5$$

Eliminating h between (4) and (5), we have $(1-k)^2 + (-1-k)^2 = (-k+7)^2/5$ or

$$9k^2 + 14k - 39 = (k+3)(9k-13) = 0 \quad \text{and} \quad k = -3,\ 13/9$$

When $k = -3$, $h = k + 5 = 2$, $r^2 = (6-2)^2 + [-1-(-3)]^2 = 20$ and the equation of the circle is $(x-2)^2 + (y+3)^2 = 20$. When $k = 13/9$, $h = k + 5 = 58/9$, $r^2 = (6 - 58/9)^2 + (-1 - 13/9)^2 = 500/81$ and the circle has equation $(x - 58/9)^2 + (y - 13/9)^2 = 500/81$.

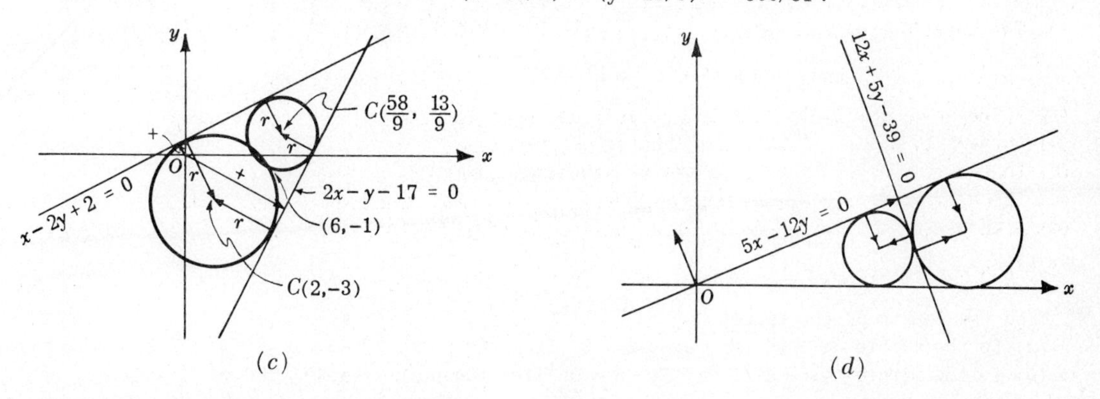

(d) Observing the directions indicated in Fig. (d) above, we have for the inscribed circle the following system of equations:

$$k = r, \quad -\frac{5h-12k}{-13} = r, \quad \text{and} \quad -\frac{12h+5k-39}{13} = r$$

Then $h = 5/2$, $k = r = 1/2$ and the equation of the circle is $(x - 5/2)^2 + (y - 1/2)^2 = 1/4$.

For the other circle, we have the following system of equations:

$$k = r, \quad -\frac{5h-12k}{-13} = r, \quad \text{and} \quad \frac{12h+5k-39}{13} = r$$

Here $h = 15/4$, $k = r = 3/4$, and the circle has equation $(x - 15/4)^2 + (y - 3/4)^2 = 9/16$.

7. Find the equations of all tangents to the circle $x^2 + y^2 - 2x + 8y - 23 = 0$:
(*a*) at the point (3,−10) on it; (*b*) having slope 3; (*c*) through the external point (8,−3).

(*a*) The center of the given circle is $C(1,-4)$. Since the tangent at any point P on the circle is perpendicular to the radius through P, the slope of the tangent is 1/3. The equation of the tangent is then

$$y + 10 = \frac{1}{3}(x-3) \qquad \text{or} \qquad x - 3y - 33 = 0$$

(*b*) Each tangent belongs to the family of lines $y = 3x + b$ or, in normal form, $\dfrac{3x - y + b}{\pm\sqrt{10}} = 0$.

The undirected distance from any tangent to the center of the circle is equal to the radius of the circle; thus

$$\left|\frac{3(1) - (-4) + b}{\pm\sqrt{10}}\right| = 2\sqrt{10} \qquad \text{or} \qquad \frac{7+b}{\sqrt{10}} = \pm 2\sqrt{10} \qquad \text{and} \qquad b = 13, -27$$

There are then two tangents having equations $y = 3x + 13$ and $y = 3x - 27$.

(*c*) The tangents to a circle through the external point (8,−3) belong to the family of lines $y + 3 = m(x-8)$ or $y = mx - 8m - 3$. When this replacement for y is made in the equation of the circle, we have

$$x^2 + (mx - 8m - 3)^2 - 2x + 8(mx - 8m - 3) - 23 = (m^2+1)x^2 + (-16m^2 + 2m - 2)x + (64m^2 - 16m - 38) = 0$$

Now this equation will have equal roots x provided the discriminant is zero, that is, provided

$$(-16m^2 + 2m - 2)^2 - 4(m^2+1)(64m^2 - 16m - 38) = -4(m-3)(9m+13) = 0$$

Then $m = 3, -13/9$ and the equations of the tangents are

$$y + 3 = 3(x-8) \quad \text{or} \quad 3x - y - 27 = 0 \qquad \text{and} \qquad y + 3 = -\frac{13}{9}(x-8) \quad \text{or} \quad 13x + 9y - 77 = 0$$

8. Prove: The length t of the tangent from the point $P_1(x_1, y_1)$ to the circle $x^2 + y^2 + 2Dx + 2Ey + F = 0$, with center $C(-D,-E)$, is given by

$$t = \sqrt{x_1^2 + y_1^2 + 2Dx_1 + 2Ey_1 + F}$$

In the figure, $(CP_1)^2 = (x_1 + D)^2 + (y_1 + E)^2$

and $(CR)^2 = D^2 + E^2 - F$

Then

$$t^2 = (RP_1)^2 = (CP_1)^2 - (CR)^2 = x_1^2 + y_1^2 + 2Dx_1 + 2Ey_1 + F$$

and

$$t = \sqrt{x_1^2 + y_1^2 + 2Dx_1 + 2Ey_1 + F}$$

9. Find the length of the tangent
(*a*) to the circle $x^2 + y^2 - 2x + 8y - 23 = 0$ from the point (8,−3)
(*b*) to the circle $4x^2 + 4y^2 - 2x + 5y - 8 = 0$ from the point (−4,4).

Denote the required length by t.

(*a*) Substituting the coordinates (8,−3) in the left member of the equation of the circle, we have

$$t^2 = (8)^2 + (-3)^2 - 2(8) + 8(-3) - 23 = 10 \qquad \text{and} \qquad t = \sqrt{10}$$

(*b*) From the general form of the equation $x^2 + y^2 - \frac{1}{2}x + \frac{5}{4}y - 2 = 0$, we find

$$t^2 = (-4)^2 + (4)^2 - \frac{1}{2}(-4) + \frac{5}{4}(4) - 2 = 37 \qquad \text{and} \qquad t = \sqrt{37}$$

10. For the circle $x^2+y^2+6x-8y = 0$, find the values of m for which the lines of the family $y = mx - 1/3$ (*a*) intersect the circle in two distinct points, (*b*) are tangent to the circle, (*c*) do not meet the circle.

Eliminating y between the two equations, we have

$$x^2 + (mx-\tfrac{1}{3})^2 + 6x - 8(mx-\tfrac{1}{3}) = (m^2+1)x^2 + (6-\tfrac{26}{3}m)x + \tfrac{25}{9} = 0$$

This equation will have two real and distinct roots, two equal roots, or two imaginary roots according as its discriminant

$$(6-\tfrac{26}{3}m)^2 - 4(m^2+1)(\tfrac{25}{9}) = \tfrac{8}{9}(72m^2-117m+28) = \tfrac{8}{9}(3m-4)(24m-7) >, =, \text{ or } < 0$$

(*a*) The lines will intersect the circle in two distinct points when $m > 4/3$ and $m < 7/24$.
(*b*) The lines will be tangent to the circle when $m = 4/3$ and $m = 7/24$.
(*c*) The lines will not meet the circle when $7/24 < m < 4/3$.

11. Write the equation of the family of circles satisfying the given conditions:
(*a*) having the common center $(-2,3)$, (*b*) having radius = 5, (*c*) with center on the x-axis.

(*a*) The equation is $(x+2)^2+(y-3)^2 = r^2$, r being a parameter.
(*b*) The equation is $(x-h)^2+(y-k)^2 = 25$, h and k being parameters.
(*c*) Let the center have coordinates $(k_1,0)$ and denote the radius by k_2. The equation of the family is $(x-k_1)^2+y^2 = k_2^2$, k_1 and k_2 being parameters.

12. Given the circles K_1: $x^2+y^2-6x+8y-16 = 0$ and K_2: $x^2+y^2+4x-2y-8 = 0$, write the equation of the family K_1+kK_2 and determine the circle of the family having the further property: (*a*) passes through $P(1,-3)$, (*b*) tangent to the line $7x-11y-40 = 0$.

The equation of the family is

$$(A) \qquad K_1+kK_2:\ x^2+y^2-6x+8y-16+k(x^2+y^2+4x-2y-8) = 0$$

or

$$(1+k)x^2 + (1+k)y^2 - (6-4k)x + (8-2k)y - (16+8k) = 0$$

(*a*) Substituting the coordinates of P in (A), we find $k = 3$. The circle of the family which passes through $(-1,3)$ has equation K_1+3K_2: $2x^2+2y^2+3x+y-20 = 0$.

(*b*) Equating the square of the radius of K_1+kK_2 to the square of the distance from the given tangent to the center and solving for k, we find

$$(-\frac{3-2k}{1+k})^2 + (\frac{4-k}{1+k})^2 + \frac{16+8k}{1+k} = \left\{\frac{7(\frac{3-2k}{1+k}) - 11(-\frac{4-k}{1+k}) - 40}{\sqrt{170}}\right\}^2$$

or $403k^2-786k-1269 = 0$ and $k = 3, -423/403$. The two circles tangent to the given line have equations

$$K_1+3K_2:\ 2x^2+2y^2+3x+y-20 = 0 \quad\text{and}\quad K_1-\tfrac{423}{403}K_2:\ 10x^2+10y^2+2055x-2035y+1532 = 0$$

13. Write the equation of the family of circles which are tangent to the circles of Problem 3 at their common point and determine the circle of the family having the property (*a*) center on $x+4y+16 = 0$, (*b*) radius is $\frac{1}{2}$.

The equation of the family is

$$K_1+kK_2:\ x^2+y^2-16x-20y+115+k(x^2+y^2+8x-10y+5) = 0$$

(*a*) The centers of K_1 and K_2 are $(8,10)$ and $(-4,5)$ respectively; the equation of their line of centers is $5x-12y+80 = 0$. This line meets $x+4y+16 = 0$ in the point $(-16,0)$, the required center.

Now (−16,0) divides the segment joining (8,10) and (−4,5) in the ratio $k:1$; thus,

$$\frac{5k+10}{k+1} = 0 \quad \text{and} \quad k = -2$$

The equation of the circle is $K_1 - 2K_2$: $x^2 + y^2 + 32x - 105 = 0$.

(*b*) Setting the square of the radius of $K_1 + kK_2$ equal to $(\frac{1}{2})^2$ and solving for k, we find

$$\left(\frac{4k-8}{1+k}\right)^2 + \left(-\frac{10+5k}{1+k}\right)^2 - \frac{115+5k}{1+k} = \frac{1}{4} \quad \text{and} \quad k = 1,\ 15/11$$

The required circles have equations

$$K_1 + K_2:\ x^2 + y^2 - 4x - 15y + 60 = 0 \quad \text{and} \quad K_1 + \frac{15}{11}K_2:\ 13x^2 + 13y^2 - 28x - 185y + 670 = 0$$

14. For the intersecting circles K_1: $x^2 + y^2 - 6x - 8y = 0$ and K_2: $x^2 + y^2 - 4x + 6y - 12 = 0$, (*a*) find the equation of the radical axis, (*b*) show that the radical axis is perpendicular to the line of centers, (*c*) find the points of intersection of the circles.

(*a*) The radical axis has equation $K_1 - K_2$: $x + 7y - 6 = 0$.

(*b*) The slope of the line of centers, joining (3,4) and (2,−3), is 7; the slope of the radical axis is −1/7. These lines are mutually perpendicular.

(*c*) Solving simultaneously the equation of the radical axis and either of the equations of the circles, the points of intersection are (6,0) and (−1,1).

15. For each pair of circles (*a*) K_1: $x^2 + y^2 - 8x - 6y = 0$, K_2: $4x^2 + 4y^2 - 10x - 10y - 13 = 0$ (*b*) K_1: $x^2 + y^2 - 12x - 16y - 125 = 0$, K_2: $3x^2 + 3y^2 - 60x - 16y + 113 = 0$

find the equation of the radical axis. Without finding the coordinates of their points of intersection, show that the circles (*a*) intersect in two distinct points while those of (*b*) are tangent internally.

(*a*) The equation of the radical axis is $K_1 - \frac{1}{4}K_2$: $22x + 14y - 13 = 0$. The undirected distance from the radical axis to the center (4,3) of K_1 $\left|\frac{22(4) + 14(3) - 13}{2\sqrt{170}}\right| = \frac{117\sqrt{170}}{340}$ is less than 5, the radius of K_1. Hence the radical axis and the circle K_2 intersect K_1 in two distinct points.

(*b*) The equation of the radical axis is $K_1 - \frac{1}{3}K_2$: $3x - 4y - 61 = 0$. The undirected distance from the radical axis to the center (6,8) of K_1 $\left|\frac{3(6) - 4(8) - 61}{5}\right| = 15$ is equal to the radius of K_1 and the circles are tangent to each other.

Since the distance between the centers of K_1 and K_2 $\sqrt{(10-6)^2 + (8/3-8)^2} = 20/3$ is equal to the difference between their radii, the circles are tangent internally.

16. Find the coordinates of the radical center R of the circles

$$K_1:\ x^2 + y^2 + 6x + 8y - 96 = 0,\quad K_2:\ x^2 + y^2 - 8x + 10y + 36 = 0,\quad \text{and} \quad K_3:\ x^2 + y^2 - 16x - 6y - 60 = 0$$

Show that the lengths of the six tangents drawn to the three circles from R are equal.

The radical axes $K_1 - K_2$: $7x - y - 66 = 0$, $K_1 - K_3$: $11x + 7y - 18 = 0$, and $K_2 - K_3$: $x + 2y + 12 = 0$ intersect in the radical center $R(8,-10)$.

The square of the length of the tangent from R to K_1 is $(8)^2 + (-10)^2 + 6(8) + 8(-10) - 96 = 36$, to K_2 is $(8)^2 + (-10)^2 - 8(8) + 10(-10) + 36 = 36$, and to K_3 is $(8)^2 + (-10)^2 - 16(8) - 6(-10) - 60 = 36$. Thus the lengths of the six tangents to the three circles from their radical center are equal.

SUPPLEMENTARY PROBLEMS

17. Write the equation of each of the following circles. (*a*) $C(0,0)$, radius 7; (*b*) $C(-4,8)$, radius 3; (*c*) $C(5,-4)$, through $(0,0)$; (*d*) $C(-4,-3)$, through $(2,1)$; (*e*) $C(-2,5)$, tangent to x-axis; (*f*) $C(-2,-5)$, tangent to $2x-y+3=0$; (*g*) tangent to both axes, radius 5; (*h*) circumscribed about the right triangle whose vertices are $(3,4)$, $(-1,-4)$, $(5,-2)$; (*i*) circumscribed about the triangle of Prob. 8, Chap. 48.
Ans. (*a*) $x^2+y^2=49$ (*c*) $x^2+y^2-10x+8y=0$ (*e*) $x^2+y^2+4x-10y+4=0$
(*b*) $x^2+y^2+8x-16y+71=0$ (*d*) $x^2+y^2+8x+6y-27=0$ (*f*) $5x^2+5y^2+20x+50y+129=0$
(*g*) $x^2+y^2\pm 10x\pm 10y+25=0$, $x^2+y^2\mp 10x\pm 10y+25=0$
(*h*) $x^2+y^2-2x-19=0$ (*i*) $56x^2+56y^2-260x-y-5451=0$

18. Find the center and radius of each of the circles.
(*a*) $x^2+y^2-6x+8y-11=0$ *Ans.* $C(3,-4)$, $r=6$
(*b*) $x^2+y^2-4x-6y-10/3=0$ *Ans.* $C(2,3)$, $r=7\sqrt{3}/3$
(*c*) $7x^2+7y^2+14x-56y-25=0$ *Ans.* $C(-1,4)$, $r=12\sqrt{7}/7$

19. Explain why any line passing through $(4,-1)$ cannot be tangent to the circle $x^2+y^2-4x+6y-12=0$.

20. (*a*) Show that the circles $x^2+y^2+6x-2y-54=0$ and $x^2+y^2-22x-8y+112=0$ do not intersect.
(*b*) Show that the circles $x^2+y^2+2x-6y+9=0$ and $x^2+y^2+8x-6y+9=0$ are tangent internally.

21. The equation of a given circle is $x^2+y^2=36$. Find (*a*) the length of the chord which lies along the line $3x+4y-15=0$, (*b*) the equation of the chord whose midpoint is $(3,2)$.
Hint: In (*a*) draw the normal to the given line. *Ans.* (*a*) $6\sqrt{3}$, (*b*) $3x+2y-13=0$

22. Find the equation of each circle satisfying the given conditions.
(*a*) Through $(6,0)$ and $(-2,-4)$, tangent to $4x+3y-25=0$.
(*b*) Tangent to $3x-4y+5=0$ at $(1,2)$, radius 5.
(*c*) Tangent to $x-2y-4=0$ and $2x-y-6=0$, passes through $(-1,2)$.
(*d*) Tangent to $2x-3y-7=0$ at $(2,-1)$; passes through $(4,1)$.
(*e*) Tangent to $3x+y+3=0$ at $(-3,6)$, tangent to $x+3y-7=0$.
Ans. (*a*) $(x-3)^2+(y+4)^2=25$, $(x-213/121)^2+(y+184/121)^2=297025/14641$
(*b*) $(x-4)^2+(y+2)^2=25$, $(x+2)^2+(y-6)^2=25$
(*c*) $x^2+y^2-2x-2y-3=0$, $x^2+y^2+118x-122y+357=0$
(*d*) $x^2+y^2+4x-10y-23=0$ (*e*) $x^2+y^2-6x-16y+33=0$, $x^2+y^2+9x-11y+48=0$

23. Find the equation of the tangent to the given circle at the given point on it.
(*a*) $x^2+y^2=169$, $(5,-12)$; (*b*) $x^2+y^2-4x+6y-37=0$, $(3,4)$.
Ans. (*a*) $5x-12y-169=0$, (*b*) $x+7y-31=0$

24. Find the equations of the tangent to each circle through the given external point.
(*a*) $x^2+y^2=25$, $(7,1)$; (*b*) $x^2+y^2-4x+2y-31=0$, $(-1,5)$
Ans. (*a*) $3x+4y-25=0$, $4x-3y-25=0$; (*b*) $y-5=0$, $4x-3y+19=0$

25. Show that the circles $x^2+y^2+4x-6y=0$ and $x^2+y^2+6x+4y=0$ are *orthogonal*, that is, that the tangents to the two circles at a point of intersection are mutually perpendicular. Also, that the square of the distance between the centers of the circles is equal to the sum of the squares of the radii.

26. Determine the equation of the circle of the family of Problem 12: (*a*) which has its center on the line $4x+3y-5=0$, (*b*) whose radius is $\sqrt{85}$.
Ans. (*a*) $x^2+y^2-16x+18y-24=0$, (*b*) $x^2+y^2+14x-12y=0$, $25x^2+25y^2-260x+310y-488=0$

27. Determine the equation of the circle of the family of Problem 13: (*a*) which passes through the point $(0,3)$, (*b*) which is tangent to the line $4x+3y-25=0$.
Ans. (*a*) $5x^2+5y^2+16x-60y+135=0$ (*b*) $x^2+y^2-40x-30y+225=0$, $8x^2+8y^2-20x-115y+425=0$

28. Prove: The radical axis of two distinct non-concentric circles is perpendicular to their line of centers.

29. Find the equation of the locus of a point from which the tangents to the circle $x^2+y^2=4$ are twice as long as the tangents to $x^2+y^2-4x=0$. *Ans.* $3x^2+3y^2-16x+4=0$

30. Show that the circles $x^2+y^2-6x-8y+16=0$, $x^2+y^2-4x+6y-20=0$, and $x^2+y^2-2x+20y+5=0$ do not have a radical center. Explain this.

CHAPTER 51

More Locus Problems

A PLANE CURVE IS THE LOCUS of all those points in the plane, and only those points, which satisfy a given geometric condition. To obtain the equation of the locus:

(*1*) Let $P(x,y)$ be any point on the locus.
(*2*) Express the given geometric condition by means of an equation in x and y.
(*3*) Simplify the equation obtained in (*2*).

Example 1. Find the equation of the locus of a point which moves in the plane so that the sum of the squares of its distances from the two lines l_1: $7x-4y-10=0$ and l_2: $4x+7y+5=0$ is always equal to 3.

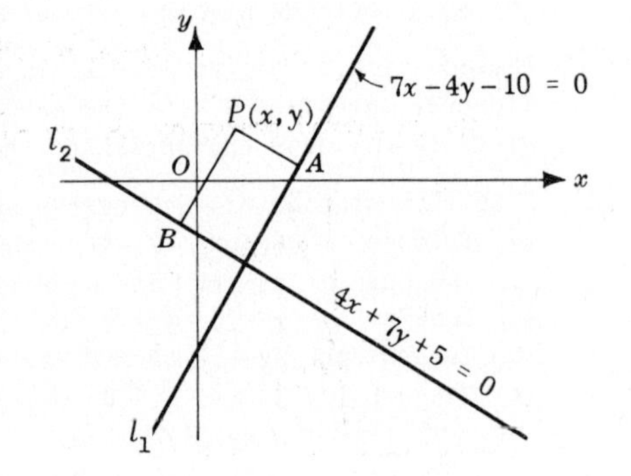

Let $P(x,y)$ be any point on the locus and drop the perpendiculars AP and BP to l_1 and l_2 respectively; then P must satisfy the condition $(AP)^2+(BP)^2 = 3$. We obtain

$$\left(\frac{7x-4y-10}{\sqrt{65}}\right)^2 + \left(\frac{4x+7y+5}{-\sqrt{65}}\right)^2 = 3$$

or $$13x^2+13y^2-20x+30y-14 = 0,$$

a circle with center at (10/13, -15/13) and radius $\sqrt{3}$. See Problems 1-7.

THE PARAMETRIC METHOD. In expressing analytically the geometric condition which defines a locus, it is often more convenient to introduce one or more auxiliary variables or parameters in addition to the coordinates of the tracing point P. Then between the relations set up, the parameters are to be eliminated. If n parameters are introduced, $(n+1)$ relations must be found in order to be able to effect their elimination. When parameters are so used to obtain the equation of the locus, we say that it has been obtained by the *parametric method.*

Example 2. Find the equation of the locus of the midpoints of the ordinates (that is, the midpoints of the vertical line segments drawn from the x-axis to points on the curve) of the circle $x^2+y^2 = r^2$.

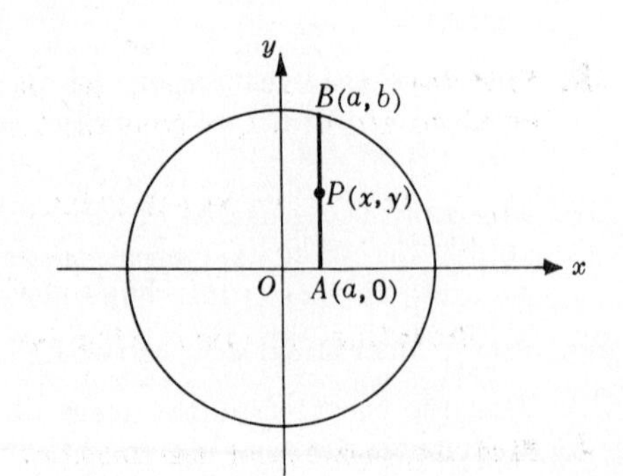

Let the vertical line drawn through $A(a,0)$ on the x-axis meet the circle in the point $B(a,b)$. If $P(x,y)$ is the tracing point of the required locus, its position on this line is at the midpoint of AB. Then

(*1*) $x = a$ and (*2*) $y = \frac{1}{2}b$

The third relation needed

$$(3) \quad a^2 + b^2 = r^2$$

expresses the condition that $B(a,b)$ be on the given circle.

To eliminate the parameters, we solve (*1*) and (*2*) for $a = x$ and $b = 2y$ respectively, and substitute in (*3*) to obtain $x^2 + 4y^2 = r^2$, the desired equation.

See Problems 8-9.

SOLVED PROBLEMS

The following refer to loci in the plane.

1. Find the equation of the locus of a point which moves so that the slope of the line joining it to $A(3,-2)$ is twice that of the line joining it to the point $B(5,8)$.

Let $P(x,y)$ be any point on the locus. Referring to Fig. (*a*) below,

$$\text{slope of } AP = 2(\text{slope of } BP) \qquad \text{or} \qquad \frac{y+2}{x-3} = 2\frac{y-8}{x-5};$$

the required equation is $xy - 18x - y + 58 = 0$. The locus is a hyperbola.

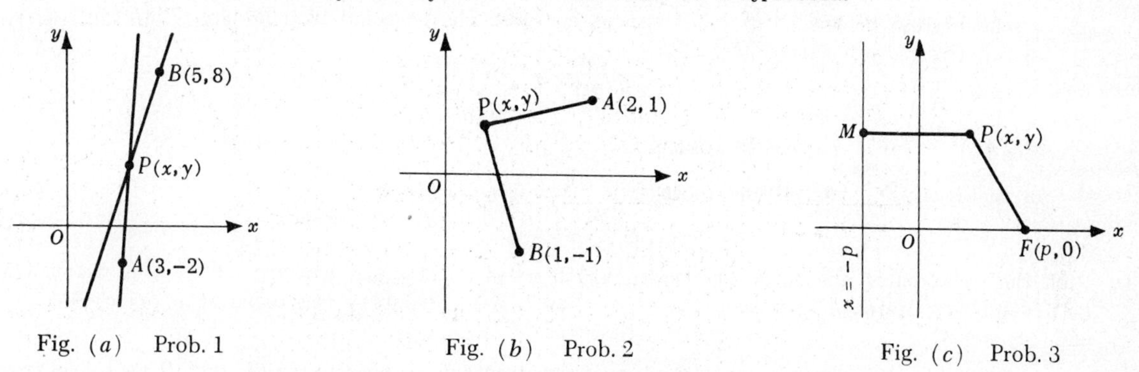

Fig. (*a*) Prob. 1 Fig. (*b*) Prob. 2 Fig. (*c*) Prob. 3

2. Find the equation of the locus of a point which moves so that the sum of the squares of its distances from the points $A(2,1)$ and $B(1,-1)$ is always equal to 3.

Let $P(x,y)$ be any point on the locus. Using Fig. (*b*) above,

$$(AP)^2 + (BP)^2 = 3, \qquad (\sqrt{(x-2)^2 + (y-1)^2})^2 + (\sqrt{(x-1)^2 + (y+1)^2})^2 = 3$$

Expanding and simplifying, we have $x^2 + y^2 - 3x + 2 = 0$, a circle with center at $(3/2, 0)$ and radius $1/2$.

3. Find the equation of the locus of a point which moves so that its distance from the point $F(p,0)$ is equal to its distance from the line d: $x = -p$. (Assume $p > 0$.)

Let $P(x,y)$ be any point on the locus and draw PM perpendicular to d, as in Fig. (*c*) above. Then since $FP = MP$,

$$\sqrt{(x-p)^2 + y^2} = x + p \qquad \text{or} \qquad (x-p)^2 + y^2 = (x+p)^2$$

Expanding and simplifying, we have $y^2 = 4px$.

4. Find the equation of the locus of a point which moves so that the sum of its distances from the points $A(5,0)$ and $B(-5,0)$ is 12.

Let $P(x,y)$ be any point on the locus, as in Fig. (*d*) below. Since $AP + BP = 12$,

$$\sqrt{(x-5)^2+y^2} + \sqrt{(x+5)^2+y^2} = 12 \quad \text{or} \quad \sqrt{(x-5)^2+y^2} = 12 - \sqrt{(x+5)^2+y^2}$$

Squaring and simplifying, we have

$$x^2 - 10x + 25 + y^2 = 144 - 24\sqrt{(x+5)^2+y^2} + x^2 + 10x + 25 + y^2 \quad \text{or} \quad 5x + 36 = 6\sqrt{(x+5)^2+y^2}$$

Squaring and simplifying, we have finally $11x^2 + 36y^2 - 396 = 0$ as the required equation. The locus is an ellipse.

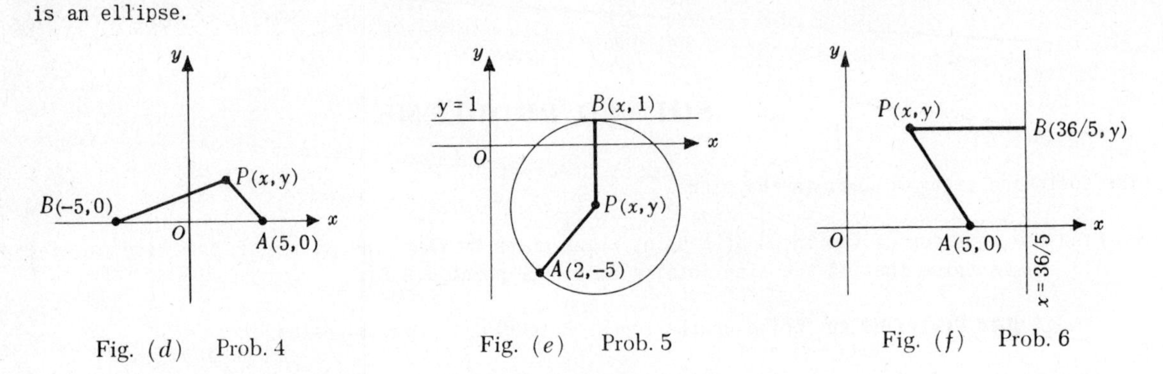

Fig. (d) Prob. 4 Fig. (e) Prob. 5 Fig. (f) Prob. 6

5. Find the equation of the locus of the center of a circle which passes through the point $A(2,-5)$ and is tangent to the line $y = 1$.

Let $P(x,y)$ be any point on the locus and drop the perpendicular BP from P to the line $y = 1$, as in Fig. (e) above. Since $AP = BP$,

$$\sqrt{(x-2)^2 + (y+5)^2} = |y-1|$$

Then $$(x-2)^2 + (y+5)^2 = (y-1)^2 \quad \text{or} \quad x^2 - 4x + 12y + 28 = 0$$

is the desired equation. The locus is a parabola.

6. Find the equation of the locus of a point which moves so that its distance from the point $A(5,0)$ is $e > 0$ times its distance from the line l_1: $x - 36/5 = 0$. Identify the locus when (a) $e = 5/6$, (b) $e = 1$ and (c) $e = 2$.

Let $P(x,y)$ be any point on the locus and drop the perpendicular BP from P to l_1, as in Fig. (f) above. Since $AP = e|BP|$,

$$\sqrt{(x-5)^2 + y^2} = e|x - 36/5|, \qquad (x-5)^2 + y^2 = e^2(x-36/5)^2,$$

and the required equation is $$(1-e^2)x^2 - (10 - 72e^2/5)x + y^2 + 25 - 1296e^2/25 = 0.$$

(a) When $e = 5/6$, the locus is the ellipse $11x^2 + 36y^2 - 396 = 0$.
(b) When $e = 1$, the locus is the parabola $25y^2 + 110x - 671 = 0$.
(c) When $e = 2$, the locus is the hyperbola $25y^2 - 75x^2 + 1190x - 4559 = 0$.

7. Find the equation of the locus of a point which moves so that the sum of its distances from the points $A(4,4)$ and $B(-4,-2)$ is 12.

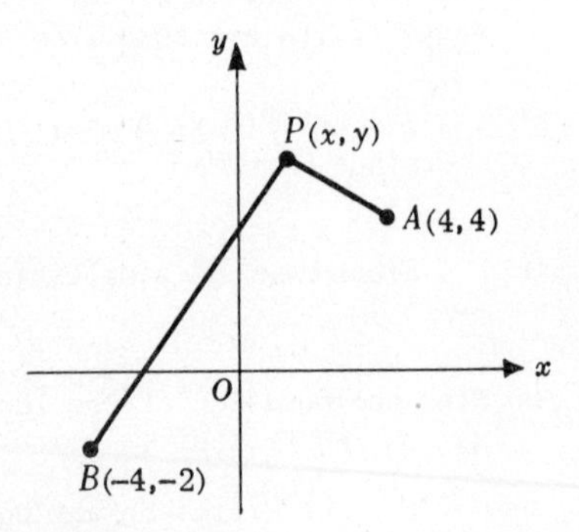

Let $P(x,y)$ be any point on the locus.

Since $AP + BP = 12$,

$$\sqrt{(x-4)^2 + (y-4)^2} + \sqrt{(x+4)^2 + (y+2)^2} = 12$$

Proceeding as in Problem 3, we obtain

$$4x + 3y + 33 = 6\sqrt{(x+4)^2 + (y+2)^2},$$

and finally $$20x^2 - 24xy + 27y^2 + 24x - 54y - 369 = 0.$$

8. A line AB of constant length $2e$ moves with the end A always on the x-axis and the end B always on the line $y = 6x$. Find the equation of the locus of the midpoint of AB.

Let A have coordinates $(a,0)$ and B have coordinates (b,c), as in Fig. (g) below. If $P(x,y)$ is the tracing point of the required locus, then

$$(1)\quad x = \tfrac{1}{2}(a+b) \qquad \text{and} \qquad (2)\quad y = \tfrac{1}{2}c$$

Also, since B is on the line $y = 6x$ and $AB = 2e$,

$$(3)\quad c = 6b \qquad \text{and} \qquad (4)\quad (a-b)^2 + c^2 = 4e^2$$

From (2), $c = 2y$; from (3), $b = c/6 = y/3$; from (1), $a = 2x - b = 2x - y/3$. Substituting in (4), we obtain the equation of the locus as

$$(2x - 2y/3)^2 + (2y)^2 = 4e^2 \qquad \text{or} \qquad 9x^2 - 6xy + 10y^2 = 9e^2$$

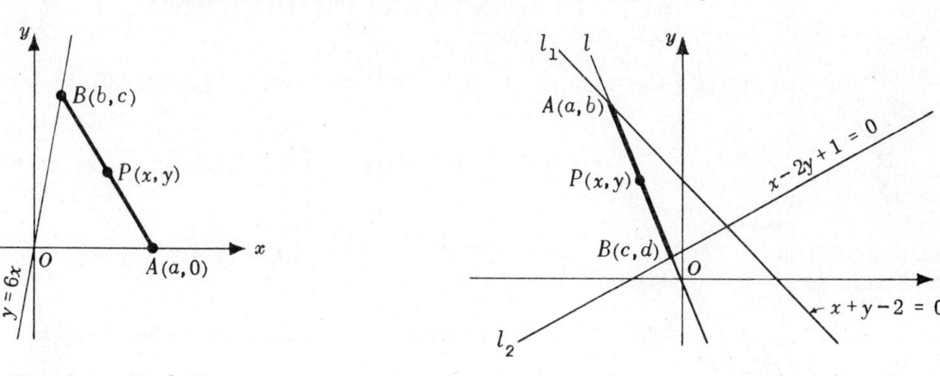

Fig. (g) Prob. 8 Fig. (h) Prob. 9

9. A line l passing through the origin intersects the line l_1: $x + y - 2 = 0$ in A and the line l_2: $x - 2y + 1 = 0$ in B. Find the locus of the midpoint of AB as l revolves about the origin. Refer to Fig. (h) above.

Let A have coordinates (a,b); then A is on l_1 and $(1)\quad a + b - 2 = 0$.

Let B have coordinates (c,d); then B is on l_2 and $(2)\quad c - 2d + 1 = 0$.

If $P(x,y)$ is the tracing point of the locus it is the midpoint of AB; then

$$(3)\quad x = \tfrac{1}{2}(a+c) \qquad \text{and} \qquad (4)\quad y = \tfrac{1}{2}(b+d)$$

Also, since A, O, B, and P are collinear, we have $(5)\quad b/a = d/c = y/x$.

To eliminate the parameters we add (1) and (2), and substitute from (3) and (4) to get

$$(a+c) + (b+d) - 3d - 1 = 0, \quad 2x + 2y - 3d - 1 = 0, \quad \text{or} \quad (6)\quad 2x + 2y - 1 = 3d$$

Next, we subtract (2) from twice (1) to get

$$2(a+c) - 3c + 2(b+d) - 5 = 0 \qquad \text{or} \qquad (7)\quad 4x + 4y - 5 = 3c$$

Finally, dividing (6) by (7) and making use of (5) we have $\dfrac{2x + 2y - 1}{4x + 4y - 5} = \dfrac{d}{c} = \dfrac{y}{x}$.

The desired equation is $2x^2 - 2xy - 4y^2 - x + 5y = 0$.

10. Find the equation of the locus of the point of intersection of any two perpendicular tangents to the circle $x^2 + y^2 = 4$.

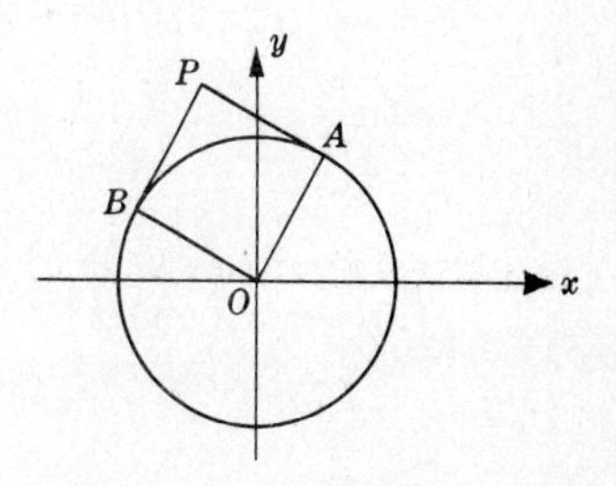

Let $P(x,y)$ be any point on the locus and let $A(a,b)$ and $B(c,d)$ be the points of tangency of the two perpendicular tangents through P. Since A and B are on the circle, we have

$$(1)\quad a^2 + b^2 = 4 \qquad \text{and} \qquad (2)\quad c^2 + d^2 = 4$$

Since $APBO$ is a rectangle: (3) $\dfrac{y-b}{x-a} = \dfrac{d}{c}$, (4) $\dfrac{y-d}{x-c} = \dfrac{b}{a}$, and (5) $\dfrac{b}{a} = -\dfrac{c}{d}$.

Using (3), (5), (4): $\dfrac{y-b}{x-a} = \dfrac{d}{c} = -\dfrac{a}{b} = -\dfrac{x-c}{y-d}$ or

$$(6)\quad x^2 + y^2 = (ax + by) + (cx + dy) - (ac + bd)$$

From $\dfrac{y-b}{x-a} = -\dfrac{a}{b}$, $ax + by = a^2 + b^2 = 4$; from $\dfrac{y-d}{x-c} = -\dfrac{c}{d}$, $cx + dy = c^2 + d^2 = 4$; from (5), $ac + bd = 0$. Then (6) becomes $x^2 + y^2 = 8$, the equation of the locus.

SUPPLEMENTARY PROBLEMS

11. Find the equation of the locus of a point whose distance from the point (3,0) is equal to its distance from the line $x + 3 = 0$. *Ans.* $y^2 - 12x = 0$

12. Find the equation of the locus of a point whose distance from the point (–3,4) is twice its distance from the line $y + 4 = 0$. *Ans.* $x^2 - 3y^2 + 6x - 40y - 39 = 0$

13. Find the equation of the locus of the center of a circle which passes through (–5,2) and is tangent to the line $x + 4 = 0$. *Ans.* $y^2 + 2x - 4y + 13 = 0$

14. Find the equation of the locus of a point the sum of whose distances from (0,4) and (0,–4) is 10. *Ans.* $25x^2 + 9y^2 - 225 = 0$

15. Find the equation of the locus of a point the difference of whose distances from (4,0) and (–4,0) is 6. *Ans.* $7x^2 - 9y^2 - 63 = 0$

16. Find the equation of the locus of a point which is twice as far from the point (3,0) as from the line $12x + 5y - 13 = 0$. *Ans.* $407x^2 + 480xy - 69y^2 - 234x - 520y - 845 = 0$

17. Find the equation of the locus of a point which is twice as far from the line $12x + 5y - 13 = 0$ as from the point (3,0). *Ans.* $532x^2 - 120xy + 651y^2 - 3744x + 130y + 5915 = 0$

18. A line is drawn from the origin to a point on the ellipse $9x^2 + 25y^2 - 225 = 0$. Find the equation of the locus of the midpoint of the line. *Ans.* $36x^2 + 100y^2 - 225 = 0$

19. A line is drawn from the point (4,0) to the hyperbola $25x^2 - 9y^2 - 225 = 0$. Find the equation of the locus of the midpoint of the line. *Ans.* $100x^2 - 36y^2 - 400x + 175 = 0$

20. A line through the origin intersects the line $2x + y - 2 = 0$ in A and the line $x - 2y + 2 = 0$ in B. Find the equation of the locus of the midpoint of AB. *Ans.* $2x^2 - 3xy - 2y^2 + x + 3y = 0$

21. A variable tangent is drawn to the circle $x^2 + y^2 - 4x = 0$. Find the equation of the locus of the foot of the perpendicular dropped from the origin to the tangent.
Hint: Let (a,b) be the point of tangency and (x,y) be the foot of the perpendicular. Then $a^2 + b^2 - 4a = 0$, $(a-2)x + by - 2a = 0$, and $b/(a-2) = y/x$. *Ans.* $(x^2 + y^2 - 2x)^2 = 4(x^2 + y^2)$

22. Find the equation of the locus of the center of a circle which is tangent to the line $x - 8 = 0$ and to the circle $x^2 + y^2 = 36$.
Hint: Let (x,y) be a point on the locus and (a,b) be the corresponding point of tangency of the two circles. Then $a^2 + b^2 = 36$, $(x-a)^2 + (y-b)^2 = (8-x)^2$, and $y/x = b/a$.
Ans. $(y^2 + 28x - 196)(y^2 + 4x - 4) = 0$

CHAPTER 52

The Parabola

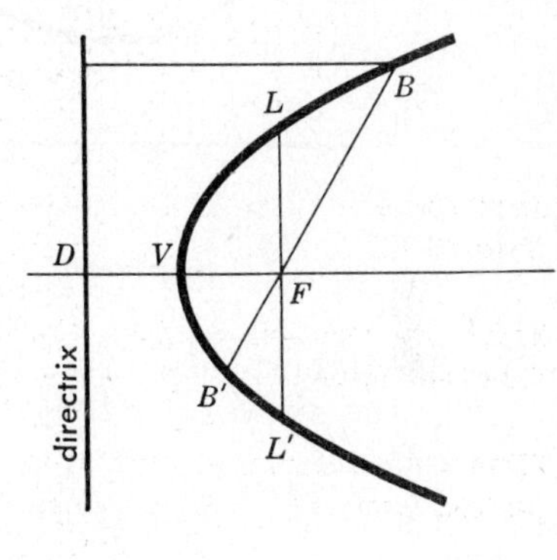

THE LOCUS OF A POINT P which moves in a plane so that its distance from a fixed line of the plane and its distance from a fixed point of the plane, not on the line, are equal is called a *parabola*.

The fixed point F is called the *focus* and the fixed line d is called the *directrix* of the parabola. The line FD through the focus and perpendicular to the directrix is called the *axis* of the parabola. The axis intersects the parabola in the point V, the midpoint of FD, called the *vertex*.

The line segment joining any two distinct points of the parabola is called a *chord*. A chord (as BB') which passes through the focus is called a *focal chord* while FB and FB' are called the *focal radii* of B and B' respectively. The focal chord LL' which is perpendicular to the axis is called the *latus rectum*.

THE EQUATION OF A PARABOLA assumes its simplest (*reduced*) form when its vertex is at the origin and its axis coincides with one of the coordinate axes.

When the vertex is at the origin and the axis coincides with the x-axis, the equation of the parabola is

$$(1) \qquad y^2 = 4px$$

Then the focus is at $F(p,0)$ and the equation of the directrix is d: $x = -p$. If $p>0$, the parabola opens to the right; if $p<0$, the parabola opens to the left.

See Problem 3, Chapter 51.

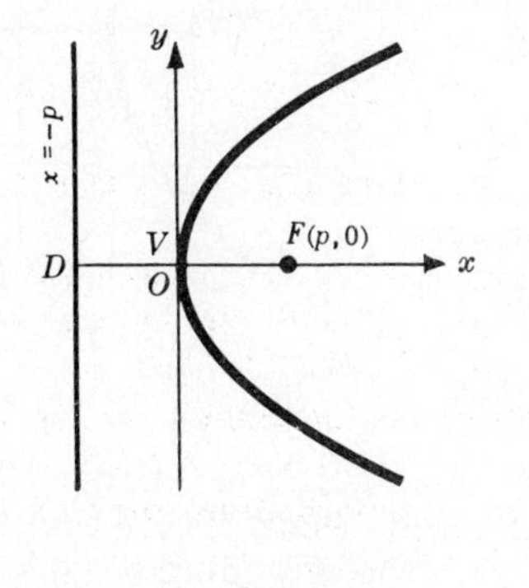

(a) $y^2 = 4px$, $p>0$

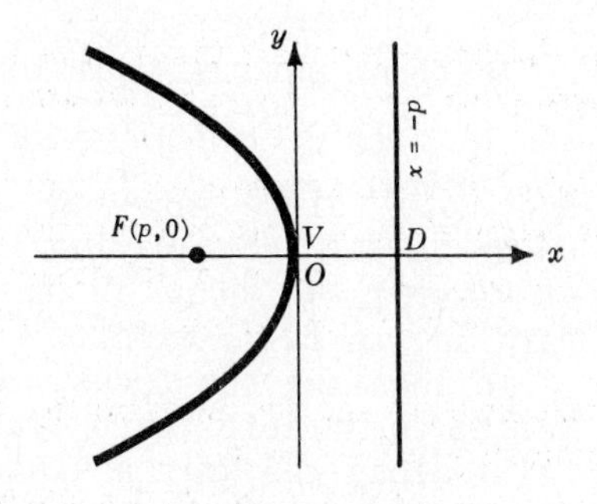

(b) $y^2 = 4px$, $p<0$

When the vertex is at the origin and the axis coincides with the y-axis, the equation of the parabola is

$$(2) \qquad x^2 = 4py$$

Then the focus is at $F(0,p)$ and the equation of the directrix is $d:\ y = -p$. If $p > 0$, the parabola opens upward; if $p < 0$, the parabola opens downward.

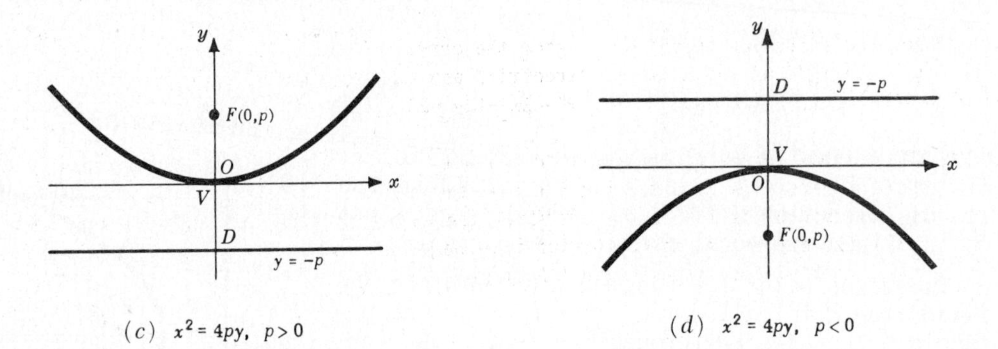

(c) $x^2 = 4py,\ p > 0$ (d) $x^2 = 4py,\ p < 0$

In either case, the distance from the directrix to the vertex and the distance from the vertex to the focus are equal to $|p|$; also, the length of the latus rectum is $|4p|$.

(Some authors define $p > 0$ and consider the four cases $y^2 = 4px$, $y^2 = -4px$, $x^2 = 4py$, and $x^2 = -4py$. Other authors label the focus $F(\frac{1}{2}p, 0)$ and directrix $d:\ x = -\frac{1}{2}p$, and obtain $y^2 = 2px$, etc.)

THE EQUATION OF A PARABOLA assumes the *semi-reduced* form

$$(1') \quad (y-k)^2 = 4p(x-h) \qquad \text{or} \qquad (2') \quad (x-h)^2 = 4p(y-k)$$

when its vertex is at the point (h, k) and its axis is parallel respectively to the x-axis or the y-axis. The distance between the directrix and vertex, the distance between the vertex and focus, and the length of the latus rectum are the same as given in the section above.

Example 1. Sketch the locus and find the coordinates of the vertex and focus, the equations of the axis and directrix, and the length of the latus rectum of the parabola

$$y^2 - 6y + 8x + 41 = 0 .$$

We first put the equation in the form

$$(1'):\ (y-3)^2 = -8(x+4)$$

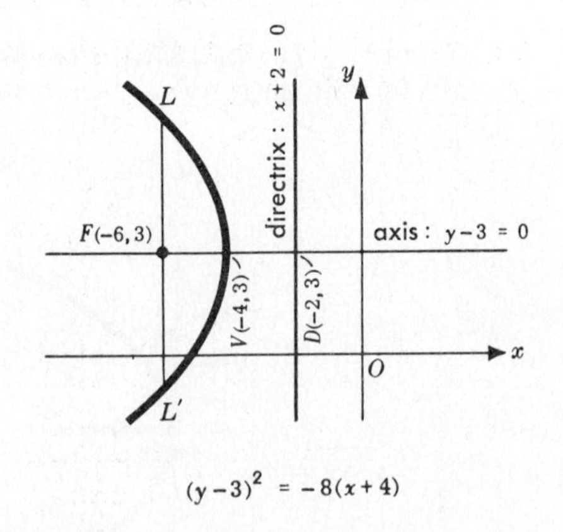

$(y-3)^2 = -8(x+4)$

and note, since $4p = -8$, that the parabola opens to the left. Having located the vertex at $V(-4,3)$, we draw in the axis through V parallel to the x-axis.

In locating the focus, we move from the vertex to the left (the parabola opens to the left) along the axis a distance $|p| = 2$ to the point $F(-6,3)$. In locating the directrix, we move from the vertex to the right (away from the focus) along the axis a distance $|p| = 2$ to the point $D(-2,3)$. The directrix passes through D and is perpendicular to the axis; its equation is $x + 2 = 0$.

The length of the latus rectum is $|4p| = 8$. We draw in the line segment LL',

through F perpendicular to the axis FV so that $FL = L'F = |2p| = 4$.
Finally, using the points L, L', and V, we sketch the parabola.

SOLVED PROBLEMS

1. For each of the following parabolas, sketch the curve and find the coordinates of the vertex and focus, the equations of the axis and directrix, and the length of the latus rectum:
(a) $y^2 = 16x$, (b) $x^2 = -9y$, (c) $x^2 - 2x - 12y + 25 = 0$, (d) $y^2 + 4y + 20x + 4 = 0$.

(a) The parabola opens to the right ($p > 0$) with vertex at $V(0,0)$. The equation of its axis is $y = 0$. Moving from V to the right along the axis a distance $|p| = 4$, we locate the focus at $F(4,0)$. Moving from V to the left along the axis a distance $|p| = 4$, we locate the point $D(-4,0)$. Since the directrix passes through D perpendicular to the axis, its equation is $x + 4 = 0$. The length of the latus rectum LL' is $|4p| = 16$.

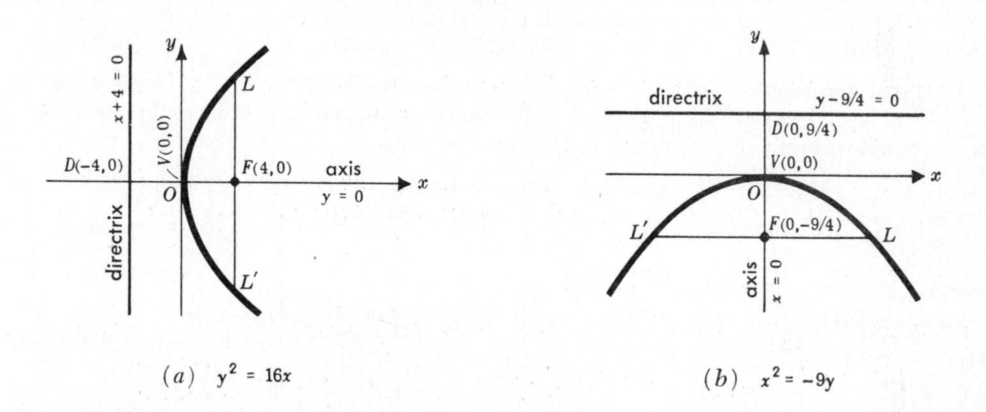

(a) $y^2 = 16x$ (b) $x^2 = -9y$

(b) The parabola opens downward ($p < 0$) with vertex at $V(0,0)$. The equation of its axis is $x = 0$. Moving from V downward along the axis a distance $|p| = 9/4$, we locate the focus at $F(0,-9/4)$. Moving from V upward along the axis a distance $|p| = 9/4$, we locate the point $D(0,9/4)$; the equation of the directrix is $4y - 9 = 0$. The length of the latus rectum is $|4p| = 9$.

(c) Here $(x-1)^2 = 12(y-2)$. The parabola opens upward ($p > 0$) with vertex at $V(1,2)$. The equation of its axis is $x - 1 = 0$. Since $|p| = 3$, the focus is at $F(1,5)$ and the equation of the directrix is $y + 1 = 0$. The length of the latus rectum is $|4p| = 12$.

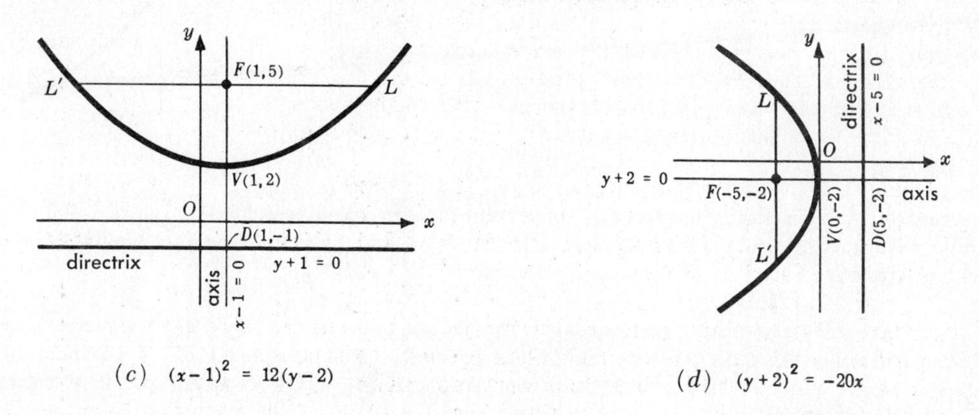

(c) $(x-1)^2 = 12(y-2)$ (d) $(y+2)^2 = -20x$

(d) Here $(y+2)^2 = -20x$. The parabola opens to the left ($p < 0$) with vertex at $V(0,-2)$. The equation of its axis is $y + 2 = 0$. Since $|p| = 5$, the focus is at $F(-5,-2)$ and the equation of the directrix is $x - 5 = 0$. The length of the latus rectum is $|4p| = 20$.

2. Find the equation of the parabola, given:

(a) $V(0,0)$; $F(0,-4)$ (c) $V(1,4)$; $F(-2,4)$

(b) $V(0,0)$; directrix: $x=-5$ (d) $F(2,3)$; directrix: $y=-1$

(e) $V(0,0)$; axis: $y=0$; passing through $(4,5)$.

(f) Axis parallel to $y=0$; passing through $(-2,4)$, $(-3,2)$, and $(-11,-2)$.
Locate the vertex, focus, and ends of the latus rectum.

(a) Since the directed distance $p = VF = -4$, the parabola opens downward.
Its equation is $x^2 = -16y$.

(b) The parabola opens to the right (away from the directrix).
Since $p = DV = 5$, the equation is $y^2 = 20x$.

(c) Here the focus lies to the left of the vertex and the parabola opens to the left.
The directed distance $p = VF = -3$ and the equation is $(y-4)^2 = -12(x-1)$.

(d) Here the focus lies above the directrix and the parabola opens upward. The axis of the parabola meets the directrix in $D(2,-1)$ and the vertex is at the midpoint $V(2,1)$ of FD.
Then $p = VF = 2$ and the equation is $(x-2)^2 = 8(y-1)$.

(e) The equation of this parabola is of the form $y^2 = 4px$.
If $(4,5)$ is a point on it then $(5)^2 = 4p(4)$, $4p = 25/4$, and the equation is $y^2 = \frac{25}{4}x$.

(f) The equation of the parabola with axis parallel to the x-axis is of the form $(y-k)^2 = 4p(x-h)$ or when expanded $x = ay^2 + by + c$. Substituting the coordinates of the given points into the latter form, we obtain the system of equations

$$-2 = 16a + 4b + c, \qquad -3 = 4a + 2b + c, \qquad -11 = 4a - 2b + c$$

having the simultaneous solution $a = -\frac{1}{4}$, $b = 2$, $c = -6$. The required equation is

$$x = -\tfrac{1}{4}y^2 + 2y - 6 \qquad \text{or} \qquad (y-4)^2 = -4(x+2)$$

The vertex is at $V(-2,4)$. Since the parabola opens to the left and $|p| = 1$, the focus is at $F(-3,4)$. Since $|2p| = 2$, the ends of the latus rectum are at $L(-3,6)$ and $L'(-3,2)$.

3. Prove: The square of the length of the perpendicular dropped from any point of a parabola to its axis is equal to the product of the length of the latus rectum and the segment of the axis between the vertex and the foot of the perpendicular.

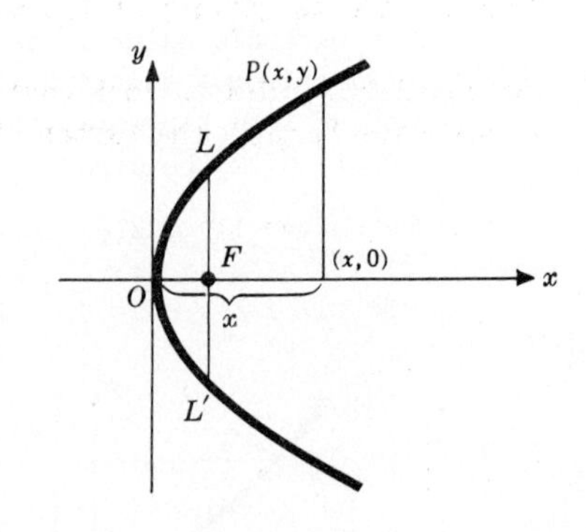

Consider the parabola $y^2 = 4px$, $p > 0$, and let $P(x,y)$ be any point on it.

The length of the perpendicular dropped from P to the axis (here, the x-axis) is y and its square is y^2. The foot of the perpendicular is x units from the vertex (origin) and the length of the latus rectum is $4p$; their product is $4px$. The equation of the parabola assures us that these two quantities are equal.

4. The cable of a suspension bridge has supporting towers which are 50 ft high and 400 ft apart. If the lowest point of the cable is 10 ft above the floor of the bridge, find the length of a supporting rod 100 ft from the center of the span.

Take the origin of coordinates at the lowest point of the cable and the positive y-axis directed upward along the axis of symmetry of the parabola. Then the equation of the parabola has the form $x^2 = 4py$. Since $(200,40)$ is a point on the parabola, $(200)^2 = 4p\cdot 40$ or $4p = 1000$, and the equation is $x^2 = 1000y$.

When $x = 100$, $(100)^2 = 1000y$, and $y = 10$ ft. The length of the supporting rod is $10 + 10 = 20$ ft.

SUPPLEMENTARY PROBLEMS

5. For each of the following parabolas, sketch the curve and find the coordinates of the vertex and focus, the equations of the axis and directrix, and the length of the latus rectum.

(a) $x^2 = 12y$ *Ans.* $V(0,0)$, $F(0,3)$; $x = 0$, $y + 3 = 0$; 12

(b) $y^2 = -10x$ *Ans.* $V(0,0)$, $F(-5/2,0)$; $y = 0$, $2x - 5 = 0$; 10

(c) $x^2 - 6x + 8y + 25 = 0$ *Ans.* $V(3,-2)$, $F(3,-4)$; $x - 3 = 0$, $y = 0$; 8

(d) $y^2 - 16x + 2y + 49 = 0$ *Ans.* $V(3,-1)$, $F(7,-1)$; $y + 1 = 0$, $x + 1 = 0$; 16

(e) $x^2 - 2x - 6y - 53 = 0$ *Ans.* $V(1,-9)$, $F(1,-15/2)$; $x - 1 = 0$, $2y + 21 = 0$; 6

(f) $y^2 + 20x + 4y - 60 = 0$ *Ans.* $V(16/5,-2)$, $F(-9/5,-2)$; $y + 2 = 0$, $5x - 41 = 0$; 20

6. Find the equation of the parabola, given:

(a) $V(0,0)$, opens to the right, latus rectum equal to 16.

(b) $V(0,0)$, $F(-2,0)$

(c) $V(0,0)$, $F(0,5)$

(d) $V(0,0)$, d: $y + 3 = 0$

(e) $V(0,0)$, F on x-axis, passes through $(-2,6)$.

(f) $V(1,3)$, $F(-1,3)$

(g) $F(3,2)$, d: $y + 4 = 0$

(h) $V(3,-2)$, latus rectum equal to 20, axis: $y + 2 = 0$.

(i) ends of latus rectum $(2,-1)$ and $(2,5)$.

Ans. (a) $y^2 = 16x$ (d) $x^2 = 12y$ (g) $x^2 - 6x - 12y - 3 = 0$

(b) $y^2 = -8x$ (e) $y^2 = -18x$ (h) $y^2 - 20x + 4y + 64 = 0$, $y^2 + 20x + 4y - 56 = 0$

(c) $x^2 = 20y$ (f) $y^2 + 8x - 6y + 1 = 0$ (i) $y^2 - 6x - 4y + 7 = 0$, $y^2 + 6x - 4y - 17 = 0$

7. Prove: The ordinate at any point $P(x,y)$ on the parabola $y^2 = 4px$ is the mean proportional between the abscissa at P and the length of the latus rectum.

8. Find the length of the focal chord of $y^2 = 8x$ parallel to $y = 2\sqrt{2}\,x + 1$. *Ans.* 9

9. Verify the following geometric construction of the parabola when the focus and the directrix are given: Draw through F the perpendicular to d meeting it in D. Locate the vertex V on DF so that $DV = VF$. Take any point A on VF on the same side of V as F and draw BA parallel to d or perpendicular to DF. With F as a center and radius equal to DA draw an arc intersecting AB at P and P'. Then P and P' are two points of the parabola.

10. Two circles are drawn, each having a focal chord of a parabola as diameter. Show that their radical axis passes through the vertex of the parabola.

CHAPTER 53

The Ellipse

THE LOCUS OF A POINT P which moves in a plane so that the sum of its distances from two fixed points in the plane is constant is called an *ellipse*.

The fixed points F and F' are called the *foci* and their midpoint C is called the *center* of the ellipse. The line FF' joining the foci intersects the ellipse in the points V and V', called the *vertices*. The segment $V'V$ intercepted on the line FF' by the ellipse is called its *major axis*; the segment $B'B$ intercepted on the line through C perpendicular to $F'F$ is called its *minor axis*.

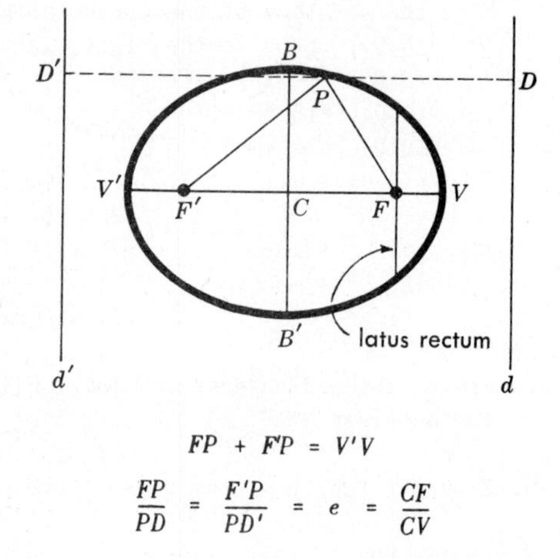

$$FP + F'P = V'V$$

$$\frac{FP}{PD} = \frac{F'P}{PD'} = e = \frac{CF}{CV}$$

A line segment whose extremities are any two points on the ellipse is called a *chord*. A chord which passes through a focus is called a *focal chord*; a focal chord perpendicular to the major axis is called a *latus rectum*.

The ellipse may also be defined as the locus of a point which moves so that the ratio of its distance from a fixed point to its distance from a fixed line is equal to $e < 1$. The fixed point is a focus F or F' and the fixed line d or d' is called a *directrix*. The ratio e is called the *eccentricity* of the ellipse.

Problems 4 and 6(*a*) of Chapter 51 illustrate the two definitions.

THE EQUATION OF AN ELLIPSE assumes its simplest (*reduced*) form when its center is at the origin and its major axis lies along one of the coordinate axes.

When the center is at the origin and the major axis lies along the x-axis, the equation of the ellipse is

$$(1) \qquad \frac{x^2}{a^2} + \frac{y^2}{b^2} = 1$$

Then the vertices are at $V(a, 0)$ and $V'(-a, 0)$ and the length of the major axis is $V'V = 2a$. The length of the minor axis is $B'B = 2b$. The foci are on the major axis at $F(c, 0)$ and $F'(-c, 0)$ where

$$c = \sqrt{a^2 - b^2}$$

When the center is at the origin and the major axis lies along the y-axis, the equation of the ellipse is

$$(2) \qquad \frac{x^2}{b^2} + \frac{y^2}{a^2} = 1$$

Then the vertices are at $V(0,a)$ and $V'(0,-a)$ and the length of the major axis is $V'V = 2a$. The length of the minor axis is $B'B = 2b$. The foci are on the major axis at $F(0,c)$ and $F'(0,-c)$, where

$$c = \sqrt{a^2 - b^2}$$

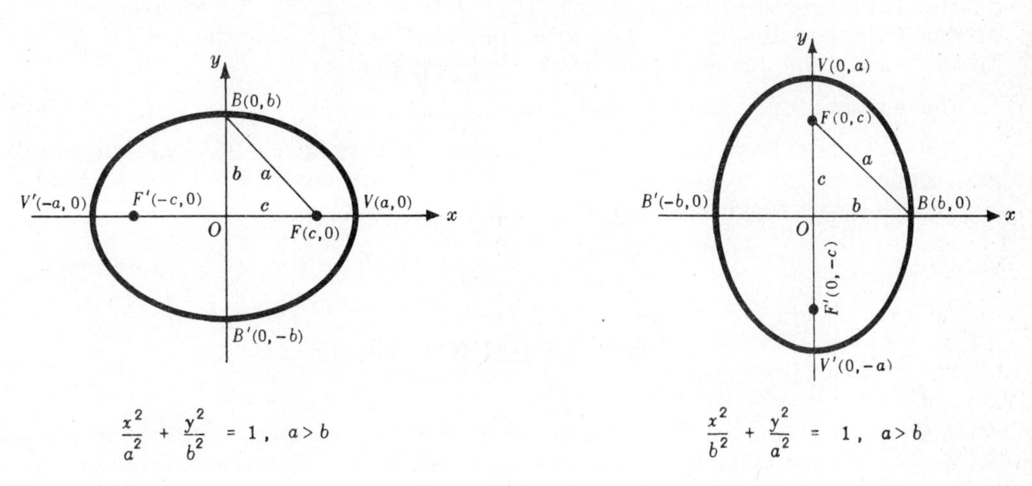

$$\frac{x^2}{a^2} + \frac{y^2}{b^2} = 1, \quad a > b \qquad\qquad \frac{x^2}{b^2} + \frac{y^2}{a^2} = 1, \quad a > b$$

In both cases, the length of a latus rectum is $2b^2/a$, the eccentricity is $e = \frac{c}{a} = \frac{\sqrt{a^2 - b^2}}{a}$, and the directrices are perpendicular to the major axis at distances $\pm a^2/c = \pm a/e$ from the center.

THE EQUATION OF AN ELLIPSE assumes the *semi-reduced* form

$$(1') \quad \frac{(x-h)^2}{a^2} + \frac{(y-k)^2}{b^2} = 1 \qquad \text{or} \qquad (2') \quad \frac{(x-h)^2}{b^2} + \frac{(y-k)^2}{a^2} = 1$$

Here the center is at the point (h,k) and the major axis is parallel respectively to the x-axis and to the y-axis. The lengths of the major and minor axes, the distance between the foci, the distance from the center to a directrix, the length of a latus rectum, and the eccentricity are as given in the section above.

Example 1. Find the coordinates of the center, vertices, and foci; the lengths of the major and minor axes; the length of a latus rectum; the eccentricity; and the equations of the directrices of the ellipse $\frac{(x-4)^2}{9} + \frac{(y+2)^2}{25} = 1$. Sketch the locus.

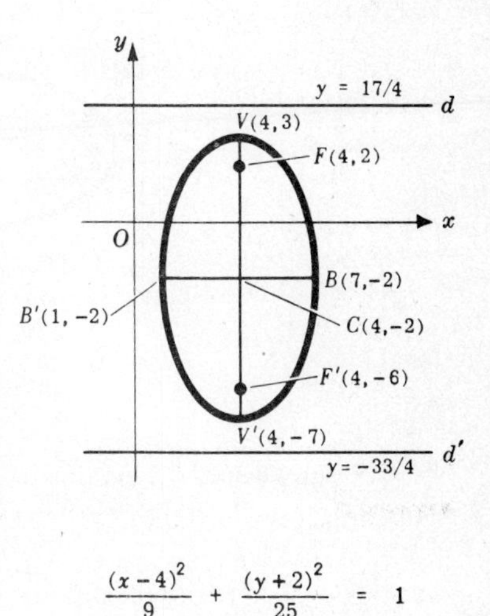

$$\frac{(x-4)^2}{9} + \frac{(y+2)^2}{25} = 1$$

The center is at the point $C(4,-2)$.

Since $a^2 > b^2$, $a^2 = 25$ and $b^2 = 9$.

Since a^2 is under the term in y, the major axis is parallel to the y-axis. To locate the vertices (the extremities of the major axis), we move from the center parallel to the y-axis a distance $a = 5$ to the points $V(4,3)$ and $V'(4,-7)$. To locate the extremities of the minor axis, we move from the center perpendicular to the major axis a distance $b = 3$ to the points $B'(1,-2)$ and $B(7,-2)$. The lengths of the major and minor axes are $2a = 10$ and $2b = 6$ respectively.

The distance from the center to a focus is $c = \sqrt{a^2 - b^2} = \sqrt{25-9} = 4$. To locate the foci, we move from the center along the major axis a distance $c = 4$ to the points $F(4,2)$ and $F'(4,-6)$.

The length of a latus rectum is $2b^2/a = 18/5$. The coordinates of the extremities of the latus rectum through F are $(11/5, 2)$ and $(29/5, 2)$, each being at a distance of one-half the length of the latus rectum from F; and the coordinates of the extremities of the latus rectum through F' are $(11/5,-6)$ and $(29/5,-6)$.

The eccentricity is $e = c/a = 4/5$.

The distance from the center to a directrix is $a^2/c = 25/4$. Since the directrices are perpendicular to the major axis, their equations are d: $y = -2 + 25/4 = 17/4$ and d': $y = -2 - 25/4 = -33/4$.

SOLVED PROBLEMS

1. For each of the following ellipses, find the coordinates of the center, vertices, and foci; the lengths of the major and minor axes; the length of a latus rectum; the eccentricity; and the equations of the directrices. Sketch the curve.
(a) $x^2/16 + y^2/4 = 1$, (b) $25x^2 + 9y^2 = 25$, (c) $x^2 + 9y^2 + 4x - 18y - 23 = 0$.

(a) Here $a^2 = 16$, $b^2 = 4$, and $c = \sqrt{a^2 - b^2} = \sqrt{16-4} = 2\sqrt{3}$.
The center is at the origin and the major axis is along the x-axis (a^2 under x^2). The vertices are on the major axis at a distance $a = 4$ from the center; their coordinates are $V(4,0)$ and $V'(-4,0)$. The minor axis is along the y-axis and its extremities, being at a distance $b = 2$ from the center, are at $B(0,2)$ and $B'(0,-2)$. The foci are on the major axis at a distance $2\sqrt{3}$ from the center; their coordinates are $F(2\sqrt{3}, 0)$ and $F'(-2\sqrt{3}, 0)$.
The lengths of the major and minor axes are $2a = 8$ and $2b = 4$ respectively.
The length of a latus rectum is $2b^2/a = 2\cdot 4/4 = 2$.
The eccentricity is $e = c/a = 2\sqrt{3}/4 = \frac{1}{2}\sqrt{3}$.
The directrices are perpendicular to the major axis and at a distance $a^2/c = 16/2\sqrt{3} = 8\sqrt{3}/3$ from the center; their equations are $x = \pm 8\sqrt{3}/3$.

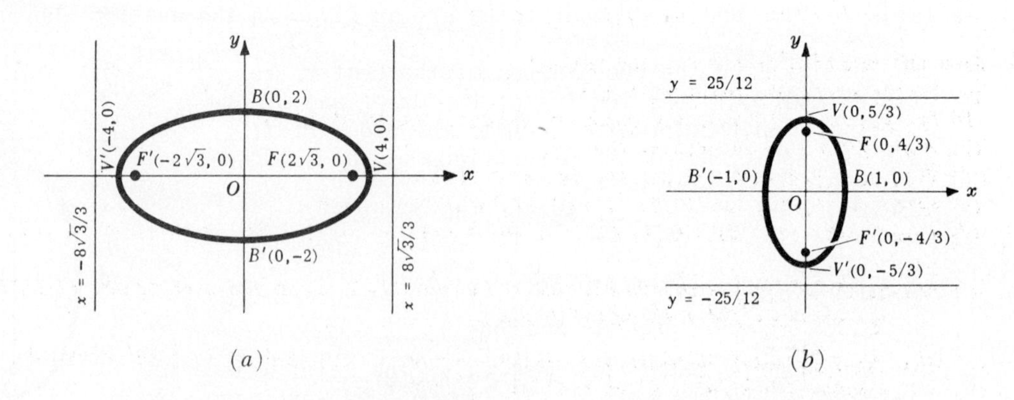

(b) When the equation is put in the form $\dfrac{x^2}{1} + \dfrac{y^2}{25/9} = 1$, we find

$$a^2 = 25/9, \quad b^2 = 1, \quad \text{and} \quad c = \sqrt{a^2 - b^2} = \sqrt{25/9 - 1} = 4/3$$

The center is at the origin and the major axis is along the y-axis (a^2 under y^2). The vertices are on the major axis at a distance $a = 5/3$ from the center; their coordinates are $V(0,5/3)$ and $V'(0,-5/3)$. The extremities of the minor axis are on the x-axis and at a distance $b = 1$ from the center; their coordinates are $B(1,0)$ and $B'(-1,0)$. The foci are on the major axis at a distance $c = 4/3$ from the center; their coordinates are $F(0,4/3)$ and $F'(0,-4/3)$.

The lengths of the major and minor axes are $2a = 10/3$ and $2b = 2$ respectively.

The length of a latus rectum is $2b^2/a = 6/5$.

The eccentricity is $e = c/a = \dfrac{4/3}{5/3} = 4/5$.

The directrices are perpendicular to the major axis and at a distance $a^2/c = 25/12$ from the center; their equations are $y = \pm 25/12$.

(*c*) When the equation is put in the form $\dfrac{(x+2)^2}{36} + \dfrac{(y-1)^2}{4} = 1$, we have $a^2 = 36$, $b^2 = 4$, and $c = \sqrt{a^2 - b^2} = 4\sqrt{2}$.

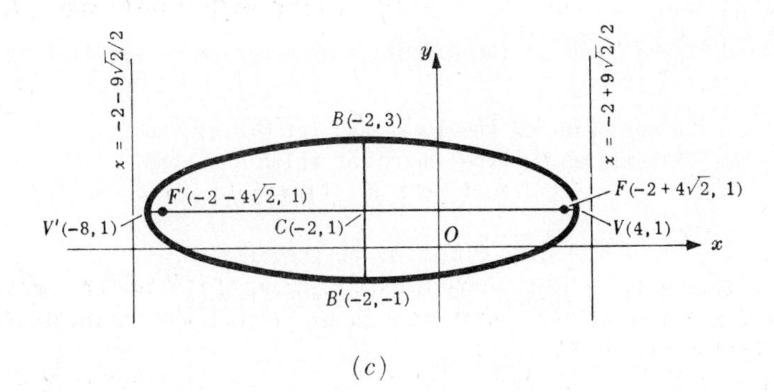

(*c*)

The center is at the point $C(-2, 1)$ and the major axis is along the line $y = 1$. The vertices are on the major axis at a distance $a = 6$ from the center; their coordinates are $V(4,1)$ and $V'(-8,1)$. The extremities of the minor axis are on the line $x = -2$ and at a distance $b = 2$ from the center; their coordinates are $B(-2,3)$ and $B'(-2,-1)$. The foci are on the major axis at a distance $c = 4\sqrt{2}$ from the center; their coordinates are $F(-2+4\sqrt{2}, 1)$ and $F'(-2-4\sqrt{2}, 1)$.

The lengths of the major and minor axes are $2a = 12$ and $2b = 4$ respectively.

The length of a latus rectum is $2b^2/a = 4/3$.

The eccentricity is $c/a = 4\sqrt{2}/6 = 2\sqrt{2}/3$.

The directrices are perpendicular to the major axis and at a distance $a^2/c = 9\sqrt{2}/2$ from the center; their equations are $x = -2 \pm \dfrac{9\sqrt{2}}{2}$.

2. Find the equation of the ellipse, given:

(*a*) Vertices $(\pm 8, 0)$, minor axis $= 6$.
(*b*) One vertex at $(0, 13)$, one focus at $(0,-12)$, center at $(0,0)$.
(*c*) Foci $(\pm 10, 0)$, eccentricity $= 5/6$.
(*d*) Vertices $(8,3)$ and $(-4,3)$, one focus at $(6,3)$.
(*e*) Vertices $(3,-10)$ and $(3,2)$, length of latus rectum $= 4$.
(*f*) Directrices $4y - 33 = 0$, $4y + 17 = 0$; major axis on $x + 1 = 0$; eccentricity $= 4/5$.

(*a*) Here $2a = V'V = 16$, $2b = 6$, and the major axis is along the x-axis. The equation of the ellipse is $x^2/a^2 + y^2/b^2 = x^2/64 + y^2/9 = 1$.

(*b*) The major axis is along the y-axis, $a = 13$, $c = 12$, and $b^2 = a^2 - c^2 = 25$. The equation of the ellipse is $x^2/b^2 + y^2/a^2 = x^2/25 + y^2/169 = 1$.

(*c*) Here the major axis is along the x-axis and $c = 10$. Since $e = c/a = 10/a = 5/6$, $a = 12$ and $b^2 = a^2 - c^2 = 44$. The equation of the ellipse is $x^2/144 + y^2/44 = 1$.

(*d*) The center is at the midpoint of $V'V$, that is, at $C(2,3)$. Then $a = CV = 6$, $c = CF = 4$, and $b^2 = a^2 - c^2 = 20$. Since the major axis is parallel to the x-axis, the equation of the ellipse is

$$\frac{(x-2)^2}{36} + \frac{(y-3)^2}{20} = 1$$

(e) The center is at $C(3,-4)$ and $a = 6$. Since $2b^2/a = 2b^2/6 = 4$, $b^2 = 12$. The major axis is parallel to the y-axis and the equation of the ellipse is $\dfrac{(x-3)^2}{12} + \dfrac{(y+4)^2}{36} = 1$.

(f) The major axis intersects the directrices in $D(-1, 33/4)$ and $D'(-1, -17/4)$. The center of the ellipse bisects $D'D$ and hence is at $C(-1,2)$. Since $CD = a/e = 25/4$ and $e = 4/5$, $a = 5$ and $c = ae = 4$. Then $b^2 = a^2 - c^2 = 9$. Since the major axis is parallel to the y-axis, the equation of the ellipse is $\dfrac{(x+1)^2}{9} + \dfrac{(y-2)^2}{25} = 1$.

3. Find the equation of the locus of the midpoints of a system of parallel chords of slope m of the ellipse $x^2/a^2 + y^2/b^2 = 1$.

Let $P(x,y)$ be any point on the locus and let $Q(p,q)$ and $R(r,s)$ be the extremities of the chord of which P is the midpoint. Then

(1) $b^2p^2 + a^2q^2 = a^2b^2$, since $Q(p,q)$ is on the ellipse,

(2) $b^2r^2 + a^2s^2 = a^2b^2$, since $R(r,s)$ is on the ellipse,

(3) $\dfrac{q-s}{p-r} = m$, since m is the slope of the chord,

(4) $x = \frac{1}{2}(p+r)$, $y = \frac{1}{2}(q+s)$, since P is the midpoint of QR.

Equating the left members of (1) and (2), we have

$$b^2p^2 + a^2q^2 = b^2r^2 + a^2s^2$$

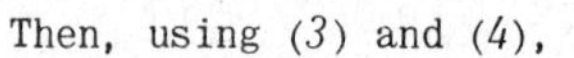

Then, using (3) and (4),

$$b^2(p^2 - r^2) = a^2(s^2 - q^2) \quad \text{and} \quad \frac{b^2}{a^2} = \frac{s^2 - q^2}{p^2 - r^2} = -\frac{q-s}{p-r}\cdot\frac{q+s}{p+r} = -m\frac{2y}{2x} = -m\frac{y}{x}$$

Thus the desired equation is $y = -\dfrac{b^2}{a^2m}x$.

The locus passes through the center $C(0,0)$ and is called a *diameter* of the ellipse.

SUPPLEMENTARY PROBLEMS

4. For each of the following ellipses, find: the coordinates of the center, vertices and foci; the lengths of the major and minor axes; the length of the latus rectum; the eccentricity; and the equations of the directrices. Sketch each curve.

(a) $4x^2 + 9y^2 = 36$

(b) $25x^2 + 16y^2 = 400$

(c) $x^2 + 4y^2 - 6x + 32y + 69 = 0$

(d) $16x^2 + 9y^2 + 32x - 36y - 92 = 0$

Ans. (a) $C(0,0)$, $V(\pm 3,0)$, $F(\pm\sqrt{5}, 0)$; 6,4; 8/3; $\sqrt{5}/3$; $x = \pm 9\sqrt{5}/5$

(b) $C(0,0)$, $V(0,\pm 5)$, $F(0,\pm 3)$; 10,8; 32/5; 3/5; $y = \pm 25/3$

(c) $C(3,-4)$, $V(5,-4)$, $V'(1,-4)$, $F(3\pm\sqrt{3}, -4)$; 4,2; 1; $\sqrt{3}/2$; $x = 3 \pm 4\sqrt{3}/3$

(d) $C(-1,2)$, $V(-1,6)$, $V'(-1,-2)$, $F(-1, 2\pm\sqrt{7})$; 8,6; 9/2; $\sqrt{7}/4$; $y = 2 \pm 16\sqrt{7}/7$

5. Find the equation of the ellipse, given:

(a) $V(\pm 13,0)$, $F(12,0)$
(b) $C(0,0)$, $a=5$, $F(0,4)$
(c) $C(0,0)$, $b=2$, d: $x=-16\sqrt{7}/7$
(d) $V(\pm 4,0)$, latus rectum = 9/2
(e) $V(7,3)$, $V'(-3,3)$, $F(6,3)$
(f) $F(5,4)$, $F'(5,-2)$, $e=\sqrt{3}/3$
(g) ends of minor axis $(-2,4)$, $(-2,2)$; d: $x=0$
(h) directrices: $y=11/5$, $y=-61/5$; major axis on $x=3$, $e=5/6$

Ans. (a) $25x^2+169y^2=4225$
(b) $25x^2+9y^2=225$
(c) $x^2+8y^2=32$, $7x^2+8y^2=32$
(d) $9x^2+16y^2=144$
(e) $9x^2+25y^2-36x-150y+36=0$
(f) $3x^2+2y^2-30x-4y+23=0$
(g) $x^2+2y^2+4x-12y+20=0$
(h) $36x^2+11y^2-216x+110y+203=0$

6. An arch in the form of a semi-ellipse is 50 ft wide at the base and 20 ft high. Find (a) its height 10 feet from the center of the base and (b) its width 10 ft above the base.
Ans. (a) $4\sqrt{21}$ ft, (b) $25\sqrt{3}$ ft

7. Justify the following construction for the ellipse $b^2x^2+a^2y^2=a^2b^2$. Through the origin draw a line meeting the circle $x^2+y^2=a^2$ in $A(p,q)$ and the circle $x^2+y^2=b^2$ in $B(r,s)$; then the point $E(p,s)$ is on the ellipse.

8. Prove: The locus of the midpoints of the chords of an ellipse drawn through one end of the major axis is an ellipse.

9. Prove: If one diameter of an ellipse bisects all chords parallel to another diameter, then the second diameter bisects all chords parallel to the first. Two such diameters are called *conjugate diameters*.

10. An ellipse whose axes are parallel to the coordinate axes passes through the points $(12,-1)$, $(-4,5)$, $(-2,-2)$, and $(10,6)$. Find its equation. *Ans.* $x^2+4y^2-8x-16y-68=0$

11. Find the equation of the locus of the center of a circle which is tangent to the circles $x^2+y^2=4$ and $x^2+y^2-8x-48=0$. *Ans.* $5x^2+9y^2-20x-25=0$, $21x^2+25y^2-84x-441=0$

CHAPTER 54

The Hyperbola

THE LOCUS OF A POINT P which moves in a plane so that the absolute value of the difference of its distances from two fixed points in the plane is constant is called a *hyperbola*. (Note that the locus consists of two distinct branches each of indefinite length.)

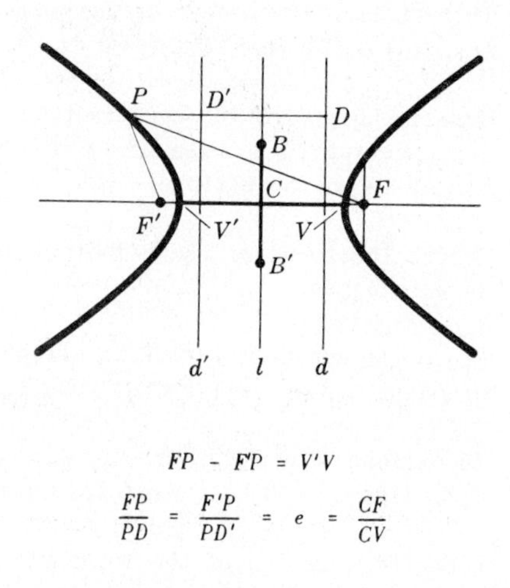

$$FP - F'P = V'V$$

$$\frac{FP}{PD} = \frac{F'P}{PD'} = e = \frac{CF}{CV}$$

The fixed points F and F' are called the *foci* and their midpoint C is called the *center* of the hyperbola. The line FF' joining the foci intersects the hyperbola in the points V and V', called the *vertices*.

The segment $V'V$ intercepted on the line FF' by the hyperbola is called its *transverse axis*. The line l through C and perpendicular to FF' does not intersect the curve but it will be found convenient to define a certain segment $B'B$ on l, having C as midpoint, as the *conjugate axis*.

A line segment whose extremities are any two points (both on the same branch or one on each branch) on the hyperbola is called a *chord*. A chord which passes through a focus is called a *focal chord*, a focal chord perpendicular to the transverse axis is called a *latus rectum*.

The hyperbola may also be defined as the locus of a point which moves so that the ratio of its distance from a fixed point and its distance from a fixed line is equal to $e > 1$. The fixed point is a focus F or F' and the fixed line d or d' is called a *directrix*. The ratio e is called the *eccentricity* of the hyperbola.

THE EQUATION OF A HYPERBOLA assumes its simplest (*reduced*) form when its center is at the origin and its transverse axis lies along one of the coordinate axes.

When the center is at the origin and the transverse axis lies along the x-axis, the equation of the hyperbola is

$$(1) \qquad \frac{x^2}{a^2} - \frac{y^2}{b^2} = 1$$

Then the vertices are at $V(a,0)$ and $V'(-a,0)$ and the length of the transverse axis is $V'V = 2a$. The extremities of the conjugate axis are $B'(0,-b)$ and $B(0,b)$ and its length is $B'B = 2b$. The foci are on the transverse axis at $F'(-c,0)$ and $F(c,0)$, where

$$c = \sqrt{a^2 + b^2}$$

When the center is at the origin and the transverse axis lies along the y-axis, the equation of the hyperbola is

$$(2) \qquad \frac{y^2}{a^2} - \frac{x^2}{b^2} = 1$$

Then the vertices are at $V(0, a)$ and $V'(0,-a)$ and the length of the transverse axis is $V'V = 2a$. The extremities of the conjugate axis are $B'(-b, 0)$ and $B(b, 0)$, and its length is $B'B = 2b$. The foci are on the transverse axis at $F(0, c)$ and $F'(0,-c)$, where

$$c = \sqrt{a^2 + b^2}$$

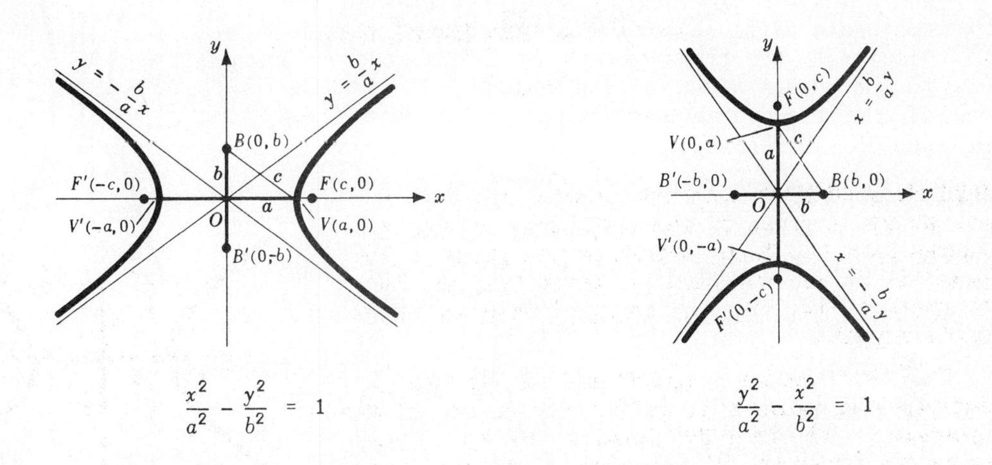

$$\frac{x^2}{a^2} - \frac{y^2}{b^2} = 1 \qquad\qquad \frac{y^2}{a^2} - \frac{x^2}{b^2} = 1$$

In both cases, the length of a latus rectum is $2b^2/a$, the eccentricity is $e = \frac{c}{a} = \frac{\sqrt{a^2 + b^2}}{a}$, and the directrices are perpendicular to the transverse axis at distances $\pm a^2/c = \pm a/e$ from the center.

THE STRAIGHT LINES $y = \pm\frac{b}{a}x$ are called the *asymptotes* of the hyperbola *(1)*, and the lines $x = \pm\frac{b}{a}y$ are called the *asymptotes* of the hyperbola (2).

These lines have the property that the perpendicular distance from a point on a hyperbola to one of them approaches zero as the point moves indefinitely far from the center.

For a proof, see Problem 4.

THE EQUATION OF A HYPERBOLA assumes the *semi-reduced* form

$$(1') \quad \frac{(x-h)^2}{a^2} - \frac{(y-k)^2}{b^2} = 1 \qquad \text{or} \qquad (2') \quad \frac{(y-k)^2}{a^2} - \frac{(x-h)^2}{b^2} = 1$$

Here the center is at the point $C(h, k)$ and the transverse axis is parallel respectively to the x-axis or to the y-axis. The lengths of the transverse and conjugate axes, the distance between the foci, the distance from the center to a directrix, the length of a latus rectum, the slope of the asymptotes, and the eccentricity are as given in the sections above.

Example 1. Find the coordinates of the center, vertices, and foci; the lengths of the transverse and conjugate axes; the length of a latus rectum; the eccentricity; and the equations of the directrices and asymptotes of the hyperbola

$$\frac{(x+3)^2}{4} - \frac{(y-1)^2}{25} = 1. \quad \text{Sketch the locus.}$$

The center is at the point $C(-3, 1)$. Since, when the equation is put in the reduced or semi-reduced form, a^2 is always in the positive term on the left,

$a^2 = 4$ and $b^2 = 25$; then $a = 2$ and $b = 5$.

The transverse axis is parallel to the x-axis (the positive term contains x). To locate the vertices, we move from the center along the transverse axis a distance $a = 2$ to the points $V(-1, 1)$ and $V'(-5, 1)$. To locate the extremities of the conjugate axis, we move from the center perpendicular to the transverse axis a distance $b = 5$ to the points $B'(-3, -4)$ and $B(-3, 6)$. The lengths of the transverse and conjugate axes are $2a = 4$ and $2b = 10$ respectively.

The distance from the center to a focus is

$$c = \sqrt{a^2 + b^2} = \sqrt{4 + 25} = \sqrt{29}$$

To locate the foci, we move from the center along the transverse axis a distance $c = \sqrt{29}$ to the points $F(-3 + \sqrt{29}, 1)$ and $F'(-3 - \sqrt{29}, 1)$.

The length of a latus rectum is $2b^2/a = 25$. The coordinates of the extremities of the latus rectum through F are $(-3 + \sqrt{29}, 27/2)$ and $(-3 + \sqrt{29}, -23/2)$, being at a distance of one-half the length of the latus rectum from F. The coordinates of the extremities of the latus rectum through F' are $(-3 - \sqrt{29}, 27/2)$ and $(-3 - \sqrt{29}, -23/2)$.

The eccentricity is $e = c/a = \sqrt{29}/2$.

The distance from the center to a directrix is $a^2/c = 4/\sqrt{29} = 4\sqrt{29}/29$. Since the directrices are perpendicular to the transverse axis, their equations are d: $x = -3 + 4\sqrt{29}/29$ and d': $x = -3 - 4\sqrt{29}/29$.

The asymptotes pass through C with slopes $\pm b/a = \pm 5/2$; thus their equations are $y - 1 = \pm \frac{5}{2}(x + 3)$. Combining these two equations, $\frac{(x+3)^2}{4} - \frac{(y-1)^2}{25} = 0$. Hence they may be obtained most readily by the simple trick of changing the right member of the equation of the hyperbola from 1 to 0.

See Problems 1-3.

THE HYPERBOLAS OF EQUATION $x^2 - y^2 = a^2$ and $y^2 - x^2 = a^2$ whose transverse and conjugate axes are of equal length $2a$ are called *equilateral hyperbolas*. Since their asymptotes $x \pm y = 0$ are mutually perpendicular, the equilateral hyperbola is also called the *rectangular hyperbola*.

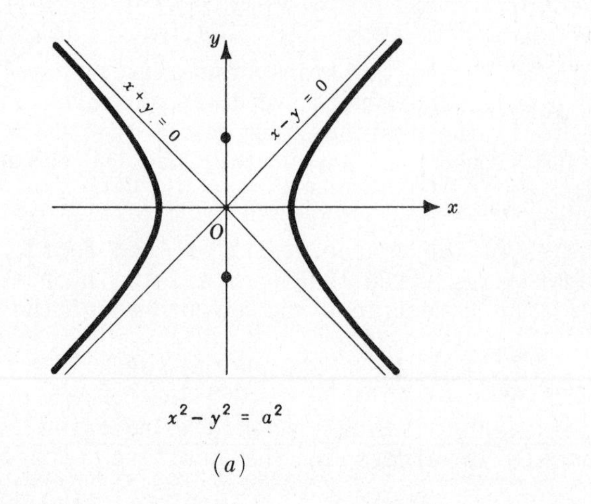

$x^2 - y^2 = a^2$

(a)

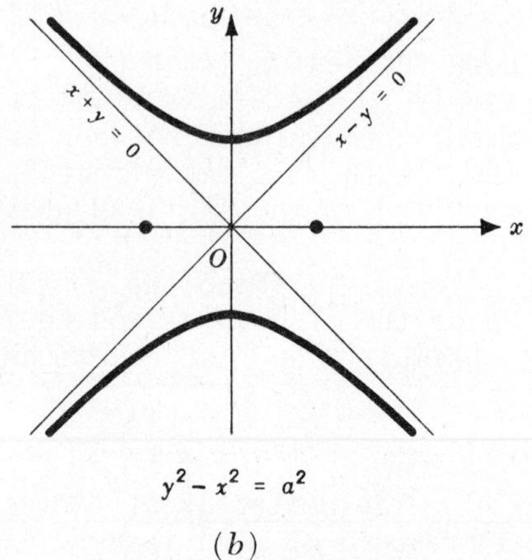

$y^2 - x^2 = a^2$

(b)

TWO HYPERBOLAS such that the transverse axis of each is the conjugate axis of the other, as

$$\frac{x^2}{16} - \frac{y^2}{9} = 1 \qquad \text{and} \qquad \frac{y^2}{9} - \frac{x^2}{16} = 1$$

are called *conjugate hyperbolas*, each being the conjugate of the other.

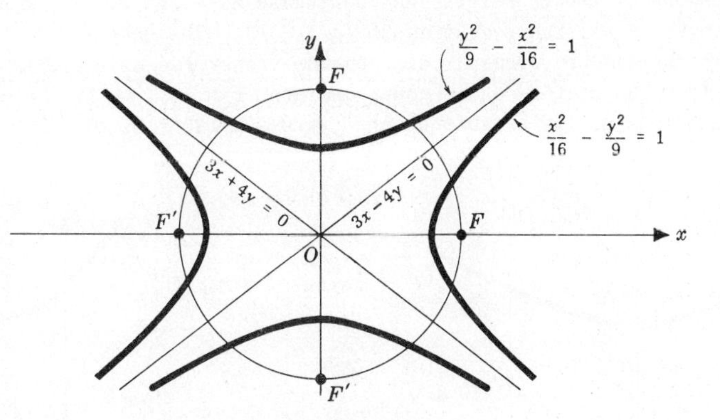

A pair of conjugate hyperbolas have the same center and the same asymptotes. Their foci lie on a circle whose center is the common center of the hyperbolas.

See Problem 5.

SOLVED PROBLEMS

1. For each of the following hyperbolas find the coordinates of the center, the vertices, and the foci; the lengths of the transverse and conjugate axes; the length of a latus rectum; the eccentricity; and the equations of the directrices and asymptotes. Sketch each locus.
(*a*) $x^2/16 - y^2/4 = 1$, (*b*) $25y^2 - 9x^2 = 225$, (*c*) $9x^2 - 4y^2 - 36x + 32y + 8 = 0$.

(*a*) Here $a^2 = 16$, $b^2 = 4$, and $c = \sqrt{a^2+b^2} = 2\sqrt{5}$.

The center is at the origin and the transverse axis is along the x-axis (a^2 under x^2). The vertices are on the transverse axis at a distance $a = 4$ from the center; their coordinates are $V(4,0)$ and $V'(-4,0)$. The extremities of the conjugate axis are on the y-axis at a distance $b = 2$ from the center; their coordinates are $B(0,2)$ and $B'(0,-2)$.

The foci are on the transverse axis at a distance $c = 2\sqrt{5}$ from the center; their coordinates are $F(2\sqrt{5},\ 0)$ and $F'(-2\sqrt{5},\ 0)$.

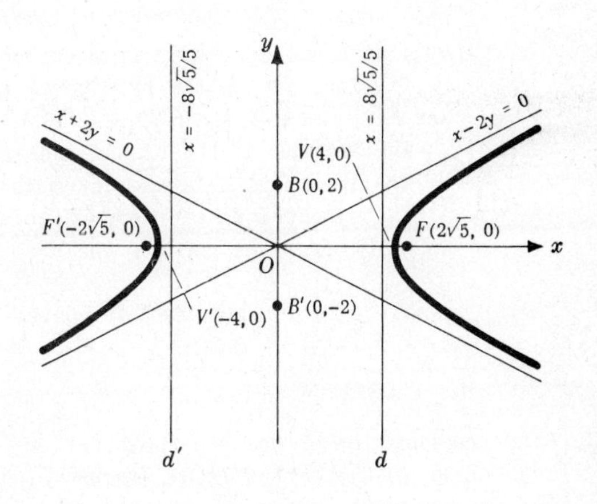

The lengths of the transverse and conjugate axes are $2a = 8$ and $2b = 4$ respectively.

The length of a latus rectum is $2b^2/a = 2$.
The eccentricity is $e = c/a = 2\sqrt{5}/4 = \frac{1}{2}\sqrt{5}$.

The directrices are perpendicular to the transverse axis and at a distance $a^2/c = 16/2\sqrt{5} = 8\sqrt{5}/5$ from the center; their equations are $x = \pm\, 8\sqrt{5}/5$.

The equations of the asymptotes are $x^2/16 - y^2/4 = 0$ or $x = \pm\, 2y$.

(*b*) When the equation is put in the form $y^2/9 - x^2/25 = 1$, we find

$$a^2 = 9,\quad b^2 = 25,\quad \text{and} \quad c = \sqrt{a^2+b^2} = \sqrt{34}.$$

The center is at the origin and the transverse axis is along the y-axis. The vertices are on the transverse axis at a distance $a = 3$ from the center; their coordinates are $V(0,3)$ and $V'(0,-3)$. The extremities of the conjugate axis are on the x-axis at a distance $b = 5$ from the center; their coordinates are $B(5,0)$ and $B'(-5,0)$.

The foci are on the transverse axis at a distance $c = \sqrt{34}$ from the center; their coordinates are $F(0, \sqrt{34})$ and $F'(0, -\sqrt{34})$.

The lengths of the transverse and conjugate axes are $2a = 6$ and $2b = 10$ respectively.

The length of a latus rectum is $2b^2/a = 50/3$. The eccentricity is $e = c/a = \sqrt{34}/3$.

The directrices are perpendicular to the transverse axis and at a distance $a^2/c = 9/\sqrt{34} = 9\sqrt{34}/34$ from the center; their equations are $y = \pm 9\sqrt{34}/34$.

The equations of the asymptotes are $y^2/9 - x^2/25 = 0$ or $5y = \pm 3x$.

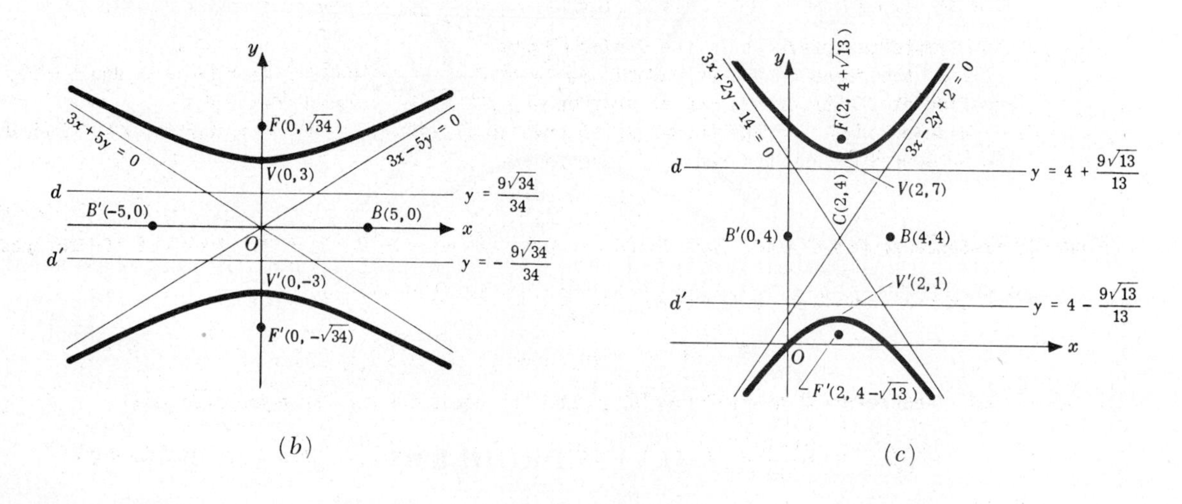

(b) (c)

(c) Putting the equation in the form $\dfrac{(y-4)^2}{9} - \dfrac{(x-2)^2}{4} = 1$, we have

$$a^2 = 9, \quad b^2 = 4, \quad \text{and} \quad c = \sqrt{a^2 + b^2} = \sqrt{13}.$$

The center is at the point $C(2,4)$ and the transverse axis is parallel to the y-axis along the line $x = 2$. The vertices are on the transverse axis at a distance $a = 3$ from the center; their coordinates are $V(2,7)$ and $V'(2,1)$. The extremities of the conjugate axis are on the line $y = 4$ at a distance $b = 2$ from the center; their coordinates are $B(4,4)$ and $B'(0,4)$.

The foci are on the transverse axis at a distance $c = \sqrt{13}$ from the center; their coordinates are $F(2, 4+\sqrt{13})$ and $F'(2, 4-\sqrt{13})$.

The lengths of the transverse and conjugate axes are $2a = 6$ and $2b = 4$ respectively.

The length of a latus rectum is $2b^2/a = 8/3$. The eccentricity is $e = c/a = \sqrt{13}/3$.

The directrices are perpendicular to the transverse axis and at a distance $a^2/c = 9/\sqrt{13} = 9\sqrt{13}/13$ from the center; their equations are $y = 4 \pm 9\sqrt{13}/13$.

The equations of the asymptotes are $\dfrac{(y-4)^2}{9} - \dfrac{(x-2)^2}{4} = 0$ or $3x - 2y + 2 = 0$ and $3x + 2y - 14 = 0$.

2. Find the equation of the hyperbola, given:

(a) Center (0,0), vertex (4,0), focus (5,0).

(b) Center (0,0), focus (0,−4), eccentricity = 2.

(c) Center (0,0), vertex (5,0), one asymptote $5y + 3x = 0$.

(d) Center (−5,4), vertex (−11,4), eccentricity = 5/3.

(e) Vertices (−11,1) and (5,1), one asymptote $x - 4y + 7 = 0$.

(f) Transverse axis parallel to the x-axis, asymptotes $3x + y - 7 = 0$ and $3x - y - 5 = 0$, passes through (4,4).

(a) Here $a = CV = 4$, $c = CF = 5$, and $b^2 = c^2 - a^2 = 25 - 16 = 9$. The transverse axis is along the x-axis and the equation of the hyperbola is $x^2/16 - y^2/9 = 1$.

(*b*) Since $c = F'C = 4$ and $e = c/a = 2$, $a = 2$ and $b^2 = c^2 - a^2 = 12$. The transverse axis is along the y-axis and the equation of the hyperbola is $y^2/4 - x^2/12 = 1$.

(*c*) The slope of the asymptote is $-b/a = -3/5$ and, since $a = CV = 5$, $b = 3$. The transverse axis is along the x-axis and the equation of the hyperbola is $x^2/25 - y^2/9 = 1$.

(*d*) Here $a = V'C = 6$ and $e = c/a = 5/3 = 10/6$; then $c = 10$ and $b^2 = c^2 - a^2 = 64$. The transverse axis is parallel to the x-axis and the equation of the hyperbola is $(x+5)^2/36 - (y-4)^2/64 = 1$.

(*e*) The center is at $(-3,1)$, the midpoint of VV'. The slope of the asymptote is $b/a = 1/4$ and, since $a = CV = 8$, $b = 2$. The transverse axis is parallel to the x-axis and the equation of the hyperbola is $(x+3)^2/64 - (y-1)^2/4 = 1$.

(*f*) The asymptotes intersect in the center $C(2,1)$.

Since the slope of the asymptote $3x - y - 5 = 0$ is $b/a = 3/1$, we take $a = m$ and $b = 3m$. The equation of the hyperbola may be written as $(x-2)^2/m^2 - (y-1)^2/9m^2 = 1$.

In order that the hyperbola pass through $(4,4)$, $4/m^2 - 9/9m^2 = 1$ and $m = \sqrt{3}$. Then $a = m = \sqrt{3}$, $b = 3m = 3\sqrt{3}$, and the required equation is $(x-2)^2/3 - (y-1)^2/27 = 1$.

Find the distance from the right hand focus of $9x^2 - 4y^2 + 54x + 16y - 79 = 0$ to one of its asymptotes.

Putting the equation in the form $\dfrac{(x+3)^2}{16} - \dfrac{(y-2)^2}{36} = 1$, we find

$$a^2 = 16, \quad b^2 = 36, \quad \text{and} \quad c = \sqrt{a^2 + b^2} = 2\sqrt{13}.$$

The right hand focus is at $F(-3 + 2\sqrt{13},\ 2)$ and the equations of the asymptotes are

$$\frac{(x+3)^2}{16} - \frac{(y-2)^2}{36} = 0 \qquad \text{or} \qquad 3x + 2y + 5 = 0 \quad \text{and} \quad 3x - 2y + 13 = 0$$

The distance from F to the first asymptote is $\left|\dfrac{3(-3 + 2\sqrt{13}) + 2\cdot 2 + 5}{-\sqrt{13}}\right| = 6$.

4. Prove: The perpendicular distance from a point on the hyperbola $x^2/a^2 - y^2/b^2 = 1$ to an asymptote $y = \pm \dfrac{b}{a}x$ approaches zero as the point moves indefinitely far from the center.

In view of the symmetry of the curve, it will suffice to prove this for the first quadrant. Let $P_1(x_1, y_1)$ be a point in the first quadrant on the hyperbola so that

$$x_1^2/a^2 - y_1^2/b^2 = 1.$$

The undirected distance of P_1 from the asymptote $y = \dfrac{b}{a}x$ is

$$\left|\frac{bx_1 - ay_1}{-\sqrt{a^2 + b^2}}\right| = \frac{a^2 b^2}{\sqrt{a^2 + b^2}} \left|\frac{1}{bx_1 + ay_1}\right| \quad \text{since}$$

$$b^2 x_1^2 - a^2 y_1^2 = (bx_1 - ay_1)(bx_1 + ay_1) = a^2 b^2 \quad \text{and}$$

$$bx_1 - ay_1 = \frac{a^2 b^2}{bx_1 + ay_1}.$$

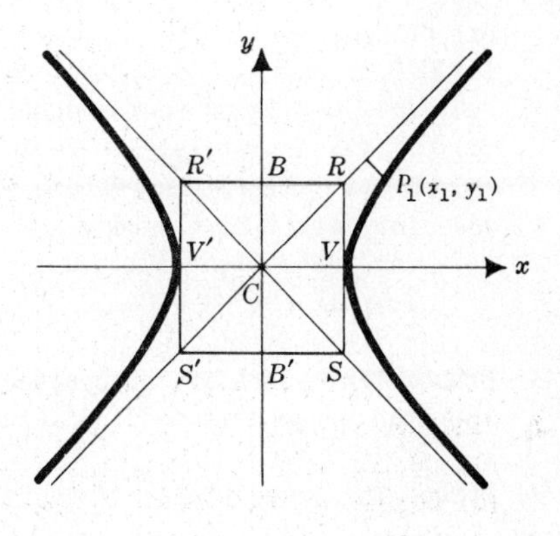

Now as P_1 moves out along the curve in the first quadrant, both x_1 and y_1 increase indefinitely and $\dfrac{1}{bx_1 + ay_1}$ approaches zero as a limit.

The asymptotes may be located by the following simple device: Through the vertices V and V' draw lines perpendicular to the transverse axis and through the extremities B and B' of the conjugate axis draw lines parallel to the transverse axis forming the rectangle $RR'S'S$. The diagonals of this rectangle are the asymptotes of the hyperbola; for example, the diagonal RS' passes through the center $C(0,0)$ with slope $\frac{VR}{CV} = \frac{CB}{CV} = \frac{b}{a}$. With the asymptotes in position as guide lines, the hyperbola may be sketched readily.

5. Write the equation of the conjugate of the hyperbola $25x^2 - 16y^2 = 400$ and sketch both curves.

The equation of the conjugate hyperbola is

$$16y^2 - 25x^2 = 400.$$

The common asymptotes have equations $y = \pm 5x/4$.

The vertices of $25x^2 - 16y^2 = 400$ are at $(\pm 4,0)$.

The vertices of $16y^2 - 25x^2 = 400$ are at $(0,\pm 5)$.

The curves are shown in the adjoining figure.

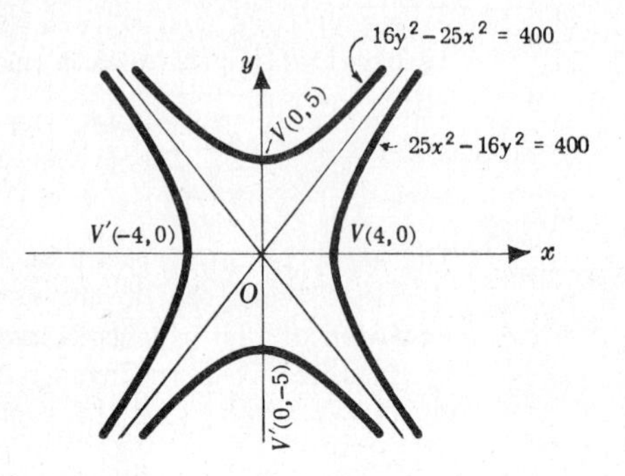

SUPPLEMENTARY PROBLEMS

6. For each of the following hyperbolas, find: the coordinates of the center, the vertices, and the foci; the lengths of the transverse and conjugate axes; the length of the latus rectum; the eccentricity; the equations of the directrices; and the equations of the asymptotes. Sketch each curve.

(a) $4x^2 - 9y^2 = 36$ (c) $x^2 - 4y^2 + 6x + 16y - 11 = 0$
(b) $16y^2 - 9x^2 = 144$ (d) $144x^2 - 25y^2 - 576x + 200y + 3776 = 0$

Ans. (a) $C(0,0)$, $V(\pm 3,0)$, $F(\pm\sqrt{13}, 0)$; 6,4; 8/3; $\sqrt{13}/3$; $x = \pm 9\sqrt{13}/13$; $2x \pm 3y = 0$.
(b) $C(0,0)$, $V(0,\pm 3)$, $F(0,\pm 5)$; 6,8; 32/3; 5/3; $y = \pm 9/5$; $3x \pm 4y = 0$.
(c) $C(-3,2)$, $V(-1,2)$, $V'(-5, 2)$, $F(-3\pm\sqrt{5}, 2)$; 4,2; 1; $\sqrt{5}/2$; $x = -3 \pm 4\sqrt{5}/5$; $x + 2y - 1 = 0$, $x - 2y + 7 = 0$.
(d) $C(2,4)$, $V(2,16)$, $V'(2,-8)$, $F(2,17)$, $F'(2,-9)$; 24,10; 25/6; 13/12; $y = 4 \pm 144/13$; $12x - 5y - 4 = 0$, $12x + 5y - 44 = 0$.

7. Find the equation of the hyperbola, given;

(a) $V(\pm 5,0)$, $F(13,0)$
(b) $C(0,0)$, $a = 5$, $F(0,6)$
(c) $V(\pm 12,0)$; latus rectum equal to 32/3
(d) $C(0,0)$; $b = 5$; d: $y = \pm 16\sqrt{41}/41$
(e) $C(2,-3)$, $V(7,-3)$, asymptote $3x - 5y - 21 = 0$
(f) $C(-3,1)$, $F(-3,5)$, eccentricity equal to 2
(g) $C(2,4)$; asymptotes $x + 2y - 10 = 0$, $x - 2y + 6 = 0$; passes through $(2,0)$
(h) $C(-3,2)$, $F(2,2)$, asymptote $4x + 3y + 6 = 0$

Ans. (a) $144x^2 - 25y^2 = 3600$
(b) $25x^2 - 11y^2 + 275 = 0$
(c) $4x^2 - 9y^2 = 576$
(d) $16x^2 - 25y^2 + 400 = 0$
(e) $9x^2 - 25y^2 - 36x - 150y - 414 = 0$
(f) $x^2 - 3y^2 + 6x + 6y + 18 = 0$
(g) $x^2 - 4y^2 - 4x + 32y + 4 = 0$
(h) $16x^2 - 9y^2 + 96x + 36y - 36 = 0$

8. Prove: The length of the conjugate axis of a hyperbola is the mean proportional between the length of the transverse axis and the latus rectum.

9. Prove: The eccentricity of every equilateral hyperbola is equal to $\sqrt{2}$.

10. Prove: The foci of a pair of conjugate hyperbolas are concyclic.

CHAPTER 55

Transformation of Coordinates

THE MOST GENERAL EQUATION of the second degree in x and y has the form

$$(1)\qquad Ax^2 + 2Bxy + Cy^2 + 2Dx + 2Ey + F = 0$$

If (1) can be factored so that we have $(ax + by + c)(dx + ey + f) = 0$, the locus consists of two straight lines; if $B = 0$, $A = C$, the locus of (1) is a circle; otherwise, the locus is one of the conics of Chapters 52-54.

The locus of the equation

$$(A)\qquad 20x^2 - 24xy + 27y^2 + 24x - 54y - 369 = 0$$

obtained in Problem 7, Chapter 51, and the locus of the equation

$$(B)\qquad 11x^2 + 36y^2 - 369 = 0$$

obtained in Problem 4, Chapter 51, are identical ellipses. The difference in equations is due to their positions with respect to the coordinate axes.

In order to make a detailed study of the loci represented by (1), say (A), it will be necessary to introduce some device to change (A) into (B). The operations, two in number, by which (A) is eventually replaced by (B), are called *transformations*. The general effect of these transformations may be interpreted as follows: Each point (x, y) of the plane remains fixed but changes its name, i.e. its coordinates, in accordance with a stated law, called the *equations of the transformation*.

TRANSLATION OF THE COORDINATE AXES. The transformation which moves the coordinate axes to a new position while keeping them always parallel to their original position is called a *translation*. In the adjoining figure, Ox and Oy are the axes and O is the origin of the original system of coordinates while $O'x'$ and $O'y'$ are the axes and O' is the origin of the new (translated) system.

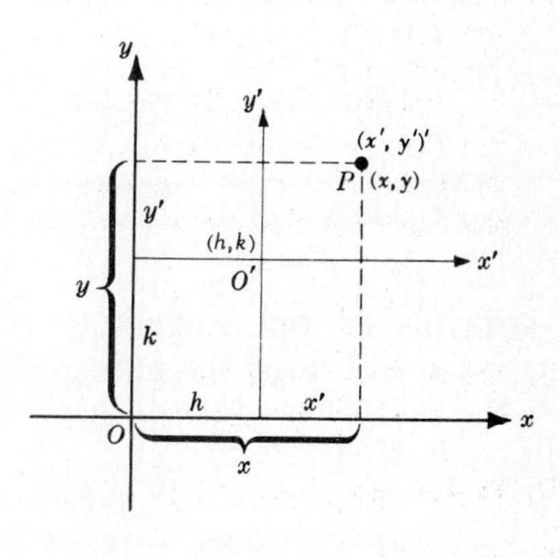

Each point in the plane will now have two sets of coordinates, the original set being the directed distances in proper order of the point from the original axes and the new set being the directed distances from the new axes. In order to avoid errors we propose to write the coordinates of a point when referred to the original system as, for example, $A(a, b)$ and the coordinates when referred to the new system as $A(c, d)'$. Also, we shall find it convenient at times to speak of the unprimed and primed systems.

If the axes with the origin O are translated to a new position with origin O' having coordinates (h, k) when referred to the original system and if the coordinates of any point are (x, y) before and $(x', y')'$ after the translation, then the equations of transformation are

$$(2) \qquad x = x' + h, \qquad y = y' + k.$$

Example 1. By means of a translation, transform $3x^2 + 4y^2 - 12x + 16y - 8 = 0$ into another equation which lacks terms of the first degree.

FIRST SOLUTION. When the values of x and y from (2) are substituted in the given equation, we obtain

$$3(x'+h)^2 + 4(y'+k)^2 - 12(x'+h) + 16(y'+k) - 8 = 0$$

or

$$(C) \qquad 3x'^2 + 4y'^2 + (6h - 12)x' + (8k + 16)y' + 3h^2 + 4k^2 - 12h + 16k - 8 = 0$$

The equation will lack terms of the first degree provided

$$6h - 12 = 0 \quad \text{and} \quad 8k + 16 = 0,$$

that is, provided $h = 2$ and $k = -2$. Thus, the translation

$$x = x' + 2, \qquad y = y' - 2$$

reduces the given equation to

$$3x'^2 + 4y'^2 - 36 = 0$$

The locus, an ellipse, together with the original and new system of coordinates are shown in the adjoining figure.

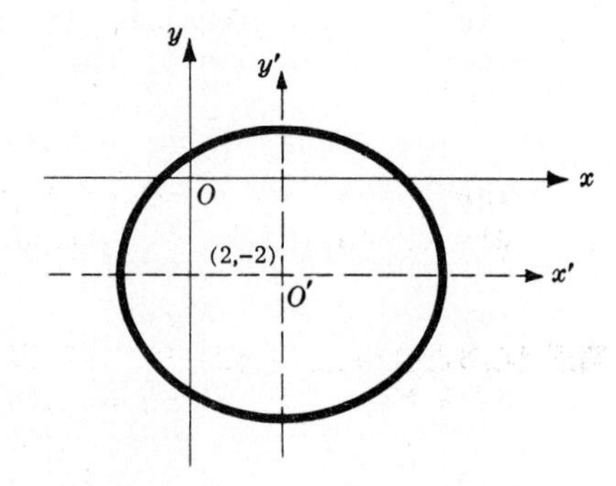

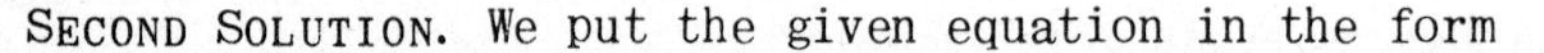

SECOND SOLUTION. We put the given equation in the form

$$3(x^2 - 4x) + 4(y^2 + 4y) = 8$$

and complete the squares to obtain

$$3(x^2 - 4x + 4) + 4(y^2 + 4y + 4) = 8 + 3(4) + 4(4) = 36$$

or

$$(D) \qquad 3(x-2)^2 + 4(y+2)^2 = 36$$

The transformation

$$x - 2 = x', \; y + 2 = y' \quad \text{or} \quad x = x' + 2, \; y = y' - 2$$

reduces (D) to $3x'^2 + 4y'^2 = 36$ as before.

See Problems 1-3.

ROTATION OF THE COORDINATE AXES. The transformation which holds the origin fixed while rotating the coordinate axes through a given angle is called a *rotation*.

If, while the origin remains fixed, the coordinate axes are rotated counterclockwise through an angle θ, and if the coordinates of any point P are (x, y) before and $(x', y')'$ after the rotation, the equations of transformation are

$$(3) \qquad \begin{aligned} x &= x' \cos\theta - y' \sin\theta \\ y &= x' \sin\theta + y' \cos\theta \end{aligned}$$

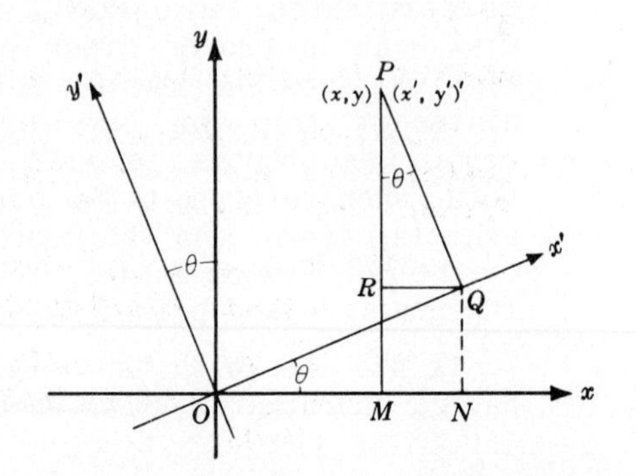

since, from the adjoining figure,

$$x = OM = ON - MN = ON - RQ = OQ\cos\theta - QP\sin\theta = x'\cos\theta - y'\sin\theta$$

and
$$y = MP = MR + RP = NQ + RP = OQ\sin\theta + QP\cos\theta = x'\sin\theta + y'\cos\theta$$

Example 2. Transform the equation

$$x^2 + \sqrt{3}\,xy + 2y^2 - 5 = 0$$

by rotating the coordinate axes through the angle 60°.

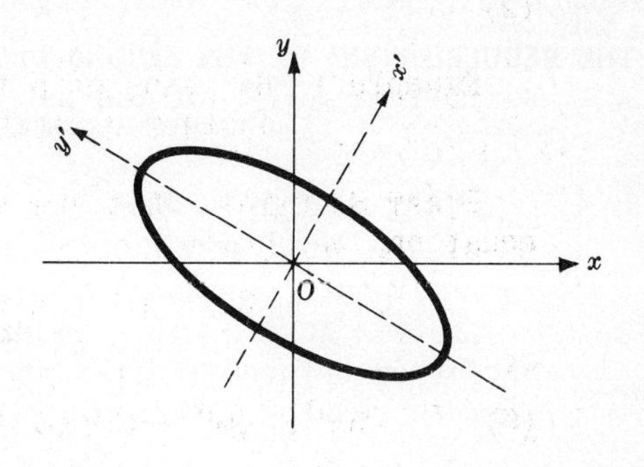

The equations of transformation are

$$x = x'\cos 60° - y'\sin 60° = \tfrac{1}{2}(x' - \sqrt{3}\,y')$$

$$y = x'\sin 60° + y'\cos 60° = \tfrac{1}{2}(\sqrt{3}\,x' + y')$$

Substituting for x and y in the given equation, we obtain

$$\tfrac{1}{4}(x' - \sqrt{3}\,y')^2 + \tfrac{1}{4}\sqrt{3}(x' - \sqrt{3}\,y')(\sqrt{3}x' + y') + \tfrac{1}{2}(\sqrt{3}\,x' + y')^2 - 5 = 0$$

$$\frac{5}{2}x'^2 + \frac{1}{2}y'^2 - 5 = 0 \qquad \text{or} \qquad 5x'^2 + y'^2 = 10$$

The locus, an ellipse, together with the original and new systems of coordinates are shown in the figure above.

See Problems 4-6.

THE SEMI-REDUCED FORM OF THE SECOND DEGREE EQUATION. Under a rotation of the coordinate axes with equations of transformation (*3*), the general equation of the second degree

$$(1) \qquad Ax^2 + 2Bxy + Cy^2 + 2Dx + 2Ey + F = 0$$

becomes

$$(1') \qquad A'x'^2 + 2B'x'y' + C'y'^2 + 2D'x' + 2E'y' + F' = 0$$

where

$$A' = A\cos^2\theta + 2B\sin\theta\cos\theta + C\sin^2\theta,$$

$$B' = (C - A)\sin\theta\cos\theta + B(\cos^2\theta - \sin^2\theta)$$

$$(4) \qquad = \tfrac{1}{2}(C - A)\sin 2\theta + B\cos 2\theta$$

$$C' = A\sin^2\theta - 2B\sin\theta\cos\theta + C\cos^2\theta,$$

$$D' = D\cos\theta + E\sin\theta, \qquad E' = E\cos\theta - D\sin\theta, \qquad F' = F.$$

If $B \neq 0$, (*1'*) will lack the term in $x'y'$ if θ is such that $0 < \theta < 90°$ and $\tan 2\theta = \dfrac{2B}{A - C}$ when $A \neq C$, and $\theta = 45°$ when $A = C$. Under this transformation, the general equation (*1*) takes the form

$$(1'') \qquad A'x'^2 + C'y'^2 + 2D'x' + 2E'y' + F' = 0$$

which will be called the *semi-reduced form* of the second degree equation.

Under *any* rotation of the coordinate axes, the quantities $A + C$ and $B^2 - AC$ are unchanged or *invariant*; that is, $A + C = A' + C'$ and $B^2 - AC = B'^2 - A'C'$. When (*1*) is transformed into (*1''*), $B' = 0$ and $B^2 - AC = -A'C'$. Then:

if $B^2 - AC < 0$, A' and C' agree in sign and (*1*) represents a real ellipse, a point ellipse or an imaginary ellipse.

if $B^2 - AC = 0$, either $A' = 0$ or $C' = 0$; now (*1''*) contains either a term in x'^2 or y'^2 (but not both), and (*1*) represents a parabola or a pair of parallel lines.

if $B^2 - AC > 0$, A' and C' differ in sign and *(1)* represents a hyperbola or a pair of intersecting lines.

THE REDUCED FORM OF THE SECOND DEGREE EQUATION. Under a suitable translation the semi-reduced form *(1″)* of the second degree equation takes the *reduced form*

$$A'x''^2 + C'y''^2 = F'', \qquad \text{when } A'C' \neq 0,$$

and

$$x''^2 = Gy'' \quad \text{or} \quad y''^2 = Hx'', \qquad \text{when } A'C' = 0.$$

If $B^2 - AC \neq 0$, it is a matter of individual preference whether the rotation is performed before or after the translation of axes; however, if $B^2 - AC = 0$, the axes *must* be rotated first.

See Problems 7-9.

SOLVED PROBLEMS

1. If the equations of translation are $x = x' - 3$, $y = y' + 4$, find
(*a*) the coordinates of $O(0,0)$ when referred to the primed system of coordinates,
(*b*) the coordinates of $O'(0,0)'$ when referred to the unprimed system,
(*c*) the coordinates of $P(5,-3)$ when referred to the primed system,
(*d*) the coordinates of $P(5,-3)'$ when referred to the unprimed system,
(*e*) the equation of l: $2x - 3y + 18 = 0$ when referred to the primed system.

(*a*) For $x = 0$, $y = 0$ the equations of transformation yield $x' = 3$, $y' = -4$; thus, in the primed system we have $O(3,-4)'$.

(*b*) For $x' = 0$, $y' = 0$ the equations of transformation yield $x = -3$, $y = 4$; thus, in the unprimed system we have $O'(-3,4)$.

(*c*) Here $x = 5$, $y = -3$; then $x' = 8$, $y' = -7$. Thus, we have $P(8,-7)'$.

(*d*) Here $x' = 5$, $y' = -3$; then $x = 2$, $y = 1$. Thus, we have $P(2,1)$.

(*e*) When the values for x and y are substituted in the given equation, we have
$$2(x'-3) - 3(y'+4) + 18 = 2x' - 3y' = 0$$
as the equation of l in the primed system. Note that the new origin was chosen on the line l.

2. Transform each of the following equations into another lacking terms of the first degree:
(*a*) $x^2 + 4y^2 - 2x - 12y + 1 = 0$, (*b*) $9x^2 - 16y^2 - 36x - 96y - 252 = 0$, (*c*) $xy + 4x - y - 8 = 0$.

(*a*) Since the given equation lacks a term in xy, we use the second method of Example 1. We have
$$(x^2 - 2x) + 4(y^2 - 3y) = -1, \qquad (x^2 - 2x + 1) + 4(y^2 - 3y + 9/4) = -1 + 1 + 4(9/4) = 9$$
and
$$(x-1)^2 + 4(y - 3/2)^2 = 9.$$
This equation takes the form $x'^2 + 4y'^2 = 9$ under the transformation
$$x - 1 = x', \ y - 3/2 = y' \qquad \text{or} \qquad x = x' + 1, \ y = y' + 3/2.$$

(*b*) We have $9(x^2 - 4x) - 16(y^2 + 6y) = 252$, $9(x^2 - 4x + 4) - 16(y^2 + 6y + 9) = 252 + 36 - 144 = 144$,
and
$$9(x-2)^2 - 16(y+3)^2 = 144.$$
This equation takes the form $9x'^2 - 16y'^2 = 144$ under the transformation
$$x - 2 = x', \ y + 3 = y' \qquad \text{or} \qquad x = x' + 2, \ y = y' - 3.$$

(c) Since the given equation contains a term in xy, we must use the first method of Example 1. We have, using the equations (2),

$$(x'+h)(y'+k) + 4(x'+h) - (y'+k) - 8 = x'y' + (k+4)x' + (h-1)y' + hk + 4h - k - 8 = 0$$

The first degree terms will disappear provided $k+4 = 0$ and $h-1 = 0$; that is, provided we take $h = 1$ and $k = -4$. For this choice, the equation becomes $x'y' - 4 = 0$.

3. By a translation of the axes, simplify each of the following:
(a) $x^2 + 6x - 4y + 1 = 0$, (b) $y^2 + 4y + 8x - 2 = 0$.

(a) Since the given equation lacks a term in y^2, it is not possible to use the second method of Example 1. Using the transformation (2), we find

$$(x'+h)^2 + 6(x'+h) - 4(y'+k) + 1 = x'^2 + 2(h+3)x' - 4y' + h^2 + 6h - 4k + 1 = 0$$

If we take $h = -3$, the term in x' disappears but it is clear that we cannot make the term in y' disappear. However, in this case, we may make the constant term

$$h^2 + 6h - 4k + 1 = (-3)^2 + 6(-3) - 4k + 1 = -8 - 4k$$

disappear by taking $k = -2$. Thus, the transformed equation becomes

$$x'^2 - 4y' = 0 \qquad \text{or} \qquad x'^2 = 4y'$$

It is now clear that this simplification may be effected by the following variation of the second method of Example 1:

$$x^2 + 6x = 4y - 1, \qquad x^2 + 6x + 9 = 4y - 1 + 9 = 4y + 8$$

or

$$(x+3)^2 = 4(y+2)$$

Then the transformation $x + 3 = x'$, $y + 2 = y'$ or $x = x' - 3$, $y = y' - 2$

reduces the equation to $x'^2 = 4y'$

(b) We have $y^2 + 4y = -8x + 2$, $y^2 + 4y + 4 = -8x + 2 + 4 = -8x + 6$, $(y+2)^2 = -8(x - 3/4)$,

and finally $y'^2 = -8x'$ under the transformation $x = x' + 3/4$, $y = y' - 2$.

4. Write the equations of transformation for a rotation of the coordinate axes through an angle of 45° and use them to find:
(a) the coordinates of $P(\sqrt{2}, 3\sqrt{2})'$ when referred to the original (unprimed) system,
(b) the coordinates of $O(0,0)$ when referred to the new (primed) system,
(c) the coordinates of $P(\sqrt{2}, 3\sqrt{2})$ when referred to the primed system,
(d) the equation of the line l: $x + y + 3\sqrt{2} = 0$ when referred to the primed system,
(e) the equation of the line l: $3x - 3y + 4 = 0$ when referred to the primed system.

The equations of transformation are

$$x = x' \cos 45° - y' \sin 45° = \frac{1}{\sqrt{2}}(x' - y')$$

$$y = x' \sin 45° + y' \cos 45° = \frac{1}{\sqrt{2}}(x' + y')$$

(a) For $x' = \sqrt{2}$, $y' = 3\sqrt{2}$, the equations of transformation yield

$$x = \frac{1}{\sqrt{2}}(\sqrt{2} - 3\sqrt{2}) = -2, \qquad y = \frac{1}{\sqrt{2}}(\sqrt{2} + 3\sqrt{2}) = 4.$$

Thus, in the unprimed system the coordinates are $P(-2, 4)$.

(b) When the equations of transformation are solved for x' and y', we have

$$x' = \frac{1}{\sqrt{2}}(x + y), \qquad y' = -\frac{1}{\sqrt{2}}(x - y).$$

For $x = 0$, $y = 0$ these equations yield $x' = 0$, $y' = 0$; thus, in the primed system, we have $O(0,0)'$. Since the coordinates are unchanged, the origin is called an *invariant point* of the transformation.

(c) For $x = \sqrt{2}$, $y = 3\sqrt{2}$ the equations of (b) yield $x' = 4$, $y' = 2$. In the primed system, we have $P(4,2)'$.

(d) When the values for x and y from the equations of transformation are substituted in the given equation of the line, we have

$$\frac{1}{\sqrt{2}}(x'-y') + \frac{1}{\sqrt{2}}(x'+y') + 3\sqrt{2} = \frac{2}{\sqrt{2}}x' + 3\sqrt{2} = 0 \qquad \text{or} \qquad x'+3 = 0.$$

Note that the x'-axis is perpendicular to the given line.

(e) Here $\dfrac{3}{\sqrt{2}}(x'-y') - \dfrac{3}{\sqrt{2}}(x'+y') + 4 = -\dfrac{6}{\sqrt{2}}y' + 4 = 0$ or $3y' - 2\sqrt{2} = 0$.

Note that the x'-axis is parallel to the given line.

5. Transform the equation $2x^2 - 4xy + 5y^2 - 18x + 12y - 24 = 0$ by rotating the coordinate axes through the angle θ where $\sin\theta = 1/\sqrt{5}$ and $\cos\theta = 2/\sqrt{5}$.

The equations of transformation are $x = \dfrac{1}{\sqrt{5}}(2x'-y')$, $y = \dfrac{1}{\sqrt{5}}(x'+2y')$ and when the values of x and y are substituted in the given equation, we find

$$x'^2 + 6y'^2 - \frac{24}{\sqrt{5}}x' + \frac{42}{\sqrt{5}}y' - 24 = 0$$

Thus, the effect of the transformation is to produce an equation in which the cross-product term $x'y'$ is missing.

6. After a rotation of axes with equations of transformation

$$x = \frac{1}{13}(12x'-5y'), \qquad y = \frac{1}{13}(5x'+12y')$$

followed by a translation with equations of transformation

$$x' = x'' + 16/13, \qquad y' = y'' - 63/13$$

a certain equation of the second degree is reduced to $y''^2 = -8x''$. Sketch the locus, showing each set of coordinate axes.

In order to distnguish between the three coordinate systems, we shall use the term *unprimed* for the original system, *primed* for the system after the rotation, and *double-primed* for the system after the translation. We begin with the original (unprimed) axes in the usual position.

Now any pair of numbers, as (12,5), which are proportional respectively to $\cos\theta = 12/13$ and $\sin\theta = 5/13$ are the coordinates of a point on the x'-axis. With the x' and y'-axes in position, we next seek the origin O'' of the double-primed system. Using first the equations of translation and then the equations of rotation, we find

$$O''(0,0)'' = O''(16/13, -63/13)' = O''(3,-4).$$

Locating $O''(3,-4)$, with reference to the original system, we draw the x''- and y''-axes through O'' parallel to the x'- and y'-axes. Finally, on this latter set of axes, we sketch the parabola $y''^2 = -8x''$.

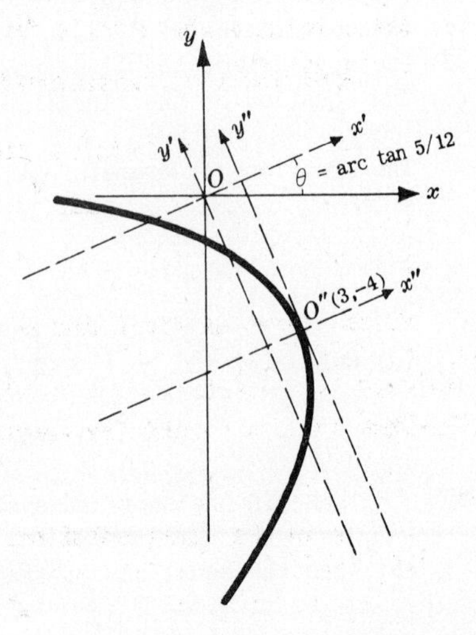

In Problems 7-9 determine the nature of the locus, obtain the reduced form of the equation, and sketch the locus showing all sets of coordinate axes.

7. $20x^2 - 24xy + 27y^2 + 24x - 54y - 369 = 0$

Since $B^2 - AC = (-12)^2 - 20\cdot 27 < 0$, the locus is an ellipse.

FIRST SOLUTION. The angle θ through which the axes must be rotated to eliminate the term in xy is given by

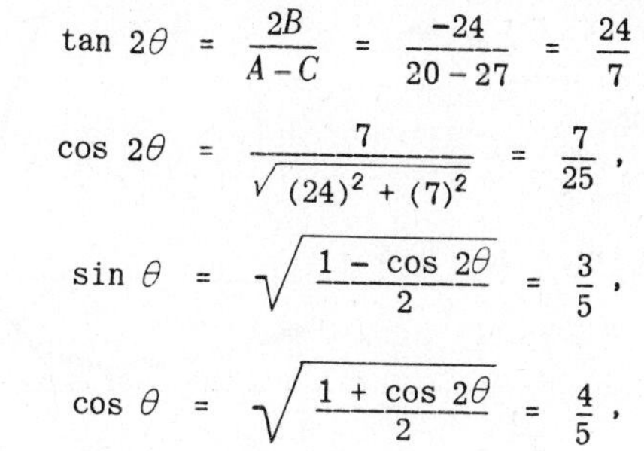

$$\tan 2\theta = \frac{2B}{A-C} = \frac{-24}{20-27} = \frac{24}{7}$$

Then $$\cos 2\theta = \frac{7}{\sqrt{(24)^2 + (7)^2}} = \frac{7}{25},$$

$$\sin\theta = \sqrt{\frac{1-\cos 2\theta}{2}} = \frac{3}{5},$$

$$\cos\theta = \sqrt{\frac{1+\cos 2\theta}{2}} = \frac{4}{5},$$

and the equations of rotation are

$$x = \frac{1}{5}(4x' - 3y'), \quad y = \frac{1}{5}(3x' + 4y')$$

When this transformation is applied to the given equation, we find

$$11x'^2 + 36y'^2 - \frac{66}{5}x' - \frac{288}{5}y' - 369 = 0 \quad \text{as the semi-reduced form.}$$

Completing squares, we have

$$11(x'^2 - \frac{6}{5}x' + 9/25) + 36(y'^2 - \frac{8}{5}y' + 16/25) = 369 + 11(\frac{9}{25}) + 36(\frac{16}{25}) = 396$$

or $$11(x' - \frac{3}{5})^2 + 36(y' - \frac{4}{5})^2 = 396$$

The translation $x' = x'' + 3/5$, $y' = y'' + 4/5$ gives the reduced form $11x''^2 + 36y''^2 = 396$.

The x'-axis passes through the point (4,3) and the coordinates of the new origin are

$$O''(0,0)'' = O''(3/5, 4/5)' = O''(0,1)$$

SECOND SOLUTION. Since some may prefer to eliminate the first degree terms before rotating the axes, we give the details for this locus.

Applying the transformation (2) to the given equation, we obtain

$$20(x'+h)^2 - 24(x'+h)(y'+k) + 27(y'+k)^2 + 24(x'+h) - 54(y'+k) - 369$$
$$= 20x'^2 - 24x'y' + 27y'^2 + (40h - 24k + 24)x' - (24h - 54k + 54)y'$$
$$+ 20h^2 - 24hk + 27k^2 + 24h - 54k - 369 = 0.$$

If the terms of first degree are to disappear, h and k must be chosen so that

$$40h - 24k + 24 = 0, \quad 24h - 54k + 54 = 0.$$

Then $h = 0$, $k = 1$ and the equation of the locus becomes

$$20x'^2 - 24x'y' + 27y'^2 - 396 = 0.$$

As in the first solution, the equations of rotation to eliminate the term in $x'y'$ are

$$x' = \frac{1}{5}(4x'' - 3y''), \quad y' = \frac{1}{5}(3x'' + 4y'').$$

When this transformation is made, we have

$$\frac{20}{25}(4x''-3y'')^2 - \frac{24}{25}(4x''-3y'')(3x''+4y'') + \frac{27}{25}(3x''+4y'')^2 - 396 = 0$$

or $$11x''^2 + 36y''^2 = 396, \quad \text{as before.}$$

8. $x^2 + 2xy + y^2 + 10\sqrt{2}\,x - 2\sqrt{2}\,y + 8 = 0$.

Here $B^2 - AC = 1 - 1\cdot 1 = 0$; the locus is either a parabola or a pair of parallel lines. Since $A = C = 1$, we rotate the axes through the angle $\theta = 45°$ to obtain the semi-reduced form. When the transformation

$$x = \frac{1}{\sqrt{2}}(x'-y'), \qquad y = \frac{1}{\sqrt{2}}(x'+y')$$

is applied to the given equation, we find

$$x'^2 + 4x' - 6y' + 4 = (x'+2)^2 - 6y' = 0$$

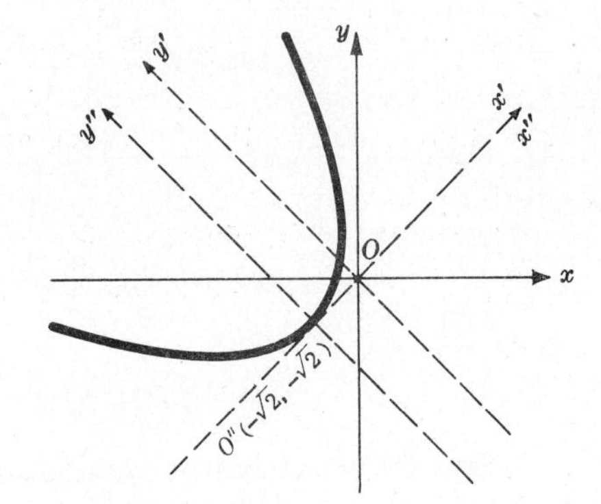

Then the translation $x' = x'' - 2$, $y' = y''$ produces the reduced form $x''^2 - 6y'' = 0$.

The locus is a parabola. The x'- axis passes through the point $(1,1)$ and

$$O''(0,0)'' = O''(-2,0)' = O''(-\sqrt{2}, -\sqrt{2}).$$

9. $27x^2 + 120xy + 77y^2 + 234x + 858y + 117 = 0$.

Here $B^2 - AC = (60)^2 - 27\cdot 77 > 0$; the locus is a hyperbola or a pair of intersecting lines.

From $\tan 2\theta = 120/(27-77) = -12/5$, we have

$$\cos 2\theta = -5/13; \text{ then}$$

$$\sin\theta = \sqrt{\tfrac{1}{2}(1+5/13)} = 3/\sqrt{13},$$

$$\cos\theta = \sqrt{\tfrac{1}{2}(1-5/13)} = 2/\sqrt{13},$$

and the equations of rotation are

$$x = \frac{1}{\sqrt{13}}(2x'-3y'), \qquad y = \frac{1}{\sqrt{13}}(3x'+2y').$$

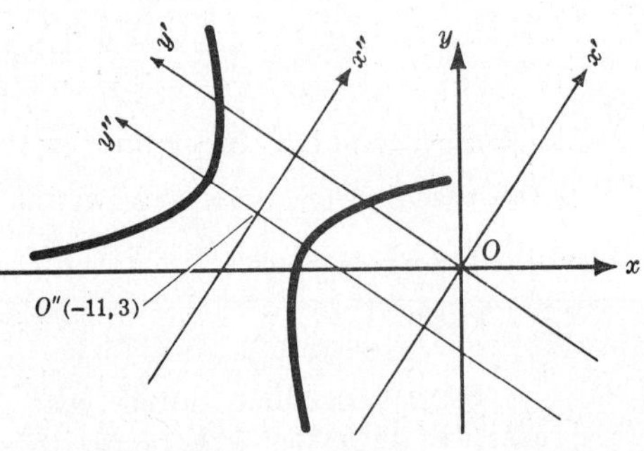

Applying the transformation, we obtain

$$9x'^2 - y'^2 + 18\sqrt{13}\,x' + 6\sqrt{13}\,y' + 9 = 0.$$

Completing the squares, we have $9(x'+\sqrt{13})^2 - (y'-3\sqrt{13})^2 + 9 = 0$ which after the translation $x' = x'' - \sqrt{13}$, $y' = y'' + 3\sqrt{13}$ becomes

$$9x''^2 - y''^2 + 9 = 0 \quad \text{or} \quad y''^2 - 9x''^2 = 9.$$

The locus is a hyperbola. The x'- axis passes through the point $(2,3)$ and

$$O''(0,0)'' = O''(-\sqrt{13}, 3\sqrt{13})' = O''(-11,3).$$

SUPPLEMENTARY PROBLEMS

10. If the equations of translation are $x = x' + 2$, $y = y' - 5$, find:
(a) the coordinates of $O(0,0)$ when referred to the primed system;
(b) the coordinates of $P(-2,4)$ when referred to the primed system;
(c) the coordinates of $P(-2,4)'$ when referred to the unprimed system;
(d) the equation of l: $5x + 2y = 0$ when referred to the primed system;
(e) the equation of l: $x - 2y + 4 = 0$ when referred to the primed system.
Ans. (a) $(-2,5)'$ (b) $(-4,9)'$ (c) $(0,-1)$ (d) $5x' + 2y' = 0$ (e) $x' - 2y' + 16 = 0$

11. Simplify each of the following equations by a suitable translation. Draw the figure showing both sets of axes.
(a) $4x^2 + y^2 - 16x + 6y - 11 = 0$ (c) $9x^2 - 4y^2 - 36x + 48y - 72 = 0$ (e) $16y^2 + 5x + 32y + 6 = 0$
(b) $9x^2 - 4y^2 - 36x + 48y - 144 = 0$ (d) $x^2 - 12x - 8y - 4 = 0$ (f) $xy - 4x + 3y + 24 = 0$
Ans. (a) $4x'^2 + y'^2 = 36$ (c) $4y'^2 - 9x'^2 = 36$ (e) $16y'^2 = -5x'$
(b) $9x'^2 - 4y'^2 = 36$ (d) $x'^2 - 8y' = 0$ (f) $x'y' + 36 = 0$

12. Simplify each of the following equations by rotating the axes through the indicated angle. Draw the figure showing both sets of axes.
(a) $x^2 - y^2 = 16$; $45°$ (c) $16x^2 + 24xy + 9y^2 + 60x - 80y = 0$; Arc cos 4/5
(b) $9x^2 + 24xy + 16y^2 = 25$; Arc cos 3/5 (d) $31x^2 - 24xy + 21y^2 = 39$; Arc cos $2/\sqrt{13}$
Ans. (a) $x'y' + 8 = 0$ (b) $x'^2 = 1$ (c) $x'^2 = 4y'$ (d) $x'^2 + 3y'^2 = 3$

13. Simplify each equation by suitable transformations and draw a figure showing all sets of axes.
(a) $3x^2 + 2xy + 3y^2 - 8x + 16y + 30 = 0$ (d) $108x^2 - 312xy + 17y^2 + 750y + 225 = 0$
(b) $13x^2 + 12xy - 3y^2 - 15x - 15y = 0$ (e) $16x^2 + 24xy + 9y^2 - 60x - 170y - 175 = 0$
(c) $25x^2 - 120xy + 144y^2 + 1300x + 1274y - 2704 = 0$ (f) $37x^2 + 32xy + 13y^2 - 42\sqrt{5}\,x - 6\sqrt{5}\,y = 0$
Ans. (a) $2x''^2 + y''^2 = 4$ (c) $y''^2 = -10x''$ (e) $x''^2 = 4y''$
(b) $30x''^2 - 10y''^2 = 3$ (d) $4x''^2 - 9y''^2 = 36$ (f) $9x''^2 + y''^2 = 18$

14. Apply the equations of transformation (2) directly to (1) and show that the first degree terms may be made to disappear provided $B^2 - AC \neq 0$.

15. Use (3) to show: (a) $A' + C' = A + C$, (b) $B'^2 - A'C' = B^2 - AC$.

16. Solve (1) to obtain $x = \dfrac{-(By + D) \pm \sqrt{H}}{A}$, where $H = (B^2 - AC)y^2 + 2(BD - AE)y + D^2 - AF$. Show that H is a perfect square when $\Delta = \begin{vmatrix} A & B & D \\ B & C & E \\ D & E & F \end{vmatrix} = 0$. Thus, prove that (1) represents a degenerate locus if and only if $\Delta = 0$.

17. Prove that Δ of Problem 16 is invariant under translation and rotation of the axes.

CHAPTER 36

Polar Coordinates

IN THE POLAR COORDINATE SYSTEM a point in the plane is located by giving its position relative to a fixed point and a fixed half-line (direction) through the fixed point. The fixed point O (see Fig. 1) is called the *pole* and the fixed half-line OA is called the *polar axis*.

Let θ denote the smallest positive angle measured counterclockwise in degrees or radians from OA to OB, and let r denote the (positively) directed distance OP. Then P is uniquely determined when r and θ are known. These two measures constitute the *polar coordinates* of P and we write $P(r,\theta)$. The quantity r is called the *radius vector* and θ is called the *vectorial angle* of P. Note that a positive direction, indicated by the arrow has been assigned on the half-line OB.

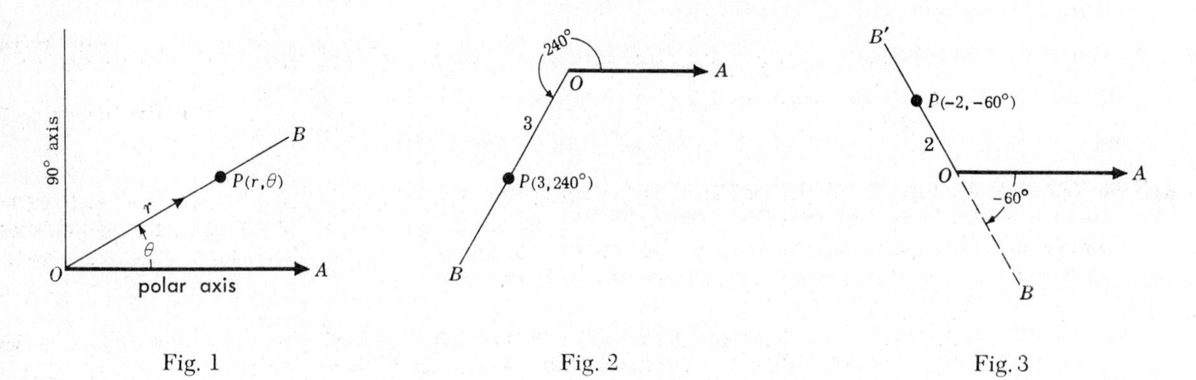

Fig. 1 Fig. 2 Fig. 3

Example 1. Locate the point $P(3,240°)$ or $P(3,4\pi/3)$. Refer to Fig. 2 above.

> Lay off the vectorial angle $\theta = \angle AOB = 240°$, measured counterclockwise from OA, and on OB locate P such that $r = OP = 3$.

In the paragraph above we have restricted r and θ so that $r \geq 0$ and $0° \leq \theta < 360°$. In general, these restrictions will be observed; however, at times it will be more convenient to permit r and θ to have positive or negative values. If θ is negative and r is positive, we lay off the angle $\theta = \angle AOB$, measured clockwise from OA, and locate P on OB so that $OP = r$. If r is negative, we lay off $\theta = \angle AOB$, extend OB through the pole to B', and locate P on OB' a distance $|r|$ from O.

Example 2. Locate the point $P(-2,-60°)$ or $P(-2,-\pi/3)$. Refer to Fig. 3 above.

> Lay off the vectorial angle $\theta = \angle AOB = -60°$, measured clockwise from OA, extend OB through the pole to B', and on OB' locate P a distance 2 units from O.

See Problems 1-2.

Although not a part of the polar system, it will be helpful at times to make use of a half-line, called the 90° axis, which issues from the pole perpendicular to the polar axis.

TRANSFORMATIONS BETWEEN POLAR AND RECTANGULAR COORDINATES. If the pole and polar axis of the polar system coincide respectively with the origin and positive x-axis of the rectangular system, and if P has rectangular coordinates (x, y) and polar coordinates (r, θ), then the following relations hold:

(1) $x = r \cos \theta$ (4) $\theta = \arctan y/x$

(2) $y = r \sin \theta$ (5) $\sin \theta = y/r$ and $\cos \theta = x/r$

(3) $r = \sqrt{x^2 + y^2}$

If relations (3)-(5) are to yield the restricted set of coordinates of the section above, θ is to be taken as the smallest positive angle satisfying (5) or, what is equivalent, θ is the smallest positive angle satisfying (4) and terminating in the quadrant in which $P(x, y)$ lies.

Example 3. Find the rectangular coordinates of $P(3, 300°)$.

Here $r = 3$ and $\theta = 300°$; then $x = r \cos\theta = 3 \cos 300° = 3(\frac{1}{2}) = 3/2$, $y = r \sin \theta = 3 \sin 300° = 3(-\frac{1}{2}\sqrt{3}) = -3\sqrt{3}/2$, and the rectangular coordinates are $(3/2, -3\sqrt{3}/2)$.

Example 4. Find the polar equation of the circle whose rectangular equation is $x^2 + y^2 - 8x + 6y - 2 = 0$.

Since $x = r \cos \theta$, $y = r \sin \theta$, and $x^2 + y^2 = r^2$, the polar equation is $r^2 - 8r \cos \theta + 6r \sin \theta - 2 = 0$.

See Problems 3-5.

CURVE TRACING IN POLAR COORDINATES. Preliminary to sketching the locus of a polar equation, we discuss symmetry, extent, etc., as in the case of rectangular equations. However, there are certain complications at times due to the fact that in polar coordinates a given curve may have more than one equation.

Example 5. Let $P(r, \theta)$ be an arbitrary point on the curve $r = 4 \cos \theta - 2$. Now P has other representations: $(-r, \theta + \pi)$, $(-r, \theta - \pi)$, $(r, \theta - 2\pi)$, ...

Since (r, θ) satisfies the equation $r = 4 \cos \theta - 2$, $(-r, \theta + \pi)$ satisfies the equation

$-r = 4 \cos(\theta + \pi) - 2 = -4 \cos\theta - 2$ or $r = 4 \cos\theta + 2$.

Thus, $r = 4\cos\theta - 2$ and $r = 4\cos\theta + 2$ are equations of the same curve. Such equations are called *equivalent*. The reader will show that $(-r, \theta - \pi)$ satisfies $r = 4\cos\theta + 2$ and $(r, \theta - 2\pi)$ satisfies $r = 4\cos\theta - 2$.

Example 6. Show that point $A(-1, \pi/6)$ is on the ellipse $r = \dfrac{3}{4 + 2\sin\theta}$.

Note that the given coordinates do not satisfy the given equation.

FIRST SOLUTION. Another set of coordinates for A is $(1, 7\pi/6)$. Since these coordinates satisfy the equation, A is on the ellipse.

SECOND SOLUTION. An equivalent equation for the ellipse is

$$-r = \frac{3}{4 + 2\sin(\theta - \pi)} \quad \text{or} \quad r = \frac{-3}{4 - 2\sin\theta}.$$

Since the given coordinates satisfy this equation, A is on the ellipse.

SYMMETRY. A locus is symmetric with respect to the polar axis if an equivalent equation is obtained when

(a) θ is replaced by $-\theta$, or

(b) θ is replaced by $\pi - \theta$ and r by $-r$ in the given equation.

A locus is symmetric with respect to the 90° axis if an equivalent equation is obtained when
(*a*) θ is replaced by $\pi - \theta$, or
(*b*) θ is replaced by $-\theta$ and r by $-r$ in the given equation.

A locus is symmetric with respect to the pole if an equivalent equation is obtained when
(*a*) θ is replaced by $\pi + \theta$, or
(*b*) r is replaced by $-r$ in the given equation.

EXTENT. The locus whose polar equation is $r = f(\theta)$ is a closed curve if r is real and finite for all values of θ, but is not a closed curve if there are values of one variable which make the other become infinite.

The equation should also be examined for values of one variable which make the other imaginary.

At times, as in the equation $r = a(1 + \sin\theta)$, the values of θ which give r its maximum values can be readily determined. Since the maximum value of $\sin\theta$ is 1, the maximum value of r is $2a$ which it assumes when $\theta = \frac{1}{2}\pi$.

DIRECTIONS AT THE POLE. Unlike all other points, the pole has infinitely many pairs of coordinates $(0,\theta)$ when θ is restricted to $0° \le \theta < 360°$. While two such pairs $(0,\theta_1)$ and $(0,\theta_2)$ define the pole, they indicate different directions (measured from the polar axis) there. Thus, the values of θ for which $r = f(\theta) = 0$ give the directions of the tangents to the locus $r = f(\theta)$ at the pole.

POINTS ON THE LOCUS. We may find as many points on a locus as desired by assigning values to θ in the given equation and solving for the corresponding values of r.

Example 7. Discuss and sketch the locus of the cardiod $r = a(1 - \sin\theta)$.

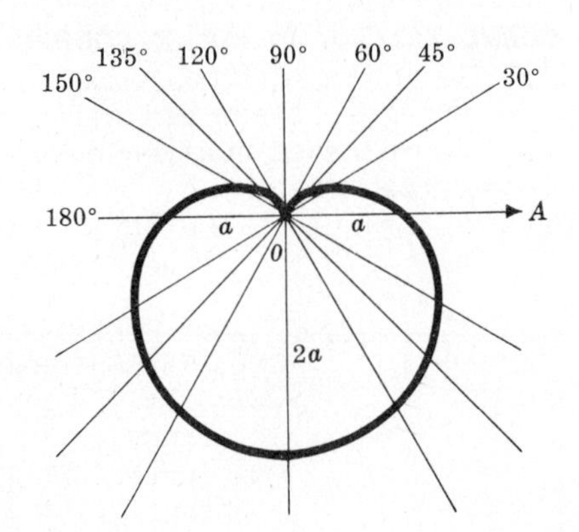

Symmetry. An equivalent equation is obtained when θ is replaced by $\pi - \theta$; the locus is symmetric with respect to the 90° axis.

Extent. Since r is real and $\le 2a$ for all values of θ, the locus is a closed curve, lying within a circle of radius $2a$ with center at the pole. Since $\sin\theta$ is of period 2π, the complete locus is described as θ varies from 0 to 2π.

Directions at the pole. When $r = 0$, $\sin\theta = 1$ and $\theta = \frac{1}{2}\pi$. Thus, the locus is tangent to the 90° axis at the pole.

After locating the following points

θ	$\frac{1}{2}\pi$	$2\pi/3$	$3\pi/4$	$5\pi/6$	π	$7\pi/6$	$5\pi/4$	$4\pi/3$	$3\pi/2$
r	0	$.13a$	$.29a$	$.5a$	a	$1.5a$	$1.71a$	$1.87a$	$2a$

and making use of symmetry of the locus with respect to the 90° axis, we obtain the required curve as shown in the above figure.

See Problems 11-17.

INTERSECTIONS OF POLAR CURVES. It is to be expected that in finding the points of intersection of two curves with polar equations $r = f_1(\theta)$ and $r = f_2(\theta)$, we set $f_1(\theta) = f_2(\theta)$ and solve for θ. However, because of the multiplicity of representations both of the coordinates of a point and the equation of a curve, this procedure will fail at times to account for all of the intersections. Thus, it is a better policy to determine from a figure the exact number of intersections before attempting to find them.

Example 8. Since each of the circles $r = 2\sin\theta$ and $r = 2\cos\theta$ passes through the pole, the circles intersect in the pole and in one other point. Since each locus is completely described on the interval 0 to π, we set $2\sin\theta = 2\cos\theta$ and solve for θ on this interval. The solution $\theta = \frac{1}{4}\pi$ yields the point $(\sqrt{2}, \frac{1}{4}\pi)$.

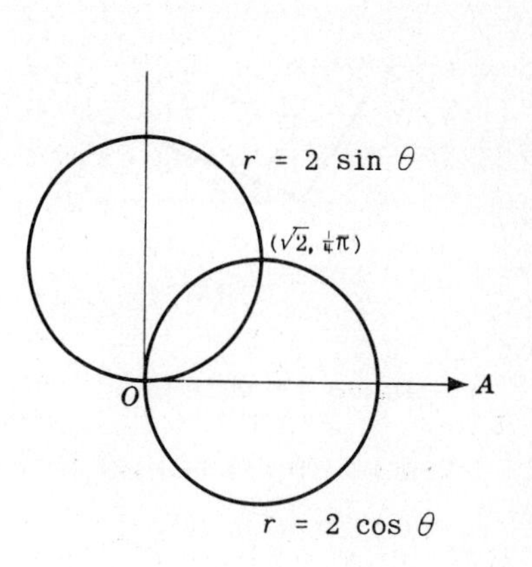

Analytically we may determine whether or not the pole is a point of intersection by setting $r = 0$ in each of the equations and solving for θ. Setting $\sin\theta = 0$ we find $\theta = 0$, and setting $\cos\theta = 0$ we find $\theta = \frac{1}{2}\pi$. Since both equations have solutions, the pole is a point of intersection. The procedure above did not yield this solution since the coordinates of the pole (0,0) satisfy $r = 2\sin\theta$ while the coordinates $(0, \frac{1}{2}\pi)$ satisfy $r = 2\cos\theta$.

See Problems 18-19.

SOLVED PROBLEMS

1. Locate the following points and determine which coincide with $P(2,150°)$ and which with $Q(2,30°)$:
(a) $A(2,750°)$ (b) $B(-2,-30°)$ (c) $C(-2,330°)$ (d) $D(-2,-150°)$ (e) $E(2,-210°)$

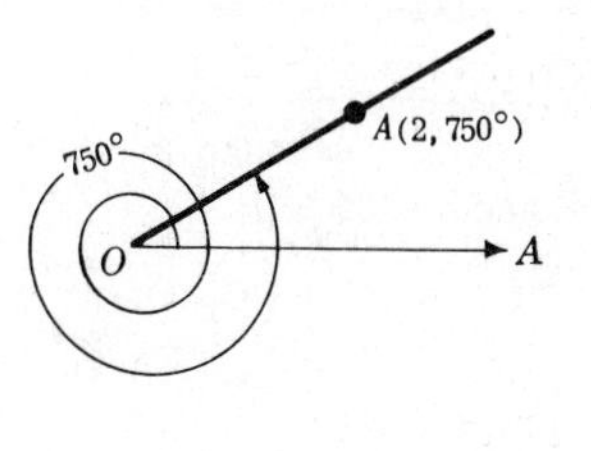

(a)

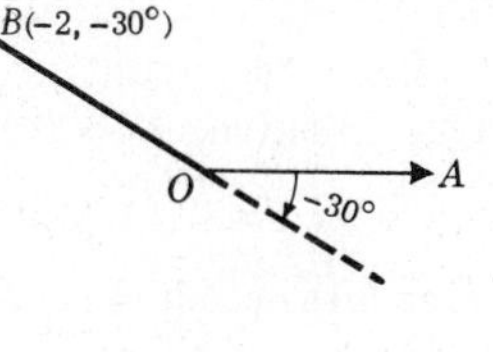

(b)

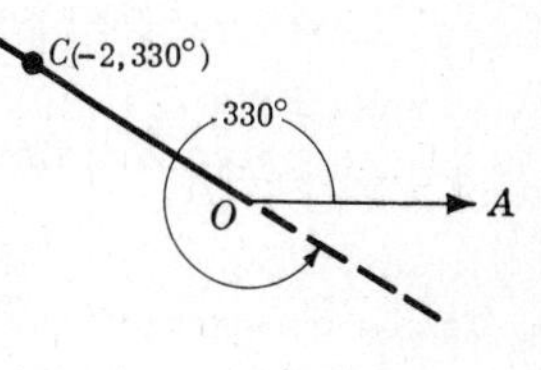

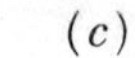

(c)

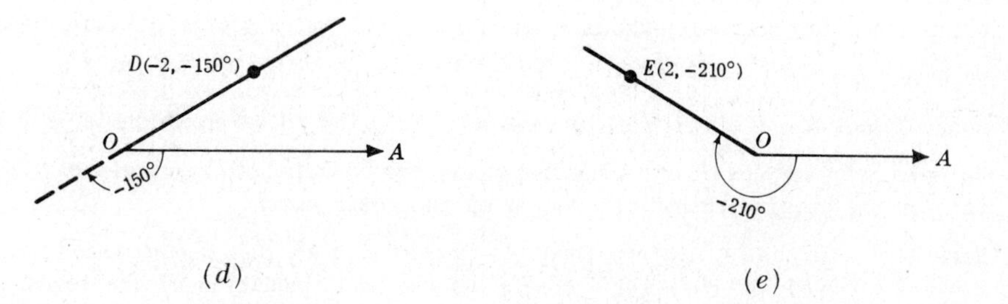

(d) (e)

The points B, C, and E coincide with P; the points A and D coincide with Q.

2. Find the distance between the points
(*a*) $P_1(5, 20°)$ and $P_2(3, 140°)$, (*b*) $P_1(4, 50°)$ and $P_2(3, 140°)$, (*c*) $P_1(r_1, \theta_1)$ and $P_2(r_2, \theta_2)$.

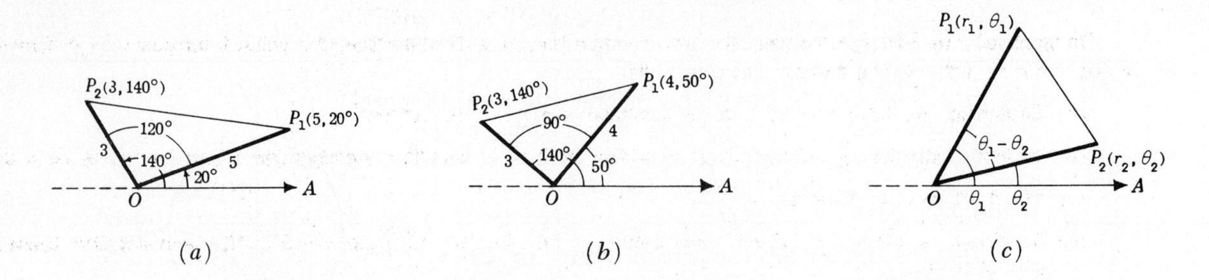

(*a*) (*b*) (*c*)

In any triangle OP_1P_2, $(P_1P_2)^2 = (OP_1)^2 + (OP_2)^2 - 2(OP_1)(OP_2) \cos \angle P_1OP_2$.

(*a*) From Fig. (*a*), $(P_1P_2)^2 = (5)^2 + (3)^2 - 2 \cdot 5 \cdot 3 \cos 120° = 49$; hence, $P_1P_2 = 7$.

(*b*) From Fig. (*b*), $P_1P_2 = \sqrt{(4)^2 + (3)^2 - 2 \cdot 4 \cdot 3 \cos 90°} = 5$.

(*c*) From Fig. (*c*), $P_1P_2 = \sqrt{r_1^2 + r_2^2 - 2r_1r_2 \cos(\theta_1 - \theta_2)}$.

3. Find the set of polar coordinates, satisfying $r \geq 0$, $0° \leq \theta < 360°$, of P whose rectangular coordinates are (*a*) $(2, -2\sqrt{3})$, (*b*) (a, a), (*c*) $(-3, 0)$, (*d*) $(0, 2)$. Find two other sets of polar coordinates for each point.

(*a*) We have $r = \sqrt{x^2 + y^2} = \sqrt{(2)^2 + (-2\sqrt{3})^2} = 4$ and $\theta = \arctan y/x = \arctan(-\sqrt{3})$. Since the point is in the fourth quadrant, we take $\theta = 300°$. The polar coordinates are $(4, 300°)$ or $(4, 5\pi/3)$. Equivalent sets of polar coordinates are $(4, -60°)$ and $(-4, 2\pi/3)$.

(*b*) Here $r = \sqrt{a^2 + a^2} = a\sqrt{2}$ and $\theta = \arctan 1$, when $a > 0$. Since the point is in the first quadrant, we take $\theta = \frac{1}{4}\pi$. The polar coordinates are $(a\sqrt{2}, \frac{1}{4}\pi)$. Equivalent sets are $(a\sqrt{2}, -7\pi/4)$ and $(-a\sqrt{2}, -3\pi/4)$.

(*c*) Here $r = \sqrt{(-3)^2 + (0)^2} = 3$. Since the point is on the negative x-axis, we take $\theta = \pi$ and the polar coordinates are $(3, \pi)$. Equivalent sets are $(-3, 0)$ and $(3, -\pi)$.

(*d*) Here $r = \sqrt{(0)^2 + (2)^2} = 2$. Since the point is on the positive y-axis, we take $\theta = \pi/2$ and the polar coordinates are $(2, \pi/2)$. Equivalent sets are $(2, -3\pi/2)$ and $(-2, 3\pi/2)$.

4. Transform each of the following rectangular equations into their polar form:

(*a*) $x^2 + y^2 = 25$ (*c*) $3x - y = 0$ (*e*) $(x^2 + y^2 - ax)^2 = a^2(x^2 + y^2)$
(*b*) $x^2 - y^2 = 4$ (*d*) $x^2 + y^2 = 4x$ (*f*) $x^3 + xy^2 + 6x^2 - 2y^2 = 0$

We make use of the transformation: $x = r \cos\theta$, $y = r \sin\theta$, $x^2 + y^2 = r^2$.

(*a*) By direct substitution we obtain $r^2 = 25$ or $r = \pm 5$. Now $r = 5$ and $r = -5$ are equivalent equations since they represent the same locus, a circle with center at the origin and radius 5.

(*b*) We have $(r\cos\theta)^2 - (r\sin\theta)^2 = r^2(\cos^2\theta - \sin^2\theta) = r^2 \cos 2\theta = 4$.

(*c*) Here $3r\cos\theta - r\sin\theta = 0$ or $\tan\theta = 3$. The polar equation is $\theta = \arctan 3$.

(*d*) We have $r^2 = 4r\cos\theta$ or $r = 4\cos\theta$ as the equation of the circle of radius 2 which passes through the origin and has its center on the polar axis.

(*e*) Here $(r^2 - ar\cos\theta)^2 = a^2r^2$; then $(r - a\cos\theta)^2 = a^2$ and $r - a\cos\theta = \pm a$. Thus we may take $r = a(1 + \cos\theta)$ or $r = -a(1 - \cos\theta)$ as the polar equation of the locus.

(*f*) Writing it as $x(x^2 + y^2) + 6x^2 - 2y^2 = 0$, we have $r^3\cos\theta + 6r^2\cos^2\theta - 2r^2\sin^2\theta = 0$. Then

$$r\cos\theta = 2\sin^2\theta - 6\cos^2\theta = 2(\sin^2\theta + \cos^2\theta) - 8\cos^2\theta = 2 - 8\cos^2\theta$$

and $r = 2(\sec\theta - 4\cos\theta)$ is the polar equation.

5. Transform each of the following equations into its rectangular form:

(a) $r = -2$ (c) $r \cos\theta = -6$ (e) $r = 4(1+\sin\theta)$ (f) $r = \dfrac{4}{2-\cos\theta}$

(b) $\theta = 3\pi/4$ (d) $r = 2\sin\theta$

In general, we attempt to put the polar equation in a form so that the substitutions x^2+y^2 for r^2, x for $r\cos\theta$ and y for $r\sin\theta$ can be made.

(a) Squaring, we have $r^2 = 4$; the rectangular equation is $x^2+y^2 = 4$.

(b) Here $\theta = \arctan\frac{y}{x} = 3\pi/4$; then $\frac{y}{x} = \tan 3\pi/4 = -1$ and the rectangular equation is $x+y = 0$.

(c) The rectangular form is $x = -6$.

(d) We first multiply the given equation by r to obtain $r^2 = 2r\sin\theta$. The rectangular form is $x^2+y^2 = 2y$.

(e) After multiplying by r, we have $r^2 = 4r + 4r\sin\theta$ or $r^2 - 4r\sin\theta = 4r$; then $(r^2 - 4r\sin\theta)^2 = 16r^2$ and the rectangular equation is $(x^2+y^2-4y)^2 = 16(x^2+y^2)$.

(f) Here $2r - r\cos\theta = 4$ or $2r = r\cos\theta + 4$; then $4r^2 = (r\cos\theta + 4)^2$ and the rectangular form of the ellipse is $4(x^2+y^2) = (x+4)^2$ or $3x^2+4y^2-8x-16 = 0$.

6. Derive the polar equation of the straight line:

(a) passing through the pole with vectorial angle k

(b) perpendicular to the polar axis and $p > 0$ units from the pole

(c) parallel to the polar axis and $p > 0$ units from the pole.

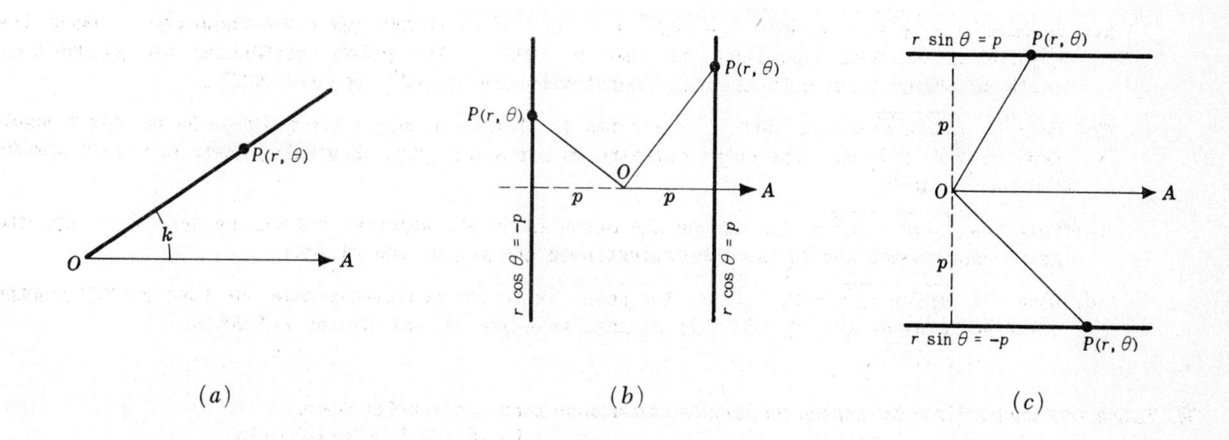

Let $P(r,\theta)$ be an arbitrary point on the line.

(a) From Fig. (a) the required equation is $\theta = k$.

(b) From Fig. (b) the equation is $r\cos\theta = p$ or $r\cos\theta = -p$ according as the line is to the right or left of the pole.

(c) From Fig. (c) the equation is $r\sin\theta = p$ or $r\sin\theta = -p$ according as the line is above or below the pole.

7. Derive the polar equivalent of the normal form of the rectangular equation of the straight line not passing through the pole.

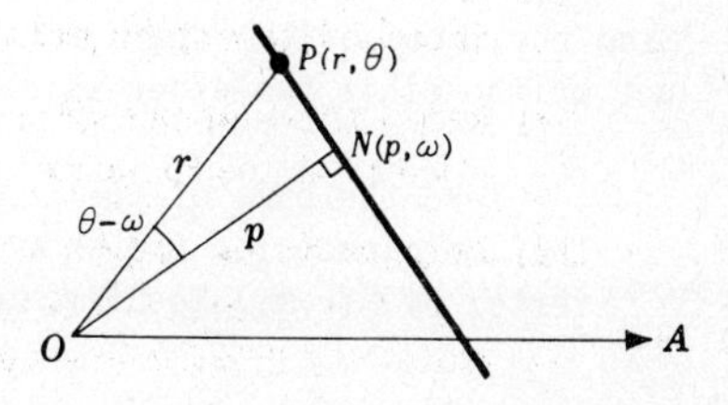

Let $P(r,\theta)$ be an arbitrary point on the line. Then the foot of the normal from the pole has coordinates $N(p,\omega)$. Using triangle ONP, the required equation is

$$r\cos(\theta-\omega) = p.$$

8. Derive the polar equation of the circle of radius a whose center is at (c,γ).

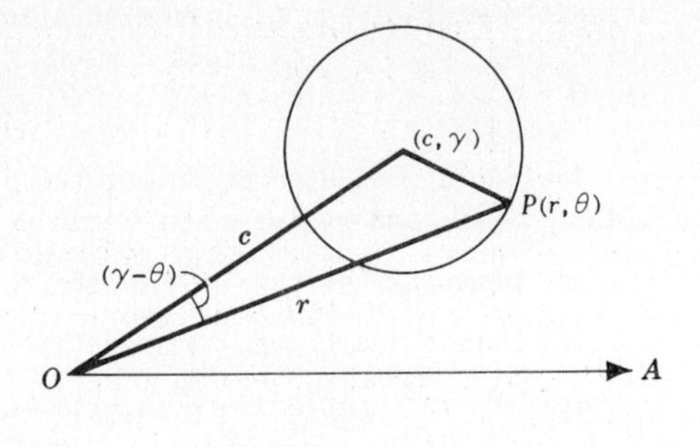

Let $P(r,\theta)$ be an arbitrary point on the circle. Then (see Problem 2(c))

$$r^2 + c^2 - 2rc\cos(\gamma-\theta) = a^2 \qquad \text{or}$$

$$(A) \qquad r^2 - 2rc\cos(\gamma-\theta) + c^2 - a^2 = 0$$

is the required equation.

The following special cases are of interest:

(a) If the center is at the pole, (A) becomes $r^2 = a^2$. Then $r = a$ or $r = -a$ is the equation of the circle of radius a with center at the pole.

(b) If $(c,\gamma) = (\pm a, 0^\circ)$, ($A$) becomes $r = \pm 2a\cos\theta$. Thus, $r = 2a\cos\theta$ is the equation of the circle of radius a passing through the pole and having its center on the polar axis; $r = -2a\cos\theta$ is the equation of the circle of radius a passing through the pole and having its center on the polar axis extended.

(c) Similarly if $(c,\gamma) = (\pm a, 90^\circ)$, we obtain $r = \pm 2a\sin\theta$ as the equation of the circle of radius a passing through the pole and having its center on the 90° axis or the 90° axis extended.

9. Derive the polar equation of a conic of eccentricity e, having a focus at the pole and p units from the corresponding directrix, when the axis on which the focus lies coincides with the polar axis.

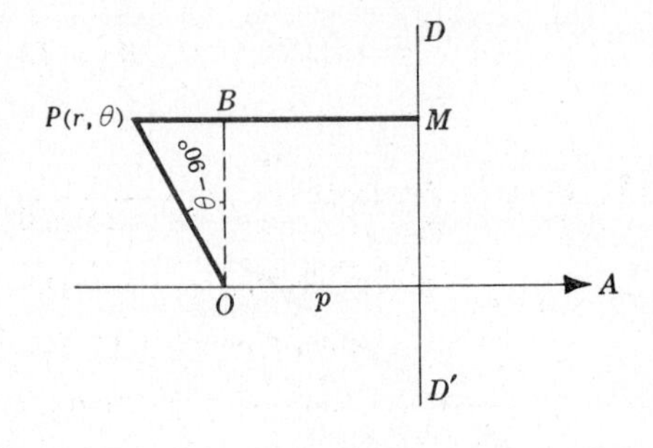

In the adjoining figure a focus is at O and the corresponding directrix DD' is to the right of O. Let $P(r,\theta)$ be an arbitrary point on the conic. Now

$$\frac{OP}{PM} = e$$

where $OP = r$ and $PM = PB + BM = r\sin(\theta - 90^\circ) + p$
$= p - r\cos\theta$.

Thus $$\frac{r}{p - r\cos\theta} = e, \qquad r(1 + e\cos\theta) = ep, \qquad \text{and} \qquad r = \frac{ep}{1 + e\cos\theta}.$$

It is left for the reader to derive the equation $r = \dfrac{ep}{1 - e\cos\theta}$ when the directrix DD' lies to the left of O.

Similarly it may be shown that the polar equation of a conic of eccentricity e, having a focus at the pole and p units from the corresponding directrix, is

$$r = \frac{ep}{1 \pm e\sin\theta}$$

where the positive sign (negative sign) is used when the directrix lies above (below) the pole.

10. Find the locus of the third vertex of a triangle whose base is a fixed line of length $2a$ and the product of the other two sides is the constant b^2.

Take the base of the triangle along the polar axis with the midpoint of the base at the pole. The coordinates of the endpoints of the base are $B(a,0)$ and $C(a,\pi)$. Denote the third (variable) vertex by $P(r,\theta)$. See Fig. (a) below.

From the triangle BOP, $(BP)^2 = r^2 + a^2 - 2ar\cos\theta$ and from the triangle COP, $(CP)^2 = r^2 + a^2 - 2ar\cos(\pi-\theta) = r^2 + a^2 + 2ar\cos\theta$. Now $(BP)(CP) = b^2$; hence

$$(r^2 + a^2 - 2ar\cos\theta)(r^2 + a^2 + 2ar\cos\theta) = (b^2)^2 = b^4.$$

Then
$$(r^2 + a^2)^2 - 4a^2r^2\cos^2\theta = b^4,$$

$$r^4 + 2a^2r^2(1 - 2\cos^2\theta) = r^4 - 2a^2r^2\cos 2\theta = b^4 - a^4,$$

$$r^4 - 2a^2r^2\cos 2\theta + a^4\cos^2 2\theta = b^4 - a^4 + a^4\cos^2 2\theta = b^4 - a^4\sin^2 2\theta,$$

and the required equation is

$$r^2 = a^2\cos 2\theta \pm \sqrt{b^4 - a^4\sin^2 2\theta}.$$

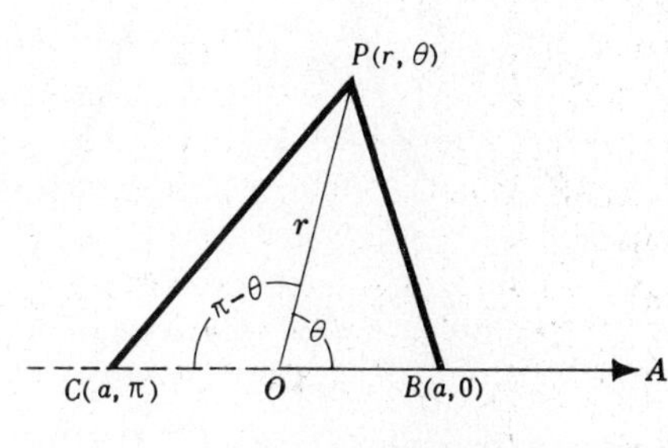

Fig. (*a*) Prob. 10

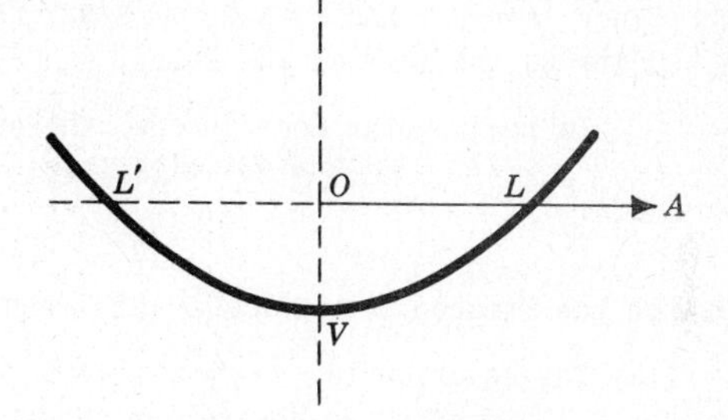

Fig. (*b*) Prob. 11

11. Sketch the conic $r = \dfrac{3}{2 - 2\sin\theta}$.

To put the equation in standard form, in which the first term in the denominator is 1, divide numerator and denominator by 2 and obtain $r = \dfrac{3/2}{1 - \sin\theta}$. The locus is a parabola $(e = 1)$ with focus at the pole. It opens upward ($\theta = \frac{1}{2}\pi$ makes r infinite).

When $\theta = 0$, $r = 3/2$; the length of the latus rectum is 3. When $\theta = 3\pi/2$, $r = 3/4$; the vertex is on the 90° axis extended 3/4 unit below the pole. With these facts the parabola may be sketched readily as in Fig. (*b*) above.

The equation in rectangular coordinates is $4x^2 = 12y + 9$.

12. Sketch the conic $r = \dfrac{18}{5 + 4\sin\theta}$.

After dividing numerator and denominator by 5, we have

$$r = \frac{18/5}{1 + (4/5)\sin\theta}.$$

The locus is an ellipse $(e = 4/5)$ with a focus at the pole.

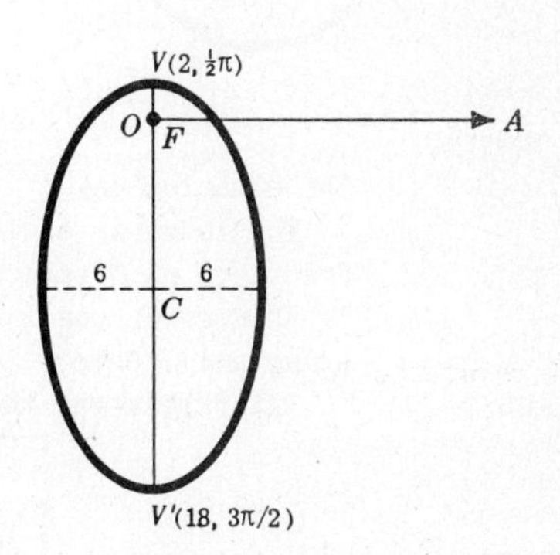

Since an equivalent equation is obtained when θ is replaced by $\pi - \theta$, the ellipse is symmetric with respect to the 90° axis; thus, the major axis is along the 90° axis. Since $ep = 18/5$ and $e = 4/5$, $p = 9/2$; the directrix is 9/2 units above the pole. When $\theta = \frac{1}{2}\pi$, $r = 2$; when $\theta = 3\pi/2$, $r = 18$. Thus the vertices are 2 units above and 18 units below the pole. Since $a = \frac{1}{2}(2 + 18) = 10$, $b = \sqrt{a^2(1 - e^2)} = 6$. With these facts the ellipse may be readily sketched as in the adjoining figure.

In rectangular coordinates, the equation is

$$25x^2 + 9y^2 + 144y - 324 = 0.$$

13. Sketch the conic $r = \dfrac{8}{3 - 5\cos\theta}$.

After dividing the numerator and denominator by 3, we have $r = \dfrac{8/3}{1 - (5/3)\cos\theta}$. The locus is a hyperbola ($e = 5/3$) with a focus at the pole.

An equivalent equation is obtained when θ is replaced by $-\theta$; hence the hyperbola is symmetric with respect to the polar axis and its transverse axis is on the polar axis. When $\theta = 0$, $r = -4$ and when $\theta = \pi$, $r = 1$; the vertices are respectively 4 units and 1 unit to the left of the pole. Then $a = \frac{1}{2}(4-1) = 3/2$ and $b = \sqrt{a^2(e^2-1)} = 2$. The asymptotes, having slopes $\pm b/a = \pm 4/3$, intersect at the center $\frac{1}{2}(1+4) = 5/2$ units to the left of the pole. Since $ep = 8/3$ and $e = 5/3$, $p = 8/5$; the directrix is 8/5 units to the left of the pole.

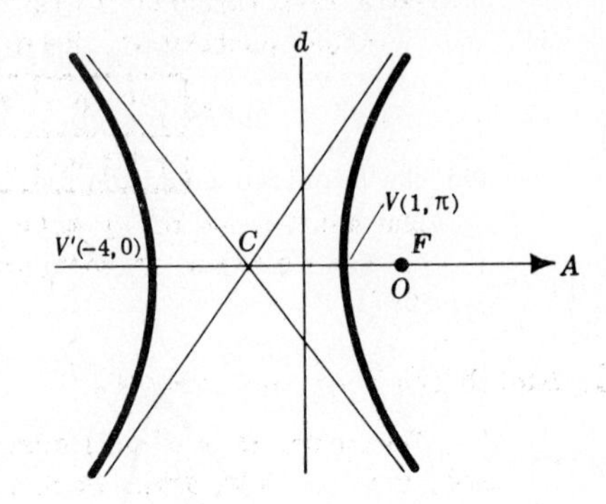

In rectangular coordinates, the equation is

$$16x^2 - 9y^2 + 80x + 64 = 0 .$$

14. Sketch the limacon $r = 2a\cos\theta + b$ when (*a*) $a = 2$, $b = 5$; (*b*) $a = 2$, $b = 4$; (*c*) $a = 2$, $b = 3$.

(*a*) The equation is $r = 4\cos\theta + 5$.

Symmetry. An equivalent equation is obtained when θ is replaced by $-\theta$; the locus is symmetric with respect to the polar axis.

Extent. Since r is real and finite for all values of θ, the locus is a closed curve. Since $\cos\theta$ is of period 2π, the complete locus is described as θ varies from 0 to 2π.

Directions at the pole. When $r = 0$, $\cos\theta = -5/4$; the locus does not pass through the pole.

After locating the following points

θ	0	$\pi/6$	$\pi/4$	$\pi/3$	$\pi/2$	$2\pi/3$	$3\pi/4$	$5\pi/6$	π
r	9.00	8.48	7.84	7.00	5.00	3.00	2.16	1.52	1.00

and making use of symmetry with respect to the polar axis, we obtain the required curve shown in Fig. (*a*) below.

The equation in rectangular coordinates is $(x^2 + y^2 - 4x)^2 = 25(x^2 + y^2)$.

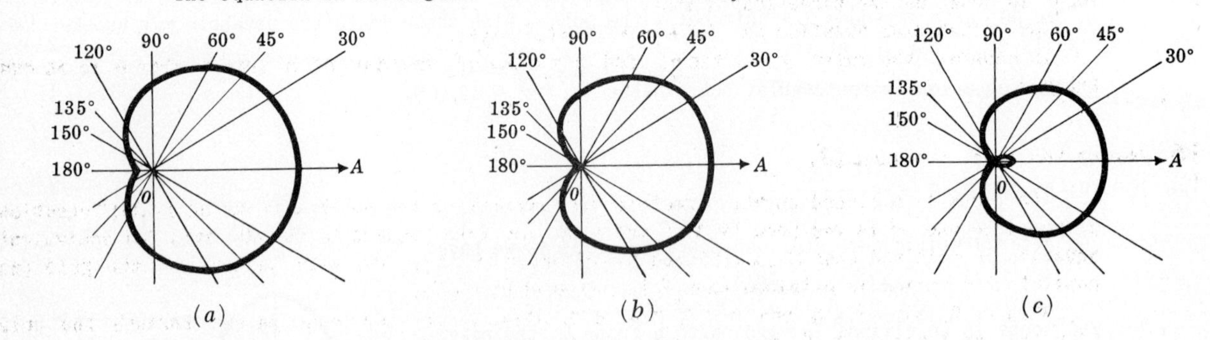

(*b*) The equation is $r = 4(1 + \cos\theta)$.

The locus is a closed curve, symmetric with respect to the polar axis, and is completely described as θ varies from 0 to 2π.

When $r = 0$, $\cos\theta = -1$ and $\theta = \pi$. The locus passes through the pole and is tangent to the polar axis there.

After locating the following points

θ	0	$\pi/6$	$\pi/4$	$\pi/3$	$\pi/2$	$2\pi/3$	$3\pi/4$	$5\pi/6$	π
r	8.00	7.48	6.84	6.00	4.00	2.00	1.16	0.52	0

and making use of symmetry, we obtain the required curve shown in Fig. (*b*) above.

In rectangular coordinates the equation of the cardiod is $(x^2 + y^2 - 4x)^2 = 16(x^2 + y^2)$.

(*c*) The equation is $r = 4\cos\theta + 3$.

The locus is a closed curve, symmetric with respect to the polar axis, and is completely described as θ varies from 0 to 2π.

When $r = 0$, $\cos\theta = -3/4 = -0.750$ and $\theta = 138°40'$, $221°20'$. The locus passes through the pole with tangents $\theta = 138°40'$ and $\theta = 221°20'$.

After putting in these tangents as guide lines, locating the following points

θ	0	$\pi/6$	$\pi/4$	$\pi/3$	$\pi/2$	$2\pi/3$	$3\pi/4$	$5\pi/6$	π
r	7.00	6.48	5.84	5.00	3.00	1.00	0.16	−0.48	−1.00

and making use of symmetry, we obtain the required curve shown in Fig. (*c*) above.

The equation in rectangular coordinates is $(x^2 + y^2 - 4x)^2 = 9(x^2 + y^2)$.

15. Sketch the rose $r = a\cos 3\theta$.

The locus is a closed curve, symmetric with respect to the polar axis. When $r = 0$, $\cos 3\theta = 0$ and $\theta = \pi/6, \pi/2, 5\pi/6, 7\pi/6, \ldots$; the locus passes through the pole with tangent lines $\theta = \pi/6$, $\theta = \pi/2$, and $\theta = 5\pi/6$ there.

The variation of r as θ changes is shown in the table.

θ	3θ	r
0 to $\pi/6$	0 to $\pi/2$	a to 0
$\pi/6$ to $\pi/3$	$\pi/2$ to π	0 to $-a$
$\pi/3$ to $\pi/2$	π to $3\pi/2$	$-a$ to 0
$\pi/2$ to $2\pi/3$	$3\pi/2$ to 2π	0 to a
$2\pi/3$ to $5\pi/6$	2π to $5\pi/2$	a to 0
$5\pi/6$ to π	$5\pi/2$ to 3π	0 to $-a$

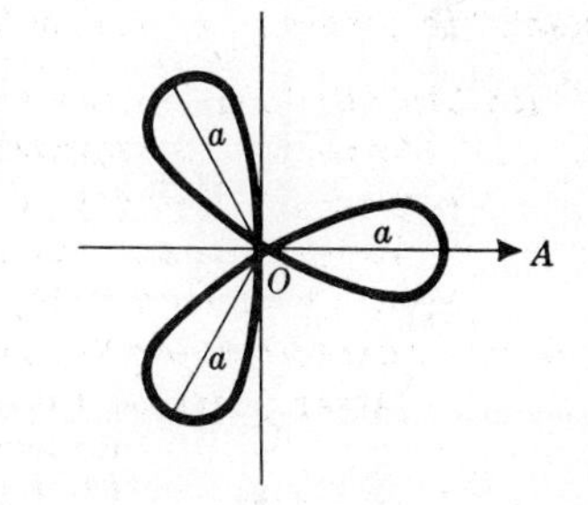

Caution: The values plotted are (r, θ) not $(r, 3\theta)$. The curve starts at a distance a to the right of the pole on the polar axis, passes through the pole tangent to the line $\theta = \pi/6$, reaches the tip of a loop when $\theta = \pi/3$, passes through the pole tangent to the line $\theta = \pi/2$, and so on. The locus is known as the three-leaved rose.

The rectangular equation is $(x^2 + y^2)^2 = ax(x^2 + 3y^2)$.

In general, the roses $r = a\sin n\theta$ and $r = a\cos n\theta$ consist of n leaves when n is an *odd* integer.

16. Sketch the rose $r = a\sin 4\theta$.

The locus is a closed curve, symmetric with respect to the polar axis (an equivalent equation is obtained when θ is replaced by $\pi - \theta$ and r by $-r$), with respect to the 90° axis (an equivalent equation is obtained when θ is replaced by $-\theta$ and r by $-r$), and with respect to the pole (an equivalent equation is obtained when θ is replaced by $\pi + \theta$).

When $r = 0$, $\sin 4\theta = 0$ and $\theta = 0, \pi/4, \pi/2, 3\pi/4, \ldots$; the locus passes through the pole with tangent lines $\theta = 0$, $\theta = \pi/4$, $\theta = \pi/2$, and $\theta = 3\pi/4$ there.

The variation of r as θ changes from 0 to $\pi/2$ is shown in the table.

θ	4θ	r
0 to $\pi/8$	0 to $\pi/2$	0 to a
$\pi/8$ to $\pi/4$	$\pi/2$ to π	a to 0
$\pi/4$ to $3\pi/8$	π to $3\pi/2$	0 to $-a$
$3\pi/8$ to $\pi/2$	$3\pi/2$ to 2π	$-a$ to 0

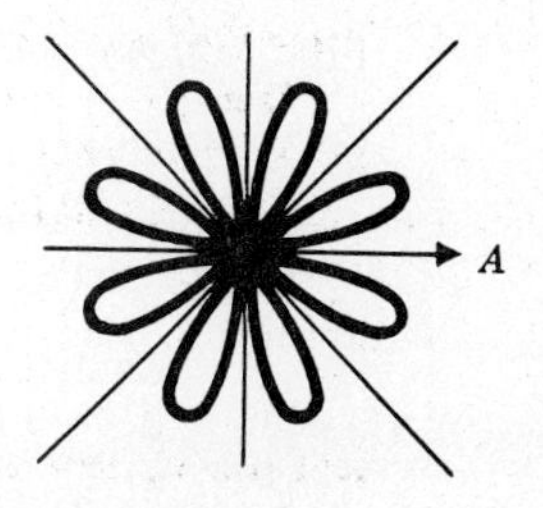

The complete curve, consisting of 8 leaves, can be traced

by making use of the symmetry.

In rectangular coordinates, the equation of the locus is $(x^2+y^2)^5 = 16a^2(x^3y-xy^3)^2$.

In general, the roses $r = a\sin n\theta$ and $r = a\cos n\theta$ consist of $2n$ leaves when n is an *even* integer.

17. Sketch the locus of $r = \cos\frac{1}{2}\theta$.

Other equations of the locus are $-r = \cos\frac{1}{2}(\theta+\pi) = -\sin\frac{1}{2}\theta$ or $r = \sin\frac{1}{2}\theta$, $-r = \cos\frac{1}{2}(\theta-\pi) = \sin\frac{1}{2}\theta$ or $r = -\sin\frac{1}{2}\theta$, and $r = \cos\frac{1}{2}(\theta-2\pi) = -\cos\frac{1}{2}\theta$.

The locus is a closed curve, symmetric with respect to the polar axis, the 90° axis, and the pole. It is completely described as θ varies from 0 to 4π.

When $r=0$, $\theta=\pi, 3\pi, \ldots$; the line $\theta=\pi$ is tangent to the locus at the pole.

The curve is traced by locating the following points and making use of symmetry.

θ	$\frac{1}{2}\theta$	r
0	0	1.00
$\pi/6$	$\pi/12$	0.97
$\pi/3$	$\pi/6$	0.87
$\pi/2$	$\pi/4$	0.71
$2\pi/3$	$\pi/3$	0.50
$5\pi/6$	$5\pi/12$	0.26
π	$\pi/2$	0.00

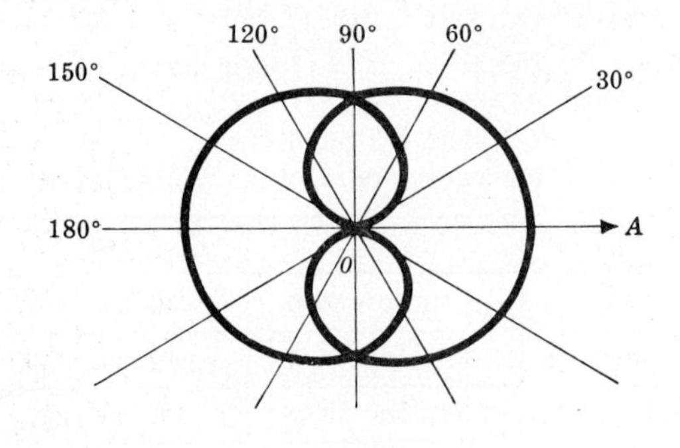

18. Find the points of intersection of the limacon $r = 2\cos\theta + 4$ and the circle $r = 8\cos\theta$.

From Fig. (c) below there are two points of intersection.

Setting $2\cos\theta + 4 = 8\cos\theta$, we obtain $\cos\theta = 2/3$; then $\theta = 48°10'$ and $311°50'$. (We solve for θ on the range $0 \le \theta < 2\pi$ since the limacon is completely described on this range.) The points of intersection are $(16/3, 48°10')$ and $(16/3, 311°50')$.

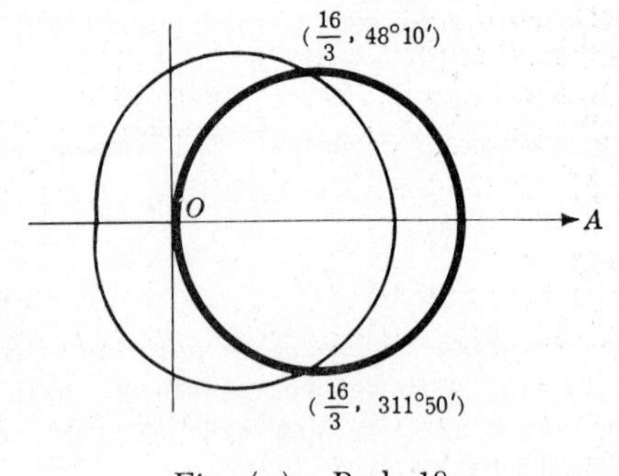

Fig. (c) Prob. 18

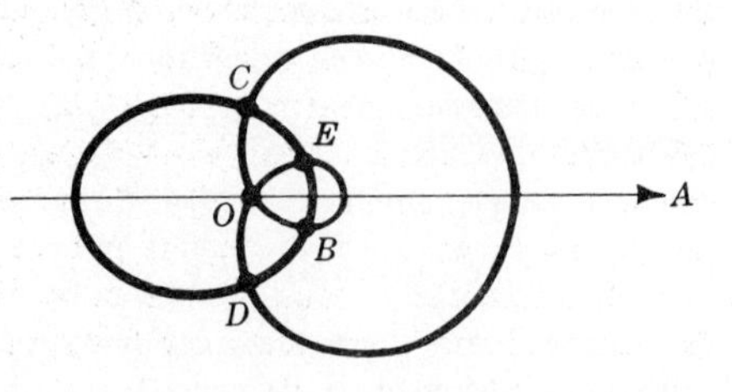

Fig. (d) Prob. 19

19. Find the points of intersection of the ellipse $r = \dfrac{4}{2+\cos\theta}$ and the limacon $r = 4\cos\theta - 2$.

From Fig. (d) above there are four points of intersection.

Setting $\dfrac{4}{2+\cos\theta} = 4\cos\theta - 2$, we have $2\cos^2\theta + 3\cos\theta - 4 = 0$. Then $\cos\theta = \dfrac{-3\pm\sqrt{41}}{4} = 0.851$ or -2.351, and $\theta = 31°40'$ and $328°20'$. The corresponding points are $E(-5+\sqrt{41}, 31°40')$ and $B(-5+\sqrt{41}, 328°20')$.

To obtain the other two points, we solve the equation of the ellipse with another equation $r = 4\cos\theta + 2$ (see Example 5) of the limacon. From $4\cos\theta + 2 = \dfrac{4}{2+\cos\theta}$, we obtain $2\cos^2\theta + 5\cos\theta = (\cos\theta)(2\cos\theta + 5) = 0$. Then $\cos\theta = 0$ and $\theta = \frac{1}{2}\pi$ and $3\pi/2$. The corresponding points are $C(2,\pi/2)$ and $D(2,3\pi/2)$. (Note. When sketching $r = 4\cos\theta - 2$, the coordinates of C were found as $(-2,3\pi/2)$ and those of D as $(-2,\pi/2)$.)

SUPPLEMENTARY PROBLEMS

20. Find the rectangular coordinates of P whose polar coordinates are:
(*a*) $(-2, 45°)$, (*b*) $(3, \pi)$, (*c*) $(2, \pi/2)$, (*d*) $(4, 2\pi/3)$.

Ans. (*a*) $(-\sqrt{2}, -\sqrt{2})$, (*b*) $(-3, 0)$, (*c*) $(0, 2)$, (*d*) $(-2, 2\sqrt{3})$.

21. Find a set of polar coordinates of P whose rectangular coordinates are:
(*a*) $(1, \sqrt{3})$, (*b*) $(0, -5)$, (*c*) $(1, -1)$, (*d*) $(-12, 5)$.

Ans. (*a*) $(2, \pi/3)$, (*b*) $(5, 3\pi/2)$, (*c*) $(\sqrt{2}, 7\pi/4)$, (*d*) $(13, \pi - \text{Arc tan } 5/12)$.

22. Transform each of the following rectangular equations into polar form:
(*a*) $x^2 + y^2 = 16$ (*b*) $y^2 - x^2 = 9$ (*c*) $x = 4$ (*d*) $y = \sqrt{3}x$ (*e*) $xy = 12$ (*f*) $(x^2 + y^2)x = 4y^2$

Ans. (*a*) $r = 4$ (*c*) $r \cos\theta = 4$ (*e*) $r^2 \sin 2\theta = 24$

(*b*) $r^2 \cos 2\theta + 9 = 0$ (*d*) $\theta = \pi/3$ (*f*) $r = 4 \tan\theta \sin\theta$

23. Transform each of the following polar equations into rectangular form:
(*a*) $r \sin\theta = -4$ (*c*) $r = 2\cos\theta$ (*e*) $r = 1 - 2\cos\theta$ (*f*) $r = \dfrac{4}{1 - 2\sin\theta}$
(*b*) $r = -4$ (*d*) $r = \sin 2\theta$

Ans. (*a*) $y = -4$ (*c*) $x^2 + y^2 - 2x = 0$ (*e*) $(x^2 + y^2 + 2x)^2 = x^2 + y^2$

(*b*) $x^2 + y^2 = 16$ (*d*) $(x^2 + y^2)^3 = 4x^2y^2$ (*f*) $x^2 - 3y^2 - 16y - 16 = 0$

24. Derive the polar equation $r = \dfrac{ep}{1 \pm e \sin\theta}$ of the conic of eccentricity e with a focus at the pole and with corresponding directrix p units from the focus.

25. Write the polar equation of each of the following:
(*a*) straight line bisecting the second and fourth quadrants;
(*b*) straight line through $(4, 2\pi/3)$ and perpendicular to the polar axis;
(*c*) straight line through $N(3, \pi/6)$ and perpendicular to the radius vector of N;
(*d*) circle with center at $C(4, 3\pi/2)$ and radius $= 4$;
(*e*) circle with center at $C(-4, 0)$ and radius $= 4$;
(*f*) circle with center at $C(4, \pi/3)$ and radius $= 4$;
(*g*) parabola with focus at the pole and directrix $r = -4 \sec\theta$;
(*h*) parabola with focus at the pole and vertex at $V(3, \pi/2)$;
(*i*) ellipse with eccentricity 3/4, one focus at the pole and the corresponding directrix 5 units above the polar axis;
(*j*) ellipse with one focus at the pole, the other focus at $(8, \pi)$ and eccentricity $= 2/3$;
(*k*) hyperbola with eccentricity $= 3/2$, one focus at the pole, and the corresponding directrix 5 units to the left of the 90° axis;
(*l*) hyperbola, conjugate axis $= 24$ parallel to and below the polar axis, transverse axis $= 10$, and one focus at the pole.

Ans. (*a*) $\theta = 3\pi/4$ (*c*) $r\cos(\theta - \pi/6) = 3$ (*e*) $r = -8\cos\theta$

(*b*) $r\cos\theta = -2$ (*d*) $r = -8\sin\theta$ (*f*) $r = 8\cos(\theta - \pi/3)$

(*g*) $r = \dfrac{4}{1 - \cos\theta}$ (*i*) $r = \dfrac{15}{4 + 3\sin\theta}$ (*k*) $r = \dfrac{15}{2 - 3\cos\theta}$

(*h*) $r = \dfrac{6}{1 + \sin\theta}$ (*j*) $r = \dfrac{10}{3 + 2\cos\theta}$ (*l*) $r = \dfrac{144}{5 - 13\sin\theta}$

26. Discuss and sketch:

(*a*) $r\sin(\theta - 45^\circ) = -2$

(*b*) $r = 10\sin\theta$

(*c*) $r = -6\cos\theta$

(*d*) $r = \dfrac{8}{2 - \sin\theta}$

(*e*) $r = \dfrac{6}{1 + 2\cos\theta}$

(*f*) $r = \dfrac{2}{1 - \cos\theta}$

(*g*) $r = 2 - 4\cos\theta$

(*h*) $r = 4 - 2\cos\theta$

(*i*) $r^2 = 9\cos 2\theta$

(*j*) $r^2 = 16\sin 2\theta$

(*k*) $r = 2\cos 2\theta$

(*l*) $r = 4\sin 2\theta$

(*m*) $r = 2a\tan\theta\sin\theta$

(*n*) $r = 4\tan^2\theta\sec\theta$

(*o*) $r = \cos\frac{3}{2}\theta$

(*p*) $r = 2\theta$

(*q*) $r = a/\theta$

27. Find the complete intersection of:

(*a*) $r = 2\cos\theta$, $r = 1$; (*b*) $r^2 = 4\cos 2\theta$, $r = 2\sqrt{2}\sin\theta$ (*c*) $r = 1 + \sin\theta$, $r = \sqrt{3}\cos\theta$

Ans. (*a*) $(1, \pi/3)$, $(1, 5\pi/3)$ (*b*) $(0,0)$, $(\sqrt{2}, \pi/6)$, $(\sqrt{2}, 5\pi/6)$ (*c*) $(0,0)$, $(3/2, \pi/6)$

28. If each radius vector through the pole to the line $r = a\sec\theta$ is increased and decreased by the constant b, the equation of the curve (conchoid of Nicomedes) thus generated is $r = a\sec\theta \pm b$. Sketch the three types corresponding to $b > a$, $b = a$, and $b < a$.

29. Let $\alpha = \angle XOM$ be any acute angle. Lay off $a = OA$ on OX and construct the perpendicular at A to OX meeting OM in C. Construct the conchoid $r = a\sec\theta + b$ with $a = OA = c\cos\alpha$ and $b = 2OC = 2c$, that is, the conchoid

$$r = c\cos\alpha\sec\theta + 2c.$$

Let the line $r = c\sin\alpha\csc\theta$ through C parallel to OX meet the conchoid in P. Show that $\theta = \angle XOP = \alpha/3$. This solution of the famous trisection problem was given by Nicomedes (circa 180 BC).

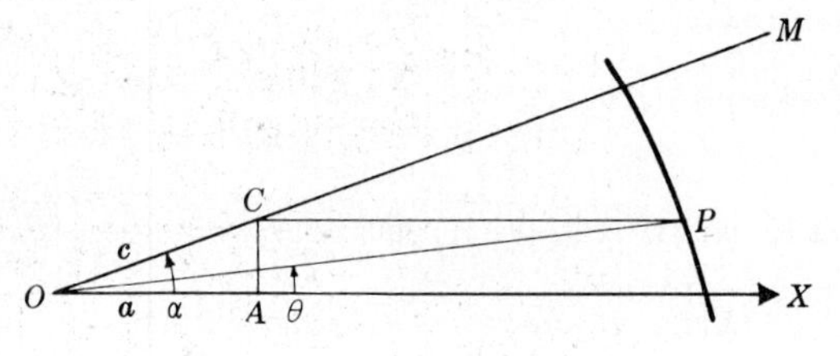

CHAPTER 57

Parametric Equations

IN THIS CHAPTER we consider the analytic representation of a plane curve by means of a pair of equations, as $x = t$, $y = 2t + 3$, in which each of the coordinates of a variable point (x, y) on the curve is expressed as a function of a third variable or *parameter*. Such equations are called *parametric equations* of the curve.

A table of values of x and y is readily obtained from the given parametric equations by assigning values to the parameter. After plotting the several points (x, y), the locus may be sketched in the usual manner.

Example. Sketch the locus of $x = t$, $y = 2t + 3$.

We form the table of values, plot the points (x, y) and join these points to obtain the straight line shown in the adjoining figure.

See Problem 1.

t	2	0	-3
x	2	0	-3
y	7	3	-3

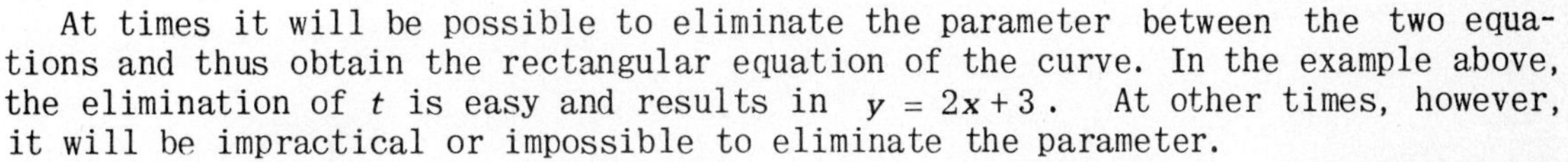

At times it will be possible to eliminate the parameter between the two equations and thus obtain the rectangular equation of the curve. In the example above, the elimination of t is easy and results in $y = 2x + 3$. At other times, however, it will be impractical or impossible to eliminate the parameter.

See Problem 2.

Parametric representation of a curve is *not* unique. For example, $x = t$, $y = 2t + 3$; $x = \frac{1}{2}u$, $y = u + 3$; $x = v - 1$, $y = 2v + 1$ are parametric representations with parameters t, u, v respectively of the straight line whose rectangular equation is $y = 2x + 3$.

See Problems 3-4.

PATH OF A PROJECTILE. If a body is projected from the origin with initial velocity v_0 ft/sec at an angle α with the positive x-axis and if all forces acting on the body after projection, excepting the force of gravity, are ignored, the coordinates of the body t seconds thereafter are given by

$$x = v_0 t \cos \alpha, \qquad y = v_0 t \sin \alpha - \tfrac{1}{2} g t^2$$

where g is the acceleration due to gravity. For convenience, we take $g = 32$ ft/sec^2.

See Problems 5-7.

SOLVED PROBLEMS

1. Sketch the locus of each of the following:
(a) $x = t,\ y = t^2$ (b) $x = 4t,\ y = 1/t$ (c) $x = 5\cos\theta,\ y = 5\sin\theta$ (d) $x = 2 + \cos\theta,\ y = \cos 2\theta$

The table of values and sketch of each are given below. In (b), $t \neq 0$; hence, we must examine the locus for values of t near 0. In (c) the complete locus is described on the interval $0 \le \theta \le 2\pi$. In (d) the complete locus is described on the interval $0 \le \theta \le \pi$. Note that only that part of the parabola below $y = 1$ is obtained; thus the complete parabola is not defined by the parametric equations.

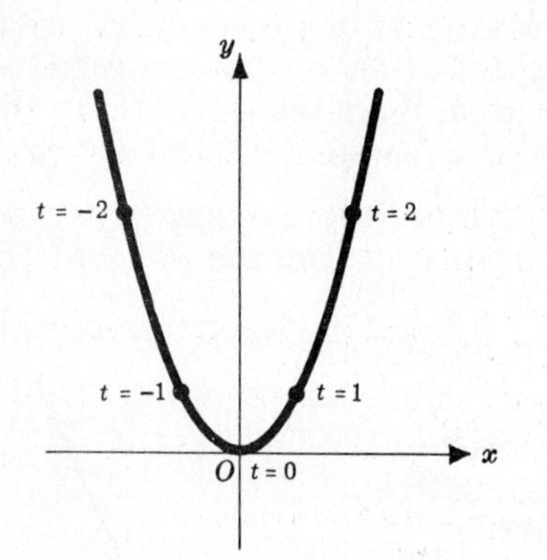

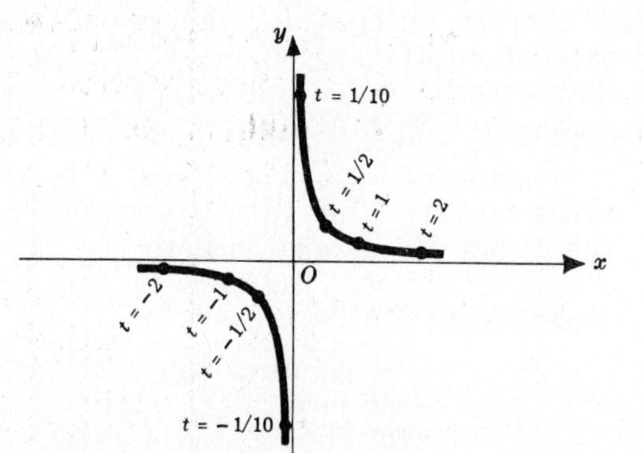

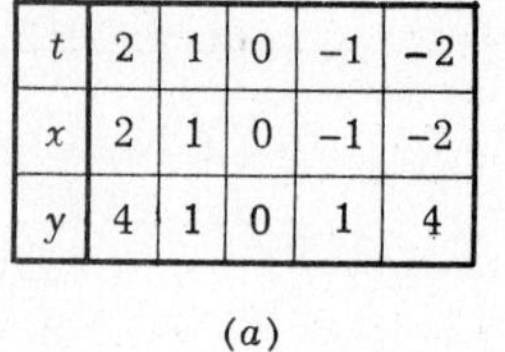

t	2	1	0	−1	−2
x	2	1	0	−1	−2
y	4	1	0	1	4

(a)

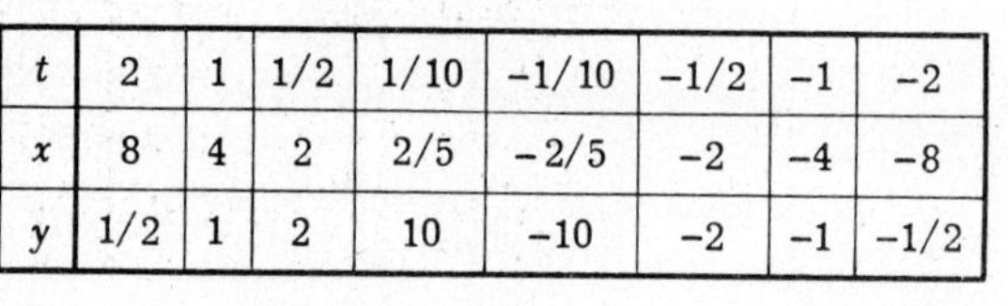

t	2	1	1/2	1/10	−1/10	−1/2	−1	−2
x	8	4	2	2/5	−2/5	−2	−4	−8
y	1/2	1	2	10	−10	−2	−1	−1/2

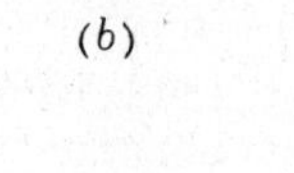

(b)

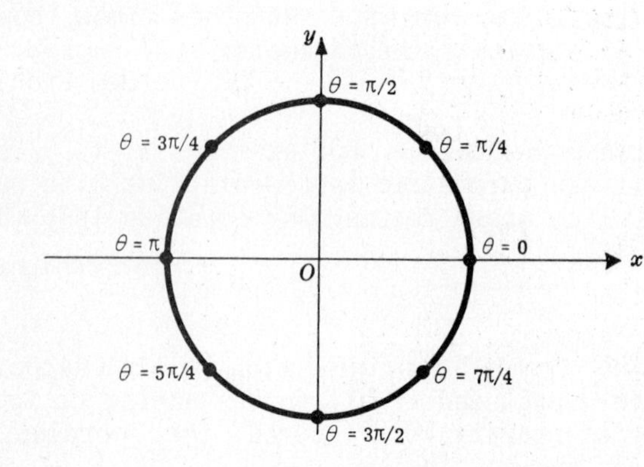

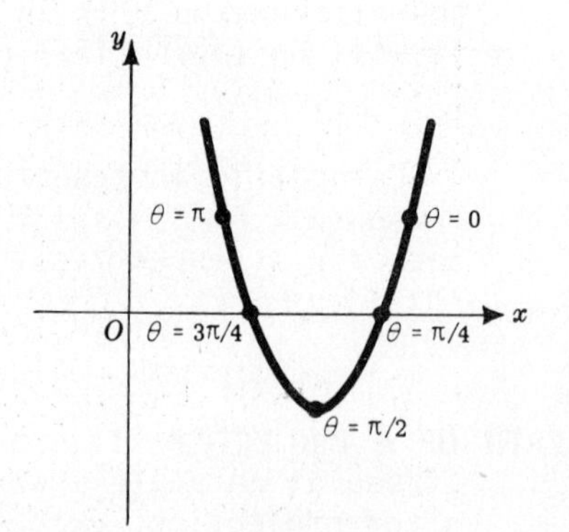

θ	0	π/4	π/2	3π/4	π	5π/4	3π/2	7π/4	2π
x	5.00	3.55	0	−3.55	−5.00	−3.55	0	3.55	5.00
y	0	3.55	5.00	3.55	0	−3.55	−5.00	−3.55	0

(c)

θ	0	π/4	π/2	3π/4	π
x	3.00	2.71	2.00	1.29	1.00
y	1.00	0	−1.00	0	1.00

(d)

2. In each of the following, eliminate the parameter to obtain the equation of the locus in rectangular coordinates: (*a*)-(*d*), Problem 1; (*e*) $x = t^2 + t$, $y = t^2 - t$; (*f*) $x = 3 \sec\phi$, $y = 2\tan\phi$; (*g*) $x = v_0(\cos\alpha)t$, $y = v_0(\sin\alpha)t - \frac{1}{2}gt^2$, t being the parameter.

(*a*) Here $t = x$ and $y = t^2 = x^2$. The required equation is $y = x^2$.
(*b*) Since $t = 1/y$, $x = 4t = 4/y$ and the equation is $xy = 4$.
(*c*) $x^2 + y^2 = (5\cos\theta)^2 + (5\sin\theta)^2 = 25(\cos^2\theta + \sin^2\theta)$. The required equation is $x^2 + y^2 = 25$.
(*d*) $\cos\theta = x - 2$ and $y = \cos 2\theta = 2\cos^2\theta - 1 = 2(x-2)^2 - 1$. The equation is $(x-2)^2 = \frac{1}{2}(y+1)$.
(*e*) Subtracting one of the equations from the other, $t = \frac{1}{2}(x-y)$. Then $x = \frac{1}{4}(x-y)^2 + \frac{1}{2}(x-y)$ and the required equation is $x^2 - 2xy + y^2 - 2x - 2y = 0$.
(*f*) $\tan\phi = \frac{1}{2}y$ and $x^2 = 9\sec^2\phi = 9(1 + \tan^2\phi) = 9(1 + \frac{1}{4}y^2)$. The equation is $4x^2 - 9y^2 = 36$.
(*g*) $t = \dfrac{x}{v_0\cos\alpha}$ and $y = \dfrac{v_0(\sin\alpha)x}{v_0\cos\alpha} - \dfrac{1}{2}\cdot\dfrac{gx^2}{(v_0\cos\alpha)^2}$. Then $y = x\tan\alpha - \dfrac{g}{2v_0^2\cos^2\alpha}x^2$.

3. Find parametric equations for each of the following, making use of the suggested substitution:
(*a*) $y^2 - 2y + 2x - 5 = 0$, $x = t + 3$
(*b*) $y^2 - 2y + 2x = 5$, $y = 1 + t$
(*c*) $9x^2 + 16y^2 = 144$, $x = 4\cos\theta$
(*d*) $x^3 + y^3 - 3axy = 0$, $y = mx$

(*a*) First write the equation as $(y-1)^2 = -2(x-3)$. Upon making the suggested substitution, we have $(y-1)^2 = -2t$ or $y = 1 \pm \sqrt{-2t}$.
We may take as parametric equations $x = t+3$, $y = 1 + \sqrt{-2t}$ or $x = t + 3$, $y = 1 - \sqrt{-2t}$.

(*b*) From $(y-1)^2 = -2(x-3)$, we obtain $t^2 = -2(x-3)$ or $x = 3 - \frac{1}{2}t^2$.
The parametric equations are $x = 3 - \frac{1}{2}t^2$, $y = 1 + t$.

(*c*) We have $9(16\cos^2\theta) + 16y^2 = 144$ or $y^2 = 9(1 - \cos^2\theta) = 9\sin^2\theta$.
The parametric equations are $x = 4\cos\theta$, $y = 3\sin\theta$ or $x = 4\cos\theta$, $y = -3\sin\theta$.

(*d*) Substituting, we have $x^3 + m^3x^3 - 3amx^2 = 0$. Dividing by x^2, we obtain $x = \dfrac{3am}{1+m^3}$. Then $y = mx = \dfrac{3am^2}{1+m^3}$ and the parametric equations are $x = \dfrac{3am}{1+m^3}$, $y = \dfrac{3am^2}{1+m^3}$.

4. (*a*) Show that $x = a\cos\theta$, $y = b\sin\theta$, $a > b$, are parametric equations of an ellipse.
(*b*) Show that these equations indicate the following method for constructing an ellipse whose major and minor axes $2a$ and $2b$ are given: With the origin as common center draw two circles having radii a and b. Through the origin draw a half-line l meeting the smaller circle in B and the larger circle in A. From A drop a perpendicular to the x-axis meeting it in R; from B drop a perpendicular to the x-axis meeting it in S and a perpendicular to RA meeting it in P. Then as l revolves about O, P describes the ellipse.

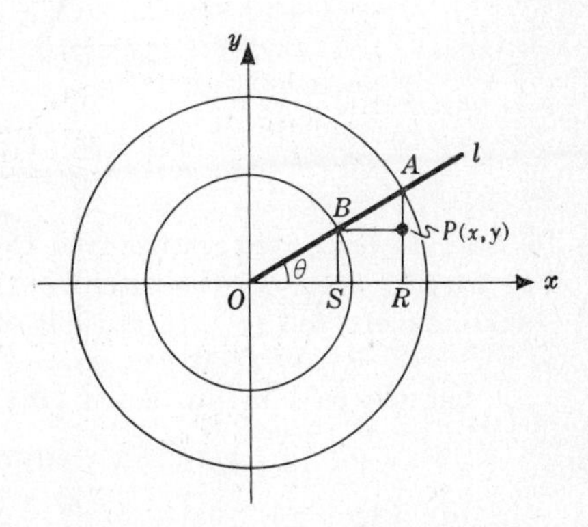

(*a*) We have $\cos\theta = x/a$ and $\sin\theta = y/b$; then
$x^2/a^2 + y^2/b^2 = \cos^2\theta + \sin^2\theta = 1$
the equation of an ellipse.

(*b*) In the adjoining diagram, let P have coordinates (x, y) and denote by θ the angle which l makes with the positive x-axis. Then

$$x = OR = OA\cos\theta = a\cos\theta$$
and
$$y = RP = SB = OB\sin\theta = b\sin\theta$$

are parametric equations of the ellipse whose major and minor axes are a and b.

5. A body is projected from the origin with initial velocity v_0 ft/sec at an angle α with the positive x-axis. Assuming that the only force acting upon the body after projection is the attraction of the earth, obtain parametric equations of the path of the body with t (the number of seconds after projection) as parameter.

If a body is released near the surface of the earth and all forces acting upon it other than gravity are neglected, the distance s ft through which it will fall in t sec is given by $s = \frac{1}{2}gt^2$, where $g = 32$ ft/sec^2 approximately.

If a body is given motion as stated in the problem and if after projection *no other force* acts on it, the motion is in a straight line (Newton's first law of motion) and after t seconds the body has coordinates $(v_0 t \cos\alpha,\ v_0 t \sin\alpha)$.

Since, when small distances are involved, the force of gravity may be assumed to act vertically, the coordinates of the projected body after t sec of motion are given by

$$(A) \qquad x = v_0 t \cos\alpha, \qquad y = v_0 t \sin\alpha - \tfrac{1}{2}gt^2.$$

In rectangular coordinates, the equation of the path is (see Problem 2(g))

$$(B) \qquad y = x \tan\alpha - \frac{g}{2v_0^2 \cos^2\alpha} x^2.$$

6. A bird is shot when flying horizontally 120 ft directly above the hunter. If its speed is 30 mi/hr, find the time during which it falls and the distance it will be from the hunter when it strikes the earth.

As in Fig. (a) below, take the bird to be at the origin when shot.
Since 30 mi/hr = $30 \cdot 5280/(60 \cdot 60) = 44$ ft/sec $= v_0$ and $\alpha = 0°$, the equations of motion are:

$$(a)\ x = v_0 t \cos\alpha = 44t \cos 0° = 44t, \qquad (b)\ y = v_0 t \sin\alpha - \tfrac{1}{2}gt^2 = -16t^2.$$

When the bird reaches the ground, its coordinates are $(x, -120)$.
From (b), $-120 = -16t^2$ and $t = \frac{1}{2}\sqrt{30}$; from (a), $x = 44(\frac{1}{2}\sqrt{30}) = 22\sqrt{30}$.
Thus the bird will fall for $\frac{1}{2}\sqrt{30}$ sec and will reach the ground $22\sqrt{30}$ ft from the hunter.

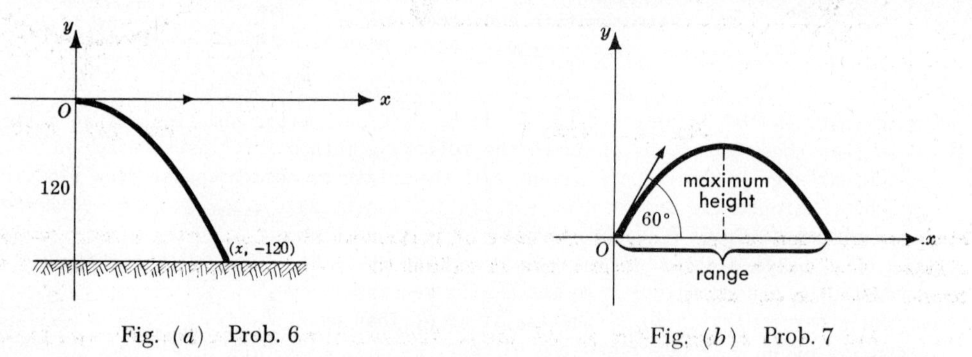

Fig. (a) Prob. 6 Fig. (b) Prob. 7

7. A ball is projected upward from the ground at an angle 60° from the horizontal with initial velocity 60 ft/sec. Find (a) the time it will be in the air, (b) its range, that is, the horizontal distance it will travel, and (c) its maximum height attained. Refer to Fig. (b) above.

Let the ball be projected from the origin; then the equations of motion are:

$$x = v_0 t \cos\alpha = 60t \cos 60° = 30t, \qquad y = v_0 t \sin\alpha - \tfrac{1}{2}gt^2 = 30t\sqrt{3} - 16t^2.$$

(a) When $y = 0$, $16t^2 - 30t\sqrt{3} = 0$ and $t = 0,\ 15\sqrt{3}/8$. Now $t = 0$ is the time when the ball was projected and $t = 15\sqrt{3}/8$ is the time when it reaches the ground again. Thus the ball was in the air for $15\sqrt{3}/8$ sec.

(b) When $t = 15\sqrt{3}/8$, $x = 30 \cdot 15\sqrt{3}/8 = 225\sqrt{3}/4$. The range is $225\sqrt{3}/4$ ft.

(c) *First Solution.* The ball will attain its maximum height when $t = \frac{1}{2}(15\sqrt{3}/8) = 15\sqrt{3}/16$. Then $y = 30t\sqrt{3} - 16t^2 = 30(15\sqrt{3}/16)\sqrt{3} - 16(15\sqrt{3}/16)^2 = 675/16$ ft, the maximum height.

Second Solution. The maximum height is attained where the horizontal distance of the ball is one-half the range, i.e., when $x = 225\sqrt{3}/8$. Using the rectangular equation $y = -\frac{4}{225}x^2 + x\sqrt{3}$, we obtain 675/16 ft as before.

8. The locus of a fixed point P on the circumference of a circle of radius a as the circle rolls without slipping along a straight line is called a *cycloid*. Obtain parametric equations of this locus.

Take the x-axis to be the line along which the circle is to roll and place the circle initially with its center C on the y-axis and P at the origin. Fig. (c) below shows the position of P after the circle has rolled through an angle θ. Drop perpendiculars PR and CS to the x-axis and PA to SC. Let P have coordinates (x,y). Then

$$x = OR = OS - RS = \text{arc } PS - PA = a\theta - a\sin\theta$$

and

$$y = RP = SA = SC - AC = a - a\cos\theta$$

Thus the equations of the locus are: $x = a(\theta - \sin\theta)$, $y = a(1 - \cos\theta)$.

The maximum height of an arch is $2a$, the diameter of the circle, and the span of an arch or the distance between two consecutive cusps is $2\pi a$, the circumference of the circle.

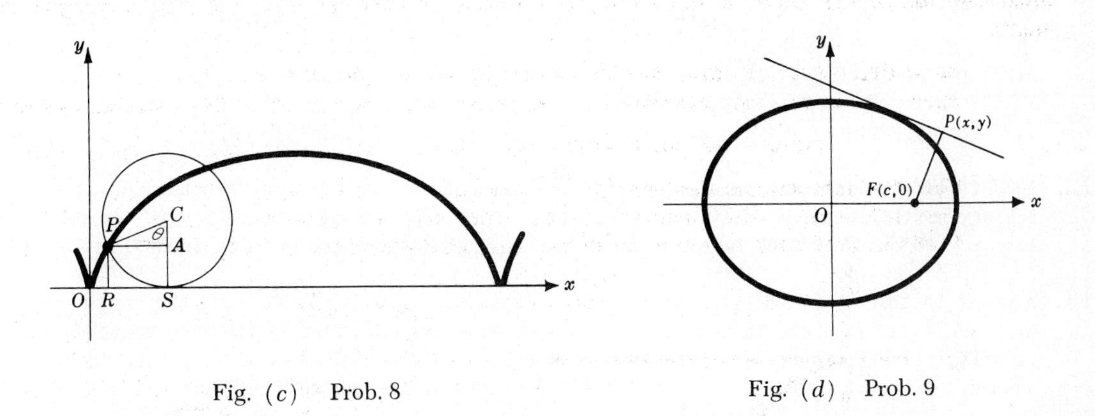

Fig. (c) Prob. 8 Fig. (d) Prob. 9

9. Find the equation of the locus of the feet of perpendiculars drawn from a focus to the tangents of the ellipse $b^2x^2 + a^2y^2 = a^2b^2$. This curve is called the *pedal curve* of the ellipse with respect to the focus. See Fig. (d) above.

Let $P(x,y)$ be any point on the locus. The equations of the tangents of slope m to the ellipse are

$$y = mx \pm \sqrt{a^2m^2 + b^2} \quad \text{or} \quad (1')\ y - mx = \pm\sqrt{a^2m^2 + b^2}.$$

The perpendicular to these tangents through $F(c,0)$ has equation

$$y = -\frac{1}{m}x + \frac{c}{m} \quad \text{or} \quad (2')\ my + x = \sqrt{a^2 - b^2}.$$

Squaring (1′) and (2′), and adding, we obtain

$$(1+m^2)y^2 + (1+m^2)x^2 = a^2(1+m^2) \quad \text{or, since } 1+m^2 \neq 0, \quad x^2 + y^2 = a^2$$

as the desired equation.

SUPPLEMENTARY PROBLEMS

10. Sketch the locus and find the rectangular equation of the curve whose parametric equations are:

(a) $x = t + 2,\ y = 3t + 5$ — *Ans.* $y = 3x - 1$

(b) $x = \tan\theta,\ y = 4\cot\theta$ — *Ans.* $xy = 4$

(c) $x = 2t^2 + 3,\ y = 3t + 2$ — *Ans.* $2y^2 - 8y - 9x + 35 = 0$

(d) $x = 3 + 2\tan\theta,\ y = -1 + 5\sec\theta$ — *Ans.* $4y^2 - 25x^2 + 8y + 150x - 321 = 0$

(e) $x = 2\sin^4\theta,\ y = 2\cos^4\theta$ — *Ans.* $(x - y)^2 - 4(x + y) + 4 = 0$

(f) $x = \tan\theta,\ y = \tan 2\theta$ — *Ans.* $x^2y - y + 2x = 0$

(g) $x = \sqrt{\cos t},\ y = \tan\frac{1}{2}t$ — *Ans.* $y^2(1 + x^2) + x^2 - 1 = 0$

(h) $x = a\cos^3\theta,\ y = a\sin^3\theta$ — *Ans.* $x^{2/3} + y^{2/3} = a^{2/3}$

(i) $x = \dfrac{2t}{1+t^2},\ y = \dfrac{1-t^2}{1+t^2}$ — *Ans.* $x^2 + y^2 = 1$

(j) $x = \dfrac{2at^2}{1+t^2},\ y = \dfrac{2at^3}{1+t^2}$ — *Ans.* $y^2(2a - x) = x^3$

11. Find parametric equations for each of the following, using the suggested value for x or y:

(a) $x^3 = 4y^2,\ x = t^2$

(b) $y = x^2 + x - 6,\ x = t + 2$

(c) $4x^2 - 9y^2 = 36,\ x = 3\sec\theta$

(d) $y = 2x^2 - 1,\ x = \cos t$

(e) $x(x^2 + y^2) = x^2 - y^2,\ y = tx$

(f) $(x^2 + 16)y = 64,\ x = 4\tan\theta$

Ans. (a) $x = t^2,\ y = \frac{1}{2}t^3$

(b) $x = t + 2,\ y = t^2 + 5t$

(c) $x = 3\sec\theta,\ y = 2\tan\theta$

(d) $x = \cos t,\ y = \cos 2t$

(e) $x = \dfrac{1-t^2}{1+t^2},\ y = \dfrac{t(1-t^2)}{1+t^2}$

(f) $x = 4\tan\theta,\ y = 4\cos^2\theta$

12. A 30 ft ladder with base on a smooth horizontal surface leans against a house. A man is standing 2/3 the way up the ladder when its foot begins to slide away from the house. Find the path of the man.
Ans. $x = 10\cos\theta,\ y = 20\sin\theta$ where θ is the angle at the foot of the ladder.

13. A stone is thrown upward with initial speed 48 ft/sec at an angle 60° with the horizontal from the top of a cliff 100 ft above the surface of a lake. Find (a) its greatest distance above the lake, (b) when it will strike the surface of the lake, and (c) its horizontal distance from the point where thrown when it strikes the surface. Hint: Take the origin at the top of the cliff.

Ans. (a) 127 ft (b) $\dfrac{3\sqrt{3} + \sqrt{127}}{4}$ sec later (c) $6(3\sqrt{3} + \sqrt{127})$ ft

14. Find the locus of the vertices of all right triangles having hypotenuse of length $2a$. Hint: Take the hypotenuse along the x-axis with its midpoint at the origin and let θ be an acute angle of the triangle. *Ans.* $x = a\cos 2\theta,\ y = a\sin 2\theta$ or $x = a\sin 2\theta,\ y = a\cos 2\theta$

15. From the two-point form of the equation of a straight line derive the parametric equations $x = x_1 + t(x_2 - x_1),\ y = y_1 + t(y_2 - y_1)$. What values of the parameter t give the points on the segment P_1P_2? Identify the points corresponding to $t = 1/2,\ 1/3,\ 2/3$.

16. Verify the following method for constructing a hyperbola with transverse axis $2a$ and conjugate axis $2b$, where $a \neq b$. With the origin as common center draw two circles having radii a and b. Through the origin pass a half line l making an angle θ with the positive x-axis and intersecting the larger circle in A. Let the tangent at A to the circle meet the x-axis in B. Through C, the intersection of the smaller circle and the positive x-axis, erect a perpendicular meeting l in D. Through D pass a line parallel to the x-axis and through B a line perpendicular to the x-axis, and denote their intersection by P. Then P is a point on the hyperbola.

17. Obtain the rectangular equation $x = a \operatorname{arc\,cos} \dfrac{a-y}{a} \mp 2ay - y^2$ of the cycloid of Problem 8.

CHAPTER 58

Points in Space

RECTANGULAR COORDINATES IN SPACE. Consider the three mutually perpendicular planes of Fig. 1. These three planes (the plane xOy or xy-plane, the plane xOz or xz-plane, the plane yOz or yz-plane) are called the *coordinate planes*; their three lines of intersection are called the *coordinate axes* (the $x'x$- or x-axis, the $y'y$- or y-axis, the $z'z$- or z-axis); and their common point O is called the *origin*. Positive direction is indicated on each axis by an arrow tip. (Note. The coordinate system of Fig. 1 is called a left-handed system. When the x- and y-axes are interchanged, the system becomes right-handed.)

The coordinate planes divide the space into eight regions, called *octants*. The octant whose edges are Ox, Oy, Oz is called the *first octant*; the other octants are usually not numbered.

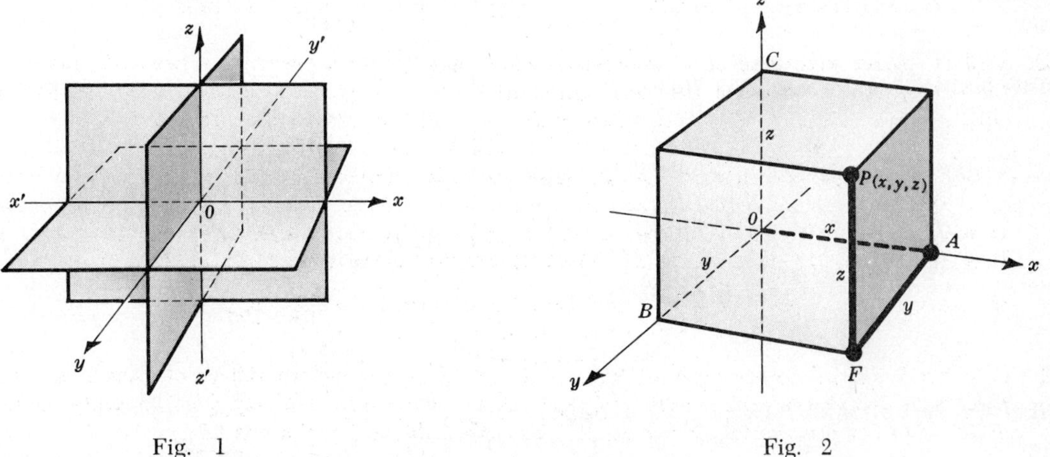

Fig. 1 Fig. 2

Let P be any point in space, not in a coordinate plane, and through P pass planes parallel to the coordinate planes meeting the coordinate axes in the points A, B, C and forming the rectangular parallelepiped of Fig. 2 above. The directed distances $x = OA$, $y = OB$, $z = OC$ are called respectively the x-coordinate, the y-coordinate, the z-coordinate of P and we write $P(x, y, z)$.

Since $AF = OB$ and $FP = OC$, it is preferable to use the three edges OA, AF, FP instead of the complete parallelepiped in locating a given point.

Example 1. Locate the points:
(a) (2, 3, 4); (b) (−2, −2, 3); (c) (2, −2, −3).

As standard procedure in representing on paper the left-handed system, we shall draw $\angle xOz = 90°$ and $\angle xOy = 135°$. Then distances on parallels to the x- and z-axes will be drawn to full scale while distances parallel to the y-axis will be drawn about 7/10 of full scale.

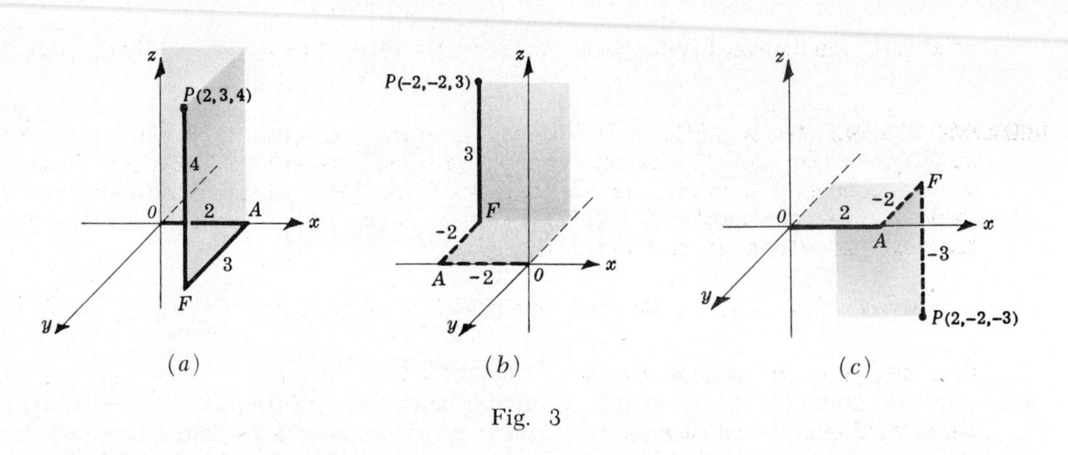

Fig. 3

(*a*) From the origin move 2 units to the right along the x-axis to $A(2,0,0)$, from A move 3 units forward parallel to the y-axis to $F(2,3,0)$, and from F move 4 units upward parallel to the z-axis to $P(2,3,4)$.
(*b*) From the origin move 2 units to the left along the x-axis to $A(-2,0,0)$, from A move 2 units backward parallel to the y-axis to $F(-2,-2,0)$, and from F move 3 units upward parallel to the z-axis to $P(-2,-2,3)$.
(*c*) From the origin move 2 units to the right along the x-axis to $A(2,0,0)$, from A move 2 units backward parallel to the y-axis to $F(2,-2,0)$, and from F move 3 units downward parallel to the z-axis to $P(2,-2,-3)$.

See Problem 1.

THE DISTANCE BETWEEN TWO POINTS $P_1(x_1, y_1, z_1)$ and $P_2(x_2, y_2, z_2)$ is from Fig. 4

$$(1) \qquad d = P_1P_2 = \sqrt{(P_1R)^2 + (RP_2)^2} = \sqrt{(P_1S)^2 + (SR)^2 + (RP_2)^2} = \sqrt{(x_2-x_1)^2 + (y_2-y_1)^2 + (z_2-z_1)^2}$$

See Problem 2.

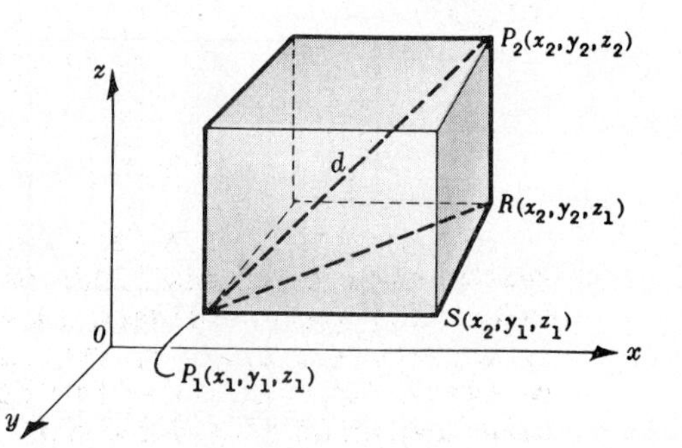

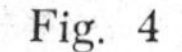
Fig. 4

IF $P_1(x_1, y_1, z_1)$ **AND** $P_2(x_2, y_2, z_2)$ are the endpoints of a line segment and if $P(x, y, z)$ divides the segment in the ratio $P_1P/PP_2 = r_1/r_2$, then

$$(2) \qquad x = \frac{r_2x_1 + r_1x_2}{r_1 + r_2}, \quad y = \frac{r_2y_1 + r_1y_2}{r_1 + r_2}, \quad z = \frac{r_2z_1 + r_1z_2}{r_1 + r_2}.$$

The coordinates of the midpoint of P_1P_2 are $(\frac{1}{2}(x_1 + x_2), \frac{1}{2}(y_1 + y_2), \frac{1}{2}(z_1 + z_2))$.

See Problems 3-4.

TWO STRAIGHT LINES IN SPACE which intersect or are parallel lie in the same plane; two lines which are not coplanar are called *skew*. By definition, the angle between two directed skew lines as b and c in Fig. 5 is the angle between any two intersecting lines as b' and

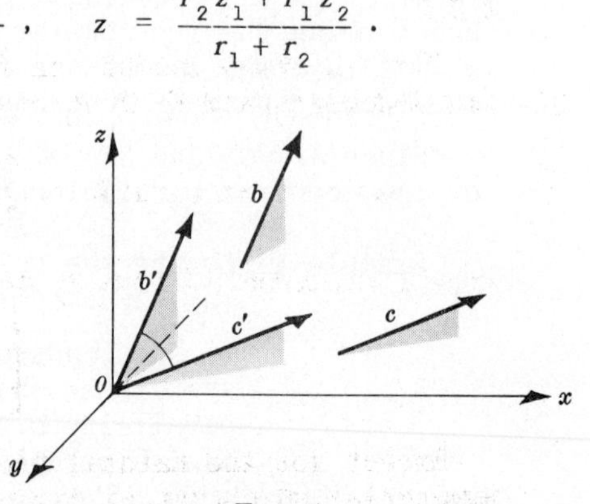

Fig. 5

c' which are respectively parallel to the skew lines and similarly directed.

DIRECTION COSINES OF A LINE. In the plane a directed line l (positive direction upward in Figures 6a, and 6b) forms the angles α and β with the positive directions on the x- and y-axes. However, in our study of the line in the plane we have favored the angle α over the angle β, calling it the angle of inclination of the line and its tangent the slope of the line.

In Fig. 6a, $\alpha + \beta = \frac{1}{2}\pi$ and $m = \tan\alpha = \dfrac{\sin\alpha}{\cos\alpha} = \dfrac{\sin(\frac{1}{2}\pi - \beta)}{\cos\alpha} = \dfrac{\cos\beta}{\cos\alpha}$; and in Fig. 6$b$, $\alpha = \frac{1}{2}\pi + \beta$ and $\tan\alpha = \dfrac{\sin(\frac{1}{2}\pi + \beta)}{\cos\alpha} = \dfrac{\cos\beta}{\cos\alpha}$. Now the angles α and β, called *direction angles* of the line, or their cosines, $\cos\alpha$ and $\cos\beta$, called *direction cosines* of the line might have been used instead of the slope to give the direction of the line l. Indeed, it will be the direction cosines which will be generalized in our study of the straight line in space.

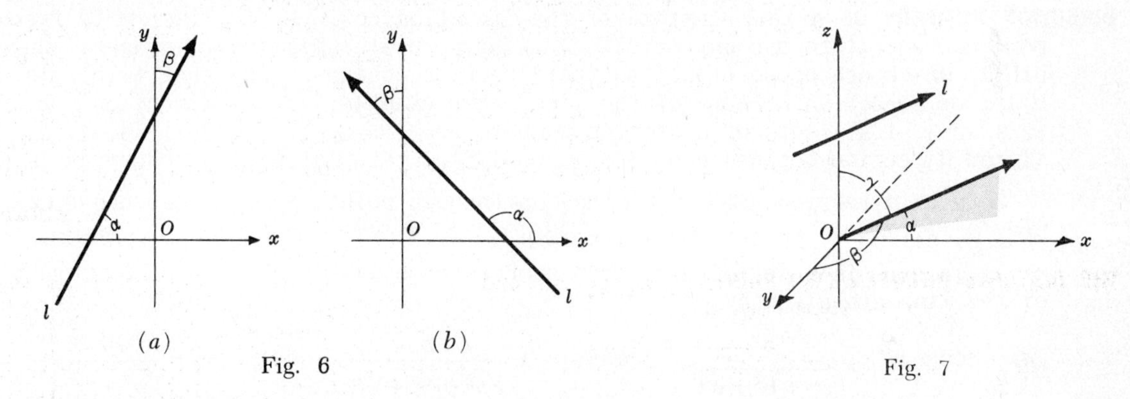

Fig. 6

Fig. 7

The direction of a line in space will be given by the three angles, called *direction angles* of the line, which it or that line through the origin parallel to it makes with the coordinate axes. If, as in Fig. 7 above, the direction angles α, β, γ, where $0 \le \alpha, \beta, \gamma < \pi$, are the respective angles between the positive directions on the x-, y-, z-axis and the directed line l (positive direction upward), the direction angles of this line when oppositely directed are $\alpha' = \pi - \alpha$, $\beta' = \pi - \beta$, $\gamma' = \pi - \gamma$. Thus, an undirected line in space has two sets of direction angles α, β, γ and $\pi - \alpha$, $\pi - \beta$, $\pi - \gamma$, and two sets of direction cosines $[\cos\alpha, \cos\beta, \cos\gamma]$ and $[-\cos\alpha, -\cos\beta, -\cos\gamma]$ since $\cos(\pi - \phi) = -\cos\phi$. To avoid confusion with the coordinates of a point, the triples of direction cosines of a line will be enclosed in a bracket. Thus we shall write l: $[A, B, C]$ to indicate the line whose direction cosines are the triple A, B, C.

The direction cosines of line l determined by points $P_1(x_1, y_1, z_1)$ and $P_2(x_2, y_2, z_2)$, and directed from P_1 to P_2 are (See Fig. 4)

$$\cos\alpha = \cos\angle P_2P_1S = \frac{P_1S}{P_1P_2} = \frac{x_2 - x_1}{d}, \quad \cos\beta = \frac{y_2 - y_1}{d}, \quad \cos\gamma = \frac{z_2 - z_1}{d}.$$

When l is directed from P_2 to P_1 the direction cosines are

$$\left[\frac{x_1 - x_2}{d}, \frac{y_1 - y_2}{d}, \frac{z_1 - z_2}{d}\right]$$

Except for the natural preference for $[1/3, 2/3, 2/3]$ over $[-1/3, -2/3, -2/3]$, it is immaterial which set of direction cosines are used when dealing with an undirected line.

Example 2. Find the two sets of direction cosines and indicate the positive direction along the line passing through the points $P_1(3, -1, 2)$ and $P_2(5, 2, -4)$.

We have $d = \sqrt{(2)^2 + (3)^2 + (-6)^2} = 7$. One set of direction cosines is

$$\left[\frac{x_2 - x_1}{d}, \frac{y_2 - y_1}{d}, \frac{z_2 - z_1}{d}\right] = \left[\frac{2}{7}, \frac{3}{7}, -\frac{6}{7}\right],$$

the positive direction being from P_1 to P_2. When the line is directed from P_2 to P_1, the direction cosines are $[-2/7, -3/7, 6/7]$.

The sum of the squares of the direction cosines of any line is equal to 1, i.e.,

$$\cos^2\alpha + \cos^2\beta + \cos^2\gamma = 1.$$

It follows immediately that at least one of the direction cosines of any line is different from 0.

DIRECTION NUMBERS OF A LINE. Instead of the direction cosines of a line, it is frequently more convenient to use any triple of numbers, preferably small integers when possible, which are proportional to the direction cosines. Any such triple is called a set of *direction numbers* of the line. For example, if the direction cosines are $[2/3, -2/3, -1/3]$, sets of direction numbers are $[2, -2, -1]$, $[-2, 2, 1]$, $[4, -4, -2]$, etc.; if the direction cosines are $[1/2, 1/\sqrt{2}, -1/2]$, a set of direction numbers is $[1, \sqrt{2}, -1]$.

Sets of direction numbers for the line through points $P_1(x_1, y_1, z_1)$ and $P_2(x_2, y_2, z_2)$ are $[x_2-x_1, y_2-y_1, z_2-z_1]$ and $[x_1-x_2, y_1-y_2, z_1-z_2]$.

If $[a, b, c]$ is a set of direction numbers of a line, then the direction cosines of the line are given by

$$(4) \quad \cos\alpha = \pm\frac{a}{\sqrt{a^2+b^2+c^2}}, \quad \cos\beta = \pm\frac{b}{\sqrt{a^2+b^2+c^2}}, \quad \cos\gamma = \pm\frac{c}{\sqrt{a^2+b^2+c^2}}$$

where the usual convention of first reading the upper signs and then the lower signs holds.

See Problems 5-8.

THE ANGLE θ BETWEEN TWO DIRECTED LINES

$$l_1: [\cos\alpha_1, \cos\beta_1, \cos\gamma_1] \quad \text{and} \quad l_2: [\cos\alpha_2, \cos\beta_2, \cos\gamma_2]$$

is given by

$$(5) \qquad \cos\theta = \cos\alpha_1\cos\alpha_2 + \cos\beta_1\cos\beta_2 + \cos\gamma_1\cos\gamma_2.$$

For a proof see Problem 9.

If the two lines are parallel then $\theta = 0$ or π, according as the lines are similarly or oppositely directed, and

$$\cos\alpha_1\cos\alpha_2 + \cos\beta_1\cos\beta_2 + \cos\gamma_1\cos\gamma_2 = \pm 1.$$

If the sign is +, then $\cos\alpha_1 = \cos\alpha_2$, $\cos\beta_1 = \cos\beta_2$, $\cos\gamma_1 = \cos\gamma_2$;
if the sign is −, then $\cos\alpha_1 = -\cos\alpha_2$, $\cos\beta_1 = -\cos\beta_2$, $\cos\gamma_1 = -\cos\gamma_2$.
Thus two undirected lines are parallel if and only if their direction cosines are the same or differ only in sign. In terms of direction numbers, *two lines are parallel if and only if corresponding direction numbers are proportional.*

If the two lines are perpendicular, then $\theta = \frac{1}{2}\pi$ or $3\pi/2$, according as the lines are similarly or oppositely directed, and

$$(6) \qquad \cos\alpha_1\cos\alpha_2 + \cos\beta_1\cos\beta_2 + \cos\gamma_1\cos\gamma_2 = 0.$$

In terms of direction numbers, *two lines with direction numbers* $[a_1, b_1, c_1]$ *and* $[a_2, b_2, c_2]$ *respectively are perpendicular if and only if*

$$(6') \qquad a_1 \cdot a_2 + b_1 \cdot b_2 + c_1 \cdot c_2 = 0 .$$

See Problems 10-12.

THE DIRECTION NUMBER DEVICE. If $l_1 : [a_1, b_1, c_1]$ and $l_2 : [a_2, b_2, c_2]$ are two non-parallel lines, then a set of direction numbers $[a, b, c]$ of any line perpendicular to both l_1 and l_2 is given by

$$a = \begin{vmatrix} b_1 & c_1 \\ b_2 & c_2 \end{vmatrix}, \qquad b = \begin{vmatrix} c_1 & a_1 \\ c_2 & a_2 \end{vmatrix}, \qquad c = \begin{vmatrix} a_1 & b_1 \\ a_2 & b_2 \end{vmatrix}.$$

These three determinants can be obtained readily as follows:

(1) Write the two sets of direction numbers in three columns $\begin{matrix} a_1 & b_1 & c_1 \\ a_2 & b_2 & c_2 \end{matrix}$.

(2) Repeat the first two columns to obtain $\begin{matrix} a_1 & b_1 & c_1 & a_1 & b_1 \\ a_2 & b_2 & c_2 & a_2 & b_2 \end{matrix}$ and strike out the first column to have $\begin{matrix} \not{a}_1 & b_1 & c_1 & a_1 & b_1 \\ \not{a}_2 & b_2 & c_2 & a_2 & b_2 \end{matrix}$.

Then a is the determinant of the first and second columns remaining, b is the determinant of the second and third columns, and c is the determinant of the third and fourth columns. This procedure will be called the *direction number device*. Note, however, that it is a mechanical procedure for obtaining one solution of two homogeneous equations in three unknowns and thus has other applications.

Example 3. Find a set of direction numbers $[a, b, c]$ of any line perpendicular to $l_1 : [2, 3, 4]$ and $l_2 : [1, -2, -3]$.

Using the direction number device, we write $\begin{matrix} \not{2} & 3 & 4 & 2 & 3 \\ \not{1} & -2 & -3 & 1 & -2 \end{matrix}$. Then

$$a = \begin{vmatrix} 3 & 4 \\ -2 & -3 \end{vmatrix} = -1 , \qquad b = \begin{vmatrix} 4 & 2 \\ -3 & 1 \end{vmatrix} = 10 , \qquad c = \begin{vmatrix} 2 & 3 \\ 1 & -2 \end{vmatrix} = -7 .$$

A set of direction numbers is $[-1, 10, -7]$ or, if preferred, $[1, -10, 7]$.

See Problem 13.

SOLVED PROBLEMS

1. What is the locus of a point:

(*a*) whose z-coordinate is always 0 ?

(*b*) whose z-coordinate is always 3 ?

(*c*) whose x-coordinate is always -5 ?

(*d*) whose x- and y-coordinates are always 0 ?

(*e*) whose x-coordinate is always 2 and whose y-coordinate is always 3 ?

(*a*) All points $(a, b, 0)$ lie in the xy-plane; the locus is that plane.

(*b*) Every point is 3 units above the xy-plane; the locus is the plane parallel to the xy-plane and 3 units above it.

(*c*) A plane parallel to the yz-plane and 5 units to the left of it.

(*d*) All points $(0, 0, c)$ lie on the z-axis; the locus is that line.

(*e*) In locating the point $P(2,3,c)$, the x- and y-coordinates are used to locate the point $F(2,3,0)$ in the xy-plane and then a distance $|c|$ is measured from F parallel to the z-axis. The locus is the line parallel to the z-axis passing through the point $(2,3,0)$ in the xy-plane.

2. (*a*) Find the distance between the points $P_1(-1,-3,3)$ and $P_2(2,-4,1)$.
(*b*) Find the perimeter of the triangle whose vertices are $A(-2,-4,-3)$, $B(1,0,9)$, $C(2,0,9)$.
(*c*) Show that the points $A(1,2,4)$, $B(4,1,6)$, and $C(-5,4,0)$ are collinear.

(*a*) Here $d = \sqrt{(x_2-x_1)^2+(y_2-y_1)^2+(z_2-z_1)^2} = \sqrt{[2-(-1)]^2+[-4-(-3)]^2+(1-3)^2} = \sqrt{14}$.

(*b*) We find $AB = \sqrt{[1-(-2)]^2+[0-(-4)]^2+[9-(-3)]^2} = 13$, $BC = \sqrt{(2-1)^2+(0-0)^2+(9-9)^2} = 1$, and $CA = \sqrt{(-2-2)^2+(-4-0)^2+(-3-9)^2} = 4\sqrt{11}$. The perimeter is $13+1+4\sqrt{11} = 14+4\sqrt{11}$.

(*c*) Here $AB = \sqrt{(3)^2+(-1)^2+(2)^2} = \sqrt{14}$, $BC = \sqrt{(-9)^2+(3)^2+(-6)^2} = 3\sqrt{14}$, and $CA = \sqrt{(6)^2+(-2)^2+(4)^2} = 2\sqrt{14}$. Since $BC = CA+AB$, the points are collinear.

3. Find the coordinates of the point P of division for each pair of points and given ratio. Find also the midpoint of the segment. (*a*) $P_1(3,2,-4)$, $P_2(6,-1,2)$; $1:2$, (*b*) $P_1(2,5,4)$, $P_2(-6,3,8)$; $-3:5$.

(*a*) Here $r_1 = 1$ and $r_2 = 2$. Then

$$x = \frac{r_2x_1+r_1x_2}{r_1+r_2} = \frac{2\cdot 3+1\cdot 6}{1+2} = 4, \quad y = \frac{r_2y_1+r_1y_2}{r_1+r_2} = \frac{2\cdot 2+1(-1)}{1+2} = 1, \quad z = \frac{r_2z_1+r_1z_2}{r_1+r_2} = -2$$

and the required point is $P(4,1,-2)$.

The midpoint has coordinates $\left(\tfrac{1}{2}(x_1+x_2), \tfrac{1}{2}(y_1+y_2), \tfrac{1}{2}(z_1+z_2)\right) = (9/2, 1/2, -1)$.

(*b*) Here $r_1 = -3$ and $r_2 = 5$. Then

$$x = \frac{5\cdot 2+(-3)(-6)}{-3+5} = 14, \quad y = \frac{5\cdot 5+(-3)3}{-3+5} = 8, \quad z = \frac{5\cdot 4+(-3)8}{-3+5} = -2$$

and the required point is $P(14,8,-2)$. The midpoint has coordinates $(-2,4,6)$.

4. Prove: The three lines joining the midpoints of the opposite edges of a tetrahedron pass through a point P which bisects each of them.

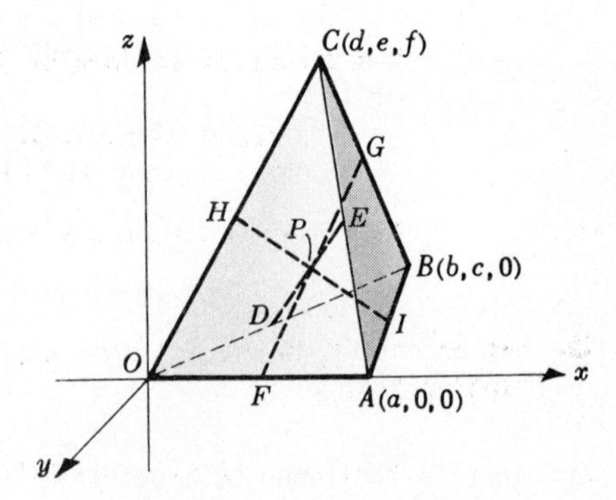

Let the tetrahedron, shown in the adjoining figure, have vertices $O(0,0,0)$, $A(a,0,0)$, $B(b,c,0)$, and $C(d,e,f)$.

The midpoints of OB and AC are respectively $D(\tfrac{1}{2}b, \tfrac{1}{2}c, 0)$ and $E(\tfrac{1}{2}(a+d), \tfrac{1}{2}e, \tfrac{1}{2}f)$, and the midpoint of DE is $P(\tfrac{1}{4}(a+b+d), \tfrac{1}{4}(c+e), \tfrac{1}{4}f)$. The midpoints of OA and BC are respectively $F(\tfrac{1}{2}a,0,0)$ and $G(\tfrac{1}{2}(b+d), \tfrac{1}{2}(c+e), \tfrac{1}{2}f)$, and the midpoint of FG is P. It is left for the reader to find the midpoints H and I of OC and AB, and show that P is the midpoint of HI.

5. Find the direction cosines of the line:
(*a*) passing through $P_1(3,4,5)$ and $P_2(-1,2,3)$ and directed from P_1 to P_2.
(*b*) passing through $P_1(2,-1,-3)$ and $P_2(-4,2,1)$ and directed from P_2 to P_1.
(*c*) passing through $O(0,0,0)$ and $P(a,b,c)$ and directed from O to P.
(*d*) passing through $P_1(4,-1,2)$ and $P_2(2,1,3)$ and directed so that γ is acute.

(a) We have $\cos\alpha = \dfrac{x_2 - x_1}{d} = \dfrac{-4}{2\sqrt{6}}$, $\cos\beta = \dfrac{y_2 - y_1}{d} = \dfrac{-2}{2\sqrt{6}}$, $\cos\gamma = \dfrac{z_2 - z_1}{d} = \dfrac{-2}{2\sqrt{6}}$.

The direction cosines are $\left[-\dfrac{2}{\sqrt{6}}, -\dfrac{1}{\sqrt{6}}, -\dfrac{1}{\sqrt{6}}\right]$.

(b) $\cos\alpha = \dfrac{x_1 - x_2}{d} = \dfrac{6}{\sqrt{61}}$, $\cos\beta = \dfrac{y_1 - y_2}{d} = \dfrac{-3}{\sqrt{61}}$, $\cos\gamma = \dfrac{z_1 - z_2}{d} = \dfrac{-4}{\sqrt{61}}$.

The direction cosines are $\left[\dfrac{6}{\sqrt{61}}, -\dfrac{3}{\sqrt{61}}, -\dfrac{4}{\sqrt{61}}\right]$.

(c) $\cos\alpha = \dfrac{a-0}{d} = \dfrac{a}{\sqrt{a^2+b^2+c^2}}$, $\cos\beta = \dfrac{b-0}{d} = \dfrac{b}{\sqrt{a^2+b^2+c^2}}$, $\cos\gamma = \dfrac{c-0}{d} = \dfrac{c}{\sqrt{a^2+b^2+c^2}}$

The direction cosines are $\left[\dfrac{a}{\sqrt{a^2+b^2+c^2}}, \dfrac{b}{\sqrt{a^2+b^2+c^2}}, \dfrac{c}{\sqrt{a^2+b^2+c^2}}\right]$.

(d) The two sets of direction cosines of the undirected line are

$$\cos\alpha = \pm\frac{2}{3}, \quad \cos\beta = \mp\frac{2}{3}, \quad \cos\gamma = \mp\frac{1}{3}$$

one set being given by the upper signs and the other by the lower signs.

When γ is acute, $\cos\gamma > 0$; hence the required set is $[-2/3, 2/3, 1/3]$.

6. Given the direction angles $\alpha = 120°$ and $\beta = 45°$, find γ if the line is directed upward.

$\cos^2\alpha + \cos^2\beta + \cos^2\gamma = \cos^2 120° + \cos^2 45° + \cos^2\gamma = (-\tfrac{1}{2})^2 + (1/\sqrt{2})^2 + \cos^2\gamma = 1$.
Then $\cos^2\gamma = \tfrac{1}{4}$ and $\cos\gamma = \pm\tfrac{1}{2}$. When the line is directed upward, $\cos\gamma = \tfrac{1}{2}$ and $\gamma = 60°$.

7. The direction numbers of a line l are given as $[2,-3,6]$. Find the direction cosines of l when directed upward.

The direction cosines of l are given by

$$\cos\alpha = \pm\frac{a}{\sqrt{a^2+b^2+c^2}} = \pm\frac{2}{7}, \quad \cos\beta = \pm\left(\frac{-3}{7}\right), \quad \cos\gamma = \pm\frac{6}{7}.$$

When γ is acute, $\cos\gamma > 0$, and the direction cosines are $[2/7, -3/7, 6/7]$.

8. Use direction numbers to show that the points $A(1,2,4)$, $B(4,1,6)$, and $C(-5,4,0)$ are collinear. (See Problem 2(c).)

A set of direction numbers for the line AB is $[3,-1,2]$, for BC is $[-9,3,-6]$.

Since the two sets are proportional, the lines are parallel; since the lines have a point in common, they are coincident and the points are collinear.

9. Prove: The angle θ between two directed lines l_1: $[\cos\alpha_1, \cos\beta_1, \cos\gamma_1]$ and l_2: $[\cos\alpha_2, \cos\beta_2, \cos\gamma_2]$ is given by $\cos\theta = \cos\alpha_1\cos\alpha_2 + \cos\beta_1\cos\beta_2 + \cos\gamma_1\cos\gamma_2$.

The angle θ is by definition the angle between two lines issuing from the origin parallel respectively to the given lines l_1 and l_2 and similarly directed.

Consider the triangle OP_1P_2, in the adjoining figure, whose vertices are the origin and the points $P_1(\cos\alpha_1, \cos\beta_1, \cos\gamma_1)$ and $P_2(\cos\alpha_2, \cos\beta_2, \cos\gamma_2)$. The line segment OP_1 is of length 1 (why?) and is parallel to l_1; similarly, OP_2 is of length 1 and is parallel to l_2. Thus, $\angle P_1OP_2 = \theta$. By the Law of Cosines,

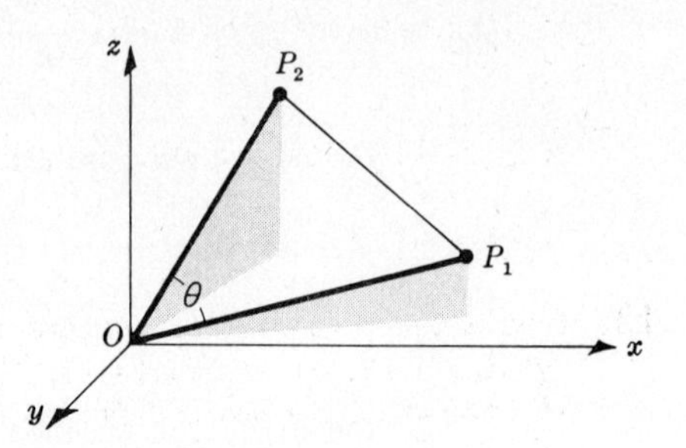

$$(P_1P_2)^2 = (OP_1)^2 + (OP_2)^2 - 2(OP_1)(OP_2)\cos\theta$$

and $\cos\theta = \cos\alpha_1\cos\alpha_2 + \cos\beta_1\cos\beta_2 + \cos\gamma_1\cos\gamma_2$.

10. (*a*) Find the angle between the directed lines l_1: [2/7, 3/7, 6/7] and l_2: [2/3, −1/3, 2/3].
(*b*) Find the acute angle between the lines l_1: [−2, 1, 2] and l_2: [2, −6, −3].
(*c*) The line l_1 passes through $A(5,-2,3)$ and $B(2,1,-4)$, and the line l_2 passes through $C(-4,1,-2)$ and $D(-3,2,3)$. Find the acute angle between them.

(*a*) We have $\cos\theta = \cos\alpha_1\cos\alpha_2 + \cos\beta_1\cos\beta_2 + \cos\gamma_1\cos\gamma_2$

$$= \frac{2}{7}\cdot\frac{2}{3} + \frac{3}{7}\left(-\frac{1}{3}\right) + \frac{6}{7}\cdot\frac{2}{3} = \frac{13}{21} = 0.619 \quad \text{and} \quad \theta = 51°50'.$$

(*b*) Since $\sqrt{(-2)^2 + (1)^2 + (2)^2} = 3$, we take [−2/3, 1/3, 2/3] as direction cosines of l_1.
Since $\sqrt{(2)^2 + (-6)^2 + (-3)^2} = 7$, we take [2/7, −6/7, −3/7] as direction cosines of l_2.

Then $\cos\theta = -\frac{2}{3}\cdot\frac{2}{7} + \frac{1}{3}\left(-\frac{6}{7}\right) + \frac{2}{3}\left(-\frac{3}{7}\right) = -\frac{16}{21} = -0.762$ and $\theta = 139°40'$.

The required angle is 40°20′.

(*c*) Take $[3/\sqrt{67}, -3/\sqrt{67}, 7/\sqrt{67}]$ as direction cosines of l_1 and $[1/3\sqrt{3}, 1/3\sqrt{3}, 5/3\sqrt{3}]$ as direction cosines of l_2.

Then $\cos\theta = \frac{1}{\sqrt{3\cdot 67}} - \frac{1}{\sqrt{3\cdot 67}} + \frac{35}{3\sqrt{3\cdot 67}} = \frac{35}{3\sqrt{201}} = 0.823$ and $\theta = 34°40'$.

11. (*a*) Show that the line joining $A(9,2,6)$ and $B(5,-3,2)$ and the line joining $C(-1,-5,-2)$ and $D(7,5,6)$ are parallel. (*b*) Show that the line joining $A(7,2,3)$ and $B(-2,5,2)$ and the line joining $C(4,10,1)$ and $D(1,2,4)$ are mutually perpendicular.

(*a*) Here [9−5, 2−(−3), 6−2] = [4, 5, 4] is a set of direction numbers of AB and [−1−7, −5−5, −2−6] = [−8, −10, −8] is a set of direction numbers of CD. Since the two sets are proportional, the two lines are parallel.
(*b*) Here [9, −3, 1] is a set of direction numbers of AB and [3, 8, −3] is a set of direction numbers of CD. Since (see Equation 6′) $9\cdot 3 + (-3)8 + 1(-3) = 0$, the lines are perpendicular.

12. Find the area of the triangle whose vertices are $A(4,2,3)$, $B(7,-2,4)$ and $C(3,-4,6)$.

The area of triangle ABC is given by $\frac{1}{2}(AB)(AC)\sin A$. We have $AB = \sqrt{26}$ and $AC = \sqrt{46}$.

To find $\sin A$, we direct the sides AB and AC away from the origin as in the adjoining figure. Then AB has direction cosines $[3/\sqrt{26}, -4/\sqrt{26}, 1/\sqrt{26}]$, AC has direction cosines $[-1/\sqrt{46}, -6/\sqrt{46}, 3/\sqrt{46}]$,

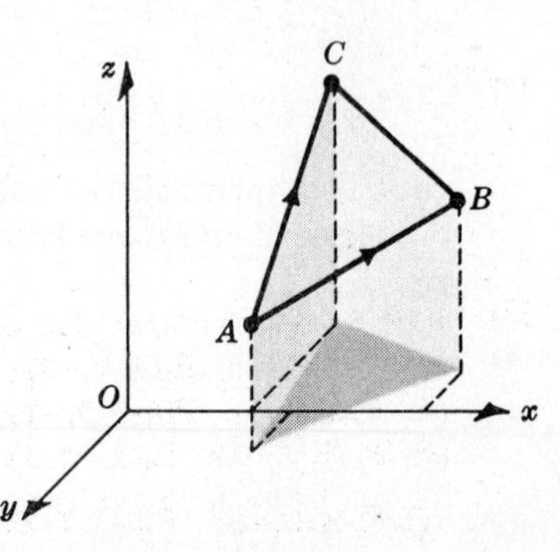

$$\cos A = \frac{3}{\sqrt{26}}\cdot\frac{-1}{\sqrt{46}} + \frac{-4}{\sqrt{26}}\cdot\frac{-6}{\sqrt{46}} + \frac{1}{\sqrt{26}}\cdot\frac{3}{\sqrt{46}} = \frac{24}{\sqrt{26}\sqrt{46}}$$

and $$\sin A = \sqrt{1-\cos^2 A} = \frac{2\sqrt{155}}{\sqrt{26}\,\sqrt{46}}.$$

The required area is $\frac{1}{2}\sqrt{26}\cdot\sqrt{46}\cdot\dfrac{2\sqrt{155}}{\sqrt{26}\,\sqrt{46}} = \sqrt{155}$.

13. Find a set of direction numbers for any line which is perpendicular to
(*a*) $l_1:[1,-2,-3]$ and $l_2:[4,-1,-5]$.
(*b*) the triangle whose vertices are $A(4,2,3)$, $B(7,-2,4)$, and $C(3,-4,6)$.

(*a*) Using the direction number device

$$\begin{matrix} 1 & -2 & -3 & 1 & -2 \\ 4 & -1 & -5 & 4 & -1 \end{matrix}, \quad \text{we obtain} \quad a = \begin{vmatrix} -2 & -3 \\ -1 & -5 \end{vmatrix} = 7, \quad b = \begin{vmatrix} -3 & 1 \\ -5 & 4 \end{vmatrix} = -7, \quad c = \begin{vmatrix} 1 & -2 \\ 4 & -1 \end{vmatrix} = 7.$$

Thus a set of direction numbers is $[7,-7,7]$; a simpler set is $[1,-1,1]$.

(*b*) Since the triangle lies in a plane determined by the lines AB and AC, we seek direction numbers for any line perpendicular to these lines. For AB and AC, respective sets of direction numbers are $[3,-4,1]$ and $[-1,-6,3]$. Using the direction number device

$$\begin{matrix} 3 & -4 & 1 & 3 & -4 \\ -1 & -6 & 3 & -1 & -6 \end{matrix}, \quad a = \begin{vmatrix} -4 & 1 \\ -6 & 3 \end{vmatrix} = -6, \quad b = \begin{vmatrix} 1 & 3 \\ 3 & -1 \end{vmatrix} = -10, \quad c = \begin{vmatrix} 3 & -4 \\ -1 & -6 \end{vmatrix} = -22.$$

Then $[-6,-10,-22]$ is a set of direction numbers and $[3,5,11]$ is a simpler set.

SUPPLEMENTARY PROBLEMS

14. Find the undirected distance between each pair of points:
(*a*) $(4,1,5)$ and $(2,-1,4)$ (*b*) $(9,7,-2)$ and $(6,5,4)$ (*c*) $(9,-2,-3)$ and $(-3,4,0)$.
Ans. (*a*) 3 (*b*) 7 (*c*) $3\sqrt{21}$

15. Find the undirected distance of each of the following points from (i) the origin, (ii) the x-axis, (iii) the y-axis, and (iv) the z-axis: (*a*) $(2,6,-3)$, (*b*) $(2,-\sqrt{3},3)$.
Ans. (*a*) 7, 2, 6, 3 (*b*) 4, 2, $\sqrt{3}$, 3

16. For each pair of points, find the coordinates of the point dividing P_1P_2 in the given ratio: find also the coordinates of the midpoint.
(*a*) $P_1(4,1,5)$, $P_2(2,-1,4)$, 3:2 *Ans.* (14/5, −1/5, 22/5), (3, 0, 9/2)
(*b*) $P_1(9,7,-2)$, $P_2(6,5,4)$, 1:4 *Ans.* (42/5, 33/5, −4/5), (15/2, 6, 1)
(*c*) $P_1(9,-2,-3)$, $P_2(-3,4,0)$, −1:3 *Ans.* (15, −5, −9/2), (3, 1, −3/2)
(*d*) $P_1(0,0,0)$, $P_2(2,3,4)$, 2:−3 *Ans.* (−4, −6, −8), (1, 3/2, 2)

17. Find the equation of the locus of a point which is (*a*) always equidistant from the points $(4,1,5)$ and $(2,-1,4)$, (*b*) always at a distance 6 units from $(4,1,5)$, (*c*) always two-thirds as far from the y-axis as from the origin.
Ans. (*a*) $4x+4y+2z-21=0$ (*b*) $x^2+y^2+z^2-8x-2y-10z+6=0$ (*c*) $5x^2-4y^2+5z^2=0$

18. Find a set of direction cosines and a set of direction numbers for the line joining P_1 and P_2, given:
(*a*) $P_1(0,0,0)$, $P_2(4,8,-8)$ *Ans.* [1/3, 2/3, −2/3], [1, 2, −2]
(*b*) $P_1(1,3,5)$, $P_2(-1,0,-1)$ *Ans.* [−2/7, −3/7, −6/7], [2, 3, 6]
(*c*) $P_1(5,6,-3)$, $P_2(1,-6,3)$ *Ans.* [−2/7, −6/7, 3/7], [2, 6, −3]
(*d*) $P_1(4,2,-6)$, $P_2(-2,1,3)$ *Ans.* $\left[\frac{6}{\sqrt{118}}, \frac{1}{\sqrt{118}}, \frac{-9}{\sqrt{118}}\right]$, [6, 1, −9]

19. Find $\cos \gamma$, given: (a) $\cos \alpha = 14/15$, $\cos \beta = -1/3$ (b) $\alpha = 60°$, $\beta = 135°$ *Ans.* (a) $\pm 2/15$ (b) $\pm \frac{1}{2}$

20. Find: (a) the acute angle between the line having direction numbers $[-4,-1,-8]$ and the line joining the points $(6,4,-1)$ and $(4,0,3)$.
(b) the interior angles of the triangle whose vertices are $A(2,-1,0)$, $B(4,1,-1)$, $C(5,-1,-4)$.
Ans. (a) $68°20'$ (b) $A = 48°10'$, $B = 95°10'$, $C = 36°40'$

21. Find the coordinates of the point P in which the line joining $A(5,-1,4)$ and $B(-5,7,0)$ pierces the yz-plane. Hint: Let P have coordinates $(0,b,c)$ and express the conditions (see Problem 8) that A,B,P be collinear. *Ans.* $P(0,3,2)$

22. Find relations which the coordinates of $P(x,y,z)$ must satisfy if P is to be collinear with $(2,3,1)$ and $(1,-2,-5)$.
Ans. $x-2 : y-3 : z-1 = 1:5:6$ or $\dfrac{x-2}{1} = \dfrac{y-3}{5} = \dfrac{z-1}{6}$

23. Find a set of direction numbers for any line perpendicular to:
(a) each of the lines $l_1 : [1,2,-4]$ and $l_2 : [2,-1,3]$,
(b) each of the lines joining $A(2,-1,5)$ to $B(-1,3,4)$ and $C(0,-5,4)$.
Ans. (a) $[2,-11,-5]$ (b) $[8,1,-20]$

24. Find the coordinates of the point P in which the line joining the points $A(4,11,18)$ and $B(-1,-4,-7)$ intersects the line joining the points $C(3,1,5)$ and $D(5,0,7)$. Hint: Let P divide AB in the ratio $1:r$ and CD in the ratio $1:s$, and obtain relations $rs - r - 4s - 6 = 0$, etc. *Ans.* $(1,2,3)$

25. Prove that the four line segments joining each vertex of a tetrahedron to the point of intersection of the medians of the opposite face have a point G in common. Prove that each of the four line segments is divided in the ratio $1:3$ by G. Note: G, the point P of Problem 4, is called the centroid of the tetrahedron.

CHAPTER 59

The Plane and Line in Space

IN THE PLANE of the rectangular coordinates (x, y) a linear equation $Ax + By + C = 0$, in which at least one of A, B is different from zero, represents a straight line in the plane. The simultaneous equations

$$A_1x + B_1y + C_1 = 0, \qquad A_2x + B_2y + C_2 = 0$$

of two non-parallel lines represent a point, the point of intersection of the lines.

In the space of the rectangular coordinates (x, y, z) a linear equation $Ax + By + Cz + D = 0$, in which at least one of A, B, C is different from zero, represents a plane of the space. The simultaneous equations

$$A_1x + B_1y + C_1z + D_1 = 0, \qquad A_2x + B_2y + C_2z + D_2 = 0$$

of two non-parallel planes represent a straight line, the line of intersection of the planes. The simultaneous equations

$$A_1x + B_1y + C_1z + D_1 = 0, \quad A_2x + B_2y + C_2z + D_2 = 0, \quad A_3x + B_3y + C_3z + D_3 = 0$$

of three planes, no two of which are parallel, represent a point, the unique point of intersection of the planes.

ANY LINE PERPENDICULAR to a plane is called a *normal* to the plane. In view of the definition of the angle between two skew lines, a normal to a plane is perpendicular to *every* line in the plane.

The equation of the plane passing through the fixed point $P_1(x_1, y_1, z_1)$ and having a line $n: [A, B, C]$ as a normal is

(1) $$A(x - x_1) + B(y - y_1) + C(z - z_1) = 0.$$

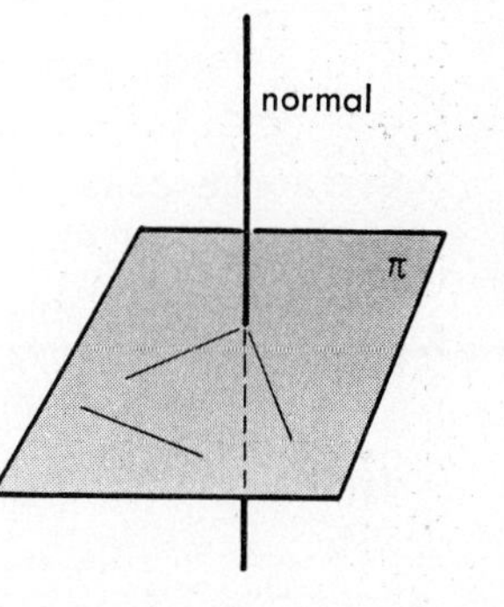

Every linear equation

(2) $$Ax + By + Cz + D = 0$$

where at least one of the coefficients A, B, C is different from zero, represents a plane whose normals have direction numbers $[A, B, C]$. Thus the equation of a plane is determined by three independent conditions.

See Problem 1.

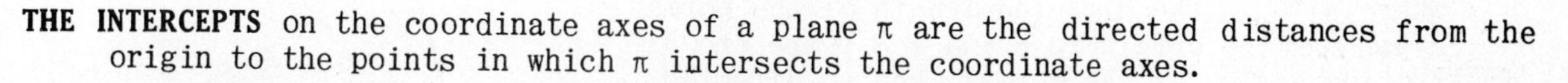

THE INTERCEPTS on the coordinate axes of a plane π are the directed distances from the origin to the points in which π intersects the coordinate axes.

If a, b, c, where $abc \neq 0$, are respectively the intercepts on the x-, y-, and z-axes, the *intercept form* of the equation of the plane is

(3) $$x/a + y/b + z/c = 1.$$

The *trace* on a coordinate plane of a plane π is the line of intersection of π and that coordinate plane.

Example 1. For the plane $\pi: 2x + 3y + 6z - 18 = 0$, find its intercepts on the coordinate axes and the equations of its traces on the coordinate planes.

Setting $y = z = 0$ in the equation of the plane and solving for x, we find the x-intercept equal to 9. In a similar manner, the y-intercept is found to be 6 and the z-intercept to be 3.

The equation of the trace in the xy-plane is

$$2x + 3y + 6z - 18 = 0, \quad z = 0$$

or, more simply,

$$2x + 3y - 18 = 0, \quad z = 0.$$

(Note that a line l may be represented by the simultaneous equations of *any* two distinct planes which contain l.)

Similarly, the equation of the trace in the xz-plane is $x + 3z - 9 = 0$, $y = 0$; and the equation of the trace in the yz-plane is $y + 2z - 6 = 0$, $x = 0$.

The intercepts and traces are shown in the figure above.

See Problems 2-3.

THE ANGLE BETWEEN TWO PLANES is defined as the angle between their respective normals. The angle θ between the planes

$$\pi_1: \; A_1x + B_1y + C_1z + D_1 = 0 \quad \text{and} \quad \pi_2: \; A_2x + B_2y + C_2z + D_2 = 0$$

is given by

$$(4) \qquad \cos\theta = \pm \frac{A_1A_2 + B_1B_2 + C_1C_2}{\sqrt{A_1^2 + B_1^2 + C_1^2}\,\sqrt{A_2^2 + B_2^2 + C_2^2}}.$$

As in the case of the corresponding formula for the angle between two lines, the two angles given by (*4*) are supplementary.

Two planes are parallel if and only if their normals are parallel, that is, if and only if $A_1/A_2 = B_1/B_2 = C_1/C_2$.

Two planes are mutually perpendicular if and only if their normals are perpendicular, that is, if and only if $A_1A_2 + B_1B_2 + C_1C_2 = 0$.

See Problems 4-5.

THE NORMAL FORM of the equation of a plane π is

$$(5) \qquad x\cos\alpha + y\cos\beta + z\cos\gamma - p = 0,$$

where $p > 0$ is the directed length of the normal drawn from the origin to π and α, β, γ are the direction angles of the normal when so directed.

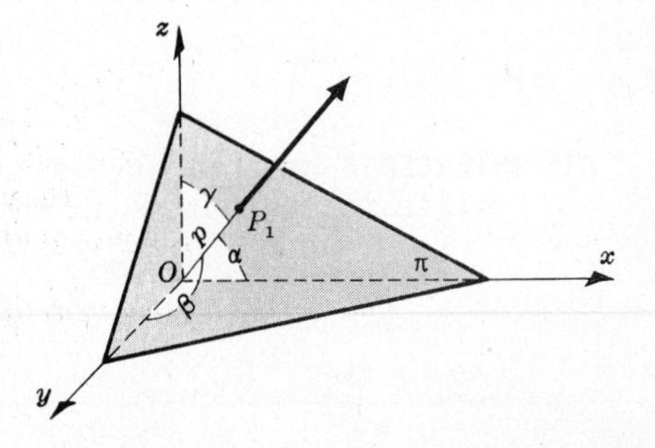

The general form (*1*) of the equation of a plane may be reduced to the normal form (*5*) by dividing it by $\pm\sqrt{A^2 + B^2 + C^2}$, choosing the sign before the radical:

(*a*) opposite to the sign of D, if $D \neq 0$,
(*b*) the same as the sign of C, if $D = 0$ and $C \neq 0$,
(*c*) the same as the sign of B, if $D = C = 0$ and $B \neq 0$,
(*d*) the same as the sign of A, if $D = C = B = 0$ and $A \neq 0$.

The directed distance *from a plane to a point* is found by writing the equation of the plane in normal form and substituting the coordinates of the point in its left member.

See Problems 7-10.

THE EQUATION OF THE FAMILY OF PLANES (also called the *pencil* of planes) passing through the intersection of the planes

$$\pi_1 :\ A_1x + B_1y + C_1z + D_1 = 0 \quad \text{and} \quad \pi_2 :\ A_2x + B_2y + C_2z + D_2 = 0$$

is

$$\pi_1 + k\pi_2 :\ A_1x + B_1y + C_1z + D_1 + k(A_2x + B_2y + C_2z + D_2) = 0$$

The common line of intersection of a family of planes is called the *axis* of the family.

See Problem 11.

THE ANALYTIC REPRESENTATION of a straight line in space is a pair of simultaneous linear equations

$$(7) \qquad \begin{aligned} \pi_1 &:\ A_1x + B_1y + C_1z + D_1 = 0 \\ \pi_2 &:\ A_2x + B_2y + C_2z + D_2 = 0 \end{aligned}$$

in which not all of the A, B, C in either equation are zero. Since the line lies in each of the planes π_1 and π_2, it is perpendicular to the normals to both. Thus a set of direction numbers for the line is obtained by the direction number device.

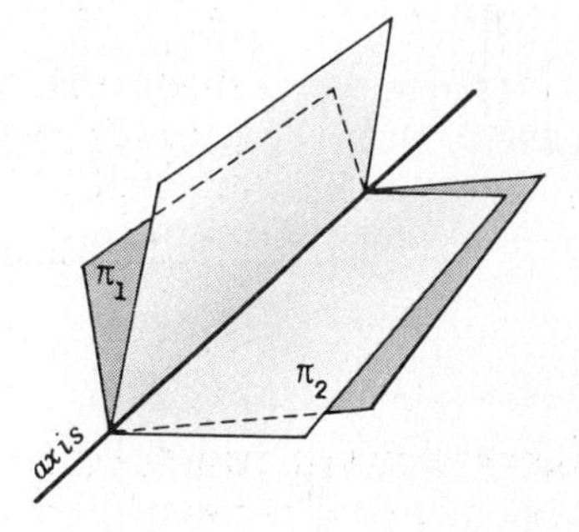

Example 2. Obtain a set of direction numbers for the line whose equations are $4x - y - z - 2 = 0$, $x + y - z + 5 = 0$.

Sets of direction numbers for the normals to the planes are respectively $[4, -1, -1]$ and $[1, 1, -1]$. Using the direction number device, a set of direction numbers of the line is $[2, 3, 5]$.

The representation (7), called the *general form* of the equations of a line, is not unique since the equations of any two distinct planes of the family $\pi_1 + k\pi_2 :$ $A_1x + B_1y + C_1z + D_1 + k(A_2x + B_2y + C_2z + D_2) = 0$ would serve equally well.

Example 3. The equations of a straight line l are $\pi_1 : 4x - y - z - 2 = 0$, $\pi_2 : x + y - z + 5 = 0$. Other representations of l are any two of the equations

(*a*) $\pi_1 + \pi_2 :\ 5x - 2z + 3 = 0$ (*b*) $\pi_1 + 2\pi_2 :\ 6x + y - 3z + 8 = 0$

(*c*) $\pi_1 - \pi_2 :\ 3x - 2y - 7 = 0$ (*d*) $\pi_1 - 4\pi_2 :\ -5y + 3z - 22 = 0$, etc.

Each of the equations (*a*), (*c*), (*d*) lacks one of the variables; hence these planes are perpendicular respectively to the *xz*-, *xy*-, and *yz*-planes. They are called respectively the *projecting planes on the xz-, xy-, and yz-planes* of the line l.

When (*a*) is put in the form $5x - 5 = 2z - 8$ or $\dfrac{x-1}{2} = \dfrac{z-4}{5}$, (*c*) is put in the form $\frac{x-1}{2} = \frac{y+2}{3}$, and (*d*) in the form $\frac{y+2}{3} = \frac{z-4}{5}$,

we may combine to have l: $\frac{x-1}{2} = \frac{y+2}{3} = \frac{z-4}{5}$. Note that from these

equations we may read a point (1,−2,4) on the line and a set of direction numbers [2,3,5] (see Example 2) of the line.

IF $P(x,y,z)$ **IS A VARIABLE POINT** and $P_1(x_1,y_1,z_1)$ is some fixed point on a line l: $[a,b,c]$, we have

$$(8) \qquad x - x_1 = at, \quad y - y_1 = bt, \quad z - z_1 = ct$$

where t is a non-zero constant.

If also $a \cdot b \cdot c \neq 0$, the equations (*8*) may be written as

$$(9) \qquad \frac{x-x_1}{a} = \frac{y-y_1}{b} = \frac{z-z_1}{c},$$

called the *symmetric form* of the equations of the line l. Again, the form (*9*) is not unique since P_1 is any known point of the line.

See Problems 12-13.

If one or more of the direction numbers are zero, (*9*) cannot be used. In such cases, we will use (*8*) with certain modifications.

See Problem 14.

SOLVED PROBLEMS

1. Write the equation of the plane
   (*a*) which passes through $P(2,3,4)$ and is perpendicular to l: $[3,-2,4]$.
   (*b*) which passes through $P(2,3,4)$ and is perpendicular to the x-axis.
   (*c*) which passes through $P(2,3,4)$ and is parallel to π: $4x - 3y + 5z - 1 = 0$.
   (*d*) which is the perpendicular bisector of the segment joining $A(2,3,1)$ and $B(6,-3,5)$.
   (*e*) which passes through the points $A(2,1,3)$, $B(3,-3,4)$, and $C(-1,1,-4)$.
   (*f*) which passes through $P(2,3,4)$ and is perpendicular to the planes π_1: $2x - y + 2z - 8 = 0$ and π_2: $x + 2y - 3z + 7 = 0$.

   (*a*) Using (*1*) with $(x_1,y_1,z_1) = (2,3,4)$ and $[A,B,C] = [3,-2,4]$, the equation is $3(x-2) - 2(y-3) + 4(z-4) = 0$ or $3x - 2y + 4z - 16 = 0$.

   (*b*) A set of direction numbers of the x-axis is $[1,0,0]$. The equation of the plane is $1(x-2) + 0(y-3) + 0(z-4) = 0$ or $x - 2 = 0$.

   (*c*) Since two parallel planes have common normals, the equation of the required plane has the form $4x - 3y + 5z + D = 0$. This plane will pass through $P(2,3,4)$ provided $4\cdot 2 - 3\cdot 3 + 5\cdot 4 + D = 0$ or $D = -19$. The desired equation is $4x - 3y + 5z - 19 = 0$.

   (*d*) A set of direction numbers of the line AB, a normal to the required plane, is $[4,-6,4]$ or $[2,-3,2]$; and the plane passes through the midpoint $(4,0,3)$ of the segment AB. The desired equation is $2(x-4) - 3(y-0) + 2(z-3) = 0$ or $2x - 3y + 2z - 14 = 0$.

   (*e*) *First Solution.* Let the equation of the plane be $Ax + By + Cz + D = 0$. Since the coordinates of each of the given points must satisfy this equation, we have

$$\begin{aligned} 2A + B + 3C + D &= 0 \\ 3A - 3B + 4C + D &= 0 \\ -A + B - 4C + D &= 0 \end{aligned}$$

Provided $D \neq 0$, we may solve for A,B,C obtaining $A = -\frac{28}{24}D$, $B = -\frac{4}{24}D$, and $C = \frac{12}{24}D$.

Then $-\frac{28}{24}Dx - \frac{4}{24}Dy + \frac{12}{24}Dz + D = 0$ or $7x + y - 3z - 6 = 0$ is the required equation.

Second Solution. The required plane passes through A and any normal is perpendicular to the lines AB and AC. Sets of direction numbers of AB and AC are respectively $[1,-4,1]$ and $[3,0,7]$. Using the direction number device, $[-28,-4,12]$ is a set of direction numbers of the normal. The equation of the plane is $-28(x-2) - 4(y-1) + 12(z-3) = 0$ or $7x + y - 3z - 6 = 0$.

(*f*) Let the equation of the required plane be $Ax + By + Cz + D = 0$.

Since the plane passes through $P(2,3,4)$, (i) $2A + 3B + 4C + D = 0$,
since it is perpendicular to π_1, (ii) $2A - B + 2C = 0$,
since it is perpendicular to π_2, (iii) $A + 2B - 3C = 0$.

A solution of (ii) and (iii) may be obtained from
$$\begin{array}{ccccc} 2 & -1 & 2 & 2 & -1 \\ 1 & 2 & -3 & 1 & 2 \end{array}$$
as $A = -1$, $B = 8$, $C = 5$. Then from (i), $D = -2A - 3B - 4C = 2 - 24 - 20 = -42$.

The desired equation is $-x + 8y + 5z - 42 = 0$ or $x - 8y - 5z + 42 = 0$.

2. For each of the given planes find the intercepts on the coordinate axes and the equations of the traces in the coordinate planes, and construct the figure.
(*a*) $x + 2y - z - 4 = 0$, (*b*) $x + 2y + 3z = 0$, (*c*) $x + 2y - 4 = 0$, (*d*) $3x - 4 = 0$.

(*a*) The x-, y-, and z-intercepts are respectively $4, 2, -4$. The equations of the traces in the xy-, xz-, and yz-planes are respectively $x + 2y - 4 = 0$, $z = 0$; $x - z - 4 = 0$, $y = 0$; $2y - z - 4 = 0$, $x = 0$.

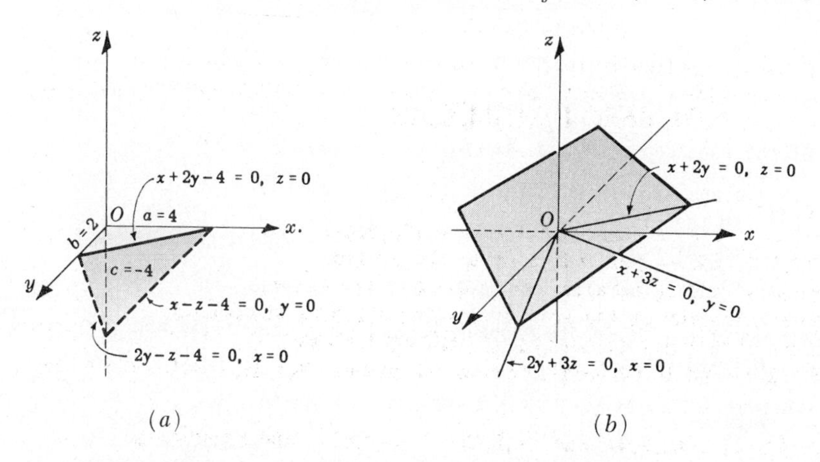

(*a*) (*b*)

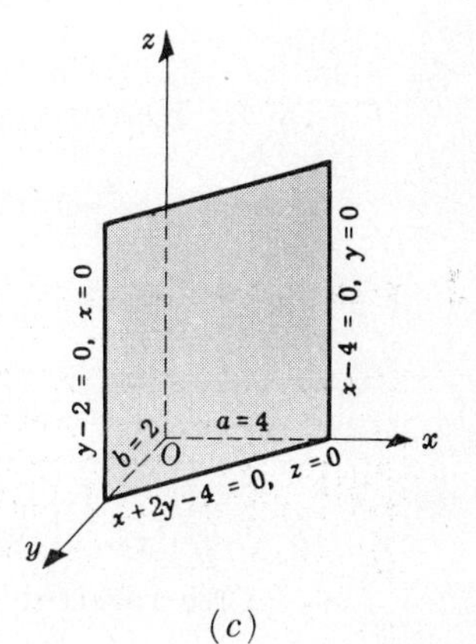

(*c*)

(*b*) Each intercept on a coordinate axis is 0. The equations of the traces are $x + 2y = 0$, $z = 0$; $x + 3z = 0$, $y = 0$; $2y + 3z = 0$, $x = 0$.

(*c*) Note that the equation lacks a term in z and the plane is thus parallel to the z-axis. The x-intercept is 4 and the y-intercept is 2; there is no z-intercept. The equations of the traces are $x + 2y - 4 = 0$, $z = 0$; $x - 4 = 0$, $y = 0$; $y - 2 = 0$, $x = 0$. The latter two traces are parallel to the z-axis.

(*d*) The x-intercept is $4/3$; there is no y- or z-intercept. Note that the equation lacks terms in y and z; being parallel to both the y- and z-axis, the plane is parallel to the yz-plane. The equations of the traces are $3x - 4 = 0$, $y = 0$ and $3x - 4 = 0$, $z = 0$. There is no trace in the yz-plane.

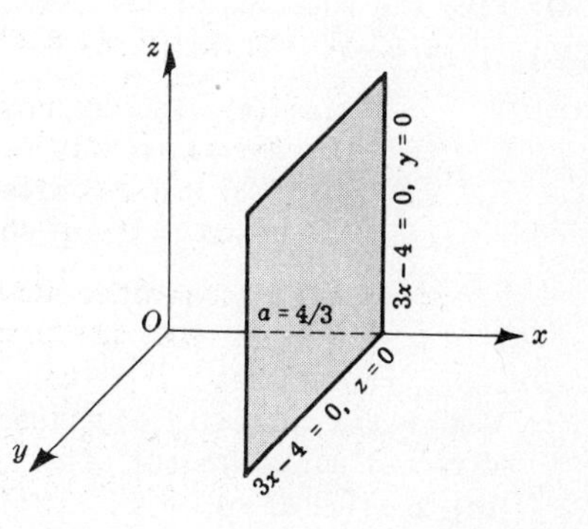

3. Write the equation of the plane
(a) having the intercepts 3,5,−2.
(b) parallel to the y-axis, x-intercept equal to −3, z-intercept equal to 4.
(c) parallel to the xz-plane, y-intercept equal to 6.

(a) Using (3), obtain $x/3 + y/5 + z/-2 = 1$ or $10x + 6y - 15z - 30 = 0$ as the required equation.

(b) Since the plane is parallel to the y-axis, its equation lacks a term in y. The desired equation is $x/-3 + z/4 = 1$ or $4x - 3z + 12 = 0$.

(c) The equation of this plane lacks terms in x and z; it is $y/6 = 1$ or $y - 6 = 0$.

4. Find the acute angle between each pair of planes:
(a) $5x - 14y + 2z - 8 = 0$ and $10x - 11y + 2z + 15 = 0$, (b) $5x - 14y + 2z - 8 = 0$ and the xy-plane.

(a) Here sets of direction numbers of normals to the given planes are respectively [5,−14,2] and [10,−11,2]. Substituting in (4) we have

$$\cos\theta = \pm\frac{5\cdot 10 + (-14)(-11) + 2\cdot 2}{\sqrt{5^2+(-14)^2+2^2}\,\sqrt{10^2+(-11)^2+2^2}} = \pm\frac{208}{225} = \pm 0.924 .$$

For the acute angle, $\cos\theta = 0.924$ and $\theta = 22°30'$.

(b) Respective sets of direction numbers of normals to the planes are [5,−14,2] and [0,0,1]. From (4), we have

$$\cos\theta = \pm\frac{5\cdot 0 + (-14)\cdot 0 + 2\cdot 1}{15\cdot 1} = \pm\frac{2}{15} = \pm 0.133 \quad\text{and}\quad \theta = 82°20'.$$

5. Find the equation of the plane perpendicular to π: $2x + 3y - 5z - 6 = 0$ and passing through the points $P(2,-1,-1)$ and $Q(1,2,3)$.

Let the equation of the required plane be $Ax + By + Cz + D = 0$.

Since this plane is perpendicular to π,	(i)	$2A + 3B - 5C = 0$,
since P is in the plane,	(ii)	$2A - B - C + D = 0$,
since Q is in the plane,	(iii)	$A + 2B + 3C + D = 0$.
Subtracting (iii) from (ii), we have	(iv)	$A - 3B - 4C = 0$.

Using the direction number device, as in Problem 1(f), a solution of (i) and (iv) is $A = -27$, $B = 3$, $C = -9$. From (ii), $D = -2A + B + C = 48$.

The required equation is $-27x + 3y - 9z + 48 = 0$ or $9x - y + 3z - 16 = 0$.

6. Find the equation of the plane π which passes through the point $A(1,-2,3)$ and is parallel to the lines l_1: [2,1,−1] and l_2: [3,6,−2].

Using (1), let the equation of π be $A(x-1) + B(y+2) + C(z-3) = 0$. Since it is parallel to l_1, a normal to π is perpendicular to l_1; then,
(i) $2A + B - C = 0$. Similarly, a normal to π is perpendicular to l_2 and
(ii) $3A + 6B - 2C = 0$. A solution of (i) and (ii) is $A = 4$, $B = 1$, $C = 9$.

The equation of π is $4(x-1) + 1(y+2) + 9(z-3) = 0$ or $4x + y + 9z - 29 = 0$.

7. Reduce the following equations to normal form and determine the length and direction angles of the directed normal to the origin:
(a) $2x - 9y - 6z + 55 = 0$, (b) $x - 2y - 2z = 0$, (c) $3x + 4y = 0$.

(a) Here $A = 2$, $B = -9$, $C = -6$, and $D = +55$. Since $D > 0$, we divide by $-\sqrt{A^2+B^2+C^2} = -11$ and obtain $-\frac{2}{11}x + \frac{9}{11}y + \frac{6}{11}z - 5 = 0$. The length of the normal is 5 and its direction angles are α = arc cos (−2/11) = 100°30′, β = arc cos 9/11 = 35°10′, γ = arc cos 6/11 = 57°0′.

(*b*) Here $D = 0$ and $C < 0$; we divide by $-\sqrt{A^2+B^2+C^2} = -3$ and obtain $-\frac{1}{3}x + \frac{2}{3}y + \frac{2}{3}z = 0$. Then $p = 0$ and the direction angles are α = arc cos (−1/3) = 109°30′, $\beta = \gamma$ = arc cos 2/3 = 48°10′.

(*c*) Here $D = C = 0$ and $B > 0$; we divide by $+\sqrt{A^2+B^2+C^2} = 5$ and obtain $\frac{3}{5}x + \frac{4}{5}y = 0$. Then p = 0 and the direction angles are α = arc cos 3/5 = 53°10′, β = arc cos 4/5 = 36°50′, and γ = arc cos 0 = 90°.

8. (*a*) Find the directed distance from the plane π: $3x - 6y + 2z + 18 = 0$ to the point $A(2,-3,-4)$.
(*b*) Find the (undirected) distance between the parallel planes π_1: $2x + y - 2z + 4 = 0$ and π_2: $2x + y - 2z - 6 = 0$.

(*a*) The normal form of the equation of π is $-\frac{3}{7}x + \frac{6}{7}y - \frac{2}{7}z - \frac{18}{7} = 0$.

Substituting into the left member of this equation the coordinates of A, we have $d = -\frac{3}{7}(2) + \frac{6}{7}(-3) - \frac{2}{7}(-4) - \frac{18}{7} = -\frac{34}{7}$. Note that the origin and A are on the same side of the plane π.

(*b*) The normal forms of the equation of the planes are

$$\pi_1:\ -\frac{2}{3}x - \frac{1}{3}y + \frac{2}{3}z - \frac{4}{3} = 0 \quad \text{and} \quad \pi_2:\ \frac{2}{3}x + \frac{1}{3}y - \frac{2}{3}z - 2 = 0.$$

These planes are on opposite sides of the origin (the coefficients in π_1 are the negatives of the corresponding coefficients in π_2) at undirected distances 4/3 and 2. The undirected distance between them is then $4/3 + 2 = 10/3$.

9. Write the equations of the planes parallel to π: $3x - 6y + 2z + 14 = 0$ and
(*a*) 6 units from the origin, (*b*) 4 units from $A(2,1,-3)$, (*c*) 5 units from π.

The required planes have equations $3x - 6y + 2z + D = 0$ or in normal form $\frac{3x - 6y + 2z + D}{\pm 7} = 0$.

(*a*) Since $D/\pm 7 = 6$ and $D = \pm 42$, the equations of the planes are $3x - 6y + 2z \pm 42 = 0$.

(*b*) Since $\frac{3(2) - 6(1) + 2(-3) + D}{\pm 7} = 4$, $D = 34$ and -22. The equations are $3x - 6y + 2z + 34 = 0$ and $3x - 6y + 2z - 22 = 0$.

(*c*) Since π is 2 units from the origin, one of the required planes is on the same side of the origin at a distance 7 from it and the other is on the opposite side of the origin at a distance 3 from it. Their equations are $3x - 6y + 2z + 49 = 0$ and $3x - 6y + 2z - 21 = 0$.

10. Find the equations of the bisectors of the dihedral angles formed by the planes π_1: $x + 2y - 2z + 6 = 0$ and π_2: $4x - y + 8z - 8 = 0$.

Let $P(x,y,z)$ be any point on a bisector. Then, as in the analogous problem in the plane, P is equidistant from π_1 and π_2: hence $\frac{x + 2y - 2z + 6}{-3} = \pm\frac{4x - y + 8z - 8}{9}$ and the required equations are $7x + 5y + 2z + 10 = 0$ and $x - 7y + 14z - 26 = 0$.

11. Write the equation of the plane passing through the line of intersection of the planes π_1: $6x + 4y + 3z + 5 = 0$ and π_2: $2x + y + z - 2 = 0$, and (*a*) passing through $A(2,-3,2)$, (*b*) 3 units from the origin, (c) perpendicular to the plane having equation $4x + y - 5z - 9 = 0$.

The required planes are members of the family $\pi_1 + k\pi_2$: $6x + 4y + 3z + 5 + k(2x + y + z - 2) = 0$ or

(i) $$(6 + 2k)x + (4 + k)y + (3 + k)z + 5 - 2k = 0$$

(a) Substituting the coordinates of A in (i), we have

$$(6 + 2k)2 + (4 + k)(-3) + (3 + k)2 + 5 - 2k = 0 \quad \text{and} \quad k = -11.$$

The equation of the plane is $-16x - 7y - 8z + 27 = 0$ or $16x + 7y + 8z - 27 = 0$.

(b) The normal form of $\pi_1 + k\pi_2$ is $\dfrac{(6 + 2k)x + (4 + k)y + (3 + k)z + 5 - 2k}{\pm\sqrt{61 + 38k + 6k^2}} = 0$.

Setting the distance from the origin equal to 3, we have $\dfrac{5 - 2k}{\pm\sqrt{61 + 38k + 6k^2}} = 3$ which, after squaring and simplifying, becomes $25k^2 + 181k + 262 = (k + 2)(25k + 131) = 0$. For $k = -2$, the plane has equation $2x + 2y + z + 9 = 0$; for $k = -131/25$, it has equation $112x + 31y + 56z - 387 = 0$.

(c) A set of direction numbers for a normal to (i) is $[6 + 2k, 4 + k, 3 + k]$; a set for the given plane is $[4, 1, -5]$. The condition for perpendicularity is $4(6 + 2k) + 1(4 + k) - 5(3 + k) = 13 + 4k = 0$. Then $k = -13/4$ and the required equation is $2x - 3y + z - 46 = 0$.

12. Write the equations (in symmetric form) of the line passing through point $P_1(2, -3, 5)$ and
(a) having direction numbers $[1, -2, 4]$.
(b) perpendicular to the plane π: $9x - 4y + 2z - 11 = 0$.
(c) parallel to the line l: $x - y + 2z + 4 = 0$, $2x + 3y + 6z - 12 = 0$.
(d) passing through the point $P_2(3, 6, -2)$.

(a) By (9), the equations are $\dfrac{x - 2}{1} = \dfrac{y + 3}{-2} = \dfrac{z - 5}{4}$.

(b) Since the line is perpendicular to π, a set of direction numbers for it is $[9, -4, 2]$. The required equations are $\dfrac{x - 2}{9} = \dfrac{y + 3}{-4} = \dfrac{z - 5}{2}$.

(c) From $\begin{matrix} 1 & -1 & 2 & 1 & -1 \\ 2 & 3 & 6 & 2 & 3 \end{matrix}$, a set of direction numbers for l is $[-12, -2, 5]$. Since the required line is parallel to l, its equations are $\dfrac{x - 2}{-12} = \dfrac{y + 3}{-2} = \dfrac{z - 5}{5}$ or $\dfrac{x - 2}{12} = \dfrac{y + 3}{2} = \dfrac{z - 5}{-5}$.

(d) A set of direction numbers for the line P_1P_2 is $[3 - 2, 6 + 3, -2 - 5] = [1, 9, -7]$. Using P_1, the equations are $\dfrac{x - 2}{1} = \dfrac{y + 3}{9} = \dfrac{z - 5}{-7}$ or, using P_2, $\dfrac{x - 3}{1} = \dfrac{y - 6}{9} = \dfrac{z + 2}{-7}$.

13. Obtain in symmetric form the equations of the following lines:
(a) π_1: $3x - 4y + 6z - 5 = 0$, π_2: $2x + 3y - 4z - 1 = 0$. (b) π_1: $x + 4y - 4z + 2 = 0$, π_2: $6x - 2y + z - 9 = 0$.

The procedure used here is to locate a point on the line by any means and to find a set of direction numbers of the line by the direction number device.

(a) Setting $z = 0$ in the equations and solving $3x - 4y - 5 = 0$ and $2x + 3y - 1 = 0$ simultaneously, we obtain the point $(19/17, -7/17, 0)$ on the line. Using the direction number device on the sets $[3, -4, 6]$ and $[2, 3, -4]$, we obtain the set of direction numbers $[-2, 24, 17]$ of the line. By (9) the required equations are $\dfrac{x - 19/17}{-2} = \dfrac{y + 7/17}{24} = \dfrac{z}{17}$.

The point $(1, 1, 1)$ is also on the line; another symmetric form is $\dfrac{x - 1}{-2} = \dfrac{y - 1}{24} = \dfrac{z - 1}{17}$.

(*b*) Setting $y = 0$ and solving $x - 4z + 2 = 0$ and $6x + z - 9 = 0$ simultaneously, we obtain the point $(34/25, 0, 21/25)$. A set of direction numbers is $[-4,-25,-26]$. Then $\frac{x - 34/25}{4} = \frac{y}{25} = \frac{z - 21/25}{26}$ are the equations of the line.

14. Write the equations of the line which passes through point $P_1(1,-2,3)$ and (*a*) has $[2,5,0]$ as direction numbers, (*b*) has $[0,1,0]$ as direction numbers, (*c*) passes through point $(3,-2,2)$.

(*a*) Using (*8*), since one direction number is 0, we obtain $x - 1 = 2t$, $y + 2 = 5t$, $z - 3 = 0$. A preferred form for these equations is $\frac{x-1}{2} = \frac{y+2}{5}$, $z - 3 = 0$ obtained by equating the values of t from the first two equations.

(*b*) Using (*8*), since two direction numbers are 0, we obtain $x - 1 = 0$, $y + 2 = t$, $z - 3 = 0$. A line is uniquely determined when the equations of two planes containing it are given; we write the equations of this line as $x - 1 = 0$, $z - 3 = 0$.

(*c*) A set of direction numbers for the line is $[2,0,-1]$. The equations are

$$\frac{x-1}{2} = \frac{z-3}{-1}, \quad y + 2 = 0 \qquad \text{or} \qquad \frac{x-3}{2} = \frac{z-2}{-1}, \quad y + 2 = 0.$$

15. Equations (*8*) when written as

$$(8') \qquad x = x_1 + at, \quad y = y_1 + bt, \quad z = z_1 + ct$$

are called *parametric equations*, t being the parameter, of the line. Use (*8′*) to find the coordinates of the point in which the line l: $\frac{x-1}{2} = \frac{y-2}{3} = \frac{z+4}{5}$ pierces the plane π: $3x - 4y + 8z - 31 = 0$.

The parametric equations of l are $x = 1 + 2t$, $y = 2 + 3t$, $z = -4 + 5t$. When these replacements are made in the equation of the plane, we have $3(1+2t) - 4(2+3t) + 8(-4+5t) - 31 = 0$ and $t = 2$. Then $x = 1 + 2t = 1 + 2\cdot 2 = 5$, $y = 2 + 3\cdot 2 = 8$, $z = -4 + 5\cdot 2 = 6$, and the point of intersection has coordinates $(5,8,6)$.

16. Write the equation of the line passing through the point $(-4,5,3)$ and

(*a*) parallel to the line l: $\frac{x+2}{3} = \frac{y-4}{-1} = \frac{z-1}{5}$.

(*b*) perpendicular to each of the lines l_1: $\frac{x-2}{3} = \frac{y+3}{2} = \frac{z-5}{5}$ and l_2: $\frac{x+1}{6} = \frac{y-2}{-4} = \frac{z+1}{3}$.

(*a*) The direction numbers of l are a set for the required line; then $\frac{x+4}{3} = \frac{y-5}{-1} = \frac{z-3}{5}$.

(*b*) Sets of direction numbers for the lines l_1 and l_2 are respectively $[3,2,5]$ and $[6,-4,3]$. A set for the required line is $[26,21,-24]$ and the equations are $\frac{x+4}{26} = \frac{y-5}{21} = \frac{z-3}{-24}$.

SUPPLEMENTARY PROBLEMS

17. For each of the following planes, find (i) a set of direction numbers of a normal and (ii) the direction cosines of a normal when directed upward:

(a) $3x+2y-6z+22=0$ *Ans.* $[3,2,-6]$, $[-3/7, -2/7, 6/7]$

(b) $4x-3y+5z-30=0$ *Ans.* $[4,-3,5]$, $[2\sqrt{2}/5, -3\sqrt{2}/10, \sqrt{2}/2]$

(c) $5y-12z+20=0$ *Ans.* $[0,5,-12]$, $[0, -5/13, 12/13]$

18. Write the equation of the plane which

(a) passes through $P(1,-3,-2)$ and is perpendicular to $l:[1,0,3]$,

(b) passes through $P(2,-3,5)$ and has the line joining $A(1,-3,-5)$ and $B(2,2,3)$ as a normal,

(c) bisects the line segment joining $(5,-2,6)$ and $(7,2,0)$,

(d) passes through $P(1,-2,-4)$ and is parallel to $7x-4y+6z+2=0$,

(e) passes through the points $(-8,6,0)$, $(0,12,0)$, $(-10,0,-9)$,

(f) passes through the points $(6,2,3)$, $(3,3,-2)$, $(2,-1,-2)$,

(g) passes through the y-axis and the point $(4,2,-3)$.

Ans. (a) $x+3z+5=0$ (b) $x+5y+8z-27=0$ (c) $x+2y-3z+3=0$ (d) $7x-4y+6z+9=0$ (e) $3x-4y+2z+48=0$ (f) $20x-5y-13z-71=0$ (g) $3x+4z=0$

19. Write the equation of the plane

(a) whose intercepts are twice those of $4x-3y+6z-24=0$,

(b) if the foot of the normal from the origin is $P(2,-1,-3)$,

(c) if the foot of the normal from the point $(3,3,0)$ is $P(2,1,1)$,

(d) which passes through $(4,1,-2)$ and is parallel to $l_1:[1,-2,3]$ and $l_2:[3,-4,0]$,

(e) which is parallel to $x-3y+4z+24=0$ and for which the sum of the intercepts on the coordinate axes is 1.

Ans. (a) $4x-3y+6z-48=0$ (b) $2x-y-3z-14=0$ (c) $x+2y-z-3=0$ (d) $12x+9y+2z-53=0$ (e) $11x-33y+44z-12=0$

20. Derive the formula for the cosine of the angle θ between the planes

$$\pi_1 : A_1x + B_1y + C_1z + D_1 = 0 \quad \text{and} \quad \pi_2 : A_2x + B_2y + C_2z + D_2 = 0 .$$

21. Find the acute angle between the pairs of planes:

(a) $x-3y+4z-8=0$ and $3x+y-4z+9=0$. *Ans.* $52°0'$

(b) $x-4y+5z+6=0$ and the yz-plane. *Ans.* $81°10'$

22. Write the equation of the plane

(a) which passes through the points $(2,-3,1)$ and $(-1,4,-1)$ and is perpendicular to $5x-3y+3z+15=0$,

(b) which passes through the point $(4,1,-3)$ and is perpendicular to each of the planes $3x-5y+2z+15=0$ and $3x-y+7z-21=0$,

(c) which passes through the points $(1,2,-4)$ and $(2,3,-5)$ and is perpendicular to the xz-plane.

Ans. (a) $15x-y-26z-7=0$ (b) $11x+5y-4z-61=0$ (c) $x+z+3=0$

23. Reduce each of the following equations to the normal form. Find the length and the direction angles of the directed normal from the origin, and the directed distance from the given plane to the point $(2,5,-4)$.

(a) $3x-4y-12z+24=0$ *Ans.* 24/13; $103°20'$, $72°0'$, $22°40'$; $-58/13$

(b) $6x-7y+6z-44=0$ *Ans.* 4; $57°0'$, $129°30'$, $57°0'$; $-91/11$

(c) $5x-12z=0$ *Ans.* 0; $112°40'$, $90°$, $22°40'$; $-58/13$

24. Find the undirected distance between each pair of parallel planes. Write the equation of the plane which is equidistant from each pair.

(a) $3x-y+z-4=0$, $3x-y+z-10=0$ *Ans.* $6\sqrt{11}/11$, $3x-y+z-7=0$

(b) $2x-3y+4z+6=0$, $2x-3y+4z-8=0$ Ans. $14\sqrt{29}/29$, $2x-3y+4z-1=0$

(c) $x-2y+2z+4=0$, $3x-6y+6z+23=0$ *Ans.* 11/9, $6x-12y+12z+35=0$

25. Determine the equation of the plane of the family $2x+ky-kz+9=0$ which

(a) passes through the point $(2,-3,-4)$. *Ans.* $2x-13y+13z+9=0$

(b) is perpendicular to the plane $4x-3y+z+8=0$. *Ans.* $2x+2y-2z+9=0$

26. Find the equations of the bisectors of the dihedral angles formed by each pair of planes:
(a) $2x-y+2z+5=0,\ 6x-7y-6z-9=0$ (b) $3x-2y+4z-8=0,\ 5x+4y+3z+6=0$
Ans. (a) $2x+5y+20z+41=0,\ 10x-8y+z+7=0$
(b) $(15\sqrt{2}\pm 5\sqrt{29})x - (10\sqrt{2}\mp 4\sqrt{29})y + (20\sqrt{2}\pm 3\sqrt{29})z - (40\sqrt{2}\mp 6\sqrt{29}) = 0$

27. Write equations of the line passing through the point (1,-2,5) and
(a) having direction numbers [2,-1,-3],
(b) having direction numbers [3,0,1],
(c) perpendicular to the plane $3x-4y-7z+13=0$,
(d) parallel to the line $\frac{x}{2}=\frac{y-1}{3}=\frac{z+2}{-4}$,
(e) parallel to the line $x+y-2z-5=0,\ 3x-2y+4z+8=0$,
(f) passing through the point (2,1,-7),
(g) perpendicular to the lines $2x-3y-2z-1=0,\ 3x-6y-2z+5=0$
and $x+2y-3z+2=0,\ 6x-y-2z+8=0$.

Ans. (a) $\frac{x-1}{2}=\frac{y+2}{-1}=\frac{z-5}{-3}$ (d) $\frac{x-3}{2}=\frac{y-1}{3}=\frac{z-1}{-4}$
(b) $\frac{x-1}{3}=\frac{z-5}{1},\ y+2=0$ (e) $\frac{y+2}{2}=\frac{z-5}{1},\ x-1=0$ (g) $\frac{x-1}{22}=\frac{y+2}{57}=\frac{z-5}{-82}$
(c) $\frac{x-1}{-3}=\frac{y+2}{4}=\frac{z-5}{7}$ (f) $\frac{x-2}{1}=\frac{y-1}{3}=\frac{z+7}{-12}$

28. Obtain in symmetric form the equations of the following lines:
(a) $2x+5y+3z-1=0,\ 6x+4y-z+10=0$
(b) $3x-2y-z-6=0,\ 4x+y-3z-1=0$
(c) $x+3y-2z+21=0,\ 3x-6y+4z-32=0$
Ans. (a) $\frac{x+4}{17}=\frac{y-3}{-20}=\frac{z+2}{22}$ (b) $\frac{x+5}{7}=\frac{y+6}{5}=\frac{z+9}{11}$ (c) $\frac{y+3}{2}=\frac{z-5}{3},\ x+2=0$

29. Find the coordinates of the point in which the given line pierces the plane $2x-3y+5z-8=0$.
(a) $\frac{x-2}{2}=\frac{y-2}{3}=\frac{z+4}{-5}$ (b) $\frac{x-7}{4}=\frac{y+2}{-3},\ z-1=0$
(c) the line joining (4,6,6) and (3,8,10) (d) the line $2x+3y+z-10=0,\ 4x+6y-3z=0$
Ans. (a) (0,-1,1) (b) (3,1,1) (c) (5,4,2) (d) (-3/2,3,4)

30. Show that the lines $\frac{x-1}{3}=\frac{y-2}{4}=\frac{z-3}{5}$ and $\frac{x+8}{6}=\frac{y+5}{8}=\frac{z+6}{10}$ are parallel and find the distance between them. Write the equation of the plane containing the two lines.
Ans. $\sqrt{11}$; $x+18y-15z+8=0$

31. If the plane $x/a+y/b+z/c=1$ is p units from the origin, show that $1/p^2=1/a^2+1/b^2+1/c^2$.

32. The shortest distance (undirected) from a given point P_1 to a given line l in space is the distance P_1P_2 where P_2 is the point in which the plane through P_1 perpendicular to l intersects it. Find the shortest distance from the point $P_1(1,-2,3)$ to the line $\frac{x-5}{4}=\frac{y+1}{2}=\frac{z}{5}$. *Ans.* $\sqrt{645}/5$

33. The shortest distance (undirected) between two skew lines l_1 and l_2 is the distance between P_1, any point on l_1, and the plane through l_2 parallel to l_1. Find the shortest distance between the two skew lines l_1: $2x+y-z-4=0,\ x+y+3z+2=0$ and l_2: $x-y-z+1=0,\ 3y+z-5=0$.
Ans. $\sqrt{6}/6$

CHAPTER 60

Surfaces and Curves in Space

THE TOTALITY OF THOSE POINTS, and only those points, whose coordinates satisfy a single equation of the form

(1) $$F(x, y, z) = 0$$

is called a *surface*. With only a few exceptions, we shall be concerned here with surfaces whose equations are of the second degree in the variables. Such surfaces are called *quadric surfaces*.

The totality of those points, and only those points, whose coordinates satisfy simultaneously a pair of equations of the form (1) $F(x,y,z) = 0$, $G(x,y,z) = 0$ is called a *space curve*. For example, $x^2 + y^2 + z^2 = 25$ is the equation of a sphere while $x^2 + y^2 + z^2 = 25$, $z = 4$ are the equations of the circle of intersection of the sphere and the plane $z = 4$.

A CYLINDER OR CYLINDRICAL SURFACE is generated by a straight line which moves so that it is always parallel to a fixed straight line and always passes through a given curve. The moving line is called the *generatrix*, the given curve is called the *directrix*, and the generatrix in any of its positions is called an *element* of the cylinder. If the directrix is a plane curve and the generatrix is perpendicular to this plane, the cylinder is called a *right cylinder*; if, in addition, the directrix is a conic, the cylinder is called a *right quadric cylinder*.

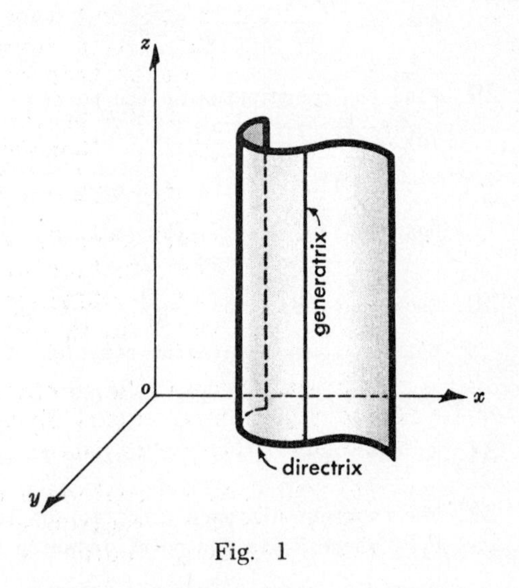

Fig. 1

We shall limit further discussion to the right cylinder with generatrix perpendicular to a coordinate plane or parallel to a coordinate axis. The equations of such cylinders have the form $f(x,y) = 0$, $g(x,z) = 0$, or $h(y,z) = 0$; in each case, the generatrix of the cylinder is parallel to the coordinate axis along which the variable missing in the equation is measured. Conversely, the locus of any equation which involves only two of the variables is a cylinder whose directrix is the curve in the plane of those variables which has the same equation and whose generatrix is parallel to the axis of the missing variable.

Example 1. Discuss the right quadric cylinder $x^2 + y^2 - 4x - 12 = 0$.

This is a right circular cylinder generated by moving a line always parallel to the z-axis (the equation lacks the variable z) and always intersecting the circle $(x-2)^2 + y^2 = 16$, $z = 0$ in the xy-plane, as shown in Fig. 2 below.

See Problem 1.

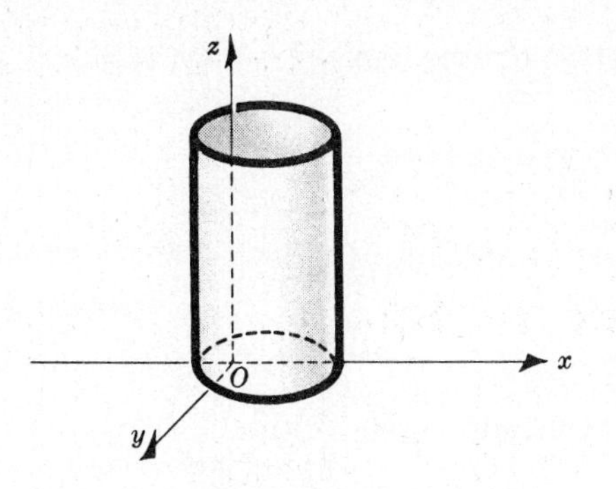

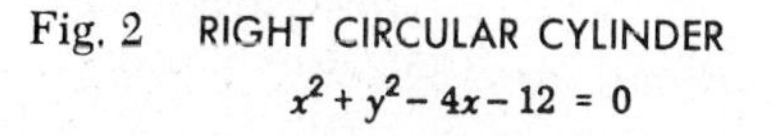

Fig. 2 RIGHT CIRCULAR CYLINDER
$x^2 + y^2 - 4x - 12 = 0$

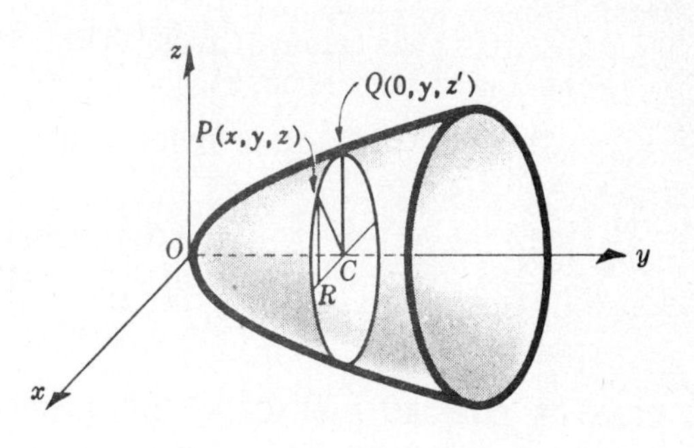

Fig. 3 PARABOLOID OF REVOLUTION
$x^2 + z^2 = 2y$

A SURFACE OF REVOLUTION is one generated by revolving a plane curve, called the *generatrix*, about a straight line (called the axis) which lies in the plane of the curve. Clearly, the section of such a surface by a plane perpendicular to the axis of revolution consists of one or more circles.

Example 2. Find the equation of the surface generated by revolving the parabola $z^2 = 2y$, $x = 0$ about the y-axis.

Although not necessary, it is a trifle more convenient in sketching the figure to choose the coordinate axes as in Fig. 3 above. Let $P(x, y, z)$ be any point on the surface. Denote by C the center of the circle cut from the surface by the plane through P perpendicular to the y-axis (axis of revolution) and by $Q(0, y, z')$ a point of intersection of this circle and the given parabola.

Let R be the foot of the perpendicular dropped from P to the xy-plane. Then $CP = CQ$ since both are radii of the same circle.

Now $CQ = z' = \sqrt{2y}$, since Q is on the parabola; and in the right triangle CRP, $CP = \sqrt{(CR)^2 + (RP)^2} = \sqrt{x^2 + z^2}$. Hence $\sqrt{x^2 + z^2} = \sqrt{2y}$ and the equation of the surface is $x^2 + z^2 = 2y$. Note that this equation can be obtained mechanically by replacing z in the equation of the parabola by $\sqrt{x^2 + z^2}$.

The equation of the surface of revolution generated by revolving a curve in one of the coordinate planes about one of the coordinate axes in that plane can be obtained as follows: If the curve is revolved about

(*a*) the x-axis, replace y or z in the equation of the curve by $\sqrt{y^2 + z^2}$;
(*b*) the y-axis, replace x or z in the equation of the curve by $\sqrt{x^2 + z^2}$;
(*c*) the z-axis, replace x or y in the equation of the curve by $\sqrt{x^2 + y^2}$.

Example 3. Write the equation of the surface of revolution generated by revolving $9x^2 + 16y^2 = 144$, $z = 0$ about the x-axis.

We replace y by $\sqrt{y^2 + z^2}$ in the equation $9x^2 + 16y^2 = 144$ and obtain $9x^2 + 16(y^2 + z^2) = 144$ or $9x^2 + 16y^2 + 16z^2 = 144$ as the equation of the surface. Since the directrix is an ellipse, the surface is called an *ellipsoid of revolution*. Note that two of the three coefficients are equal.

See Problems 2-3.

A SPHERE OR SPHERICAL SURFACE is a surface of revolution obtained by revolving a circle about one of its diameters. It is also the locus of a point which moves in space so that it is always at a fixed distance (radius of the sphere) from a fixed point (center of the sphere).

The equation of a sphere whose center is at the origin and whose radius is r is $x^2 + y^2 + z^2 = r^2$.

The equation of the sphere whose center is at the point $C(a, b, c)$ and whose radius is r is $(x-a)^2 + (y-b)^2 + (z-c)^2 = r^2$. See Problems 4-5.

A CONE OR CONICAL SURFACE is one generated by a straight line (called the *generatrix*) which moves so that it always intersects a given curve (called the *directrix*) and always passes through a given point (called the vertex). The generatrix in any position is called an *element* of the cone. The vertex divides the surface into two distinct pieces, called *nappes*.

When the vertex of the cone is at the origin (see Fig. 4 below), its equation is homogeneous in the three variables, i.e., if $f(x, y, z) = 0$ of degree n is the equation of the cone, then $f(kx, ky, kz) = k^n f(x, y, z)$.

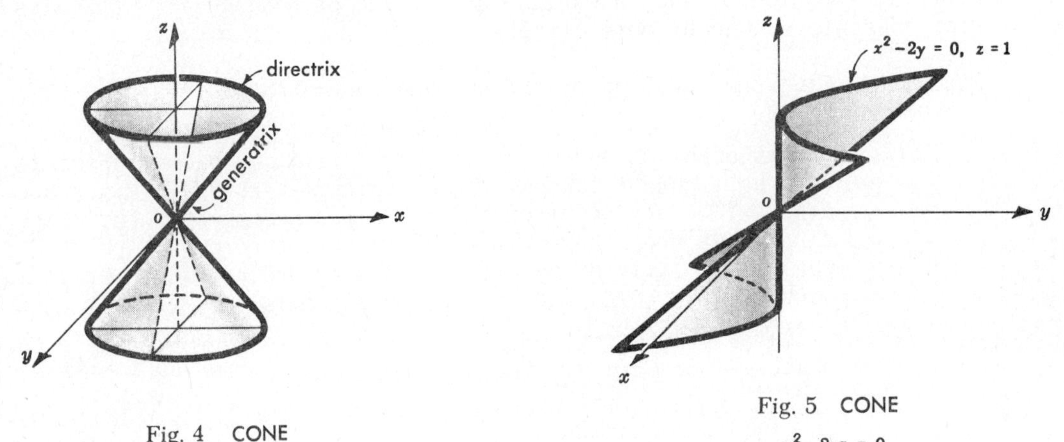

Fig. 4 CONE

Fig. 5 CONE
$x^2 - 2yz = 0$

Example 4. Identify and construct the surface having equation $x^2 - 2yz = 0$. See Fig. 5 above.

Let $f(x, y, z) = x^2 - 2yz$; then $f(kx, ky, kz) = (kx)^2 - 2(ky)(kz) = k^2(x^2 - 2yz) = k^2 f(x, y, z)$. The equation is homogeneous; hence the locus is a quadric cone (degree of equation is 2) with vertex at the origin.

In generating the surface, we may use as directrix the parabola $x^2 = 2y$, $z = 1$ lying in the plane $z = 1$. After drawing in the parabola and several of the elements (lines joining the origin with points on the parabola) we obtain a satisfactory figure. See Problem 6.

THE GENERAL EQUATION OF THE SECOND DEGREE IN THREE VARIABLES,

$$(2) \qquad Ax^2 + By^2 + Cz^2 + Dxy + Exz + Fyz + Gx + Hy + Iz + K = 0,$$

where at least one of the coefficients A, B, C, D, E, F is different from zero, represents a quadric surface.

As in the case of the general equation of the second degree in two variables, a properly chosen rotation and translation of the coordinate axes transform (2) into its reduced form. We propose now to study briefly each of the quadric surfaces not already covered above using in each case the reduced form of the equation. In this study, we shall make use of symmetry, intercepts, and extent as in the case of the conics. However, the nature of a surface is best revealed by studying its sections by planes parallel to the coordinate planes.

A SURFACE IS SYMMETRIC with respect to a coordinate plane if its equation is unchanged when the variable measured from that plane is changed in sign.

A surface is symmetric with respect to a coordinate axis if its equation is unchanged when the variables not measured along that axis are changed in sign.

A surface is symmetric with respect to the origin if its equation is unchanged when the three variables are changed in sign.

Example 5. The surface of equation $x^2 + 4y^2 + 3z^2 - 4z + 5 = 0$ is symmetric with respect to the yz-plane since its equation is unchanged when x is replaced by $-x$. It is symmetric with respect to the xz-plane since its equation is unchanged when y is replaced by $-y$. It is not symmetric with respect to the xy-plane since the equation is changed when z is replaced by $-z$.

The surface is symmetric with respect to the z-axis since its equation is unchanged when x and y are replaced respectively by $-x$ and $-y$. It is not symmetric with respect to either the x- or y-axis.

The surface is not symmetric with respect to the origin since its equation is changed into $x^2 + 4y^2 + 3z^2 + 4z + 5 = 0$, when the replacements $-x, -y, -z$ for x, y, z respectively are made in the given equation.

THE INTERCEPTS ON THE COORDINATE AXES of a surface are the directed distances from the origin to the points in which the coordinate axes pierce the surface. The intercepts on the coordinate axes are found by setting the variables equal to zero in pairs and solving for the remaining variable.

The *trace* in a coordinate plane of a surface is the curve in which the coordinate plane intersects the surface. The traces on the coordinate planes are found by setting in turn one of the variables equal to zero.

Example 6. Find the intercepts on the coordinate axes and the traces in the coordinate planes of the surface $x^2 + 4y^2 - 8z = 16$.

Setting $y = z = 0$, we have $x^2 = 16$; then the x-intercepts are ± 4. Setting $x = z = 0$, we have $4y^2 = 16$; the y-intercepts are ± 2. Setting $x = y = 0$, the z-intercept is -2.

Setting $z = 0$, the trace in the xy-plane is the ellipse $x^2 + 4y^2 = 16$, $z = 0$. The trace in the xz-plane is the parabola $x^2 - 8z = 16$, $y = 0$; and the trace on the yz-plane is the parabola $y^2 - 2z = 4$, $x = 0$.

THE LOCUS OF THE EQUATION $\dfrac{x^2}{a^2} + \dfrac{y^2}{b^2} + \dfrac{z^2}{c^2} = 1$ is called an *ellipsoid*. If at least two of a, b, c are equal, the locus is called an ellipsoid of revolution; if $a = b = c$, the locus is a sphere.

The ellipsoid is symmetric with respect to the coordinate planes, the coordinate axes, and the origin. The traces in the coordinate planes

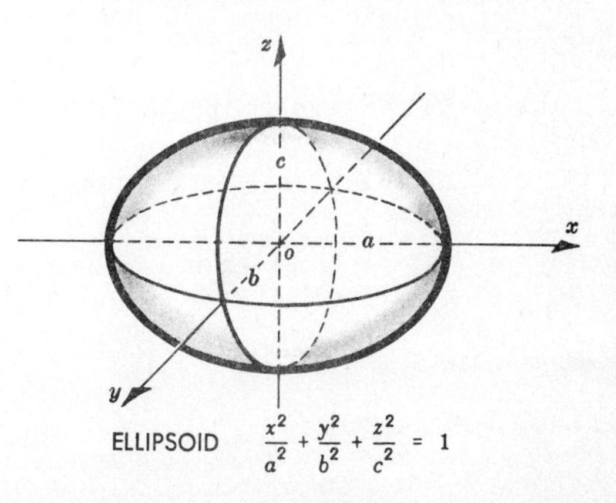

ELLIPSOID $\dfrac{x^2}{a^2} + \dfrac{y^2}{b^2} + \dfrac{z^2}{c^2} = 1$

$$\frac{x^2}{a^2} + \frac{y^2}{b^2} = 1,\ z = 0;\qquad \frac{x^2}{a^2} + \frac{z^2}{c^2} = 1,\ y = 0;\qquad \frac{y^2}{b^2} + \frac{z^2}{c^2} = 1,\ x = 0$$

are ellipses or circles.

A section by the plane $z = k$ is the ellipse (or circle) $\frac{x^2}{a^2} + \frac{y^2}{b^2} = 1 - \frac{k^2}{c^2}$. The ellipses decrease in size as the plane recedes from the xy-plane. The ellipse reduces to a point when $k = c$ and is imaginary when $k > c$. Similar statements may be made regarding the sections by the planes $y = k$ and $x = k$.

THE LOCUS OF THE EQUATION $\frac{x^2}{a^2} + \frac{y^2}{b^2} - \frac{z^2}{c^2} = 1$ is called a *hyperboloid of one sheet.* (If $a = b$, the locus is a hyperboloid of rotation.) It is symmetric with respect to the coordinate planes, the coordinate axes, and the origin.

The trace in the xy-plane is the ellipse $\frac{x^2}{a^2} + \frac{y^2}{b^2} = 1,\ z = 0$; the traces in the xz- and yz-planes are the hyperbolas $\frac{x^2}{a^2} - \frac{z^2}{c^2} = 1,\ y = 0$ and $\frac{y^2}{b^2} + \frac{z^2}{c^2} = 1,\ x = 0$.

A section by the plane $z = k$ is an ellipse, the size increasing as the plane recedes from the xy-plane. Sections by the planes $y = k$ and $x = k$ are hyperbolas.

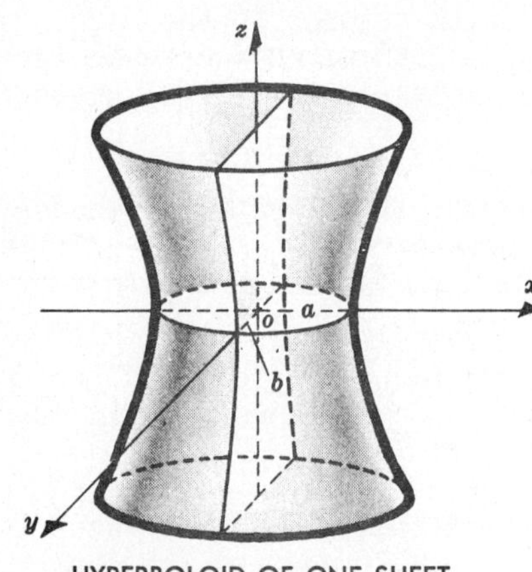

HYPERBOLOID OF ONE SHEET

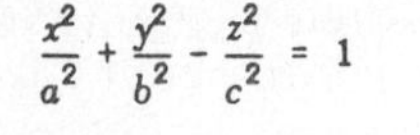

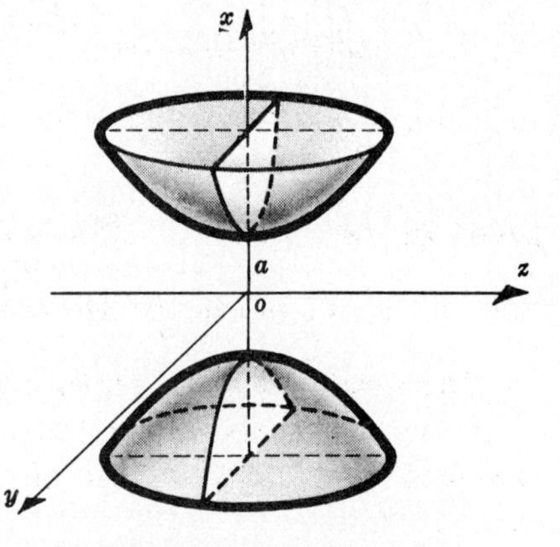

HYPERBOLOID OF TWO SHEETS

$$\frac{x^2}{a^2} - \frac{y^2}{b^2} - \frac{z^2}{c^2} = 1$$

THE LOCUS OF THE EQUATION $\frac{x^2}{a^2} - \frac{y^2}{b^2} - \frac{z^2}{c^2} = 1$ is called a *hyperboloid of two sheets.* If $b = c$, the locus is a hyperboloid of revolution. It is symmetric with respect to the coordinate planes, the coordinate axes, and the origin.

The traces in the xy- and xz-planes are the hyperbolas $\frac{x^2}{a^2} - \frac{y^2}{b^2} = 1,\ z = 0$ and $\frac{x^2}{a^2} - \frac{z^2}{c^2} = 1,\ y = 0$; the trace in the yz-plane is imaginary.

The sections by the planes $y = k$ and $z = k$ are hyperbolas; the section by the plane $x = k$ is imaginary when $|k| < a$, is a point when $|k| = a$, and an ellipse (or circle) when $|k| > a$.

THE LOCUS OF THE EQUATION $\frac{x^2}{a^2} + \frac{y^2}{b^2} = cz$ is called an *elliptic paraboloid.* If $a = b$, the locus is a paraboloid of revolution. It is symmetric with respect to the xz- and yz-planes, and the z-axis. If $c > 0$, the surface lies entirely above the xy-plane; if $c < 0$, the surface lies entirely below the xy-plane.

The traces in the xz- and yz-planes are the parabolas $\frac{x^2}{a^2} = cz$, $y = 0$ and $\frac{y^2}{b^2} = cz$, $x = 0$; the trace in the xy-plane is the origin.

The sections by the planes $x = k$ and $y = k$ are parabolas; the section by the plane $z = k$ is imaginary when $kc < 0$, a point when $k = 0$, an ellipse when $kc > 0$.

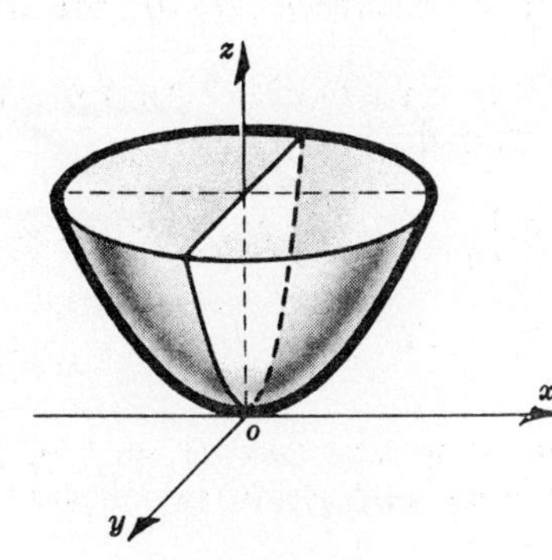

ELLIPTIC PARABOLOID

$$\frac{x^2}{a^2} + \frac{y^2}{b^2} = cz$$

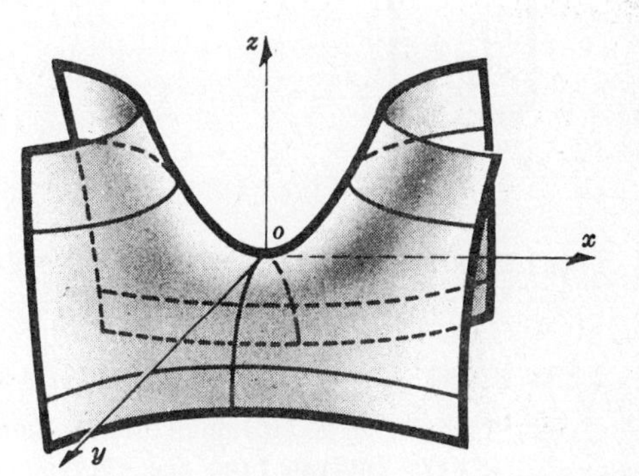

HYPERBOLIC PARABOLOID

$$\frac{x^2}{a^2} - \frac{y^2}{b^2} = cz$$

THE LOCUS OF THE EQUATION $\frac{x^2}{a^2} - \frac{y^2}{b^2} = cz$ is called a *hyperbolic paraboloid.* It is symmetric with respect to the xz- and yz-planes, and the z-axis.

The trace in the xy-plane is the pair of lines $\frac{x}{a} \pm \frac{y}{b} = 0$; the traces in the xz- and yz-planes are the parabolas $\frac{x^2}{a^2} = cz$ and $\frac{y^2}{b^2} = -cz$.

The section by the plane $z = k$ is a hyperbola except when $k = 0$, when it is a pair of lines as already noted. The sections by the planes $x = k$ and $y = k$ are parabolas.

IN ADDITION TO THE SURFACES DESCRIBED there are certain degenerate loci consisting of a pair of distinct planes, one plane counted twice, one line (a circular cylinder of radius 0), and one point.

See Problem 7.

SOLVED PROBLEMS

1. Discuss and sketch each of the following right cylinders:
(a) $x^2 + 4y^2 = 16$, (b) $y^2 = 4z - 8$, (c) $xz = -12$.

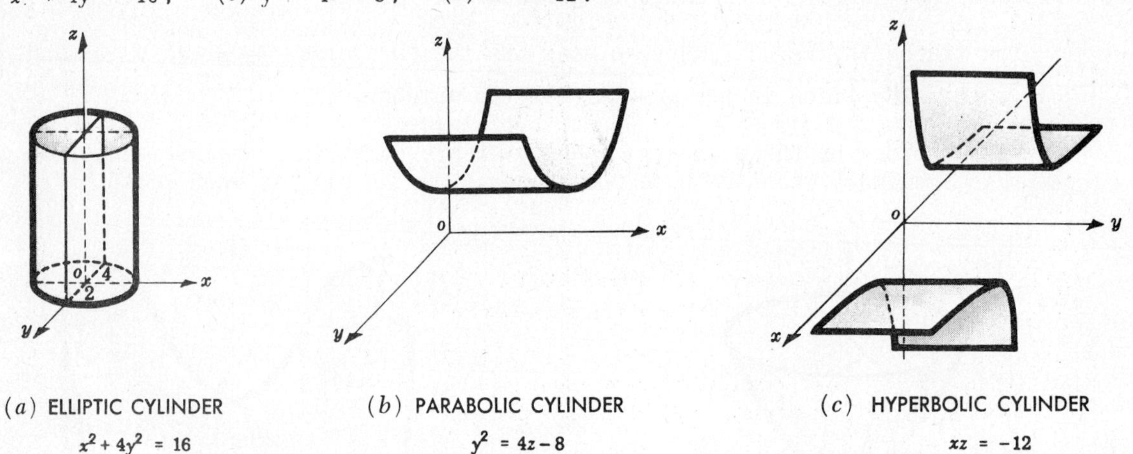

(a) ELLIPTIC CYLINDER $x^2 + 4y^2 = 16$ (b) PARABOLIC CYLINDER $y^2 = 4z - 8$ (c) HYPERBOLIC CYLINDER $xz = -12$

(a) This is an elliptic cylinder, generated by a straight line moving parallel to the z-axis and always intersecting the ellipse $x^2 + 4y^2 = 16,\ z = 0$.

(b) This is a parabolic cylinder, generated by a straight line moving parallel to the x-axis and always intersecting the parabola $y^2 = 4z - 8,\ x = 0$.

(c) This is a hyperbolic cylinder, generated by a straight line moving parallel to the y-axis and always intersecting the hyperbola $xz = -12,\ y = 0$.

2. Write the equation of the surface of revolution generated by revolving the given curve about the indicated axis.
(a) $x^2 + y^2 = 4,\ z = 0$; about the x-axis.
(b) $9x^2 - 4z^2 = 36,\ y = 0$; about the z-axis.
(c) $y + 2z + 4 = 0,\ x = 0$; about the y-axis.

(a) Replacing y by $\sqrt{y^2 + z^2}$, we obtain $x^2 + y^2 + z^2 = 4$. This is a *sphere*.

(b) Replacing x by $\sqrt{x^2 + y^2}$, we obtain $9x^2 + 9y^2 - 4z^2 = 36$. This is a *hyperboloid of revolution*.

(c) Replacing z by $\sqrt{x^2 + z^2}$, $y + 2\sqrt{x^2 + z^2} + 4 = 0$. When this is written as $y + 4 = -2\sqrt{x^2 + z^2}$ and both members are squared, we obtain $4x^2 + 4z^2 - (y + 4)^2 = 0$. This is a *cone*.

3. Identify and sketch:
(a) $x^2 + y^2 + z^2 = 9$, (b) $x^2 + 4y^2 + z^2 = 4$, (c) $z^2 - 4x^2 - 4y^2 = 4$, (d) $x^2 + z^2 - 8y = 0$.

(a) The locus is a sphere, generated by revolving the circle $x^2 + y^2 = 9,\ z = 0$ about the x- or the y-axis, or by revolving the circle $x^2 + z^2 = 9,\ y = 0$ about the x- or the z-axis, or by revolving the circle $y^2 + z^2 = 9,\ x = 0$ about the y-axis or the z-axis. Refer to the adjoining figure.

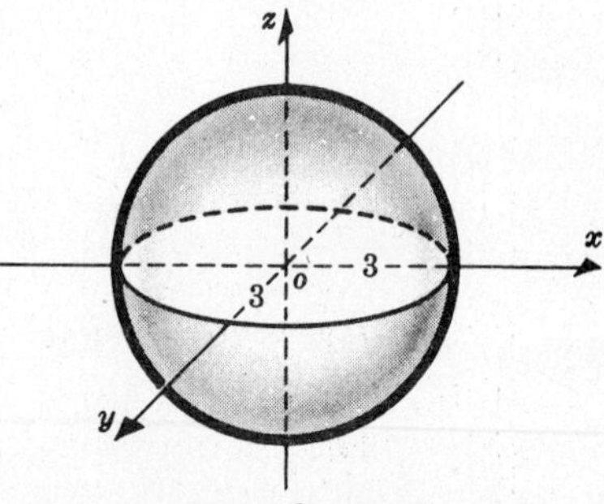

(a) SPHERE $x^2 + y^2 + z^2 = 9$

(b) The locus is an ellipsoid of revolution, generated by revolving the ellipse $x^2 + 4y^2 = 4,\ z = 0$ about the y-axis or by revolving the ellipse $4y^2 + z^2 = 4,\ x = 0$ about the y-axis. Refer to Fig. (b) below.

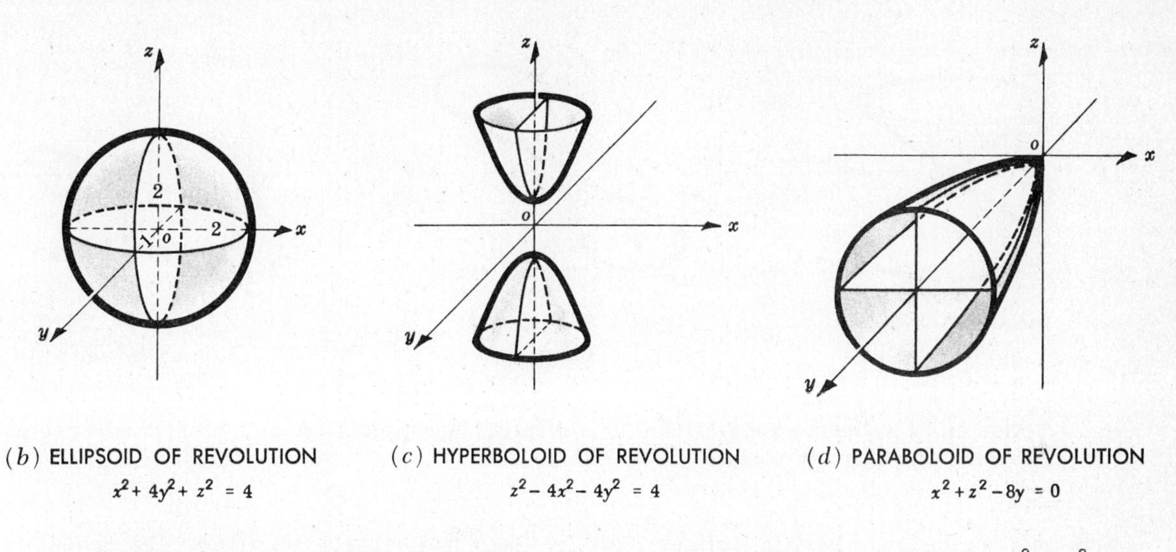

(b) ELLIPSOID OF REVOLUTION $x^2+4y^2+z^2=4$

(c) HYPERBOLOID OF REVOLUTION $z^2-4x^2-4y^2=4$

(d) PARABOLOID OF REVOLUTION $x^2+z^2-8y=0$

(c) The locus is a hyperboloid of revolution, generated by revolving the hyperbola $z^2-4x^2=4$, $y=0$ about the z-axis or by revolving the hyperbola $z^2-4y^2=4$, $x=0$ about the z-axis.

(d) The locus is a paraboloid of revolution generated by revolving the parabola $x^2-8y=0$, $z=0$ about the y-axis or by revolving the parabola $z^2-8y=0$, $x=0$ about the y-axis.

4. Write the equation of each of the following spheres: (a) $C(2,-3,-4)$, $r=5$; (b) $C(1,-2,3)$, tangent to π: $x+4y-5z+2=0$; (c) center on the x-axis, passes through $A(2,3,5)$ and $B(6,-3,3)$.

(a) The equation is $(x-2)^2+(y+3)^2+(z+4)^2 = 25$.

(b) Since C is $\dfrac{1+4(-2)-5\cdot3+2}{-\sqrt{42}} = \dfrac{20}{\sqrt{42}}$ units from the given plane, the radius of the sphere is $\dfrac{20}{\sqrt{42}}$ and its equation is $(x-1)^2+(y+2)^2+(z-3)^2 = 400/42 = 200/21$.

(c) Let the center have coordinates $(a,0,0)$. Now $(CA)^2=(CB)^2$ or $(a-2)^2+9+25 = (a-6)^2+9+9$ and then $a=2$. The center of the sphere is at $C(2,0,0)$ and the square of its radius is $r^2 = (a-2)^2+9+25 = 34$. The required equation is $(x-2)^2+y^2+z^2 = 34$.

5. Find the coordinates of the center and the radius of the spheres:
(a) $(x-2)^2+(y-3)^2+(z+4)^2 = 36$, ($b$) $x^2+y^2+z^2-6x+8y-10z+25 = 0$

(a) The center is at $C(2,3,-4)$ and the radius is $\sqrt{36}=6$.

(b) Completing the squares, we have $(x-3)^2+(y+4)^2+(z-5)^2 = -25+9+16+25 = 25$.
The center is at $C(3,-4,5)$ and the radius is 5.

6. Identify and construct each of the following surfaces:
(a) $x^2+y^2-4z^2=0$, (b) $x^2-2y^2+4z^2=0$, (c) $x^2-2y^2-4z^2=0$.

(a) This is a circular cone generated by a line through the origin which moves so as to always intersect the circle $x^2+y^2=4$, $z=1$ or $x^2+y^2=16$, $z=-2$, etc.
We draw in the first named circle and through its points pass lines intersecting at the origin. See Fig. (a) below.

(b) This is an *elliptic cone* generated by a straight line through the origin which moves so as to always intersect the ellipse $x^2+4z^2=2$, $y=1$. See Fig. (b) below.

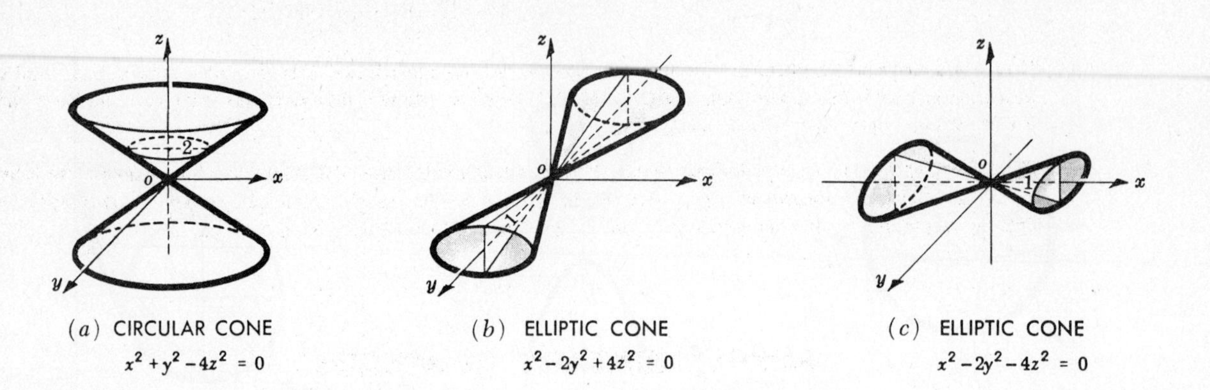

(*a*) CIRCULAR CONE $x^2+y^2-4z^2=0$

(*b*) ELLIPTIC CONE $x^2-2y^2+4z^2=0$

(*c*) ELLIPTIC CONE $x^2-2y^2-4z^2=0$

(*c*) This is an elliptic cone generated by a straight line through the origin which moves so as to always intersect the ellipse $2y^2+4z^2=1,\ x=1$.

7. Construct the following quadric surfaces: (*a*) $4x^2+9y^2+16z^2=144$, (*b*) $x^2+4y^2-9z^2=36$, (*c*) $x^2-4y^2-9z^2=36$, (*d*) $4y^2+9z^2=36x$, (*e*) $4x^2-9y^2=72z$.

(*a*) This is an ellipsoid whose traces in the coordinate planes are the ellipses $4x^2+9y^2=144,\ z=0$; $x^2+4z^2=36,\ y=0$; and $9y^2+16z^2=144,\ x=0$. These traces are sufficient to indicate the surface.

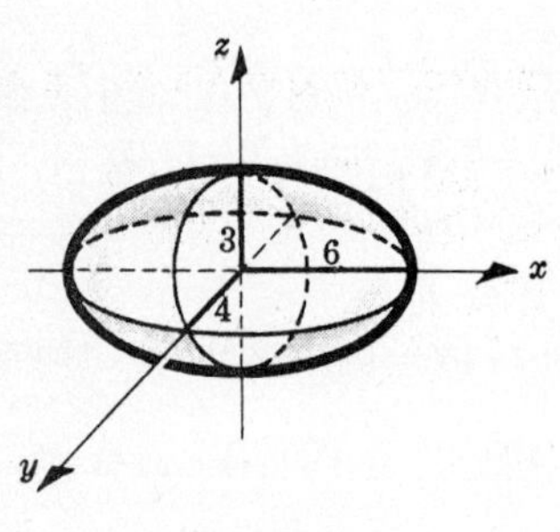

(*a*) ELLIPSOID

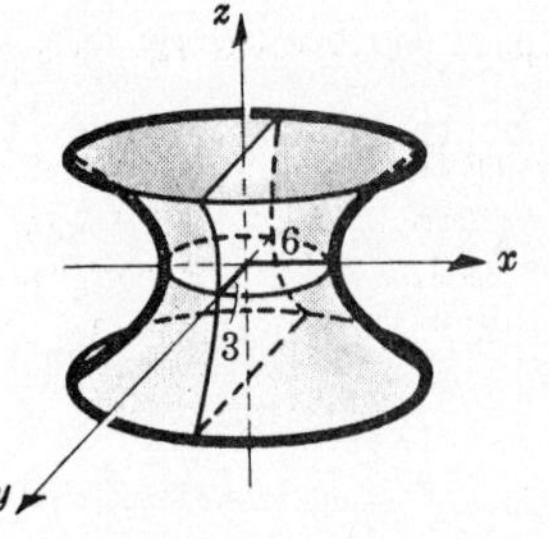

(*b*) HYPERBOLOID OF ONE SHEET

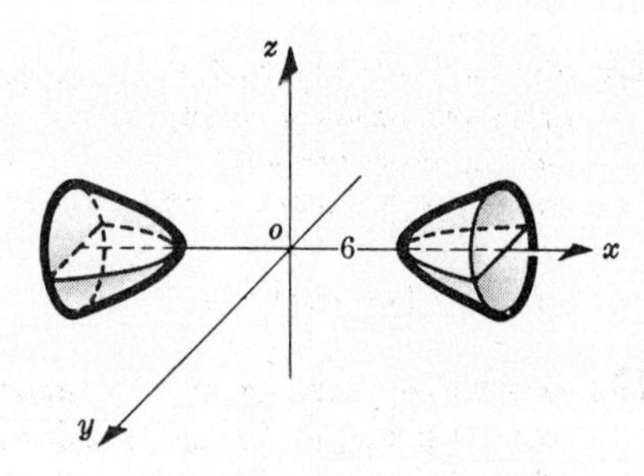

(*c*) HYPERBOLOID OF TWO SHEETS

(*b*) This is a hyperboloid of one sheet whose traces in the coordinate planes are the ellipse $x^2+4y^2=36,\ z=0$ and the hyperbolas $x^2-9z^2=36,\ y=0$ and $4y^2-9z^2=36,\ x=0$. Fig. (*b*) above, shows these traces together with the sections $x^2+4y^2=180,\ z=\pm 4$.

(*c*) This is a hyperboloid of two sheets whose real traces in the coordinate planes are the hyperbolas $x^2-4y^2=36,\ z=0$ and $x^2-9z^2=36,\ y=0$. Fig. (*c*) above, shows these traces together with the sections $4y^2+9z^2=108,\ x=\pm 12$.

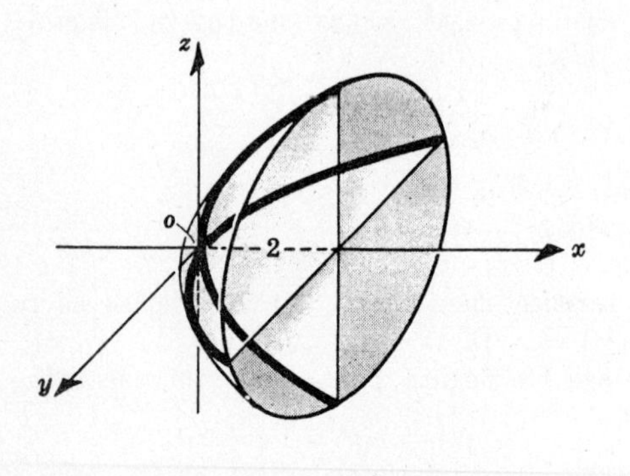

(*d*) ELLIPTIC PARABOLOID

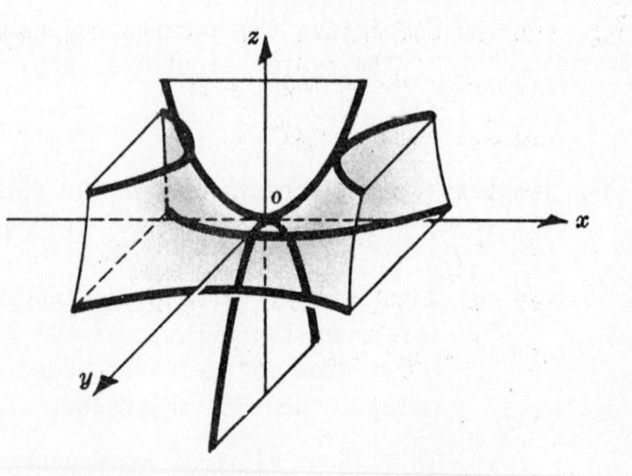

(*e*) HYPERBOLIC PARABOLOID

(*d*) This is an elliptic paraboloid whose traces in the coordinate planes are the origin and the parabolas $y^2 = 9x,\ z = 0$ and $z^2 = 4x,\ y = 0$. Fig. (*d*) above shows these traces together with the section $4y^2 + 9z^2 = 72,\ x = 2$.

(*e*) This is a hyperbolic paraboloid whose traces in the coordinate planes are the lines $2x \pm 3y = 0,\ z = 0$ and the parabolas $x^2 = 18z,\ y = 0$ and $y^2 = -8z,\ x = 0$. Fig. (*e*) above, indicates these traces together with the sections $4x^2 - 9y^2 = 72,\ z = 1$ and $9y^2 - 4x^2 = 72,\ z = -1$.

SUPPLEMENTARY PROBLEMS

8. Discuss and construct each of the following right cylinders.
(*a*) $y^2 + z^2 = 16$ (*b*) $4x^2 + 9y^2 = 36$ (*c*) $x^2 - 4z^2 = 36$ (*d*) $x^2 = 8y - 24$

9. Write the equation of the surface of revolution generated by revolving the given curve about the indicated axis.
(*a*) $x^2 + y^2 = 4,\ z = 0$; about the *y*-axis. *Ans.* $x^2 + y^2 + z^2 = 4$
(*b*) $x^2 - 4z^2 = 16,\ y = 0$; about the *z*-axis. *Ans.* $x^2 + y^2 - 4z^2 = 16$
(*c*) $y = 2x,\ z = 0$; about the *x*-axis. *Ans.* $4x^2 - y^2 - z^2 = 0$
(*d*) $x^2 + 3y = 6,\ z = 0$; about the *y*-axis. *Ans.* $x^2 + z^2 + 3y - 6 = 0$

10. For each of the following, find the axis of revolution and the equations of the generatrix in the coordinate plane containing the axis.
(*a*) $9x^2 + y^2 + z^2 - 36 = 0$ *Ans.* *x*-axis; $9x^2 + y^2 = 36,\ z = 0$ or $9x^2 + z^2 = 36,\ y = 0$
(*b*) $2x^2 + 3y^2 + 2z^2 = 12$ *Ans.* *y*-axis; $2x^2 + 3y^2 = 12,\ z = 0$ or $3y^2 + 2z^2 = 12,\ x = 0$
(*c*) $x^2 + y^2 = 4$ *Ans.* *z*-axis; $x = 2,\ y = 0$ or $y = 2,\ x = 0$
(*d*) $x^2 - 3y^2 - 3z^2 = 9$ *Ans.* *x*-axis; $x^2 - 3y^2 = 9,\ z = 0$ or $x^2 - 3z^2 = 9,\ y = 0$

11. Write the equation of the sphere
(*a*) with center $(1,2,-3)$ and radius 2.
(*b*) with center $(2,-1,1)$ and passing through $(5,2,-3)$.
(*c*) with center $(3,2,4)$ and tangent to $2x + y + 2z - 31 = 0$.
(*d*) passing through $(3,5,4)$, $(4,4,-8)$, and $(-5,0,1)$.
Ans. (*a*) $x^2 + y^2 + z^2 - 2x - 4y + 6z + 10 = 0$ (*c*) $x^2 + y^2 + z^2 - 6x - 4y - 8z + 4 = 0$
(*b*) $x^2 + y^2 + z^2 - 4x + 2y - 2z - 28 = 0$ (*d*) $x^2 + y^2 + z^2 - 2x - 4y + 4z - 40 = 0$

12. Find the coordinates of the center and the radius of each of the following spheres.
(*a*) $x^2 + y^2 + z^2 + 6x - 2y - 8z + 10 = 0$ *Ans.* $C(-3,1,4)$; $r = 4$
(*b*) $x^2 + y^2 + z^2 - 4x + 6y - 12 = 0$ *Ans.* $C(2,-3,0)$; $r = 5$
(*c*) $4x^2 + 4y^2 + 4z^2 - 4x - 12y - 16z + 10 = 0$ *Ans.* $C(1/2,\ 3/2,\ 2)$; $r = 2$

13. Discuss and sketch each of the following.
(*a*) $4x^2 + 9y^2 + 36z^2 = 36$ (*g*) $x^2 + 4z^2 = 16$
(*b*) $4x^2 + 4y^2 - 25z^2 = 100$ (*h*) $x^2 + y^2 + 4z^2 = 0$
(*c*) $36x^2 + 9y^2 - 4z^2 = 36$ (*i*) $x^2 + 4y^2 + 4z = 0$
(*d*) $x^2 + y^2 + z^2 - 8x + 6y = 0$ (*j*) $y^2 = 4xz$
(*e*) $4x^2 - 16y^2 - 25z^2 = 400$ (*k*) $x^2 + 4y^2 - 9z^2 = 0$
(*f*) $x^2 - 4y^2 - 4z = 0$ (*l*) $x^{1/2} + y^{1/2} = a^{1/2}$

CHAPTER 61

The Derivative

IN THIS AND SUBSEQUENT CHAPTERS, it will be understood that number refers always to a real number, that the range of any variable (as x) is a set of real numbers, and that a function of one variable (as $f(x)$) is a single-valued function.

In this chapter a procedure is given by which from a given function $y = f(x)$ another function, denoted by y' or $f'(x)$ and called the derivative of y or of $f(x)$ with respect to x, is obtained. Depending upon the quantities denoted by x and $y = f(x)$, the derivative may be interpreted as the slope of a tangent line to a curve, as velocity, as acceleration, etc.

LIMIT OF A FUNCTION. A given function $f(x)$ is said to have a *limit* M as x approaches c (in symbols, $\lim_{x \to c} f(x) = M$) if $f(x)$ is as near to M as we please for all values of $x \neq c$ but sufficiently near to c.

Example 1. Consider $f(x) = x^2 - 2$ for x near 3.

If x is near to 3, say $2.99 < x < 3.01$, then

$(2.99)^2 - 2 < f(x) < (3.01)^2 - 2$ or $6.9401 < f(x) < 7.0601$.

If x is nearer to 3, say $2.999 < x < 3.001$, then

$(2.999)^2 - 2 < f(x) < (3.001)^2 - 2$ or $6.994001 < f(x) < 7.006001$.

If x is still nearer to 3, say $2.9999 < x < 3.0001$, then

$(2.9999)^2 - 2 < f(x) < (3.0001)^2 - 2$ or $6.99940001 < f(x) < 7.00060001$.

It appears reasonable to conclude that as x is taken in a smaller and smaller interval about 3, the corresponding $f(x)$ will lie in a smaller and smaller interval about 7. Conversely, it seems reasonable to conclude that if we demand that $f(x)$ have values lying in smaller and smaller intervals about 7, we need only to choose x in sufficiently smaller and smaller intervals about 3. Thus we conclude

$$\lim_{x \to 3} (x^2 - 2) = 7.$$

Example 2. Consider $f(x) = \dfrac{x^2 - x - 6}{x - 3}$ for x near 3.

When $x \neq 3$, $f(x) = \dfrac{x^2 - x - 6}{x - 3} = x + 2$. Thus, for x near 3, $x + 2$ is near to 5

and
$$\lim_{x \to 3} \frac{x^2 - x - 6}{x - 3} = 5.$$

THEOREMS ON LIMITS. If $\lim_{x\to c} f(x) = M$ and $\lim_{x\to c} g(x) = N$, then

I. $\lim_{x\to c} \{f(x) \pm g(x)\} = \lim_{x\to c} f(x) \pm \lim_{x\to c} g(x) = M \pm N$.

II. $\lim_{x\to c} \{kf(x)\} = k \lim_{x\to c} f(x) = kM$, where k is a constant.

III. $\lim_{x\to c} \{f(x)\cdot g(x)\} = \lim_{x\to c} f(x) \cdot \lim_{x\to c} g(x) = MN$.

IV. $\lim_{x\to c} \dfrac{f(x)}{g(x)} = \dfrac{\lim_{x\to c} f(x)}{\lim_{x\to c} g(x)} = \dfrac{M}{N}$, provided $N \neq 0$.

See Problem 2.

CONTINUOUS FUNCTIONS. A function $f(x)$ is called continuous at $x = c$, provided

(*1*) $f(c)$ is defined, (*2*) $\lim_{x\to c} f(x)$ exists, (*3*) $\lim_{x\to c} f(x) = f(c)$.

Example 3. (*a*) The function $f(x) = x^2 - 2$ of Example 1 is continuous at $x = 3$ since (*1*) $f(3) = 7$, (*2*) $\lim_{x\to 3} (x^2 - 2) = 7$, (*3*) $\lim_{x\to 3} (x^2 - 2) = f(3)$.

(*b*) The function $f(x) = \dfrac{x^2 - x - 6}{x - 3}$ of Example 2 is not continuous at $x = 3$ since $f(3)$ is not defined.

A function $f(x)$ is said to be *continuous* on the interval (a, b) if it is continuous for every value of x of the interval. A polynomial in x is continuous for all values of x. A rational function in x, $f(x) = \dfrac{P(x)}{Q(x)}$, where $P(x)$ and $Q(x)$ are polynomials, is continuous for all values of x except those for which $Q(x) = 0$. Thus

$f(x) = \dfrac{x^2 + x + 1}{(x - 1)(x^2 + 2)}$ is continuous for all values of x, except $x = 1$.

INCREMENTS. Let x_0 and x_1 be two distinct values of x. It is customary to denote their difference $x_1 - x_0$ by Δx (read, delta x) and to write $x_0 + \Delta x$ for x_1.

Now if $y = f(x)$ and if x changes in value from $x = x_0$ to $x = x_0 + \Delta x$, y will change in value from $y_0 = f(x_0)$ to $y_0 + \Delta y = f(x_0 + \Delta x)$. The change in y due to a change in x from $x = x_0$ to $x = x_0 + \Delta x$ is $\Delta y = f(x_0 + \Delta x) - f(x_0)$.

Example 4. Compute the change in $y = f(x) = x^2 - 2x + 5$ when x changes in value from (*a*) $x = 3$ to $x = 3.2$, (*b*) $x = 3$ to $x = 2.9$.

(*a*) Take $x_0 = 3$ and $\Delta x = 0.2$. Then $y_0 = f(x_0) = f(3) = 8$, $y_0 + \Delta y = f(x_0 + \Delta x) = f(3.2) = 8.84$, and $\Delta y = 8.84 - 8 = 0.84$.

(*b*) Take $x_0 = 3$ and $\Delta x = -0.1$. Then $y_0 = f(3) = 8$, $y_0 + \Delta y = f(2.9) = 7.61$, and $\Delta y = 7.61 - 8 = -0.39$.

See Problems 3-4.

THE DERIVATIVE. The *derivative* of $y = f(x)$ at $x = x_0$ is

$$\lim_{\Delta x\to 0} \frac{\Delta y}{\Delta x} = \lim_{\Delta x\to 0} \frac{f(x_0 + \Delta x) - f(x_0)}{\Delta x}, \quad \text{provided the limit exists.}$$

In finding derivatives, we shall use the following 5-step rule:

(1) Write $y_0 = f(x_0)$.

(2) Write $y_0 + \Delta y = f(x_0 + \Delta x)$.

(3) Obtain $\Delta y = f(x_0 + \Delta x) - f(x_0)$.

(4) Obtain $\Delta y/\Delta x$.

(5) Evaluate $\lim_{\Delta x \to 0} \frac{\Delta y}{\Delta x}$.

Example 5. Find the derivative of $y = f(x) = 2x^2 - 3x + 5$ at $x = x_0$.

(1) $y_0 = f(x_0) = 2x_0^2 - 3x_0 + 5$

(2) $y_0 + \Delta y = f(x_0 + \Delta x) = 2(x_0 + \Delta x)^2 - 3(x_0 + \Delta x) + 5$
$= 2x_0^2 + 4x_0 \cdot \Delta x + 2(\Delta x)^2 - 3x_0 - 3 \cdot \Delta x + 5$

(3) $\Delta y = f(x_0 + \Delta x) - f(x_0) = 4x_0 \cdot \Delta x - 3 \cdot \Delta x + 2(\Delta x)^2$

(4) $\frac{\Delta y}{\Delta x} = 4x_0 - 3 + 2 \cdot \Delta x$

(5) $\lim_{\Delta x \to 0} \frac{\Delta y}{\Delta x} = \lim_{\Delta x \to 0} (4x_0 - 3 + 2 \cdot \Delta x) = 4x_0 - 3$.

If in the example above the subscript 0 is deleted, the 5-step rule yields a function of x (here, $4x - 3$) called the derivative with respect to x of the given function. The derivative with respect to x of the function $y = f(x)$ is denoted by some one of the symbols y', $\frac{dy}{dx}$, $f'(x)$, or $D_x y$.

Provided it exists, the value of the derivative for any given value of x, say x_0, will be denoted by $y'\Big|_{x=x_0}$, $\frac{dy}{dx}\Big|_{x=x_0}$, or $f'(x_0)$.

See Problems 5-13.

HIGHER ORDER DERIVATIVES. The process of finding the derivative of a given function is called *differentiation*.

By differentiation, we obtain from a given function $y = f(x)$ another function $y' = f'(x)$ which will now be called the *first derivative* of y or of $f(x)$ with respect to x. If, in turn, $y' = f'(x)$ is differentiated with respect to x, another function $y'' = f''(x)$, called the second derivative of y or of $f(x)$ is obtained. Similarly, a third derivative may be found, and so on.

Example 7. Let $y = f(x) = x^4 - 3x^2 + 8x + 6$. Then $y' = f'(x) = 4x^3 - 6x + 8$, $y'' = f''(x) = 12x^2 - 6$, and $y''' = f'''(x) = 24x$.

See Problem 14.

SOLVED PROBLEMS

1. Investigate $f(x) = 1/x$ for values of x near $x = 0$.

If x is near 0, say $-.01 < x < .01$, then $\frac{1}{-.01} < \frac{1}{x} < \frac{1}{.01}$ or $-100 < \frac{1}{x} < 100$.

If x is nearer to 0, say $-.0001 < x < .0001$, then $\frac{1}{-.0001} < \frac{1}{x} < \frac{1}{.0001}$ or $-10000 < \frac{1}{x} < 10000$.

It is clear that as x is taken in smaller and smaller intervals about 0, the corresponding $f(x)$ does *not* lie in smaller and smaller intervals about any number M. Hence, $\lim_{x\to 0}(1/x)$ does not exist.

2. Evaluate when possible:

(*a*) $\lim_{x\to 2}(4x^2 - 5x)$, (*b*) $\lim_{x\to 1}(x^2 - 4x + 10)$, (*c*) $\lim_{x\to 2}\frac{x^2+6x+5}{x^2-2x-3}$, (*d*) $\lim_{x\to 3}\frac{x^2+6x+5}{x^2-2x-3}$, (*e*) $\lim_{x\to -1}\frac{x^2+6x+5}{x^2-2x-3}$.

(*a*) $\lim_{x\to 2}(4x^2-5x) = \lim_{x\to 2}4x^2 - \lim_{x\to 2}5x = 4\lim_{x\to 2}x^2 - 5\lim_{x\to 2}x = 4\cdot 4 - 5\cdot 2 = 6$.

(*b*) $\lim_{x\to 1}(x^2-4x+10) = (1)^2 - 4\cdot 1 + 10 = 7$.

(*c*) $\lim_{x\to 2}(x^2+6x+5) = 21$ and $\lim_{x\to 2}(x^2-2x-3) = -3$; hence

$$\lim_{x\to 2}\frac{x^2+6x+5}{x^2-2x-3} = \frac{\lim_{x\to 2}(x^2+6x+5)}{\lim_{x\to 2}(x^2-2x-3)} = \frac{21}{-3} = -7.$$

(*d*) $\lim_{x\to 3}(x^2+6x+5) = 32$ and $\lim_{x\to 3}(x^2-2x-3) = 0$; hence $\lim_{x\to 3}\frac{x^2+6x+5}{x^2-2x-3}$ does not exist.

(*e*) $\lim_{x\to -1}(x^2+6x+5) = 0$ and $\lim_{x\to -1}(x^2-2x-3) = 0$. Then, when $x \neq -1$,

$$\frac{x^2+6x+5}{x^2-2x-3} = \frac{(x+5)(x+1)}{(x-3)(x+1)} = \frac{x+5}{x-3} \quad\text{and}\quad \lim_{x\to -1}\frac{x^2+6x+5}{x^2-2x-3} = \lim_{x\to -1}\frac{x+5}{x-3} = \frac{4}{-4} = -1.$$

3. Let $P(x_0, y_0)$ and $Q(x_0+\Delta x, y_0+\Delta y)$ be two distinct points on the parabola $y = x^2 - 3$. Compute $\Delta y/\Delta x$ and interpret.

Here
$$\begin{aligned} y_0 &= x_0^2 - 3 \\ y_0 + \Delta y &= (x_0+\Delta x)^2 - 3 = x_0^2 + 2x_0\cdot\Delta x + (\Delta x)^2 - 3 \\ \Delta y &= [x_0^2 + 2x_0\cdot\Delta x + (\Delta x)^2 - 3] - [x_0^2 - 3] \\ &= 2x_0\cdot\Delta x + (\Delta x)^2 \end{aligned}$$

and
$$\frac{\Delta y}{\Delta x} = 2x_0 + \Delta x.$$

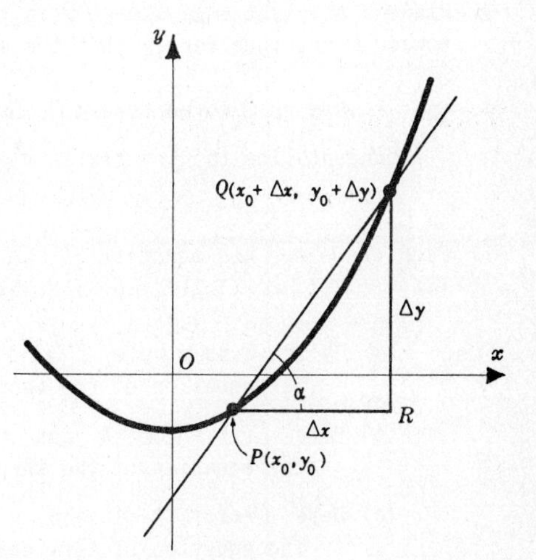

In the adjacent figure, PR is parallel to the x-axis and QR is parallel to the y-axis. If α denotes the inclination of the secant line PQ, $\tan\alpha = \frac{\Delta y}{\Delta x}$; thus, $\frac{\Delta y}{\Delta x}$ is the slope of the secant line PQ.

4. If $s = 3t^2 + 10$ is the distance a body moving in a straight line is from a fixed point O of the line at time t, (*a*) find the change Δs in s when t changes from $t = t_0$ to $t = t_0 + \Delta t$, (*b*) find $\Delta s/\Delta t$ and interpret.

(a) Here $s_0 = 3t_0^2 + 10$.
Then $s_0 + \Delta s = 3(t_0 + \Delta t)^2 + 10 = 3t_0^2 + 6t_0 \cdot \Delta t + 3(\Delta t)^2 + 10$ and $\Delta s = 6t_0 \cdot \Delta t + 3(\Delta t)^2$.

(b) $\dfrac{\Delta s}{\Delta t} = \dfrac{6t_0 \cdot \Delta t + 3(\Delta t)^2}{\Delta t} = 6t_0 + 3\Delta t$. Since Δs is the distance the body moves in time Δt, $\dfrac{\Delta s}{\Delta t}$ is the average rate of change of distance with respect to time or the average velocity of the body in the interval t_0 to $t_0 + \Delta t$.

5. Find: (a) $g'(x)$, given $g(x) = 5$ (c) $k'(x)$, given $k(x) = 4x^2$
(b) $h'(x)$, given $h(x) = 3x$ (d) $f'(x)$, given $f(x) = 4x^2 + 3x + 5$.
Thus verify: If $f(x) = k(x) + h(x) + g(x)$, then $f'(x) = k'(x) + h'(x) + g'(x)$.

(a)
$$y = g(x) = 5$$
$$y + \Delta y = g(x + \Delta x) = 5$$
$$\Delta y = 0$$
$$\frac{\Delta y}{\Delta x} = 0$$
$$g'(x) = \lim_{\Delta x \to 0} 0 = 0.$$

(b)
$$y = h(x) = 3x$$
$$y + \Delta y = 3(x + \Delta x) = 3x + 3\Delta x$$
$$\Delta y = 3\Delta x$$
$$\frac{\Delta y}{\Delta x} = 3$$
$$h'(x) = \lim_{\Delta x \to 0} 3 = 3.$$

(c)
$$y = k(x) = 4x^2$$
$$y + \Delta y = 4(x + \Delta x)^2 = 4x^2 + 8x \cdot \Delta x + 4(\Delta x)^2$$
$$\Delta y = 8x \cdot \Delta x + 4(\Delta x)^2$$
$$\frac{\Delta y}{\Delta x} = 8x + 4\Delta x$$
$$k'(x) = \lim_{\Delta x \to 0} (8x + 4\Delta x) = 8x.$$

(d)
$$y = f(x) = 4x^2 + 3x + 5$$
$$y + \Delta y = 4(x + \Delta x)^2 + 3(x + \Delta x) + 5 = 4x^2 + 8x \cdot \Delta x + 4(\Delta x)^2 + 3x + 3\Delta x + 5$$
$$\Delta y = 8x \cdot \Delta x + 3\Delta x + 4(\Delta x)^2$$
$$\frac{\Delta y}{\Delta x} = 8x + 3 + 4\Delta x$$
$$f'(x) = \lim_{\Delta x \to 0} (8x + 3 + 4\Delta x) = 8x + 3.$$

Thus $f'(x) = k'(x) + h'(x) + g'(x) = 8x + 3 + 0$.

6. Place a straight edge along PQ in the figure of Problem 3. Keeping P fixed, let Q move along the curve toward P and thus verify that the straight edge approaches the tangent line PT as limiting position.

Now as Q moves toward P, $\Delta x \to 0$ and $\displaystyle\lim_{\Delta x \to 0} \frac{\Delta y}{\Delta x} = \lim_{\Delta x \to 0} (2x_0 + \Delta x) = 2x_0$. Thus the slope of the tangent line to $y = f(x) = x^2 - 3$ at the point $P(x_0, y_0)$ is $m = f'(x_0) = 2x_0$.

7. Find the slope and equation of the tangent line to the given curve $y = f(x)$ at the given point:
(a) $y = 2x^3$ at $(1,2)$, (b) $y = -3x^2 + 4x + 5$ at $(3,-10)$, (c) $y = x^2 - 4x + 3$ at $(2,-1)$.

(a) By the 5-step rule, $f'(x) = 6x^2$; then the slope $m = f'(1) = 6$.
The equation of the tangent line at $(1,2)$ is $y - 2 = 6(x - 1)$ or $6x - y - 4 = 0$.

(b) Here $f'(x) = -6x + 4$ and $m = f'(3) = -14$.
The equation of the tangent line at $(3,-10)$ is $14x + y - 32 = 0$.

(c) Here $f'(x) = 2x - 4$ and $m = f'(2) = 0$.
The equation of the tangent line at $(2,-1)$ is $y + 1 = 0$. Identify the given point on the parabola.

8. Find the equation of the tangent line to the parabola $y^2 = 8x$ at (a) the point $(2,4)$, (b) the point $(2,-4)$.

Let $P(x,y)$ and $Q(x+\Delta x,\ y+\Delta y)$ be two nearby points on the parabola. Then

$$(1)\qquad y^2 = 8x$$

$$(2)\qquad (y+\Delta y)^2 = 8(x+\Delta x)$$

or
$$y^2 + 2y\cdot\Delta y + (\Delta y)^2 = 8x + 8\cdot\Delta x$$

Subtracting (1) from (2), $2y\cdot\Delta y + (\Delta y)^2 = 8\cdot\Delta x$, $\Delta y(2y+\Delta y) = 8\cdot\Delta x$, and $\dfrac{\Delta y}{\Delta x} = \dfrac{8}{2y+\Delta y}$.

Now as Q moves along the curve toward P, $\Delta x \to 0$ and $\Delta y \to 0$. Thus

$$m = \lim_{\Delta x\to 0}\frac{\Delta y}{\Delta x} = \lim_{\Delta y\to 0}\frac{8}{2y+\Delta y} = \frac{8}{2y} = \frac{4}{y}$$

(*a*) At point (2,4) the slope of the tangent line (also called the slope of the curve) is $m = 4/4 = 1$ and the equation of the tangent line is $x - y + 2 = 0$.

(*b*) At point (2,−4) the slope of the tangent line is $m = -1$ and the equation is $x + y + 2 = 0$.

9. Find the equation of the tangent line to the ellipse $4x^2 + 9y^2 = 25$ at (*a*) the point (2,1), (*b*) the point (0,5/3).

Let $P(x,y)$ and $Q(x+\Delta x,\ y+\Delta y)$ be two nearby points on the ellipse. (Why should P not be taken at an extremity of the major axis?) Then

$$(1)\qquad 4x^2 + 9y^2 = 25$$

$$(2)\qquad 4x^2 + 8x\cdot\Delta x + 4(\Delta x)^2 + 9y^2 + 18y\cdot\Delta y + 9(\Delta y)^2 = 25$$

Subtracting (1) from (2), $8x\cdot\Delta x + 4(\Delta x)^2 + 18y\cdot\Delta y + 9(\Delta y)^2 = 0$. Then

$$\Delta y(18y + 9\Delta y) = -\Delta x(8x + 4\cdot\Delta x) \quad\text{and}\quad \frac{\Delta y}{\Delta x} = -\frac{8x + 4\Delta y}{18y + 9\Delta y}$$

When Q moves along the curve toward P, $\Delta x \to 0$ and $\Delta y \to 0$. Then $m = \lim\limits_{\Delta x\to 0}\dfrac{\Delta y}{\Delta x} = -\dfrac{4x}{9y}$.

(*a*) At point (2,1), $m = -8/9$ and the equation of the tangent line is $8x + 9y - 25 = 0$.

(*b*) At point (0,5/3), $m = 0$ and the equation of the tangent line is $y - 5/3 = 0$.

10. The normal line to a curve at a point P on it is perpendicular to the tangent line at P. Find the equation of the normal line to the given curve at the given point of (*a*) Problem 7(*b*), (*b*) Problem 9.

(*a*) The slope of the tangent line is −14; the slope of the normal line is 1/14. The equation of the normal line is $y + 10 = \frac{1}{14}(x-3)$ or $x - 14y - 143 = 0$.

(*b*) The slope of the tangent line at (2,1) is −8/9; the slope of the normal line is 9/8. The equation of the normal line is $9x - 8y - 10 = 0$.
At the point (0,5/3) the normal line is vertical. Its equation is $x = 0$.

11. If $s = f(t)$ is the distance of a body, moving in a straight line, is from a fixed point O of the line at time t, then (see Problem 4) $\dfrac{\Delta s}{\Delta t} = \dfrac{f(t+\Delta t) - f(t)}{\Delta t}$ is the average velocity of the body in the interval of time t to $t+\Delta t$ and

$$v = s' = \lim_{\Delta t\to 0}\frac{\Delta s}{\Delta t} = \lim_{\Delta t\to 0}\frac{f(t+\Delta t) - f(t)}{\Delta t}$$

is the *instantaneous velocity of the body at time t*. For $s = 3t^2 + 10$ of Problem 4, find the (instantaneous) velocity of the body at time (a) $t=0$, (b) $t=4$.

Here $v = s' = 6t$. (a) When $t=0$, $v=0$. (b) When $t=4$, $v=24$.

12. The height above the ground of a bullet shot vertically upward with initial velocity of 1152 ft/sec is given by $s = 1152t - 16t^2$. Find (a) the velocity of the bullet 20 sec after it was fired, (b) the time required for the bullet to reach its maximum height and the maximum height attained.

Here $v = 1152 - 32t$.

(a) When $t = 20$, $v = 1152 - 32(20) = 512$ ft/sec.

(b) At its maximum height, the velocity of the bullet is 0 ft/sec. When $v = 1152 - 32t = 0$, $t = 36$ sec. When $t = 36$, $s = 1152(36) - 16(36)^2 = 20,736$ ft, the maximum height.

13. Find the derivative of each of the following polynomials:
(a) $f(x) = 3x^2 - 6x + 5$, (b) $f(x) = 2x^3 - 8x + 4$, (c) $f(x) = (x-2)^2(x-3)^2$.

(a) $f'(x) = 3\cdot 2x^{2-1} - 6x^{1-1} + 0 = 6x - 6$.

(b) $f'(x) = 2\cdot 3x^{3-1} - 8x^{1-1} + 0 = 6x^2 - 8$.

(c) Here $f(x) = x^4 - 10x^3 + 37x^2 - 60x + 36$. Then $f'(x) = 4x^3 - 30x^2 + 74x - 60 = (x-2)(x-3)(4x-10)$.

14. For each of the following functions, find $f'(x)$, $f''(x)$, and $f'''(x)$:
(a) $f(x) = 2x^2 + 7x - 5$, (b) $f(x) = x^3 - 6x^2$, (c) $f(x) = x^5 - x^3 + 3x$.

(a) $f'(x) = 4x + 7$, $f''(x) = 4$, $f'''(x) = 0$

(b) $f'(x) = 3x^2 - 12x$, $f''(x) = 6x - 12$, $f'''(x) = 6$

(c) $f'(x) = 5x^4 - 3x^2 + 3$, $f''(x) = 20x^3 - 6x$, $f'''(x) = 60x^2 - 6$

SUPPLEMENTARY PROBLEMS

15. Find all (real) values of x for which each of the following is defined.

(a) $x^2 - 3x + 4$ (d) $\dfrac{1}{(x-2)(x+3)}$ (g) $\dfrac{1}{x^2+4}$

(b) $\dfrac{1}{x^2}$ (e) $\dfrac{1}{x^2 - 4x + 3}$ (h) $\dfrac{x^2-9}{x-3}$

(c) $\dfrac{1}{x-2}$ (f) $\dfrac{1}{x^2-4}$ (i) $\dfrac{x-3}{x^2-9}$

Ans. (a) all x (b) $x \neq 0$ (c) $x \neq 2$ (d) $x \neq 2, -3$ (e) $x \neq 1, 3$ (f) $x \neq \pm 2$ (g) all x (h) $x \neq 3$ (i) $x \neq \pm 3$

16. For each function $f(x)$ of Problem 15 evaluate $\lim_{x\to 1} f(x)$, when it exists.

Ans. (a) 2 (b) 1 (c) −1 (d) −1/4 (f) −1/3 (g) 1/5 (h) 4 (i) 1/4

17. For each function $f(x)$ of Problem 15 evaluate $\lim_{x\to 3} f(x)$, when it exists.

Ans. (a) 4 (b) 1/9 (c) 1 (d) 1/6 (f) 1/5 (g) 1/13 (h) 6 (i) 1/6

18. Use the 5-step rule to obtain $f'(x)$ or $f'(t)$, given:
(a) $f(x) = 3x+5$ (b) $f(x) = x^2-3x$ (c) $f(t) = 2t^2+8t+9$ (d) $f(t) = 2t^3-12t^2+20t+3$
Ans. (a) 3 (b) $2x-3$ (c) $4t+8$ (d) $6t^2-24t+20$

19. Find the equations of the tangent and normal to each curve at the given point on it.
(a) $y = x^2+2$, $P(1,3)$ *Ans.* $2x-y+1 = 0$, $x+2y-7 = 0$
(b) $y = 2x^2-3x$, $P(1,-1)$ *Ans.* $x-y-2 = 0$, $x+y = 0$
(c) $y = x^2-4x+5$, $P(1,2)$ *Ans.* $2x+y-4 = 0$, $x-2y+3 = 0$
(d) $y = x^2+3x-10$, $P(2,0)$ *Ans.* $7x-y-14 = 0$, $x+7y-2 = 0$
(e) $x^2+y^2 = 25$, $P(4,3)$ *Ans.* $4x+3y-25 = 0$, $3x-4y = 0$
(f) $y^2 = 4x-8$, $P(3,-2)$ *Ans.* $x+y-1 = 0$, $x-y-5 = 0$
(g) $x^2+4y^2 = 8$, $P(-2,-1)$ *Ans.* $x+2y+4 = 0$, $2x-y+3 = 0$
(h) $2x^2-y^2 = 9$, $P(-3,3)$ *Ans.* $2x+y+3 = 0$, $x-2y+9 = 0$

20. A particle moves along the x-axis according to the law $s = 2t^2 + 8t + 9$ (see Problem 18c), where s (ft) is the directed distance of the particle from the origin O at time t (sec). Locate the particle and find its velocity when (a) $t = 0$, (b) $t = 1$.
Ans. (a) 9 ft to the right of O, $v = 8$ ft/sec (b) 19 ft to the right of O, $v = 12$ ft/sec

21. A particle moves along the x-axis according to the law $s = 2t^3 - 12t^2 + 20t + 3$ (see Problem 18d), where s is defined as in Problem 20.
(a) Locate the particle and find its velocity when $t = 2$. (b) Locate the particle when $v = 2$ ft/sec.
Ans. (a) 11 ft to the right of O, $v = -4$ ft/sec
(b) $t = 1$, 13 ft to the right of O; $t = 3$, 9 ft to the right of O

22. The height (s ft) of a bullet shot vertically upward is given by $s = 1280t - 16t^2$, with t measured in seconds. (a) What is the initial velocity? (b) For how long will it rise? (c) How high will it rise?
Ans. (a) 1280 ft/sec (b) 40 sec (c) 25,600 ft

23. Find the coordinates of the points for which the slope of the tangent to $y = x^3-12x+1$ is 0.
Ans. (2,−15), (−2,17)

24. At what point on $y = \frac{1}{2}x^2-2x+3$ is the tangent perpendicular to that at the point (1,0)?
Ans. (3, 3/2)

25. Show that the equation of the tangent to the conic $Ax^2+2Bxy+Cy^2+2Dx+2Ey+F = 0$ at the point $P_1(x_1, y_1)$ on it is given by $Ax_1x + B(x_1y+y_1x) + Cy_1y + D(x_1+x) + E(y_1+y) + F = 0$. Use this as a formula to solve Problem 19.

26. Show that the tangents at the extremities of the latus rectum of the parabola $y^2 = 4px$ (a) are mutually perpendicular, (b) intersect on the directrix.

27. Show that the tangent of slope $m \neq 0$ to the parabola $y^2 = 4px$ has equation $y = mx+p/m$.

28. Show that the slope of the tangent at either end of either latus rectum of the ellipse $b^2x^2+a^2y^2 = a^2b^2$ is equal numerically to its eccentricity. Investigate the case of the hyperbola.

CHAPTER 62

Differentiation of Algebraic Expressions

DIFFERENTIATION FORMULAS.

I. If $y = f(x) = kx^n$, where k and n are constants, then $y' = f'(x) = knx^{n-1}$.

For a verification, see Problem 1.

II. If $y = k \cdot u^n$, where k and n are constants and u is a function of x, then $y' = knu^{n-1} \cdot u'$, provided u' exists.

For a verification, see Problem 2.

Example 1. Find y', given (a) $y = 8x^{5/4}$, (b) $y = (x^2 + 4x - 1)^{3/2}$.

(a) Here $k = 8$, $n = 5/4$. Then $y' = knx^{n-1} = 8 \cdot \frac{5}{4} x^{5/4-1} = 10x^{1/4}$.

(b) Let $u = x^2 + 4x - 1$ so that $y = u^{3/2}$.
Then differentiating with respect to x, $u' = 2x + 4$ and

$$y' = \frac{3}{2} u^{1/2} \cdot u' = \frac{3}{2}\sqrt{x^2 + 4x - 1}\,(2x + 4) = 3(x + 2)\sqrt{x^2 + 4x - 1}$$

See Problem 3.

III. If $y = f(x) \cdot g(x)$, then $y' = f(x) \cdot g'(x) + g(x) \cdot f'(x)$, provided $f'(x)$ and $g'(x)$ exist.

For the derivation, see Problem 4.

Example 2. Find y' when $y = (x^3 + 3x^2 + 1)(x^2 + 2)$.

Take $f(x) = x^3 + 3x^2 + 1$ and $g(x) = x^2 + 2$.
Then $f'(x) = 3x^2 + 6x$, $g'(x) = 2x$, and

$$\begin{aligned} y' &= f(x) \cdot g'(x) + g(x) \cdot f'(x) \\ &= (x^3 + 3x^2 + 1)(2x) + (x^2 + 2)(3x^2 + 6x) \\ &= 5x^4 + 12x^3 + 6x^2 + 14x \end{aligned}$$

IV. If $y = \dfrac{f(x)}{g(x)}$, then $y' = \dfrac{g(x) \cdot f'(x) - f(x) \cdot g'(x)}{\{g(x)\}^2}$, when $f'(x)$ and $g'(x)$ exist and $g(x) \neq 0$.

For a derivation, see Problem 6.

Example 3. Find y', given $y = \dfrac{x + 1}{x^2 + 1}$.

Take $f(x) = x + 1$ and $g(x) = x^2 + 1$. Then

$$y' = \frac{g(x) \cdot f'(x) - f(x) \cdot g'(x)}{\{g(x)\}^2} = \frac{(x^2 + 1)(1) - (x + 1)(2x)}{(x^2 + 1)^2}$$

$$= \frac{1 - 2x - x^2}{(x^2 + 1)^2}.$$

SOLVED PROBLEMS

1. Use the 5-step rule to obtain y' when $y = 6x^{3/2}$.

We have

$$y = 6x^{3/2}$$
$$y + \Delta y = 6(x+\Delta x)^{3/2}$$
$$\Delta y = 6(x+\Delta x)^{3/2} - 6x^{3/2} = 6[(x+\Delta x)^{3/2} - x^{3/2}]$$

and

$$\frac{\Delta y}{\Delta x} = 6\cdot\frac{(x+\Delta x)^{3/2} - x^{3/2}}{\Delta x} = 6\cdot\frac{(x+\Delta x)^{3/2} - x^{3/2}}{\Delta x}\cdot\frac{(x+\Delta x)^{3/2} + x^{3/2}}{(x+\Delta x)^{3/2} + x^{3/2}}$$

$$= 6\cdot\frac{(x+\Delta x)^3 - x^3}{\Delta x[(x+\Delta x)^{3/2} + x^{3/2}]} = 6\cdot\frac{3x^2 + 3x\cdot\Delta x + (\Delta x)^2}{(x+\Delta x)^{3/2} + x^{3/2}}.$$

Then

$$y' = \lim_{\Delta x\to 0} 6\cdot\frac{3x^2 + 3x\cdot\Delta x + (\Delta x)^2}{(x+\Delta x)^{3/2} + x^{3/2}} = 6\cdot\frac{3x^2}{2x^{3/2}} = 9x^{1/2}.$$

Note. By Formula I, with $k = 6$ and $n = 3/2$, we find $y' = knx^{n-1} = 6\cdot\frac{3}{2}x^{1/2} = 9x^{1/2}$.

2. Use the 5-step rule to find y' when $y = (x^2+4)^{1/2}$. Solve also by using Formula II.

We have

$$y = (x^2+4)^{1/2}$$
$$y + \Delta y = [(x+\Delta x)^2 + 4]^{1/2}$$
$$\Delta y = [(x+\Delta x)^2 + 4]^{1/2} - (x^2+4)^{1/2}$$

and

$$\frac{\Delta y}{\Delta x} = \frac{[(x+\Delta x)^2 + 4]^{1/2} - (x^2+4)^{1/2}}{\Delta x}\cdot\frac{[(x+\Delta x)^2 + 4]^{1/2} + (x^2+4)^{1/2}}{[(x+\Delta x)^2 + 4]^{1/2} + (x^2+4)^{1/2}}$$

$$= \frac{(x+\Delta x)^2 + 4 - (x^2+4)}{\Delta x\{[(x+\Delta x)^2 + 4]^{1/2} + (x^2+4)^{1/2}\}} = \frac{2x + \Delta x}{[(x+\Delta x)^2 + 4]^{1/2} + (x^2+4)^{1/2}}.$$

Then

$$y' = \lim_{\Delta x\to 0}\frac{2x + \Delta x}{[(x+\Delta x)^2 + 4]^{1/2} + (x^2+4)^{1/2}} = \frac{2x}{2(x^2+4)^{1/2}} = \frac{x}{(x^2+4)^{1/2}}.$$

Let $u = x^2 + 4$ so that $y = u^{1/2}$. Then $u' = 2x$ and $y' = \frac{1}{2}u^{-1/2}\cdot u' = \frac{1}{2}(x^2+4)^{-1/2}\cdot 2x = \dfrac{x}{\sqrt{x^2+4}}$.

3. Find y', given: (*a*) $y = (2x-5)^3$, (*b*) $y = \frac{2}{3}(x^6 + 4x^3 + 5)^2$, (*c*) $y = \dfrac{4}{x^2}$, (*d*) $y = \dfrac{1}{\sqrt{x}}$, (*e*) $y = 2(3x^2+2)^{1/2}$

(*a*) Let $u = 2x - 5$ so that $y = u^3$. Then, differentiating with respect to x, $u' = 2$ and

$$y' = 3u^2\cdot u' = 3(2x-5)^2\cdot 2 = 6(2x-5)^2.$$

(*b*) Let $u = x^6 + 4x^3 + 5$ so that $y = \frac{2}{3}u^2$. Diferentiating with respect to x, $u' = 6x^5 + 12x^2$ and

$$y' = \frac{4}{3}u\cdot u' = \frac{4}{3}(x^6 + 4x^3 + 5)(6x^5 + 12x^2) = 8(x^6 + 4x^3 + 5)(x^5 + 2x^2).$$

(*c*) Here $y = 4x^{-2}$ and $y' = 4(-2)x^{-3} = -8/x^3$.

(*d*) Here $y = x^{-1/2}$ and $y' = (-\frac{1}{2})x^{-3/2} = -1/(2x\sqrt{x})$.

(*e*) Since $y = 2(3x^2+2)^{1/2}$, $y' = 2(\frac{1}{2})(3x^2+2)^{-1/2}(6x) = 6x/(3x^2+2)^{1/2}$.

4. Derive: If $y = f(x)\cdot g(x)$, then $y' = f(x)\cdot g'(x) + g(x)\cdot f'(x)$, provided $f'(x)$ and $g'(x)$ exist.

Let $u = f(x)$ and $v = g(x)$ so that $y = u\cdot v$.
As x changes to $x+\Delta x$, let u change to $u+\Delta u$, v change to $v+\Delta v$, and y change to $y+\Delta y$. Then

$$y + \Delta y = (u+\Delta u)(v+\Delta v) = uv + u\cdot\Delta v + v\cdot\Delta u + \Delta u\cdot\Delta v$$
$$\Delta y = u\cdot\Delta v + v\cdot\Delta u + \Delta u\cdot\Delta v$$

and
$$\frac{\Delta y}{\Delta x} = u\cdot\frac{\Delta v}{\Delta x} + v\cdot\frac{\Delta u}{\Delta x} + \Delta u\cdot\frac{\Delta v}{\Delta x}.$$

Then
$$y' = \lim_{\Delta x\to 0}\left(u\cdot\frac{\Delta v}{\Delta x} + v\cdot\frac{\Delta u}{\Delta x} + \Delta u\cdot\frac{\Delta v}{\Delta x}\right)$$
$$= u\cdot v' + v\cdot u' + 0\cdot v' = u\cdot v' + v\cdot u' = f(x)\cdot g'(x) + g(x)\cdot f'(x).$$

5. Find y', given: (a) $y = x^5(1-x^2)^4$, (b) $y = x^2\sqrt{x^2+4}$, (c) $y = (3x+1)^2(2x^3-3)^{1/3}$.

(a) Set $f(x) = x^5$ and $g(x) = (1-x^2)^4$. Then $f'(x) = 5x^4$, $g'(x) = 4(1-x^2)^3(-2x)$, and

$$y' = f(x)\cdot g'(x) + g(x)\cdot f'(x) = x^5\cdot 4(1-x^2)^3(-2x) + (1-x^2)^4\cdot 5x^4$$
$$= x^4(1-x^2)^3[-8x^2 + 5(1-x^2)] = x^4(1-x^2)^3(5-13x^2).$$

(b)
$$y = x^2\cdot\tfrac{1}{2}(x^2+4)^{-1/2}\cdot 2x + (x^2+4)^{1/2}\cdot 2x$$
$$= x^3(x^2+4)^{-1/2} + 2x(x^2+4)^{1/2} = \frac{x^3 + 2x(x^2+4)}{(x^2+4)^{1/2}} = \frac{3x^3+8x}{\sqrt{x^2+4}}.$$

(c) Here $y = (3x+1)^2(2x^3-3)^{1/3}$ and

$$y' = (3x+1)^2\cdot\frac{1}{3}(2x^3-3)^{-2/3}\cdot 6x^2 + (2x^3-3)^{1/3}\cdot 2(3x+1)\cdot 3$$
$$= 2x^2(3x+1)^2(2x^3-3)^{-2/3} + 6(3x+1)(2x^3-3)^{1/3} = \frac{2x^2(3x+1)^2 + 6(3x+1)(2x^3-3)}{(2x^2-3)^{2/3}}$$
$$= \frac{2(3x+1)[x^2(3x+1) + 3(2x^3-3)]}{(2x^3-3)^{2/3}} = \frac{2(3x+1)(9x^3+x^2-9)}{(2x^3-3)^{2/3}}.$$

6. Prove: If $y = \dfrac{f(x)}{g(x)}$, if $f'(x)$ and $g'(x)$ exist, and if $g(x)\neq 0$, then $y' = \dfrac{g(x)\cdot f'(x) - f(x)\cdot g'(x)}{\{g(x)\}^2}$.

Let $u = f(x)$ and $v = g(x)$ so that $y = \dfrac{u}{v}$. Then

$$y + \Delta y = \frac{u+\Delta u}{v+\Delta v}, \qquad \Delta y = \frac{u+\Delta u}{v+\Delta v} - \frac{u}{v} = \frac{v\cdot\Delta u - u\cdot\Delta v}{v(v+\Delta v)}$$

and
$$\frac{\Delta y}{\Delta x} = \frac{v\cdot\Delta u - u\cdot\Delta v}{\Delta x\cdot v(v+\Delta v)} = \frac{v\cdot\dfrac{\Delta u}{\Delta x} - u\cdot\dfrac{\Delta v}{\Delta x}}{v(v+\Delta v)}.$$

Then
$$y' = \lim_{\Delta x\to 0}\frac{v\cdot\dfrac{\Delta u}{\Delta x} - u\cdot\dfrac{\Delta v}{\Delta x}}{v(v+\Delta v)} = \frac{v\cdot u' - u\cdot v'}{v^2} = \frac{g(x)\cdot f'(x) - f(x)\cdot g'(x)}{\{g(x)\}^2}.$$

7. Find y', given: (a) $y = 1/x^3$, (b) $y = \dfrac{2x}{x-3}$, (c) $y = \dfrac{x+5}{x^2-1}$, (d) $y = \dfrac{x^3}{\sqrt{4-x^2}}$, (e) $y = \dfrac{\sqrt{2x-3x^2}}{x+1}$.

(a) Take $f(x) = 1$ and $g(x) = x^3$; then

$$y' = \frac{g(x)\cdot f'(x) - f(x)\cdot g'(x)}{\{g(x)\}^2} = \frac{x^3\cdot 0 - 1\cdot 3x^2}{(x^3)^2} = -\frac{3}{x^4}.$$

Note that it is simpler here to write $y = x^{-3}$ and $y' = -3x^{-4} = -3/x^4$.

(*b*) Take $f(x) = 2x$ and $g(x) = x-3$; then

$$y' = \frac{g(x)\cdot f'(x) - f(x)\cdot g'(x)}{\{g(x)\}^2} = \frac{(x-3)\cdot 2 - 2x\cdot 1}{(x-3)^2} = \frac{-6}{(x-3)^2}.$$

Note that $y = \dfrac{2x}{x-3} = 2 + \dfrac{6}{x-3} = 2 + 6(x-3)^{-1}$ and $y' = 6(-1)(x-3)^{-2} = -6/(x-3)^2$.

(*c*) $$y' = \frac{(x^2-1)(1) - (x+5)(2x)}{(x^2-1)^2} = -\frac{1+10x+x^2}{(x^2-1)^2}.$$

(*d*) $$y' = \frac{(4-x^2)^{1/2}\cdot 3x^2 - x^3\cdot\frac{1}{2}(4-x^2)^{-1/2}(-2x)}{4-x^2} = \frac{(4-x^2)(3x^2) + x^3\cdot x}{(4-x^2)^{3/2}} = \frac{12x^2-2x^4}{(4-x^2)^{3/2}}.$$

The derivative exists for $-2 < x < 2$.

(*e*) $$y' = \frac{(x+1)\cdot\frac{1}{2}(2x-3x^2)^{-1/2}(2-6x) - (2x-3x^2)^{1/2}(1)}{(x+1)^2}$$

$$= \frac{(x+1)(1-3x) - (2x-3x^2)}{(x+1)^2(2x-3x^2)^{1/2}} = \frac{1-4x}{(x+1)^2(2x-3x^2)^{1/2}}.$$

The derivative exists for $0 < x < 2/3$.

8. Find y', y'', y''', given: (*a*) $y = 1/x$, (*b*) $y = \dfrac{1}{x-1} + \dfrac{1}{x+1}$, (*c*) $y = \dfrac{2}{x^2-1}$.

(*a*) Here $y = x^{-1}$; then $y' = -1\cdot x^{-2} = -x^{-2}$, $y'' = 2x^{-3}$, $y''' = -6x^{-4}$ or

$$y' = -1/x^2, \quad y'' = 2/x^3, \quad y''' = -6/x^4.$$

(*b*) Here $y = (x-1)^{-1} + (x+1)^{-1}$; then

$$y' = -1(x-1)^{-2} + (-1)(x+1)^{-2} = -\frac{1}{(x-1)^2} - \frac{1}{(x+1)^2}$$

$$y'' = 2(x-1)^{-3} + 2(x+1)^{-3} = \frac{2}{(x-1)^3} + \frac{2}{(x+1)^3}$$

$$y''' = -6(x-1)^{-4} - 6(x+1)^{-4} = -\frac{6}{(x-1)^4} - \frac{6}{(x+1)^4}.$$

(*c*) Here $y = 2(x^2-1)^{-1}$; then

$$y' = 2(-1)(x^2-1)^{-2}\cdot(2x) = -4x(x^2-1)^{-2} = -4x/(x^2-1)^2$$

$$y'' = -4(x^2-1)^{-2} - 4x(-2)(x^2-1)^{-3}(2x) = -4(x^2-1)^{-2} + 16x^2(x^2-1)^{-3} = \frac{12x^2+4}{(x^2-1)^3}$$

$$y''' = \frac{(x^2-1)^3(24x) - (12x^2+4)\cdot 3(x^2-1)^2(2x)}{(x^2-1)^6} = \frac{(x^2-1)(24x) - 6x(12x^2+4)}{(x^2-1)^4} = \frac{-48x(x^2+1)}{(x^2-1)^4}.$$

SUPPLEMENTARY PROBLEMS

9. Use the differentiation formulas to find y', given:

(a) $y = 2x^3 + 4x^2 - 5x + 8$ *Ans.* $y' = 6x^2 + 8x - 5$

(b) $y = -5 + 3x - \frac{3}{2}x^2 - 7x^3$ *Ans.* $y' = 3 - 3x - 21x^2$

(c) $y = (x-2)^4$ *Ans.* $y' = 4(x-2)^3$

(d) $y = (x^2+2)^3$ *Ans.* $y' = 6x(x^2+2)^2$

(e) $y = (4-x^2)^{10}$ *Ans.* $y' = -20x(4-x^2)^9$

(f) $y = (2x^2+4x-5)^6$ *Ans.* $y' = 24(x+1)(2x^2+4x-5)^5$

(g) $y = \frac{1}{5}x^{5/2} + \frac{1}{3}x^{3/2}$ *Ans.* $y' = \frac{1}{2}x^{1/2}(x+1)$

(h) $y = (x^2-4)^{3/2}$ *Ans.* $y' = 3x(x^2-4)^{1/2}$

(i) $y = (1-x^2)^{1/2}$ *Ans.* $y' = -\dfrac{x}{(1-x^2)^{1/2}}$

(j) $y = \dfrac{6}{x} + \dfrac{4}{x^2} - \dfrac{3}{x^3}$ *Ans.* $y' = -\dfrac{6}{x^2} - \dfrac{8}{x^3} + \dfrac{9}{x^4}$

(k) $y = x^3(x+1)^2$ *Ans.* $y' = x^2(x+1)(5x+3)$

(l) $y = (x+1)^3(x-3)^2$ *Ans.* $y' = (x+1)^2(x-3)(5x-7)$

(m) $y = (x+2)^2(2-x)^3$ *Ans.* $y' = -(x+2)(2-x)^2(5x+2)$

(n) $y = \dfrac{x+1}{x-1}$ *Ans.* $y' = -\dfrac{2}{(x-1)^2}$

(o) $y = \dfrac{x^2+2x-3}{x^2}$ *Ans.* $y' = \dfrac{6-2x}{x^3}$

(p) $y = \dfrac{x^2+1}{x^2+2}$ *Ans.* $y' = \dfrac{2x}{(x^2+2)^2}$

(q) $y = \dfrac{1}{(2x+1)^3}$ *Ans.* $y' = -\dfrac{6}{(2x+1)^4}$

(r) $y = \dfrac{1}{(x^2-9)^{1/2}}$ *Ans.* $y' = -\dfrac{x}{(x^2-9)^{3/2}}$

(s) $y = \dfrac{1}{(16-x^2)^{1/2}}$ *Ans.* $y' = \dfrac{x}{(16-x^2)^{3/2}}$

(t) $y = \dfrac{x}{(x+1)^{1/2}}$ *Ans.* $y' = \dfrac{x+2}{2(x+1)^{3/2}}$

(u) $y = \dfrac{(x^2+2)^{1/2}}{x}$ *Ans.* $y' = \dfrac{-2}{x^2(x^2+2)^{1/2}}$

10. For each of the following, find $f'(x)$, $f''(x)$, and $f'''(x)$.

(a) $f(x) = 3x^4 - 8x^3 + 12x^2 + 5$ (c) $f(x) = \dfrac{1}{4-x}$

(b) $f(x) = x^3 - 6x^2 + 9x + 18$ (d) $f(x) = (1-x^2)^{3/2}$

Ans. (a) $f'(x) = 12x(x^2 - 2x + 2)$, $f''(x) = 12(3x^2 - 4x + 2)$, $f'''(x) = 24(3x - 2)$

(b) $f'(x) = 3(x^2 - 4x + 3)$, $f''(x) = 6(x-2)$, $f'''(x) = 6$

(c) $f'(x) = \dfrac{1}{(4-x)^2}$, $f''(x) = \dfrac{2}{(4-x)^3}$, $f'''(x) = \dfrac{6}{(4-x)^4}$

(d) $f'(x) = -3x(1-x^2)^{1/2}$, $f''(x) = \dfrac{3(2x^2-1)}{(1-x^2)^{1/2}}$, $f'''(x) = \dfrac{3x(3-2x^2)}{(1-x^2)^{3/2}}$

11. In each of the following state the values of x for which $f(x)$ is continuous; also find $f'(x)$ and state the values of x for which it is defined.

(a) $f(x) = \dfrac{1}{x^2}$ (b) $f(x) = \dfrac{1}{x-2}$ (c) $f(x) = (x-2)^{4/3}$ (d) $f(x) = (x-2)^{1/3}$

Ans. (a) $x \neq 0$; $f'(x) = \dfrac{-2}{x^3}$, $x \neq 0$ (c) all x; $f'(x) = \frac{4}{3}(x-2)^{1/3}$; all x

(b) $x \neq 2$; $f'(x) = \dfrac{-1}{(x-2)^2}$, $x \neq 2$ (d) all x; $f'(x) = \dfrac{1}{3(x-2)^{2/3}}$, $x \neq 2$

Note. Parts (a) and (b) verify: If $f(x)$ is not continuous at $x = x_0$ then $f'(x)$ does not exist at $x = x_0$.

Parts (c) and (d) verify: If $f(x)$ is continuous at $x = x_0$, its derivative $f'(x)$ may or may not exist at $x = x_0$.

CHAPTER 63

Applications of Derivatives

INCREASING AND DECREASING FUNCTIONS. A function $y=f(x)$ is said to be an *increasing function* if y *increases* as x increases, and a *decreasing function* if y *decreases* as x increases.

Let the graph of $y=f(x)$ be as shown in Fig. 1. Clearly $y=f(x)$ is an increasing function from A to B and from C to D, and is a decreasing function from B to C and from D to E. At any point of the curve between A and B (also between C and D), the inclination θ of the tangent line to the curve is acute; hence, $f'(x) = \tan\theta > 0$. At any point of the curve between B and C (also, between D and E), the inclination θ of the tangent line is obtuse; hence, $f'(x) = \tan\theta < 0$.

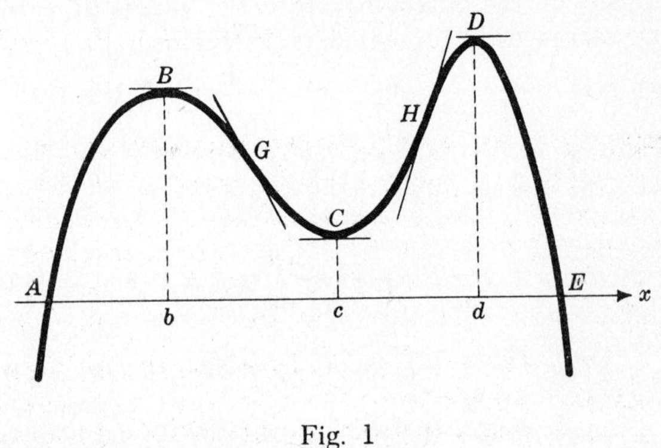

Fig. 1

Thus for values of x for which $f'(x)>0$, the function $f(x)$ is an increasing function; for values of x for which $f'(x)<0$, the function is a decreasing function.

When $x=b$, $x=c$, and $x=d$, the function is neither increasing nor decreasing since $f'(x)=0$. Such values of x are called *critical values* for the function $f(x)$.

Example 1. For the function $f(x) = x^2-6x+8$, $f'(x) = 2x-6$.

Setting $f'(x)=0$, we find the critical value $x=3$.
Now $f'(x)<0$ when $x<3$, and $f'(x)>0$ when $x>3$.
Thus, $f(x) = x^2-6x+8$ is a decreasing function when $x<3$ and an increasing function when $x>3$.

See Problem 1.

RELATIVE MAXIMUM AND MINIMUM VALUES. Let the curve of Fig. 1 be traced from left to right. Leaving A, the tracing point rises to B and then begins to fall. At B the ordinate $f(b)$ is greater than at any point of the curve near to B. We say that the point $B(b, f(b))$ is a *relative maximum point* of the curve or that the function $f(x)$ has a *relative maximum* ($=f(b)$) when $x=b$. By the same argument $D(d, f(d))$ is also a relative maximum point of the curve or $f(x)$ has a relative maximum ($=f(d)$) when $x=d$.

Leaving B, the tracing point falls to C and then begins to rise. At C the ordinate $f(c)$ is smaller than at any point of the curve near to C. We say that the point $C(c, f(c))$ is a *relative minimum point* of the curve or that $f(x)$ has a *relative minimum* ($=f(c)$) when $x=c$.

Note that the relative maximum and minimum of this function occur at the critical values. While not true for all functions, the above statement is true for all of the functions considered in this chapter.

Test for relative maximum: If $x=a$ is a critical value for $y=f(x)$ and if $f'(x) > 0$ for all values of x less than but near to $x=a$ while $f'(x)<0$ for all values of x greater than but near to $x=a$, then $f(a)$ is a relative maximum value of the function.

Test for relative minimum: If $x=a$ is a critical value for $y=f(x)$ and if $f'(x) < 0$ for all values of x less than but near to $x=a$ while $f'(x)>0$ for all values of x greater than but near to $x=a$, then $f(a)$ is a relative minimum value of the function.

If as x increases in value through a critical value, $x=a$, $f'(x)$ does not change sign, then $f(a)$ is neither a relative maximum nor a relative minimum value of the function.

Example 2. For the function of Example 1, the critical value is $x=3$.

Since $f'(x) = 2(x-3) < 0$ for $x<3$ and $f'(x)>0$ for $x>3$, the given function has a relative minimum value $f(3)=-1$.

In geometric terms, the point $(3,-1)$ is a relative minimum point of the curve $y = x^2-6x+8$.

See Problem 2.

ANOTHER TEST FOR MAXIMUM AND MINIMUM VALUES. At A on the curve of Fig. 1, the inclination θ of the tangent line is acute. As the tracing point moves from A to B, θ decreases; thus $f'(x) = \tan\theta$ is a decreasing function. At B, $f'(x)=0$. As the tracing point moves from B to G, θ is obtuse and decreasing; thus $f'(x) = \tan\theta$ is a decreasing function. Hence from A to G, $f'(x)$ is a decreasing function and its derivative $f''(x)<0$. In particular, $f''(b)<0$. Similarly, $f''(d)<0$.

As the tracing point moves from G to C, θ is obtuse and increasing; thus $f'(x)$ is an increasing function. At C, $f'(x)=0$. As the tracing point moves from C to H, θ is acute and increasing; thus $f'(x)$ is an increasing function. Hence, from G to H, $f'(x)$ is an increasing function and $f''(x)>0$. In particular, $f''(c)>0$.

Test for relative maximum: If $x=a$ is a critical value for $y=f(x)$ and if $f''(a) < 0$, then $f(a)$ is a relative maximum value of the function $f(x)$.

Test for relative minimum: If $x=a$ is a critical value for $y=f(x)$ and if $f''(a) > 0$, then $f(a)$ is a relative minimum value of the function $f(x)$.

The test fails when $f''(a)=0$. When this occurs, the tests of the preceding section must be used.

See Problem 3.

INFLECTION POINT OF A CURVE. If at $x=a$, not necessarily a critical value for $f(x)$, $f''(a)=0$ while $f'''(a)\neq 0$, the point $(a, f(a))$ is called an *inflection point* of the curve $y=f(x)$.

In Fig. 1, G and H are inflection points of the curve. Note that at points between A and G the tangent lines to the curve lie above the curve, at points between G and H the tangent lines lie below the curve, and at points between H and E the tangent lines lie above the curve. At G and H, the points of inflection, the tangent line *crosses* the curve.

Example 3. For the function $f(x) = x^2-6x+8$ of Example 1, $f'(x) = 2x-6$ and $f''(x)=2$. At the critical value $x=3$, $f''(x)>0$; hence $f(3)=-1$ is a relative minimum value. Since $f''(x)=2\neq 0$, the parabola $y = x^2-6x+8$ has no inflection point.

See Problem 4.

VELOCITY AND ACCELERATION. Let a particle move along a horizontal line and let its distance (in feet) at time $t \geq 0$ (in seconds) from a fixed point O of the line be given by $s = f(t)$. Let positive direction on the line be to the right (that is, the direction of increasing s). A complete description of the motion may be obtained by examining $f(t)$, $f'(t)$, and $f''(t)$. It was noted in Chapter 61 that $f'(t)$ gives the velocity v of the particle. The *acceleration* of the particle is given by $a = f''(t)$.

Example 4. Discuss the motion of a particle which moves along a horizontal line according to the equation $s = t^3 - 6t^2 + 9t + 2$.

When $t = 0$, $s = f(0) = 2$. The particle begins its motion from $A(s = 2)$.

Direction of Motion. Here $v = f'(t) = 3t^2 - 12t + 9 = 3(t-1)(t-3)$. When $t = 0$, $v = f'(0) = 9$; the particle leaves A with initial velocity 9 ft/sec.

Now $v = 0$ when $t = 1$ and $t = 3$. Thus, the particle moves (from A) for 1 sec, stops momentarily ($v = 0$, when $t = 1$), moves off for two more seconds, stops momentarily, and then moves off indefinitely.

On the interval $0 < t < 1$, $v > 0$. Now $v > 0$ indicates that s is increasing; thus the body leaves A with initial velocity 9 ft/sec and moves to the right for 1 sec to B ($s = f(1) = 6$) where it stops momentarily.

On the interval $1 < t < 3$, $v < 0$. Now $v < 0$ indicates that s is decreasing; thus the particle leaves B and moves to the left for 2 sec to A ($s = f(3) = 2$) where it stops momentarily.

On the interval $t > 3$, $v > 0$. The particle leaves A for the second time and moves to the right indefinitely.

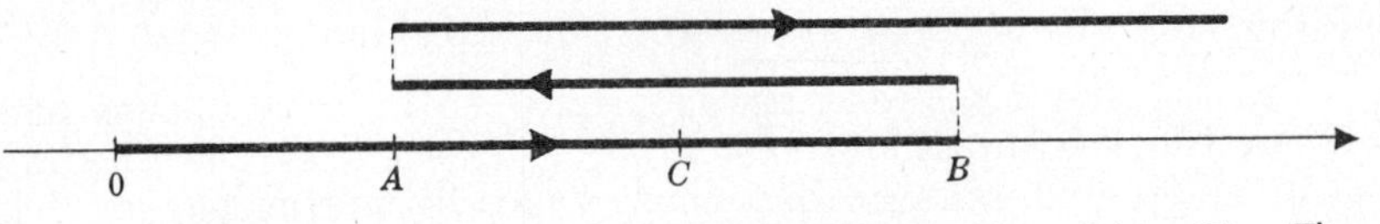

Velocity and Speed. We have $a = f''(t) = 6t - 12 = 6(t-2)$. The acceleration is 0 when $t = 2$.

On the interval $0 < t < 2$, $a < 0$. Now $a < 0$ indicates that v is decreasing; thus the particle moves for the first two seconds with decreasing velocity. For the first second (from A to B) the velocity decreases from $v = 9$ to $v = 0$. The speed (numerical value of the velocity) decreases from 9 to 0, that is, the particle "slows up". When $t = 2$, $f(t) = 4$ (the particle is at C) and $f'(t) = -3$. Thus from B to C ($t = 1$ to $t = 2$), the velocity decreases from $v = 0$ to $v = -3$. On the other hand, the speed increases from 0 to 3, that is, the particle "speeds up".

On the interval $t > 2$, $a > 0$; thus the velocity is increasing. From C to A ($t = 2$ to $t = 3$) the velocity increases from $v = -3$ to $v = 0$ while the speed decreases from 3 to 0. Thereafter ($t > 3$) both the velocity and speed increase indefinitely.

See Problem 9.

DIFFERENTIALS. Let $y = f(x)$. Define dx (read, differential x) by the relation $dx = \Delta x$ and define dy (read, differential y) by the relation $dy = f'(x) \cdot dx$. Note $dy \neq \Delta y$.

Example 5. If $y = f(x) = x^3$, then

$$\Delta y = (x + \Delta x)^3 - x^3 = 3x^2 \cdot \Delta x + 3x(\Delta x)^2 + (\Delta x)^3$$
$$= 3x^2 \cdot dx + 3x(dx)^2 + (dx)^3$$

while $dy = f'(x) \cdot dx = 3x^2 \cdot dx$. Thus if dx is small numerically, dy is a fairly good approximation of Δy and simpler to compute.

Suppose now that $x = 10$ and $dx = \Delta x = .01$. Then for the function above,

$$\Delta y = 3(10)^2(.01) + 3(10)(.01)^2 + (.01)^3 = 3.0031 \quad \text{while}$$
$$dy = 3(10)^2(.01) = 3.$$

SOLVED PROBLEMS

1. Determine the intervals on which each of the following is an increasing function and the intervals on which it is a decreasing function:

(a) $f(x) = x^2 - 8x$ (c) $f(x) = -x^3 + 3x^2 + 9x + 5$ (e) $f(x) = (x-2)^3$

(b) $f(x) = 2x^3 - 24x + 5$ (d) $f(x) = x^3 + 3x$ (f) $f(x) = (x-1)^3(x-2)$

(a) Here $f'(x) = 2(x-4)$. Setting this equal to 0 and solving, we find the critical value to be $x = 4$. We locate the point $x = 4$ on the x-axis and find that $f'(x) < 0$ for $x < 4$, and $f'(x) > 0$

$f'(x) < 0$ | 4 | $f'(x) > 0$ → x

when $x > 4$. Thus, $f(x) = x^2 - 8x$ is an increasing function when $x > 4$, and is a decreasing function when $x < 4$.

(b) $f'(x) = 6x^2 - 24 = 6(x+2)(x-2)$; the critical values are $x = -2$ and $x = 2$. Locating these points and determining the sign of $f'(x)$ on each of the intervals $x < -2$, $-2 < x < 2$, and $x > 2$, we find

$f'(x) > 0$ | -2 | $f'(x) < 0$ | 2 | $f'(x) > 0$ → x

that $f(x) = 2x^3 - 24x + 5$ is an increasing function on the intervals $x < -2$ and $x > 2$, and is a decreasing function on the interval $-2 < x < 2$.

(c) $f'(x) = -3x^2 + 6x + 9 = -3(x+1)(x-3)$; the critical values are $x = -1$ and $x = 3$. Then $f(x)$ is an

$f'(x) < 0$ | -1 | $f'(x) > 0$ | 3 | $f'(x) < 0$ → x

increasing function on the interval $-1 < x < 3$, and a decreasing function on the intervals $x < -1$ and $x > 3$.

(d) $f'(x) = 3x^2 + 3 = 3(x^2 + 1)$; there are no critical values. Since $f'(x) > 0$ for all values of x, $f(x)$ is everywhere an increasing function.

(e) $f'(x) = 3(x-2)^2$; the critical value is $x = 2$. Then $f(x)$ is an increasing function on the in-

$f'(x) > 0$ | 2 | $f'(x) > 0$ → x

tervals $x < 2$ and $x > 2$.

(f) $f'(x) = (x-1)^2(4x-7)$; the critical values are $x = 1$ and $x = 7/4$. Then $f(x)$ is an increasing

$f'(x) < 0$ | 1 | $f'(x) < 0$ | 7/4 | $f'(x) > 0$ → x

function on the interval $x > 7/4$ and is a decreasing function on the intervals $x < 1$ and $1 < x < 7/4$.

2. Find the relative maximum and minimum values of the functions of Problem 1.

(a) The critical value is $x = 4$. Since $f'(x) < 0$ for $x < 4$ and $f'(x) > 0$ for $x > 4$, the function has a relative minimum value $f(4) = -16$.

(b) The critical values are $x = -2$ and $x = 2$. Since $f'(x) > 0$ for $x < -2$ and $f'(x) < 0$ for $-2 < x < 2$, the function has a relative maximum value $f(-2) = 37$. Since $f'(x) < 0$ for $-2 < x < 2$ and $f'(x) > 0$ for $x > 2$, the function has a relative minimum value $f(2) = -27$.

(c) The critical values are $x = -1$ and $x = 3$. Since $f'(x) < 0$ for $x < -1$ and $f'(x) > 0$ for $-1 < x < 3$, $f(x)$ has a relative minimum value $f(-1) = 0$. Since $f'(x) > 0$ for $-1 < x < 3$ and $f'(x) < 0$ for $x > 3$, the function has a relative maximum value $f(3) = 32$.

(d) The function has neither a relative maximum nor a relative minimum value.

(e) The critical value is $x = 2$. Since $f'(x) > 0$ for $x < 2$ and $f'(x) > 0$ for $x > 2$, the function has neither a relative maximum nor minimum value.

(f) The critical values are $x = 1$ and $x = 7/4$. The function has a relative minimum value $f(7/4) = -27/256$. The critical value $x = 1$ yields neither a relative maximum nor minimum value.

3. Find the relative maximum and minimum values of the functions of Problem 1, using the second derivative test.

(a) $f(x) = x^2 - 8x$, $f'(x) = 2x - 8$, $f''(x) = 2$.

The critical value is $x = 4$. Since $f''(4) = 2 \neq 0$, $f(4) = -16$ is a relative minimum value of the function.

(b) $f(x) = 2x^3 - 24x + 5$, $f'(x) = 6x^2 - 24$, $f''(x) = 12x$.

The critical values are $x = -2$ and $x = 2$. Since $f''(-2) = -24 < 0$, $f(-2) = 37$ is a relative maximum value of the function; since $f''(2) = 24 > 0$, $f(2) = -27$ is a relative minimum value.

(c) $f(x) = -x^3 + 3x^2 + 9x + 5$, $f'(x) = -3x^2 + 6x + 9$, $f''(x) = -6x + 6$.

The critical values are $x = -1$ and $x = 3$. Since $f''(-1) > 0$, $f(-1) = 0$ is a relative minimum value of the function; since $f''(3) < 0$, $f(3) = 32$ is a relative maximum value.

(d) $f(x) = x^3 + 3x$, $f'(x) = 3x^2 + 3$, $f''(x) = 6x$.

There are no critical values; hence the function has neither a relative minimum nor a relative maximum value.

(e) $f(x) = (x-2)^3$, $f'(x) = 3(x-2)^2$, $f''(x) = 6(x-2)$.

The critical value is $x = 2$. Since $f''(2) = 0$, the test fails. The test of Problem 2 shows that the function has neither a relative maximum nor a relative minimum value.

(f) $f(x) = (x-1)^3(x-2)$, $f'(x) = (x-1)^2(4x-7)$, $f''(x) = 6(2x-3)(x-1)$.

The critical values are $x = 1$ and $x = 7/4$. Since $f''(1) = 0$, the test fails; the test of Problem 2 shows that $f(1)$ is neither a relative maximum nor a relative minimum value of the function. Since $f''(7/4) > 0$, $f(7/4) = -27/256$ is a relative minimum value.

4. Find the inflection points and plot the graph of each of the given curves. In sketching the graph, locate the x- and y-intercepts when they can be found, the relative maximum and minimum points (see Problem 2), and the inflection points, if any. Additional points may be found if necessary.

(a) $y = f(x) = x^2 - 8x$
(b) $y = f(x) = 2x^3 - 24x + 5$
(c) $y = f(x) = -x^3 + 3x^2 + 9x + 5$
(d) $y = f(x) = x^3 + 3x$
(e) $y = f(x) = (x-2)^3$
(f) $y = f(x) = (x-1)^3(x-2)$

(a) Since $f''(x) = 2$, the parabola does not have an inflection point.

The x- and y-intercepts are $x = 0$, $x = 8$, and $y = 0$; $(4,-16)$ is a relative minimum point. See Fig. (a) below.

(b) $f''(x) = 12x$ and $f'''(x) = 12$. Since $f''(x) = 0$ when $x = 0$ and $f'''(0) = 12 \neq 0$, $(0,5)$ is an inflection point.

The y-intercept is $y = 5$, the x-intercepts cannot be determined; $(-2,37)$ is a relative maximum point, $(2,-27)$ is a relative minimum point; $(0,5)$ is an inflection point. See Fig. (b) below.

(c) $f''(x) = -6x + 6$ and $f'''(x) = -6$. Since $f''(x) = 0$ when $x = 1$ and $f'''(1) = -6 \neq 0$, $(1,16)$ is an inflection point.

The x- and y-intercepts are $x = -1$, $x = 5$, and $y = 5$; $(-1,0)$ is a relative minimum point and $(3,32)$ is a relative maximum point; $(1,16)$ is an inflection point. See Fig. (c) below.

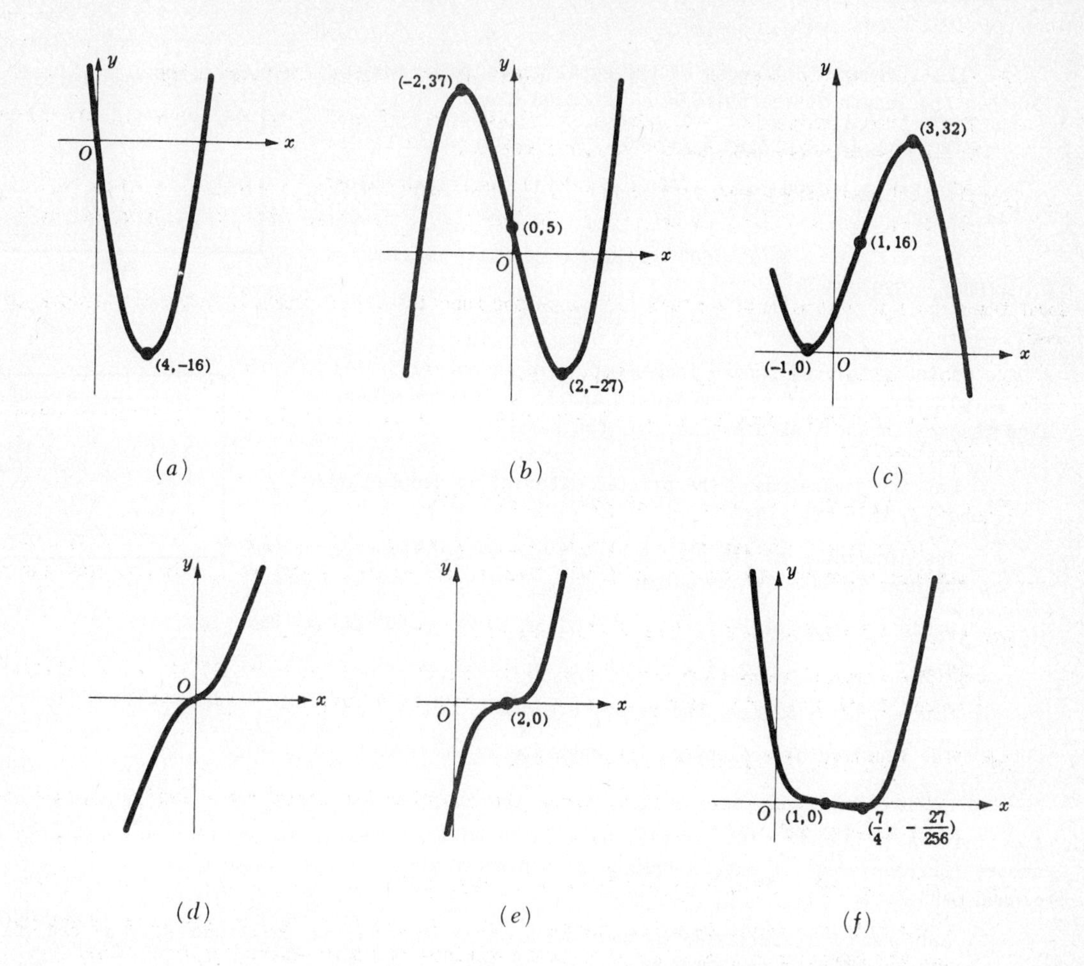

(*d*) $f''(x) = 6x$ and $f'''(x) = 6$. The point (0,0) is an inflection point.

The x- and y-intercepts are $x = 0$, $y = 0$; (0,0) is an inflection point. The curve can be sketched after locating the points (1,4), (2,14), (−1,−4), and (−2,−14). See Fig. (*d*) above.

(*e*) $f''(x) = 6(x-2)$ and $f'''(x) = 6$. The point (2,0) is an inflection point.

The x- and y-intercepts are $x = 2$, $y = -8$; (2,0) is an inflection point. The curve can be sketched after locating the points (3,1), (4,8), and (1,−1). See Fig. (*e*) above.

(*f*) $f''(x) = 6(2x-3)(x-1)$ and $f'''(x) = 6(4x-5)$. The inflection points are (1,0) and (3/2,−1/16).

The x- and y-intercepts are $x = 1$, $x = 2$, and $y = 2$; (7/4,−27/256) is a relative minimum point; (1,0) and (3/2,−1/16) are inflection points. For the graph, see Fig. (*f*) above.

5. Find two integers whose sum is 12 and whose product is a maximum.

Let x and $12-x$ be the integers; their product is $P = f(x) = x(12-x) = 12x - x^2$.

Since $f'(x) = 12 - 2x = 2(6-x)$, $x = 6$ is the critical value. Now $f''(x) = -2$; hence $f''(6) = -2 < 0$ and $x = 6$ yields a relative maximum. The integers are 6 and 6.

Note that we have, in effect, proved that the rectangle of given perimeter has maximum area when it is a square.

6. A farmer wishes to enclose a rectangular plot for a pasture, using a wire fence on three sides and a hedge row as the fourth side. If he has 2400 ft of wiring, what is the greatest area he can fence off?

Let x denote the length of the equal sides to be wired; then the length of the third side is $2400 - 2x$.

The area is $A = f(x) = x(2400 - 2x) = 2400x - 2x^2$.

Now $f'(x) = 2400 - 4x = 4(600 - x)$ and the critical value is $x = 600$.

Since $f''(x) = -4$, $x = 600$ yields a relative maximum $= f(600) = 720{,}000 \text{ ft}^2$.

7. A page is to contain 54 square inches of printed material. If the margins are 1 in. at top and bottom and $1\frac{1}{2}$ in. at the sides, find the most economical dimensions of the page.

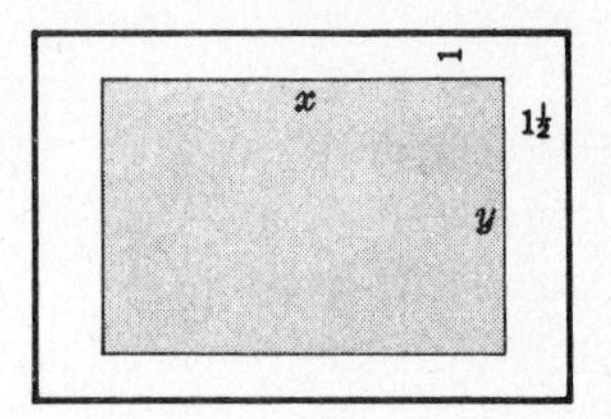

Let the dimensions of the printed material be denoted by x and y; then $xy = 54$.

The dimensions of the page are $x+3$ and $y+2$; the area of the page is $A = (x+3)(y+2)$.

Since $y = 54/x$, $A = f(x) = (x+3)(54/x + 2) = 60 + 162/x + 2x$.

Then $f'(x) = -162/x^2 + 2$ and the critical values are $x = \pm 9$.

Since $f''(x) = 324/x^3$, the relative minimum is given by $x = 9$.

The required dimensions of the page are 12 in. wide and 8 in. high.

8. A cylindrical container with circular base is to hold 64 cubic inches. Find the dimensions so that the amount (surface area) of metal required is a minimum when (*a*) the container is an open cup and (*b*) a closed can.

Let r and h respectively be the radius of the base and height in inches, V be the volume of the container, and A be the surface area of the metal required.

(*a*) $V = \pi r^2 h = 64$ and $A = 2\pi rh + \pi r^2$. Solving for $h = 64/\pi r^2$ in the first relation and substituting in the second, we have $A = 2\pi r\left(\frac{64}{\pi r^2}\right) + \pi r^2 = \frac{128}{r} + \pi r^2$.

Then $\frac{dA}{dr} = -\frac{128}{r^2} + 2\pi r = \frac{2(\pi r^3 - 64)}{r^2}$ and the critical value is $r = \frac{4}{\sqrt[3]{\pi}}$.

Now $h = \frac{64}{\pi r^2} = \frac{4}{\sqrt[3]{\pi}}$; thus, $r = h = \frac{4}{\sqrt[3]{\pi}}$ in.

(*b*) $V = \pi r^2 h = 64$ and $A = 2\pi rh + 2\pi r^2 = 2\pi r\left(\frac{64}{\pi r^2}\right) + 2\pi r^2 = \frac{128}{r} + 2\pi r^2$.

Then $\frac{dA}{dr} = -\frac{128}{r^2} + 4\pi r = \frac{4(\pi r^3 - 32)}{r^2}$ and the critical value is $r = 2\sqrt[3]{\frac{4}{\pi}}$.

Now $h = \frac{64}{\pi r^2} = 4\sqrt[3]{\frac{4}{\pi}}$; thus, $h = 2r = 4\sqrt[3]{\frac{4}{\pi}}$ in.

9. Study the motion of a particle which moves along a horizontal line in accordance with:
(*a*) $s = t^3 - 6t^2 + 3$, (*b*) $s = t^3 - 5t^2 + 7t - 3$, (*c*) $v = (t-1)^2(t-4)$, (*d*) $v = (t-1)^4$.

(*a*) Here $v = 3t^2 - 12t = 3t(t-4) = 0$ when $t = 0$ and $t = 4$;
$a = 6t - 12 = 6(t-2) = 0$ when $t = 2$.

The particle leaves $A(s = 3)$ with velocity 0 and moves to $B(-29)$ where it stops momentarily; thereafter it moves to the right.

The intervals of increasing and decreasing speed are given below:

$v<0,\ a<0$ | $v<0,\ a>0$ | $v>0,\ a>0$

0 *velocity dec.* *speed inc.* 2 *velocity inc.* *speed dec.* 4 *velocity inc.* *speed inc.* t

(*b*) Here $v = 3t^2 - 10t + 7 = (t-1)(3t-7) = 0$ when $t = 1$ and $t = 7/3$;
$a = 6t - 10 = 2(3t-5) = 0$ when $t = 5/3$.

The particle leaves $A(s = -3)$ with velocity 7 ft/sec and moves to 0 where it stops momentarily, then it moves to $B(-32/27)$ where it stops momentarily. Thereafter it moves to the right.

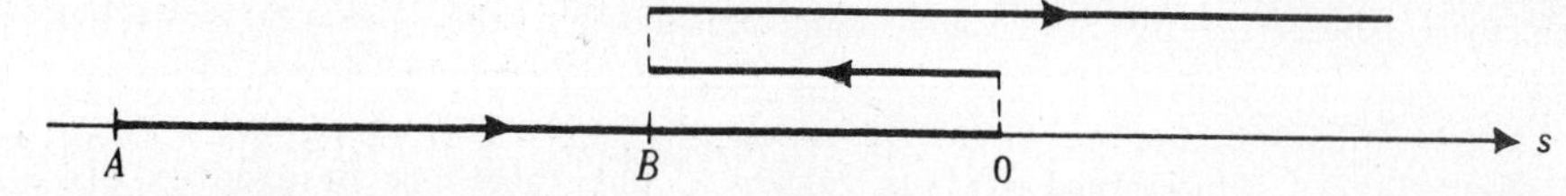

The intervals of increasing and decreasing speed are given below:

$v>0,\ a<0$ | $v<0,\ a<0$ | $v<0,\ a>0$ | $v>0,\ a>0$

0 *velocity dec.* *speed dec.* 1 *vel. dec.* *speed inc.* 5/3 *vel. inc.* *speed dec.* 7/3 *velocity inc.* *speed inc.* t

(*c*) Here $v = 0$ when $t = 1$ and $t = 4$. Also $a = 3t^2 - 12t + 9 = 3(t-1)(t-3) = 0$ when $t = 1$ and $t = 3$. The intervals of increasing and decreasing speed are

$v<0,\ a>0$ | $v<0,\ a<0$ | $v<0,\ a>0$ | $v>0,\ a>0$

0 *vel. inc.* *speed dec.* 1 *velocity dec.* *speed inc.* 3 *vel. inc.* *speed dec.* 4 *velocity inc.* *speed inc.* t

Note that the particle stops momentarily at the end of one second but does not then reverse it direction of motion as, for example, in (*b*).

(*d*) Here $a = 4(t-1)^3$ and $v = a = 0$ when $t = 1$. The intervals of increasing and decreasing speed are:

$v>0,\ a<0$ | $v>0,\ a>0$

0 *velocity dec.* *speed dec.* 1 *velocity inc.* *speed inc.* t

10. Find dy in terms of x and dx, given:
(*a*) $y = f(x) = x^2 + 5x + 6$, (*b*) $y = f(x) = x^4 - 4x^3 + 8$, (*c*) $y = f(x) = x^2 + 1/x^2$.

(*a*) Since $f'(x) = 2x + 5$, $dy = f'(x)\,dx = (2x+5)dx$.
(*b*) Since $f'(x) = 4x^3 - 12x^2$, $dy = (4x^3 - 12x^2)\,dx$.
(*c*) Since $f'(x) = 2x - 2/x^3$, $dy = (2x - 2/x^3)\,dx$.

11. Find the approximate displacement of a particle moving along the x-axis in accordance with the law $s = t^4 - t^2$, from the time $t = 1.99$ to $t = 2$.

Here $ds = (4t^3 - 2t)dt$. We take $t = 2$ and $dt = -0.01$. Then $ds = (4\cdot 8 - 2\cdot 2)(-.01) = -0.28$ and the displacement is 0.28 units.

12. Find using differentials the approximate area of a square whose side is 3.01 in.

Here $A = x^2$ and $dA = 2x\,dx$. Taking $x = 3$ and $dx = 0.01$, we find $dA = 2\cdot 3(.01) = 0.06\ \text{in}^2$. Now the area ($9\ \text{in}^2$) of a square 3 in. on a side is increased approximately $0.06\ \text{in}^2$ when the side is increased to 3.01 in. Hence the approximate area is $9.06\ \text{in}^2$. The true area is $9.0601\ \text{in}^2$.

SUPPLEMENTARY PROBLEMS

13. Determine the intervals on which each of the following is an increasing function and the intervals on which it is a decreasing function.

(a) $f(x) = x^2$ — *Ans.* Dec. for $x < 0$; inc. for $x > 0$

(b) $f(x) = 4 - x^2$ — *Ans.* Inc. for $x < 0$; dec. for $x > 0$

(c) $f(x) = x^2 + 6x - 5$ — *Ans.* Dec. for $x < -3$; inc. for $x > -3$

(d) $f(x) = 3x^2 + 6x + 18$ — *Ans.* Dec. for $x < -1$; inc. for $x > -1$

(e) $f(x) = (x-2)^4$ — *Ans.* Dec. for $x < 2$; inc. for $x > 2$

(f) $f(x) = (x-1)^3(x+2)^2$ — *Ans.* Inc. for $x < -2$; dec. for $-2 < x < -4/5$; inc. for $-4/5 < x < 1$ and for $x > 1$

14. Find the relative maximum and minimum values of the functions of Problem 13.

Ans. (a) Min. = 0 (b) Max. = 4 (c) Min. = −14 (d) Min. = 15 (e) Min. = 0 (f) Max. = 0, Min. = −26244/3125

15. Investigate for relative maximum (minimum) points and points of inflection. Sketch each locus.

(a) $y = x^2 - 4x + 8$ (b) $y = (x-1)^3 + 5$ (c) $y = x^4 + 32x + 40$ (d) $y = x^3 - 3x^2 - 9x + 6$

Ans. (a) Min. (2,4) (b) I.P. (1,5) (c) Min. (−2,−8) (d) Max. (−1,11), Min. (3,−21), I.P. (1,−5)

16. The sum of two positive numbers is 12. Find the numbers

(a) if the sum of their squares is a minimum,

(b) if the product of one and the square of the other is a maximum,

(c) if the product of one and the cube of the other is a maximum.

Ans. (a) 6 and 6 (b) 4 and 8 (c) 3 and 9

17. Find the dimensions of the largest open box which can be made from a sheet of tin 24 inches square by cutting equal squares from the corners and turning up the sides. *Ans.* $16 \times 16 \times 4$ in.

18. Find the dimensions of the largest open box which can be made from a sheet of tin 60 in. by 28 in. by cutting equal squares from the corners and turning up the sides. *Ans.* $48 \times 16 \times 6$ in.

19. A rectangular field is to be enclosed by a fence and divided into two smaller plots by a fence parallel to one of the sides. Find the dimensions of the largest such field which can be enclosed by 1200 ft of fencing. *Ans.* 200×300 ft

20. If a farmer harvests his crop today he will have 1200 bushels worth \$2.00 per bushel. Every week he waits, the crop increases by 100 bu but the price drops 10¢ per bu. When should he harvest the crop? *Ans.* 4 weeks from today

21. The base of an isoscles triangle is 20 ft and its altitude is 40 ft. Find the dimensions of the largest inscribed rectangle if two of the vertices are on the base of the triangle. *Ans.* 10×20 ft

22. For each of the following compute Δy, dy, and $\Delta y - dy$.

(a) $y = \frac{1}{2}x^2 + x$; $x = 2$, $\Delta x = \frac{1}{4}$ — *Ans.* $\Delta y = 25/32$, $dy = 3/4$, $\Delta y - dy = 1/32$

(b) $y = x^2 - x$; $x = 3$, $\Delta x = .01$ — *Ans.* $\Delta y = .0501$, $dy = .05$, $\Delta y - dy = .0001$

23. Approximate using differentials the volume of a cube whose side is 3.005 in. *Ans.* 27.135 in^3

24. Approximate using differentials the area of a circular ring whose inner radius is 5 in. and whose width is 1/8 in. *Ans.* 1.25π in^2

CHAPTER 64

Integration

IF $F(x)$ IS A FUNCTION whose derivative $F'(x) = f(x)$, then $F(x)$ is called an *integral* of $f(x)$.

For example, $F(x) = x^3$ is an integral of $f(x) = 3x^2$ since $F'(x) = 3x^2 = f(x)$. Also, $G(x) = x^3 + 5$ and $H(x) = x^3 - 6$ are integrals of $f(x) = 3x^2$. Why ?

If $F(x)$ and $G(x)$ are two distinct integrals of $f(x)$, then $F(x) = G(x) + C$, where C is a constant.

See Problem 1.

THE INDEFINITE INTEGRAL of $f(x)$, denoted by $\int f(x)\,dx$, is the most general integral of $f(x)$, that is,

$$\int f(x)\,dx = F(x) + C$$

where $F(x)$ is any function such that $F'(x) = f(x)$ and C is an arbitrary constant. Thus the indefinite integral of $f(x) = 3x^2$ is $\int 3x^2\,dx = x^3 + C$.

We shall use the following integration formulas:

I. $\int x^n\,dx = \dfrac{x^{n+1}}{n+1} + C$, where $n \neq -1$

II. $\int c\,f(x)\,dx = c\int f(x)\,dx$, where c is a constant

III. $\int \{f(x) + g(x)\}\,dx = \int f(x)\,dx + \int g(x)\,dx$

Example 1. (a) $\int x^5\,dx = \dfrac{x^{5+1}}{5+1} + C = \dfrac{x^6}{6} + C$

(b) $\int 4x^3\,dx = 4\int x^3\,dx = 4\cdot\dfrac{x^4}{4} + C = x^4 + C$

(c) $\int 3x\,dx = 3\int x\,dx = 3\cdot\dfrac{x^2}{2} + C = \dfrac{3}{2}x^2 + C$

(d) $\int \dfrac{dx}{x^3} = \int x^{-3}\,dx = \dfrac{x^{-2}}{-2} + C = -\dfrac{1}{2x^2} + C$

(e) $\int (x^5 + 4x^3 + 3x)\,dx = \int x^5\,dx + \int 4x^3\,dx + \int 3x\,dx$

$= x^6/6 + x^4 - 3x^2/2 + C$

See Problems 2-8.

AREA BY INTEGRATION. Consider the area $A(x)$ bounded by the curve $y = f(x)$, the x-axis, the fixed ordinate $x = a$, and a variable ordinate $x = x$. (For simplicity, we shall assume that $y \geq 0$ at all times.) Let ΔA denote the increase (shaded area) in $A(x)$ when x is increased from x to $x+\Delta x$. From the figure below, it is clear that

area rect. $MPRN \leq \Delta A \leq$ area rect. $MSQN$

or $\qquad y \cdot \Delta x \leq \Delta A \leq (y+\Delta y)\Delta x.$

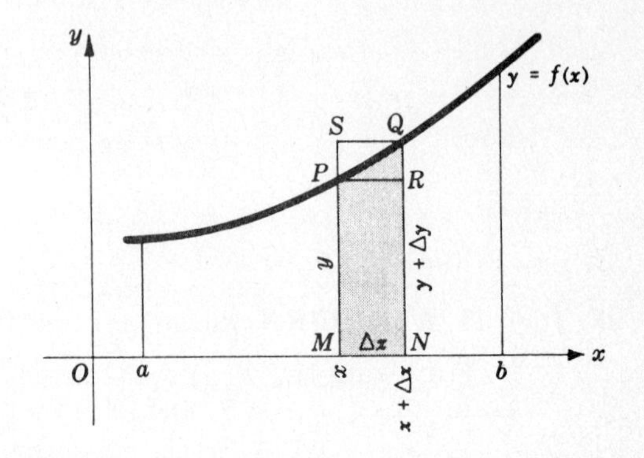

Dividing by $\Delta x \neq 0$, we have

$$y \leq \frac{\Delta A}{\Delta x} \leq y + \Delta y.$$

Let $\Delta x \to 0$; then $\Delta y \to 0$ and $\Delta A \to 0$, and

$$y \leq \lim_{\Delta x \to 0} \frac{\Delta A}{\Delta x} \leq y \quad \text{or} \quad y \leq \frac{dA}{dx} \leq y$$

so that $\qquad \dfrac{dA}{dx} = y = f(x).$

Then $\qquad dA = f(x)\,dx \quad$ and $\quad A(x) = \int f(x)\,dx = F(x) + C,$

where $F'(x) = f(x)$. To determine C, we note that when $x = a$, that is, when the bounding ordinates coincide, there is no area and hence $A(a) = 0$. Now $A(a) = F(a) + C$; thus $C = -F(a)$ and

$$A(x) = F(x) - F(a)$$

Let it now be required to find the area A between the curve $y = f(x)$, the x-axis, and the two fixed ordinates $x = a$ and $x = b$, where $b > a$. Then

$$A = A(b) = F(b) - F(a)$$

Example 2. Determine the area bounded by the parabola $y = x^2$, the x-axis, and the ordinates $x = 1$ and $x = 3$.

Since $\dfrac{dA}{dx} = y = x^2, \quad A(x) = \int x^2\,dx = \dfrac{x^3}{3} + C.$

When $x = 1$, $A(x) = 0$; then $0 = \dfrac{1}{3} + C$ and $C = -\dfrac{1}{3}$.

Thus $A(x) = \dfrac{x^3}{3} - \dfrac{1}{3}$ and $A = A(3) = 9 - 1/3 = 26/3$ sq units.

We may simplify this procedure as follows: Since $F(x) = x^3/3$ is an integral of $f(x) = x^2$, $A = F(3) - F(1) = 9 - 1/3 = 26/3$ sq units.

IF WE DEFINE $\int_a^b f(x)\,dx$ (read, the *definite integral* of $f(x)$ between $x = a$ and $x = b$) as follows

$$\int_a^b f(x)\,dx = F(x)\Big|_a^b = F(b) - F(a), \qquad (F'(x) = f(x))$$

then the area bounded by $y = f(x) \geq 0$, the x-axis, and the ordinates $x = a$ and $x = b$, $(b > a)$, is given by

$$A = \int_a^b f(x)\,dx$$

See Problems 9-12.

SOLVED PROBLEMS

1. Prove: If $F(x)$ and $G(x)$ are distinct integrals of $f(x)$ then $F(x) = G(x) + C$, where C is a constant.

Since $F(x)$ and $G(x)$ are integrals of $f(x)$, $F'(x) = G'(x) = f(x)$.

Suppose $F(x) - G(x) = H(x)$; differentiating with respect to x, $F'(x) - G'(x) = H'(x)$ and $H'(x) = 0$. Thus $H(x)$ is a constant, say C, and $F(x) = G(x) + C$.

2. (a) $\int \sqrt{x}\, dx = \int x^{1/2}\, dx = \dfrac{x^{3/2}}{3/2} + C = \dfrac{2}{3}x^{3/2} + C$

(b) $\int (3x^2 + 5)dx = x^3 + 5x + C$

(c) $\int (5x^6 + 2x^3 - 4x + 3)dx = \dfrac{5}{7}x^7 + \dfrac{1}{2}x^4 - 2x^2 + 3x + C$

(d) $\int (80x^{19} - 32x^{15} - 12x^{-3})dx = 4x^{20} - 2x^{16} + 6/x^2 + C$

3. At every point (x,y) of a certain curve, the slope is equal to 8 times the abscissa. Find the equation of the curve if it passes through $(1,3)$.

Since $m = \dfrac{dy}{dx} = 8x$ we have $dy = 8x\, dx$. Then $y = \int 8x\, dx = 4x^2 + C$, a family of parabolas. We seek the equation of the parabola of this family which passes through the point $(1,3)$.

Then $3 = 4(1)^2 + C$ and $C = -1$. The curve has equation $y = 4x^2 - 1$.

4. For a certain curve $y'' = 6x - 10$, Find its equation if it passes through point $(1,1)$ with slope -1.

Since $y'' = 6x - 10$, $y' = 3x^2 - 10x + C_1$; since $y' = -1$ when $x = 1$, we have $-1 = 3 - 10 + C_1$ and $C_1 = 6$. Then $y' = 3x^2 - 10x + 6$.

Now $y = x^3 - 5x^2 + 6x + C_2$ and since $y = 1$ when $x = 1$, $1 = 1 - 5 + 6 + C_2$ and $C_2 = -1$. Thus the equation of the curve is $y = x^3 - 5x^2 + 6x - 1$.

5. The velocity at time t of a particle moving along the x-axis is given by $v = x' = 2t + 5$. Find the position of the particle at time t, if $x = 2$ when $t = 0$.

Select a point on the x-axis as origin and assume positive direction to the right. Then at the beginning of the motion $(t = 0)$ the particle is 2 units to the right of the origin.

Since $v = \dfrac{dx}{dt} = 2t + 5$, $dx = (2t + 5)dt$. Then $x = \int (2t + 5)dt = t^2 + 5t + C$.

Substituting $x = 2$ and $t = 0$, we have $2 = 0 + 0 + C$ so that $C = 2$. Thus the position of the particle at time t is given by $x = t^2 + 5t + 2$.

6. A body moving in a straight line has an acceleration equal to $6t^2$, where time (t) is measured in seconds and distance s is measured in feet. If the body starts from rest, how far will it move during the first two seconds?

Let the body start from the origin; then it is given that when $t = 0$, $v = 0$ and $s = 0$.

Since $a = \dfrac{dv}{dt} = 6t^2$, $dv = 6t^2\, dt$. Then $v = \int 6t^2\, dt = 2t^3 + C_1$. When $t = 0$, $v = 0$; then $0 = 2\cdot 0 + C_1$ and $C_1 = 0$. Thus $v = 2t^3$.

Now $v = \frac{ds}{dt} = 2t^3$; then $ds = 2t^3\,dt$ and $s = \int 2t^3\,dt = \frac{1}{2}t^4 + C_2$. When $t = 0$, $s = 0$; then $C_2 = 0$ and $s = \frac{1}{2}t^4$. When $t = 2$, $s = \frac{1}{2}(2)^4 = 8$. The body moves 8 ft during the first two sec.

7. A ball is thrown upward from the top of a building 320 ft high with initial velocity 128 ft/sec. Determine the velocity with which the ball will strike the street below. (Assume acceleration is 32 ft/sec^2, directed downward.)

First we choose an origin from which all distances are to be measured and a direction (upward or downward) which will be called positive.

First Solution. Take the origin at the top of the building and positive direction as upward. Then

$$a = \frac{dv}{dt} = -32 \qquad \text{and} \qquad v = -32t + C_1$$

When the ball is released, $t = 0$ and $v = 128$; then $128 = -32(0) + C_1$ and $C_1 = 128$.

Now $v = \frac{ds}{dt} = -32t + 128$ and $s = -16t^2 + 128t + C_2$. When the ball is released, $t = 0$ and $s = 0$; then $C_2 = 0$ and $s = -16t^2 + 128t$.

When the ball strikes the street it is 320 ft below the origin, that is, $s = -320$; hence

$$-320 = -16t^2 + 128t, \quad t^2 - 8t - 20 = (t+2)(t-10) = 0 \quad \text{and} \quad t = 10$$

Finally, when $t = 10$, $v = -32(10) + 128 = -192$ ft/sec.

Second Solution. Take the origin on the street and positive direction as before. Then $a = \frac{dv}{dt} = -32$ and $v = -32t + 128$ as in the first solution.

Now $s = -16t^2 + 128t + C_2$ but when $t = 0$, $s = 320$. Thus $C_2 = 320$ and $s = -16t^2 + 128t + 320$. When the ball strikes the street $s = 0$; then $t = 10$ and $v = -192$ ft/sec as before.

8. A ball was dropped from a balloon 640 ft above the ground. If the balloon was rising at the rate of 48 ft/sec, find (*a*) the greatest distance above the ground attained by the ball, (*b*) the time the ball was in the air, (*c*) the speed of the ball when it struck the ground.

Assume the origin at the point where the ball strikes the ground and positive distance to be directed upward, Then

$$a = \frac{dv}{dt} = -32 \qquad \text{and} \qquad v = -32t + C_1$$

When $t = 0$, $v = 48$; hence $C_1 = 48$. Then $v = \frac{ds}{dt} = -32t + 48$ and $s = -16t^2 + 48t + C_2$.
When $t = 0$, $s = 640$; hence $C_2 = 640$ and $s = -16t^2 + 48t + 640$.

(*a*) When $v = 0$, $t = 3/2$ and $s = -16(3/2)^2 + 48(3/2) + 640 = 676$. The greatest height attained by the ball was 676 ft.

(*b*) When $s = 0$, $-16t^2 + 48t + 640 = 0$ and $t = -5, 8$. The ball was in the air for 8 sec.

(*c*) When $t = 8$, $v = -32(8) + 48 = -208$. The ball struck the ground with speed 208 ft/sec.

9. Find the area bounded by the line $y = 4x$, the x-axis, and the ordinates $x = 0$ and $x = 5$.

Here $y \geq 0$ on the interval $0 \leq x \leq 5$. Then

$$A = \int_0^5 4x\,dx = 2x^2\Big|_0^5 = 50 \text{ sq units}$$

Note that we have found the area of a right triangle whose legs are 5 and 20 units. The area is $\frac{1}{2}(5)(20) = 50$ sq units.

10. Find the area bounded by the parabola $y = 8 + 2x - x^2$ and the x-axis.

The x-intercepts are $x = -2$ and $x = 4$; $y \geq 0$ on the interval $-2 \leq x \leq 4$. Hence

$$A = \int_{-2}^{4} (8 + 2x - x^2)\,dx = (8x + x^2 - \frac{x^3}{3})\Big|_{-2}^{4}$$

$$= (8\cdot 4 + 4^2 - \frac{4^3}{3}) - [8(-2) + (-2)^2 - \frac{(-2)^3}{3}] = 36 \text{ sq units.}$$

11. Find the area bounded by the parabola $y = x^2 + 2x - 3$, the x-axis, and the ordinates $x = -2$ and $x = 0$.

On the interval $-2 \leq x \leq 0$, $y \leq 0$. Here

$$A = \int_{-2}^{0} (x^2 + 2x - 3)\,dx = (\frac{x^3}{3} + x^2 - 3x)\Big|_{-2}^{0} = 0 - \{\frac{(-2)^3}{3} + (-2)^2 - 3(-2)\} = -\frac{22}{3}.$$

The negative sign indicates that the area lies entirely below the x-axis. The area is 22/3 sq units. See Fig. (a) below.

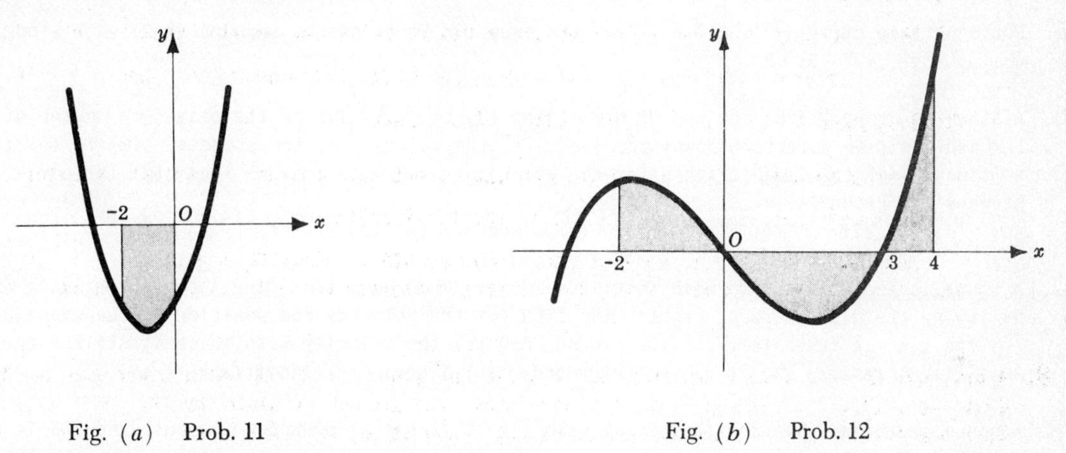

Fig. (a) Prob. 11

Fig. (b) Prob. 12

12. Find the area bounded by the curve $y = x^3 - 9x$, the x-axis, and the ordinates $x = -2$ and $x = 4$.

The purpose of this problem is to show that the required area is *not* given by $\int_{-2}^{4} (x^3 - 9x)\,dx$

From Fig. (b) above, we note that y changes sign at $x = 0$ and at $x = 3$. The required area consists of three pieces, the individual areas being given, apart from sign, by

$$A_1 = \int_{-2}^{0} (x^3 - 9x)\,dx = (\frac{1}{4}x^4 - \frac{9}{2}x^2)\Big|_{-2}^{0} = 0 - (4 - 18) = 14$$

$$A_2 = \int_{0}^{3} (x^3 - 9x)\,dx = (\frac{1}{4}x^4 - \frac{9}{2}x^2)\Big|_{0}^{3} = (\frac{81}{4} - \frac{81}{2}) - 0 = -\frac{81}{4}$$

$$A_3 = \int_{3}^{4} (x^3 - 9x)\,dx = (\frac{1}{4}x^4 - \frac{9}{2}x^2)\Big|_{3}^{4} = (64 - 72) - (\frac{81}{4} - \frac{81}{2}) = \frac{49}{4}.$$

Thus $A = A_1 - A_2 + A_3 = 14 + \frac{81}{4} + \frac{49}{4} = \frac{93}{2}$ sq units.

Note that $\int_{-2}^{4} (x^3 - 9x)\,dx = 6 < A_1$, an absurd result.

SUPPLEMENTARY PROBLEMS

13. Work out the following indefinite integrals.

(a) $4\int dx = 4x + C$

(b) $\int \frac{1}{2}x\,dx = \frac{1}{4}x^2 + C$

(c) $\int (3x^2 + 4x - 5)\,dx = x^3 + 2x^2 - 5x + C$

(d) $\int x(1-x)\,dx = \frac{1}{2}x^2 - \frac{1}{3}x^3 + C$

(e) $\int 3(x+1)^2\,dx = (x+1)^3 + C$

(f) $\int (x-1)(x+2)\,dx = \frac{1}{3}x^3 + \frac{1}{2}x^2 - 2x + C$

(g) $\int \frac{dx}{x^2} = -\frac{1}{x} + C$

(h) $\int \frac{x^2-2}{x^2}\,dx = x + \frac{2}{x} + C$

14. Find the equation of the family of curves whose slope is the given function of x. Find also the equation of the curve of the family passing through the given point.

(a) $m = 1$, $(1,-2)$ *Ans.* $y = x + C$, $y = x - 3$

(b) $m = -6x$, $(0,0)$ *Ans.* $y = -3x^2 + C$, $y = -3x^2$

(c) $m = 3x^2 + 2x$, $(1,-3)$ *Ans.* $y = x^3 + x^2 + C$, $y = x^3 + x^2 - 5$

(d) $m = 6x^2$, $(0,1)$ *Ans.* $y = 2x^3 + C$, $y = 2x^3 + 1$

15. For a certain curve $y'' = 6x + 8$. Find its equation if it passes through $(1,2)$ with slope $m = 6$.
Ans. $y = x^3 + 4x^2 - 5x + 2$

16. A stone is dropped from the top of a building 400 ft high. Taking the origin at the top of the building and positive direction downward, find (a) the velocity of the stone at time t, (b) the position at time t, (c) the time it takes for the stone to reach the ground, and (d) the velocity when it strikes the ground.
Ans. (a) $v = 32t$ (b) $s = 16t^2$ (c) 5 sec (d) 160 ft/sec

17. A stone is thrown downward with initial velocity 20 ft/sec from the top of a building 336 ft high. Following the directions of Problem 16, find (a) the velocity and position of the stone 2 sec later, (b) the time it takes to reach the ground, and (c) the velocity with which it strikes the ground.
Ans. (a) 84 ft/sec, 232 ft above the ground (b) 4 sec (c) 148 ft/sec

18. A stone is thrown upward with initial velocity 16 ft/sec from the top of a building 192 ft high. Find (a) the greatest height attained by the stone, (b) the total time in motion, and (c) the speed with which the stone strikes the ground.
Ans. (a) 196 ft (b) 4 sec (c) 112 ft/sec

19. A boy on top a building 192 ft high throws a rock straight down. What initial velocity did he give it if it strikes the ground after 3 sec? *Ans.* 16 ft/sec

20. Find the area bounded by the x-axis, the given curve, and the indicated ordinates.

(a) $y = x^2$ between $x = 2$ and $x = 4$ *Ans.* 56/3 square units

(b) $y = 4 - 3x^2$ between $x = -1$ and $x = 1$ *Ans.* 6 square units

(c) $y = x^{1/2}$ between $x = 0$ and $x = 9$ *Ans.* 18 square units

(d) $y = x^2 - x - 6$ between $x = 0$ and $x = 2$ *Ans.* 34/3 square units

(e) $y = x^3$ between $x = -2$ and $x = 4$ *Ans.* 68 square units

(f) $y = x^3 - x$ between $x = -1$ and $x = 1$ *Ans.* 1/2 square unit

CHAPTER 65

Summation and the Definite Integral

AREA BY SUMMATION. Consider the area A bounded by the curve $y = f(x) \geq 0$, the x-axis, and the ordinates $x = a$ and $x = b$, where $b > a$.

Let the interval $a \leq x \leq b$ be divided into n equal parts each of length Δx. At each point of subdivision construct the ordinate thus dividing the area into n strips, as in Fig. 1(a). Since the area of the strips are unknown, we propose to approximate each strip by a rectangle whose area can be found. In Fig. 1(b) a representative strip and its approximating rectangle are shown.

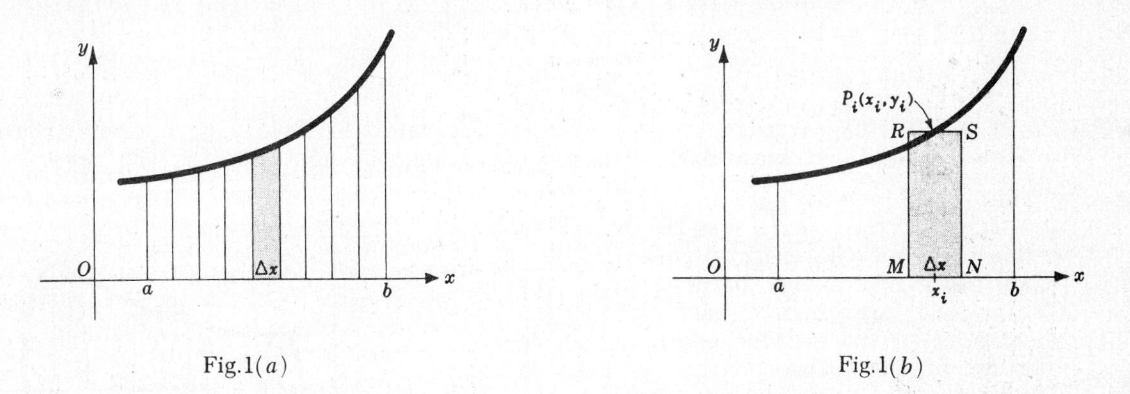

Fig.1(a) Fig.1(b)

Suppose the representative strip is the ith strip counting from the left, and let $x = x_i$ be the coordinate of the midpoint of its base. Denote by $y_i = f(x_i)$ the ordinate of the point P_i (on the curve) whose abscissa is x_i. Through P_i pass a line parallel to the x-axis and complete the rectangle $MRSN$. This rectangle of area $y_i \Delta x$ is the approximating rectangle of the ith strip. When each strip is treated similarly, it seems reasonable to take

$$y_1 \Delta x + y_2 \Delta x + y_3 \Delta x + \ldots + y_n \Delta x = \sum_{i=1}^{n} y_i \Delta x$$

as an approximation of the area sought.

See Problem 1.

Now suppose that the number of strips (with approximating rectangles) is indefinitely increased so that $\Delta x \to 0$. It is evident from the figure that by so increasing the number of approximating rectangles the sum of their areas more nearly approximates the area sought, that is,

$$A = \lim_{n \to \infty} \sum_{i=1}^{n} y_i \Delta x = \int_a^b y \, dx = \int_a^b f(x) \, dx$$

Example 1. Find the area between the curves $y = x^2$ and $y = 2x + 8$.

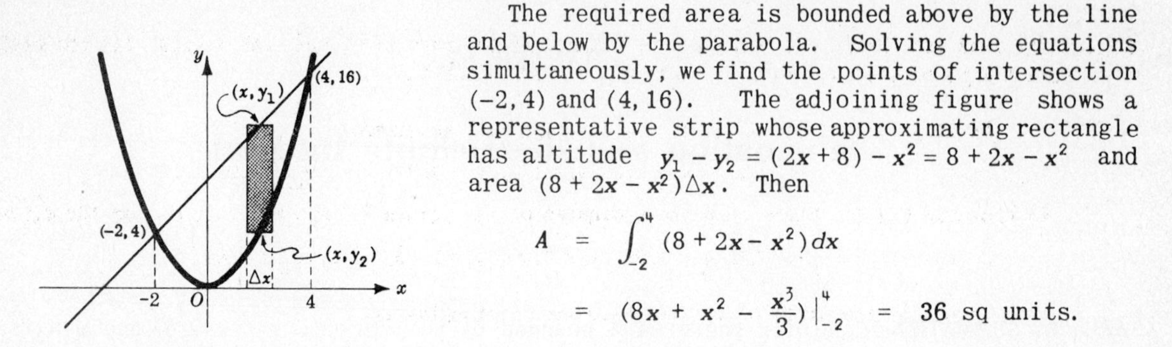

The required area is bounded above by the line and below by the parabola. Solving the equations simultaneously, we find the points of intersection (-2, 4) and (4, 16). The adjoining figure shows a representative strip whose approximating rectangle has altitude $y_1 - y_2 = (2x+8) - x^2 = 8 + 2x - x^2$ and area $(8 + 2x - x^2)\Delta x$. Then

$$A = \int_{-2}^{4} (8 + 2x - x^2)\,dx$$

$$= (8x + x^2 - \frac{x^3}{3})\Big|_{-2}^{4} = 36 \text{ sq units.}$$

Note. This problem could have been solved in Chapter 64 as follows. Denote by A_1 the area bounded by the line $y = 2x + 8$, the x-axis, and the ordinates $x = -2$ and $x = 4$; denote by A_2 the area bounded by the parabola $y = x^2$, the x-axis, and the ordinates $x = -2$ and $x = 4$. Then the required area A is given by

$$A = A_1 - A_2 = \int_{-2}^{4} (8 + 2x)\,dx - \int_{-2}^{4} x^2\,dx = \int_{-2}^{4} (8 + 2x - x^2)\,dx$$

See Problems 2-3.

VOLUMES OF REVOLUTION. A solid of revolution is generated by revolving a plane area about a line (the axis of rotation) lying in the plane but not crossing the area.

Let the plane area be bounded by the curve $y = f(x) \geq 0$, the x-axis, and the ordinates $x = a$ and $x = b$; let the axis of rotation be the x-axis. We first slice the area into vertical strips and approximate each strip by a rectangle as in the section above. As an approximation of the volume sought, we take the sum of the volumes generated by revolving the several rectangles about the axis of rotation. Since the axis of rotation is part of the boundary of each rectangle, the solid of revolution in each case is a cylindrical disk.

For the representative disk of the adjoining figure, the radius is y_i, the altitude is Δx, and the volume is $\pi y_i^2 \Delta x$. Thus the volume sought is approximated by

$$\pi y_1^2\,\Delta x + \pi y_2^2\,\Delta x + \ldots + \pi y_n^2\,\Delta x = \pi \sum_{i=1}^{n} y_i^2\,\Delta x$$

By increasing the number of strips indefinitely and passing to the limit, the required volume is given by

$$V = \lim_{n \to \infty} \pi \sum_{i=1}^{n} y_i^2\,\Delta x = \pi \int_a^b y^2\,dx$$

See Problems 4-7.

SOLVED PROBLEMS

1. Let the area (see Example 2, Chapter 64) bounded by the parabola $y = x^2$, the x-axis, and the ordinates $x = 1$ and $x = 3$ be sliced into 10 strips each of width $x = 0.2$.

The midpoints of the bases of these strips have abscissas

$$x_1 = 1.1,\ x_2 = 1.3,\ x_3 = 1.5,\ x_4 = 1.7,\ \ldots,\ x_{10} = 2.9$$

as shown in Fig. (a) below, and the ordinates of the points P_1, P_2, P_3, ..., P_{10} on the curve are

$$y_1 = 1.21,\ y_2 = 1.69,\ y_3 = 2.25,\ y_4 = 2.89,\ \ldots,\ y_{10} = 8.41.$$

The sum of the areas of the approximating rectangles is

$$\sum_{i=1}^{10} y_i\,\Delta x = (1.21 + 1.69 + 2.25 + 2.89 + \ldots + 8.41)(0.2) = (43.30)(0.2) = 8.66 \text{ sq units.}$$

This is a close approximation of the required area = 8 2/3 sq units.

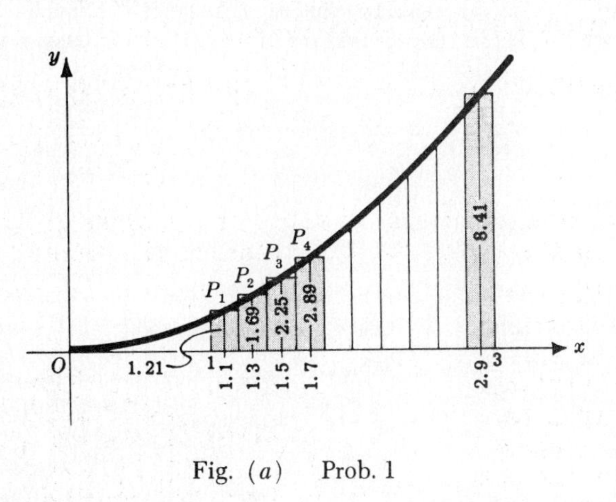

Fig. (a) Prob. 1

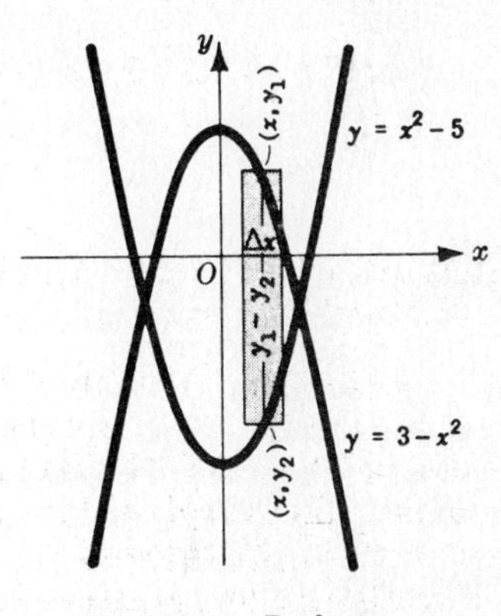

Fig. (b) Prob. 2

2. Find the area bounded by the parabolas $y = x^2 - 5$ and $y = 3 - x^2$.

Solving the equations simultaneously, we find the points of intersection (−2, −1) and (2, −1). Fig. (b) above, shows a representative strip whose approximating rectangle has altitude $y_1 - y_2 = (3 - x^2) - (x^2 - 5) = 8 - 2x^2$ and area $(8 - 2x^2)\Delta x$. The required area is given by

$$A = \int_{-2}^{2} (8 - 2x^2)\,dx = (8x - \frac{2}{3}x^3)\Big|_{-2}^{2} = \frac{64}{3} \text{ sq units.}$$

Note 1. Since the y-axis is an axis of symmetry, the required area is twice that lying to the right of the y-axis; thus

$$A = 2\int_{0}^{2} (8 - 2x^2)\,dx = 2(\frac{32}{3}) = \frac{64}{3} \text{ sq units.}$$

Note. 2. Unlike Example 2, this problem cannot be solved simply by the procedure of Chapter 64.

3. Find the area bounded by the parabola $x = 4 - y^2$ and the y-axis.

The x-intercept of the parabola is 4 and the y-intercepts are $y = \pm 2$.

First Solution. We slice the area into strips by equally spaced vertical lines. Fig. (c) below shows a representative strip whose approximating rectangle has altitude $2y = 2\sqrt{4-x}$ and area $2\sqrt{4-x}\,\Delta x$. Then

$$A = \int_0^4 2\sqrt{4-x}\,dx = -2\frac{2}{3}(4-x)^{3/2}\Big|_0^4 = -\frac{4}{3}(0-8) = \frac{32}{3} \text{ square units.}$$

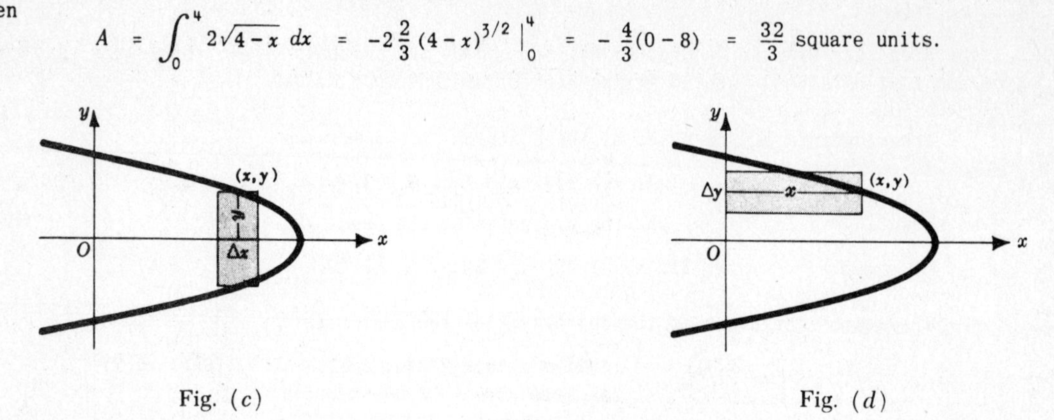

Fig. (c) Fig. (d)

Second Solution. We slice the area into strips by equally spaced horizontal lines. Fig. (d) above shows a representative strip of width Δy whose approximating rectangle has base $x = 4-y^2$ and area $(4-y^2)\Delta y$. Then

$$A = \int_{-2}^{2}(4-y^2)dy = (4y - \frac{y^3}{3})\Big|_{-2}^{2} = \frac{32}{3} \text{ square units.}$$

4. Find the volume of a sphere of radius r.

Consider the sphere as generated by revolving the upper half of the circle $x^2+y^2=r^2$ about the x-axis. When the representative rectangle of Fig. (e) below is revolved about the x-axis the volume of the disk generated is $\pi y^2\Delta x = \pi(r^2-x^2)\Delta x$. Then

$$V = \int_{-r}^{r}(r^2-x^2)dx = (r^2x - \frac{x^3}{3})\Big|_{-r}^{r} = \frac{4}{3}\pi r^3 \text{ cubic units.}$$

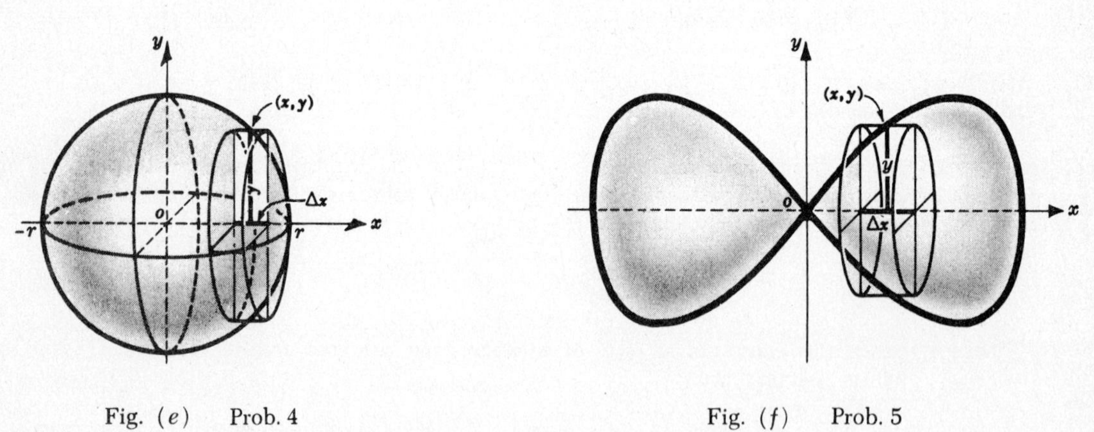

Fig. (e) Prob. 4 Fig. (f) Prob. 5

5. Find the volume generated by revolving the curve $y^2 = x^2 - x^4$ about the x-axis.

From Fig. (f) above it is clear that the required volume is twice that generated by revolving the first quadrant area about the x-axis. The representative rectangle generates a disk of radius y and volume $\pi y^2\Delta x = \pi(x^2-x^4)\Delta x$. Then

$$V = 2\pi\int_0^1(x^2-x^4)dx = 2\pi(\frac{x^3}{3} - \frac{x^5}{5})\Big|_0^1 = 4\pi/15 \text{ cubic units.}$$

6. Find the volume generated by revolving the area bounded by the parabola $y^2 = 12x$ and its latus rectum ($x = 3$) about the latus rectum.

Divide the area by equally spaced horizontal lines. When the approximating rectangle of Fig. (g) below is revolved about the latus rectum a disk of radius $3 - x$, altitude Δy, and volume $\pi(3-x)^2\Delta y = \pi(3 - \frac{y^2}{12})^2\Delta y$ is generated. Then

$$V = 2\pi\int_0^6 (3 - \frac{y^2}{12})^2 dy = 2\pi\int_0^6 (9 - \frac{y^2}{2} + \frac{y^4}{144})dy = 2\pi(9y - \frac{y^3}{6} + \frac{y^5}{720})\Big|_0^6 = \frac{288\pi}{5} \text{ cubic units.}$$

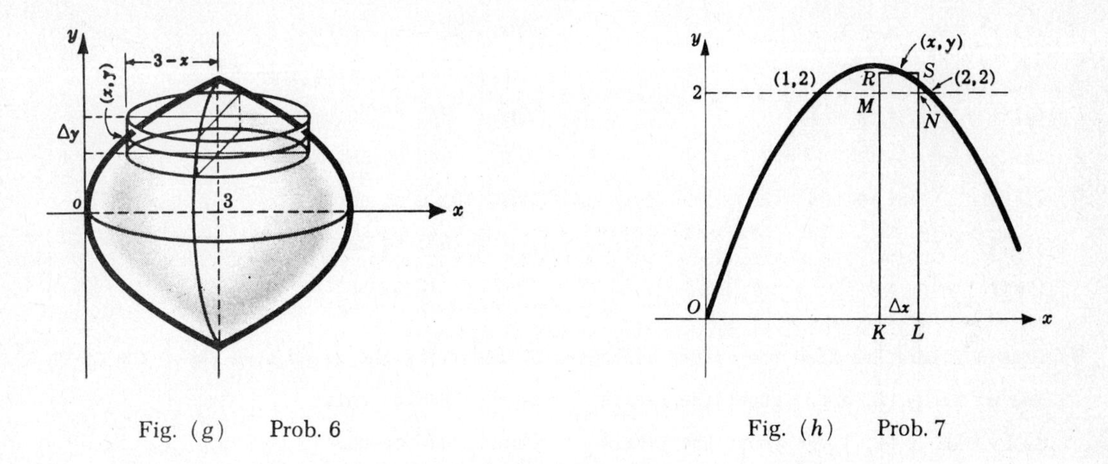

Fig. (g) Prob. 6

Fig. (h) Prob. 7

7. Find the volume generated by revolving the portion of the parabola $y = 3x - x^2$ cut off by the line $y = 2$ about the x-axis.

The points of intersection of the line and the parabola are (1,2) and (2,2). When the approximating rectangle $MRSN$ of Fig. (h) above is revolved about the x-axis a washer is generated. The volume of this washer is equal to the volume generated by revolving $KRSL$ about the x-axis minus the volume generated by revolving $KMNL$ about the x-axis =

$$\pi\int_1^2 (3x - x^2)^2 dx - \pi\int_1^2 2^2 dx = \pi\int_1^2 (9x^2 - 6x^3 + x^4 - 4)dx$$

$$= \pi(3x^3 - 3x^4/2 + x^5/5 - 4x)\Big|_1^2 = 7\pi/10 \text{ cubic units.}$$

SUPPLEMENTARY PROBLEMS

8. Draw a figure and find the area bounded as follows.

(*a*) $y = x^2,\ y = 0,\ x = 3$ *Ans.* 9 sq units

(*b*) $y = x^2,\ y = 0,\ x = -2,\ x = 3$ *Ans.* 35/3 sq units

(*c*) $y = x^2,\ y = 9$ *Ans.* 36 sq units

(*d*) $y^2 = x,\ y = 0,\ x = 9$ *Ans.* 18 sq units

(*e*) $y^2 = x,\ x = 4$ *Ans.* 32/3 sq units

(*f*) $y^2 = x,\ x = 0,\ y = -2$ *Ans.* 8/3 sq units

(*g*) $y = x^2,\ x + y = 2$ *Ans.* 9/2 sq units

(*h*) $y = x^2 - 5x + 6,\ y = 2$ *Ans.* 9/2 sq units

(*i*) $y = x^2 - 5x + 6,\ y = 2x$ *Ans.* 125/6 sq units

(*j*) $4y = x^2,\ 4x = y^2$ *Ans.* 16/3 sq units

(*k*) $y = x^3,\ x = 0,\ y = 8$ *Ans.* 12 sq units

9. Draw a figure and find the volume generated by revolving the given area about the given line.

(*a*) $y = 2x,\ y = 0,\ x = 3$ about the x-axis *Ans.* 36π cu units

(*b*) $y = 2x,\ x = 0,\ y = 6$ about the y-axis *Ans.* 18π cu units

(*c*) $y = x^2,\ y = 0,\ x = 5$ about the x-axis *Ans.* 625π cu units

(*d*) $y = x^2,\ x = 0,\ y = 9$ about the y-axis *Ans.* $\frac{81}{2}\pi$ cu units

(*e*) $\frac{x^2}{4} + y^2 = 1$ about the x-axis *Ans.* $\frac{8}{3}\pi$ cu units

(*f*) $\frac{x^2}{4} + y^2 = 1$ about the y-axis *Ans.* $\frac{16}{3}\pi$ cu units

10. For the first quadrant area bounded by $y^2 = 8x^3,\ y = 0,\ x = 2$ find the volume generated by revolving it about:
(*a*) the x-axis (*b*) the y-axis (*c*) the line $x = 2$ (*d*) the line $x = 4$

Ans. (*a*) 32π cu units (*b*) $\frac{128}{7}\pi$ cu units (*c*) $\frac{256}{35}\pi$ cu units (*d*) $\frac{1152}{35}\pi$ cu units

N	0	1	2	3	4	5	6	7	8	9
10	0000	0043	0086	0128	0170	0212	0253	0294	0334	0374
11	0414	0453	0492	0531	0569	0607	0645	0682	0719	0755
12	0792	0828	0864	0899	0934	0969	1004	1038	1072	1106
13	1139	1173	1206	1239	1271	1303	1335	1367	1399	1430
14	1461	1492	1523	1553	1584	1614	1644	1673	1703	1732
15	1761	1790	1818	1847	1875	1903	1931	1959	1987	2014
16	2041	2068	2095	2122	2148	2175	2201	2227	2253	2279
17	2304	2330	2355	2380	2405	2430	2455	2480	2504	2529
18	2553	2577	2601	2625	2648	2672	2695	2718	2742	2765
19	2788	2810	2833	2856	2878	2900	2923	2945	2967	2989
20	3010	3032	3054	3075	3096	3118	3139	3160	3181	3201
21	3222	3243	3263	3284	3304	3324	3345	3365	3385	3404
22	3424	3444	3464	3483	3502	3522	3541	3560	3579	3598
23	3617	3636	3655	3674	3692	3711	3729	3747	3766	3784
24	3802	3820	3838	3856	3874	3892	3909	3927	3945	3962
25	3979	3997	4014	4031	4048	4065	4082	4099	4116	4133
26	4150	4166	4183	4200	4216	4232	4249	4265	4281	4298
27	4314	4330	4346	4362	4378	4393	4409	4425	4440	4456
28	4472	4487	4502	4518	4533	4548	4564	4579	4594	4609
29	4624	4639	4654	4669	4683	4698	4713	4728	4742	4757
30	4771	4786	4800	4814	4829	4843	4857	4871	4886	4900
31	4914	4928	4942	4955	4969	4983	4997	5011	5024	5038
32	5051	5065	5079	5092	5105	5119	5132	5145	5159	5172
33	5185	5198	5211	5224	5237	5250	5263	5276	5289	5302
34	5315	5328	5340	5353	5366	5378	5391	5403	5416	5428
35	5441	5453	5465	5478	5490	5502	5514	5527	5539	5551
36	5563	5575	5587	5599	5611	5623	5635	5647	5658	5670
37	5682	5694	5705	5717	5729	5740	5752	5763	5775	5786
38	5798	5809	5821	5832	5843	5855	5866	5877	5888	5899
39	5911	5922	5933	5944	5955	5966	5977	5988	5999	6010
40	6021	6031	6042	6053	6064	6075	6085	6096	6107	6117
41	6128	6138	6149	6160	6170	6180	6191	6201	6212	6222
42	6232	6243	6253	6263	6274	6284	6294	6304	6314	6325
43	6335	6345	6355	6365	6375	6385	6395	6405	6415	6425
44	6435	6444	6454	6464	6474	6484	6493	6503	6513	6522
45	6532	6542	6551	6561	6571	6580	6590	6599	6609	6618
46	6628	6637	6646	6656	6665	6675	6684	6693	6702	6712
47	6721	6730	6739	6749	6758	6767	6776	6785	6794	6803
48	6812	6821	6830	6839	6848	6857	6866	6875	6884	6893
49	6902	6911	6920	6928	6937	6946	6955	6964	6972	6981
50	6990	6998	7007	7016	7024	7033	7042	7050	7059	7067
51	7076	7084	7093	7101	7110	7118	7126	7135	7143	7152
52	7160	7168	7177	7185	7193	7202	7210	7218	7226	7235
53	7243	7251	7259	7267	7275	7284	7292	7300	7308	7316
54	7324	7332	7340	7348	7356	7364	7372	7380	7388	7396
N	0	1	2	3	4	5	6	7	8	9

N	0	1	2	3	4	5	6	7	8	9
55	7404	7412	7419	7427	7435	7443	7451	7459	7466	7474
56	7482	7490	7497	7505	7513	7520	7528	7536	7543	7551
57	7559	7566	7574	7582	7589	7597	7604	7612	7619	7627
58	7634	7642	7649	7657	7664	7672	7679	7686	7694	7701
59	7709	7716	7723	7731	7738	7745	7752	7760	7767	7774
60	7782	7789	7796	7803	7810	7818	7825	7832	7839	7846
61	7853	7860	7868	7875	7882	7889	7896	7903	7910	7917
62	7924	7931	7938	7945	7952	7959	7966	7973	7980	7987
63	7993	8000	8007	8014	8021	8028	8035	8041	8048	8055
64	8062	8069	8075	8082	8089	8096	8102	8109	8116	8122
65	8129	8136	8142	8149	8156	8162	8169	8176	8182	8189
66	8195	8202	8209	8215	8222	8228	8235	8241	8248	8254
67	8261	8267	8274	8280	8287	8293	8299	8306	8312	8319
68	8325	8331	8338	8344	8351	8357	8363	8370	8376	8382
69	8388	8395	8401	8407	8414	8420	8426	8432	8439	8445
70	8451	8457	8463	8470	8476	8482	8488	8494	8500	8506
71	8513	8519	8525	8531	8537	8543	8549	8555	8561	8567
72	8573	8579	8585	8591	8597	8603	8609	8615	8621	8627
73	8633	8639	8645	8651	8657	8663	8669	8675	8681	8686
74	8692	8698	8704	8710	8716	8722	8727	8733	8739	8745
75	8751	8756	8762	8768	8774	8779	8785	8791	8797	8802
76	8808	8814	8820	8825	8831	8837	8842	8848	8854	8859
77	8865	8871	8876	8882	8887	8893	8899	8904	8910	8915
78	8921	8927	8932	8938	8943	8949	8954	8960	8965	8971
79	8976	8982	8987	8993	8998	9004	9009	9015	9020	9025
80	9031	9036	9042	9047	9053	9058	9063	9069	9074	9079
81	9085	9090	9096	9101	9106	9112	9117	9122	9128	9133
82	9138	9143	9149	9154	9159	9165	9170	9175	9180	9186
83	9191	9196	9201	9206	9212	9217	9222	9227	9232	9238
84	9243	9248	9253	9258	9263	9269	9274	9279	9284	9289
85	9294	9299	9304	9309	9315	9320	9325	9330	9335	9340
86	9345	9350	9355	9360	9365	9370	9375	9380	9385	9390
87	9395	9400	9405	9410	9415	9420	9425	9430	9435	9440
88	9445	9450	9455	9460	9465	9469	9474	9479	9484	9489
89	9494	9499	9504	9509	9513	9518	9523	9528	9533	9538
90	9542	9547	9552	9557	9562	9566	9571	9576	9581	9586
91	9590	9595	9600	9605	9609	9614	9619	9624	9628	9633
92	9638	9643	9647	9652	9657	9661	9666	9671	9675	9680
93	9685	9689	9694	9699	9703	9708	9713	9717	9722	9727
94	9731	9736	9741	9745	9750	9754	9759	9763	9768	9773
95	9777	9782	9786	9791	9795	9800	9805	9809	9814	9818
96	9823	9827	9832	9836	9841	9845	9850	9854	9859	9863
97	9868	9872	9877	9881	9886	9890	9894	9899	9903	9908
98	9912	9917	9921	9926	9930	9934	9939	9943	9948	9952
99	9956	9961	9965	9969	9974	9978	9983	9987	9991	9996
N	0	1	2	3	4	5	6	7	8	9

Angle	Sin	Tan	Cot	Cos	
0°00′	.0000	.0000	—	1.0000	90°00′
10	.0029	.0029	343.77	1.0000	50
20	.0058	.0058	171.89	1.0000	40
30	.0087	.0087	114.59	1.0000	30
40	.0116	.0116	85.940	.9999	20
50	.0145	.0145	68.750	.9999	10
1°00′	.0175	.0175	57.290	.9998	89°00′
10	.0204	.0204	49.104	.9998	50
20	.0233	.0233	42.964	.9997	40
30	.0262	.0262	38.188	.9997	30
40	.0291	.0291	34.368	.9996	20
50	.0320	.0320	31.242	.9995	10
2°00′	.0349	.0349	28.636	.9994	88°00′
10	.0378	.0378	26.432	.9993	50
20	.0407	.0407	24.542	.9992	40
30	.0436	.0437	22.904	.9990	30
40	.0465	.0466	21.470	.9989	20
50	.0494	.0495	20.206	.9988	10
3°00′	.0523	.0524	19.081	.9986	87°00′
10	.0552	.0553	18.075	.9985	50
20	.0581	.0582	17.169	.9983	40
30	.0610	.0612	16.350	.9981	30
40	.0640	.0641	15.605	.9980	20
50	.0669	.0670	14.924	.9978	10
4°00′	.0698	.0699	14.301	.9976	86°00′
10	.0727	.0729	13.727	.9974	50
20	.0756	.0758	13.197	.9971	40
30	.0785	.0787	12.706	.9969	30
40	.0814	.0816	12.251	.9967	20
50	.0843	.0846	11.826	.9964	10
5°00′	.0872	.0875	11.430	.9962	85°00′
10	.0901	.0904	11.059	.9959	50
20	.0929	.0934	10.712	.9957	40
30	.0958	.0963	10.385	.9954	30
40	.0987	.0992	10.078	.9951	20
50	.1016	.1022	9.7882	.9948	10
6°00′	.1045	.1051	9.5144	.9945	84°00′
10	.1074	.1080	9.2553	.9942	50
20	.1103	.1110	9.0098	.9939	40
30	.1132	.1139	8.7769	.9936	30
40	.1161	.1169	8.5555	.9932	20
50	.1190	.1198	8.3450	.9929	10
7°00′	.1219	.1228	8.1443	.9925	83°00′
10	.1248	.1257	7.9530	.9922	50
20	.1276	.1287	7.7704	.9918	40
30	.1305	.1317	7.5958	.9914	30
40	.1334	.1346	7.4287	.9911	20
50	.1363	.1376	7.2687	.9907	10
8°00′	.1392	.1405	7.1154	.9903	82°00′
10	.1421	.1435	6.9682	.9899	50
20	.1449	.1465	6.8269	.9894	40
30	.1478	.1495	6.6912	.9890	30
40	.1507	.1524	6.5606	.9886	20
50	.1536	.1554	6.4348	.9881	10
9°00′	.1564	.1584	6.3138	.9877	81°00′
	Cos	Cot	Tan	Sin	Angle

Angle	Sin	Tan	Cot	Cos	
9°00′	.1564	.1584	6.3138	.9877	81°00′
10	.1593	.1614	6.1970	.9872	50
20	.1622	.1644	6.0844	.9868	40
30	.1650	.1673	5.9758	.9863	30
40	.1679	.1703	5.8708	.9858	20
50	.1708	.1733	5.7694	.9853	10
10°00′	.1736	.1763	5.6713	.9848	80°00′
10	.1765	.1793	5.5764	.9843	50
20	.1794	.1823	5.4845	.9838	40
30	.1822	.1853	5.3955	.9833	30
40	.1851	.1883	5.3093	.9827	20
50	.1880	.1914	5.2257	.9822	10
11°00′	.1908	.1944	5.1446	.9816	79°00′
10	.1937	.1974	5.0658	.9811	50
20	.1965	.2004	4.9894	.9805	40
30	.1994	.2035	4.9152	.9799	30
40	.2022	.2065	4.8430	.9793	20
50	.2051	.2095	4.7729	.9787	10
12°00′	.2079	.2126	4.7046	.9781	78°00′
10	.2108	.2156	4.6382	.9775	50
20	.2136	.2186	4.5736	.9769	40
30	.2164	.2217	4.5107	.9763	30
40	.2193	.2247	4.4494	.9757	20
50	.2221	.2278	4.3897	.9750	10
13°00′	.2250	.2309	4.3315	.9744	77°00′
10	.2278	.2339	4.2747	.9737	50
20	.2306	.2370	4.2193	.9730	40
30	.2334	.2401	4.1653	.9724	30
40	.2363	.2432	4.1126	.9717	20
50	.2391	.2462	4.0611	.9710	10
14°00′	.2419	.2493	4.0108	.9703	76°00′
10	.2447	.2524	3.9617	.9696	50
20	.2476	.2555	3.9136	.9689	40
30	.2504	.2586	3.8667	.9681	30
40	.2532	.2617	3.8208	.9674	20
50	.2560	.2648	3.7760	.9667	10
15°00′	.2588	.2679	3.7321	.9659	75°00′
10	.2616	.2711	3.6891	.9652	50
20	.2644	.2742	3.6470	.9644	40
30	.2672	.2773	3.6059	.9636	30
40	.2700	.2805	3.5656	.9628	20
50	.2728	.2836	3.5261	.9621	10
16°00′	.2756	.2867	3.4874	.9613	74°00′
10	.2784	.2899	3.4495	.9605	50
20	.2812	.2931	3.4124	.9596	40
30	.2840	.2962	3.3759	.9588	30
40	.2868	.2994	3.3402	.9580	20
50	.2896	.3026	3.3052	.9572	10
17°00′	.2924	.3057	3.2709	.9563	73°00′
10	.2952	.3089	3.2371	.9555	50
20	.2979	.3121	3.2041	.9546	40
30	.3007	.3153	3.1716	.9537	30
40	.3035	.3185	3.1397	.9528	20
50	.3062	.3217	3.1084	.9520	10
18°00′	.3090	.3249	3.0777	.9511	72°00′
	Cos	Cot	Tan	Sin	Angle

Angle	Sin	Tan	Cot	Cos	
18°00′	.3090	.3249	3.0777	.9511	72°00′
10	.3118	.3281	3.0475	.9502	50
20	.3145	.3314	3.0178	.9492	40
30	.3173	.3346	2.9887	.9483	30
40	.3201	.3378	2.9600	.9474	20
50	.3228	.3411	2.9319	.9465	10
19°00′	.3256	.3443	2.9042	.9455	71°00′
10	.3283	.3476	2.8770	.9446	50
20	.3311	.3508	2.8502	.9436	40
30	.3338	.3541	2.8239	.9426	30
40	.3365	.3574	2.7980	.9417	20
50	.3393	.3607	2.7725	.9407	10
20°00′	.3420	.3640	2.7475	.9397	70°00′
10	.3448	.3673	2.7228	.9387	50
20	.3475	.3706	2.6985	.9377	40
30	.3502	.3739	2.6746	.9367	30
40	.3529	.3772	2.6511	.9356	20
50	.3557	.3805	2.6279	.9346	10
21°00′	.3584	.3839	2.6051	.9336	69°00′
10	.3611	.3872	2.5826	.9325	50
20	.3638	.3906	2.5605	.9315	40
30	.3665	.3939	2.5386	.9304	30
40	.3692	.3973	2.5172	.9293	20
50	.3719	.4006	2.4960	.9283	10
22°00′	.3746	.4040	2.4751	.9272	68°00′
10	.3773	.4074	2.4545	.9261	50
20	.3800	.4108	2.4342	.9250	40
30	.3827	.4142	2.4142	.9239	30
40	.3854	.4176	2.3945	.9228	20
50	.3881	.4210	2.3750	.9216	10
23°00′	.3907	.4245	2.3559	.9205	67°00′
10	.3934	.4279	2.3369	.9194	50
20	.3961	.4314	2.3183	.9182	40
30	.3987	.4348	2.2998	.9171	30
40	.4014	.4383	2.2817	.9159	20
50	.4041	.4417	2.2637	.9147	10
24°00′	.4067	.4452	2.2460	.9135	66°00′
10	.4094	.4487	2.2286	.9124	50
20	.4120	.4522	2.2113	.9112	40
30	.4147	.4557	2.1943	.9100	30
40	.4173	.4592	2.1775	.9088	20
50	.4200	.4628	2.1609	.9075	10
25°00′	.4226	.4663	2.1445	.9063	65°00′
10	.4253	.4699	2.1283	.9051	50
20	.4279	.4734	2.1123	.9038	40
30	.4305	.4770	2.0965	.9026	30
40	.4331	.4806	2.0809	.9013	20
50	.4358	.4841	2.0655	.9001	10
26°00′	.4384	.4877	2.0503	.8988	64°00′
10	.4410	.4913	2.0353	.8975	50
20	.4436	.4950	2.0204	.8962	40
30	.4462	.4986	2.0057	.8949	30
40	.4488	.5022	1.9912	.8936	20
50	.4514	.5059	1.9768	.8923	10
27°00′	.4540	.5095	1.9626	.8910	63°00′
	Cos	Cot	Tan	Sin	Angle

Angle	Sin	Tan	Cot	Cos	
27°00′	.4540	.5095	1.9626	.8910	63°00′
10	.4566	.5132	1.9486	.8897	50
20	.4592	.5169	1.9347	.8884	40
30	.4617	.5206	1.9210	.8870	30
40	.4643	.5243	1.9074	.8857	20
50	.4669	.5280	1.8940	.8843	10
28°00′	.4695	.5317	1.8807	.8829	62°00′
10	.4720	.5354	1.8676	.8816	50
20	.4746	.5392	1.8546	.8802	40
30	.4772	.5430	1.8418	.8788	30
40	.4797	.5467	1.8291	.8774	20
50	.4823	.5505	1.8165	.8760	10
29°00′	.4848	.5543	1.8040	.8746	61°00′
10	.4874	.5581	1.7917	.8732	50
20	.4899	.5619	1.7796	.8718	40
30	.4924	.5658	1.7675	.8704	30
40	.4950	.5696	1.7556	.8689	20
50	.4975	.5735	1.7437	.8675	10
30°00′	.5000	.5774	1.7321	.8660	60°00′
10	.5025	.5812	1.7205	.8646	50
20	.5050	.5851	1.7090	.8631	40
30	.5075	.5890	1.6977	.8616	30
40	.5100	.5930	1.6864	.8601	20
50	.5125	.5969	1.6753	.8587	10
31°00′	.5150	.6009	1.6643	.8572	59°00′
10	.5175	.6048	1.6534	.8557	50
20	.5200	.6088	1.6426	.8542	40
30	.5225	.6128	1.6319	.8526	30
40	.5250	.6168	1.6212	.8511	20
50	.5275	.6208	1.6107	.8496	10
32°00′	.5299	.6249	1.6003	.8480	58°00′
10	.5324	.6289	1.5900	.8465	50
20	.5348	.6330	1.5798	.8450	40
30	.5373	.6371	1.5697	.8434	30
40	.5398	.6412	1.5597	.8418	20
50	.5422	.6453	1.5497	.8403	10
33°00′	.5446	.6494	1.5399	.8387	57°00′
10	.5471	.6536	1.5301	.8371	50
20	.5495	.6577	1.5204	.8355	40
30	.5519	.6619	1.5108	.8339	30
40	.5544	.6661	1.5013	.8323	20
50	.5568	.6703	1.4919	.8307	10
34°00′	.5592	.6745	1.4826	.8290	56°00′
10	.5616	.6787	1.4733	.8274	50
20	.5640	.6830	1.4641	.8258	40
30	.5664	.6873	1.4550	.8241	30
40	.5688	.6916	1.4460	.8225	20
50	.5712	.6959	1.4370	.8208	10
35°00′	.5736	.7002	1.4281	.8192	55°00′
10	.5760	.7046	1.4193	.8175	50
20	.5783	.7089	1.4106	.8158	40
30	.5807	.7133	1.4019	.8141	30
40	.5831	.7177	1.3934	.8124	20
50	.5854	.7221	1.3848	.8107	10
36°00′	.5878	.7265	1.3764	.8090	54°00′
	Cos	Cot	Tan	Sin	Angle

Angle	Sin	Tan	Cot	Cos	
36°00′	.5878	.7265	1.3764	.8090	54°00′
10	.5901	.7310	1.3680	.8073	50
20	.5925	.7355	1.3597	.8056	40
30	.5948	.7400	1.3514	.8039	30
40	.5972	.7445	1.3432	.8021	20
50	.5995	.7490	1.3351	.8004	10
37°00′	.6018	.7536	1.3270	.7986	53°00′
10	.6041	.7581	1.3190	.7969	50
20	.6065	.7627	1.3111	.7951	40
30	.6088	.7673	1.3032	.7934	30
40	.6111	.7720	1.2954	.7916	20
50	.6134	.7766	1.2876	.7898	10
38°00′	.6157	.7813	1.2799	.7880	52°00′
10	.6180	.7860	1.2723	.7862	50
20	.6202	.7907	1.2647	.7844	40
30	.6225	.7954	1.2572	.7826	30
40	.6248	.8002	1.2497	.7808	20
50	.6271	.8050	1.2423	.7790	10
39°00′	.6293	.8098	1.2349	.7771	51°00′
10	.6316	.8146	1.2276	.7753	50
20	.6338	.8195	1.2203	.7735	40
30	.6361	.8243	1.2131	.7716	30
40	.6383	.8292	1.2059	.7698	20
50	.6406	.8342	1.1988	.7679	10
40°00′	.6428	.8391	1.1918	.7660	50°00′
10	.6450	.8441	1.1847	.7642	50
20	.6472	.8491	1.1778	.7623	40
30	.6494	.8541	1.1708	.7604	30
40	.6517	.8591	1.1640	.7585	20
50	.6539	.8642	1.1571	.7566	10
41°00′	.6561	.8693	1.1504	.7547	49°00′
10	.6583	.8744	1.1436	.7528	50
20	.6604	.8796	1.1369	.7509	40
30	.6626	.8847	1.1303	.7490	30
40	.6648	.8899	1.1237	.7470	20
50	.6670	.8952	1.1171	.7451	10
42°00′	.6691	.9004	1.1106	.7431	48°00′
10	.6713	.9057	1.1041	.7412	50
20	.6734	.9110	1.0977	.7392	40
30	.6756	.9163	1.0913	.7373	30
40	.6777	.9217	1.0850	.7353	20
50	.6799	.9271	1.0786	.7333	10
43°00′	.6820	.9325	1.0724	.7314	47°00′
10	.6841	.9380	1.0661	.7294	50
20	.6862	.9435	1.0599	.7274	40
30	.6884	.9490	1.0538	.7254	30
40	.6905	.9545	1.0477	.7234	20
50	.6926	.9601	1.0416	.7214	10
44°00′	.6947	.9657	1.0355	.7193	46°00′
10	.6967	.9713	1.0295	.7173	50
20	.6988	.9770	1.0235	.7153	40
30	.7009	.9827	1.0176	.7133	30
40	.7030	.9884	1.0117	.7112	20
50	.7050	.9942	1.0058	.7092	10
45°00′	.7071	1.0000	1.0000	.7071	45°00′
	Cos	Cot	Tan	Sin	Angle

Angle	L Sin	L Tan	L Cot	L Cos	
0°00′	—	—	—	10.0000	90°00′
10	7.4637	7.4637	12.5363	10.0000	50
20	7.7648	7.7648	12.2352	10.0000	40
30	7.9408	7.9409	12.0591	10.0000	30
40	8.0658	8.0658	11.9342	10.0000	20
50	8.1627	8.1627	11.8373	10.0000	10
1°00′	8.2419	8.2419	11.7581	9.9999	89°00′
10	8.3088	8.3089	11.6911	9.9999	50
20	8.3668	8.3669	11.6331	9.9999	40
30	8.4179	8.4181	11.5819	9.9999	30
40	8.4637	8.4638	11.5362	9.9998	20
50	8.5050	8.5053	11.4947	9.9998	10
2°00′	8.5428	8.5431	11.4569	9.9997	88°00′
10	8.5776	8.5779	11.4221	9.9997	50
20	8.6097	8.6101	11.3899	9.9996	40
30	8.6397	8.6401	11.3599	9.9996	30
40	8.6677	8.6682	11.3318	9.9995	20
50	8.6940	8.6945	11.3055	9.9995	10
3°00′	8.7188	8.7194	11.2806	9.9994	87°00′
10	8.7423	8.7429	11.2571	9.9993	50
20	8.7645	8.7652	11.2348	9.9993	40
30	8.7857	8.7865	11.2135	9.9992	30
40	8.8059	8.8067	11.1933	9.9991	20
50	8.8251	8.8261	11.1739	9.9990	10
4°00′	8.8436	8.8446	11.1554	9.9989	86°00′
10	8.8613	8.8624	11.1376	9.9989	50
20	8.8783	8.8795	11.1205	9.9988	40
30	8.8946	8.8960	11.1040	9.9987	30
40	8.9104	8.9118	11.0882	9.9986	20
50	8.9256	8.9272	11.0728	9.9985	10
5°00′	8.9403	8.9420	11.0580	9.9983	85°00′
10	8.9545	8.9563	11.0437	9.9982	50
20	8.9682	8.9701	11.0299	9.9981	40
30	8.9816	8.9836	11.0164	9.9980	30
40	8.9945	8.9966	11.0034	9.9979	20
50	9.0070	9.0093	10.9907	9.9977	10
6°00′	9.0192	9.0216	10.9784	9.9976	84°00′
10	9.0311	9.0336	10.9664	9.9975	50
20	9.0426	9.0453	10.9547	9.9973	40
30	9.0539	9.0567	10.9433	9.9972	30
40	9.0648	9.0678	10.9322	9.9971	20
50	9.0755	9.0786	10.9214	9.9969	10
7°00′	9.0859	9.0891	10.9109	9.9968	83°00′
10	9.0961	9.0995	10.9005	9.9966	50
20	9.1060	9.1096	10.8904	9.9964	40
30	9.1157	9.1194	10.8806	9.9963	30
40	9.1252	9.1291	10.8709	9.9961	20
50	9.1345	9.1385	10.8615	9.9959	10
8°00′	9.1436	9.1478	10.8522	9.9958	82°00′
10	9.1525	9.1569	10.8431	9.9956	50
20	9.1612	9.1658	10.8342	9.9954	40
30	9.1697	9.1745	10.8255	9.9952	30
40	9.1781	9.1831	10.8169	9.9950	20
50	9.1863	9.1915	10.8085	9.9948	10
9°00′	9.1943	9.1997	10.8003	9.9946	81°00′
*Add −10	L Cos	L Cot	L Tan	L Sin	Angle

Angle	L Sin	L Tan	L Cot	L Cos	
9°00′	9.1943	9.1997	10.8003	9.9946	81°00′
10	9.2022	9.2078	10.7922	9.9944	50
20	9.2100	9.2158	10.7842	9.9942	40
30	9.2176	9.2236	10.7764	9.9940	30
40	9.2251	9.2313	10.7687	9.9938	20
50	9.2324	9.2389	10.7611	9.9936	10
10°00′	9.2397	9.2463	10.7537	9.9934	80°00′
10	9.2468	9.2536	10.7464	9.9931	50
20	9.2538	9.2609	10.7391	9.9929	40
30	9.2606	9.2680	10.7320	9.9927	30
40	9.2674	9.2750	10.7250	9.9924	20
50	9.2740	9.2819	10.7181	9.9922	10
11°00′	9.2806	9.2887	10.7113	9.9919	79°00′
10	9.2870	9.2953	10.7047	9.9917	50
20	9.2934	9.3020	10.6980	9.9914	40
30	9.2997	9.3085	10.6915	9.9912	30
40	9.3058	9.3149	10.6851	9.9909	20
50	9.3119	9.3212	10.6788	9.9907	10
12°00′	9.3179	9.3275	10.6725	9.9904	78°00′
10	9.3238	9.3336	10.6664	9.9901	50
20	9.3296	9.3397	10.6603	9.9899	40
30	9.3353	9.3458	10.6542	9.9896	30
40	9.3410	9.3517	10.6483	9.9893	20
50	9.3466	9.3576	10.6424	9.9890	10
13°00′	9.3521	9.3634	10.6366	9.9887	77°00′
10	9.3575	9.3691	10.6309	9.9884	50
20	9.3629	9.3748	10.6252	9.9881	40
30	9.3682	9.3804	10.6196	9.9878	30
40	9.3734	9.3859	10.6141	9.9875	20
50	9.3786	9.3914	10.6086	9.9872	10
14°00′	9.3837	9.3968	10.6032	9.9869	76°00′
10	9.3887	9.4021	10.5979	9.9866	50
20	9.3937	9.4074	10.5926	9.9863	40
30	9.3986	9.4127	10.5873	9.9859	30
40	9.4035	9.4178	10.5822	9.9856	20
50	9.4083	9.4230	10.5770	9.9853	10
15°00′	9.4130	9.4281	10.5719	9.9849	75°00′
10	9.4177	9.4331	10.5669	9.9846	50
20	9.4223	9.4381	10.5619	9.9843	40
30	9.4269	9.4430	10.5570	9.9839	30
40	9.4314	9.4479	10.5521	9.9836	20
50	9.4359	9.4527	10.5473	9.9832	10
16°00′	9.4403	9.4575	10.5425	9.9828	74°00′
10	9.4447	9.4622	10.5378	9.9825	50
20	9.4491	9.4669	10.5331	9.9821	40
30	9.4533	9.4716	10.5284	9.9817	30
40	9.4576	9.4762	10.5238	9.9814	20
50	9.4618	9.4808	10.5192	9.9810	10
17°00′	9.4659	9.4853	10.5147	9.9806	73°00′
10	9.4700	9.4898	10.5102	9.9802	50
20	9.4741	9.4943	10.5057	9.9798	40
30	9.4781	9.4987	10.5013	9.9794	30
40	9.4821	9.5031	10.4969	9.9790	20
50	9.4861	9.5075	10.4925	9.9786	10
18°00′	9.4900	9.5118	10.4882	9.9782	72°00′
*Add −10	L Cos	L Cot	L Tan	L Sin	Angle

Angle	L Sin	L Tan	L Cot	L Cos	
18°00′	9.4900	9.5118	10.4882	9.9782	72°00′
10	9.4939	9.5161	10.4839	9.9778	50
20	9.4977	9.5203	10.4797	9.9774	40
30	9.5015	9.5245	10.4755	9.9770	30
40	9.5052	9.5287	10.4713	9.9765	20
50	9.5090	9.5329	10.4671	9.9761	10
19°00′	9.5126	9.5370	10.4630	9.9757	71°00′
10	9.5163	9.5411	10.4589	9.9752	50
20	9.5199	9.5451	10.4549	9.9748	40
30	9.5235	9.5491	10.4509	9.9743	30
40	9.5270	9.5531	10.4469	9.9739	20
50	9.5306	9.5571	10.4429	9.9734	10
20°00′	9.5341	9.5611	10.4389	9.9730	70°00′
10	9.5375	9.5650	10.4350	9.9725	50
20	9.5409	9.5689	10.4311	9.9721	40
30	9.5443	9.5727	10.4273	9.9716	30
40	9.5477	9.5766	10.4234	9.9711	20
50	9.5510	9.5804	10.4196	9.9706	10
21°00′	9.5543	9.5842	10.4158	9.9702	69°00′
10	9.5576	9.5879	10.4121	9.9697	50
20	9.5609	9.5917	10.4083	9.9692	40
30	9.5641	9.5954	10.4046	9.9687	30
40	9.5673	9.5991	10.4009	9.9682	20
50	9.5704	9.6028	10.3972	9.9677	10
22°00′	9.5736	9.6064	10.3936	9.9672	68°00′
10	9.5767	9.6100	10.3900	9.9667	50
20	9.5798	9.6136	10.3864	9.9661	40
30	9.5828	9.6172	10.3828	9.9656	30
40	9.5859	9.6208	10.3792	9.9651	20
50	9.5889	9.6243	10.3757	9.9646	10
23°00′	9.5919	9.6279	10.3721	9.9640	67°00′
10	9.5948	9.6314	10.3686	9.9635	50
20	9.5978	9.6348	10.3652	9.9629	40
30	9.6007	9.6383	10.3617	9.9624	30
40	9.6036	9.6417	10.3583	9.9618	20
50	9.6065	9.6452	10.3548	9.9613	10
24°00′	9.6093	9.6486	10.3514	9.9607	66°00′
10	9.6121	9.6520	10.3480	9.9602	50
20	9.6149	9.6553	10.3447	9.9596	40
30	9.6177	9.6587	10.3413	9.9590	30
40	9.6205	9.6620	10.3380	9.9584	20
50	9.6232	9.6654	10.3346	9.9579	10
25°00′	9.6259	9.6687	10.3313	9.9573	65°00′
10	9.6286	9.6720	10.3280	9.9567	50
20	9.6313	9.6752	10.3248	9.9561	40
30	9.6340	9.6785	10.3215	9.9555	30
40	9.6366	9.6817	10.3183	9.9549	20
50	9.6392	9.6850	10.3150	9.9543	10
26°00′	9.6418	9.6882	10.3118	9.9537	64°00′
10	9.6444	9.6914	10.3086	9.9530	50
20	9.6470	9.6946	10.3054	9.9524	40
30	9.6495	9.6977	10.3023	9.9518	30
40	9.6521	9.7009	10.2991	9.9512	20
50	9.6546	9.7040	10.2960	9.9505	10
27°00′	9.6570	9.7072	10.2928	9.9499	63°00′
*Add −10	L Cos	L Cot	L Tan	L Sin	Angle

Angle	L Sin	L Tan	L Cot	L Cos	
27°00′	9.6570	9.7072	10.2928	9.9499	63°00′
10	9.6595	9.7103	10.2897	9.9492	50
20	9.6620	9.7134	10.2866	9.9486	40
30	9.6644	9.7165	10.2835	9.9479	30
40	9.6668	9.7196	10.2804	9.9473	20
50	9.6692	9.7226	10.2774	9.9466	10
28°00′	9.6716	9.7257	10.2743	9.9459	62°00′
10	9.6740	9.7287	10.2713	9.9453	50
20	9.6763	9.7317	10.2683	9.9446	40
30	9.6787	9.7348	10.2652	9.9439	30
40	9.6810	9.7378	10.2622	9.9432	20
50	9.6833	9.7408	10.2592	9.9425	10
29°00′	9.6856	9.7438	10.2562	9.9418	61°00′
10	9.6878	9.7467	10.2533	9.9411	50
20	9.6901	9.7497	10.2503	9.9404	40
30	9.6923	9.7526	10.2474	9.9397	30
40	9.6946	9.7556	10.2444	9.9390	20
50	9.6968	9.7585	10.2415	9.9383	10
30°00′	9.6990	9.7614	10.2386	9.9375	60°00′
10	9.7012	9.7644	10.2356	9.9368	50
20	9.7033	9.7673	10.2327	9.9361	40
30	9.7055	9.7701	10.2299	9.9353	30
40	9.7076	9.7730	10.2270	9.9346	20
50	9.7097	9.7759	10.2241	9.9338	10
31°00′	9.7118	9.7788	10.2212	9.9331	59°00′
10	9.7139	9.7816	10.2184	9.9323	50
20	9.7160	9.7845	10.2155	9.9315	40
30	9.7181	9.7873	10.2127	9.9308	30
40	9.7201	9.7902	10.2098	9.9300	20
50	9.7222	9.7930	10.2070	9.9292	10
32°00′	9.7242	9.7958	10.2042	9.9284	58°00′
10	9.7262	9.7986	10.2014	9.9276	50
20	9.7282	9.8014	10.1986	9.9268	40
30	9.7302	9.8042	10.1958	9.9260	30
40	9.7322	9.8070	10.1930	9.9252	20
50	9.7342	9.8097	10.1903	9.9244	10
33°00′	9.7361	9.8125	10.1875	9.9236	57°00′
10	9.7380	9.8153	10.1847	9.9228	50
20	9.7400	9.8180	10.1820	9.9219	40
30	9.7419	9.8208	10.1792	9.9211	30
40	9.7438	9.8235	10.1765	9.9203	20
50	9.7457	9.8263	10.1737	9.9194	10
34°00′	9.7476	9.8290	10.1710	9.9186	56°00′
10	9.7494	9.8317	10.1683	9.9177	50
20	9.7513	9.8344	10.1656	9.9169	40
30	9.7531	9.8371	10.1629	9.9160	30
40	9.7550	9.8398	10.1602	9.9151	20
50	9.7568	9.8425	10.1575	9.9142	10
35°00′	9.7586	9.8452	10.1548	9.9134	55°00′
10	9.7604	9.8479	10.1521	9.9125	50
20	9.7622	9.8506	10.1494	9.9116	40
30	9.7640	9.8533	10.1467	9.9107	30
40	9.7657	9.8559	10.1441	9.9098	20
50	9.7675	9.8586	10.1414	9.9089	10
36°00′	9.7692	9.8613	10.1387	9.9080	54°00′
*Add −10	L Cos	L Cot	L Tan	L Sin	Angle

Angle	L Sin	L Tan	L Cot	L Cos	
36°00′	9.7692	9.8613	10.1387	9.9080	54°00′
10	9.7710	9.8639	10.1361	9.9070	50
20	9.7727	9.8666	10.1334	9.9061	40
30	9.7744	9.8692	10.1308	9.9052	30
40	9.7761	9.8718	10.1282	9.9042	20
50	9.7778	9.8745	10.1255	9.9033	10
37°00′	9.7795	9.8771	10.1229	9.9023	53°00′
10	9.7811	9.8797	10.1203	9.9014	50
20	9.7828	9.8824	10.1176	9.9004	40
30	9.7844	9.8850	10.1150	9.8995	30
40	9.7861	9.8876	10.1124	9.8985	20
50	9.7877	9.8902	10.1098	9.8975	10
38°00′	9.7893	9.8928	10.1072	9.8965	52°00′
10	9.7910	9.8954	10.1046	9.8955	50
20	9.7926	9.8980	10.1020	9.8945	40
30	9.7941	9.9006	10.0994	9.8935	30
40	9.7957	9.9032	10.0968	9.8925	20
50	9.7973	9.9058	10.0942	9.8915	10
39°00′	9.7989	9.9084	10.0916	9.8905	51°00′
10	9.8004	9.9110	10.0890	9.8895	50
20	9.8020	9.9135	10.0865	9.8884	40
30	9.8035	9.9161	10.0839	9.8874	30
40	9.8050	9.9187	10.0813	9.8864	20
50	9.8066	9.9212	10.0788	9.8853	10
40°00′	9.8081	9.9238	10.0762	9.8843	50°00′
10	9.8096	9.9264	10.0736	9.8832	50
20	9.8111	9.9289	10.0711	9.8821	40
30	9.8125	9.9315	10.0685	9.8810	30
40	9.8140	9.9341	10.0659	9.8800	20
50	9.8155	9.9366	10.0634	9.8789	10
41°00′	9.8169	9.9392	10.0608	9.8778	49°00′
10	9.8184	9.9417	10.0583	9.8767	50
20	9.8198	9.9443	10.0557	9.8756	40
30	9.8213	9.9468	10.0532	9.8745	30
40	9.8227	9.9494	10.0506	9.8733	20
50	9.8241	9.9519	10.0481	9.8722	10
42°00′	9.8255	9.9544	10.0456	9.8711	48°00′
10	9.8269	9.9570	10.0430	9.8699	50
20	9.8283	9.9595	10.0405	9.8688	40
30	9.8297	9.9621	10.0379	9.8676	30
40	9.8311	9.9646	10.0354	9.8665	20
50	9.8324	9.9671	10.0329	9.8653	10
43°00′	9.8338	9.9697	10.0303	9.8641	47°00′
10	9.8351	9.9722	10.0278	9.8629	50
20	9.8365	9.9747	10.0253	9.8618	40
30	9.8378	9.9772	10.0228	9.8606	30
40	9.8391	9.9798	10.0202	9.8594	20
50	9.8405	9.9823	10.0177	9.8582	10
44°00′	9.8418	9.9848	10.0152	9.8569	46°00′
10	9.8431	9.9874	10.0126	9.8557	50
20	9.8444	9.9899	10.0101	9.8545	40
30	9.8457	9.9924	10.0076	9.8532	30
40	9.8469	9.9949	10.0051	9.8520	20
50	9.8482	9.9975	10.0025	9.8507	10
45°00′	9.8495	10.0000	10.0000	9.8495	45°00′
*Add −10	L Cos	L Cot	L Tan	L Sin	Angle

INDEX